Neurobiology
of Learning
and Memory

Neurobiology of Learning and Memory

SECOND EDITION

Edited by

RAYMOND P. KESNER

Department of Psychology
University of Utah
Salt Lake City, Utah

JOE L. MARTINEZ, JR.

Division of Life Sciences
The University of Texas at San Antonio
San Antonio, Texas

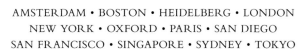

AMSTERDAM • BOSTON • HEIDELBERG • LONDON
NEW YORK • OXFORD • PARIS • SAN DIEGO
SAN FRANCISCO • SINGAPORE • SYDNEY • TOKYO

ELSEVIER

Academic Press is an imprint of Elsevier

Publisher: Nikki Levy
Senior Developmental Editor: Barbara Makinster
Marketing Manager: Patricia Howard
Project Manager: Jeff Freeland
Cover Designer: Eric DeCicco
Compositor: SNP Best-set Typesetter Ltd., Hong Kong
Printer/Binder: Hing Yip Printing Co., Ltd.

Academic Press is an imprint of Elsevier
30 Corporate Drive, Suite 400, Burlington, MA 01803, USA
Linacre House, Jordan Hill, Oxford OX2 8DP, UK

 Recognizing the importance of preserving what has been written, Elsevier prints its books on acid-free paper whenever possible.

Library of Congress Cataloging-in-Publication Data
Neurobiology of learning and memory / edited by Raymond P. Kesner, Joe L. Martinez, Jr. — 2nd ed.
    p. cm.
    Includes bibliographical references (p.).
    ISBN-13: 978-0-12-372540-0 (alk. paper)
    ISBN-10: 0-12-372540-2 (alk. paper)
    1. Learning — Physiological aspects.   2. Memory — Physiological aspects.   3. Neurobiology.
    I. Kesner, Raymond P.   II. Martinez, Joe L.
    QP408.N492 2007

    612.8′2 — dc22

                                                                                    2006051913

British Library Cataloguing-in-Publication Data
A catalogue record for this book is available from the British Library.

ISBN: 978-0-12-372540-0

For information on all Academic Press publications
visit our Web site at www.books.elsevier.com

Printed in China
07 08 09 10 11 12   10 9 8 7 6 5 4 3 2 1

This book is dedicated to our wives,
Drs. Laya Kesner and Kimberly Smith-Martinez.

Contents

Part I Approaches to Understanding the Neurobiological Basis of Learning and Memory

1 Historical Perspective

Mark Rosenzweig

2 Developmental Approaches to the Memory Process

Julie A. Markham, James E. Black, and William T. Greenough

3 Genetics in Learning and Memory

Yalin Wang, Josh Dubnau, Tim Tully, and Yi Zhong

4 Gene Expression in Learning and Memory

Joe L. Martinez, Jr., Kenira J. Thompson, and Angela M. Sikorski

5 Mnemonic Contributions of Hippocampal Place Cells

Sherri J. Y. Mizumori, D. M. Smith, and C. B. Puryear

Part II The Contribution of Neural Systems in Mediating Learning and Memory

12 *Neural Systems Involved in Fear and Anxiety Based on the Fear-Potentiated Startle Test*

Michael Davis

13 *Cerebellar Learning*

Tatsuya Ohyama and Michael D. Mauk

Part III Applications of the Importance of Learning and Memory to Applied Issues

17 Enhancement of Learning and Memory Performance: Modality-Specific Mechanisms of Action

Stephen C. Heinrichs

Contributors

Numbers in parentheses indicate the chapter contributed by authors.

Carol A. Barnes (15)
Arizona Research Labs
Division of Neural Systems, Memory
   and Aging
University of Arizona
Tucson, AZ 85724

James E. Black (2)
Department of Psychiatry
Southern Illinois School of Medicine
Springfield, IL 62794

Timothy J. Buschman (10)
Picower Institute for Learning and Memory
RIKEN-MIT Neuroscience Research
   Center
Department of Brain and Cognitive
   Sciences
Massachusetts Institute of Technology
Cambridge, MA 02142

Michael Davis (12)
Department of Psychiatry
Emory University School of Medicine
Atlanta, GA 30322

Josh T. Dubnau (3)
Cold Spring Harbor Laboratory
Cold Spring Harbor, NY 11724

Paul E. Gold (7)
Departments of Psychology and
   Psychiatry
Neuroscience Program
Institute for Genomic Biology
University of Illinois
Champaign, IL 61820

William T. Greenough (2)
Beckman Institute
University of Illinois
Urbana, IL 61801

Stephen C. Heinrichs (17)
Department of Psychology
Boston College
Chestnut Hill, MA 02467

Peter W. Kalivas (14)
Department of Physiology
Medical University of South Carolina
Charleston, SC 29425

Raymond P. Kesner (8)
Psychology Department
University of Utah
Salt Lake City, UT 84112

Donna L. Korol (7)
Department of Psychology
Neuroscience Program
Institute for Genomic Biology
University of Illinois
Champaign, IL 61820

Ryan T. LaLumiere (14)
Department of Neurosciences
Medical University of South Carolina
Charleston, SC 29425

Julie A. Markham (2)
Beckman Institute
University of Illinois
Urbana, IL 61801

Joe L. Martinez, Jr. (4)
Biology Department
University of Texas
San Antonio, TX 78249

Michael D. Mauk (13)
Department of Neurobiology and
    Anatomy
University of Texas Medical School
Houston, TX 77030

Earl K. Miller (10)
Picower Institute for Learning and
    Memory
Riken-MIT Neuroscience Research
    Center
Department of Brain and Cognitive
    Sciences
Massachusetts Institute of Technology
Cambridge, MA 02139

Sherri J. Y. Mizumori (5)
Department of Psychology
University of Washington
Seattle, WA 98195

Tatsuya Ohyama (13)
Department of Neurobiology and
    Anatomy
University of Texas Medical School
Houston, TX 77030

Marsha R. Penner (15)
Division of Neural Systems, Memory
    and Aging
University of Arizona
Tucson, AZ 85724

Alison R. Preston (9)
Department of Psychology
Neurosciences Program
Stanford University
Stanford, CA 94305

Corey B. Puryear (5)
Department of Psychology
University of Washington
Seattle, WA 98195

Michael E. Ragozzino (11)
Department of Psychology
University of Illinois
Chicago, IL 60607

Edmund T. Rolls (6)
Department of Experimental Psychology
University of Oxford
Oxford OX1 3UD, United Kingdom

Mark R. Rosenzweig (1)
Department of Psychology
University of California
Berkeley, CA 94720

Angela M. Sikorski (4)
Department of Biology
University of Texas
San Antonio, TX 78249

D. M. Smith (5)
Department of Psychology
University of Washington
Seattle, WA 98195

Kenira J. Thompson (4)
Department of Physiology
Ponce School of Medicine
Ponce, PR 00732

Tim Tully (3)
Cold Spring Harbor Laboratory
Cold Spring Harbor, NY 11724

Anthony D. Wagner (9)
Department of Psychology
Neurosciences Program
Stanford University
Stanford, CA 94305

Yalin Wang (3)
Cold Spring Harbor Laboratory
Cold Spring Harbor, NY 11724

Gary L. Wenk (16)
Department of Psychology
Ohio State University
Columbus, OH 43210

Yi Zhong (3)
Cold Spring Harbor Laboratory
Cold Spring Harbor, NY 11724

Preface

Graduate students interested in the neurosciences with a special interest in behavior are the intended audience. The major aim of this book is to present up-to-date information on the neurobiology of learning and memory based on multiple levels of analysis, contributions of multiple brain regions, systems that modulate memory and applications to aging, drugs of abuse, neurodegenerative diseases, and models of enhancement of memory. The emphasis will be on both animal and human studies.

The first section of the book covers different approaches to understanding the neurobiological basis of learning and memory. More specifically, there is an excellent introduction to the history of the neurobiology of learning and memory incorporating information from all of the contributing authors (Rosenzweig). With three chapters the book covers the developmental and genetic contributions to memory. This topic is becoming very important with the discovery of a variety of genetic tools to examine the role of specific genes and their contribution to learning, memory formation, and memory storage and retrieval (Markham, Black, and Greenough; Wang, Dubnau, Tully, and Zhong; and Martinez, Thompson, and Sikorski). Plasticity as it relates to memory has played a critical role in delineating the cellular properties of neurons that can maintain information over time (Mizumori and Smith). There is one chapter that emphasizes the role of place cells in the hippocampus and interconnected neural circuits based primarily on an electrophysiological analysis of cellular changes associated with learning and memory (Mizumori and Smith). There is one chapter that will cover a new area of theoretical importance for all of the different brain regions, namely the use of computational models to aid in providing a new theoretical approach to understand the processes that subserve memory (Rolls). Finally, in this section there is a chapter that covers the influence of hormonal processes on learning and memory (Korol and Gold).

The second section of the book covers the contribution of neural systems in mediating learning and memory. Since there are many brain regions associated with the processing of information of importance to learning and memory, six chapters outline a multiple system and multiple processes approach to understanding the complexity of information processing resulting in memory encoding, storage, and retrieval. The chapters deal with the following neural substrates, namely the medial temporal lobe, the frontal lobes, amygdala, basal ganglia, and cerebellum and cover experimental results and theoretical ideas based on research with humans, monkeys, and rats. Multiple approaches and techniques aimed at studying these brain regions are presented including, neuroanatomy, electrophysiology, lesion, pharmacology, fMRI, behavior, and cognitive analysis (Kesner; Preston and Wagner; Miller and Buschman; Ragozzino; Davis; Ohyama and Mauk).

The third section of the book emphasizes applications of the importance of learning and memory to applied issues. There are four chapters that provide a connection between all the previous chapters and important applications of the basic empirical findings to real world issues. The chapters cover issues of reward and drugs of abuse, the effects of aging on memory, the importance of studying neurodegenerative diseases from both the molecular and treatment approaches to memory and a final look at our ability to enhance memory (Balmier and Kalivas; Barnes and Penner; Wenk; Heinrichs).

The emphasis of each chapter will be on the presentation of the latest and most important research on the topic, the development of a theoretical perspective, and providing an outline that will aid a student in understanding the most important concepts presented in each chapter.

Ray Kesner

Joe Martinez

Part I

Approches to Understanding the Neurobiological Basis of Learning and Memory

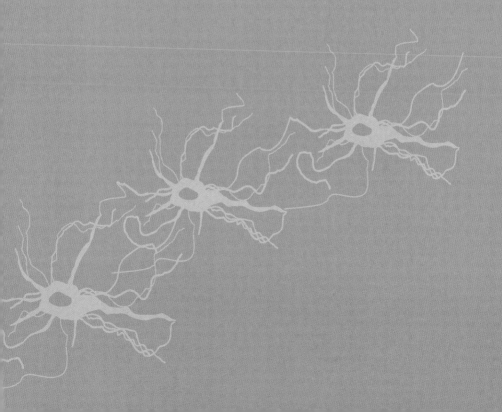

CHAPTER 1

Historical Perspective

Mark R. Rosenzweig

Department of Psychology, University of California, Berkeley, CA 94720

I. INTRODUCTION

The following chapters review recent and current research on many important aspects of the neurobiology of learning and memory. This chapter gives some historical perspective to this active field. Having participated in this research for half a century, I am happy to share information, interpretations, and insights about this productive multidisciplinary area.

In antiquity, speculation about mechanisms of memory took the form of metaphors, and metaphors of memory continue to be proposed in the present day. By the last quarter of the nineteenth century, scientific hypotheses and investigations of memory and its mechanisms began to be made, and it appeared that progress would be rapid. Research on neurobiological mechanisms of memory appeared to stall, however, and by the middle of the twentieth century, some thinkers despaired about the possibility of progress in this apparently intractable field. But shortly after midcentury, research and theory took off again, and rapid progress has continued to this day.

II. METAPHORS OF MEMORY

Concern about memory and its mechanisms goes far back in recorded history. An ancient Egyptian legend, related by the Greek philosopher Plato (427–347 BCE) in his *Phaedrus*, told that Thoth, the god of knowledge, offered the gift of writing to King Thamus of Egypt. The king was reluctant to accept the gift, expressing the fear that writing would cause forgetfulness because people would no longer exercise their memories but tend to rely instead on external written characters.

Thinkers in antiquity speculated about the mechanisms of memory and suggested metaphors for them. A widespread metaphor for memory was writing on a tablet coated with wax. The god Thoth was often depicted writing on such a tablet. As Draaisma notes in his book *Metaphors of Memory: A History of Ideas About the Mind* (2000), the classic passage on the wax tablet as the metaphor for memory appears in Plato's (1987) *Theaetetus*. In this dialogue, Socrates suggests:

> [O]ur minds contain a wax block, which may vary in size, cleanliness and consistency in different individuals, but in some people is just right. . . . [W]henever we want to remember something we've seen or heard or conceived on our own, we subject the block to the perception or the idea and stamp the impression into it. . . . We remember and know anything imprinted, as long as the impression remains in the block; but we forget and do not know anything which is erased or cannot be imprinted (pp. 99–100).

This wax tablet, wrote Plato, was a gift of Mnemosyne, the goddess of memory in the Greek pantheon and mother of the muses. We still acknowledge this goddess when we speak of mnemonic devices.

The metaphor of the wax tablet returned at greater length and in greater detail in the work of Aristotle (384–322 BCE), the pupil of Plato. Aristotle suggested that in the case of illness that affected memory, the consistency of the wax would be too loose, so no clear image could be stamped on it, just as no impression would be formed if a seal were to impinge on running water. He proposed that this is also why young children and old people have poor memories. They are in a state of flux, the former because of their growth, the latter because of their decay (see ref. in Draaisma, 2000, p. 46). The close association of memory and writing appears from Latin through French to English. The Latin word *memoria* meant both "memory" and "memoir." In French, *la memoire* means "memory" and *le memoire* means "memoir." And English has the related words "memory" and "memoir," derived from French.

Throughout the centuries, a series of metaphors was proposed for mechanisms of memory, each in keeping with current practices and technology. Here are some examples: The metaphor of a dovecote or aviary was long used; we still refer to this when we speak of placing a memory in a mental pigeonhole.

In the Middle Ages, books as well as libraries provided metaphors of memory. In the nineteenth century, the rapid progress of technologies for recording and transmitting information provided a series of metaphors for memory. Photography (from the 1830s) was one; think of the expression "a photographic memory." Telegraphy, also from the 1830s, provided another metaphor. The telephone system, with its switchboard, offered a more flexible system in the 1870s. In 1877 came the phonograph, which provided a mechanical memory for sound. Early phonograph records inscribed the sound on wax-covered cylinders, thus updating the ancient technology of writing on wax tablets. Even in the late twentieth century, while research in the neurobiology of memory prospered, metaphors of memory based on recent technologies continued to be proposed, such as the digital computer and the hologram.

III. ADVANCES IN THE LAST QUARTER OF THE NINETEENTH CENTURY

By the last quarter of the nineteenth century, sufficient progress had been made in psychology and neurobiology for scientific research to begin in memory and its neural mechanisms. Psychology was becoming established as an independent academic discipline and as a laboratory science in Europe and North America. Wilhelm Wundt, a professor of philosophy with a doctorate in medicine, had founded the first formal laboratory of psychology at the University of Leipzig in 1879. William James, also a professor of philosophy with a medical degree, began teaching physiological psychology at Harvard University in 1875, and he had an informal laboratory of psychology.

The decade of the 1880s saw major advances in research on learning and memory. French psychologist Théodule Ribot published an important book, *The Diseases of Memory* (1881), in which he described and discussed impairments of memory as consequences of brain lesions and brain diseases. From his study of published reports, Ribot proposed that more recent memories were more likely to be impaired than were older memories. This formulation became known as "Ribot's law," and it was verified by experimental research a century later. In his book, Ribot wrote that he regretted that it was not possible to state impairments of memory in quantitative terms. Only a few years later, German psychologist Hermann Ebbinghaus showed how memory could be measured in his pathbreaking book *On Memory* (1885). This book inaugurated the experimental investigation of learning and memory, a field that soon expanded rapidly.

Contemporaries and immediate successors of Ebbinghaus soon enlarged the work he had started, emphasizing controlled research on memory in a laboratory setting. Although Ebbinghaus' research obviously encouraged others, they were ready to move in this direction, as was shown in a review by Postman:

> Ebbinghaus' paradigm did not dominate or constrain the development of the field in its early years. Not only were many new methods of measurement and types of materials introduced in rapid succession, but the kinds of questions that were asked about memory soon began to move in different directions (Postman, 1985, p. 127).

An important monograph on studies of verbal memory was published by Müller and Pilzecker in 1900. In it they put forth the *perseveration–consolidation hypothesis*, which engendered much further research. This hypothesis held that neural activity initiated by a learning trial continues and recurs for some time after the original stimulation has ceased and that this perseveration aids the consolidation of a stable memory trace. In reviewing this book, William McDougall (1901) pointed out that the perseveration–consolidation hypothesis could be used to account for retrograde amnesia following head injury.

A. William James (1890) on the Physical Basis of Habit and Memory

In his major textbook *Principles of Psychology* (1890), James devoted separate chapters to habit, association, and memory. James asserted that habit, memory, and other aspects of behavior are based on physiological properties of the brain, even when he could not specify those properties very clearly. Thus James stated that the cerebral hemispheres seem to be the chief seat of memory (p. 98). James devoted Chapter 4 to habit and Chapter 16 to memory; a related chapter, 14, was devoted to association. The separation of the chapters on habit and on memory can be seen as a precursor to the distinction made in the 1980s between nondeclarative and declarative memories. Habits, according to James, reflected the "plasticity of the organic material [of the nervous system]" (p. 105). Neural activity could either "deepen old paths or . . . make new ones" (p. 107). James admitted that it was not yet possible to define in a detailed way what happens in the nervous system when habits are formed or changed, but he was confident that scientific research would find the answers (1890, p. 107):

> [O]ur usual scientific custom of interpreting hidden molecular events after the analogy of visible massive ones enables us to frame easily an abstract and general scheme of processes which the physical changes in question *may* be like. And when once the possibility of *some* kind of mechanical[1] interpretation is established, Mechanical Science, in her present mood, will not hesitate to set her brand of ownership upon the matter, feeling sure that it is only a question of time when the exact mechanical explanation of the case shall be found out.

[1]James used *mechanical* here in the sense of mechanistic, that is, interpreting and explaining phenomena by referring to causally determined material forces.

James gave lessons on how to form habits effectively. And he drew an ethical lesson, with a molecular basis:

> Could the young but realize how soon they will become mere walking bundles of habits, they would give more heed to their conduct while in the plastic state. . . . Every smallest stroke of virtue or vice leaves its never-so-little scar. The drunken Rip Van Winkle, in Jefferson's play, excuses himself for every fresh dereliction by saying, "I won't count this time!" Well! he may not count it, and a kind Heaven may not count it; but it is being counted none the less. Down among his nerve cells and fibres the molecules are counting it, registering and storing it up to be used against him when the next temptation comes (1890, p. 127).

James distinguished between what later came to be called short-term and long-term memories, referring to them as "primary" and "secondary" memories (1890, p. 670). Concerning the tendency of emotionally exciting experiences to be remembered well, James wrote, "An impression may be so exciting emotionally as to almost leave a *scar* on the cerebral tissues" (1890, p. 670).

James devoted three pages (pp. 676–678) to the experiments of Ebbinghaus (1885) under the heading "Exact Measurements of Memory." Considering Ebbinghaus' curve of forgetting, James commented, "The nature of this result might have been anticipated, but hardly its numerical proportions" (p. 677). James praised Ebbinghaus especially for his novel and successful attempt to test experimentally between two opposed hypotheses: This referred to Ebbinghaus' evidence that serial learning involves not only direct associations between adjacent items but also the formation of remote associations between nonadjacent items. James commented that the fact of these remote associations

> ought to make us careful, when we speak of nervous "paths," to use the word in no restricted sense. They add one more fact to the set of facts which prove that association is subtler than consciousness, and that a nerve-process may, without producing consciousness, be effective in the same way in which consciousness would have seemed to be effective if it had been there (p. 678).

As of 1890 there were few techniques available to study neural processes that might occur during learning and memory formation or ways of studying possible effects of memory on brain anatomy or neurochemistry. The development and use of such techniques characterized the research of the twentieth century, but speculation about neural junctions as sites of change in learning were already prevalent in the late nineteenth century, as we note next.

B. Neural Junctions as Sites of Change in Learning

In the 1890s, several scientists speculated that changes at neural junctions might account for memory. This was anticipated, as Finger (1994) points out, by associationist philosopher Alexander Bain (1872), who suggested that memory formation involves growth of what we now call synaptic junctions: "For every act of memory, every exercise of bodily aptitude, every habit, recollection, train of ideas, there is a specific grouping or coordination of sensations and movements, by virtue of specific growths in the cell junctions" (p. 91).

Such speculations were put on a firmer basis when neuroanatomist Wilhelm von Waldeyer (Waldeyer-Hartz, 1891) enunciated the neuron doctrine, largely based on the research of Santiago Ramón y Cajal. Neurologist Eugenio Tanzi (1893) proposed the hypothesis that the plastic changes involved in learning probably take place at the junctions between neurons. He expressed confidence that investigators would soon be able to test by direct inspection the junctional changes he hypothesized to occur with development and training. About 80 years were to elapse, however, before the first results of this sort were announced.

Ramón y Cajal, apparently independent of Tanzi, went somewhat further in his Croonian lecture to the Royal Society of London (Cajal, 1894). He stated that the higher one looked in the vertebrate scale, the more the neural terminals and collaterals ramified. During development of the individual, neural branching increased, probably up to adulthood. And he held it likely that mental exercise also leads to greater growth of neural branches, as he stated with a colorful set of metaphors:

> The theory of free arborization of cellular branches capable of growing seems not only to be very probable but also most encouraging. A continuous preestablished network — a sort of system of telegraphic wires with no possibility for new stations or new lines — is something rigid and unmodifiable that clashes with our impression that the organ of thought is, within certain limits, malleable and perfectible by well-directed mental exercise, especially during the developmental period. If we are not worried about putting forth analogies, we could say that the cerebral cortex is like a garden planted with innumerable trees — the pyramidal cells — which, thanks to intelligent cultivation, can multiply their branches and sink their roots deeper, producing fruits and flowers of ever greater variety and quality (Cajal, 1894, pp. 467–468).

But Ramón y Cajal then considered an obvious objection to his hypothesis:

> You may well ask how the volume of the brain can remain constant if there is a greater branching and even formation of new terminals of the neurons. To meet this objection we may hypothesize either a reciprocal diminution of the cell bodies or a shrinkage of other areas of the brain whose function is not directly related to intelligence (p. 467).

We will return later to this assumption of constancy of brain volume and Ramón y Cajal's hypotheses to permit constancy in the face of increased neuronal ramification.

The neural junctions didn't have a specific name when Tanzi and Ramón y Cajal wrote early in the 1890s, but a few years later neurophysiologist Charles Sherrington (Foster and Sherrington, 1897) gave them the name *synapse*. Sherrington also stated that the synapse was likely to be strategic for learning, putting it in this picturesque way:

> Shut off from all opportunities of reproducing itself and adding to its number by mitosis or otherwise, the nerve cell directs its pent-up energy towards amplifying its connections with its fellows, in response to the events which stir it up. Hence, it is capable of an education unknown to other tissues. (p. 1117).

During the first half of the twentieth century, psychologists and other scientists proposed memory hypotheses involving either the growth of neural fibrils toward one another to narrow the synaptic gap or more subtle chemical changes at synapses (see review in Finger, 1994). But the techniques then available allowed little progress on this issue.

C. Introduction of Research on Learning in Animal Subjects

Research on learning and memory was extended to animal subjects independently by psychologist Edward L. Thorndike and physiologist Ivan P. Pavlov. Thorndike demonstrated in his doctoral thesis (1898), conducted under the supervision of William James, how learning and memory can be measured in animal subjects, using cats, dogs, and chicks. This research led to the concept of trial-and-error learning and, later, to the "law of effect" (Thorndike, 1911). The field Thorndike opened with this research was quickly entered by others (Hilgard and Marquis, 1940, p. 6).

In 1902, American psychologist Shepard I. Franz opened a further line in animal research on learning and memory. He sought to determine the site of learning in the brain by combining Thorndike's methods of training and testing animals with the technique of localized brain lesions. Franz later recruited Karl S. Lashley, and through Lashley many others, to research on this topic.

In contrast to Thorndike's planned study of animal learning, Pavlov came upon the concept of conditioning from observations on salivary responses, made during his Nobel Prize–winning research on secretions of the alimentary tract. His initial contribution to the study of learning has been dated anywhere from 1897 to 1904 or even 1906. The *American Psychologist* [1997, *52*(9)] and the *European Psychologist* [1997, *2*(3)] published parallel sections in 1997 to commemorate the centenary of Pavlov's book, in Russian, *Lectures on the Work of the Principal Digestive Glands* (Pavlov, 1897). Pavlov's book included observations on *psychic secretion*, which foreshadowed his later research on conditioning. The first published use of the term *conditioned reflex* (actually *conditional reflex*) was in a report by I.F. Tolotschinoff (Tolochinov), one of Pavlov's associates, at the Congress of Natural Sciences in Helsinki in 1902. Pavlov discussed conditioning in his Nobel Prize lecture in 1904, although the main subject of the lecture was the research on the digestive glands, for which the Nobel Prize was awarded. Pavlov's first paper in English on salivary conditioning was his 1906 Huxley lecture, "The scientific investigation of the psychical faculties or processes in the higher animals," which was published in both *The Lancet* and *Science*. Even this review did not, however, "lead to any immediate repetitions of Pavlov's work in America, so far as published records reveal" (Hilgard and Marquis, 1940, p. 10).

Conditioning is now such a widely used technique — including in the research reviewed in several chapters in this volume — that it is interesting to note that it did not gain acceptance rapidly. Only after the presidential address of John B. Watson to the American Psychological Association in 1915, "The place of the conditioned reflex in psychology" (Watson, 1916), did conditioning begin to gain a prominent place in textbooks, and its place in the laboratory lagged behind still further. The publication in 1927 and 1928 of translations of books by Pavlov, revealing the wealth of facts discovered by Pavlov and his colleagues during more than a quarter of a century of research on salivary conditioning in dogs, stimulated a series of replications and extensions to conditioning in other species.

1. Earlier Observations of "Psychical Secretion"

In evaluating Pavlov's contributions, it is important to note that Pavlov, as he stated in his 1904 Nobel Prize lecture, was not the first to observe that secretions of the salivary and gastric glands can be evoked by "psychic" (i.e., non-gustatory) stimuli. Although Pavlov did not feel it necessary to name his predecessors in this respect, several medical or physiological investigators recorded such observations in the eighteenth and nineteenth centuries, and many more must have seen this phenomenon. One of the earliest such reports I have seen is that of Robert Whytt in his book *An Essay on the Vital and Other Involuntary Motions of Animals* (1763, p. 280):

> We consider, that not only an irritation of the muscles of animals, or parts nearly connected with them, is followed by convulsive motions; but that the remembrance or *idea* of substances, formerly applied to different parts of the body, produces almost the same effect, as if these substances were really present. Thus the sight, or even the recalled *idea* of grateful food causes an uncommon flow of spittle into the mouth of a hungry person; and the seeing of a lemon cut produces the same effect in many people. . . . The sight of a medicine that has often provocked [sic] vomiting, nay, the very mention of its name, will in many delicate persons raise a nausea.

Note that in the last sentence, Whytt also anticipated Garcia's (1990) *bait-shyness* learning. Further descriptions of salivary responses presumably elicited by learned stimuli were made by Erasmus Darwin (the grandfather of Charles Darwin) in 1796, French physiologist C.-L. Dumas (1803), Claude Bernard

(1872), and others, as I have documented elsewhere (Rosenzweig, 1959, 1960).

Pavlov's contribution was not to discover this phenomenon but to investigate it. He was the first to demonstrate that salivation could be evoked by a previously neutral stimulus after this had been paired with an effective stimulus. And he investigated carefully and skillfully both the conditions under which such acquisition occurs and conditions that do not lead to acquisition even though stimuli have been paired. This is one of many instances in the history of the field in which a casual observation has been exploited to lead to an important advance in knowledge.

2. Pavlov's Physiological Theory

The fundamental concepts in Pavlov's physiological theory, as summarized by Hilgard and Marquis (1940), were excitation and inhibition, conceived as states or processes located in the cerebral cortex. Afferent stimulation by an originally neutral stimulus caused an excitatory process to be initiated at a particular point A on the cortex, from whence it spread or irradiated over the cortex. The irradiating excitation

> will be concentrated at any other focus of excitation, such as that aroused by an unconditioned stimulus. After a number of repetitions of the two stimuli, the excitation aroused by the neutral stimulus is drawn to the locus of the unconditioned stimulus in sufficient intensity to elicit the unconditioned response. The direction of the drainage of excitation is from the weaker to the stronger or more dominant focus of excitation (Hilgard and Marquis, 1940, p. 310).

These concepts were elaborated by Pavlov to account for such phenomena as conditioning, generalization, and extinction and also for sleep, hypnosis, and neurosis.

In spite of the tempting simplicity and scope of Pavlov's conception of cortical physiology, Hilgard and Marquis (1940) noted that it did not attract any wide degree of acceptance. Two of the primary objections they summarized are these:

1. Concepts of cortical physiology should be based on direct measures of cortical function, but Pavlov's "excitation" and "inhibition" were purely inferential concepts based on overt movements or amounts of saliva secreted (Hilgard and Marquis, 1940, p. 312).

2. Pavlov's physiological conceptions are explicitly based on the premise that conditioning is exclusively a cortical function. Recent experimentation . . . demonstrates, however, that conditioning is possible at a subcortical level. . . . The two-dimensional character of Pavlov's irradiation concept does not easily permit extension of the theory to embrace the integrated functioning of cortical and subcortical centers" (p. 313).

IV. PESSIMISM IN MIDCENTURY, THEN RAPID GAINS

By 1950, the search for neural mechanisms of learning and memory seemed to have reached an impasse. During graduate studies at Harvard in the late 1940s, I heard mainly pessimistic opinions about prospects for the field. For example, when Karl S. Lashley surveyed the literature on possible synaptic changes as a result of training, he concluded that there was no solid evidence to support any of the "growth" theories. Specifically, Lashley offered these criticisms: (a) Neural cell growth appears to be too slow to account for the rapidity with which some learning can take place. (We return to this point later.) (b) Because he was unable to localize the memory trace, Lashley held there was no warrant to look for localized changes.

Edwin G. Boring, the historian of psychology, also testified in 1950 to the lack of progress in this area: "Where or how does the brain store its memories? That is the great mystery. . . . The physiology of memory has been so baffling a problem that most psychologists in facing it have gone positivistic, being content with hypothesized intervening variables or with empty correlations" (1950, p. 670).

In other words, we were still at the level of the ancient Egyptians and Greeks, having metaphors but no neurobiological mechanisms for memory. At the end of his chapter on the history of research on brain functions, Boring gave his view about what was needed for further progress:

> In general it seems safe to say that progress in this field is held back, not by lack of interest, ability, or industry, but by the absence of some one of the other essentials for scientific progress. Knowledge of the nature of the nerve impulse waited upon the discovery of electric currents and galvanometer of several kinds. Knowledge in psychoacoustics seemed to get nowhere until electronics developed. The truth about how the brain functions may eventually yield to a technique that comes from some new field remote from either physiology or psychology. Genius waits on insight, but insight may wait on the discovery of new concrete factual knowledge (1950, p. 688).

A few years later, Hans-Lukas Teuber stated, in an *Annual Review of Psychology* chapter on physiological psychology, that

> the absence of any convincing physiological correlate of learning is the greatest gap in physiological psychology. Apparently, the best we can do with learning is to prevent it from occurring, by intercurrent stimulation through implanted electrodes, . . . by cerebral ablation, . . . or by depriving otherwise intact organisms, early in life, of normal sensory influx (Teuber, 1955, p. 267).

In fact, some major advances were beginning to occur in research on the neural mechanisms of learning and memory. Some of these resulted from the application of recently developed techniques, such as single-cell electrophysiological recording and electron microscopy, and the use of new neurochemical methods, as we review shortly. Another major influence encouraging research on neural mechanisms of learning and memory was Donald O. Hebb's 1949 monograph, *The Organization of Behavior*. I had the good fortune to be exposed to Hebb's optimistic perspective in a seminar he gave at Harvard in the summer of 1947, using as a text a mimeographed version of his book that was published in 1949. Hebb (1949) was more positive about possible synaptic changes in learning than his colleague Lashley. Hebb noted some evidence for neural changes and did not let the absence of conclusive evidence deter him from reviving hypotheses about the conditions that could lead to the formation of new synaptic junctions and that underlie memory. In essence, Hebb's hypothesis of synaptic change underlying learning resembled James' formulation: "When two elementary brain processes have been active together or in immediate succession, one of them, on recurring, tends to propagate its excitement into the other" (James, 1890, p. 566). Hebb's *dual-trace hypothesis* also resembled the *consolidation-perseveration* hypothesis of Müller and Pilzecker. Much current neuroscience research concerns properties of what are now known as Hebbian synapses. Hebb was wryly amused that his name was connected to this resurrected hypothesis rather than to concepts he considered original (Milner, 1993, p. 127). The emphasis on the synaptic hypothesis reflects the fact, noted by Gallistel (1990, p. 570), that most neuroscientists have been more concerned with how synaptic changes can store information than with how neural networks can compute memories. By the 1990s the idea that correlated activity could lead to new neural connections was so well accepted that it could be epitomized in six words: Neurons that fire together wire together (Löwel and Singer, 1992, p. 211).

V. NEUROCHEMICAL AND NEUROANATOMICAL EFFECTS OF TRAINING AND EXPERIENCE

Ten years after Hebb's 1949 book was published, his postulate of use-dependent neural plasticity had still not been demonstrated experimentally. It seemed to many that it would not be possible, with available techniques, to find changes in the brain induced by training or experience. In fact, some neurobiologists spoke of a catch-22 in trying to find neurochemical changes as a result of training in an extract of whole brain: If a change is detected, it can probably be ruled out as being a result of training — any changes observed can more reasonably be attributed to grosser and less specific concomitants of learning such as stress, attentiveness, and so on (Agranoff, Burrell, Dokas, and Springer, 1978, p. 628). At a symposium in 1957 my colleagues and I proposed that an approach to this problem would be to make neurochemical analyses of specific regions of trained and untrained brains. This might be able to integrate and permit measurement of small changes taking place over many thousands of neural units. If such changes were found within a region, then subsequent analyses might be able to focus down more closely (Rosenzweig, Krech, and Bennett, 1958, p. 338).

In the early 1960s, two experimental programs announced findings demonstrating that the brain can be altered measurably by training or differential experience. First was the demonstration by our group at Berkeley that both formal training and informal experience in varied environments led to measurable changes in neurochemistry and neuroanatomy of the rodent brain (Krech, Rosenzweig, and Bennett, 1960; Rosenzweig, Krech, and Bennett, 1961; Rosenzweig, Krech, Bennett, and Diamond, 1962). Soon after came the report of Hubel and Wiesel that occluding one eye of a kitten led to reduction in the number of cortical cells responding to that eye (Wiesel and Hubel, 1963; Hubel and Wiesel, 1965; Wiesel and Hubel, 1965).

The original clues for the discovery of the Berkeley group came from data on rats given formal training in a variety of problems in order to examine possible relations between individual differences in brain chemistry and problem-solving ability. We did obtain significant correlations between levels of activity of the enzyme acetylcholinesterase (AChE) in the cerebral cortex and the ability to solve spatial problems (e.g., Krech, Rosenzweig, and Bennett, 1956; Rosenzweig, Krech, and Bennett, 1958). When we tested the generality of this finding over six different behavioral tests, we found a surprise: As we reported at a 1959 symposium, total AChE activity was higher in the cerebral cortex of groups that had been trained and tested on more difficult problems than in those given easier problems, and all the tested groups measured higher in total cortical AChE activity than groups given no training and testing (Rosenzweig, Krech, and Bennett, 1961, p. 102 and Fig. 4). It appeared that training could alter the AChE activity of the cortex! To test this further, we conducted an experiment in which littermates were either trained on a difficult problem or left untrained; the trained rats developed significantly higher total cortical AChE activity than their untrained littermates (Rosenzweig, Krech, and Bennett, 1961, p. 103). Control experiments showed that the results could not be attributed to the fact that the trained rats were underfed to increase their motivation or were handled.

Instead of continuing to train rats in problem-solving tests, a time-consuming and expensive procedure, we decided to house the animals in different environments that provided differential opportunities for informal learning. Measures made at the end of the experiment showed that informal enriched experience led to increased cortical AChE activity (Krech, Rosenzweig, and Bennett, 1960). The discovery that formal training or differential experience caused changes in cortical chemistry was soon followed by the even more surprising finding that enriched experience increased the *weights* of regions of the neocortex (Rosenzweig et al., 1962).

Work by students of Hebb (e.g., Forgays and Forgays, 1952) provided the models for the environments used in these experiments. Typically, we assigned littermates of the same sex by a random procedure among various laboratory environments, the three most common being these: (a) a large cage containing a group of 10–12 animals and a variety of stimulus objects, which were changed daily, called the *enriched condition* (EC) because it provided greater opportunities for informal learning than did the other conditions; (b) the *standard colony* or *social condition* (SC), with three animals in a standard laboratory cage; (c) SC-size cages housing single animals, called the *impoverished condition* or *isolated condition* (IC). All three conditions provided food and water *ad libitum*.

Over the next several years, replications and extensions by us (e.g., Bennett, Diamond, Krech, and Rosenzweig, 1964a) and by others (e.g., Altman and Das, 1964; Geller, Yuweiler, and Zolman, 1965; Greenough and Volkmar, 1973) added to the evidence that training or differential experience could produce measurable changes in the brain. Control experiments demonstrated that the cerebral differences could not be attributed to differential handling, locomotor activity, or diet. The brain-weight differences caused by differential experience were extremely reliable, although small in percentage terms. Moreover, these differences were not uniformly distributed throughout the cerebral cortex. They were almost invariably largest in the occipital cortex and smallest in the adjacent somesthetic cortex; the rest of the brain outside the cerebral cortex tended to show very little effect (Bennett et al., 1964a; Bennett, Krech, and Rosenzweig, 1964b). Thus the experience caused changes in specific cortical regions and not undifferentiated growth of brain. Later work also showed effects of differential experience in other parts of the brain that have been implicated in learning and formation of memory — the cerebellar cortex (Pysh and Weiss, 1979) and the hippocampal dentate gyrus (Juraska, Fitch, Henderson, and Rivers, 1985; Juraska, Fitch, and Washburne, 1989).

Further early studies revealed experience-induced changes in other measures, especially in the occipital cortex. These measures included not only cortical thickness (Diamond, Krech, and Rosenzweig, 1964) but also detailed cellular measures: sizes of neuronal cell bodies and nuclei (Diamond, 1967), size of synaptic contact areas (West and Greenough, 1972), an increase of 10% in the numbers of dendritic spines per unit of length of basal dendrites (Globus et al., 1973), an increase in the extent and branching of dendrites (Holloway, 1966) amounting to 25% or more (Greenough and Volkmar, 1973), and a parallel increase in the numbers of synapses per neuron (Turner and Greenough, 1985). Mainly because of the increase in dendritic branching, the neuronal cell bodies were spaced farther apart in the cortex of EC rats than in IC rats. These effects indicated substantial increases in cortical volume and intracortical connections; they suggested greater processing capacity of the cortical region concerned. They contradicted the speculation of Ramón y Cajal (Cajal, 1894), noted earlier, that with training, neural cell bodies would shrink in order to allow neural arborizations to grow, thus allowing brain volume to remain constant.

Instead, increased arborization requires *larger* cell bodies to maintain them, and the volume of the cortex increases as cell bodies and dendrites grow.

These experimental reports indicated growth of number and/or size of synaptic connections as results of training or enriched experience. Some workers declared for one or the other of these possibilities, as when neurophysiologist John C. Eccles (1965, p. 97) stated his belief that learning and memory storage involve "growth just of bigger and better synapses that are already there, not growth of new connections." But Rosenzweig et al. (1972) reviewed findings and theoretical discussions suggesting that negative as well as positive synaptic changes may store memory. Depending on where one measures and the kind of training or differential experience, one may find an increase in the number of synapses, an increase in their size, a decrease in the number, or decrease in the size.

Did the discovery of neurochemical and neuroanatomical effects of training or experience require novel experimental techniques? Yes and no. Accurate measurement of AChE activity in large numbers of tissue samples became practical only in the early 1950s, when we began our research. We first used a newly devised "pHstat" to titrate automatically the rate of hydrolysis and liberation of acid catalyzed by AChE. Then, when the Beckman UV spectrophotometer became available, we used it. On the other hand, most of the neuroanatomical effects of training or experience could have been discovered decades earlier, if anyone had had a reason to look for them; these were not findings that required technical advances for their discovery.

Skepticism or frank disbelief was the initial reaction to our reports that significant changes in the brain were caused by relatively simple and benign exposure of animals to differential experience. By the early 1970s, some neurobiologists began to accept these results. Thus neurobiologist B.G. Cragg (1972, p. 42) wrote,

> Initial incredulity that such differences in social and psychological conditions could give rise to significant differences in brain weight, cortical thickness, and glial numbers seems to have been overcome by the continued series of papers from Berkeley reporting consistent results. Some independent confirmation by workers elsewhere has also been obtained.

Soon after the early publications of neurochemical and anatomical plasticity came another kind of evidence of cortical plasticity — the announcement by Hubel and Wiesel that depriving one eye of light in a young animal, starting at the age at which the eyes open, reduced the number of cortical cells responding to stimulation of that eye (Wiesel and Hubel 1963; Hubel and Wiesel 1965; Wiesel and Hubel 1965). Depriving an eye of light is a rather severe and pathological condition. In contrast, giving animals different amounts of experience without depriving them of any sensory modality is a rather mild and natural treatment, yet it leads to measurable changes of neurochemistry and neuroanatomy, and it has significant effects on problem-solving ability. The report of Wiesel and Hubel (1965) that changes can be induced in the visual system only during a critical period early in the life of the kitten served to solidify the belief of many neurobiologists that neural connections in the adult brain are fixed and do not vary as a result of training.

Greenough (see Chapter 2 of this volume and Black and Greenough, 1998) follows Piaget (1980) in distinguishing between two kinds of information acquired from the environment: (1) general information acquired by all members of a species from common features of their environments (i.e., "expected" information), and (2) idiosyncratic information that the individual uses to adapt to its unique environment (i.e., "unexpected" information). As Black and Greenough (1998) point out, exposure to light and visual pattern stimulation provides general, "expected" information, whereas exposure to a complex environment provides idiosyncratic information. Exposure to "expected" stimulation usually occurs early in development, and it may be important in preparing the animal to respond adequately to idiosyncratic information.

A. Differential Experience Produces Cerebral Changes Throughout the Life Span and Rather Rapidly

Further experiments revealed that significant cerebral effects of enriched versus impoverished experience could be induced at any part of the life span and with relatively short periods of exposure. In contrast, Hubel and Wiesel had reported that depriving an eye of light altered cortical responses only if the eye was occluded during a critical period early in life. Later, however, investigators found that modifying sensory experience in adult animals — especially in the modalities of touch and hearing — could alter both receptive fields of cells and cortical maps, as reviewed by Kaas (1991) and Weinberger (1995).

Initially we supposed that cerebral plasticity might be restricted to the early part of the life span, so we assigned animals to differential environments at weaning (about 25 days of age) and kept them there for 80 days. Later, members of our group obtained similar effects in rats assigned to the differential environments for 30 days as juveniles at 50 days of age (Zolman and Morimoto, 1962) and as young adults at 105 days of age (Rosenzweig, Bennett, and Krech, 1964; Bennett, Diamond, Krech, and Rosenzweig, 1964a). Riege (1971), in our laboratory, found that similar effects occurred in rats assigned to the differential environments at 285 days of age and kept there for periods of 30, 60, or 90 days. Two hours a day in the differential environments for a period of 30 or 54 days produced similar cerebral effects to 24-hr exposure for the same periods (Rosenzweig, Love, and Bennett, 1968). Four days of differential housing produced clear effects on cortical weights (Bennett, Rosenzweig, and Diamond, 1970) and on dendritic branching (Kilman et al., 1988); Ferchmin and Eterovic (1986) reported that four 10-min daily sessions in EC significantly altered cortical RNA concentrations.

The fact that differential experience can cause cerebral changes throughout the life span and relatively rapidly was consistent with our interpretation of these effects as due to learning. Recall also that our original observation of differences in cortical neurochemistry came from experiments on formal training. Later, Chang and Greenough (1982) reported that formal visual training confined to one eye of rats caused increased dendritic branching in the visual cortex contralateral to the open eye. Also, single-trial peck-avoidance training in chicks was found to result in changes in density of dendritic spines (Lowndes and Stewart, 1994).

Although the capacity for these plastic changes of the nervous system and for learning remains in older subjects, the cerebral effects of differential environmental experience develop somewhat more rapidly in younger than in older animals, and the magnitude of the effects is often larger in the younger animals. Also, continuing plasticity does not hold for all brain systems and types of experience. As noted earlier, changes in responses of cortical cells to an occluded eye are normally restricted to early development, as Wiesel and Hubel (1963) found. But this restriction may itself be modifiable. Baer and Singer (1986) reported that plasticity of the adult visual cortex could be restored by infusing acetylcholine and noradrenaline. Further work showed that the plastic response of the young kitten brain to occlusion of one eye also depends on glutamate transmission, because treating the striate cortex with an inhibitor of the glutamate NMDA receptor prevented the changes (Kleinschmidt, Baer, and Singer, 1987). Thus, the extent to which the brain shows plastic changes in response to a particular kind of experience depends on the age of the subject, the brain region, and the kind of experience and also on special circumstances or treatments that enhance or impair plasticity. The factor of age is reviewed in Chapter 15 of this volume, by Barnes and Penner.

B. Enriched Experience Improves Ability to Learn and Solve Problems

Hebb (1949, pp. 298–299) reported briefly that when he allowed laboratory rats to explore his home for some weeks as pets of his children and then returned the rats to the laboratory, they showed better problem-solving ability than rats that had remained in the laboratory throughout. Furthermore, they maintained their superiority or even increased it during a series of tests. (Whether Hebb's children showed better problem-solving ability after having rats as household pets was apparently not investigated.) Hebb concluded that *"the richer experience of the pet group during development made them better able to profit by new experience at maturity* — one of the characteristics of the 'intelligent' human being" (pp. 298–299, italics in the original). Moreover the results seemed to show a *permanent* effect of early experience on problem solving at maturity.

We and others have found that experience in an enriched laboratory environment improves learning and problem-solving ability on a wide variety of tests, although such differences have not been found invariably. One general finding is that the more complex the task, the more likely it is that animals with EC experience will perform better than animals from SC or IC groups (see review and different explanations offered for this effect: Renner and Rosenzweig, 1987, pp. 46–48).

We were unable, however, to replicate an important aspect of Hebb's report — that over a series of tests, EC rats maintain or increase their superiority over IC rats. On the contrary, we found that IC rats tend to catch up with EC rats over a series of trials; this occurred with each of three different tests, including the Hebb–Williams mazes (Rosenzweig, 1971, p. 321). Thus we did not find that early deprivation of experience caused a permanent deficit, at least for rats tested on spatial problems. Also, decreases in cortical weights induced by 300 days in the IC (versus the EC) environment could be overcome by a few weeks of training and testing in the Hebb–Williams mazes (Cummins, Walsh, Budtz-Olsen, Konstantinos, and Horsfal, 1973). Later we will see a similar effect in birds.

C. Similar Neuroanatomical Effects of Training and Experience Occur in All Species Tested to Date

Experiments with several strains of rats showed similar effects of EC versus IC experience on both brain values and problem-solving behavior, as reviewed by Renner and Rosenzweig (1987, pp. 53–54). Similar effects on brain measures have been found in several species of mammals — mice, gerbils, ground squirrels, cats, and monkeys (reviewed by Renner and Rosenzweig, 1987, pp. 54–59); and effects of training on brain values of birds have also been found. Thus the cerebral effects of experience that were surprising when first found in rats have now been generalized to several mammalian and avian species. Anatomical effects of training or differential experience have been measured in specific brain regions of *Drosophila* (R. Davis, 1993; Heisenberg et al., 1995). Synaptic changes with training have also been found in the nervous systems of the molluscs *Aplysia* and *Hermissenda*, as reviewed by Krasne and Glanzman (1995). In *Aplysia*, long-term habituation led to decreased numbers of synaptic sites, whereas long-term sensitization led to an increase (Bailey and Chen, 1983); this is a case where either a decrease or an increase in synaptic numbers stores memory. Thus, as noted by Greenough, Withers, and Wallace (1990, p. 164), "experience-dependent synaptic plasticity is more widely reported, in terms of species, than any other putative memory mechanisms." Thus the cerebral effects of experience that were surprising when first reported for rats in the early 1960s are now seen to occur widely in the animal kingdom, "from flies to philosophers" (Mohammed, 2001).

D. Experience May Be Necessary for Full Growth of Brain and of Behavioral Potential

Sufficiently rich experience may be necessary for full growth of species-specific brain characteristics and behavioral potential. This was seen in research on differential experience conducted with different species of the crow family. Species that cache food in a variety of locations for future use are found to have significantly larger hippocampal formations than related species that do not cache food (Krebs et al., 1989; Sherry et al., 1989). But the difference in hippocampal size is not found in young birds who are still in the nest; it appears only after food storing has started, a few weeks after the birds have left the nest (Healy and Krebs, 1993). Even more interesting is the finding that this species-typical difference in hippocampal size depends on experience; it does not appear in birds that have not had the opportunity to cache food (Clayton and Krebs, 1994). Different groups of hand-raised birds were given experience in storing food at three different ages: either 35–59 days posthatch, 60–83 days, or 115–138 days. Experience at each of these periods led to increased hippocampal size, much as we had found for measures of occipital cortex in the rat. Thus, both birds and rats appear to retain considerable potential for experience-induced brain growth if it does not occur at the usual early age.

VI. GENETIC STUDIES OF LEARNING ABILITY: FROM SELECTION TO MOLECULAR BIOLOGY

A major advance at midcentury was the discovery of the structure of DNA by Francis Crick and James D. Watson in 1953. This soon led to understanding the genetic code and major advances in molecular biology, including progress in the neurobiology of learning and memory, as reviewed in Chapter 8. Because of the importance of their discovery, Crick and Watson were awarded the Nobel Prize in Physiology or Medicine in 1962. As a key to their discovery, Crick and Watson relied on X-ray photographs of DNA made by chemist Rosalind Franklin; some scientists believe that if she had not died prematurely in 1958, Franklin might have shared the Nobel Prize (Maddox, 2002, 2003). In fact, genetic research on learning ability began long before the work of Crick and Watson, as we review next.

A. Genetic Selection for Learning Ability

In the late nineteenth century, Francis Galton was convinced that intelligence and learning ability are inherited, and he had thought of breeding dogs for intelligence, as he reported later: "[I]t would be a most interesting occupation . . . to pick the cleverest dogs [one] could hear of, and mate them together, generation after generation — breeding purely for intellectual power" (Galton, 1909, p. 319). Although he believed that the costs of such an experiment could largely be covered by selling the superior animals that would result, Galton never undertook this project, nor was he able to persuade others to do so. I believe the first experiment to breed animals for learning ability was that conducted by Edward C. Tolman (1924) with rats. This successful preliminary work was then extended by Tolman's former student Robert C. Tryon (1940, 1942), using a 17-unit automatic maze developed by Tolman, Tryon, and Jeffress (1929). Tryon started by testing a large number of male and female rats of heterogeneous stocks. Males and females with low error scores were then bred together, and so were males and females with high error scores. Among the offspring of the low-error parents, those who themselves made few errors were kept for breeding. Similarly, in the other group, those who made many errors were mated. By the seventh generation, there was very little overlap of scores between the "bright" and "dull" lines. Further selective breeding did not increase the separation.

Why did experimental selection for learning ability wait for the 1920s when Galton had conceived of such an experiment by the end of the nineteenth century? Factors that made the experiment feasible by the 1920s but not when Galton originally conceived of it include the following:

1. Choice of the laboratory rat, rather than the dog, as the main subject for experiments on learning made such selection experiments economically feasible. This was especially the case for an experiment of selective breeding, since the generation time for rats is considerably shorter than for dogs.

2. By the 1920s there were animal laboratories in university departments of psychology supported by academic budgets. Galton would have had to undertake such an experiment with his own means, and even though he thought that eventually some of the costs could be recouped by sale of intelligent dogs, there would have been important start-up costs.

3. There was also the conceptual question of measuring intelligence. Galton did not indicate how he would measure the intelligence of dogs other than by observation and rating. By the 1920s there was a considerable background of experience and theory for testing the learning ability of animals. Tolman and his students improved the feasibility of testing large numbers of animals

over several successive generations by devising a multiple-unit automatically recording maze (Tolman, Tryon, and Jeffress, 1929).

Beginning in midcentury, however, behavior geneticists began to believe that Tryon's strains differed mainly in motivation rather than in learning per se. This came about when Tryon's student Lloyd V. Searle (1941, 1949) attempted to determine whether Tryon's maze-bright rats were generally superior in learning to the maze-dulls or whether their superiority was confined to the test employed in the selection program. Searle used 10 maze-bright, 10 maze-dull, and 15 animals of a crossed line, giving them a variety of tests of learning, activity, and emotional behavior. He (1949, p. 323) concluded:

> No evidence was found that a difference exists between the Brights and Dulls in the learning capacity per se. A detailed study of the behavior profiles indicated that the Brights are characteristically food-driven, economical of distance, . . . and timid in response to open space. Dulls are disinterested in food . . . and timid of mechanical apparatus features. It is concluded that brightness and dullness in the original Tryon Maze may be accounted for in large part by such motivational and emotional patterns. Although indications exist that the two strains may be differentiated with reference to certain basic "cognitive" tendencies, the procedures followed in this experiment were not sufficiently analytical to indicate their nature.

In fact, Tryon had investigated the same question, earlier than Searle and with a more complete experiment. About 1940, Tryon sought to test the possibility that motivational differences between the strains might account for the difference in their error scores in the maze. To do this, he ran an experiment with animals of the 22nd generation, using the following groups: (a) 71 maze-bright rats with "normal" hunger motivation, i.e., given the standard ration throughout the experiment; (b) 43 maze-bright rats that had been satiated with extra rations; (c) 71 maze-dulls with "normal" hunger motivation; and (d) 57 maze-dulls whose motivation was heightened by reduced rations. The results showed that the level of hunger motivation affected running speed but did not affect mean error scores of the groups. Whether normally hungry or satiated, the maze-brights made only about a third as many errors as the normally hungry or strongly hungry maze-dulls. Tryon never published these results, but about 20 years later he gave them to me to include in a paper on the effects of heredity and environment on brain measures and learning ability in the rat (Rosenzweig, 1964).

Tryon concluded, unlike Searle, that error scores were practically independent of food motivation in both strains. But Searle's results, having been published, convinced many readers that Tryon had selected for motivation and emotion rather than for learning ability. In the 1960s, my colleagues and I found that descendants of the maze-bright rats made significantly fewer errors than descendants of the maze-dulls on the Hebb–Williams maze, the Dashiell checkerboard maze, and the Lashley III maze (Rosenzweig, 1964), thus indicating some generality for Tryon's conclusions.

B. Effects of Mutations on Learning Ability

Geneticists have employed the fruit fly, *Drosophila melanogaster*, as a favorite subject since the early 1900s. Modern neurogenetic dissection of *Drosophila* behavior was pioneered by Seymour Benzer (1967, 1973). The application of this approach to the study of mechanisms of learning and memory became possible only after Benzer and his colleagues demonstrated that *Drosophila* can learn (Quinn et al., 1979). As reviewed by Dudai (1989), two methods were then employed to isolate mutations that affect learning or memory specifically without affecting other factors, such as perception and motivation. In the first, mutants previously isolated by a variety of criteria — morphological, developmental, biochemical, or physiological — are subjected to tests of learning and memory. Because these mutants have salient abnormalities, the specificity of any defect in learning or memory must be tested with special care. Several previously identified mutants have been reported to show relatively specific impairments in learning (e.g., Tempel, Livingstone, and Quinn, 1984; Heisenberg et al., 1985; Cowan and Siegel, 1986).

The second method is more straightforward. Here one treats flies with a mutagen and screens the progeny for defects in learning and/or memory. If such effects are found, the mutants must also be tested for defects in factors, such as perception and motivation, that might account for impaired performance on tests of learning or memory. Several mutants for learning have been isolated in this way (e.g., Dudai et al., 1976; Quinn, Sziber, and Booker, 1979; Aceves-Pina, Booker, Duerr, Livingstone, et al., 1983).

C. Molecular Biological Studies of Learning and Memory Formation

In addition to studying mutations, methods of molecular biology have made it possible to affect genes in a number of ways that have been applied to research on the mechanisms of learning and memory. Chapter 3 in this book, by Wang, Dubnau, Tully, and Zhong, discusses in detail the main methods and the results obtained to date with them. Mutations and gene knockouts are genetic lesions, and the results of such treatments are subject to the problems and criticisms that beset lesion techniques in general. Moreover, at first genetic techniques were rather blunt instruments with which to perform lesions. That is, they were not restricted in time or location — they affected animals throughout their development and throughout the body. Fortunately, techniques were later developed to restrict the changes to specific times and to certain brain regions. The importance of the gene-modification techniques for research in learning and memory is reflected by the fact that Martinez, Thompson, and Sikorski (Chapter 4 of this volume) devote substantial sections of their chapter to gene expression in learning and memory, and Wang, Dubnau, Tully, and Zhong (Chapter 3 of this volume) discuss the role of genetic manipulation in *Drosophila* on learning and memory.

VII. CHANGING CONCEPTS OF LEARNING AND MEMORY FORMATION

Research on the neurobiology of learning and memory has been influenced in the second half of the twentieth century by changing ideas about learning and memory formation. One important idea has been the variety of forms of learning and of memory mechanisms. Another has been the distinction between direct and modulatory processes in memory formation.

A. Variety of Forms of Learning and of Memory Mechanisms

Most theorists of learning in the first half of the twentieth century, such as Clark Hull (1943), E. R. Guthrie (1935), and B. F. Skinner (1938), attempted to explain learning by means of a single set of rules. Edward C. Tolman (1949), however, was convinced that there is more than one kind of learning and that different kinds might follow different laws. Findings in the second half of the twentieth century bore out Tolman's insight as a variety of kinds of learning were described and as different neural mechanisms were discovered. Some of the different kinds of learning and memory stores were these: (a) short-term versus long-term memory, a distinction that had been anticipated by William James in 1890; (b) declarative versus procedural (or nondeclarative) memory, distinguished in 1980 (Cohen and Squire, 1980); (c) storage of different attributes of memory, distinguished by Underwood (1969) and Spear (1976), as processed by different regions of the brain (Kesner, 1980 and Chapter 8 of this volume; McDonald and White, 1993); (d) processing of spatial memory as accomplished by certain cells in the hippocampus (O'Keefe and Dostrovsky, 1971; see Chapter 5 of this volume, by Mizumori, Smith, and Puryear); circuits in the cerebellum suffice for formation of so-called delay eye-blink conditioning when little or no time elapses between the end of the conditioned stimulus (CS) and the onset of the unconditioned stimulus (US), but the hippocampus is required for formation of so-called trace conditioning, when time intervals between the US and CS are longer (see Chapter 13 of this volume, by Ohyama and Mauk).

The discovery of different forms of learning and memory was dependent on elaboration of more specific behavioral tests that were capable of discriminating among the varieties of learning and memory. Lashley's pessimistic evaluation of theories of memory formation, mentioned earlier, in Section IV, came from his failure to discriminate between short-term and long-term memory stores and his use of spatial mazes that could be solved with inputs from any of several sensory modalities.

An important concept in testing types of memory and their neural substrates is *double dissociation*. Lashley (1952) had criticized ablation studies that purported to show localization even of sensory function. He pointed out that failure of an animal to continue to make a trained visual object discrimination after lesion of a temporal cortical area might instead mean impairment of comparison behavior or of comprehension of the training situation. This might be overcome, he suggested, by testing whether the lesion left intact the ability to discriminate in another modality, such as somesthesis. In an important review, Hans-Lukas Teuber (1955, p. 283) countered that more was needed to resolve this question than simply to show that the lesion did not impair discrimination in another modality. Such a "simple dissociation" might only mean that visual discrimination was more vulnerable to temporal cortical lesions than was tactile discrimination. What was needed for conclusive proof, Teuber argued, was *double dissociation*, that is, evidence that lesion of one cortical area impaired visual object discriminations without loss on comparable tactile tasks, while lesion of another cortical area impaired tactile discriminations without loss on the visual tasks, and that the impairments in the two tasks be comparable in severity. Subsequent investigators took up the challenge of finding double dissociations and also extended it to studies of the localization of brain regions involved in learning and memory. At the same time, investigators devised learning tasks intended to involve rather specific processes, and they abandoned earlier, rather nonspecific learning paradigms, such as Thorndike's puzzle boxes and Lashley's mazes, that could be solved in a variety of ways.

Knowlton, Mangels, and Squire (1996) started a review article by stating, "Students of brain and behavior have long recognized that double dissociations [references to Teuber (1955) and later authors] provide the strongest evidence for separating the functions of brain systems" (p. 1399). They presented evidence for a double dissociation between human brain regions and kinds of memory: Amnesic patients, with damage to the limbic-diencephalic regions, show impaired formation of declarative but not of nondeclarative memories, whereas patients with Parkinson's disease, who have damage to the neostriatum (caudate nucleus and putamen), show impaired formation of habits but not of declarative memories. Kesner, in a series of studies summarized in his chapter in this volume (Chapter 8), has found evidence of double and triple dissociations between brain regions involved in working memory for different attributes of the learning situation. A similar research project, by McDonald and White (1993), also found a triple dissociation, using three different problems, all run in the radial maze: (1) A neural system that includes the hippocampus acquires information about relationships among stimuli and events (declarative memories); (2) a different system that includes the dorsal striatum (mainly the caudate nucleus) mediates the formation of reinforced stimulus–response associations (habits, or nondeclarative memories); (3) a third system that includes the amygdala mediates rapid acquisition of behaviors based on biologically significant events with affective properties.

B. Distinction Between Direct and Modulatory Processes in the Formation of Memory

Beginning in the 1970s, a distinction began to be drawn between systems that might provide the substrates for memory and systems whose manipulations could modulate memory formation (e.g., Gold and McGaugh, 1975). This distinction became widely used, as shown by a 1984 symposium on the neurobiology of learning and memory (Lynch, McGaugh, and Weinberger, 1984) that included a section on modulation of memory, with four papers and five commentaries. The first of these papers stated: "Formation of the memory trace involves both necessary or basic processes (sometime called intrinsic) and also modulatory (or extrinsic) influences that affect the rate or level of the direct processes" (Rosenzweig and Bennett, 1984, p. 265). As examples, Rosenzweig and Bennett noted that synthesis of proteins is required for formation of long-term memory, and this direct process can be modulated by several sorts of processes and treatments, including the level of arousal, excitant or depressant agents, and drugs that affect the cholinergic system. Making this distinction between direct and modulatory processes helps to clarify the roles of the many factors that affect memory formation.

In the present book, direct processes are considered in Chapters 8, 9, 10, and 13, by Kesner, Preston and Wagner, Miller and Buschman, and Ohyama and Mauk, respectively, among others. Modulatory processes are considered in Chapters 7 and 16, by Korol and Gold and by Wenk, respectively, among others.

VIII. NEUROCHEMICAL MECHANISMS OF LEARNING AND MEMORY

Research on neurochemical mechanisms of learning and memory has become a prominent line of investigation since the 1960s. In part this was encouraged by the findings of the effects of enriched experience and formal training on brain chemistry. Some investigators suggested that memory might be encoded in one or another brain chemical. Another source of this research was interest in mechanisms of consolidation of memory.

A. Tests of the Hypothesis That Protein Synthesis Is Required for Memory Storage

By what processes does enriched experience or formal training lead to plastic changes in cerebral neurochemistry and neuroanatomy? We found early on that enriched experience causes increased rates of protein synthesis and increased amounts of protein in the cortex (Bennett et al., 1964a). Later, training (imprinting) was reported to increase the rates of incorporation of precursors into RNA and protein in the forebrain of the chick (Haywood, Rose, and Bateson, 1970), and enriched experience in rats was found to lead to increased amounts of RNA (Ferchmin, Eterovic, and Caputto, 1970; Bennett, 1976) and increased expression of RNA in rat brain (Grouse, Schrier, Bennett, Rosenzweig, and Nelson, 1978). Maze training led to increased ratios of RNA to DNA in rat cortex (Bennett, Rosenzweig, Morimoto, and Hebert, 1979). We viewed these findings in the light of the hypothesis, perhaps first enunciated by Katz and Halstead (1950), that protein synthesis is required for memory storage.

Tests of the protein-synthesis hypothesis of memory formation were initiated by Flexner and associates in the early 1960s (e.g., Flexner, Flexner, Stellar, de la Haba, and Roberts, 1962; Flexner, Flexner, de la Haba, and Roberts, 1965), and much research followed their design: (1) giving animal subjects brief training that, without further treatment, would yield evidence of retention at a test a few days later; (2) administering to experimental subjects an inhibitor of protein synthesis at various times close to training while control subjects received an inactive substance; and (3) comparing test performance of experimental and control subjects. By the early 1970s, considerable evidence indicated that protein synthesis during or soon after training is necessary for the formation of long-term memory (LTM), but the interpretation of the findings was clouded by serious problems, including the following. (1) The inhibitors of protein synthesis then available for research (such as puromycin and cycloheximide) were rather toxic, which impeded experiments and complicated interpretation. (2) It appeared that inhibition of protein synthesis could prevent memory formation after weak training but not after strong training (e.g., Barondes, 1970).

A newly discovered protein-synthesis inhibitor, anisomycin (ANI), helped to overcome these problems. Schwartz, Castellucci, and Kandel (1971) reported that ANI did not prevent an electrophysiological correlate of short-term habituation or sensitization in an isolated ganglion of *Aplysia*, but they did not investigate whether ANI can prevent long-term effects. The discovery by Bennett, Orme, and Hebert (1972) that ANI is an effective amnestic agent in rodents opened the way to resolving the main challenges to the protein-synthesis hypothesis of the formation of LTM. ANI is much less toxic than other protein-synthesis inhibitors, and giving doses repeatedly at 2-hr intervals can prolong the duration of cerebral inhibition at amnestic levels. By varying the duration of amnestic levels of inhibition in this way, we found that the stronger the training, the longer protein synthesis had to be inhibited to prevent formation of LTM (Flood, Bennett, Orme, and Rosenzweig, 1975; Flood, Bennett, Rosenzweig, and Orme, 1973). We also found that protein must be synthesized in the cortex soon after training if LTM is to be formed; short-term memory (STM) and intermediate-term memory (ITM) do not require protein synthesis (e.g., Bennett, Orme, and Hebert, 1972; Mizumori, Rosenzweig, and Bennett, 1985; Mizumori, Sakai, Rosenzweig, Bennett, and Wittreich, 1987).

B. Neurochemistry of Short-Term and Intermediate-Term Memories

Further studies were then designed to find the neurochemical processes that underlie formation of STM and ITM. Lashley's concern, mentioned earlier, that some kinds of memory appear to be formed too quickly to allow growth of neural connections, ignored the distinction between STM and LTM, even though William James (1890) had already distinguished between these stores (although under different names). Observing this distinction was necessary if one was to look for different mechanisms of the two kinds of memory traces that Hebb distinguished: transient, labile memory traces, on the one hand, and stable, structural traces, on the other.

Much research on the neurochemistry of STM and ITM has been done with chicks, which have several advantages for this research: Chicks can be trained rapidly in a one-trial peck-avoidance paradigm and can be tested within seconds after training or hours or days later. Large numbers of chicks can be studied in a single run, so one can compare different agents, doses, and times of administration within the same batch of subjects. Unlike invertebrate preparations, the chick system can be used to study the roles of different vertebrate brain structures and to investigate questions of cerebral asymmetry in learning and memory. The chick system permits study of learning and memory in the intact animal. The successive neurochemical stages occur more slowly in the chick than in the rat, thus allowing them to be separated more clearly. Further advantages have been stated elsewhere (e.g., Rosenzweig, 1990; Rosenzweig, Bennett, Martinez, Colombo, et al., 1992).

Although some amnestic agents, such as ANI, diffuse readily throughout the brain, others affect only a restricted volume of tissue at amnestic concentrations (Patterson, Alvarado, Warner, Rosenzweig, and Bennett, 1986). Such agents can be used to reveal the roles of different brain structures in different stages of memory formation (e.g., Patterson, Alvarado, et al., 1986; Serrano, Rodriguez, Bennett, and Rosenzweig, 1995).

C. Both Enriched Experience and Formal Training Evoke Similar Neurochemical Cascades

Using the chick system, several investigators have traced parts of a cascade of neurochemical events from initial stimulation to synthesis of protein and structural changes (e.g., Gibbs & Ng, 1977; Ng and Gibbs, 1991; Rose 1992a, 1992b; Rosenzweig, Bennett, Martinez, Colombo, et al., 1992). At some, if not all, stages, parallel processes occur. Briefly, here are some of the stages. The cascade is initiated when sensory stimulation activates receptor organs, which stimulate afferent neurons by using various synaptic transmitters, such as acetylcholine (ACh) and glutamate. Inhibitors of ACh synaptic activity, such as scopolamine and pirenzepine, can prevent STM. So can inhibitors of glutamate receptors, including both the NMDA and AMPA receptors. Alteration of regulation of ion channels in the neuronal membrane can inhibit STM formation, as seen in effects of lanthanum chloride on calcium channels and of ouabain on sodium and potassium channels. Inhibition of second messengers is also amnestic, for example, inhibition of adenylate cyclase by forskolin or of diacylglycerol by bradykinin. These second messengers can activate protein kinases — enzymes that catalyze the addition of phosphate molecules to proteins. We found that two kinds of protein kinases are important in the formation, respectively, of ITM and LTM. Agents that inhibit calcium-calmodulin protein kinases (CaM kinases) prevent formation of ITM, whereas agents that do *not* inhibit CaM kinases but *do* inhibit protein kinase A (PKA) or protein kinase C (PKC) prevent formation of LTM (Rosenzweig, Bennett, Martinez, Colombo, et al., 1992; Serrano, Beniston, Oxonian, Rodriguez, Rosenzweig, and Bennett, 1994). From this research, Serrano et al. (1995) were able to predict for a newly available inhibitor of PKC its effective amnestic dose and how long after training it would cause memory to decline. One-trial training leads to an increase of immediate early gene messenger RNA in the chick forebrain (Anokhin and Rose, 1991) and to an increase in the density of dendritic spines (Lowndes and Stewart, 1994). Many of these effects occur only in the left hemisphere of the chick or are more prominent in the left than in the right hemisphere. Thus, learning in the chick system permits study of many steps that lead from sensory stimulation to formation of neuronal structures involved in memory.

The neurochemical cascade involved in formation of memory in the chick is similar to the cascades found in long-term potentiation (LTP) in the mammalian brain (e.g., Colley and Routtenberg, 1993) and in the nervous systems of invertebrates (e.g., Krasne and Glanzman, 1995).

Many of the steps in the formation of memory in the chick can also be modulated by opioids and other substances. Opioid agonists tend to impair, and opioid antagonists to enhance, memory formation. Different opioids appear to modulate formation of different stages of memory (e.g., Colombo, Martinez, Bennett, and Rosenzweig, 1992; Colombo, Thompson, Martinez, Bennett, and Rosenzweig, 1993; Patterson, Schulteis, Alvarado, Martinez, Bennett, Rosenzweig, and Hruby, 1989; Rosenzweig et al., 1992).

Several groups of investigators have sought to determine which proteins must be synthesized to hold LTP or LTM; only a few examples will be mentioned here. Routtenberg and his colleagues produced evidence that synthesis of protein F1 (GAP-43) is involved in LTP (Meberg, Valcourt, and Routtenberg, 1995). Studies with *Aplysia* suggested that different proteins are involved in LTM (e.g., Kennedy et al., 1992; Kuhl et al., 1992).

D. Can Parts of the Neurochemical Cascade Be Related to Different Stages of Memory Formation?

Some of the difficulty in attempting to relate parts of the neurochemical cascade to different stages of memory formation have come from problems of defining stages of memory, as discussed more fully elsewhere (Rosenzweig, Bennett, Colombo, Lee, and Serrano, 1993). Consider, for example, some very different attempts to state the duration of STM. Early investigators of human STM (Brown, 1958; Peterson and Peterson, 1959) reported that it lasts only about 30 sec if rehearsal is prevented. Agranoff, Davis, and Brink (1966) reported that in goldfish, if formation of LTM is prevented by an inhibitor of protein synthesis, STM can last up to three days, although normally LTM forms within an hour after training. Kandel et al. (1987) wrote that in *Aplysia*, "A single training trial produces short-term sensitization that lasts from minutes to hours" (p. 17) and that long-term memory is "memory that lasts more than one day" (p. 35). Rose (1995) suggested that, in the chick, memories that persist only a few hours involve a first wave of glycoprotein synthesis, whereas "true long-term memory" requires a second wave of glycoprotein synthesis, occurring about 6 hr after training.

Instead of considering that STM can last several hours or even a day or more, others posited one or more intermediate-term memory (ITM) stages occurring between STM and LTM (e.g., McGaugh 1966, 1968). Thus, Gibbs and Ng (1977) referred to a "labile" stage occurring between STM and LTM and later called this intermediate-term memory (e.g., Gibbs and Ng, 1984; Ng and Gibbs, 1991). My coworkers and I have discussed mechanisms of STM, ITM, and LTM in a series of papers (e.g., Rosenzweig and Bennett, 1984; Rosenzweig, Bennett, Martinez, and Colombo, 1992, 1993; Mizumori, Sakai, et al., 1987; Patterson, Alvarado, Rosenzweig, and Bennett, 1988). In investigating effects of protein kinase inhibitors (PKIs) on memory formation in chicks, we reported that those agents that inhibit CaM kinase activity disrupted formation of what some workers with chicks identified as ITM (lasting from about 15 min to about 60 min posttraining); those agents that inhibit PKC, PKA, or PKG (protein kinase G) but do not inhibit CaM kinase disrupted the formation of LTM (Rosenzweig, Bennett, Martinez, Colombo, et al., 1992; Serrano et al., 1994). Other investigators preferred to refer to different phases or stages of LTM rather than use the term ITM. Thus, studying the LTP analog to memory in slices of rat hippocampus, Huang and Kandel (1994) reported findings similar to those of Rosenzweig et al. (1992) and Serrano et al. (1994) with regard to the roles of two classes of protein kinases: Inhibitors of CaM kinase activity disrupted what Huang and Kandel called a transient, early phase of LTP (E-LTP), evoked by moderately strong stimuli and lasting from 1 hr to less than 3 hr after induction of LTP. Agents that inhibit PKA but do not inhibit CaM kinase disrupt the formation of what they called a later, more enduring phase of LTP (L-LTP), evoked by strong stimulation and lasting at least 6–10 hr. Weak stimuli evoke only short-term potentiation (STP), lasting only 20–30 min. As already mentioned, Rose (1995) suggested that, in the chick, a kind of LTM that lasts a few hours involves a first wave of glyco-protein synthesis, whereas "true long-term memory" requires a second wave of glycoprotein synthesis, occurring about 6 hr after training. Rather than call the memory associated with Rose's first 6-hr-long wave a form of LTM, it may be better to think of it as ITM and to note that there is an earlier STM, lasting only a few minutes, as has been shown in many experiments with the chick.

These and other findings supported the hypothesis of at least three sequentially dependent stages of memory formation, each dependent on different neurochemical processes. These results are important, not only for investigators of the neurochemistry of memory, but also for neuropsychologists and others who work with patients who suffer from memory disorders. A review by Kopelman (1992, pp. 136–138) found mixed results in attempts to distinguish losses of ITM and LTM in Korsakoff's and Alzheimer's patients. If it becomes possible to distinguish patients with disorders of ITM from those with impairment of STM or LTM, then perhaps their deficits can be traced to different disorders of the nervous system. If we can identify the neurochemical processes underlying each stage of memory formation, this could lead to attempts at rational pharmacological treatments. If investigators could then understand the genetics involved, they might eventually find genetic treatments for some memory defects.

E. Is Memory Encoded in Brain Chemicals?

Beginning early in the 1960s, several investigators proposed that memory is encoded in one or another brain chemical. Among the chemicals proposed as repositories of memory were RNA, glucocorticoids, and peptides such as "scotophobin." Later, investigators attempted to set up guidelines to evaluate such proposals.

1. Proposed "Memory Molecules"

Reports of "memory molecules" were prominent for a time, not only in scientific circles but also in the popular press. A book by Louis Irwin (2006) presents a lively account of this topic and offers biographical sketches of several of the contributors. One of the first investigators in this field was Swedish neurochemist Holger Hydén. In 1959 Hydén hypothesized that any particular pattern of neuronal activation would alter the sequence of bases in molecules of RNA. Then in 1962, Hydén and Egyházi sampled RNA from the vestibular nucleus of rats taught to climb a wire to obtain food reward. They reported that the training caused changes in concentration of RNA and in the base composition of RNA. This report set off a flurry of speculation that memory could be encoded in RNA.

Hydén's reports suggested to psychologist James McConnell a mechanism for results he had been obtaining with planaria. He had been training planaria to turn left or right at a choice point and claimed that if a naive planarian cannibalized a trained planarian, the cannibal was then likely to choose the direction to which its prey had been trained. When these results were published, training planaria soon became a popular topic for high school science fairs, where positive results were often reported, although university laboratories had difficulty replicating them. A witticism of the time was that students might gain knowledge more efficiently by cannibalizing their professors than by studying. Biochemists at Berkeley were impressed enough by the possibility of finding a chemical code for memory that they decided to try to replicate the work with planaria. Edward Bennett was in charge of this project, but he could not obtain clear evidence of learning by planarians.

Attention then moved to "transfer of memory" in rats, when psychologist Allan Jacobson and coworkers reported that when they extracted RNA from the brains of trained rats and injected it into naive rats, the naive rats displayed operant-training responses acquired by the trained rats (Babich, Jacobson, et al., 1965). Many laboratories sought to replicate or to extend this exciting report. Our laboratory attempted to replicate the study exactly, even ordering rats from the small supplier Jacobson used and getting his technician to extract RNA by the method used in their report. We were unable to obtain positive results, and other labs also reported failure to replicate Jacobson's transfer of training. In 1966, *Science* published a report by 23 authors from eight different laboratories, including ours, announcing failure to obtain transfer of a variety of tasks with RNA (Byrne, Samuel, et al., 1966).

Meanwhile biochemist Georges Ungar had reported that trained responses in rats could be transferred to naive animals by peptides extracted from the brain (Ungar and Oceguera-Navarro, 1965). The best known of these peptides became one that Ungar called *scotophobin* because it appeared to encode learned fear of the dark; it was extracted from brains of rats who learned to avoid the dark side of an enclosure because they were given foot shock there (Ungar et al., 1972). As with Jacobson's report of transfer, most laboratories that tried to replicate Ungar's report failed to do so. Later, David Malin (1976), who had earlier collaborated with Ungar, reported that when scotophobin was injected into mice forced to remain in a black box, the mice developed elevated blood corticosteroid level, while mice in a lighted box did not. Thus scotophobin was apparently interacting with a particular environmental stimulus to elevate stress and cause the animal to flee the stress-inducing situation. In other words, scotophobin may have been a modulatory agent, a concept discussed earlier, in Section VII-B, rather than encoding memory.

2. Criteria for Neurochemistry of Memory

As evidence accumulated that learning and experience induce chemical changes in the brain and that inhibiting some chemical processes around the time of learning blocks formation of memory, some investigators tried to devise guidelines and criteria to judge whether such changes and processes are necessary and sufficient for formation of memory. Of course, reports of many studies stated one or more criteria against which to test their findings, but Entingh, Dunn, Wilson, Glassman, and Hogan (1975) and Rose (1981) tried to list several guidelines or criteria that would be applicable to a variety of studies. Some of these criteria are the following. (a) There must be changes in the quantity of the system or substance or its rate of production or turnover in some localized region of the brain during memory formation. (b) The amount of change should be related to the strength or amount of training, up to a limit. (c) Stress, motor activity, or other processes that accompany learning must not, in the absence of memory formation, result in the structural or biochemical changes. (d) If the cellular or biochemical changes are inhibited during the period over which memory formation would normally occur, then memory formation should be prevented and the animal should be amnesic. (However, Flood, Bennett, Rosenzweig, and Orme, 1973, found cases in which the protein synthesis required for LTM formation was only postponed by inhibition of protein synthesis and occurred later than usual, after the inhibition wore off.) Research on learning and memory, chiefly with chicks, showed that some neurochemical processes appeared to fulfill all the stated criteria, as I have discussed elsewhere (Rosenzweig, 1996, pp. 18–19).

In Chapter 4 of this volume, Martinez, Thompson, and Sikorski discuss whether LTP and long-term depression (LTD) — which involve neurochemical, electrophysiological, and neuroanatomical changes — are memory mechanisms. Most of the research on LTP has been done on rodent hippocampal preparations. It is generally believed that the hippocampus does not store memories for the long term, because ablation of the hippocampus does not destroy long-term memories; rather, the hippocampus appears to help to process information for long-term storage elsewhere in the brain. Experiments in which hippocampal lesions were made at different numbers of days after training in rodents showed that memory was not impaired if the lesions were made more than two or three days after training (Kim and Fanselow, 1992; Winocur, 1990). Because of such findings, it was not clear what purpose would be served by a hippocampal mechanism for holding memory more than a few days in the rodent. Thus, some theorists considered the hippocampus to be a "temporary memory store" (Rawlins, 1985) or an "intermediate-term buffer store" (Treves and Rolls, 1994).

While conceding that convincing proof does not exist that LTP and LTD are involved in learning and memory, Martinez, Thompson, and Sikorski (Chapter 4 of this volume) believe that after many years of research, dating back to the initial discovery of LTP by Bliss and Lomo (1973), LTP and LTD remain the best candidates for a cellular process of synaptic change that underlies learning and memory in the vertebrate brain. They review findings of a cascade of neurochemical events underlying LTP that is similar to those found in research on memory formation.

IX. ELECTROPHYSIOLOGICAL STUDIES OF LEARNING AND MEMORY

A lucky accident led to the first electrophysiological observations of training with a human subject. The French neurophysiologists Gustav Durup and Alfred Fessard (1935) were studying how the alpha rhythm is blocked when a person's field of vision is illuminated. One day, after switching on the light several times and seeing the subject's alpha rhythm disappear from the record each time, the experimenter again threw the light switch, but the bulb failed and the room remained dark — nevertheless the alpha rhythm again disappeared! Seeking to explain this puzzling occurrence, the investigators hypothesized that the sound of the switch became a conditioned stimulus predicting the appearance of light and thus caused the EEG to respond as if light were present. Tests with other subjects soon demonstrated that the sound of the switch did not block the alpha rhythm in naive subjects but came to do so after pairings of sound and light.

This research became more widely known after the end of World War II, and many investigators took up studies of EEG correlates of conditioning in the late 1940s and the 1950s. But precise localization of EEG activity in the human cortex is difficult because of the overlying skull and tissue. Besides, the critical events might not be occurring in the cortex but in deeper brain structures. So the focus of research shifted to recording from the brains of alert, behaving animal subjects, often with indwelling electrodes. With the invention of microelectrodes around 1950 it became possible to record the activity of single neurons during training. We will see this technique applied to investigating cellular activity during conditioning of a variety of animals, including relatively simple mollusks.

A. Sites of Synaptic Plasticity in the Nervous System of *Aplysia*

Observations on nonassociative learning in relatively simple invertebrates date back at least to the first decade of the twentieth century (e.g., Jennings, 1906). The relative simplicity of the central nervous systems of some invertebrates led several investigators to try to find in them the neural circuits necessary and sufficient for learning, with the goal of studying plastic synaptic changes in these circuits. Invertebrate preparations, such as the large sea slug *Aplysia californica*, appeared to offer the following advantages for this research, although we will see that some of these were overestimated.

1. The number of nerve cells in an *Aplysia* ganglion is relatively small compared to that in a mammalian brain or even a brain region, although the number in an *Aplysia* ganglion is still of the order of 1,000.

2. In the ganglia of mollusks such as *Aplysia*, the cell bodies form the outside and the dendritic processes are on the inside. This arrangement, the opposite of that in the mammal, made it easy to identify and record from cells of such invertebrates.

3. Many individual cells in molluskan ganglia can be recognized both because of their shapes and sizes and because the cellular structure of the ganglion is uniform from individual to individual. Thus it was possible to identify certain cells and to trace their sensory and motor connections. The neurotransmitters in some of the large identifiable cells were also known.

Because of such advantages, J.W. Davis (1986, p. 268) stated, in the first edition of this book, that "invertebrates offer the promise of immediate and comprehensive understanding of the physiological processes underlying associative learning, which may in turn provide insights into mammalian learning." But research in the decades since 1986 showed Davis' prediction to have been overoptimistic.

A well-known example of such research is the program initiated by Eric Kandel that has investigated sites and mechanisms of plasticity for both nonassociative and associative learning in *Aplysia* (e.g., Kandel, Schacher, Castellucci, and Goelet, 1987; Kandel, Schwartz, and Jessell, 1995). Kandel's research indicated that conditioning of the gill-withdrawal reflex took place within a straight-through sensory-motor chain that controls the behavior being studied. Many interesting results have been reported from this program, although some investigators have voiced reservations about the some of the methods and findings.

Although Kandel concluded the gill-withdrawal response is a simple unitary reflex and is controlled only by cells in the abdominal ganglion, other investigators have challenged both of these conclusions. In fact, the gill-withdrawal response was found to occur even when the central nervous system of *Aplysia* has been inactivated or removed.

Colebrook and Lukowiak (1988) further pointed out that in experiments on conditioning *Aplysia* no one had recorded both the electrical activity of the motor neurons and the gill responses in the *same* animals. When they carried out such an experiment, they found that although both the neural responses and the gill-withdrawal amplitudes to the CS showed mean increases as a result of conditioning, over one-third of the animals showed an increase in one but not in the other measure! That is, the behavioral response and its supposed neural cause did not necessarily act in the same way. Colebrook and Lukowiak (1988) concluded that many loci and neural mechanisms are likely to be involved in conditioning of the gill-withdrawal response, with both the ganglia and peripheral sites combining their effects.

Kandel and his associates then began to move in this direction, using a reduced preparation for simultaneous behavioral and cellular studies of plasticity of the gill-withdrawal response. They published preliminary reports on nonassociative learning with this preparation (Cohen, Henzi, Kandel, and Hawkins, 1991; Hawkins, Cohen, and Kandel, 1992). To investigate the role of different motor neurons in the ganglion, they inactivated one or another neuron by hyperpolarization; they reported that one motor neuron is responsible for about 70% of the gill-withdrawal response. They then recorded responses of this neuron during habituation, dishabituation, and sensitization. The "results suggest that habituation in this preparation is largely due to depression at central synapses, whereas dishabituation and sensitization are due to central and peripheral facilitation with different time courses" (Hawkins, Cohen, and Kandel, 1992, p. 360).

Further work on the sites involved in conditioning has not yet appeared, but it is likely that conditioning as well as sensitization involves the peripheral as well as the central nervous system of *Aplysia*. Thus the results of more recent research challenged the earlier conclusions that plasticity is located exclusively in the ganglia of *Aplysia*. However, there seems to be no reason to question that synaptic mechanisms of plasticity occur at some large neurons in the abdominal ganglion. Support for Kandel's neurochemical hypotheses came from research with *Drosophila* mutants that were impaired in learning and memory, as Kandel and associates pointed out in a review (Kandel, Schacter, et al., 1987, p. 26). These mutants were found to have deficiencies in some of the neurochemical steps identified by Kandel and his associates as being important for learning, and this provided independent support for the generality of their hypotheses.

But even in the ganglion, the story was far from complete, because a single touch to the siphon was found to activate electrical responses in about 150 different neurons (Zecevic, Wu, Cohen, London, Hopp, and Falk, 1989), and many of these probably played roles in the complex gill movements. Other investigators reported that approximately 200 abdominal ganglion neurons are involved in the gill-withdrawal response, and most of them were also involved in respiratory movements (Wu, Cohen, and Falk, 1994). Study of the different kinds of responses mediated by these neurons suggested that the different behaviors are generated by altered activities of a single, large distributed network rather than by separate small networks, each dedicated to a particular response. Wu, Cohen, and Falk (1994) reported that the large motor neurons probably contributed less than 10% to the gill-withdrawal response. Beyond these problems at the *central* sites, the mechanisms of plasticity at *peripheral* neural sites in *Aplysia* have not yet been studied, so there is still much to learn about the mechanisms of learning, even in what some investigators hoped would be a "simple" kind of learning in a "simple" organism.

B. Conditioning in *Hermissenda*

Another marine mollusk, *Hermissenda crassicornia*, has been the subject of extensive research on mechanisms of conditioning, by Daniel Alkon and his colleagues (e.g., Alkon, 1975, 1989, 1992; Farley and Alkon, 1985). In the laboratory, pairing light with rotation on a turntable caused conditioned suppression of the tendency to approach the light. The plasticity in this system occurred in the eyes of *Hermissenda*, which contain only five photoreceptor cells.

The work with *Hermissenda*, which found the important changes with training to occur in the neuronal membrane, afforded quite a different picture of basic mechanisms of conditioning from that furnished by the research with *Aplysia*, which focused on changes that occur at the presynaptic side of the synaptic junction. Similarities as well as differences of the neurochemical mechanisms of learning in *Aplysia* and *Hermissenda* emerged from a comparison by Clark and Schuman (1992). After noting important similarities, they pointed out some distinctions:

> Compared with plasticity in *Aplysia* siphon sensory cells, plasticity in *Hermissenda* Type B photoreceptors involves a different sensory modality (light rather than touch), different types of potassium conductances (I_A and I_{K-Ca}, rather than I_S), primarily a different second-messenger system (protein kinase C, rather than CAMP-dependent kinase), and an inhibitory rather than an excitatory synaptic potential, among other differences. These are meaningful distinctions, and their existence suggests that each preparation will provide unique insights into cellular mechanisms of learning (Clark and Schuman, 1992, p. 598).

It was proposed that similar research with other species of invertebrates and vertebrates might show whether these are only two of a wide variety of possible mechanisms of learning or whether either would prove to be general over a number of species.

C. The Mammalian Cerebellum Houses the Brain Circuit for a Simple Conditioned Reflex

While many investigators studied learning in the apparently simpler nervous systems of invertebrates, others tried to define a circuit for learning in intact mammals. Thus, psychologist Richard F. Thompson and his colleagues have been studying the neural circuitry of eyelid conditioning since the 1970s (Thompson, 1990; Lavond, Kim, and Thompson, 1993). Prior behavioral research had produced a great deal of knowledge about how the eye-blink reflex of the rabbit became conditioned when an air puff to the cornea (US) follows an acoustic tone (CS). A stable conditioned response (CR) developed rather rapidly, and this is similar to eyelid conditioning in humans. The basic circuit of the eye-blink reflex is simple, involving two cranial nerves and some interneurons that connect their nuclei.

Early in their work, Thompson and his colleagues found that during conditioning the hippocampus developed neural responses whose temporal patterns resembled closely those of the eyelid responses. Although the hippocampal activity closely paralleled the course of conditioning and did so better than the activity of other limbic structures, this result did not prove that the hippocampus is required for conditioning to occur. In fact, destruction of the hippocampus had little effect on acquisition or retention of the conditioned eyelid response in rabbits (Lockhart and Moore, 1975). Therefore the hippocampus is not *required* for this conditioning. It may, however, participate in the conditioning, as indicated by the finding that abnormal hippocampal activity can disrupt the acquisition of conditioning.

Thompson and his coworkers then searched further, mapping in detail the brain structures where neurons were active electrically during conditioning. They found that learning-related increases in activity of individual neurons were prominent in the cerebellum, both in its cortex and deep nuclei and in certain nuclei in the pons.

In the cerebellum, there were only negligible responses to CS and US before the stimuli were paired, but a neuronal replica of the learned behavioral response emerged during conditioning. These responses, which preceded the behavioral eye-blink responses by 50 milliseconds or more, were found in the deep cerebellar nuclei ipsilateral to the eye that was trained. The interpositus nucleus appeared to be particularly involved (McCormick and Thompson, 1984). Lesion experiments were then undertaken to find whether the cerebellar responses were required for conditioning or whether, like the hippocampal responses, they only correlated with the CRs. In an animal that had already been conditioned, destruction of the ipsilateral interpositus nucleus abolished the CR. The CR could not be relearned on the ipsilateral side, but the contralateral eye could still be conditioned normally. In a naive animal, prior destruction of the interpositus nucleus on one side prevented conditioning on that side. The effect of the cerebellar lesions could not be attributed to interference with sensory or motor tracts because the animal still showed a normal unconditioned blink when an air puff was delivered to its eye.

The circuit of the conditioned reflex was then mapped in further detail using a combination of methods: electrophysiological recording, localized lesions, localized stimulation of neurons, localized infusion of small amounts of drugs, and tracing of fiber pathways (Krupa, Thompson, and Thompson, 1993). Based on these experiments, Thompson proposed a schematic circuit for the conditioned eye-blink response (Thompson and Krupa, 1994, Fig. 1, p. 536).

Since the main input to the deep cerebellar nuclei comes from the cerebellar cortex, lesions of the cortex would be expected to abolish the eyelid CR, just as lesions of the deep nuclei do. Such a finding was reported by a group working in England (Yeo, Hardiman, and Glickstein, 1985). Thompson and his associates, however, did not find lesions of the cerebellar cortex to interfere with the CR unless the lesions were very large. Perrett, Ruiz, and Mauk (1993) reported that lesions of the anterior cerebellar cortex prevented rabbits from acquiring accurate timing of the CR. They proposed that motor learning involves two sites of plasticity in the cerebellum: the CS–US association occurs at synapses between the mossy fibers and the deep cerebellar nuclei, whereas temporal discrimination is mediated by synapses between granule cells and Purkinje cells in the cerebellar cortex. More recent work on this question is reviewed in Chapter 13 of this volume, by Ohyama and Mauk.

The role of the cerebellum in conditioning is not restricted to eye-blink conditioning. The cerebellum is also needed for conditioning of leg flexion; in this task, an animal learns to withdraw its leg when a tone sounds in order to avoid a shock to the paw (Donegan, Foy, and Thompson, 1983; Voneida, 1990). On the other hand, the cerebellum is not required for all forms of conditioning of skeletal muscular responses. Thompson and his colleagues found that cerebellar lesions did not prevent operant conditioning of a treadle-press response in the rabbit (Holt, Mauk, and Thompson, unpublished, cited in Lavond, Kim, and Thompson, 1993, p. 328).

Studies with human subjects were consistent with the animal research. Patients with unilateral cerebellar lesions (usually caused by a stroke) show normal eyeblink reflexes with both eyes, but they can acquire a conditioned eyeblink responses only on the side where the cerebellum is intact (Papka, Ivry, and Woodruff-Pak, 1994). A PET study found that when humans received paired tone–airpuff training, several regions of the cerebellum and other brain structures showed increased glucose metabolism (Logan and Grafton, 1995). During the first, control session of the experiment, PET scans were taken while subjects received unpaired tone and right-eye airpuff stimuli. In the second session, one to six days later, subjects were given paired tone–airpuff trials. In the third session, two to seven days after the first, PET scans were made while the subjects received paired trials. Comparison of the scans showed increased session-three activity in several regions of the cerebellum and also in other brain regions: right inferior thalamus/red nucleus, right hippocampal formation, right and left ventral striatum, right cortical middle temporal gyrus, left cortex occipitotemporal fissure. Thus the neural network involved in human eyelid conditioning includes not only the cerebellar and brainstem regions found by Thompson and his colleagues but also the hippocampus, the ventral striatum, and regions of the cerebral cortex.

D. Sleep and Memory Consolidation

After electrical recording helped to define the stages of sleep, beginning with the report of Aserinsky and Kleitman (1953), some investigators began to study the relation of stages of sleep to memory consolidation. Leconte and Bloch (1970) found that depriving rats of rapid-eye-movement (REM) sleep in the hours after learning impaired retention for avoidance conditioning. In a related experiment, the percentage of REM sleep to total sleep increased after a session of avoidance conditioning (Leconte, Hennevin, and Bloch, 1973). These results suggested that processing of newly acquired information continues during sleep as well as during waking. Bloch (1976) referred in this regard to the perseveration–consolidation hypothesis of Müller and Pilzecker (1900), which we cited earlier (Section IV-B). Some of Müller and Pilzecker's subjects in verbal learning experiments reported that, although instructed not to rehearse material between experimental sessions scheduled days apart, they found the material coming back to mind without their trying to recall it.

Attempts to relate learning in humans to REM sleep have usually yielded negative results, according to a review by sleep researcher Peretz Lavie (1996, p. 140). He did, however, note some positive findings: During intensive learning of a new language, young people show increases in REM sleep, and so do people who recover language after becoming aphasic because of brain damage. Most significant, however, was the discovery by Lavie and colleagues of a young man who showed virtually no REM sleep — some nights showed no REM and overall only 2–5% of his total sleep was spent in REM, whereas healthy people of his age have 20–25% REM sleep (Lavie et al., 1984). The patient had been injured when fragments of a shell entered his brain; one splinter was lodged in the pons in the region believed to control activation of REM sleep. After recovery from the main effects of his injuries, this man had completed high school and then law school, so the great reduction in REM sleep had not impaired his learning or memory.

Indications that brain activity during sleep is related to memory formation continued to appear. By monitoring the electrical activity of neurons in the hippocampus during sleep, Bruce McNaughton and colleagues may have observed such consolidation in process. The neurons in question appeared to be "place cells"; that is, while a rat learns its way in a maze, certain hippocampal cells come to favor firing when the rat is in a particular place in the maze, some cells firing when the rat is in one place, other cells firing more commonly when it is in another place. In such studies, large, prominent landmarks are placed around the room containing the maze so that the rat can use them to keep track of its position. The firing of a particular hippocampal cell indicates not where the rat actually is, but the rat's perceived location in the maze.

Wilson and McNaughton (1994) simultaneously recorded the activity of many (>50) hippocampal cells before, during, and after rats learned a new maze. The activity of two cells that developed very different place preferences in the course of learning the maze was uncorrelated throughout the experiment. But hippocampal cells that developed place preferences for neighboring portions of the maze came to fire together. So while the activity of these cells was uncorrelated before learning the maze, their firing was positively correlated by the end of the task. When the scientists examined the records made before the maze was learned, they found, as expected, that the activity of the cell pairs that was uncorrelated before maze learning during waking was also uncorrelated during prior sleep episodes. But after the maze was learned and the hippocampal neurons developed a correlation in their discharge, that correlated discharge was also seen during slow-wave sleep (SWS). They were able to find such neuron pairs (with uncorrelated discharge before learning and correlated discharge after learning) only by monitoring so many neurons at once. Although these results were intriguing, it was not clear how to integrate them with results of other studies, most of which related memory formation to rapid-eye-movement (REM) sleep rather than to SWS. But the facts that hippocampal neurons were active during sleep in these rats and that the postlearning sleep activity reflected the modified discharge learned in the maze indicated that some active process was going on. It was almost as if the postlearning sleep reinforced the new relationships between the cells in their firing. A further study reported that even the order in which various hippocampal cells fired during the training session was reflected in the order in which they fired during sleep afterwards (Skaggs and McNaughton, 1996). Whether this electrical activity during the posttraining sleep period in fact helped the rats remember the maze in subsequent trials remained to be seen. Because posttraining sleep has been shown to improve memory retention and because the hippocampus seems to be important for at least some kinds of memory formation, these observations of neuronal activity seemed to resemble tantalizingly sleep-consolidation of memory.

X. MEMORY DURING AGING

As we have noted, thinkers in antiquity (e.g., Aristotle) commented on the decline of memory with old age and tried to account for it, and this decline was mentioned over the ages. But when scientists began in the second half of the twentieth century to measure memory as a function of age and to test possible mechanisms for its changes, the subject became more complicated. Some kinds of memory appeared to start their decline at relatively early ages, whereas others remained relatively intact over age. The capacity of older people to learn and remember has become a topic of heightened interest, in part because of the growing proportion of elderly people in the populations of developed countries and concern for reductions in performance that accompany normal aging. It also reflects attention to pathological forms of cognitive impairment that are more likely to affect older people, such as Alzheimer's disease.

Accurate comparisons of learning and memory in people of different ages were impeded by confounding factors. For example, apparent differences in learning ability and memory formation may be influenced by such factors as educational level, how recently subjects had experienced formal training, and motivation. One way to avoid such confounds was to employ animal subjects. And work with animals showed declines with age in some forms of learning and memory (e.g., Kubanis and Zornetzer, 1981). After experimenters allowed for confounding factors with human subjects, differences related to age were found to occur with some tasks but not with all (e.g., Anderson and Craik, 2000; Balota et al., 2000).

What kinds of tasks were found to be more likely than others to show decrements with aging? Elderly people in normal health were found to show some memory impairment in tasks of conscious recollection that required effort (Hasher and Zacks, 1979) and that relied primarily on internal generation of the memory rather than on external cues (Craik, 1985). Giving elderly subjects easily organized task structures, or cues, often raised their performance to the level of the young. Thus the type of task helped to determine whether impairment was observed. A number of different mechanisms appeared to be involved in the changes in learning and memory with age, as reviewed by Barnes and Penner in Chapter 15 of this volume.

XI. HOW TO IMPROVE MEMORY

As far back as classical antiquity, thinkers were concerned with how to improve memory. Thus, to be able to remember the successive parts of a speech in correct order, the Roman orator Cicero (106–43 BCE) advocated use of the method of loci, which he attributed to the Greek poet Simonides. In this mnemonic method, one imagines the successive parts as being located at distinctive points along a path through a familiar house or along a familiar street. What has research contributed to this question?

In the first experimental study of memory, Ebbinghaus (1885) investigated the effects of repetition in improving memory for lists of words. In some earlier sections of this chapter, we noted methods that improved learning and memory. Thus, in Section V-B we saw that enriched experience improves ability to learn. In Section VI we reviewed genetic methods of improving learning ability. In Section X we noted that providing easily organized task structures can improve memory. Heinrichs brings this topic up to date in Chapter 17 of this volume.

Research on neurochemical processes in memory formation in the last quarter of the twentieth century led to efforts to find drugs that could improve learning and memory. Many experiments with animals showed beneficial effects of certain drugs on learning and/or on memory performance, but they were far from finding a drug that met three major criteria: (1) The drug should be able to improve several of the many kinds of learning and memory. (2) The drug should have a broad effective dose range so that determining the correct dose for an individual would not be a time-consuming and expensive procedure. (3) Use of the drug over long periods of time should produce no important negative side effects. In 1992 I listened to a debate on this topic between two experts on the effects of drugs on memory. The optimist (if that is the right term in this case) made the provocative prediction that within 18 months a drug would be found that could improve memory for large numbers of people. The pessimist predicted that it would be 10 years before such a drug would be discovered. In fact, even the 10-year span was too short; no such drug has yet been discovered.

In this area, as in many others, people tend to be gullible and look for "quick fixes." Not only individuals, but large businesses and government agencies, including the armed forces, have spent enormous amounts of money on unproved and dubious systems to improve learning and memory. In the 1980s the National Research Council of the United States appointed a committee to evaluate several popular systems alleged to improve learning and memory. Their report, a book entitled *Enhancing Human Performance* (Druckman and Swets, 1988), found that none of those systems had proved merit, and most of them had no plausible scientific basis. The committee also spelled out procedures and criteria that a business or organization should use to evaluate a system claimed to improve learning and memory before deciding whether to adopt it.

Although it will take time, more effective systems of training and effective memory drugs are likely to be discovered, so it is important to make adequate preparations for a drug that could significantly improve the ability of large numbers of people to learn better and to remember more surely. Such a "smart pill" would effectively increase intelligence by making more information immediately available to individuals and by cutting the time needed to learn and remember. But would we be ready for the social effects of a drug that would significantly increase memory and intelligence? For one thing, who would benefit first from such a drug, both within nations and among nations? Would not such a drug be likely to increase the disparities between the wealthier and poorer social groups, both nationally and internationally? What would be the effects on society if students could complete the curriculum significantly more rapidly and enter the labor market earlier? Although such a drug would promise many benefits in the long run, it would cause a host of difficult problems during the transition to the new era of generally higher intelligence.

These problems were raised in the 1960s by Prof. René Cassin, one of the authors of the *International Declaration of Human Rights*. He was awarded the Nobel Peace Prize in 1968 for his work in promoting binding international treaties on human rights. In his 80s, Cassin toured the world speaking to groups of biological scientists, educators, and jurists and warning them of this problem as well as of others that could arise from modern biological research. I have raised these problems at three International Congresses of Psychology (Rosenzweig, 1981, 1989, 1994) and I do so again here, because I believe that psychologists and neurobiologists should be aware of the problems as well as the benefits that could arise from this research. We have a responsibility, along with others, to plan for the beneficial use of this research and to anticipate and try to prevent negative effects. We must help to decide how to set up adequate planning for the time when our research can lead to significant improvements in learning and memory.

XII. CONCLUSIONS

1. Concern about memory and its mechanisms goes far back in recorded history. In antiquity, speculation about mechanisms of memory took the form of metaphors, and metaphors of memory continue to be proposed in the present.

2. Formal research on learning and memory extends back to the late nineteenth century. Early work has often been neglected, and older ideas have sometimes later been presented as new.

3. Hermann Ebbinghaus' 1885 book initiated empirical research on human learning and memory. Psychology was ready for this, as shown by the fact that other investigators soon joined this field, using a variety of methods and materials.

4. Laboratory research on learning in animal subjects was begun independently by Edward L. Thorndike (1898), on the basis of planned experiments, and Ivan P. Pavlov (1897–1906), on the basis of unexpected observations during research on alimentary secretions. Thorndike's research was followed up rapidly by other investigators. Although Pavlov conducted an extensive program of

research on conditioning, few other laboratories took up this research until the late 1920s.

5. Training and differential experience were demonstrated by the early 1960s to cause measurable changes in neurochemistry and neuroanatomy of weanling rats (e.g., Krech, Rosenzweig, and Bennett, 1960; Rosenzweig, Krech, Bennett, and Diamond, 1962). Similar results were later shown in adult animals. Later research found a variety of neurochemical and neuroanatomical changes.

6. Depriving one eye of visual experience in young kittens was found to cause changes in the electrical responses of cells in the visual cortex (Wiesel and Hubel, 1963). Wiesel and Hubel reported that such effects could not be induced in kittens more than a few months of age, but later workers found changes in cortical responses of adult animals as a result of experience.

7. Formation of long-term memory was demonstrated in the early 1970s to require synthesis of proteins in the hours following training (e.g., Bennett, Orme, and Hebert, 1972; Flood, Bennett, Rosenzweig, and Orme, 1973).

8. Formation of short-term, intermediate-term, and long-term memories has been found to require a cascade of neurochemical events, and rather similar sequences have been found in birds, mammals, and invertebrates.

9. Different kinds of learning and memory follow different rules, involve different brain sites, and require somewhat different cascades of neurochemical events. Because of these differences, generalizing about learning and memory and their biological mechanisms must be done with great care.

10. Newer biomedical and behavioral techniques are adding to knowledge of the neural mechanisms of learning and memory. The biomedical techniques include noninvasive brain imaging and molecular biological approaches. The behavioral techniques include tests for implicit as well as explicit memory, and tests for various kinds of declarative and nondeclarative memories. Biomedical as well as behavioral investigators use computational models.

11. Double dissociation of brain regions and cognitive processes is a strong method of localizing regions particularly involved in specific cognitive processes.

12. Although it may be simpler to conduct behavioral tests on one set of subjects and neurobiological tests on others, more powerful experimental designs perform both behavioral and neurobiological measures on the same set of subjects.

13. Increasingly, investigators are using multiple tests of hypotheses and seeking converging evidence to establish conclusions.

14. Changes in learning and memory formation with aging are diverse and have been hard to characterize. A number of neurobiological changes in the brain with aging have been found to correlate with changes in performance.

15. Many methods have been found to improve learning and memory, but discovery of a "smart pill" has proved elusive. Critics have pointed out that

discovery of such a drug would pose social problems and that investigators should share responsibility in preparing for this eventuality.

REFERENCES

Aceves-Pina, E.O., Booker, R., Duerr, J.S., Livingstone, M.S., Quinn, W.G., Smith, R.F., Sziber, P.P., Tempel, B.L., and Tully, T.P. (1983). Learning and memory in *Drosophila*, studied with mutants. *Cold Spring Harbor Symposium in Quantitative Biology, 48,* 831–840.

Agranoff, B.W., Burrell, H.R., Dokas, L.A., and Springer, A.D. (1978). Progress in biochemical approaches to learning and memory. In M.A. Lipton, A. DiMascio, and K.F. Killam (eds.), *Psychopharmacology: A Generation of Progress* (pp. 623–635). New York: Raven Press.

Agranoff, B.W., Davis, R.E., and Brink, J.J. (1966). Chemical studies on memory fixation in goldfish. *Brain Research, 1,* 303–309.

Alkon, D.L. (1975). Neural correlates of associative training in Hermissenda. *Journal of General Physiology, 65,* 46–56.

Alkon, D.L. (1989). Memory storage and neural systems. *Scientific American, 260,* 42–50.

Alkon, D.L. (1992). *Memory's Voice: Deciphering the Mind–Brain Code.* New York: HarperCollins.

Altman J., and Das G.D. (1964). Autoradiographic examination of the effects of enriched environment on the rate of glial multiplication in the adult rat brain. *Nature, 204,* 1161–1163.

Anokhin, K.V., and Rose, S.P.R. (1991). Learning-induced increase of early immediate gene messenger RNA in the chick forebrain. *European Journal of Neuroscience, 3,* 162–167.

Aristotle (1951). *The Works of Aristotle.* Oxford, U.K.: Oxford University Press.

Aserinsky, E., and Kleitman, N. (1953). Regularly occurring periods of eye motility, and concomitant phenomena during sleep. *Science, 118,* 273–274.

Babich, F.R., Jacobson, A.L., Bubash, S., and Jacobson, A. (1965). Transfer of a response to naive rats by injection of ribonucleic acid from trained rats. *Science, 149,* 656–657.

Baer, M.F., and Singer, W. (1986). Modulation of visual cortical plasticity by acetylcholine and noradrenaline. *Nature, 320,* 172–176.

Bailey, C.H., and Chen, M. (1983). Morphological basis of long-term habituation and sensitization in *Aplysia. Science, 220,* 91–93.

Bain, A. (1872). *Mind and Body: The Theories of Their Relation.* London: Henry S. King.

Barondes, S.H. (1970). Some critical variables in studies of the effect of inhibitors of protein synthesis on memory. In W.L. Byrne (ed.), *Molecular Approaches to Learning and Memory* (pp. 27–34). New York: Academic Press.

Bennett, E.L. (1976). Cerebral effects of differential experience and training. In M.R. Rosenzweig and E.L. Bennett (eds.), *Neural Mechanisms of Learning and Memory* (pp. 279–287). Cambridge, MA: MIT Press.

Bennett, E.L., Diamond, M.C., Krech, D., and Rosenzweig, M.R. (1964a). Chemical and anatomical plasticity of brain. *Science, 146,* 610–619.

Bennett, E.L., Krech, D., and Rosenzweig, M.R. (1964b). Reliability and regional specificity of cerebral effects of environmental complexity and training. *Journal of Comparative and Physiological Psychology, 57,* 440–441.

Bennett, E.L., Orme, A.E., and Hebert, M. (1972). Cerebral protein synthesis inhibition and amnesia produced by scopolamine, cycloheximide, streptovitacin A, anisomycin, and emetine in rat. *Federation Proceedings, 31,* 838.

Bennett, E. L., Rosenzweig, M.R., and Diamond, M.C. (1970). Time courses of effects on differential experience on brain measures and behavior of rats. In W.L. Byrne (ed.), *Molecular Approaches to Learning and Memory* (pp. 69–85). New York: Academic Press.

Bennett, E.L., Rosenzweig, M.R., Morimoto, H., and Hebert, M. (1979). Maze training alters brain anatomy and cortical RNA/DNA ratios. *Behavioral and Neural Biology, 26*, 1–22.

Benzer, S. (1967). Behavioral mutants of *Drosophila* isolated by countercurrent distribution. *Proceedings of the National Academy of Sciences, U.S.A., 58*, 1112–1119.

Benzer, S. (1973). Genetic dissection of behavior. *Scientific American, 229*(12), 24–37.

Black, J.E., and Greenough, W.T. (1998). Developmental approaches to the memory process. In J.L. Martinez and R.P. Kesner (eds.), *Neurobiology of Learning and Memory* (pp. 55–88). San Diego: Academic Press.

Bliss, T.V., and Lomo, T. (1973). Long-lasting potentiation of synaptic transmission in the dentate area of the anaesthetized rabbit following stimulation of the perforant path. *Journal of Physiology (London), 232*, 331–356.

Bloch, V. (1976). Brain activation and memory consolidation. In M.R. Rosenzweig and E.L. Bennett (eds.), *Neural Mechanisms of Learning and Memory* (pp. 583–590). Cambridge, MA: MIT Press.

Boring, E.G. (1950). *A History of Experimental Psychology* (2nd ed.). New York: Appleton-Century-Crofts.

Brown. J. (1958). Some tests of the decay theory of immediate memory. *Quarterly Journal of Expimental Psychology, 10*, 12–21.

Byrne, W.L., Samuel, D., Bennett, E.L., Rosenzweig, M.R., Wasserman, E. et al., (1966). Memory transfer. *Science, 153*, 658–659.

Cajal, R.S. (1894). La fine structure des centres nerveux. *Proceedings of the Royal Society, London, 55*, 444–468.

Chang, F.-L.F., and Greenough, W.T. (1982). Lateralized effects of monocular training on dendritic branching in adult split-brain rats. *Brain Research, 232*, 283–292.

Clark, G.C., and Schuman, E.M. (1992). Snails' tales: Initial comparisons of synaptic plasticity underlying learning in *Hermissenda* and *Aplysia*. In L.R. Squire and N. Butters (eds.), *Neuropsychology of Learning and Memory* (pp. 588–602). New York: Guilford Press.

Clayton, N.S., and Krebs, J.R. (1994). Hippocampal growth and attrition in birds affected by experience. *Proceedings of the National Academy of Sciences, U.S.A., 91*, 7410–7414.

Cohen, N.J., and Squire, L.R. (1980). Preserved learning and retention of pattern-analyzing skill in amnesia: Dissociation of knowing how and knowing what. *Science, 210*, 207–210.

Cohen, T.E., Henzi, V., Kandel, E.R., and Hawkins, R.D. (1991). Further behavioral and cellular studies of dishabituation and sensitization in *Aplysia*. *Society for Neuroscience Abstracts, 17*, 1302.

Colebrook, E., and Lukowiak, K. (1988). Learning by the *Aplysia* model system: Lack of correspondence between gill and gill motor neurone responses. *Journal of Experimental Biology, 135*, 411–429.

Colley, P.A., and Routtenberg, A. (1993). Long-term potentiation as synaptic dialogue. *Brain Research Review, 18*, 115–122.

Colombo, P.J., Martinez, J.L., Bennett, E.L., and Rosenzweig, M.R. (1992). Kappa opioid receptor activity modulates memory for peck-avoidance training in the 2-day-old chick. *Psychopharmacology, 108*, 235–240.

Colombo, P.J., Thompson, K.R., Martinez, J.L. Jr., Bennett, E.L., and Rosenzweig, M.R. (1993). Dynorphin (1–13) impairs memory formation for aversive and appetitive learning in chicks. *Peptides, 14*, 1165–1170.

Cowan, T.M., and Siegel, R.W. (1986). *Drosophila* mutations that alter ionic conduction disrupt acquisition and retention of a conditioned odor avoidance response. *Journal of Neurogenetics, 3*, 187–201.

Cragg, B.G. (1972). Plasticity of synapses. In G.H. Bourne (ed.), *The Structure and Function of Nervous Tissue* (vol. 4, pp. 2–60). New York: Academic Press.

Cummins, R.A., Walsh, R.N., Budtz-Olsen, O.E., Konstantinos, T., and Horsfall, C.R. (1973). Environmentally induced changes in the brains of elderly rats. *Nature, 243*, 516–518.

Davis, J.W. (1986). Invertebrate model systems. In J.L. Martinez, Jr., and R.P. Kesner (eds.), *Learning and Memory: A Biological View.* Orlando, FL: Academic Press.

Davis, R. (1993). Mushroom bodies and *Drosophila* learning. *Neuron, 11*, 1–14.

Diamond, M.C. (1967). Extensive cortical depth measurements and neuron size increases in the cortex of environmentally enriched rats. *Journal of Comparative Neurology, 131*, 357–364.

Diamond, M.C., Krech, D., and Rosenzweig, M.R. (1964). The effects of an enriched environment on the histology of the rat cerebral cortex. *Journal of Comparative Neurology, 123*, 111–119.

Donegan, N.H., Foy, M.R., and Thompson, R.F. (1983). Neuronal responses of the rabbit cerebellar cortex during performance of the classically conditioned eyelid response. *Society for Neuroscience Abstracts, 11*, 835.

Draaisma, D. (2000). *Metaphors of Memory: A History of Ideas About the Mind.* Cambridge, U.K.: Cambridge University Press.

Druckman, D., and Swets, J.A. (1988). *Enhancing human performance.* Washington, D.C.: National Academy Press.

Dudai, Y. (1989). *The Neurobiology of Memory.* Oxford, U.K.: Oxford University Press.

Dudai, Y., Jan, Y.-N., Beyrs, D., Quinn, W.G., and Benzer, S. (1976). *Dunce*, a mutant of *Drosophila* deficient in learning. *Proceedings of the National Academy of Sciences, U.S.A., 73*, 1684–1688.

Durup, G., and Fessard, A. (1935). L'électroencéphalogramme de l'homme. *L'Année Psychologique, 36*, 1–32.

Ebbinghaus, H. (1885). *Ueber das Gedächtnis* [On memory]. Leipzig: Dunker and Humbolt.

Eccles, J.C. (1965). Comment. In D.P. Kimble (ed.), *The anatomy of memory* (p. 97). Palo Alto, CA: Science and Behavior Books.

Entingh, D., Dunn, A., Wilson, J.E., Glassman, E., and Hogan, E. (1975). Biochemical approaches to the biological basis of memory. In M.S. Gazzaniga and C. Blakemore (eds.), *Handbook of Psychobiology* (pp. 201–238). New York: Academic Press.

Farley, J., and Alkon, D.L. (1985). Cellular mechanisms of learning, memory, and information storage. *Annual Review of Psychology, 36*, 419–494.

Ferchmin, P., and Eterovic, V. (1986). Forty minutes of experience increase the weight and RNA content of cerebral cortex in periadolescent rats. *Developmental Psychobiology, 19*, 511–519.

Ferchmin, P., Eterovic. V., and Caputto R. (1970). Studies on brain weight and RNA content after short periods of exposure to environmental complexity. *Brain Research, 20*, 49–57.

Finger, S. (1994). Chap. 23. The nature of the memory trace (pp. 332–348). Chap. 24. The neuropathology of memory (pp. 349–368). *Origins of Neuroscience: A History of Explorations into Brain Function.* New York: Oxford University Press.

Flexner, J.B., Flexner, L.B., de la Haba, G., and Roberts R.B. (1965). Loss of memory as related to inhibition of cerebral protein synthesis. *Journal of Neurochemistry, 12*, 535–541.

Flexner, J.B., Flexner, L.B., Stellar, E., de la Haba, G., and Roberts, R.B. (1962). Inhibition of protein synthesis in brain and learning and memory following puromycin. *Journal of Neurochemistry, 9*, 595–605.

Flood, J.F., Bennett, E.L., Orme, A.E., and Rosenzweig, M.R. (1975). Relation of memory formation to controlled amounts of brain protein synthesis. *Physiology and Behavior, 15*, 97–102.

Flood, J.F., Bennett, E.L., Rosenzweig, M.R., and Orme, A.E. (1973). The influence of duration of protein synthesis inhibition on memory. *Physiology and Behavior, 10*, 555–562.

Forgays, D.G., and Forgays, J.W. (1952). The nature of the effect of free-environmental experience on the rat. *Journal of Comparative and Physiological Psychology, 45*, 747–750.

Foster, M., and Sherrington, C.S. (1897). *A Textbook of Physiology. Part III. The Central Nervous System.* London: Macmillan.

Franz, S.I. (1902). On the functions of the cerebrum. I. The frontal lobes in the production and retention of simple sensory-motor habits. *American Journal of Physiology, 8,* 1–22.

Gallistel, C.R. (1990). *The Organization of Learning.* Cambridge, MA: MIT Press.

Galton, F. (1909). *Memories of My Life* (3rd ed.). London: Methuen.

Garcia, J. (1990). Learning without memory. *Journal of Cognitive Neuroscience, 2,* 287–305.

Geller, E., Yuwiler, A., and Zolman, J.F. (1965). Effects of environmental complexity on constituents of brain and liver. *Journal of Neurochemistry, 12,* 949–955.

Gibbs, M.E., and Ng, K.T. (1977). Psychobiology of memory: Towards a model of memory formation. *Biobehavioral Reviews, 1,* 13–36.

Gibbs, M.E., and Ng, K.T. (1984). Diphenylhydantoin extension of short-term and intermediate stages of memory. *Behavioural Brain Research, 11,* 103–108.

Globus, A., Rosenzweig, M.R., Bennett, E.L., and Diamond, M.C. (1973). Effects of differential experience on dendritic spine counts in rat cerebral cortex. *Journal of Comparative and Physiological Psychology, 82,* 175–181.

Gold, P.E., and McGaugh, J.L. (1976). A single-trace, two-process view of memory storage processes. In D. Deutsch and J.A. Deutsch (Eds.) *Short-term memory* (pp. 355–378). New York: Academic.

Greenough, W.T., and Volkmar, F.R. (1973). Pattern of dendritic branching in occipital cortex of rats reared in complex environments. *Experimental Neurology, 40,* 491–504.

Greenough, W.T., Withers, G.S., and Wallace, C.S. (1990). Morphological changes in the nervous system arising from behavioral experience: What is the evidence they are involved in learning and memory? In L.R. Squire and E. Lindenlaub (eds.), *The Biology of Memory* (pp. 159–185). Stuttgart, Germany: F.K. Schattauer Verlag.

Grouse, L.D., Schrier, B.K., Bennett, E.L., Rosenzweig, M.R., and Nelson, P.G. (1978). Sequence diversity studies of rat brain RNA: Effects of environmental complexity on rat brain RNA diversity. *Journal of Neurochemistry, 30,* 191–203.

Guthrie, E.R. (1943). *The psychology of learning.* New York: Harper.

Hawkins, R.D., Cohen. T.E., and Kandel, E.R. (1992). Motor neuron correlates of dishabituation and sensitization of the gill-withdrawal reflex in *Aplysia. Society for Neuroscience Abstracts, 18,* 360.

Haywood, J., Rose, S.P.R., and Bateson, P.P.G. (1970). Effects of an imprinting procedure on RNA polymerase activity in the chick brain. *Nature, 288,* 373–374.

Healy, S.D., and Krebs, J.R. (1993). Development of hippocampal specialisation in a food-storing bird. *Behavioural Brain Research, 53,* 127–130.

Hebb, D.O. (1949). *The Organization of Behavior: A Neuropsychological Theory.* New York: Wiley.

Heisenberg, M., Borst, A., Wagner, S., and Byers, D. (1985). *Drosophila* mushroom body mutants are deficient in olfactory learning. *Journal of Neurogenetics, 2,* 1–30.

Heisenberg, M., Heusipp, M., and Wanke, C. (1995). Structural plasticity in the *Drosophila* brain. *Journal of Neuroscience, 15,* 1951–1960.

Hilgard, E.R., and Marquis, D.G. (1940). *Conditioning and Learning.* New York: Appleton-Century.

Holloway, R.L. (1966). Dendritic branching: Some preliminary results of training and complexity in rat visual cortex. *Brain Research, 2,* 393–396.

Huang, Y.Y., and Kandel, E.R. (1994). Recruitment of long-lasting and protein kinase A–dependent long-term potentiation in the CA1 region of hippocampus requires repeated tetanization. *Learning & Memory, 1,* 74–82.

Hubel, D.H., and Wiesel, T.N. (1965). Binocular interaction in striate cortex of kittens reared with artificial squint. *Journal of Neurophysiology, 28,* 1041–1059.

Hull, C.J. (1943). *Principles of behavior.* New York: Appleton-Century-Crofts.

Hydén, H. (1959). Biochemical changes in glial cells and nerve cells. In F. Brücke (Ed.), Proceedings of the Fourth International Congress of Biochemistry (pp. 49–89). New York: Pergamon.

Hydén, H., and Egyhazi, E. (1962). Nuclear RNA changes of nerve cells during a learning experiment in rats. *Proceedings of the National Academy of Sciences, U.S.A.*, *48*, 1366–1373.

Irwin, L.N. (2006). *Scotophobin: Darkness at the Dawn of the Search for Memory Molecules*. New York: Hamilton Books.

James, W. (1890). *Principles of Psychology*. New York: Henry Holt.

Jennings, H.S. (1906). *Behavior of the Lower Organisms*. New York: Columbia University Press.

Juraska, J.M,, Fitch, J.M,, Henderson, C., and Rivers, N. (1985). Sex differences in dendritic branching of dentate granule cells following differential experience. *Brain Research*, *333*, 73–80.

Juraska, J.M., Fitch, J.M., and Washburne, D.L. (1989). The dendritic morphology of neurons in the rat hippocampus CA3 area. II. Effects of gender and the environment. *Brain Research*, *479*, 115–119.

Kaas, J.H. (1991). Plasticity of sensory and motor maps in adult animals. *Annual Review of Neuroscience*, *14*, 137–167.

Kandel, E.R., Schacher, S., Castelluci, V.F., and Goelet, P. (1987). The long and short of memory in *Aplysia*: A molecular perspective. In *Fidia Research Foundation Neuroscience Award Lectures*. Padua, Italy: Liviana Press.

Kandel, E.R., Schwartz, J.H., and Jessell, T.M. (1995). *Essentials of Neural Science*. Norwalk, CT: Appleton & Lange.

Katz, J.J., and Halstead, W.G. (1950). Protein organization and mental function. *Comparative Psychology Monographs*, *20*, 1–38.

Kennedy, T.E., Kuhl, D., Barzilai, A., Sweatt, J.D., and Kandel, E.R. (1992). Long-term sensitization training in *Aplysia* leads to an increase in calreticulin, a major presynaptic calcium-binding protein. *Neuron*, *9*, 1013–1024.

Kilman, V.L., Wallace, C.S., Withers, G.S., and Greenough, W.T. (1988). Four days of differential housing alters dendritic morphology of weanling rats. *Society for Neuroscience Abstracts*, *14*, 1135.

Kim, J.J., and Fanselow, M.S. (1992). Modality-specific retrograde amnesia of fear. *Science*, *256*, 675–677.

Kleinschmidt, A., Baer, M.F., and Singer, W. (1987). Blockade of NMDA receptors disrupts experience-dependent plasticity of kitten striate cortex. *Science*, *238*, 355–358.

Knowlton, B.J., Mangels, J.A., and Squire, L.R. (1996). A neostriatal habit learning system in humans. *Science*, *273*, 1399–1402.

Kopelman, M.D. (1992). The "new" and the "old": Components of the anterograde and retrograde memory loss in Korsakoff and Alzheimer patients. In L.R. Squire and N. Butters (eds.), *Neuropsychology of Memory* (2nd ed.) (pp. 130–146). New York: Guilford Press.

Krasne, F.B., and Glanzman, D.L. (1995). What we can learn from invertebrate learning. *Annual Review of Psychology*, *46*, 585–624.

Krebs, J.R., Sherry, D.F., Healy, S.D., Perry, V.H., and Vaccarino, A.L. (1989). Hippocampal specialization of food-storing birds. *Proceedings of the National Academy of Sciences, U.S.A.*, *86*, 1388–1392.

Krech, D., Rosenzweig, M.R., and Bennett, E.L. (1956). Dimensions of discrimination and level of cholinesterase activity in the cerebral cortex of the rat. *Journal of Comparative and Physiological Psychology*, *82*, 261–268.

Krech, D., Rosenzweig, M.R., and Bennett, E.L. (1960). Effects of environmental complexity and training on brain chemistry. *Journal of Comparative and Physiological Psychology*, *53*, 509–519.

Krupa, D.J., Thompson, J.K., and Thompson, R.F. (1993). Localization of a memory trace in the mammalian brain. *Science*, *260*, 989–991.

Kubanis, P., and Zornetzer, S.F. (1981). Age-related behavioral and neurobiological changes: A review with emphasis on memory. *Behavioral and neural biology, 31,* 115–172.

Kuhl, D., Kennedy, T.E., Barzilai, A., and Kandel, E.R. (1992). Long-term sensitization training in *Aplysia* leads to an increase in the expression of *BiP*, the major protein chaperone of the ER. *Journal of Cell Biology, 119,* 1069–1076.

Lashley K.S. (1950). In search of the engram. *Symposia of the Society for Experimental Biology, 4,* 454–482.

Lashley K.S. (1952). Functional interpretation of anatomic patterns. *Research Publications of the Association for Nervous and mental Disease, 30,* 529–547.

Lavie, P. (1996). *The Enchanted World of Sleep.* New Haven, CT: Yale University Press.

Lavie, P., Pratt, H., Scharf, B., Peled, R., and Brown, J. (1984). Localized pontine lesion: Nearly total absence of REM sleep. *Neurology, 34,* 118–120.

Lavond, D., Kim, J.J., and Thompson, R.F. (1993). Mammalian brain substrates of aversive conditioning. *Annual Review of Psychology, 44,* 317–342.

Leconte, P., and Bloch, V. (1970). Déficit de la rétention d'un conditionnement après privation de sommeil paradoxal chez le rat [Impairment in retention for conditioning after deprivation of paradoxical sleep in the rat]. *Comptes Rendus de l'Académie des Sciences (Paris), 271D,* 226–229.

Leconte, P., Hennevin, E., and Bloch, V. (1973). Analyse des effets d'un apprentissage et de son niveau d'acquisition sur le sommeil paradoxal consécutif [Analysis of the effects of learning and its strength on the paradoxical sleep that follows]. *Brain Research, 49,* 367–379.

Lockhart, M., and Moore, J.W. (1975). Classical differential and operant conditioning in rabbits (*Orycytolagus cuniculus*) with septal lesions. *Journal of Comparative and Physiological Psychology, 88,* 147–154.

Logan, C.G., and Grafton, S.T. (1995). Functional anatomy of human eye-blink conditioning determined with regional cerebral glucose metabolism and positron emission tomography, *Proceedings of the National Academy of Sciences, U.S.A., 92,* 7500–7504.

Löwel, S., and Singer, W. (1992). Selection of intrinsic horizontal connections in the visual cortex by correlated neuronal activity. *Science, 255,* 209–212.

Lowndes, M., and Stewart, M.G. (1994). Dendritic spine density in the lobus parolfactorius of the domestic chick is increased 24 h after one-trial passive avoidance training. *Brain Research, 654,* 129–136.

Lynch, G., McGaugh, J.L., and Weinberger, N. (Eds.), *Neurobiology of learning and memory.* New York: Guilford.

Maddox, B. (2002). *Rosalind Franklin: The dark lady of DNA.* London: Harper-Collins.

Maddox, B. (2003). The double helix and the "wronged heroine." *Nature, 421,* 407–408.

McCormick, D.A., and Thompson, R.F. (1984). Cerebellum: Essential involvement in the classically conditioned eyelid response. *Science, 223,* 296–299.

McDonald, R.J., and White, N.M. (1993). A triple dissociation of memory systems: Hippocampus, amygdala, and dorsal striatum. *Behavioral Neuroscience, 107,* 3–22.

McDougall, W. (1901). Experimentelle Beiträge zur Lehre vom Gedächtnis, by G.E. Müller and A. Pilzecker [Review of "Experimental research on memory," by G.E. Müller and A. Pilzecker]. *Mind, 10,* 388–394.

McGaugh, J.L. (1966). Time-dependent processes in memory storage. *Science, 153,* 1351–1358.

McGaugh, J.L. (1968). A multi-trace view of memory storage processes. In D. Bovet, F. Bovet-Nitti and A. Oliverio (Eds.) *Attuali orientamenti dela ricerca sull'apprendimento e la memoria* (pp. 13–24). Rome: Accademia Nazionale dei Lincei.

Meberg, P.J., Valcourt, E.G., and Routtenberg, A. (1995). Protein F1/GAP-43 and *PKC* gene expression patterns in hippocampus are altered 1–2 h after LTP. *Brain Research. Molecular Brain Research, 34,* 343–346.

Milner, P.M. (1993). The mind and Donald O. Hebb. *Scientific American, 268*(1), 124–129.

Mizumori, S.J.Y., Rosenzweig, M.R., and Bennett, E.L. (1985). Long-term working memory in the rat: Effects of hippocampally applied anisomycin. *Behavioral Neuroscience, 99,* 220–232.

Mizumori, S.J.Y., Sakai, D.H., Rosenzweig, M.R,, Bennett, E.L., and Wittreich, P. (1987). Investigations into the neuropharmacological basis of temporal stages of memory formation in mice trained in an active avoidance task. *Behavioral Brain Research, 23,* 239–250.

Mohammed, A. (2001). Enrichment and the brain. Plasticity in the adult brain. From genes to neurotherapy. 22nd International Summer School of Brain Research. Amsterdam, Netherlands.

Müller, G.E., and Pilzecker, A. (1900). Experimentale Beitrage zur Lehre vom Gedächtnis [Experimental research on memory]. *Zeitschrift für Psychologie,* Suppl., 1–288.

Ng, K.T., and Gibbs, M.E. (1991). Stages in memory formation: A review. In R.J. Andrew (ed.), *Neural and Behavioural Plasticity: The Use of the Domestic Chick as a Model* (pp. 351–369). Oxford, U.K.: Oxford University Press.

O'Keefe, J., and Dostrovsky, J. (1971). The hippocampus as a spatial map. Preliminary evidence from unit activity in the freely-moving rat. *Brain Research, 34,* 171–175.

Papka, M., Ivry, R., and Woodruff-Pak, D.S. (1994). Eyeblink classical conditioning and time production in patients with cerebellar damage. *Society for Neuroscience Abstracts, 20,* 360.

Patterson, T.A., Alvarado, M.C., Rosenzweig, M.R., and Bennett, E.L. (1988). Time courses of amnesia development in two areas of the chick forebrain. *Neurochemical Research, 13,* 643–647.

Patterson, T.A., Alvarado, M.C., Warner, I.T., Rosenzweig, M.R., and Bennett, E.L. (1986). Memory stages and brain asymmetry in chick learning. *Behavioral Neuroscience, 100,* 856–865.

Patterson, T.A., Schulteis, G., Alvarado, M.C., Martinez, J.L., Bennett, E.L., Rosenzweig, M.R., and Hruby, V.J. (1989). Influence of opioid peptides on learning and memory processes in the chick. *Behavioral Neuroscience, 103,* 429–437.

Pavlov, I.P. (1897). *Lekstii o Rabote Glavnykh Pishchevaritel'nykh Zhelez* [Lectures on the Work of the Principal Digestive Glands]. St. Petersburg, Russia: Typographiia Ministerstva Putei Soobsheniia.

Pavlov, I.P. (1906). The scientific investigation of the psychical faculties or processes in the higher animals. *Science, 24,* 613–619. (Also in *The Lancet, 2,* 911–915.)

Pavlov, I.P. (1927). *Conditioned Reflexes* (translated by G.V. Anrep). London: Oxford University Press.

Pavlov, I.P. (1928). *Lectures on Conditioned Reflexes* (translated by W.H. Gantt). New York: International.

Perret, S.P., Ruiz, B.P., and Mauk, M.D. (1993). Cerebellar cortex lesions disrupt learning-dependent timing of conditioned eyelid responses. *Journal of Neuroscience, 13,* 1708–1718.

Peterson, L.R., and Peterson, M.J. (1959). Short-term retention of individual verbal items. *Journal of Experimental Psychology, 58,* 193–198.

Piaget, J. (1980). *Adaptation and intelligence: Organic selection and phenocopy.* (Translator S.S. Earnes). Chicago: University of Chicago Press.

Plato (1987). *Theaetus.* (translated by R.A.H. Waterfield). Harmondsworth, U.K.: Penguin Books.

Postman, L. (1985). Human learning and memory. In G. Kimble and K. Schlessinger (eds.), *Topics in the History of Psychology.* Hillsdale, NJ: Erlbaum.

Pysh, J.J., and Weiss, M. (1979). Exercise during development induces an increase in Purkinje cell dendritic tree size. *Science, 206,* 230–232.

Quinn, W.G., Sziber, P.P., and Booker, R. (1979). The *Drosophila* memory mutant *amnesiac. Nature, 277,* 212–214.

Rawlins, J.N. (1985). Associations across time: The hippocampus as a temporary memory store. *Behavioral & Brain Sciences, 8,* 479–528.

Renner, M.J., and Rosenzweig, M.R. (1987). *Enriched and Impoverished Environments: Effects on Brain and Behavior.* New York: Springer-Verlag.

Ribot, T. (1881). *Les maladies de la mémoire.* Paris: J.B. Ballière. Translated by J. Fitzgerald as *The Diseases of Memory,* New York: Humboldt Library of Popular Science Literature (Vol. 46), 453–500, 1883.

Riege, W.H. (1971). Environmental influences on brain and behavior of old rats. *Developmental Psychobiology, 4,* 157–167.

Rose, S.P.R. (1981). What should a biochemistry of learning and memory be about? *Neuroscience, 6,* 811–821.

Rose, S.P.R. (1992a). *The Making of Memory.* New York: Doubleday.

Rose, S.P.R. (1992b). Of chicks and Rosetta stones. In L.R. Squire and N. Butters (eds.), *Neuropsychology of Memory* (2nd ed.) (pp. 547–556). New York: Guilford Press.

Rose, S.P.R. (1995). Glycoproteins and memory formation. *Behavioural Brain Research, 66,* 73–78.

Rosenzweig, M.R. (1959). Salivary conditioning before Pavlov. *American Journal of Psychology, 72,* 628–633.

Rosenzweig, M.R. (1960). Pavlov, Bechterev, and Twitmeyer on conditioning. *American Journal of Psychology, 73,* 312–316.

Rosenzweig, M.R. (1964). Effects of heredity and environment on brain chemistry, brain anatomy, and learning ability in the rat. *Kansas Studies in Education, 14*(3), 3–34.

Rosenzweig, M.R. (1971). Effects of environment on development of brain and behavior. In E. Tobach (ed.), *Biopsychology of Development* (pp. 303–342). New York: Academic Press.

Rosenzweig, M.R. (1981). Brain mechanisms of learning and memory: Research and applications. *Proceedings of XXII International Congress of Psychology, Leipzig 1980* pp. 200–207.

Rosenweig, M.R. (1989). Neural and cognitive processes in learning and memory: Findings and implications for education and human rights. In N. Bond and D. Siddle (Eds.), *Proceedings of the XXIV International Congress of Psychology. Vol. 6. Psychobiology: Issues and Applications* (pp. 3–22). Amsterdam: Elsevier.

Rosenzweig, M.R. (1990). The chick as a model system for studying neural processes in learning and memory. In L. Erinoff (ed.), *Behavior as an Indicator of Neuropharmacological Events: Learning and Memory* (pp. 1–20). Washington, DC: NIDA Research Monographs.

Rosenzweig, M.R. (1992). Research on the neural bases of learning and memory. In M.R. Rosenzweig (ed.), *International Psychological Science: Progress, Problems, and Prospects* (pp. 103–136). Washington, DC: American Psychological Association.

Rosenzweig, M.R. (1994). Some surprising findings about memory and its biological bases. In P. Bertelson, P. Eelen, and G. d'Ydewalle (Eds.), *Current Advances in Psychological Science* (pp. 1–23). Hove, England: Lawrence Erlbaum Associates, Ltd. (Presidential Address, XXV International Congress of Psychology, 1992).

Rosenzweig, M.R. (1996). Aspects of the search for neural mechanisms of memory. *Annual Review of Psychology, 47,* 1–32.

Rosenzweig, M.R., and Bennett, E.L. (1984). Studying memory formation with chicks and mice. In N. Butters and L.R. Squire (eds.), *The Neuropsychology of Memory* (pp. 555–565). New York: Guilford Press.

Rosenzweig, M.R., Bennett, E.L., Colombo, P.J., Lee, D.W., and Serrano, P.A. (1993). Short-term, intermediate-term, and long-term memories. *Behavioural Brain Research, 57,* 193–198.

Rosenzweig, M.R., Bennett, E.L., and Diamond, M.C. (1972). Brain changes in response to experience. *Scientific American, 226,* 22–29.

Rosenzweig, M.R., Bennett, E.L., and Krech, D. (1964). Cerebral effects of environmental complexity and training among adult rats. *Journal of Comparative and Physiological Psychology, 57,* 438–439.

Rosenzweig, M.R., Bennett, E.L., Martinez, J.L. Jr., Colombo, P.J., Lee, D.W., and Serrano, P.A. (1992). Studying stages of memory formation with chicks. In L.R. Squire and N. Butters (eds.), *Neuropsychology of Memory* (2nd ed.) (pp. 533–546). New York: Guilford Press.

Rosenzweig, M.R., Krech, D., and Bennett, E.L. (1958). Brain chemistry and adaptive behavior. In H.F. Harlow and C.N. Woolsey (eds.), *Biological and Biochemical Bases of Behavior* (pp. 367–400). Madison, WI: University of Wisconsin Press.

Rosenzweig, M.R., Krech, D., and Bennett, E.L. (1961). Heredity, environment, brain biochemistry, and learning. In *Current Trends in Psychological Theory* (pp. 87–110). Pittsburgh: University of Pittsburgh Press.

Rosenzweig, M.R., Krech, D., Bennett, E.L., and Diamond, M.C. (1962). Effects of environmental complexity and training on brain chemistry and anatomy: A replication and extension. *Journal of Comparative and Physiological Psychology, 55*, 429–437.

Rosenzweig, M.R., Love, W., and Bennett, E.L. (1968). Effects of a few hours of enriched experience on brain chemistry and brain weights. *Physiology and Behavior, 3*, 819–825.

Schwartz, J.H., Castellucci, V.F., and Kandel, E.R. (1971). Functioning of identified neurons and synapses in abdominal ganglia of *Aplysia* in absence of protein synthesis. *Journal of Neurophysiology, 34*, 939–963.

Searle, L.V. (1941). A study of the generality of inherited maze-brightness and maze-dullness. *Psychological Bulletin, 38*, 742.

Searle, L.V. (1949). The organization of hereditary maze-brightness and maze-dullness. *Genetic Psychology Monographs, 39*, 279–325.

Serrano P.A., Beniston, D.S., Oxonian, M.G., Rodriguez, W.A., Rosenzweig, M.R., and Bennett, E.L. (1994). Differential effects of protein kinase inhibitors and activators on memory formation in the 2-day-old chick. *Behavioral and Neural Biology, 61*, 60–72.

Serrano, P.A., Rodriguez, W.A., Bennett, E.L., and Rosenzweig, M.R. (1995). Protein kinase inhibitors disrupt memory formation in two chick brain regions. *Pharmacology Biochemistry and Behavior, 52*(3), 547–554.

Sherry, D.F., Vaccarino, A.L., Buckenham, K., and Herz, R.S. (1989). The hippocampal complex of food-storing birds. *Brain, Behavior and Evolution, 34*, 308–317.

Skaggs, W.E., and McNaughton, B.L. (1996) Replay of neuronal firing sequences in rat hippocampus during sleep following spatial experience. *Science, 271*, 1870–1873.

Skinner, B.F. (1938). *The behavior of organisms.* New York: Appleton-Century-Crofts.

Spear, N.F. (1974). Retrieval of memories: A psychobiological approach. In W.K. Estes (Ed.), *Handbook of learning and cognitive processes. Vol. 4. Attention and memory* (pp. 17–90). Hillsdale, N.J.: Erlbaum.

Tanzi, E. (1893). I fatti e le induzioni nell'odierna isiologia del sistema nervoso. *Revista Sperimentale di Freniatria e di Medicina Legale, 19*, 419–472.

Tempel, B.L., Livingstone, M.S., and Quinn, W.G. (1984). Mutations on the dopa decarboxylase gene affect learning in *Drosophila. Proceedings of the National Academy of Sciences, U.S.A., 81*, 3577–3581.

Teuber, H.-L. (1955). Physiological psychology. *Annual Review of Psychology, 6*, 267–296.

Thompson, R.F. (1990). Neural mechanisms of classical conditioning in mammals. *Philosophical Transactions of the Royal Society, London, Series B, 329*, 161–170.

Thompson, R.F., and Krupa, D.J. (1994). Organization of memory traces in the mammalian brain. *Annual Review of Neuroscience, 17*, 519–549.

Thorndike, E.L. (1898). Animal intelligence: An experimental study of the associative processes in animals. *Psychological Monographs,* No. 8, 1–109.

Thorndike, E.L. (1911). *Animal Intelligence: Experimental Studies.* New York: Macmillan.

Tolman, E.C. (1924). The inheritance of maze-learning ability in rats. *Journal of Comparative Psychology, 4*, 1–18.

Tolman, E.C. (1949). There is more than one kind of learning. *Psychological Review, 56*, 144–155.

Tolman, E.C., Tryon, R.C., and Jeffress, L.A. (1929). A self-recording maze with an automatic-delivery table. *University of California Publications in Psychology, 4*, 99–112.

Tolotschinoff, I.F. (1902). Contribution à l'étude de la physiologie et de la psychologie des glandes salivaires [Contribution to the study of the physiology and psychology of the salivary glands]. *Forhandlinga vid Nordiska naturforskare och lakermotet*, 42–46. Helsinki, Finland.

Treves, A., and Rolls, E.T. (1994). Computational analysis of the role of the hippocampus in memory. *Hippocampus, 4*, 374–391.

Tryon, R.C. (1940). Genetic differences in maze learning ability in rats. *Yearbook of the National Society for Studies in Education, 39*, Part I, 111–119.

Tryon, R.C. (1942). Individual differences. In F.A. Moss (ed.), *Comparative Psychology* (rev. ed.) (pp. 330–365). New York: Prentice-Hall.

Turner, A.M., and Greenough, W.T. (1985). Differential rearing effects on rat visual cortex synapses. I. Synaptic and neuronal density and synapses per neuron. *Brain Research, 329*, 195–203.

Underwood, B.J. (1969). Attributes of memory. *Psychological Review, 76*, 559–573.

Ungar, G., and Oceguera-Navarro, C. (1965). Transfer of habituation by material extracted from brain. *Nature, 207*, 301–302.

Ungar, G., Desiderio, D., and Parr, W. (1972). Isolation, identification and synthesis of a specific behaviour-inducing brain peptide. *Nature, 238*, 198–202.

Voneida, T.J. (1990). The effect of rubrospinal tractotomy on a conditioned limb response in the cat. *Society for Neuroscience Abstracts, 16*, 279.

Waldeyer-Hartz, W. von (1891). Ueber einige neuere Forschungen im Gebiete der Anatomie des Centralnervensystems. *Deutsche Medizinische Wochenschrift, 17*, 1213–1218, 1244–1246, 1267–1269, 1287–1289, 1331–1332, 1352–1356.

Watson, J.B. (1916). The place of the conditioned reflex in psychology. *Psychological Review, 23*, 89–116.

Weinberger, N.M. (1995). Dynamic regulation of receptive fields and maps in the adult sensory cortex. *Annual Review of Neuroscience, 18*, 129–158.

West, R.W., and Greenough, W.T. (1972). Effects of environmental complexity on cortical synapses of rats: Preliminary results. *Behavioral Biology, 7*, 279–284.

Whytt, R. (1763). *An Essay on the Vital and Other Involuntary Motions of Animals* (2nd ed.). Edinburgh, Scotland: John Balfour.

Wiesel, T.N., and Hubel, D.H. (1963). Single-cell responses in striate cortex of kittens deprived of vision in one eye. *Journal of Neurophysiology, 26*, 1003–1017.

Wiesel, T.N., and Hubel, D.H. (1965). Comparison of the effects of unilateral and bilateral eye closure on cortical unit responses in kittens. *Journal of Neurophysiology, 28*, 1029–1040.

Wilson, M.A., and McNaughton, B.L. (1994). Reactivation of hippocampal ensemble memories during sleep. *Science, 265*, 676–679.

Winocur, G. (1990). Anterograde and retrograde amnesia in rats with dorsal hippocampal or dorsomedial thalamic lesions. *Behavioural Brain Research, 38*, 145–154.

Wu, J.Y., Cohen, L.B., and Falk, C.X. (1994). Neuronal activity during different behaviors in *Aplysia*: A distributed organization? *Science, 263*, 820–823.

Yeo, C.H., Hardiman, M.J., and Glickstein, M. (1985). Classical conditioning of the nictitating membrane response of the rabbit. II. Lesions of the cerebellar cortex. *Experimental Brain Research, 60*, 99–113.

Zecevic, D., Wu, J.Y., Cohen, L.B., London, J.A., Hopp, H.P., and Falk, C.X. (1989). Hundreds of neurons in the *Aplysia* abdominal ganglion are active during the gill-withdrawal reflex. *Journal of Neuroscience, 9*, 3681–3689.

Zolman, J.F., and Morimoto, H. (1962). Effects of age of training on cholinesterase activity in the brains of maze-bright rats. *Journal of Comparative and Physiological Psychology, 55*, 794–800.

CHAPTER 2

Developmental Approaches to the Memory Process

Julie A. Markham

Beckman Institute, University of Illinois, Urbana, IL 61801

James E. Black

Department of Psychiatry, Southern Illinois School of Medicine, Springfield, IL 62794

William T. Greenough

Beckman Institute, University of Illinois, Urbana, IL 61801

I. INTRODUCTION

In the last half of the nineteenth century, scientists in fields ranging from histology to psychiatry were proposing that memory and development were intimately linked to subtle changes in neural processes. Because synapses were not visible with the microscopic techniques available at the time, these connections between neurons became the focus of much speculation. Ramon y Cajal (1893) suggested that learning might involve the formation of new synaptic connections between neurons. Tanzi (1893), noting that the resistance to transmission between neurons might vary with the size of the connection, proposed that frequent use of a synapse might produce growth similar to that produced by exercising a muscle, thereby strengthening preexisting connections. Both theories assumed that the structural plasticity seen in development extended into adulthood, a concept that was not demonstrated until nearly a century later.

Little theoretical or empirical progress was made along this line until Hebb (1949) suggested how experience could be represented in new or modified neural organizations. Rejecting the "one memory — one neuron" concept,

Hebb proposed that memory involved large structures of interconnected neurons, termed *cell assemblies*. Memory thus became a process more than a place, since encoding and retrieval depended on the cooperation of many neurons rather than a small subset. Hebb, who also felt that developmental plasticity and adult memory might share mechanisms, proposed what is now termed the *Hebb synapse*, a model synapse with a rule that concurrent pre- and postsynaptic activity increases synaptic efficacy. This basic concept of a cooperative set of modifiable connections as the basis of learning and memory continues to have substantial influence on neural network theory (McClelland and Rumelhart, 1988; Kesner and Rolls, 2001). Initially, alterations in neuronal structure were the focus of investigation; more recently, however, it has become clear that other nervous system components, such as macroglial cells and cerebrovasculature, also exhibit robust plasticity in response to experience. Regardless of the particular brain substrate under investigation, the goal of much scientific effort is to understand the degree to which adult memory processes are related to developmental plasticity (Yuste and Bonhoeffer, 2001).

II. EXPERIENCE-EXPECTANT AND EXPERIENCE-DEPENDENT NEURAL PLASTICITY

The organization of functional brain circuits is governed by the interplay of an individual's genetic makeup and his or her experiences. In Piaget's view (1980), an organism's interaction with its environment results in two types of experiences: those that are commonly shared by all members of a species (general information, or expected environmental features), and individual-specific experiences (unexpected environmental features), which allow the organism to adapt its behavior to its own unique surroundings. It has been proposed that these represent two types of brain information storage, termed *experience-expectant* and *experience-dependent* processes, respectively, and that they are supported by different brain mechanisms (Black and Greenough, 1986).

On the one hand, experience-expectant processes mean that the amount of information carried by the genome can be greatly simplified — if an individual member of a given species will almost certainly encounter particular features of the environment during its development, then the neural circuit can be genetically programmed to respond to those features, allowing experience of those features thenceforth to drive development of the system. In this case, the genome is programmed to anticipate particular experiences, which themselves must occur in order for proper development to take place. Thus experience-expectant information storage occurs during a brief window of time during development (termed a *critical* or *sensitive period*), which is both species and system specific, when the organism is optimally primed to respond to a particular type of information. Experience-dependent information, on the other hand, is idiosyncratic for each individual and thus is not able to be anticipated by evolution, so storage of this type of information is not programmed by the genome. Experience-dependent information storage, which includes what is commonly regarded as learning, can happen at any time across the life span and so does not involve a critical period. The brain substrates of experience-expectant and experience-dependent information storage may differ, although it seems likely that some mechanisms would be shared between them. We discuss what is known about these mechanisms following a brief description of the methods used in this research.

III. QUANTITATIVE METHODS IN DEVELOPMENTAL NEUROBIOLOGY

Nerve cell bodies and processes, glial cells, and the brain's vasculature are tightly packed and intertwined in all brain regions. Figure 2-1A is a drawing of a thin section of rat visual cortex in which all tissue components were stained. While larger dendrites, somata, nuclei, and some other features can be observed, it is impossible to identify other parts of the tissue that are associated with any particular cell. Stains such as that used in Figure 2-1A are useful

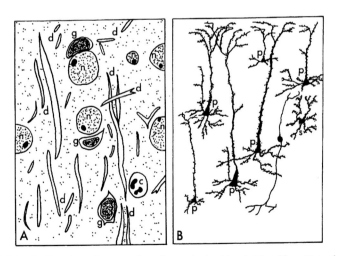

FIGURE 2-1   **A.** Drawing of a cortical section stained with toluidine blue. Note the tangled packing of glial somata (g), neuronal somata (n), dendrites (d), and capillaries (c). The density and size of the neuronal somata allow the cortex to be divided into six layers. **B.** Drawing at lower magnification of a Golgi-stained cortical section. Fewer pyramidal (p) and stellate (s) neurons are stained, with complete dendrites reaching across cortical layers. Unstained tissue is relatively transparent.

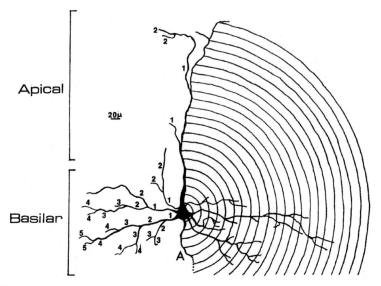

FIGURE 2-2 The amount of dendritic material for cortical neurons can be analyzed in two ways: (1) The intersections between dendrites and a superimposed series of concentric rings are counted. Here there are six basilar intersections at the fourth ring on the right. (2) The number and length of dendrites at each order are measured. Here there are six basilar segments of third order on the left. The apical dendrite is that emanating from the top, or apex, of the pyramid-like cell body of a pyramidal neuron. Basilar dendrites radiate from its base. A (axon).

for quantifying numbers of neurons, glial cells, blood vessels, etc.; however, since all of them are stained, they are all visible. Figure 2-1B is a drawing from a much thicker section stained via the Golgi method, named after its inventor, the turn-of-the-century anatomist Camillo Golgi, which stains only a few neurons in a region and thus allows their dendrites (and their axons to varying degrees) to be viewed. To obtain accurate light microscopic measurement of neuron morphology, the dendritic field must be traced completely without becoming lost in the tangle of other neural processes. A camera lucida can be used to collect data by superimposing the slide's image onto a two-dimensional drawing, or a computer-aided microscope can record the three-dimensional coordinates of points on the dendritic branches, storing a mathematical representation of the neuron.

Dendritic branches are commonly described in terms of order of bifurcation, as indicated in Figure 2-2. A first-order segment is defined as originating from the soma (cell body), and a second-order segment has its root in the forked end of a first-order segment. Another method of measuring dendritic trees uses a two-dimensional transparent overlay of concentric rings (Sholl, 1956) or concentric spheres for three-dimensional computer-microscope data. The fre-

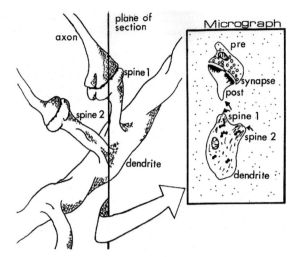

FIGURE 2-3   Drawings of a synapse before sectioning and of the electron microscope image of the tissue at the plane of section. One can count and classify synapses from micrographs as well as measure the synaptic cleft, postsynaptic thickening, and pre- or postsynaptic areas. Note how difficult it would be to identify which axon and dendrite a synapse belongs to by examining a single electron micrograph.

quency of ring intersections indicates dendritic volume distribution. Neurons stained using the Golgi method also allow analysis of dendritic spine numbers, their sizes, and their shapes. Spines are small postsynaptic extensions on the dendrites of many kinds of neurons (see Figure 2-1B and Figure 2-3). The size and shape of spines may affect their conductive properties with regard to the effectiveness of transmission from one neuron to another, and hence the possible modification of spine structure by learning has been investigated in several studies (Chen et al., 2000).

Electron microscopic studies are useful for measurement of the number of synapses and their size and other characteristics. Figure 2-3 illustrates the appearance of synapses in an electron micrograph. Since the thin electron microscopy section cuts through synapses at arbitrary locations, statistical corrections of size and shape must be made. Corrections of apparent synapse density (counts on photomicrographs) must also be made to account for the synapse size distribution, for failure to recognize synapses just barely in the section plane, and for other distorting factors. Mathematically unbiased procedures termed *stereological methods* must be used to provide accurate estimates of synapse numbers. The result of these corrections, an accurate density estimate (e.g., number of synapses per unit volume), may still be misleading, since various manipulations may change the overall size of the brain region being investigated, such as the visual cortex. When this so-called "reference volume" cannot easily be determined, for comparison with Golgi data the best measure using electron micrographs is probably the number of synapses per neuron (obtained using the ratio of synapses per unit volume to neurons per unit volume). Electron microscopic estimates of the number of synapses per neuron correspond well to estimates obtained from quantitative studies of Golgi material (Turner and Greenough, 1985; Sirevaag and Greenough, 1987). The electron microscope can also be used as a tool to investigate developmental and experience-induced changes in astrocytes, oligodendrocytes, myelin, and cerebrovasculature.

The technique of two-photon laser scanning fluorescent microscopy was first developed by Denk, Strickler, and Webb (1990). In recent years a number of investigators have employed this technique to engage in the repeated imaging of dendritic spines in the live rodent. Two-photon imaging involves the excitation of a fluorescent marker (termed a *fluorochrome*) by two photons of longer-wavelength light, rather than a single photon of shorter wavelength (as in one-photon fluorescence imaging). The value of the technique lies in the fact that the excitation of the fluorochrome is limited to a small focal point (rather than the entire sample), thereby greatly reducing the photobleaching and photodamage that have been the major drawbacks of one-photon fluorescence imaging, preventing longitudinal studies. Additionally, the longer wavelength scatters less, allowing for much deeper optical sections to be obtained through the tissue than was previously possible. Cells in cultured brain slices can be labeled with a special protein, such as GFP (green fluorescent protein), which will fluoresce when excited by the proper wavelength (Mainen et al., 1999). To achieve imaging in the live animal, neurons can be labeled in the cortical region(s) of interest while the animal is anesthetized or, alternatively, transgenic mice that express GFP in a small percentage of neurons can be used. In either case, neurons residing in layer II/III or V are typically labeled, because their apical dendrites extend up to layer I (layer I is conveniently located on the outer surface of the cortex but does not itself contain pyramidal neurons), where they can be viewed through a coverslip carefully secured in place of a portion of removed skull (Chen et al., 2000).

IV. NEUROBIOLOGICAL CORRELATES OF THE LEARNING PROCESS

Use of these techniques following early sensory deprivation, general enrichment of the environment, and training tasks has provided evidence that changes in both numbers of synapses and structural characteristics of synapses, as well as changes in nonneuronal components of the nervous system, may be involved in the storage of experiential information. Remarkably similar but still distinct effects have been seen with manipulations directed at the experience-expectant and experience-dependent types of information storage.

A. Early Sensory Deprivation

Because experience-expectant information storage occurs during a sensitive developmental time window, or critical period, the processes underlying such storage can be investigated by withholding the relevant experience during this window. For instance, behavioral and structural effects of early light and visual pattern deprivation have been reported in most mammalian species tested. The effects are most pronounced with monocular deprivation in species with binocularly overlapping visual systems, such as cats and monkeys, but significant effects also occur in largely nonoverlapping species, such as rats.

Animals raised in total darkness are impaired in behavior requiring vision. For example, Walk and Walters (1973) showed that dark-reared animals have long-lasting deficits on a shallow visual cliff, where animals can unwisely choose to step off a small platform. Cats and rodents reared in darkness or unpatterned illumination are slower to learn complex visual discriminations, such as an X versus an N (Riesen, 1965; Tees, 1968). This effect is related to difficulties in visual processing rather than to some general learning disability, since dark-reared animals have no problem associating two auditory stimuli but do have difficulty with an auditory-visual stimulus pair (Tees and Cartwright, 1972). These studies suggest that visual learning of adult animals is impaired if early visual experience has not established effective information-processing schemes.

Depriving a cat of light to one eye (monocular deprivation) during the critical period for visual cortical development results in structural and functional changes in the visual cortex (Wiesel and Hubel, 1965a, 1965b). If this deprivation is maintained for long enough, then even after the animal's eye is opened it will remain blind in that eye, even though its retina is still capable of transmitting visual information. How does this happen? Prior to deprivation, there is an equal number of cells in the visual cortex that "expect," or are programmed to respond to, visual input from the left and right eyes. However, for an animal that is monocularly deprived, the expected experience of visual input from both eyes does not occur. Visual cortical organization thus gradually begins to shift in favor of cells responding to the nondeprived eye, and synapses carrying information from the nondeprived eye are maintained, at the expense of synapses from the deprived eye. By the time the deprived eye is allowed to open, few neurons remain capable of responding to the information it is sending.

1. Synapses

What are the neurobiological correlates of impaired visual processing observed in light-deprived animals? Monocular deprivation studies in animals with extensive binocular overlap of the visual fields have indicated involvement of competitive processes in establishing synaptic connections (LeVay et al., 1980).

For instance, cells that respond to input from the left versus the right eye are spatially segregated into neighboring "ocular dominance columns" in the adult animal. When an animal experiences monocular deprivation during the critical period for the development of these columns, however, the functional blindness of the deprived eye is associated with a narrowing of its cortical columns, whereas the cortical columns of the nondeprived eye expand in width (LeVay et al., 1980). This is due to the enhanced regression of axon terminals in columns of the deprived eye coupled with attenuated regression of axon terminals in the columns for the other eye (Antonini and Stryker, 1993). Furthermore, the average axon from a deprived eye has fewer synaptic terminals and smaller pre- and postsynaptic components than those from an experienced eye (Tieman, 1984, 1985). Findings such as these have been attributed to competition between the axon terminal fields, with the more active connections winning out over the deprived. Activity-dependent synaptic competition also underlies the normal development of ocular dominance columns, as visual cortical neurons initially receive inputs from both eyes, and the columns develop over time (LeVay et al., 1978; LeVay et al., 1980). Thus inappropriate connections are routinely made early in development of the visual system that, with continued normal visual experience, are eventually pruned (reviewed in Katz and Shatz, 1996).

Other changes in the visual cortex of light-deprived animals include a reduction in the dendritic fields of pyramidal neurons (Coleman and Riesen, 1968; Valverde, 1971) and potentially delayed development of dendritic spines. Freire (1978) examined the spines of apical dendrites in layer IV of occipital cortex of 19-day-old dark-reared and normal mice with serial-section electron microscopy. The three-dimensional reconstructions indicated that spine development progressed from small spines with no spine apparati (a characteristic structure typically more common in larger spines) to large spines with extensive spine apparati. Dark-reared mice had more of the small-type spines, while normally reared animals had more of the large-type spines. Electron microscopy studies also suggest substantial changes in the number of synapses after visual deprivation. Cragg (1975) found that light-experienced kittens had about 40% more synapses per neuron than binocularly deprived kittens. These findings are compatible with the aforementioned Golgi studies, which suggest a reduction in the number of synapses per neuron with visual deprivation. It is possible that deprivation causes a maturational lag in the formation of synapses for which compensation can later occur. For example, Winfield (1981) showed that binocularly sutured cats eventually catch up with normally reared cats in synapses per neuron, and others have similarly shown that differences between dark- and normal-reared animals become smaller as the animals get older (Cragg, 1967; Valverde, 1971).

Using a more selective type of visual deprivation, it has been found that kittens exposed only to horizontal or vertical stripes during development have visual cortex neurons that respond selectively to visual stimuli of the exposure orientation (Hirsch and Spinelli, 1970). Layer IV stellate cells of visual cortex from horizontal- and vertical-stripe-raised kittens were not found to differ in dendritic length or number of branches, but the angular distribution of distal dendritic segments were at approximatly 90° from each other, just as their stimuli were at right angles (Coleman et al., 1981). Tieman and Hirsch (1982) similarly reported that stripe-rearing modifies the dendrite orientation of Layer III pyramidal cells of kitten visual cortex — horizontal-stripe-reared cats and vertical-stripe-reared cats had approximately perpendicular distributions of dendritic orientation, suggesting a specific relationship between the morphology of Layer III pyramidal cells, their physiological orientation, and their early experience.

Visual deprivation studies collectively indicate that early visual experience substantially affects experience-expectant neural plasticity and that these effects impair later experience-dependent visual learning. Functional and neurobiological changes in the brain in response to sensory deprivation are not limited to the visual system. In fact, it appears that competition for survival between synapses is the rule rather than the exception in brain development and that such competition is what drives the initial formation of mature neural circuits in the normal individual. In addition to the work conducted in the auditory and olfactory systems (e.g., Coss et al., 1980; Feng and Rogowski, 1980), a large literature also exists using somatosensory deprivation as a model system in which to investigate experience-expectant plasticity, primarily using the whisker system of rodents and the corresponding barrel cortex (in which each whisker is represented by a highly specialized functional unit, termed a *barrel*) (Fox, 2002). The topographic map linking one whisker to a single barrel in the somatosensory cortex that is maximally responsive to it develops early, perhaps before postnatal day 5, and is resistant to disruptions after this time (O'Leary et al., 1994; Shepherd et al., 2003). Other aspects of barrel cortical development continue to mature at later time points; in fact, most of the synaptic circuitry in this cortical region occurs after the barrels are formed (see Shepherd et al., 2003). For instance, synaptic responses of layer II/III neurons to whisker deflection are undetectable at postnatal day 12; however, two days later, on day 14, receptive fields of these cells are of a mature organization (Stern et al., 2001). Sensory deprivation prior to day 14 disrupts receptive field organization. The same study reported that receptive field properties of layer IV neurons, on the other hand, are mature by day 12 and are actually unaffected by deprivation at this time. This example illustrates that multiple critical periods exist in the barrel cortex, the timing of which can be cortical layer dependent.

Recently, it has become possible to monitor changes in dendritic spine shape and mobility over relatively long periods of time following sensory deprivation, both in the live animal and in cultured tissue, through the use of two-photon fluorescent imaging. On the whole, these studies confirm the findings obtained using traditional neuroanatomical quantification techniques regarding deprivation-induced neuronal plasticity. Just as with the studies discussed earlier, when approaching this literature it is important to keep in mind when the deprivation occurred — whether it was prior to, during, or after the critical period for the development of that system (whether visual or somatosensory).

In the mouse visual cortex, Grutzendler and colleagues (2002) have repeatedly imaged the same dendritic segments over a number of weeks and found that highly transitory, filopodia-like dendritic protrusions are abundant in young animals but nearly absent in the adult. During the critical period for visual cortical development, the majority of observed changes were due to spine elimination, consistent with activity-dependent refinement of circuitry, which is known to occur during this time. In contrast, the overwhelming majority (about 96%) of spines in adult mice were stable over the one-month period. Similarly, Majewska and Sur (2003) found that spine motility was high at young ages, decreased between postnatal days 21–28 [the height of the critical period (Gordon and Stryker, 1996)], and then remained stable through day 42. Binocular deprivation before eye opening (which occurs on day 13) greatly increased the motility of spines during the critical period (day 28) but not at the beginning of (day 21) or after (day 42) the critical period. On the other hand, they did not observe high rates of spine turnover during the study's two hours of observation, and there was no difference in this measurement between groups, although it is not clear if the trend toward increased turnover in deprived mice at the critical period might have reached statistical significance had the imaging observation window exceeded two hours (Majewska and Sur, 2003).

Similar to the pruning of spines during the critical period in the visual cortex (Majewska and Sur, 2003), a recent study using in vivo imaging of spines also found spine turnover on the apical tufts of layer II/III somatosensory neurons, but a net loss of spines between postnatal days 16–25 (Holtmaat et al., 2005). The degree of spine turnover gradually reduced over development, and the proportion of stable spines (those with a lifetime of 8 days or more) gradually increased over development (Holtmaat et al., 2005). Zuo et al. (2005b) confirmed this net loss of spines (13–20% reduction compared to 5–8% formation) over two weeks in the barrel cortex and also in motor and frontal cortices, indicating that synaptic pruning occurs in many different cortical areas during this time. In comparison, only 3–5% of spines were either eliminated or formed over a two-week period during adulthood; and after an amazing 18 months, only 26% were observed to be eliminated and 19% formed in adult barrel cortex (Zuo et al., 2005b). It is important to remember that these mice were housed in relatively uninteresting environments during their lifetimes and did not experience any sensory deprivation; manipulations of either sort would be expected to alter spine numbers considerably over time.

Interestingly, the results of a recent in vivo imaging study suggest that sensory deprivation may impair activity-dependent pruning in the somatosensory cortex. In adolescent mice, whisker trimming prevented the loss of spines on layer V apical dendrites, which normally occurs during this time period, by reducing the rate of spine elimination (rather than increasing the formation of new spines) (Zuo et al., 2005a). In another study, Trachtenberg and colleagues (2002) repeatedly imaged dendritic spines on layer V pyramidal neurons of mice aged 34–74 days, both prior to and following selective unilateral whisker trimming. Deprivation increased the proportion of spines that were transient (present for a single day or less) and decreased the proportion of spines that remained stable over several days, indicating that, in addition to spine motility, spine turnover is heavily influenced by sensory experience. At the conclusion of the imaging portion of the experiment, they processed the brain tissue for electron microscopy and confirmed that the observed growth of new spines corresponded to synaptogenesis and that the retraction of spines corresponded to synapse elimination (Trachtenberg et al., 2002). The authors interpreted the deprivation-induced increased turnover of spines as a response to the destabilization of previously stable synapses.

Collectively, these findings indicate that spine motility is greatest when input is either immature or not present and that synaptic activity stabilizes spines. In support of this, in vitro work has shown that spines are stabilized by calcium influx into the spine head following synaptic activation by either AMPA or NMDA receptors (Fischer et al., 2000; Korkotian and Segal, 2001). It is important to note, however, that not all studies find that deprivation increases spine motility. For instance, Lendvai and colleagues (2000) also found that spines and filopodia of layer II/III neurons imaged in vivo were highly motile. However, in this case, trimming of all major whiskers (on one side) between postnatal days 11–13 was found to *reduce* the motility of spines by approximately 40% but was without effect in younger (days 8–10) or older (days 14–16) animals. The differences between Lendvai et al. (2000) and Trachtenberg et al. (2002) may be due to methodological discrepancies: Different cortical layers were imaged, different ages were examined, and the pattern of whisker deprivation differed, with Lendvai et al. trimming all major whiskers and Trachtenberg et al. trimming approximately half of the whiskers, in a "chessboard" pattern of deprivation. Future studies are necessary to resolve these differences. Regardless of the details, the in vivo imaging studies to date are generally compatible with static anatomy results, in that they suggest activity-dependent mechanisms regulate aspects of synapse development, especially during a critical period. Static anatomical measurements do appear to have underestimated the degree of synapse elimination/turnover that may be involved in generating and maintaining such activity-dependent connections. Nevertheless, findings resulting from both traditional neuroanatomical techniques and in vivo imaging support a model in which synaptogenesis early in

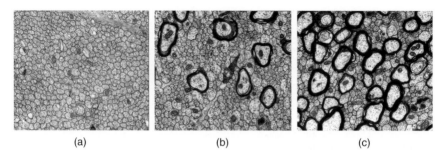

(a)   (b)   (c)

FIGURE 2-4   Electron micrographs taken from the splenium of the rat corpus callosum (bar = 0.6 μm). At postnatal day 15 (a), just prior to the onset of myelination, many unmyelinated axons can be observed. At day 25 (b), myelination is under way and myelin rings can be observed around several large axons here. By two months of age (c), which corresponds to young adulthood, the amount of area occupied by myelinated axons rivals or exceeds that occupied by unmyelinated axons, although the difference is due to the greater size of myelinated axons rather than number. [From Kim and Juraska (1997). Reprinted with permission from Developmental Brain Research.]

development is followed by activity-dependent pruning of connections, with spine synapses become permanent through transformations that are evident in morphology.

2. Glia and Cerebrovasculature

In addition to synaptic changes, there is good evidence that the myelination process, whereby axons are covered in a fatty insulating sheath, is also sensitive to experience during critical periods of development. Myelinated axons can conduct action potentials at rates that are 50–100 times greater than those accomplished by their unmyelinated counterparts; thus proper myelination is an important part of information processing in the nervous system. The number of myelinated axons increases greatly over development and continues (at a slower rate) into adulthood (Figure 2-4), raising the possibility that the myelination process may be sensitive to experience across the life span. Visual deprivation results in reduced myelination of the optic nerve (Gyllensten and Malmfors, 1963), whereas, conversely, premature eyelid opening accelerates the onset of myelination in this structure (Tauber et al., 1980). Although not all studies have found a relationship between early visual experience and myelination in the optic nerve (see Moore et al., 1976; Fukui et al., 1991), there is additional evidence using other techniques for a relationship between axonal activity and myelination. For example, Demerens and colleagues (1996) have shown that experimentally increasing axonal activity promotes optic nerve myelination, and blocking of action potentials with tetrodotoxin inhibits myelination. Furthermore, cortical axon branch formation as well as proliferation of precursors to oligodendrocytes (the glial cells responsible for myelin formation) are both activity-dependent processes (Barres and Raff, 1993; Uesaka et al., 2005). A number of signaling molecules are known to be released by axons; however, the molecular mechanisms responsible for inducing activity-dependent myelination are essentially unknown. Potential activity-dependent initiators of myelination include the neurotransmitters glutamate and γ-aminobutyric acid (GABA), receptors for which are expressed by oligodendrocytes (Butt and Tutton, 1992), adenosine triphosphate (ATP) (Fields and Stevens, 2000), and neuregulin (Taveggia et al., 2005). Regardless of the mechanism, the fact that early experience can influence myelination indicates that oligodendroglia are sensitive to manipulations of expected features of the environment during a critical window of development.

Maturation of astrocytes, another type of macroglial cell in the brain, is also disrupted in response to sensory deprivation during a critical period. Dark rearing reduces the size and number of astrocytes in the visual cortex (Gabbott et al., 1986; Stewart et al., 1986; Muller, 1990; Argandona et al., 2003). Hawrylak and Greenough (1995) also found that monocular deprivation during the critical period reduces the surface density of astrocytic processes as well the ratio of these processes to neurons; however, deprivation during adulthood was without effect on these measures. Although the plasticity of astrocytes in response to visual deprivation has received less attention than that of neurons, astrocytic plasticity in fact appears to play an important role in the formation of ocular dominance columns. The evidence for this comes from studies in which astrocytic function was impaired, either by a suppression of metabolic activity of these cells (Imamura et al., 1993) or by a disruption of the astrocytic-specific protein S-100β (Muller et al., 1993), both resulting in an unusually high proportion of cells that stubbornly maintained binocular responses following monocular deprivation, indicative of a failure in activity-dependent selection of connections. Similarly, astrocyte–neuron interactions appear to be crucial to the maturation of the somatosensory system. For instance, whisker stimulation–induced neuronal uptake of glucose (an indication of activity) is greatly impaired in the barrel cortex of mice lacking a functional gene for either of the two glial glutamate transporters (GLT-1 or GLAST knockout mice) (Voutsinos-Porche et al., 2003a). Glial glutamate transport is involved in triggering enhanced glucose utilization by neurons and in other aspects of metabolic crosstalk between neurons and glia (Voutsinos-Porche et al., 2003b).

Finally, there are other changes that occur in response to the reduction in metabolic demands that accompany the loss of sensory experience in the brains of light-deprived or whisker-trimmed animals. Argandona and Lafuente (1996) have observed that dark rearing reduces the density of cerebrovascular elements in the visual cortex. They also report that normal vascular development in the visual cortex involves the early presence of a number of vertically oriented vascular "trunks," the density of which is greatly decreased after birth in light-reared rats but remains elevated in dark-reared rats (Argandona and Lafuente, 1996). More recently, these authors have employed unbiased stereological techniques to determine that, despite the reduced density of blood vessels, the ratio of vessels per neuron to vascular area per neuron remains stable in the visual cortex of dark-reared animals (neuronal density was also reduced in the dark-reared animals, possibly due to increased apoptosis in these animals) (Argandona and Lafuente, 2000). Thus the changes in the vasculature in the visual cortex in response to deprivation are consistent with reduced metabolic demand.

It is clear that experience-expectant mechanisms have a complex role. For example, in the mammalian visual system, the simple presence of light does not trigger complete maturation of the visual system. Rather, coherent experience of a range of features and relationships is required. The requirement for early coherent experience ultimately affects experience-*dependent* processes, since, as discussed earlier, young animals deprived of sensation in a given modality are impaired as adults in learning situations requiring that modality.

B. Manipulation of the Complexity of the Environment

The initial overproduction of synapses that occurs in many brain areas during development and that is part of the brain's experience-expectant plasticity allows for subsequent experience-dependent remodeling of synaptic architecture that is exquisitely tailored to allow optimal functioning in the individual's own unique environment. A large body of work has been conducted to elucidate the morphological changes in the brain that accompany such experience-dependent information storage. Much of this work has come from studies conducted on rodents raised in a complex (also called "enriched," relative to standard lab cages), stimulating environment. The complex-environment paradigm, pioneered by Donald Hebb and his students (e.g., Hebb, 1949; Forgays and Forgays, 1952; Hymovitch, 1952), involves housing a group of animals together in a large cage containing numerous toys, such as balls, tunnels, and ladders, which are changed daily to provide a continuously stimulating environment (Figure 2-5). Comparisons of both behavior and brain structure are then made between animals housed in a complex environment (EC), those housed in a normal cage in isolation (IC), and those housed in a normal cage but with other animals (SC, or social condition). Importantly, as we will discuss, components of the nervous system other than synapses (such as glial cells and vasculature) are also sensitive to experience.

The behavioral effects of differential rearing are profound. It was Hebb (1949) who first reported that rats raised as pets at home were cognitively superior to laboratory rats. In general, rats raised in complex laboratory envi-

FIGURE 2-5   In a complex environment, each rat has many opportunities to interact socially with other rats, explore novel arrangements of barriers and ramps, and play with toys that roll, make noises, smell, etc. [From Black and Greenough (1986). Copyright 1986, reprinted with permission from Lawrence Earlbaum Assoc.]

ronments have been found superior to isolated or socially raised rats on memory tests, including the Hebb–Williams maze (Mohammed et al., 1986; Galani et al., 1997), the Morris water maze (Whishaw et al., 1984; Mohammed et al., 1990; Leggio et al., 2005), and the radial arm maze (Galani et al., 1998). Greenough, Wood, and Madden (1972) argued that the information-processing capability of EC mice was superior to that of IC or SC mice because they were uniquely capable of mastering the difficult Lashley III maze when trials were run immediately after one another. While learning a Hebb–Williams maze, ECs make fewer errors than IC rats. But that superiority vanishes if the maze is rotated, effectively disrupting extramaze cues (Hymovitch, 1952; Brown, 1968). Ravizza and Herschberger (1966), however, found EC rats better at maze learning even if extramaze cues were hidden by a curtain, suggesting that intramaze cues can also be better utilized by ECs in the absence of extramaze cues. Interestingly, the superior performance of EC animals is not due simply to greater visual experience, since Krech, Rosenzweig, and Bennett (1962) found blinded EC rats superior to blind IC rats in maze performance.

1. Synapses

Bennett, Diamond, Krech, and Rosenzweig (1964) first reported that several cortical regions were heavier and thicker in EC rats than in IC rats, particularly

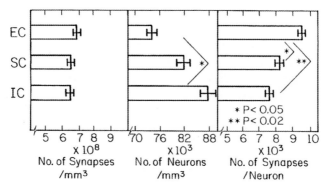

FIGURE 2-6   Numerical density of synapses and neuronal nuclei and ratio of synapses to neurons in upper occipital cortex (layers I–IV) of rats reared from 23 to 55 days of age in environmental complexity (EC), in pairs in social cages (SC), or in individual cages (IC). Synapses were counted in conventionally stained (osmium–uranyl–lead) electron micrographs and the counts corrected for differences in size using stereological formulae. Neuronal nuclei were estimated by point counting in toluidine blue–stained light microscopic sections. Reduced neuronal density in more experienced animals reflects the greater volume of neuronal processes, glia, vasculature, neuronal somata, etc. that accompanies the new synapses. [Data from Sirevaag and Greenough (1985); figure from Turner and Greenough (1985). Copyright 1985, Elsevier Science Publishers, reprinted with permission.]

the occipital cortex. Since then it has been shown that animals raised in EC have greater dendritic arborization, increased dendritic spine density, and more synapses per neuron in a number of brain areas as compared with IC animals (reviewed by Markham & Greenough, 2004). Holloway (1966) first reported that ring analysis (Figure 2-2) of visual cortical neurons indicated larger dendritic fields in rats reared in EC. Greenough and Volkmar (1973) placed rats in the EC, SC, and IC conditions at 23–25 days of age for 30 days. Pyramidal neurons from Layers II, IV, and V and Layer IV stellates in visual cortex had more ring intersections in ECs than in ICs. These effects were most pronounced in higher-order branches — the outer part of the dendritic field. To confirm that these dendritic differences reflected synapse number differences, Turner and Greenough used electron microscopy to show that EC rats exceeded ICs in synapses per neuron in upper visual cortex by roughly the amount predicted from the Golgi studies, with SCs intermediate but somewhat closer to the ICs (Figure 2-6) (Turner and Greenough, 1983; 1985). Thus quantitative Golgi procedures appear to accurately indicate differences in synaptic numbers. Dendritic elaboration (e.g., Greenough et al., 1973; Kolb et al., 2003; Leggio et al., 2005) and increased spine density (Moser et al., 1994) as a result of EC also occurs in other neocortical areas and in the dentate gyrus and area CA3 of the hippocampus, although interestingly the direction of the changes in dendritic arbor in the latter has been found to vary by sex (Juraska et al., 1985, 1989). The increase in dendritic length can be detected after as few as four days in EC in the visual cortex (Wallace et al., 1992), and it contributes to the greater thickness of the visual cortex among EC animals that was initially reported (Bennett et al., 1964). The fact that EC induces plasticity in nonvisual cortical areas indicates that the effects do not merely result from visual stimulation. On the other hand, EC-induced dendritic elaboration is not a ubiquitous phenomenon, for Greenough and colleagues (Greenough et al., 1973) did not find this effect in frontolateral (sensorimotor) cortex (but did in visual and auditory cortices), suggesting that general hormonal or metabolic factors, which would be expected to affect all cortical areas, do not play a significant role in mediating EC-induced neuronal plasticity. The lateralized effects of training on the cortex (discussed later) also argue against general hormonal and metabolic effects. (For a more comprehensive review of this topic, see Grossman et al., 2002.) Finally, reports of EC–IC differences in cerebellar cortex (Floeter and Greenough, 1979; Pysh and Weiss, 1979), superior colliculus (Fuchs et al., 1990), and striatum (Comery et al., 1995) indicate that the experience-dependent effects of rearing complexity are not restricted to phylogenetically newer brain structures.

In addition to inducing the formation of new synapses, manipulations of environmental complexity can also modify the morphology of existing synapses (or, alternatively, induce the formation and/or loss of synapses exhibiting particular characteristics) (reviewed by Greenough and Chang, 1988). The morphology of dendritic spines and synapses has been shown to be important for their function, including synaptic efficacy, conducive properties, and biochemical compartmentalization (Sorra and Harris, 2000; Tsay and Yuste, 2004; Noguchi et al., 2005); thus interest in these factors stems from the idea that they may ultimately play a role in learning and memory. EC animals have, on average, larger pre- and postsynaptic components, including postsynaptic densities (PSD) and cross-sectional area of presynaptic vesicle aggregate profiles (West and Greenough, 1972; Diamond et al., 1975; Sirevaag and Greenough, 1985, 1987; Turner and Greenough, 1985). Reflective of the maturational state of a spine on a cortical neuron, spine shape changes in similar ways (from the initial sessile shape, to exhibiting a clearly discernible head or neck, and finally to the large mushroom shape with a mature spine apparatus) over development (Galofre and Ferrer, 1987) and in response to EC (Sirevaag and Greenough, 1985) and long-term potentiation (LTP, a synaptic model for memory) (Chang and Greenough, 1984). The proportion of perforated synapses (those in which the PSD has enlarged and assumed a more complex shape, such as a horseshoe or doughnut) also increases in response to EC (Greenough et al., 1978; Jones and Calverley, 1991) as well as in the hippocampus in response to kindling or LTP induction (Geinisman et al., 1990, 1991). Perforated synapses are characterized by the incorporation of greater numbers of glutamate receptors (both AMPA and NMDA subtypes) in the PSD, which may enable them to evoke larger postsynaptic responses (relative to nonperforated synapses) and thereby enhance synaptic plasticity (Ganeshina et al., 2004a, 2004b).

On the presynaptic side, boutons in animals exposed to EC are more concave than those in IC rats, with SC rats being intermediate (Wesa et al., 1982). EC also increases the number of multiple synapse boutons (MSBs; two postsynaptic contacts innervated by the same presynaptic varicosity). Comery et al. (1996) reported 60% greater density of multiple-headed dendritic spines on spiny neurons in the striatum of EC as compared to IC rats. Similarly, the number of MSBs per neuron that contacted both a dendritic spine and a dendritic shaft were greatly increased in layer IV of the visual cortex of rats exposed to EC for 60 days as compared to either SC or IC controls (Jones et al., 1997). From these examples it is clear that the formation of novel dendritic contacts onto existing axonal boutons or varicosities is a common form of experience-driven synaptic plasticity, one that would seem to alter the efficacy of a preexisting pathway rather than creating novel connections (further discussed later, in subsection C: Skill Learning).

2. Glia and Cerebrovasculature

The differences in synaptic number and morphology in response to a complex environment are accompanied by differences in supportive tissue components, such as glial cells and blood vessels. Some early studies indicated that EC resulted in changes in astroyctic morphology (Diamond et al., 1964; Szeligo and Leblond, 1977); since that time, EC-induced increases in astrocytic cell size (hypertrophy) and number (hyperplasia) have been confirmed using unbiased stereological techniques (Sirevaag and Greenough, 1987, 1991) (reviewed in Jones, 2002). In general, it has been found that morphological plasticity of astrocytes in response to EC occurs on a time scale that is comparable to neuronal changes observed in this paradigm (Jones et al., 1996; Sirevaag and Greenough, 1985; Jones and Greenough, 1996), raising the possibility that experience-dependent changes in nonneuronal components of the nervous system could contribute to the behavioral changes observed in animals exposed to EC.

Plasticity of astrocytes is gaining more attention because there is increasing evidence for their role in synaptic function (see Volterra et al., 2002). Interestingly, the degree of synaptic ensheathement by astrocytic processes is increased by exposure to EC (Figure 2-7) (Jones and Greenough, 1996). This experience-dependent enhancement of astrocytic-synaptic communication is an important finding in light of the fact that perisynaptic astrocytes modulate synaptic transmission in response to synaptically released neurotransmitters and, in fact, themselves release neurotransmitters (Oliet et al., 2001; Zhang

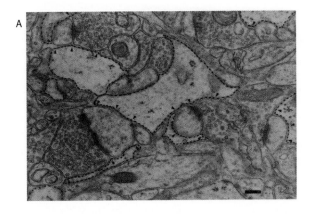

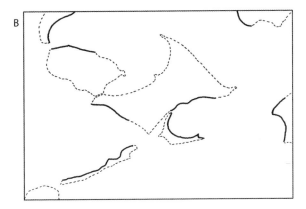

Figure 2-7   Astrocytic processes (dashed outline) within layer IV of the rat visual cortex as revealed by electron microscopy (A) and a tracing of these processes (B). Processes in direct apposition to synaptic elements are indicated (arrows, A; solid lines, B). Scale bar, 0.2 μm. [Reprinted with permission from T.A. Jones and Greenough (1996).]

and Haydon, 2005). Astrocytes are also involved in GABA and glutamate reuptake and metabolism (Schousboe et al., 1992; Bezzi et al., 1999) and can conduct excitation via propagated Ca^{2+} waves, which can directly influence neuronal activity (reviewed by Zhang and Haydon, 2005). Astrocytic coverage of synapses may thus also serve to enhance the input specificity of information, thus facilitating learning and memory.

Szeligo and Leblond (1977), who were the first to examine the influence of rearing environment on brain fiber tracts, found increases in oligodendrocytes in the visual cortex of EC rats. Subsequently, Sirevaag and Greenough (1987) also found the volume fraction of oligodendrocyte nuclei in the visual cortex to be greater among EC-raised rats. The influence of developmental experience on oligodendrocytes is not limited to the visual cortex. An electron microscopic study (Juraska and Kopcik, 1988) found that raising rats in EC increases the number of myelinated axons in the splenial portion of the corpus callosum, which contains axons of visual cortical neurons that carry information between the two cerebral hemispheres. The positive effect of a complex rearing environment on the size of the corpus callosum (size of a fiber tract is typically correlated with the degree of myelination) has also been demonstrated in rhesus monkeys (Sanchez et al., 1998).

Finally, the brain's vasculature is also quite responsive to manipulations of environmental complexity. Animals raised in EC have larger and more elaborately branched capillaries in the visual cortex, compared to both IC- and SC-raised animals (Black et al., 1987; Sirevaag et al., 1988; Black et al., 1991). Also, the plasticity of cerebrovasculature in response to behavioral demands appears to be far greater than that of synapses — volume fraction of capillaries (which combines diameter and density effects) nearly doubles following EC exposure (Black et al., 1987; Sirevaag et al., 1988). This effect is likely more related to satisfying the increased metabolic demands of neurons (undergoing dendritic arborization and activity-dependent synaptogenesis) and glial cells (undergoing hypertrophy, hyperplasia, and, in some cases, myelin synthesis) in response to new behavioral demands than to accomplishing new memories per se. The permanence of EC-induced angiogenesis has not been examined, but the effect is likely to be transient, for exercise-induced alterations in the brain's vasculature do not persist for very long beyond the period of increased physical activity (Rhyu et al., 2003). Because, like exercise, experimentally induced hypoxia induces rapid angiogenesis (Harik et al., 1995), indicators of blood oxygen levels and/or a related metabolic demand may serve as physiological signals that trigger vascular proliferation. Although the precise signals are unknown, it is known that, similar to synaptogenesis, angiogenesis in response to experience is greatest during development, also occurs during adulthood, and remains present, although diminished, during aging (Black et al., 1989).

3. Plasticity in the Adult Brain

The neocortex retains considerable structural plasticity in response to differential housing into adulthood, as we would expect for experience-dependent mechanisms designed for learning. EC/IC branching differences of 10% or more, nearly equivalent in magnitude to those described in animals exposed as weanlings, have been found in the adult visual cortex (Greenough and Volkmar, 1973; Uylings et al., 1978; Juraska et al., 1980). Briones et al. (2004) have observed that rats exposed as adults to EC for either 30 or 60 days had significantly more synapses per neuron in layer IV of the visual cortex than did IC animals of the same age, as revealed by electron microscopy (Figure 2-8). The increased synapse number was not diminished by a subsequent

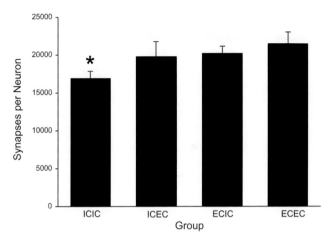

FIGURE 2-8   The EC-induced increase in the number of synapses per neuron in the adult rat visual cortex persists for at least 30 days after animals are removed from a complex environment (EC). ICIC animals were individually caged (IC) for 60 days and were significantly different (*, $p < 0.05$) from each of the three other groups: ICEC animals (housed in IC for 30 days followed by EC housing for 30 days), ECIC animals (housed in EC for 30 days followed by IC housing for 30 days), and ECEC animals (housed in EC for 60 days). [Modified from Briones et al. (2004), with permission.]

period of 30 days of IC housing. Increases in dendritic branching, synapse number, and number of synapses per neuron have also been demonstrated to occur in response to EC in aging rats (Green et al., 1983; Greenough et al., 1986). Thus, in contrast to the greatly diminished effects of sensory deprivation during adulthood, the environmental complexity studies in adult animals clearly suggest that this kind of experience alters the neurons in the adult neocortex in a similar way to that seen in young animals. On the other hand, the production of new blood vessels appears to be substantially impaired in middle-aged rats, and continuing progressive failure of new blood vessel production may restrict their capacity for storing information in the form of new synapses when they reach old age (Black et al., 1989).

4. Adult Neurogenesis

Most neurons in the brain proliferate during gestation, and, until recently, the notion that neurogenesis does not occur in the adult mammalian brain (outside of the olfactory bulb) was part of neuroscience dogma. Although there were earlier indications to the contrary (Altman, 1962, 1963; Kaplan, 1981), these were largely ignored until several key studies were published within the last decade. These studies confirmed the occurrence of adult neurogenesis in the dentate gyrus of the hippocampus of both rodents and primates (Kuhn et al., 1996; Kempermann et al., 1997; Eriksson et al., 1998; Kornack and Rakic, 1999). Whether neurogenesis occurs in the adult neocortex remains controversial, with some reports of substantial neurogenesis (Gould et al., 1999a, 2001; Dayer et al., 2005) and others reporting little or no detectable production of new neurons, although proliferation of astrocytes and other non-neuronal cell types was detected (Kornack and Rakic, 2001). Although the number of neurons added to the adult brain is small in comparison to both total neuron number and glial cell genesis, several environmental factors have been shown to influence this process. In general, stress, glucocorticoids, and alcohol exposure — experienced either during pre- or postnatal development or during adulthood — and the aging process all decrease the number of new neurons added to the adult brain, whereas antidepressants, estrogen, environmental stimulation, and exercise all increase it (Kuhn et al., 1996; Gould et al., 1997; Cameron et al., 1998; Tanapat et al., 1999; van Praag et al., 1999; Malberg et al., 2000; Mirescu et al., 2004; Crews et al., 2006; Redila et al., 2006; Wong and Herbert, 2006).

Housing in a complex environment is one experience that can robustly induce neurogenesis in the adult dentate gyrus. Kempermann, Kuhn, and Gage (1997) were the first to demonstrate that adult mice housed in EC have an approximately 15% increase in the number of new cells in the dentate gyrus of the hippocampus (an effect that is reflected in an increased size of the granule cell layer in this structure). This finding has since been replicated by this laboratory and others in both mice (Kempermann et al., 1998b; Kempermann and Gage, 1999; J. Brown et al., 2003) and rats (Nilsson et al., 1999; Bruel-Jungerman et al., 2005). Although a decline in neurogenesis occurs during normal aging (Kuhn et al., 1996), the dentate gyrus remains sensitive to environmental stimulation, and EC housing can still increase the number of new hippocampal neurons during senescence (Kempermann et al., 1998a). Conversely, social isolation (IC) during adulthood reduces basal levels of neurogenesis and prevents the normal exercise-induced increase in neurogenesis (Lu et al., 2003; Stranahan et al., 2006). Importantly, the mechanisms by which environmental factors result in greater numbers of new cells added to the dentate gyrus of the adult rodent can vary: Some factors, such as exercise, tend to increase the rate of neurogenesis directly, whereas exposure to EC and learning tend to increase the survival of newly generated cells rather than affecting the rate at which cells are born (Kempermann et al., 1998a; van Praag et al., 1999; Ambrogini et al., 2000; Olson et al., 2006).

Because rodents exposed to EC exhibit superior learning and memory ability as compared to IC rats, and because EC exposure during adulthood increases neurogenesis in the dentate gyrus of the hippocampus but not in the olfactory bulb (J. Brown et al., 2003), a learning-specific role for neurons added to the adult brain is suggested. There is evidence that adult neurogenesis and learning are indeed correlated. IC animals, in which hippocampal neurogenesis is reduced, also show a reduction in LTP in the hippocampus and impaired performance on Morris water maze spatial learning (which relies on the hippocampus); all three phenotypes are reversible by subsequent group housing (Lu et al., 2003). Conversely, EC-induced neurogenesis is associated with an improvement in water maze learning (Nilsson et al., 1999). In aged rats, a reduced survival of newly proliferated hippocampal neurons is associated with impairment in another form of learning that relies heavily on the integrity of the hippocampus, called *contextual fear conditioning* (Wati et al., 2006). Similarly, performance of aged rats on the spatial version of the Morris water maze is predictive of the level of hippocampal neurogenesis (Drapeau et al., 2003). Removal of the olfactory bulbs or developmental lead exposure both result in decreased adult hippocampal neurogenesis and impaired contextual fear conditioning (Jaako-Movits and Zharkovsky, 2005; Jaako-Movits et al., 2005). Sleep deprivation, which (in rats, as in humans) disrupts memory, has also been shown to profoundly reduce adult hippocampal neurogenesis (Guzman-Marin et al., 2005). In fact, sleep deprivation can prevent neurogenesis from being induced by spatial water maze learning (Hairston et al., 2005). Interestingly, the same study found that performance on the spatial version of the task suffered as a result of sleep deprivation, whereas learning a version of the task that is not dependent on the hippocampus was not affected.

More direct evidence for a link between adult neurogenesis and learning comes from studies that induce hippocampal neurogenesis by training on hippocampus-dependent learning tasks as well as from those that prevent or reduce neurogenesis in the hippocampus directly and then observe the consequences of this for memory. An important study by Gould and colleagues (1999b) found that training on associative learning tasks that require the hippocampal formation, but not training on hippocampal-independent tasks, increases the number of new neurons in the dentate gyrus. Subsequently, this group found that hippocampal-dependent learning enhanced the survival of newly born cells in the adult dentate gyrus long beyond the time when the hippocampus was required for learning of the task (Leuner et al., 2004). Ambrogini and colleagues (2000) have also found that training on a hippocampal-dependent learning task (spatial version of the Morris water maze) increased the survival of newly generated cells in the adult dentate gyrus. Rampon's group confirmed the benefit conferred on both memory performance and hippocampal neurogenesis by EC housing and furthermore reported that blocking adult neurogenesis (using the antimitotic agent methylazoxymethanol acetate) abolished the EC-induced improvement in hippocampal-dependent memory (Bruel-Jungerman et al., 2005). Mild irradiation, which inhibits adult neurogenesis, has also been found to impair hippocampal-dependent fear conditioning and performance on a delayed nonmatch-to-sample task in which longer delays were imposed (also hippocampal dependent) (Winocur et al., 2006). Two other groups have also found irradiation to reduce neurogenesis and to impair performance on hippocampus-dependent (but not hippocampus-independent) learning tasks (Madsen et al., 2003; Raber et al., 2004; Rola et al., 2004), further strengthening the link between hippocampal neurogenesis and hippocampus-dependent learning. Interestingly, it may be the neurons born prior to the learning experience, and not those generated by the learning experience itself, that are critical for memory performance. Irradiation disrupted performance on the spatial (hippocampal-dependent) version of the Morris water maze (but was without effect on performance of the hippocampal-independent, visible platform version of the maze) when administered 4–28 days prior to maze training but not when administered just prior to or immediately following maze training (Snyder et al., 2005). This finding is perhaps not surprising in light of the fact that the brain must rely on past experiences to predict future ones. Thus cells may be added to the adult hippocampus in anticipation of their need to mediate the acquisition, storage, and/or consolidation of future memories. Whatever the subtleties may be, what is clear is that adult hippocampal neurogenesis is an additional form of experience-dependent brain plasticity that appears to contribute to the learning process.

5. Conclusions from Environmental Complexity Studies

The complexity of the rearing environment can profoundly affect the structure of the brain. However, the changes resulting from differential rearing complexity are probably not a simple extension of those found in the visual deprivation studies. While the deprivation studies demonstrated that a drastic but simple manipulation of experience can modify connectivity and subsequent learning ability, visual experience is definitely "expected" during ontogeny. The types of visual experience of which the animals are deprived are normally quite uniform, for all species members, in their timing (i.e., after eye opening) and quality (e.g., all visual angles present). Visual deprivation at later ages, once the animals have had experience, has minimal lasting effect. On the other hand, the modification of experience in the environmental complexity research has a character that is much less "expected," from the phylogenetic perspective. The timing and character of individual experience in the EC environment cannot be uniformly predicted for all species members, such that synaptic plasticity must remain able to capture new information from experience whenever it becomes available.

The connectivity modifications observed in the EC animals appear more related to how neural activity is processed than to how much is processed. For example, both EC and IC animals use approximately the same amount of light (average intensity on the retina) quite differently, one with self-initiated activity and its visual consequences, the other with dull routine. The importance of active involvement is evident in the finding that there is essentially no brain effect of rearing rats within a small cage inside the EC environment (Ferchmin and Bennett, 1975). The fact that enhanced visual experience does not solely explain the effects is further highlighted by the finding that differences between blinded EC rats and blinded IC rats are comparable in magnitude to those between intact EC versus IC rats (Krech et al., 1962). Combined with the findings of widespread similarity of effects noted in the brains of EC animals and those trained in learning paradigms (discussed later), the available evidence to date indicates that EC-induced brain plasticity is related specifically to the learning process.

C. Skill Learning

1. Synapses

If the dendritic and synaptic alterations seen after EC experience are related to experience-dependent mechanisms such as learning, then we would expect to observe similar structural changes after training on traditional psychological learning tasks. Greenough et al. (1979) used the Hebb–Williams maze, which has movable barriers, allowing a large variety of problems. Adult rats received extensive training for 25–26 days on a new problem plus several old problems each day for water reward. Littermate control rats were allowed to drink water several times daily while held by the investigator. Trained animals had more dendrite branches along distal apical dendrites of Layers IV and V pyramidal cells in occipital cortex. Similarly, Bennett, Rosenzweig, Morimoto, and Hebert (1979) exposed rats to complex mazes in their cages, which were changed daily for 30 days, while their littermates were kept in IC. The maze-reared animals had heavier visual cortices. The complexity of the environment is a factor in the effect, since rats housed with a single, simple maze for 30 days had brain weights between those of EC and IC rats. Another group similarly found that extensive Hebb–Williams maze training of 500-day-old rats otherwise kept in IC increased forebrain weight and cortical area relative to baseline groups remaining in the IC cages (Cummins et al., 1973). However, littermates from an EC condition showed no effect of maze training, suggesting that additional effects were small or were obscured by the neural effects of EC exposure.

To further examine the specificity of training effects, Chang and Greenough (1982) studied monocular maze training effects on visual cortex of split-brain rats. Since about 90% of visual afferents in the rat cross to the contralateral cortex, use of an opaque contact lens over one eye of a split-brain rat can effectively isolate one occipital area from the other. Split-brain littermate triplets were assigned to one of four groups: (1) the left and right eyes were occluded on alternate days during successive training periods, (2) the same eye was occluded during all training periods, and (3) alternating or (4) unilateral occlusion of the eyes with no training at all. The occluders were worn for about four hours daily during the training period. There was no effect of unilateral versus alternating occluder position in the nontrained rats (group 3 versus group 4), indicating that occluder insertion alone did not affect the brain measures. The apical dendrites of cells in trained visual cortex were more extensive both within the fixed occluder rats (group 2, comparing adjacent hemispheres) and between the alternating occluder and nontrained rats (group 1 versus groups 3 and 4, comparing two trained hemispheres to two nontrained). This indicates that the effects of training are relatively restricted to the side of the brain most involved in learning the task. Thus the effects appear to be related to where memory for the task may be stored rather than to general metabolic or other activity.

A similar interpretation arises from experiments in which rats are trained to reach food pellets through a thin slot in their cage using one or both forepaws. Dendritic changes occurred in several neuronal populations in the sensory motor cortex region that governs forelimb activity (Greenough et al., 1985; Withers and Greenough, 1989). Using this skilled reaching task, Kleim and colleagues have continued to study the structural and functional correlates of motor learning. They found that, in response to motor skill learning, the area of motor cortex controlling forelimb movements expands and that in this region but not adjacent regions, the number of synapses per neuron increases (Kleim et al., 1998, 2002). Upon further testing they found that the addition of synapses precedes expansion of these motor maps and that both occur during the late phase of training (Kleim et al., 2004), prompting the authors to speculate that synaptogenesis and its functional correlates may play a role in the consolidation of motor learning. Some of the changes mentioned in the foregoing studies were lateralized with respect to the particular forelimb that had been trained, while others were more general, occurring on both sides of the brain, even in unilaterally trained animals. There are at least two possible interpretations of this result: (1) Some changes (bilateral) could be due to general activation of the tissue by the training experience (e.g., motor activity, sensory input), while others (unilateral) reflect changes associated with memory storage. (2) All of the changes reflect memory-associated brain reorganization, which must occur on both sides of the brain for the task to be learned.

In a series of studies designed to tease apart morphological changes associated with learning from those associated with general physical activity, a group of adult female rats that had been trained on a motor skill–learning task (using a challenging "acrobatic" course) were compared with animals allowed to exercise freely (on a treadmill) but with minimal opportunity for learning. The rats trained on the acrobat course had more synapses per neuron (Black et al., 1990; Kleim et al., 1996) (Figure 2-9), more perforated synapses (Jones, 1999), and more multiple synapse boutons (MSBs; see earlier)

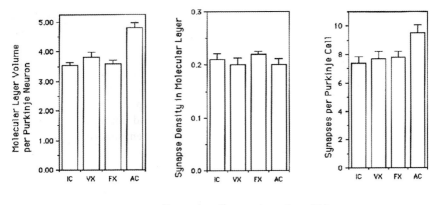

Exercise/Learning Condition

FIGURE 2-9   Data from cerebellar cortex indicate that new patterns of neural activity associated with motor learning, rather than the stereotyped patterns associated with repetitive exercise, affect synaptic connectivity. The paramedian lobule of cerebellar cortex was examined in four groups of adult rats: acrobatic conditioning (AC), which were trained to traverse a complicated elevated obstacle course that became progressively more difficult over the 30-day training period; voluntary exercise (VX), which had free access to a running wheel; forced exercise (FX), which were subjected to a treadmill exercise routine; and individual condition (IC), which were kept in standard cages without opportunity for additional exercise. In terms of distance traveled, the AC group covered much less than the two exercise groups; whereas in terms of opportunity for learning, the AC condition offered new skills to be acquired each day while the exercise conditions quickly became dull routine. Relative to the other three groups, the AC group had a lower density of Purkinje somata. Since the density of molecular layer synapses did not differ, the number of synapses per Purkinje neuron was substantially higher in the learning group than in the others. [From Black et al. (1990).]

(Federmeier et al., 2002) in regions of the brain involved in the control of fine motor movements, such as the motor cortex and the cerebellum, as compared to animals that exercised without the opportunity for learning and as compared to animals that were sedentary during the course of the experiment. The learning-induced changes in synapse number have been found to persist for at least four weeks after training has finished (Figure 2-10A) (Kleim et al., 1997). Motor skill learning (and not exercise) also increases the number of parallel fiber synapses and climbing fiber synapses per unit of cerebellar Purkinje cell reference volume (Anderson et al., 1996). Purkinje cell morphology is also altered by classical eye-blink conditioning (Anderson et al., 1999). Olfactory learning has also been found to increase spine density along apical dendrites of pyramidal neurons in the piriform (olfactory) cortex (Knafo et al., 2001). Finally, still others have found that associative memory formation induces synaptogenesis and the formation of multiple synapse boutons in the

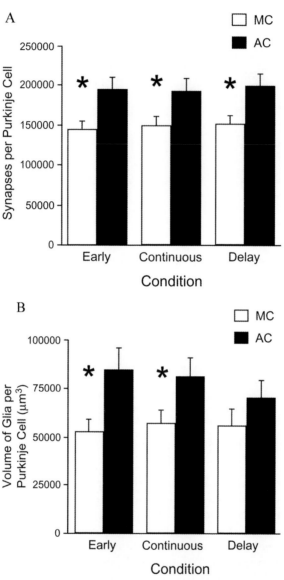

FIGURE 2-10   The increase in synapses per neuron in the motor cortex is stable in the absence of continued training. AC (acrobat) rats were trained on a motor skill learning task, whereas MC (motor control) animals ran on a treadmill but were not given an opportunity for learning. Animals in the Early group participated in training (AC) or exercised (MC) for 10 days, animals in the Continuous group participated for 38 days, and animals in the Delay group participated for 10 days and then training (or exercise) was discontinued for the following 28 days before histological examination. ★ indicates $p < 0.05$ for the comparison between the MC and AC animals of a particular group (Early, Continuous, or Delay). [Modified from Kleim et al. (1997b) and Kleim et al. (in press) with permission.]

hippocampus, a region with a known major role in this type of memory (O'Malley et al., 2000; Geinisman et al., 2001; Leuner et al., 2003). Thus it appears that learning, and not merely a general enhancement of brain activity, is required to induce synaptogenesis.

2. Glia and Cerebrovasculature

Astrocytic hypertrophy has also been found to accompany skill learning (but not physical exercise in the absence of learning). And because astrocytic and synaptic changes in the cerebellar cortex are correlated on an animal-by-animal basis, increased astrocytic volume can be inferred to arise in association with learning-specific synaptogenesis and not merely to constitute a response to a general increase in neural activity (Anderson et al., 1994). There is some evidence that learning-induced changes in astrocytes may be transient. When animals were first trained on a motor skill learning task for 10 days and then left idle for the following 28 days, synaptogenesis that had occurred during learning remained clearly evident in these "delay" animals, whereas training-induced effects on astrocytes were reduced and no longer statistically significant (Figure 2-10A and B) (Kleim et al., 1997; Kleim et al., in press). Therefore it is tempting to speculate that astrocytic changes may be necessary to induce and enhance, but not to maintain, adaptive changes in the brain's "wiring diagram" in response to experience. Learning-induced responses of oligodendrocytes and/or the myelination process have not been examined. Motor skill learning induces angiogenesis. However, blood vessel changes are also induced by physical exercise in the absence of learning (Black et al., 1990), further indicating that changes in cerebrovasculature are driven more by the repeated performance of movements rather than by the learning process itself.

D. Experience-Expectant and Experience-Dependent Plasticity in the Human Brain

The first demonstration of developmental overproduction and subsequent pruning of synaptic connections in a primate was accomplished by Boothe et al. (1979), who showed that synaptogenesis early in monkey visual cortical development was followed several months later by a reduction to adult levels of connectivity. Although technically more difficult to accomplish in human subjects, influential early work conducted by Huttenlocher (Huttenlocher, 1979; Huttenlocher and Dabholkar, 1997) suggested that a similar pattern of synaptogenesis followed by synaptic pruning occurs during human development. Potentially reflective of this underlying cellular pattern, volumetric increases in cortical gray matter followed by postpubertal decreases to adult values are observed in pediatric neuroimaging studies (e.g., Giedd et al., 1999). Although scientists cannot ethically conduct experiments aimed at manipulating the experience-expectant and experience-dependent processes that undoubtedly are at work here, there are some studies that nevertheless strongly suggest that the human brain responds with similar plasticity to early sensory deprivation, manipulations of environmental complexity, and skill learning.

Critical periods in human development have been characterized. For example, children who have amblyopia due to a misalignment of one eye will permanently have impaired vision if surgical correction does not occur prior to the age of 7; surgical correction of the eye's alignment is without effect on the brain later in development, so vision remains impaired (Daw, 1998). In congenitally blind individuals, the occipital cortex reorganizes in the absence of visual information and, far from being functionally silent, appears to actually respond to nonvisual information (Theoret et al., 2004). In fact, such reorganization in response to visual deprivation may underlie the superior auditory and tactile processing of blind individuals (Van Boven et al., 2000; Doucet et al., 2005). Although some changes result if blindness occurs during adulthood, they are far less than what is observed in the congenitally blind (Sathian, 2005). A similar phenomenon has been observed for those born congenitally deaf — in order to maximally benefit from a cochlear implant, in terms of both nonlinguistic sounds and the development of normal speech, this intervention must occur at a very young age (Harrison et al., 2005). In contrast, individuals who have developed normal hearing and speech prior to deafness can benefit from cochlear implantation technology even in adulthood (Harrison et al., 2005). The development of phoneme distinction is another example of an early window of plasticity: Infants initially respond to a large number of distinct phonemes, but within the first year of life their responsiveness to phonemes present in their native language is enhanced, whereas their responsiveness to phonemes not present in their native language is reduced. This effect was first observed behaviorally (Kuhl et al., 1992) and has since been observed in event-related potentials recorded from the brain (Cheour et al., 1998). Such findings indicate that, similar to the animal models discussed earlier, the human brain has critical periods of maximal responsiveness to visual, auditory, and linguistic experience.

Studies of Romanian orphans have made it clear that the human brain is terribly sensitive to deficiencies in environmental complexity. Due to social policies and economic problems in Romania in the 1980s, these children were subjected to early, global, and serious environmental deprivation. Neuropsychological assessment of these children revealed that they suffer from remarkable deficits in cognitive and social functioning (Kaler and Freeman, 1994). Furthermore, a neuroimaging study revealed that regional cerebral blood flow in these children was greatly below that of normal children's in areas crucial for learning and memory, including the prefrontal cortex, the amygdala, and the hippocampus (Chugani et al., 2001). These studies are reminiscent of the impaired mnemonic abilities of IC animals in comparison to EC-raised animals. Fortunately, just as in rodent models, childhood enrichment programs are able to produce a long-lasting increase in IQ, especially among children from disadvantaged backgrounds (Campbell et al., 2001).

Experience-dependent remodeling of the brain has also been shown to occur in primates. For instance, training a monkey to use a specific fingertip increases the area devoted to representation of that finger in the somatosensory cortex (Recanzone et al., 1992). Interestingly, the same phenomenon of altered cortical representation occurs for the fingers used by Braille readers (Pascual-Leone and Torres, 1993) and for the left hand of string musicians (Pantev et al., 2003). Such experience-dependent brain reorganization occurs in other sensory modalities; for example, enlarged cortical representation of tones of the musical scale (as compared to pure tones) were found in skilled musicians. Furthermore, the degree of enlargement is correlated with the age at which musicians began to practice (reviewed in Pantev et al., 2003). Very recently it was found that the structure of white matter (measured using diffusion tensor imaging and believed to correlate with degree of myelination) in the brain of adult pianists is correlated with the amount of time spent practicing or playing the piano during adolescence (Bengtsson et al., 2005). Studies such as these indicate that both neuronal and glial components of the human brain can reorganize in an experience-dependent fashion. These studies validate the continued use of animal models of experience-dependent plasticity.

V. IMPLICATIONS FOR THE NEUROBIOLOGICAL STUDY OF MEMORY

The evidence is strong for learning-induced changes in the number of synapses, glial cell morphology, and neuron–glia communication in many brain regions of both young and mature animals. The fact that such effects (1) are seen across a number of species, including nonmammalian (e.g., Coss et al., 1980; Patel and Stewart, 1988; Withers et al., 1995), and across types of experience, from light deprivation to maze learning; and (2) are correlated with changes in neuronal function and behavior suggests that evolution may not have established one simple mechanism for developmental plasticity and another for memory. Rather, there may be a set of cellular mechanisms on which the organism can draw for the incorporation of information from a wide variety of experiences, and it seems likely that we have not discovered all of the cellular mechanisms involved in encoding experience. Some mechanisms may not even be structurally detectable, at least with currently available techniques. At this point, however, there are consistently reported effects of various experiences on the number, and presumably the pattern, of synaptic connections, changes in glial cells and cerebrovasculature, and alterations in the number of new neurons and glial cells added to the adult brain. The consistency of these effects across visual deprivation, environmental complexity, and training paradigms may lead one to question whether our distinction between experience-expectant and experience-dependent neural plasticity is biologically meaningful, despite differences in the timing of susceptibility, response pattern across brain regions, and relative magnitude of effects. We argue here that the distinction is meaningful and that different neural mechanisms underlie these two forms of neural plasticity.

Specifically, we suggest that species survival may be facilitated by information storage processes anticipating an experience with identical timing and features for all juvenile members. A structural correlate of "expectation" may be a temporary overproduction of synapses during the sensitive period, with a subsequent pruning back of inappropriate synapses. The neuromodulatory event that triggers this synapse overproduction may be under maturational control or may be activity dependent (as after eye opening), but it is diffuse and pervasive. The expected experience produces patterned activity of neurons, effectively targeting which synapses will be selected. For example, Cragg (1975) reported that the number of synapses per neuron in cat visual cortex reached a peak at about 5 weeks of age and then fell to lower levels in adulthood. Similarly, Boothe and colleagues (1979) reported that spine frequency on some types of monkey visual cortex neurons reached an early peak and later declined to adult values. The peak values are reached, in both cases, at about the time that sensitivity to gross manipulations of visual experience, such as monocular pattern deprivation, is also maximal. In both species, afferent axonal terminal fields overlap during early development, segregating through the elimination of overlapping synapses as development progresses (e.g., LeVay et al., 1980); and in both species, occlusion of one eye causes more of its connections to be lost and more of the open eye's connections to be preserved (LeVay et al., 1980; Tieman, 1984). Similar phenomena are seen in other developing sensory projection systems, and entire dendritic or axonal branches regress in some cases (Falls and Gobel, 1979; Feng and Rogowski, 1980; Mariani and Changeux, 1981a, 1981b; Greenough and Chang, 1988). This overproduction of synapses followed by pruning appears to occur during normal development of the human cortex as well (Huttenlocher, 1979). It thus appears that the nervous system may become ready for expected experience by overproducing connections on a sensory-systemwide basis, such that experience-related neural activity can select a functionally appropriate subset of them. If not confirmed or stabilized, these synapses regress according to a developmental schedule and/or due to competition from confirmed synapses. Even when an aggregate overproduction is not observed (e.g., Valverde, 1971), it is quite possible that the process is being masked by the concomitant generation of some synapses and the loss of others in approximately equal magnitude.

The neural basis of experience-expectant information storage thus appears to be (1) the overproduction of potentially permanent synaptic connections paired with (2) the selective survival of connections deemed valuable by use during a critical window in development.

In contrast, for experience-dependent neural information storage, which includes traditional forms of learning, neither the timing nor the specific nature of the information can be anticipated by the nervous system, and there is little evidence for systemwide overproduction in these cases [although low-level systemwide turnover cannot be ruled out (Greenough and Chang, 1988)]. Because the phylogenetic adaptations of neural plasticity cannot anticipate the timing or specific features of such idiosyncratic experience, synapses are generated locally, on demand. Metabolically, the most efficient way to generate synapses locally within the system would be for activity to trigger local synaptogenesis, from which activity-dependent stabilization might further select an appropriate subset. Thus new synapses would be formed only when they were needed for incorporation of new information. A wonderful illustration of the specificity of experience-dependent synaptic plasticity was provided by Knott and colleagues (2002), who showed that, in adult mice, a single 24-hour period of whisker stimulation was sufficient to increase synaptic density by 36% in the barrel corresponding to the stimulated whisker, whereas there was no change in synapse density in a neighboring barrel corresponding to an unstimulated whisker. We suggest that active participation by the animal is necessary for it to obtain the necessary coherent relationship between responses and their consequences. This experience-dependent localized shaping of connectivity suggests that very general experience (as in EC) would produce a widespread increase in synaptic frequency but that relatively specific experience (as in training tasks) would produce more localized increases. In describing information storage mechanisms, we have tried to break down some of the distinctions between maturation, experience-sensitive development, and learning. Some aspects of experience (such as juvenile EC/IC rearing) may influence both experience-expectant and experience-dependent processes. In fact, these processes probably cannot be entirely isolated, since they have substantial interactive consequences for how the brain processes information and they share mechanisms at the cellular level. Thus we propose that learning results in the formation of new synapses as well as new neurons (in the dentate gyurs) that are involved in the permanent encoding of the memory as alterations in the circuitry of the brain systems in which the memory is stored.

ACKNOWLEDGMENTS

This work was supported by NIMH 35321, AG10154, and AA09838. JAM was supported by NIMHD07333

REFERENCES

Altman J (1962) Are new neurons formed in the brains of adult mammals? *Science* 135:1127–1128.

Altman J (1963) Autoradiographic investigation of cell proliferation in the brains of rats and cats. *Anat Rec* 145:573–591.

Ambrogini P, Cuppini R, Cuppini C, Ciaroni S, Cecchini T, Ferri P, Sartini S, and Del Grande P (2000) Spatial learning affects immature granule cell survival in adult rat dentate gyrus. *Neurosci Lett* 286:21–24.

Anderson BJ, Li X, Alcantara AA, Isaacs KR, Black JE, and Greenough WT (1994) Glial hypertrophy is associated with synaptogenesis following motor-skill learning, but not with angiogenesis following exercise. *Glia* 11:73–80.

Anderson BJ, Alcantara AA, and Greenough WT (1996) Motor-skill learning: Changes in synaptic organization of the rat cerebellar cortex. *Neurobiol Learn Mem* 66:221–229.

Anderson BJ, Relucio K, Haglund K, Logan C, Knowlton B, Thompson J, Steinmetz JE, Thompson RF, and Greenough WT (1999) Effects of paired and unpaired eye-blink conditioning on Purkinje cell morphology. *Learn Mem* 6:128–137.

Antonini A, and Stryker MP (1993) Rapid remodeling of axonal arbors in the visual cortex. *Science* 260:1819–1821.

Argandona EG, and Lafuente JV (1996) Effects of dark-rearing on the vascularization of the developmental rat visual cortex. *Brain Res* 732:43–51.

Argandona EG, and Lafuente JV (2000) Influence of visual experience deprivation on the postnatal development of the microvascular bed in layer IV of the rat visual cortex. *Brain Res* 855:137–142.

Argandona EG, Rossi ML, and Lafuente JV (2003) Visual deprivation effects on the s100beta positive astrocytic population in the developing rat visual cortex: A quantitative study. *Brain Res Dev Brain Res* 141:63–69.

Barres BA, and Raff MC (1993) Proliferation of oligodendrocyte precursor cells depends on electrical activity in axons. *Nature* 361:258–260.

Bengtsson SL, Nagy Z, Skare S, Forsman L, Forssberg H, and Ullen F (2005) Extensive piano practicing has regionally specific effects on white matter development. *Nat Neurosci* 8:1148–1150.

Bennett EL, Diamond MC, Krech D, and Rosenzweig MR (1964) Chemical and anatomical plasticity of the brain. *Science* 146:610–619.

Bennett EL, Rosenzweig MR, Morimoto H, and Hebert M (1979) Maze training alters brain weights and cortical RDA/DNA ratios. *Behav Neural Biol* 26:1–22.

Bezzi P, Vesce S, Panzarasa P, and Volterra A (1999) Astrocytes as active participants of glutamatergic function and regulators of its homeostasis. *Adv Exp Med Biol* 468:69–80.

Black JE, and Greenough WT (1986) Induction of pattern in neural structure by experience: Implications for cognitive development. In: *Advances in Developmental Psychology*, Volume 4 (Lamb ME, Brown AL, Rogoff B, eds), pp. 1–50. Hillsdale, NJ: Erlbaum.

Black JE, Isaacs KR, Anderson BJ, Alcantara AA, and Greenough WT (1990a) Learning causes synaptogenesis, whereas motor activity causes angiogenesis, in cerebellar cortex of adult rats. *Proc Natl Acad Sci USA* 87:5568–5572.

Black JE, Polinsky M, and Greenough WT (1989) Progressive failure of cerebral angiogenesis supporting neural plasticity in aging rats. *Neurobiol Aging* 10:353–358.

Black JE, Sirevaag AM, and Greenough WT (1987) Complex experience promotes capillary formation in young rat visual cortex. *Neurosci Lett* 83:351–355.

Black JE, Zelazny AM, and Greenough WT (1991) Capillary and mitochondrial support of neural plasticity in adult rat visual cortex. *Exp Neurol* 111:204–209.

Boothe RG, Greenough WT, Lund JS, and Wrege K (1979) A quantitative investigation of spine and dendrite development of neurons in visual cortex (area 17) of *Macaca nemestrina* monkeys. *J Comp Neurol* 186:473–489.

Briones TL, Klintsova AY, and Greenough WT (2004) Stability of synaptic plasticity in the adult rat visual cortex induced by complex environment exposure. *Brain Res* 1018:130–135.

Brown J, Cooper-Kuhn CM, Kempermann G, Van Praag H, Winkler J, Gage FH, and Kuhn HG (2003) Enriched environment and physical activity stimulate hippocampal but not olfactory bulb neurogenesis. *Eur J Neurosci* 17:2042–2046.

Brown RT (1968) Early experience and problem-solving ability. *J Comp Physiol Psychol* 65:433–440.

Bruel-Jungerman E, Laroche S, and Rampon C (2005) New neurons in the dentate gyrus are involved in the expression of enhanced long-term memory following environmental enrichment. *Eur J Neurosci* 21:513–521.

Butt AM, and Tutton M (1992) Response of oligodendrocytes to glutamate and gamma-aminobutyric acid in the intact mouse optic nerve. *Neurosci Lett* 146:108–110.

Cameron HA, Tanapat P, and Gould E (1998) Adrenal steroids and N-methyl-D-aspartate receptor activation regulate neurogenesis in the dentate gyrus of adult rats through a common pathway. *Neuroscience* 82:349–354.

Campbell FA, Pungello EP, Miller-Johnson S, Burchinal M, and Ramey CT (2001) The development of cognitive and academic abilities: Growth curves from an early childhood educational experiment. *Dev Psychol* 37:231–242.

Chang FL, and Greenough WT (1982) Lateralized effects of monocular training on dendritic branching in adult split-brain rats. *Brain Res* 232:283–292.

Chang FL, and Greenough WT (1984) Transient and enduring morphological correlates of synaptic activity and efficacy change in the rat hippocampal slice. *Brain Res* 309:35–46.

Chen BE, Lendvai B, Nimchinsky EA, Burbach B, Fox K, and Svoboda K (2000) Imaging high-resolution structure of GFP-expressing neurons in neocortex in vivo. *Learn Mem* 7:433–441.

Cheour M, Ceponiene R, Lehtokoski A, Luuk A, Allik J, Alho K, and Naatanen R (1998) Development of language-specific phoneme representations in the infant brain. *Nat Neurosci* 1:351–353.

Chugani HT, Behen ME, Muzik O, Juhasz C, Nagy F, and Chugani DC (2001) Local brain functional activity following early deprivation: A study of postinstitutionalized Romanian orphans. *Neuroimage* 14:1290–1301.

Coleman PD, and Riesen AH (1968) Evironmental effects on cortical dendritic fields. I. Rearing in the dark. *J Anat* 102:363–374.

Coleman PD, Flood DG, Whitehead MC, and Emerson RC (1981) Spatial sampling by dendritic trees in visual cortex. *Brain Res* 214:1–21.

Comery TA, Shah R, and Greenough WT (1995) Differential rearing alters spine density on medium-sized spiny neurons in the rat corpus striatum: Evidence for association of morphological plasticity with early response gene expression. *Neurobiol Learn Mem* 63:217–219.

Comery TA, Stamoudis CX, Irwin SA, and Greenough WT (1996) Increased density of multiple-head dendritic spines on medium-sized spiny neurons of the striatum in rats reared in a complex environment. *Neurobiol Learn Mem* 66:93–96.

Coss RG, Brandon JG, and Globus A (1980) Changes in morphology of dendritic spines on honeybee calycal interneurons associated with cumulative nursing and foraging experiences. *Brain Res* 192:49–59.

Cragg BG (1967) Changes in visual cortex on first exposure of rats to light. Effect on synaptic dimensions. *Nature* 215:251–253.

Cragg BG (1975) The development of synapses in kitten visual cortex during visual deprivation. *Exp Neurol* 46:445–451.

Crews FT, Mdzinarishvili A, Kim D, He J, and Nixon K (2006) Neurogenesis in adolescent brain is potently inhibited by ethanol. *Neuroscience* 137:437–445.

Cummins RA, Walsh RN, Budtz-Olsen OE, Konstantinos T, and Horsfall CR (1973) Environmentally induced changes in the brains of elderly rats. *Nature* 243:516–518.

Daw NW (1998) Critical periods and amblyopia. *Arch Ophthalmol* 116:502–505.

Dayer AG, Cleaver KM, Abouantoun T, and Cameron HA (2005) New GABAergic interneurons in the adult neocortex and striatum are generated from different precursors. *J Cell Biol* 168:415–427.

Demerens C, Stankoff B, Logak M, Anglade P, Allinquant B, Couraud F, Zalc B, and Lubetzki C (1996) Induction of myelination in the central nervous system by electrical activity. *Proc Natl Acad Sci USA* 93:9887–9892.

Denk W, Strickler JH, and Webb WW (1990) Two-photon laser scanning fluorescence microscopy. *Science* 248:73–76.

Diamond MC, Krech D, and Rosenzweig MR (1964) The effects of an enriched environment on the histology of the rat cerebral cortex. *J Comp Neurol* 123:111–120.

Diamond MC, Lindner B, Johnson R, Bennett EL, and Rosenzweig MR (1975) Differences in occipital cortical synapses from environmentally enriched, impoverished, and standard colony rats. *J Neurosci Res* 1:109–119.

Doucet ME, Guillemot JP, Lassonde M, Gagne JP, Leclerc C, and Lepore F (2005) Blind subjects process auditory spectral cues more efficiently than sighted individuals. *Exp Brain Res* 160:194–202.

Drapeau E, Mayo W, Aurousseau C, Le Moal M, Piazza PV, and Abrous DN (2003) Spatial memory performances of aged rats in the water maze predict levels of hippocampal neurogenesis. *Proc Natl Acad Sci USA* 100:14385–14390.

Eriksson PS, Perfilieva E, Bjork-Eriksson T, Alborn AM, Nordborg C, Peterson DA, and Gage FH (1998) Neurogenesis in the adult human hippocampus. *Nat Med* 4:1313–1317.

Falls W, and Gobel S (1979) Golgi and EM studies of the formation of dendritic and axonal arbors: The interneurons of the substantia gelatinosa of *Rolando* in newborn kittens. *J Comp Neurol* 187:1–18.

Federmeier KD, Kleim JA, and Greenough WT (2002) Learning-induced multiple synapse formation in rat cerebellar cortex. *Neurosci Lett* 332:180–184.

Feng AS, and Rogowski BA (1980) Effects of monaural and binaural occlusion on the morphology of neurons in the medial superior olivary nucleus of the rat. *Brain Res* 189:530–534.

Ferchmin PA, and Bennett EL (1975) Direct contact with enriched environment is required to alter cerebral weights in rats. *J Comp Physiol Psychol* 88:360–367.

Fields RD, and Stevens B (2000) ATP: An extracellular signaling molecule between neurons and glia. *Trends Neurosci* 23:625–633.

Fischer M, Kaech S, Wagner U, Brinkhaus H, and Matus A (2000) Glutamate receptors regulate actin-based plasticity in dendritic spines. *Nat Neurosci* 3:887–894.

Floeter MK, and Greenough WT (1979) Cerebellar plasticity: Modification of Purkinje cell structure by differential rearing in monkeys. *Science* 206:227–229.

Forgays DG, and Forgays JW (1952) The nature of the effect of free-environmental experience in the rat. *J Comp Physiol Psychol* 45:322–328.

Fox K (2002) Anatomical pathways and molecular mechanisms for plasticity in the barrel cortex. *Neuroscience* 111:799–814.

Freire M (1978) Effects of dark rearing on dendritic spines in layer IV of the mouse visual cortex. A quantitative electron microscopical study. *J Anat* 126:193–201.

Fuchs JL, Montemayor M, Greenough WT (1990) Effect of environmental complexity on size of the superior colliculus. *Behav Neural Biol* 54:198–203.

Fukui Y, Hayasaka S, Bedi KS, Ozaki HS, and Takeuchi Y (1991) Quantitative study of the development of the optic nerve in rats reared in the dark during early postnatal life. *J Anat* 174:37–47.

Gabbott PL, Stewart MG, and Rose SP (1986) The quantitative effects of dark-rearing and light exposure on the laminar composition and depth distribution of neurons and glia in the visual cortex (area 17) of the rat. *Exp Brain Res* 64:225–232.

Galani R, Coutureau E, and Kelche C (1998) Effects of enriched postoperative housing conditions on spatial memory deficits in rats with selective lesions of either the hippocampus, subiculum or entorhinal cortex. *Restor Neurol Neurosci* 13:173–184.

Galani R, Jarrard LE, Will BE, and Kelche C (1997) Effects of postoperative housing conditions on functional recovery in rats with lesions of the hippocampus, subiculum, or entorhinal cortex. *Neurobiol Learn Mem* 67:43–56.

Galofre E, and Ferrer I (1987) Development of dendritic spines in the Vth's layer pyramidal neurons of the rat's somatosensory cortex. A qualitative and quantitative study with the Golgi method. *J Hirnforsch* 28:653–659.

Ganeshina O, Berry RW, Petralia RS, Nicholson DA, and Geinisman Y (2004a) Synapses with a segmented, completely partitioned postsynaptic density express more AMPA receptors than other axospinous synaptic junctions. *Neuroscience* 125:615–623.

Ganeshina O, Berry RW, Petralia RS, Nicholson DA, and Geinisman Y (2004b) Differences in the expression of AMPA and NMDA receptors between axospinous perforated and nonperforated synapses are related to the configuration and size of postsynaptic densities. *J Comp Neurol* 468:86–95.

Geinisman Y, Berry RW, Disterhoft JF, Power JM, and Van der Zee EA (2001) Associative learning elicits the formation of multiple-synapse boutons. *J Neurosci* 21:5568–5573.

Geinisman Y, Morrell F, and de Toledo-Morrell L (1990) Increase in the relative proportion of perforated axospinous synapses following hippocampal kindling is specific for the synaptic field of stimulated axons. *Brain Res* 507:325–331.

Geinisman Y, de Toledo-Morrell L, and Morrell F (1991) Induction of long-term potentiation is associated with an increase in the number of axospinous synapses with segmented postsynaptic densities. *Brain Res* 566:77–88.

Giedd JN, Blumenthal J, Jeffries NO, Castellanos FX, Liu H, Zijdenbos A, Paus T, Evans AC, and Rapoport JL (1999) Brain development during childhood and adolescence: A longitudinal MRI study. *Nat Neurosci* 2:861–863.

Gordon JA, and Stryker MP (1996) Experience-dependent plasticity of binocular responses in the primary visual cortex of the mouse. *J Neurosci* 16:3274–3286.

Gould E, Beylin A, Tanapat P, Reeves A, and Shors TJ (1999b) Learning enhances adult neurogenesis in the hippocampal formation. *Nat Neurosci* 2:260–265.

Gould E, McEwen BS, Tanapat P, Galea LA, and Fuchs E (1997) Neurogenesis in the dentate gyrus of the adult tree shrew is regulated by psychosocial stress and NMDA receptor activation. *J Neurosci* 17:2492–2498.

Gould E, Reeves AJ, Graziano MS, and Gross CG (1999a) Neurogenesis in the neocortex of adult primates. *Science* 286:548–552.

Gould E, Vail N, Wagers M, and Gross CG (2001) Adult-generated hippocampal and neocortical neurons in macaques have a transient existence. *Proc Natl Acad Sci USA* 98:10910–10917.

Green EJ, Greenough WT, and Schlumpf BE (1983) Effects of complex or isolated environments on cortical dendrites of middle-aged rats. *Brain Res* 264:233–240.

Greenough WT, and Chang FL (1988) Plasticity of synapse structure and pattern in the cerebral cortex. In: *Cerebral Cortex: Development and Maturation of Cerebral Cortex* (Peters A, Jones EG, eds). New York: Plenum Press.

Greenough WT, Juraska JM, and Volkmar FR (1979) Maze-training effects on dendritic branching in occipital cortex of adult rats. *Behav Neurol Biol* 26:287–297.

Greenough WT, Larson JR, and Withers GS (1985) Effects of unilateral and bilateral training in a reaching task on dendritic branching of neurons in the rat motor-sensory forelimb cortex. *Behav Neurol Biol* 44:301–314.

Greenough WT, McDonald JW, Parnisari RM, and Camel JE (1986) Environmental conditions modulate degeneration and new dendrite growth in cerebellum of senescent rats. *Brain Res* 380:136–143.

Greenough WT, and Volkmar FR (1973) Pattern of dendritic branching in occipital cortex of rats reared in complex environments. *Exp Neurol* 40:491–504.

Greenough WT, Volkmar FR, and Juraska JM (1973) Effects of rearing complexity on dendritic branching in frontolateral and temporal cortex of the rat. *Exp Neurol* 41:371–378.

Greenough WT, West RW, and DeVoogd TJ (1978) Subsynaptic plate perforations: Changes with age and experience in the rat. *Science* 202:1096–1098.

Greenough WT, Wood WE, and Madden TC (1972) Possible memory storage differences among mice reared in environments varying in complexity. *Behav Biol* 7:717–722.

Grossman AW, Churchill JD, Bates KE, Kleim JA, and Greenough WT (2002) A brain adaptation view of plasticity: Is synaptic plasticity an overly limited concept? *Prog Brain Res* 138:91–108.

Grutzendler J, Kasthuri N, and Gan WB (2002) Long-term dendritic spine stability in the adult cortex. *Nature* 420:812–816.

Guzman-Marin R, Suntsova N, Methippara M, Greiffenstein R, Szymusiak R, and McGinty D (2005) Sleep deprivation suppresses neurogenesis in the adult hippocampus of rats. *Eur J Neurosci* 22:2111–2116.

Gyllensten L, and Malmfors T (1963) Myelinization of the optic nerve and its dependence on visual function — a quantitative investigation in mice. *J Embryol Exp Morphol* 11:255–266.

Hairston IS, Little MT, Scanlon MD, Barakat MT, Palmer TD, Sapolsky RM, and Heller HC (2005) Sleep restriction suppresses neurogenesis induced by hippocampus-dependent learning. *J Neurophysiol* 94:4224–4233.

Harik SI, Hritz MA, and LaManna JC (1995) Hypoxia-induced brain angiogenesis in the adult rat. *J Physiol* 485 (Pt 2):525–530.

Harrison RV, Gordon KA, and Mount RJ (2005) Is there a critical period for cochlear implantation in congenitally deaf children? Analyses of hearing and speech perception performance after implantation. *Dev Psychobiol* 46:252–261.

Hawrylak N, and Greenough WT (1995) Monocular deprivation alters the morphology of glial fibrillary acidic protein-immunoreactive astrocytes in the rat visual cortex. *Brain Res* 683:187–199.

Hebb DO (1949) *The Organization of Behavior.* New York: Wiley.

Hirsch HV, and Spinelli DN (1970) Visual experience modifies distribution of horizontally and vertically oriented receptive fields in cats. *Science* 168:869–871.

Holloway RL, Jr (1966) Dendritic branching: Some preliminary results of training and complexity in rat visual cortex. *Brain Res* 2:393–396.

Holtmaat AJ, Trachtenberg JT, Wilbrecht L, Shepherd GM, Zhang X, Knott GW, and Svoboda K (2005) Transient and persistent dendritic spines in the neocortex in vivo. *Neuron* 45:279–291.

Huttenlocher PR (1979) Synaptic density in human frontal cortex — developmental changes and effects of aging. *Brain Res* 163:195–205.

Huttenlocher PR, and Dabholkar AS (1997) Regional differences in synaptogenesis in human cerebral cortex. *J Comp Neurol* 387:167–178.

Hymovitch B (1952) The effects of experimental variations on problem solving in the rat. *J Comp Physiol Psychol* 45:313–321.

Imamura K, Mataga N, and Watanabe Y (1993) Gliotoxin-induced suppression of ocular dominance plasticity in kitten visual cortex. *Neurosci Res* 16:117–124.

Jaako-Movits K, and Zharkovsky A (2005) Impaired fear memory and decreased hippocampal neurogenesis following olfactory bulbectomy in rats. *Eur J Neurosci* 22:2871–2878.

Jaako-Movits K, Zharkovsky T, Romantchik O, Jurgenson M, Merisalu E, Heidmets LT, and Zharkovsky A (2005) Developmental lead exposure impairs contextual fear conditioning and reduces adult hippocampal neurogenesis in the rat brain. *Int J Dev Neurosci* 23:627–635.

Jones DG, and Calverley RK (1991) Frequency of occurrence of perforated synapses in developing rat neocortex. *Neurosci Lett* 129:189–192.

Jones TA (1999) Multiple synapse formation in the motor cortex opposite unilateral sensorimotor cortex lesions in adult rats. *J Comp Neurol* 414:57–66.

Jones TA (2002) Behavioural experience-dependent plasticity of glial-neuronal interactions. In: *The Tripartite Synapse: Glia in Synaptic Transmission* (Volterra A, Magistretti P, and Hayden PG, eds.), pp. 248–265. Oxford, UK: Oxford University Press.

Jones TA, and Greenough WT (1996) Ultrastructural evidence for increased contact between astrocytes and synapses in rats reared in a complex environment. *Neurobiol Learn Mem* 65:48–56.

Jones TA, Hawrylak N, and Greenough WT (1996) Rapid laminar-dependent changes in GFAP immunoreactive astrocytes in the visual cortex of rats reared in a complex environment. *Psychoneuroendocrinology* 21:189–201.

Jones TA, Klintsova AY, Kilman VL, Sirevaag AM, and Greenough WT (1997) Induction of multiple synapses by experience in the visual cortex of adult rats. *Neurobiol Learn Mem* 68:13–20.

Juraska JM, Fitch JM, Henderson C, and Rivers N (1985) Sex differences in the dendritic branching of dentate granule cells following differential experience. *Brain Res* 333:73–80.

Juraska JM, Fitch JM, and Washburne DL (1989) The dendritic morphology of pyramidal neurons in the rat hippocampal CA3 area. II. Effects of gender and the environment. *Brain Res* 479:115–119.

Juraska JM, Greenough WT, Elliott C, Mack KJ, and Berkowitz R (1980) Plasticity in adult rat visual cortex: An examination of several cell populations after differential rearing. *Behav Neural Biol* 29:157–167.

Juraska JM, and Kopcik JR (1988) Sex and environmental influences on the size and ultrastructure of the rat corpus callosum. *Brain Res* 450:1–8.

Kaler SR, and Freeman BJ (1994) Analysis of environmental deprivation: Cognitive and social development in Romanian orphans. *J Child Psychol Psychiatry* 35:769–781.

Kaplan MS (1981) Neurogenesis in the 3-month-old rat visual cortex. *J Comp Neurol* 195:323–338.

Katz LC, and Shatz CJ (1996) Synaptic activity and the construction of cortical circuits. *Science* 274:1133–1138.

Kempermann G, Brandon EP, and Gage FH (1998b) Environmental stimulation of 129/SvJ mice causes increased cell proliferation and neurogenesis in the adult dentate gyrus. *Curr Biol* 8:939–942.

Kempermann G, and Gage FH (1999) Experience-dependent regulation of adult hippocampal neurogenesis: Effects of long-term stimulation and stimulus withdrawal. *Hippocampus* 9:321–332.

Kempermann G, Kuhn HG, and Gage FH (1997) More hippocampal neurons in adult mice living in an enriched environment. *Nature* 386:493–495.

Kempermann G, Kuhn HG, and Gage FH (1998a) Experience-induced neurogenesis in the senescent dentate gyrus. *J Neurosci* 18:3206–3212.

Kesner RP, and Rolls ET (2001) Role of long-term synaptic modification in short-term memory. *Hippocampus* 11:240–250.

Kim JH, and Juraska JM (1997) Sex differences in the development of axon number in the splenium of the rat corpus callosum from postnatal day 15 through 60. *Brain Res Dev Brain Res* 102:77–85.

Kleim JA, Barbay S, Cooper NR, Hogg TM, Reidel CN, Remple MS, and Nudo RJ (2002) Motor learning-dependent synaptogenesis is localized to functionally reorganized motor cortex. *Neurobiol Learn Mem* 77:63–77.

Kleim JA, Barbay S, and Nudo RJ (1998) Functional reorganization of the rat motor cortex following motor skill learning. *J Neurophysiol* 80:3321–3325.

Kleim JA, Hogg TM, VandenBerg PM, Cooper NR, Bruneau R, and Remple M (2004) Cortical synaptogenesis and motor map reorganization occur during late, but not early, phase of motor skill learning. *J Neurosci* 24:628–633.

Kleim JA, Lussnig E, Schwarz ER, Comery TA, and Greenough WT (1996) Synaptogenesis and *Fos* expression in the motor cortex of the adult rat after motor skill learning. *J Neurosci* 16:4529–4535.

Kleim JA, Markham JA, Kapil V, Kelly JL, Ballard DH, and Greenough WT (in press). Motor learning induces astrocytic hypertrophy in the cerebellar cortex. *Behavioural Brain Research*.

Kleim JA, Vij K, Ballard DH, and Greenough WT (1997) Learning-dependent synaptic modifications in the cerebellar cortex of the adult rat persist for at least four weeks. *J Neurosci* 17:717–721.

Knafo S, Grossman Y, Barkai E, and Benshalom G (2001) Olfactory learning is associated with increased spine density along apical dendrites of pyramidal neurons in the rat piriform cortex. *Eur J Neurosci* 13:633–638.

Knott GW, Quairiaux C, Genoud C, and Welker E (2002) Formation of dendritic spines with GABAergic synapses induced by whisker stimulation in adult mice. *Neuron* 34:265–273.

Kolb B, Gorny G, Soderpalm AH, and Robinson TE (2003) Environmental complexity has different effects on the structure of neurons in the prefrontal cortex versus the parietal cortex or nucleus accumbens. *Synapse* 48:149–153.

Korkotian E, and Segal M (2001) Regulation of dendritic spine motility in cultured hippocampal neurons. *J Neurosci* 21:6115–6124.

Kornack DR, and Rakic P (1999) Continuation of neurogenesis in the hippocampus of the adult macaque monkey. *Proc Natl Acad Sci USA* 96:5768–5773.

Kornack DR, and Rakic P (2001) Cell proliferation without neurogenesis in adult primate neocortex. *Science* 294:2127–2130.

Krech D, Rosenzweig MR, and Bennett EL (1962) Relations between chemistry and problem-solving among rats raised in enriched and impoverished environments. *J Comp Physiol Psychol* 55:801–807.

Kuhl PK, Williams KA, Lacerda F, Stevens KN, and Lindblom B (1992) Linguistic experience alters phonetic perception in infants by 6 months of age. *Science* 255:606–608.

Kuhn HG, Dickinson-Anson H, and Gage FH (1996) Neurogenesis in the dentate gyrus of the adult rat: Age-related decrease of neuronal progenitor proliferation. *J Neurosci* 16:2027–2033.

Leggio MG, Mandolesi L, Federico F, Spirito F, Ricci B, Gelfo F, and Petrosini L (2005) Environmental enrichment promotes improved spatial abilities and enhanced dendritic growth in the rat. *Behav Brain Res* 163:78–90.

Lendvai B, Stern EA, Chen B, and Svoboda K (2000) Experience-dependent plasticity of dendritic spines in the developing rat barrel cortex in vivo. *Nature* 404:876–881.

Leuner B, Falduto J, and Shors TJ (2003) Associative memory formation increases the observation of dendritic spines in the hippocampus. *J Neurosci* 23:659–665.

Leuner B, Mendolia-Loffredo S, Kozorovitskiy Y, Samburg D, Gould E, and Shors TJ (2004) Learning enhances the survival of new neurons beyond the time when the hippocampus is required for memory. *J Neurosci* 24:7477–7481.

LeVay S, Stryker MP, and Shatz CJ (1978) Ocular dominance columns and their development in layer IV of the cat's visual cortex: A quantitative study. *J Comp Neurol* 179:223–244.

LeVay S, Wiesel TN, and Hubel DH (1980) The development of ocular dominance columns in normal and visually deprived monkeys. *J Comp Neurol* 191:1–51.

Lu L, Bao G, Chen H, Xia P, Fan X, Zhang J, Pei G, and Ma L (2003) Modification of hippocampal neurogenesis and neuroplasticity by social environments. *Exp Neurol* 183:600–609.

Madsen TM, Kristjansen PE, Bolwig TG, and Wortwein G (2003) Arrested neuronal proliferation and impaired hippocampal function following fractionated brain irradiation in the adult rat. *Neuroscience* 119:635–642.

Mainen ZF, Maletic-Savatic M, Shi SH, Hayashi Y, Malinow R, and Svoboda K (1999) Two-photon imaging in living brain slices. *Methods* 18:231–239, 181.

Majewska A, and Sur M (2003) Motility of dendritic spines in visual cortex in vivo: Changes during the critical period and effects of visual deprivation. *Proc Natl Acad Sci USA* 100:16024–16029.

Malberg JE, Eisch AJ, Nestler EJ, and Duman RS (2000) Chronic antidepressant treatment increases neurogenesis in adult rat hippocampus. *J Neurosci* 20:9104–9110.

Mariani J, and Changeux JP (1981a) Ontogenesis of olivocerebellar relationships. I. Studies by intracellular recordings of the multiple innervation of Purkinje cells by climbing fibers in the developing rat cerebellum. *J Neurosci* 1:696–702.

Mariani J, and Changeux JP (1981b) Ontogenesis of olivocerebellar relationships. II. Spontaneous activity of inferior olivary neurons and climbing fibermediated activity of cerebellar Purkinje cells in developing rats. *J Neurosci* 1:703–709.

Markham JA, and Greenough WT (2004) Experience-driven brain plasticity: beyond the synapse. *Neuron Glia Biology* 1:351–363.

McClelland JL, and Rumelhart DE (1988) *Parallel Distributed Processing: Explorations in the Microstructure of Cognition. Volume 2: Psychological and Biological Models.* Cambridge, MA: MIT Press.

Mirescu C, Peters JD, and Gould E (2004) Early life experience alters response of adult neurogenesis to stress. *Nat Neurosci* 7:841–846.

Mohammed AK, Jonsson G, and Archer T (1986) Selective lesioning of forebrain noradrenaline neurons at birth abolishes the improved maze learning performance induced by rearing in complex environment. *Brain Res* 398:6–10.

Mohammed AK, Winblad B, Ebendal T, and Larkfors L (1990) Environmental influence on behavior and nerve growth factor in the brain. *Brain Res* 528:62–72.

Moore CL, Kalil R, and Richards W (1976) Development of myelination in optic tract of the cat. *J Comp Neurol* 165:125–136.

Moser MB, Trommald M, and Andersen P (1994) An increase in dendritic spine density on hippocampal CA1 pyramidal cells following spatial learning in adult rats suggests the formation of new synapses. *Proc Natl Acad Sci USA* 91:12673–12675.

Muller CM (1990) Dark-rearing retards the maturation of astrocytes in restricted layers of cat visual cortex. *Glia* 3:487–494.

Muller CM, Akhavan AC, and Bette M (1993) Possible role of S-100 in glia-neuronal signalling involved in activity-dependent plasticity in the developing mammalian cortex. *J Chem Neuroanat* 6:215–227.

Nilsson M, Perfilieva E, Johansson U, Orwar O, and Eriksson PS (1999) Enriched environment increases neurogenesis in the adult rat dentate gyrus and improves spatial memory. *J Neurobiol* 39:569–578.

Noguchi J, Matsuzaki M, Ellis-Davies GC, and Kasai H (2005) Spine-neck geometry determines NMDA receptor-dependent Ca^{2+} signaling in dendrites. *Neuron* 46:609–622.

O'Leary DD, Ruff NL, and Dyck RH (1994) Development, critical period plasticity, and adult reorganizations of mammalian somatosensory systems. *Curr Opin Neurobiol* 4:535–544.

Oliet SH, Piet R, and Poulain DA (2001) Control of glutamate clearance and synaptic efficacy by glial coverage of neurons. *Science* 292:923–926.

Olson AK, Eadie BD, Ernst C, and Christie BR (2006) Environmental enrichment and voluntary exercise massively increase neurogenesis in the adult hippocampus via dissociable pathways. *Hippocampus* 16:250–260.

O'Malley A, O'Connell C, Murphy KJ, and Regan CM (2000) Transient spine density increases in the mid-molecular layer of hippocampal dentate gyrus accompany consolidation of a spatial learning task in the rodent. *Neuroscience* 99:229–232.

Pantev C, Ross B, Fujioka T, Trainor LJ, Schulte M, and Schulz M (2003) Music and learning-induced cortical plasticity. *Ann NY Acad Sci* 999:438–450.

Pascual-Leone A, and Torres F (1993) Plasticity of the sensorimotor cortex representation of the reading finger in Braille readers. *Brain* 116 (Pt 1):39–52.

Patel SN, and Stewart MG (1988) Changes in the number and structure of dendritic spines 25 hours after passive avoidance training in the domestic chick, *Gallus domesticus*. *Brain Res* 449:34–46.

Piaget J (1980) *Adaptation and Intelligence: Organic Selection and Phenocopy* (S.S. Eames, Trans.). Chicago: University of Chicago Press.

Pysh JJ, and Weiss GM (1979) Exercise during development induces an increase in Purkinje cell dendritic tree size. *Science* 206:230–232.

Raber J, Rola R, LeFevour A, Morhardt D, Curley J, Mizumatsu S, VandenBerg SR, and Fike JR (2004) Radiation-induced cognitive impairments are associated with changes in indicators of hippocampal neurogenesis. *Radiat Res* 162:39–47.

Ramony Y, and Cajal S (1893) Neue Darstellung vom histologischen Bau des Centralnerven-system. [New Findings About the Histological Structure of the Central Nervous System]. *Arch Anat Physiol (Anat)* 319–428.

Ravizza RJ, and Herschberger AC (1966) The effect of prolonged motor restriction upon later behavior of the rat. *Psychol Rec* 16:73–80.

Recanzone GH, Merzenich MM, Jenkins WM, Grajski KA, and Dinse HR (1992) Topographic reorganization of the hand representation in cortical area 3b owl monkeys trained in a frequency-discrimination task. *J Neurophysiol* 67:1031–1056.

Redila VA, Olson AK, Swann SE, Mohades G, Webber AJ, Weinberg J, and hristie BR (2006) Hippocampal cell proliferation is reduced following prenatal ethanol exposure but can be rescued with voluntary exercise. *Hippocampus* 16:305–311.

Rhyu IJ, Boklewski J, Ferguson B, Lee KJ, Lange H, Bytheway J, Lamb J, McCormick K, Williams N, Cameron J, and Greenough WT (2003) Exercise training is associated with increased cortical vascularization in adult female cynomolgus monkeys. *Soc Neurosci Abstracts* 33:920–921.

Riesen AH (1965) Effects of visual deprivation on perceptual function and the neural substrate. In: *Symposium bel air II, Desafferentation experimentale et Clinique* (DeAjuriaguerra J, ed), pp. 47–66. Geneva: George and Cie.

Rola R, Raber J, Rizk A, Otsuka S, VandenBerg SR, Morhardt DR, and Fike JR (2004) Radiation-induced impairment of hippocampal neurogenesis is associated with cognitive deficits in young mice. *Exp Neurol* 188:316–330.

Sanchez MM, Hearn EF, Do D, Rilling JK, and Herndon JG (1998) Differential rearing affects corpus callosum size and cognitive function of rhesus monkeys. *Brain Res* 812:38–49.

Sathian K (2005) Visual cortical activity during tactile perception in the sighted and the visually deprived. *Dev Psychobiol* 46:279–286.

Schousboe A, Westergaard N, Sonnewald U, Petersen SB, Yu AC, and Hertz L (1992) Regulatory role of astrocytes for neuronal biosynthesis and homeostasis of glutamate and GABA. *Prog Brain Res* 94:199–211.

Shepherd GM, Pologruto TA, and Svoboda K (2003) Circuit analysis of experience-dependent plasticity in the developing rat barrel cortex. *Neuron* 38:277–289.

Sholl DA (1956) The measurable parameters of the cerebral cortex and their significance in its organization. *Prog Neurobiol* 2:324–333.

Sirevaag AM, Black JE, Shafron D, and Greenough WT (1988) Direct evidence that complex experience increases capillary branching and surface area in visual cortex of young rats. *Brain Res* 471:299–304.

Sirevaag AM, and Greenough WT (1985) Differential rearing effects on rat visual cortex synapses. II. Synaptic morphometry. *Brain Res* 351:215–226.

Sirevaag AM, and Greenough WT (1987) Differential rearing effects on rat visual cortex synapses. III. Neuronal and glial nuclei, boutons, dendrites, and capillaries. *Brain Res* 424:320–332.

Sirevaag AM, and Greenough WT (1991) Plasticity of GFAP-immunoreactive astrocyte size and number in visual cortex of rats reared in complex environments. *Brain Res* 540:273–278.

Snyder JS, Hong NS, McDonald RJ, and Wojtowicz JM (2005) A role for adult neurogenesis in spatial long-term memory. *Neuroscience* 130:843–852.

Sorra KE, and Harris KM (2000) Overview on the structure, composition, function, development, and plasticity of hippocampal dendritic spines. *Hippocampus* 10:501–511.

Stern EA, Maravall M, and Svoboda K (2001) Rapid development and plasticity of layer $\frac{2}{3}$ maps in rat barrel cortex in vivo. *Neuron* 31:305–315.

Stewart MG, Bourne RC, and Gabbott PL (1986) Decreased levels of an astrocytic marker, glial fibrillary acidic protein, in the visual cortex of dark-reared rats: Measurement by enzyme-linked immunosorbent assay. *Neurosci Lett* 63:147–152.

Stranahan AM, Khalil D, and Gould E (2006) Social isolation delays the positive effects of running on adult neurogenesis. *Nat Neurosci* 9:526–533.

Szeligo F, and Leblond CP (1977) Response of the three main types of glial cells of cortex and corpus callosum in rats handled during suckling or exposed to enriched, control and impoverished environments following weaning. *J Comp Neurol* 172:247–263.

Tanapat P, Hastings NB, Reeves AJ, and Gould E (1999) Estrogen stimulates a transient increase in the number of new neurons in the dentate gyrus of the adult female rat. *J Neurosci* 19:5792–5801.

Tanzi E (1893) I fatti e le induzioni nell'odierna istologia del sistema nervoso. [The facts and the inductions in current histology of the nervous system.]. *Riv Sperimen Fren Med LegMent Alien* 19:419–472.

Tauber H, Waehneldt TV, and Neuhoff V (1980) Myelination in rabbit optic nerves is accelerated by artificial eye opening. *Neurosci Lett* 16:235–238.

Taveggia C, Zanazzi G, Petrylak A, Yano H, Rosenbluth J, Einheber S, Xu X, Esper RM, Loeb JA, Shrager P, Chao MV, Falls DL, Role L, and Salzer JL (2005) Neuregulin-1 type III determines the ensheathment fate of axons. *Neuron* 47:681–694.

Tees RC (1968) Effect of early restriction on later form discrimination in the rat. *Can J Psychol* 22:294–301.

Tees RC, and Cartwright J (1972) Sensory preconditioning in rats following early visual deprivation. *J Comp Physiol Psychol* 81:12–20.

Theoret H, Merabet L, and Pascual-Leone A (2004) Behavioral and neuroplastic changes in the blind: Evidence for functionally relevant cross-modal interactions. *J Physiol Paris* 98:221–233.

Tieman SB (1984) Effects of monocular deprivation on geniculocortical synapses in the cat. *J Comp Neurol* 222:166–176.

Tieman SB (1985) The anatomy of geniculocortical connections in monocularly deprived cats. *Cell Mol Neurobiol* 5:35–45.

Tieman SB, and Hirsch HV (1982) Exposure to lines of only one orientation modifies dendritic morphology of cells in the visual cortex of the cat. *J Comp Neurol* 211:353–362.

Trachtenberg JT, Chen BE, Knott GW, Feng G, Sanes JR, Welker E, and Svoboda K (2002) Long-term in vivo imaging of experience-dependent synaptic plasticity in adult cortex. *Nature* 420:788–794.

Tsay D, and Yuste R (2004) On the electrical function of dendritic spines. *Trends Neurosci* 27:77–83.

Turner AM, and Greenough WT (1983) Synapses per neuron and synaptic dimensions in occipital cortex of rats reared in complex, social, or isolation housing. *Acta Stereol 2* (Suppl. I):239–244.

Turner AM, and Greenough WT (1985) Differential rearing effects on rat visual cortex synapses. I. Synaptic and neuronal density and synapses per neuron. *Brain Res* 329:195–203.

Uesaka N, Hirai S, Maruyama T, Ruthazer ES, and Yamamoto N (2005) Activity dependence of cortical axon branch formation: A morphological and electrophysiological study using organotypic slice cultures. *J Neurosci* 25:1–9.

Uylings HB, Kuypers K, Diamond MC, and Veltman WA (1978) Effects of differential environments on plasticity of dendrites of cortical pyramidal neurons in adult rats. *Exp Neurol* 62:658–677.

Valverde F (1971) Rate and extent of recovery from dark rearing in the visual cortex of the mouse. *Brain Res* 33:1–11.

Van Boven RW, Hamilton RH, Kauffman T, Keenan JP, and Pascual-Leone A (2000) Tactile spatial resolution in blind braille readers (1). *Am J Ophthalmol* 130:542.

van Praag H, Kempermann G, and Gage FH (1999) Running increases cell proliferation and neurogenesis in the adult mouse dentate gyrus. *Nat Neurosci* 2:266–270.

Volterra A, Magistretti P, and Haydon PG (2002) *The Tripartite Synapse: Glia in Synaptic Transmission.* Oxford, UK: Oxford University Press.

Voutsinos-Porche B, Bonvento G, Tanaka K, Steiner P, Welker E, Chatton JY, Magistretti PJ, and Pellerin L (2003b) Glial glutamate transporters mediate a functional metabolic crosstalk between neurons and astrocytes in the mouse developing cortex. *Neuron* 37:275–286.

Voutsinos-Porche B, Knott G, Tanaka K, Quairiaux C, Welker E, and Bonvento G (2003a) Glial glutamate transporters and maturation of the mouse somatosensory cortex. *Cereb Cortex* 13:1110–1121.

Walk RD, and Walters CP (1973) Effect of visual deprivation on depth discrimination of hooded rats. *J Comp Physiol Psychol* 85:559–563.

Wallace CS, and Kilman VL, Withers GS, and Greenough WT (1992) Increases in dendritic length in occipital cortex after 4 days of differential housing in weanling rats. *Behav Neural Biol* 58:64–68.

Wati H, Kudo K, Qiao C, Kuroki T, and Kanba S (2006) A decreased survival of proliferated cells in the hippocampus is associated with a decline in spatial memory in aged rats. *Neurosci Lett.*

Wesa JM, Chang FL, Greenough WT, and West RW (1982) Synaptic contact curvature: Effects of differential rearing on rat occipital cortex. *Brain Res* 256:253–257.

West RW, and Greenough WT (1972) Effect of environmental complexity on cortical synapses of rats: Preliminary results. *Behav Biol* 7:279–284.

Whishaw IQ, Zaborowski JA, and Kolb B (1984) Postsurgical enrichment aids adult hemidecorticate rats on a spatial navigation task. *Behav Neural Biol* 42:183–190.

Wiesel TN, and Hubel DH (1965a) Extent of recovery from the effects of visual deprivation in kittens. *J Neurophysiol* 28:1060–1072.

Wiesel TN, and Hubel DH (1965b) Comparison of the effects of unilateral and bilateral eye closure on cortical unit responses in kittens. *J Neurophysiol* 28:1029–1040.

Winfield DA (1981) The postnatal development of synapses in the visual cortex of the cat and the effects of eyelid closure. *Brain Res* 206:166–171.

Winocur G, Wojtowicz JM, Sekeres M, Snyder JS, and Wang S (2006) Inhibition of neurogenesis interferes with hippocampus-dependent memory function. *Hippocampus* 16:296–304.

Withers GS, Fahrbach SE, and Robinson GE (1995) Effects of experience and juvenile hormone on the organization of the mushroom bodies of honey bees. *J Neurobiol* 26:130–144.

Withers GS, and Greenough WT (1989) Reach training selectively alters dendritic branching in subpopulations of layer II–III pyramids in rat motor-somatosensory forelimb cortex. *Neuropsychologia* 27:61–69.

Wong EY, and Herbert J (2006) Raised circulating corticosterone inhibits neuronal differentiation of progenitor cells in the adult hippocampus. *Neuroscience* 137:83–92.

Yuste R, and Bonhoeffer T (2001) Morphological changes in dendritic spines associated with long-term synaptic plasticity. *Annu Rev Neurosci* 24:1071–1089.

Zhang Q, and Haydon PG (2005) Roles for gliotransmission in the nervous system. *J Neural Transm* 112:121–125.

Zuo Y, Lin A, Chang P, and Gan WB (2005b) Development of long-term dendritic spine stability in diverse regions of cerebral cortex. *Neuron* 46:181–189.

Zuo Y, Yang G, Kwon E, and Gan WB (2005a) Long-term sensory deprivation prevents dendritic spine loss in primary somatosensory cortex. *Nature* 436:261–265.

CHAPTER 3

Genetics in Learning and Memory

Yalin Wang, Josh Dubnau, Tim Tully, and Yi Zhong

Cold Spring Harbor Laboratory, Cold Spring Harbor, NY 11724

I. INTRODUCTION

The power of genetics has been clearly demonstrated in studies of embryonic development. Systematic searches for mutant phenotypes during embyogenesis in *Drosophila* (Nusslein-Volhard and Wieschaus, 1980; Johnston and Nusslein-Volhard, 1992; Wieschaus, 1996) and zebrafish (Grunwald and Eisen, 2002) have uncovered most of the genes controlling embryonic pattern formation. The number of genes involved is limited, and they can be categorized into specific functional groups according to their phenotypic effects. Many of the corresponding genes have been characterized molecularly, thereby identifying components of gene regulation and cell signaling pathways. Genes first discovered in flies have subsequently been identified in vertebrates, and basic genetic mechanisms controlling early development of invertebrates have been shown to be conserved in vertebrate systems. Thus, a driving force behind analysis of development has been gene discovery in flies, which in turn has provided experimental and conceptual tools to dissect mechanisms of morphogenesis in both invertebrate and vertebrate species.

Neurobiology of Learning and Memory, Second Edition

For centuries it has been accepted that heredity contributes to learning and memory. The experimental evidence came in the mid-twentieth century when Tryon (Tryon, 1940) used bidirectional selection experiments to breed "bright" and "dull" strains of rats. Bright and dull rats were selected based on their abilities to learn to navigate a maze for a food reward and were mated among themselves. After several generations, rats from the "maze-bright" strain learned quickly, whereas those from the "maze-dull" strain learned much more slowly. Similarly in the blowfly *Phormia regina*, Hirsch and coworkers (T.R. McGuire and Hirsch, 1977; Zawistowski and Hirsch, 1984) bred flies showing high or low performance during classical conditioning of the proboscis extension reflex (PER). They were able to generate bright and dull strains of flies that had significantly different learning scores for classical conditioning.

These experiments established genetic components in learning and memory. The application of genetics in studying learning and memory started when Seymour Benzer at the California Institute of Technology brought to the field of *Drosophila* behavioral genetics the single-gene mutant approach in the early 1970s. He suggested that most genes involved with a complex trait might be identified by direct chemical mutagenesis to isolate mutations one gene at a time (Tully, 1996) — in much the same way that genetic screens were being used to study simpler processes. He and colleagues carried out genetic screening for learning and memory mutants in *Drosophila melanogaster*. Although greeted in the beginning by skepticism, this approach has led to some fundamental insights into mechanisms of learning and memory. The findings in *Drosophila* have converged with studies of *Aplysia* in identifying the cAMP-mediated signal transduction as a central pathway for learning and memory. The genetic studies of learning and memory have also been extended to other model systems, most notably, mice.

With modern techniques of genetics, many genes involved in learning and memory have been discovered in *Drosophila*, characterizations of which have given us considerable understanding of learning and memory mechanisms. The success of genetics in embryonic development is being played out in studies of learning and memory. In this chapter, we review how genetic approaches have been applied to dissect the processes of learning and memory in *Drosophila melanogaster* and use them to exemplify the role of genetics in learning and memory research.

II. GENETIC SCREENING OF LEARNING AND MEMORY MUTANTS

A. Behavior Measures of Learning and Memory

As important as a readily identifiable trait is for developmental studies, a simple and reliable behavioral assay is critical for identifying abnormal learning and/or

FIGURE 3-1   The T-maze assay for Pavlovian olfactory learning and memory in *Drosophila*.
A. *Training*: About 100 flies are sequestered in a training tube whose inner surface is covered by
an electrifiable grid. Odor currents can be drawn through the tube using a vacuum source
attached to chamber **a** in the central elevator. The flies are first exposed to odor 1 while electric
current is delivered to the grid and then to odor 2 with the current switched off. Such training
can be repeated without an interval (massed training) or with an interval (spaced training). **B.**
Testing: After training, the flies are transferred to chamber **b** in the central elevator to a choice
point where currents of odor 1 and 2 are drawn in through two opposite testing tubes by vacuum
attached to chamber **b**. If they have learned, the flies will avoid odor 1, which is associated with
electric shock. Such performance can be quantified by the distribution of flies in the two testing
tubes. Performance measured immediately after training is regarded as learning and at later time
points as memory.

memory defects in large populations of flies. This challenge was met by Benzer
and his colleagues Quinn and Harris at the California Institute of Technology,
who developed an olfactory behavior test that paired electric shock with an
odor cue, allowing assessment of associative learning in flies. Subsequently,
Tully and Quinn (1985), then at Princeton University, developed a Pavlovian
olfactory learning assay (Fig. 3-1) that allows direct comparisons of the
behavioral properties of learning and memory among *Drosophila* and other
invertebrate and vertebrate species. In this assay, flies are exposed to two odors.
During exposure to one of them, the flies are given an electric shock. Such
training results in a strong avoidance of the odor that is accompanied by electric shock and, consequently, robust memory retention. Flies can remember up
to one day after a single training session, and four days after 10 successive
training sessions without any interval. The memory can be further extended
to a week if the flies are trained for 10 sessions with proper intervals (Tully
et al., 1994).

B. Chemical Mutagenesis

The initial screening was carried out by feeding flies the chemical mutagen
ethylmethane sulfonate (EMS). EMS is an alkylating agent that adds an alkyl
group to the guanine bases of DNA. On replication, the modified guanine
(G) base will pair with thymine (T) instead of the preferred cytosine (C) base,
leading to point mutations from G:C to A:T. Other chemical mutagens,

including N-ethyl-N-nitrosourea (ENU), also directly modify DNA bases, resulting in point mutations. With such an approach, Benzer's group at Caltech identified the first single-gene mutant for associative learning, *dunce* (*dnc*) (Dudai et al., 1976). Continuing efforts by Quinn's group at Princeton isolated *rutabaga* (*rut*) (Livingstone et al., 1984), *amnesiac* (*amn*) (Quinn et al., 1979), *radish* (*rsh*) (Folkers et al., 1993), *cabbage* (*cab*) (Aceves-Pina et al., 1979), and *turnip* (*tur*) (Choi et al., 1991).

Chemical mutagenesis offers a high mutation rate and a broad target range. Chemical mutagens usually have low target specificity, and therefore the whole genome can be targeted for their mutational effects. The point mutations induced by chemical mutagens generate a diverse range of alleles, sometimes with conditional alleles (for example, thermosensitive alleles), which can provide more informative insights into the functions of a gene. One limiting factor for this approach is that the affected genes are hard to clone.

C. Transposon Mutagenesis

The transposon mutagenesis is achieved by random transposition of transposons, mobile DNA elements (P-elements in *Drosophila*). Stripped of their autonomous ability to transpose, engineered P-elements can be stably integrated into the *Drosophila* genome and remobilized simply by crossing the P-element-carrying fly with one possessing a transgenic transposase activity (Rubin and Spradling, 1982; Cooley et al., 1988). The mobilized P-element can insert into random chromosomal loci. Genes neighboring the P-element insertion sites are often disrupted. Compared to chemical mutagenesis, affected genes in transposon mutagenesis can be easily identified with the help of the P-element tag (Cooley et al., 1988). The limitation is that in transposon mutagenesis, complete loss-of-function alleles are typically generated and the mutations are biased by the sequence specificity of the transposon. With this method, three additional learning mutants were isolated: *latheo* (*lat*) (Boynton and Tully, 1992), *linotte* (*lio*) (Dura et al., 1993), and *nalyot* (*nal*) (DeZazzo et al., 2000).

A variant of the transposon approach utilizes the enhancer-trap technique (O'Kane and Gehring, 1987), which in addition to producing mutagenesis, allows quick identification of the expression pattern of the affected gene. It incorporates in the P-element a reporter gene, *lacZ*, whose expression generally reflects the nearby gene and which can be revealed by β-galactosidase staining. Screening was performed using the enhancer-detection method to identify genes showing preferential expression the mushroom body (MB), a brain structure critical for olfactory learning and memory (Han et al., 1996). Among these genes, *leonardo* (*leo*) (Skoulakis and Davis, 1996), *volado* (*vol*) (Grotewiel et al., 1998), and *fasciclin II* (*fasII*) (Cheng et al., 2001) are found to be involved in olfactory learning and memory. This approach also revealed that *dnc*, *rut*, and *DC0* (see later) are preferentially expressed in the MB. A cautionary note is that preferential expression does not necessarily imply the region of relevant gene function. For example, *rut* is also required for visual memory in another structure of the brain (Liu et al., 2006), although its expression is enriched in the MB.

Most recently, Tully's group at Cold Spring Harbor Laboratory used the enhancer-trap approach and screened for mutants with deficits in one-day memory after spaced training (Dubnau et al., 2003b). Sixty mutant strains were identified, and the P-element insertion sites were defined for 58 of those. It is revealed that in 28 mutant strains, the P-element landed in 25 known transcriptional units (three of these genes were hit twice). One of these transcriptional units is *pumilio* (*pum*), a transcript-specific translational repressor. Notably, *pumilio* was also uncovered in a DNA microarray screening (see Section D). A similar screening effort for identifying long-term-memory mutants uncovered *crammer* (*cer*), which might function as a transinhibitor of cathepsins (Comas et al., 2004).

D. DNA Microarray Screening

Instead of working on one gene at a time, DNA microarray offers the opportunity to monitor a whole genome simultaneously on a chip. Dubnau's group at the Cold Spring Harbor Laboratory used this technology and identified genes that are transcriptionally regulated during long-term-memory formation in normal flies (Dubnau et al., 2003b). Among those are *pumilio*, which is also identified in the aforementioned mutagenesis screening, and a few others, including *staufen* (*stau*), *orb*, and *eIF-5C*, which are known to be involved in local control of mRNA translation.

E. Anatomical Screening

Heisenberg's group at the University of Würzburg, Germany, screened for single-gene mutants with gross anatomical defects in various regions of the adult brain. *Minibrain*, a mutation that reduces the size of the adult brain, and several mutations affecting the mushroom bodies (MBs) or central complex (CC) — namely, *mushroom body miniature*, *mushroom body deranged*, *central body defect*, *ellipsoid body open*, *central complex deranged*, *central complex broad*, and *no bridge* — also show olfactory learning defects (Heisenberg et al., 1985; R.L. Davis, 1996; de Belle and Heisenberg, 1996).

F. Reverse Genetics to Identify Learning and Memory Mutants

While forward genetics starts from a phenotype and moves toward identification of the mutated gene, reverse genetics goes in the opposite direction. It starts with the knowledge of a gene sequence. Then the gene is disrupted or modified through targeted mutagenesis, followed by examination of the ensuing phenotypes to determine the gene's function. In *Drosophila*, several systematic gene-disruption projects using P-element insertion have generated thousands of stocks that each harbor a single P-element construct inserted at a known location in the genome. In addition, mutations in a known DNA sequence can be generated by homologous recombination (Rong and Golic, 2000; Rong et al., 2002) or by combining chemical mutagenesis and methods that detect single-nucleotide polymorphism (Stemple, 2004; Winkler et al., 2005). Aided with knowledge derived from molecular identification of the learning and memory mutants generated from mutagenesis screenings, gene discovery can be facilitated in reverse-genetics approaches by examining ready-made stocks with mutations in genes of interest for learning and memory defects. For example, biochemical analysis of *dunce* and *rutabaga* suggested that the cAMP signal transduction cascade is involved with olfactory learning. This discovery prompted a focused analysis of learning in extant mutants of *DC0* (the catalytic subunit of the cAMP-dependent protein kinase, PKA), *Su-var(3)* (a type I protein phosphatase PP1), and *Shaker* (R.L. Davis, 1996; Waddell and Quinn, 2001). Moreover, reverse-genetic disruptions of *Gα*, *PKA-RI* (a type I regulatory subunit of PKA) and *dCREB2* (cAMP-responsive element binding transcription factor) revealed roles for these genes in olfactory associative learning and memory (Waddell and Quinn, 2001; Dubnau et al., 2003a).

Other reverse-genetics strategies include direct manipulation of the gene of interest. The reverse-genetics approach from studies of human neurodegenerative disorders has led to identification of *neurofibromatosis 1* (*NF1*) and *nebula* (*nla*) as learning and memory genes (Guo et al., 2000; Chang et al., 2003). Neurofibromatosis type 1 (NF1) is a common human genetic disorder that exhibits tumors in the nervous system and learning and memory defects. It is caused by mutations in the *NF1* gene (Cawthon et al., 1990; Viskochil et al., 1990; Wallace et al., 1990). Down's syndrome (DS) is the most common cause of mental retardation. Studies in DS patients have revealed a segment of chromosome 21, the DS region, that is closely related to phenotypic features of the DS. One of the genes in the DS region is the DS critical region 1 (*DSCR1*) gene. The cognitive defects in both disorders prompted interests in examining possible roles of *NF1* and *DSCR1* in learning and memory in *Drosophila*. It has turned out that both genes are involved in learning and memory. *Drosophila* *NF1* mutants show learning defects that could be rescued by conditioned expression of the *NF1* transgene (Guo et al., 2000). *Nebula* is a *Drosophila* homolog of DSCR1. Mutations in *nla* produce learning and long-term memory defects that can be rescued by conditioned expression of an *nla* transgene (Chang et al., 2003).

G. Essential Controls

Critical to the characterization of single-gene mutations involved with learning is proper assessment of sensorimotor responses. This issue arises from the fact that learning per se is not observed directly but is inferred from a change in behavioral response after exposure to a stimulus. Hence, alterations in factors such as perception of the stimuli, fatigue, and motivational state also can affect behavioral responses. Single-gene mutants often have pleiotropic effects on many different behavioral responses, which also can make it difficult to determine if a given gene is involved with learning per se. A resolution to these problems lies first in determining which behavioral responses are pertinent to a given learning task and then in developing task-relevant assays for the sensorimotor responses required for normal performance. For Pavlovian olfactory conditioning, olfactory acuity and shock reactivity assays quantitate the flies' abilities to sense and respond to the odors and shock stimuli used in the conditioning procedure. When a particular mutant fails these sensorimotor assays, one cannot conclude that poor performance in conditioning experiments results from a defect in learning per se. Conversely, when a learning mutant passes these sensorimotor tests, one gains confidence that the corresponding gene is involved in learning, regardless of any pleiotropic effects it may have on other behaviors. Of the genes mentioned earlier, seven (six affecting CC anatomy and *tur*) fail the task-relevant assays of sensorimotor responses required for Pavlovian olfactory learning. This highlights the importance of assaying performance in task-relevant sensorimotor responses.

III. GENETIC MANIPULATION OF CANDIDATE LEARNING AND MEMORY GENES

Unlike anatomical screens, behavioral screens for learning mutants can identify genes involved in both the development of neuronal structures and the biochemical pathways underlying behavioral plasticity of adults. Distinguishing these roles cannot be accomplished simply by analyzing gross neuroanatomical structure in the adult. Furthermore, to understand the underlying neural circuits of learning and memory, it is critical to know when and where in the brain the function of a candidate gene is required. Therefore, a candidate gene needs to be manipulated both in time, to rule out possible developmental effects and to determine its temporal requirement, and in space, to identify

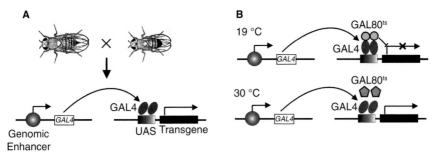

FIGURE 3-2   Transgene targeting systems. **A.** *The GAL4-UAS system*: In the GAL4 line, the yeast transcriptional activator GAL4 is inserted into the fly genome. In a separate line, a transgene placed next to an upstream activating sequence (UAS), which contains multiple GAL4 sites, is also inserted into the fly genome. After the two lines are crossed, transgene expression is activated in the progeny when GAL4 binds UAS. Regional specificity is conferred by GAL4 expression, which is driven by its proximity to a specific genomic enhancer, thus reflecting the expression pattern of a nearby gene. **B.** *The TARGET system*: It adds a temporal control over the GAL4-UAS system. The temporal specificity comes from GAL80^{ts}, the expression of which is controlled by a ubiquitous tubulin 1a (Tub) promoter. A fly carrying GAL80^{ts} and a GAL4 that is expressed in a tissue of interest is crossed with a fly carrying a transgene. At permissive temperatures, GAL80^{ts} binds to GAL4 and suppresses transgene expression in progeny flies, whereas at restrictive temperatures, GAL80^{ts} becomes inactivated and allows GAL4-activated transgene expression to go forward. Therefore, by raising F1 flies at different temperatures at specific times, transgene expression can be controlled temporally and spatially. (Adapted from S.E. McGuire et al., 2003.)

the neural correlates of learning and memory. The temporal manipulation may be accomplished if fortuitously there exist conditional alleles of the gene of interest. This is usually through a temperature-sensitive mutant allele, whose function is disrupted at restrictive temperatures. Generally, the temporal and spatial controls over expression of a cloned gene can be achieved separately by the heat-inducible system and the GAL4-UAS system.

The temporal control in the heat-inducible system is conferred by a promoter from the *Drosophila heat-shock protein 70 (hsp70)* gene. The heat-inducible hsp70 promoter is cloned into a P-element vector upstream of a transgene to be expressed, and together they are incorporated into the fly genome following germ-line transformation. Expression of the transgene is controlled by the hsp70 promoter, which can be induced by heat. Hsp70-driven expression is global without any spatial specificity.

The GAL4-UAS system (Brand and Perrimon, 1993), which is based on the yeast GAL4 transcription factor and its upstream activating sequence (UAS), provides excellent spatial specificity for transgene expression (Fig. 3-2A). Analogous to an enhancer element in multicellular eukaryotes, UAS is essential for the transcriptional activation of GAL4-regulated genes. GAL4 does not activate native *Drosophila* genes but can drive expression of a transgene placed downstream of UAS. GAL4, constructed in the same way as in an enhancer trap, respond to neighboring enhancers, and its expression often reflects that of the neighboring gene. Thus, the GAL4-driven transgene can be expressed in the same cell population or tissue as an endogenous *Drosophila* gene next to the GAL4 insertion site. Many fly lines with specific expression of GAL4 in various tissues and cell types have been generated. Additionally, lines carrying various transgenes fused with a UAS promoter have been established.

However, the GAL4–UAS system does not provide a temporal control. This can be achieved with a GAL4 under the control of the hsp70 promoter (Brand et al., 1994), but the spatial specificity is lost. To capitalize on the great collection of GAL4 and UAS lines and at the same time allow a temporal control, several systems have been developed with modifications of the GAL4-UAS system. They include the GeneSwitch, the tetracycline-inducible system, and TARGET (Osterwalder et al., 2001; Roman et al., 2001; Stebbins and Yin, 2001; S.E. McGuire et al., 2003). In all three systems, the spatial control is conferred by the GAL4-UAS system, but the eventual expression of the transgene is dependent on either feeding/withdrawing of specific substances, such as in GeneSwitch and the tetracycline-inducible system, or on elevation of temperature, such as in TARGET (Fig. 3-2B), and therefore a temporal handle is added onto the GAL4-UAS system. However, these systems usually take a day or days to attain the behavioral effect of transgene expression. So far, the heat-shock-induced transgene expression remains the best choice for the temporal control.

With these tools in hand, a gene of interest can be manipulated in many ways. Pertinent to the demonstration of a gene's involvement in learning and memory, one would naturally like to disrupt its function in time and space and examine the impact on learning and memory. This type of manipulation can be achieved by expression of a transgene dominant-negative to the endogenous gene or by RNA interference (RNAi). While the dominant-negative trangene product will compete in vivo with the endogenous protein and interrupt its function, transgenic expression of double-stranded RNA triggers an enzymatic degradation of the homologous endogenous mRNA, thus producing gene-specific posttranscriptional silencing (Lam and Thummel, 2000; Zamore et al., 2000). Another type of manipulation involves reintroducing a wild-type transgene into the mutant fly strain. Rescue of the mutant learning and memory defect by the expression of the transgene can provide ultimate evidence that the right gene has been identified from the mutant. Further insight into the involvement of the gene in learning and memory can be gained from the temporal and spatial requirement of the transgene expression for the rescue.

IV. GENETIC DISSECTION OF LEARNING AND MEMORY

Genetic tools provide powerful means to study learning and memory. Genetic screening has enabled unbiased discovery of gene mutations affecting learning and memory. The identification of these genes and subsequent reverse-genetics manipulation have allowed dissection of learning and memory at the molecular, synaptic, and systems levels.

A. Dissection of Biochemical Pathways

The behavioral and molecular studies of the mutants *dnc, rut,* and *amn* have suggested that signal transduction mediated by cAMP plays a key role for learning and memory in the fly. *dnc, rut,* and *amn* loci encode, respectively, a phosphodiesterase that degrades cAMP, an adenylyl cyclase (AC) that converts ATP to cAMP, and a peptide transmitter that stimulates AC. Reverse genetics has shown that other players in the cAMP pathway all affect learning and memory.

For *dnc, rut, DC0,* and *dCREB2,* inducible transgenes and conditional alleles have been used to demonstrate an acute role for these genes in the biochemistry of adult associative learning (Drain et al., 1991; Yin et al., 1994; Dauwalder and Davis, 1995; Yin et al., 1995; Li et al., 1996; McGuire et al., 2003). Induced expression of a transgene encoding a *dnc⁺* cDNA yields partial rescue of the learning defect in *dnc^{M14}* mutants, revealing a role for *dnc* during adult associative learning. In the absence of complete rescue, however, it remains possible that the residual learning defect results from anatomical defects that arise during development. An acute role for *rut* in memory is demonstrated with the TARGET system (S.E. McGuire et al., 2003). Transient expression of the *rut*-encoded AC in the adult is necessary and sufficient to rescue the *rut* mutant memory defect, while expression during development is not.

DC0, the catalytic subunit of PKA, also has been shown to function in learning and memory in the adult. Pavlovian olfactory learning is reduced in heteroallelic combinations of *DC0^{581}* and *DC0^{B10}* (Skoulakis et al., 1993); memory retention three hours after training, however, is normal in these mutants. Induced expression of transgenes encoding either the catalytic subunit of PKA, a peptide inhibitor of PKA, or a truncated mammalian type II regulatory subunit (RII) of PKA (in which the cAMP binding site was removed) causes a decrease in Pavlovian olfactory learning (Drain et al., 1991). In contrast, transgenes expressing full-length RII subunit have no effect. Task-relevant sensorimotor responses are normal before or after heat shock in each of these transgenic lines and in the heteroalleleic mutants. These findings demonstrate an acute role for PKA in adult olfactory learning.

Evidence for an acute requirement for PKA activity during memory formation after Pavlovian olfactory learning comes from studies of DCO^{X4}, a conditional lethal allele. At the permissive temperature, DCO^{X4}/Df hemizygous mutants show a reduction in learning, but they have normal memory decay. At the restrictive temperature, however, DCO^{X4}/Df hemizygous mutants show the same learning defect and a disruption of middle-term memory (Li et al., 1996). This study suggests that different thresholds of PKA activity may be required for development, learning, and memory formation.

More insights into cAMP signaling have come from analyses of a neuropeptide known as pituitary AC activating peptide (PACAP) and the *Drosophila* homolog of the *human neurofibromatosis type 1* (*NF1*) gene. In vertebrates, PACAP activates AC through a G protein–coupled receptor. In *Drosophila*, application of mammalian PACAP38 to the neuromuscular junction (NMJ) results in a slow inward current lasting tens of seconds, followed by an enhancement of outward K^+ current. This PACAP-induced response is impaired in *rut* mutants (Zhong, 1995; Zhong and Pena, 1995), suggesting activation of AC by PACAP. Concomitant activation of the Ras and cAMP signaling pathways is required to mediate PACAP function.

In mammals, *NF1* is believed to function as a GTPase-activating protein for Ras (Ras-GAP) and therefore as a negative regulator of Ras (Ballester et al., 1990; Buchberg et al., 1990; Martin et al., 1990; Xu et al., 1990a, 1990b). In flies, mutations in *NF1* eliminate the PACAP response. Application of either cAMP analogs or forskolin (which stimulates G protein-coupled AC) restores the PACAP response in *NF1* mutant flies, suggesting that *NF1* regulates *rutabaga*-encoded AC (Guo et al., 1997). G protein–stimulated AC activity is reduced in *NF1* mutants and can be restored by both *Drosophila* and human *NF1* transgenes (Guo et al., 2000; Tong et al., 2002). Moreover, over-expression of PKA is sufficient to rescue developmental defects of *NF1* mutant flies (The et al., 1997).

Three distinct AC signaling pathways in *Drosophila* have now been identified, including a novel growth factor–activated NF1/Ras-dependent AC pathway, as well as two separate neurotransmitter-stimulated AC pathways (Fig. 3-3) (Hannan et al., 2006). The growth factor–stimulated AC pathway can be disrupted by mutations in the epidermal growth factor receptor (EGFR), NF1, and Ras but not $G\alpha_s$. The second AC pathway is stimulated by serotonin and histamine requiring NF1 and $G\alpha_s$. The third is a classical $G\alpha_s$-dependent AC pathway, which is stimulated by Phe-Met-Arg-Phe-amide (FMRFamide) and dopamine. The NF1-dependent Rut-AC is required for learning (Guo et al., 2000).

Further study of the cAMP signaling cascade components, along with information about other learning/memory genes, suggests the involvement of additional biochemical pathways. Amorphic alleles of *rut*, for instance, reduce but do not eliminate olfactory learning (Livingstone et al., 1984). This suggests

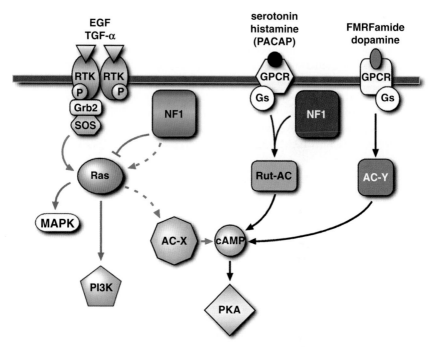

FIGURE 3-3   Three putatively distinct pathways that can activate AC. The first is mediated by a receptor tyrosin kinase (RTK) pathway that is stimulated by growth factors such as EGF and TGFα and is dependent on both NF1 and Ras. This pathway activates an unidentified AC (AC-X) and does not involve Gα$_s$. The second is stimulated by serotonin and histamine, and possibly PACAP38. It acts through *rutabaga*-encoded AC (Rut-AC) and depends on both NF1 and Gα$_s$ but not Ras. The third is stimulated by FMRFamide and dopamine, which activate an unidentified AC (AC-Y). It requires Gα$_s$ but not NF1 or Ras. (Adapted from Hannan et al., 2006.)

the involvement of additional cyclases or of novel signaling pathways. The *rsh* mutant is deficient in anesthesia-resistant memory, a form of long-term memory in *Drosophila*, and it has been recently found that *rsh* encodes a protein with 23 predicted PKA phosphorylation sites (Folkers et al., 2006). In addition to genes involved in the cAMP signaling cascade, mutations in other pathways have been identified. In the learning mutant *tur*, protein kinase C (PKC) activity is significantly reduced compared with activity in wild-type flies (Choi et al., 1991). Moreover, an atypical PKC is required for a specific phase of memory (Drier et al., 2002). Ca^{2+}-CaM-dependent protein kinase II (CaMKII) is involved in courtship learning (Griffith et al., 1993). The learning mutant gene *leo* encodes a *Drosophila* homolog of the vertebrate 14-3-3ζ isoform, a protein that is known to interact with diverse signaling proteins, including RAF-1, in the mitogen-activated protein kinase (MAPK) pathway and PKC (Skoulakis and Davis, 1996). A role for cell adhesion is demonstrated by the discovery of two mutants of cell adhesion molecules, fasciclin II and integrin,

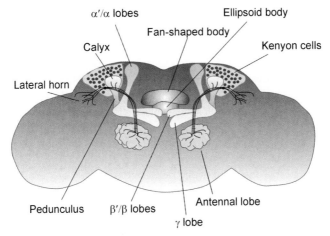

FIGURE 3-4   Cross section of the *Drosophila melanogaster* brain, ventral view. Olfactory information is relayed from the antennal lobe by projection neurons in the inner antennocerebral tract to the MB and further to the lateral horn of the protocerebrum. The MB also receives the US information for the aversive and appetitive olfactory learning through innervations by dopaminergic and octopaminergic neurons, respectively. Axonal projections of the MB neurons (Kenyon cells) extend anteriorly and then divide into α'/α and $\beta'/\beta/\gamma$ lobes. The central complex (CC), a motor output control center, receives inputs from a variety of brain centers, including prominent visual input. The CC comprises four substructures, namely the ellipsoid body and the fan-shaped body (depicted here) and the protocerebral bridge and nodulii (not shown). (Adapted from Dubnau and Tully, 1998.)

that are defective in learning and memory (Grotewiel et al., 1998; Cheng et al., 2001). The characterization of additional learning and memory genes, such as *nal* and *lat*, which respectively encode the Adf1 transcription factor (DeZazzo et al., 2000) and a subunit of the origin recognition complex that is involved in DNA replication (Pinto et al., 1999), and findings of requirement of *Notch* in long-term memory (Ge et al., 2004; Presente et al., 2004) suggest that still more biochemical pathways may be involved. These multiple signaling cascades may act in series or in parallel; they may act in the same neurons or in different brain structures; and they may contribute to different aspects of the learning and memory process.

B. Dissection of Neuroanatomical Pathways

The insect brain contains several prominent neuropillar structures (Fig. 3-4). MBs are large bilaterally symmetric structures that are positioned in the dorsal/posterior region of the central brain. In adult *Drosophila*, MBs consist of ~2,500 neurons (Kenyon cells) in each brain hemisphere. The MB neurons can be divided into three types, the γ, α'/β', and α/β neurons, depending on their birth orders, and they give rise to separate axonal bundles (Crittenden et al., 1998; Lee and Luo, 1999). Axons of the γ neurons form the horizontal γ lobe, while those of α'/β' and α/β neurons bifurcate and extend into vertical α' and α and horizontal β' and β lobes, respectively. The primary olfactory inputs are conveyed from the antennal lobe (AL) along the inner antennal-cerrebral tract (iACT) to the MB and to the lateral horn (LH) of the protocerebrum (Stocker et al., 1990). The LH also receives direct input from the AL along the outer and medial ACT (oACT, mACT). In addition to the cholinergic input that conveys olfactory information to the MB, there are dopaminergic and octopaminergic inputs to MB neurons. The dopaminergic neurons, which innervate both the calyx (dendritic field) and lobes (axon terminals) of the MB, are required for association of odor with electric shock, while octopaminergic innervation of MB underlies association of odors with sugar in an appetitive conditioning procedure (Schwaerzel et al., 2003). The dopaminergic and octopaminergic neurons may carry the US information in the associative learning. This is supported by the finding that dopaminergic neurons respond to electric shock experienced during learning (Riemensperger et al., 2005).

Convergent data demonstrate that MBs play an important role in olfactory associative learning in *Drosophila* (Heisenberg, 2003). Analyses of learning in mutants with defects in brain structure indicated that the MBs are required for olfactory (but not visual) learning (Heisenberg et al., 1985; de Belle and Heisenberg, 1996). Chemical ablation of the MB or targeted disruption of the cAMP signaling cascade with a constitutively active $G\alpha_s$ ($G\alpha_s{}^*$) in the MB abolishes olfactory learning with no effects on the "task-relevant" sensorimotor responses (de Belle and Heisenberg, 1994; Connolly et al., 1996). Associative learning is normal when activated $G\alpha_s{}^*$ is expressed in CC (Connolly et al., 1996). Expression of the *rut*-encoded AC in the MB rescues the *rut* memory defect (Zars et al., 2000; S.E. McGuire et al., 2003), further demonstrating a critical role for cAMP signaling within the MB, at least in early olfactory memory. This is consistent with the finding that multiple components of the cAMP signaling cascade (AC, PDE, PKA catalytic subunit, and RI regulatory subunit) are expressed at high levels in MBs (Han et al., 1996). Temporary disruption of MB neuron output with expression of a temperature-sensitive dominant-negative *dynamin* transgene, *shibirie*, blocks memory retrieval (Dubnau et al., 2001; S.E. McGuire et al., 2001). Taken together, these findings support the idea that MB Kenyon cells are part of an anatomical circuit underlying olfactory learning and memory. In particular, the anatomical "circuit diagram" as well as the genetic perturbations of cAMP signaling in MB support a model in which MB Kenyon cells are at least one anatomical site where the US (electric shock) and CS (odors) are associated.

Results from studies of the *amnesiac* mutant, which is defective in memory but normal in learning, indicates that neurons outside the MB may also be part of the neural circuits underlying learning and memory (Waddell et al., 2000). *Amn* encodes a PACAP-like peptide, and its expression is restricted to the dorsal paired medial (DPM) neurons, a pair of neurons that innervate the axon terminals of MB neurons. Developmental expression of an *amn* transgene in DPM neurons rescues (DeZazzo et al., 1999; Waddell et al., 2000), whereas transiently blocking DPM neuron output with the temperature-sensitive *shibirie* mimics the *amn* mutant memory defect without affecting learning, establishing a link between the DPM neurons and *amn*-dependent memory (Waddell et al., 2000).

A recent forward enhancer-trap mutagenesis screen for genes affecting long-term memory has also revealed the involvement of structures other than the MBs in long-term memory (Dubnau et al., 2003b). Some 60 new transposon-insertion mutants with one-day memory defect after spaced training are identified in this screening. Not surprisingly, about 60% of the mutants yield reporter expression in the MB. But more interestingly, a significant number of the mutants show expression exclusively outside the MB. One such mutant, *murashka*, shows restricted expression in a few neurons that innervate the MB calyx. Genetic manipulation of these newly found candidate genes in both time and space will certainly help us to construct the complete neural circuitry underlying olfactory learning and memory.

The CC is another prominent brain structure situated in the center of the supraesophageal ganglia. It comprises four substructures: the ellipsoid body, the fan-shaped body, the nodulii, and the protocerebral bridge (Fig. 3-4). The CC forms intricate connections to a variety of brain centers, is believed to be a control center for many different behavioral responses (motor output), and receives prominent, but not exclusive, visual input (Strauss, 2002). CC participates in visual memory as mutants with CC structural abnormalities are impaired in visual pattern memory (Liu et al., 2006). Intact *rut* gene is also required for visual-pattern memory. Expression of the wild-type *rut* cDNA (UAS–*rut*$^+$) in CC rescues the visual memory deficit in the mutant *rut* flies. Strikingly, memories of different visual features — the angular orientation of a bar and its vertical position on the fly's retinal field — require *rut*-AC in different subsets of CC neurons.

C. Dissection of Synaptic Plasticity

Because of the remarkable biochemical and pharmacological similarities between mechanisms of cellular and behavioral plasticity, modifications of synaptic strength and structure are widely believed to underlie learning and memory (Kandel and Abel, 1995). Physiological and neuroanatomical studies of synaptic plasticity in learning and memory mutants have strengthened this notion considerably.

Several mutants show defective synaptic structure and function at the larval NMJ. Mutations in either *eag* or *Sh* (which encode K^+ channel subunits), for instance, produce an altered K^+ conductance, and *eag-Sh* double mutants show higher baseline activity and evoked hyperexcitability and decreased expression of the cell adhesion molecule, fasiclin II, resulting in increased synaptic arborization (Budnik et al., 1990; Zhong and Wu, 1991; Zhong et al., 1992; Schuster et al., 1996). *dnc* mutants also show increased neuronal activity and synaptic arborization, and these defects are enhanced in *dnc-Sh* double mutants (Zhong et al., 1992). Moreover, the synaptic defects in *dnc-Sh* and *dnc-eag* double mutants are suppressed by *rut* in triple mutant combinations. At the larval NMJ, then, cAMP signaling modulates synaptic structure in an activity-dependent manner.

Both mutant *Vol* (encodes a–integrin protein) and *lat* are impaired in multiple forms of synaptic plasticity at the NMJ (Rohrbough et al., 1999, 2000). In addition to functional abnormalities, *Vol* mutant synaptic arbors are structurally enlarged, suggesting Volado negatively regulates developmental synaptic sprouting and growth. In the *leo* mutant NMJ, basal synaptic transmission is reduced by 30% and transmission amplitude, fidelity, and fatigue-resistance properties are reduced at elevated stimulation frequencies and in low external $[Ca^{2+}]$ (Broadie et al., 1997). Moreover, transmission augmentation and posttetanic potentiation (PTP) are disrupted in the mutant. These results suggest that Leonardo plays a role in the regulation of synaptic vesicle dynamics.

Overexpression of CREB repressor suppresses the increase in neuronal activity but has no effect on the increased arborization, produced by the *dnc* mutation (G.W. Davis et al., 1996). Moreover, overexpression of CREB activator alone does not appear to alter structure or function at the larval NMJ but does produce an increase in synaptic function in *Fas* II mutants, which normally show increased arborization without an increase in activity. On the other hand, reduced expression of *Adf1* in the mutant *nal* and heat-shock-induced overexpression of the wild-type *Adf1* gene have opposite effects on synaptic structure, but neither affects synaptic function (DeZazzo et al., 2000). These data suggest that activity-dependent modulation can be dissected into a *CREB*-mediated functional pathway and an *Adf1*-mediated structural pathway. AP-1, a heterodimeric transcription factor composed of *Fos* and *Jun*, regulates both synaptic strength and synapse number (Sanyal et al., 2002). Overexpression of *dnc* phosphodiesterase fails to block the influence of AP-1 on synapse number. Conditioned overexpression of AP-1 results in robust increase in CREB mRNA levels. These observations indicate that AP-1 acts upstream of CREB.

NMDA receptor (NMDAR) plays an important role in experience-dependent synaptic plasticity, including long-term potentiation and long-term depression (Bear, 1996). Because of its unique property of voltage-dependent activation by ligand, NMDAR has been suggested as a "Hebbian coincidence detector" underlying associative learning. Mutations of the *Drosophila* NMDAR gene cause learning defects that can be restored by wild-type transgenes (Xia et al., 2005). Acute disruption of NMDAR with an NMDAR antisense RNA transgene impairs learning and long-term memory formation, providing direct evidence for acute involvement of NMDAR in associative learning and memory.

D. Dissection of Memory Formation

Memory formation also represents a signaling pathway of sorts — from learning to long-term memory. Behavior-genetic analyses of this process have suggested five distinct temporal phases: long-term memory (LTM), anesthesia-resistant memory (ARM), middle-term memory (MTM), short-term memory (STM), and acquisition [or learning (LRN)]. Single-gene-mutant analyses indicate that memory processing is sequential from LRN to MTM but that consolidation of ARM and LTM then occurs in parallel (Fig. 3-5).

1. LTM

LTM is protein synthesis dependent, appears within 24 hr after spaced training (10 training sessions with a 15-min rest interval between each), and lasts for more than one week. In contrast, LTM is not produced after massed training (10 training sessions with no rest interval between each) (Tully et al., 1994). LTM is bloked by protein synthesis inhibitors and depends critically on gene expression mediated by Adf1, CREB transcription factors, and Notch signaling (Yin et al., 1994, 1995; DeZazzo et al., 2000; Ge et al., 2004; Presente et al., 2004).

2. ARM

Memory is initially labile but eventually can be consolidated into more stable forms. Memory soon after Pavlovian olfactory learning, for example, is disrupted by cold shock. During the first two hours after training, however, a cold-shock-insensitive form of memory (or ARM) appears (Tully et al., 1994). ARM is the longest-lasting memory produced by massed training, appears to be protein synthesis independent, and decays away within four days. Although multiple training cycles do produce increasingly higher levels of ARM, 10 massed and 10 spaced training sessions both yield similar maximal levels of ARM (Tully et al., 1994). ARM is unaffected by inhibition of protein synthesis or by disruptions of Adf1, CREB, or Notch transcription factors that block LTM (Yin et al., 1994; DeZazzo et al., 2000; Ge et al., 2004; Presente et al., 2004), but it is disrupted in *rsh* mutants (Tully et al., 1994). In contrast,

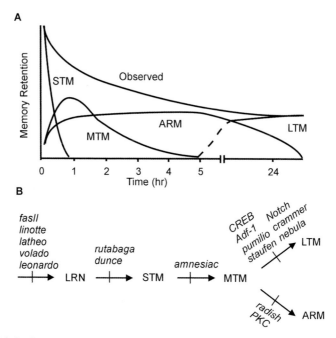

FIGURE 3-5   Distinct memory phases. **A.** Four functionally distinct memory phases have been detected underlying the observed memory retention scores. Short-term memory (STM), middle-term memory (MTM), anesthesia-resistant memory (ARM), and long-term memory (LTM) appear sequentially, and each has a progressively slower rate of decay. **B.** A genetic model of information flow showing where in the pathway single-gene mutants have their primary effects. The pathway is sequential from learning (LRN) to MTM, but then it branches into two independent paths — one leading to ARM and the other leading to LTM. (Reproduced with modifications from Dubnau and Tully, 1998.)

LTM in *rsh* mutants is normal. Thus, ARM and LTM are genetically and functionally independent forms of long-lasting memory that exist in parallel for several days after spaced training. A study on mutant flies devoid of MB vertical lobes (*ala*) suggests that ARM and LTM are exclusive to each other (Isabel et al., 2004). It has to be cautioned that *ala* is a developmental mutant and that the abnormal memory formation in *ala* may be a result of aberrant wiring of the MB (Margulies et al., 2005).

3. MTM

Two hours after training, when ARM is maximal, approximately 50% of observed memory is still cold-shock sensitive. This early memory is also resistant to protein synthesis inhibitors and can be further decomposed into STM and MTM by the *amn* mutation. Initial learning in *amn* is near normal, as is

7-hr memory retention. Memory retention at intermediate time points, however, is reduced. This observation first suggested the existence of MTM.

Evidence for MTM in wild-type flies emerged from reversal retention experiments (Tully et al., 1990, 1996). During the first training session, odor A (the CS⁺) is paired with electroshock, and odor B (the CS⁻) is not. In a second training session, this arrangement is reversed: Odor B becomes the CS⁺, and odor A becomes the CS⁻. Different groups of flies are then subjected to reversal learning at various time points after the first training session. In each case, however, conditioned responses are quantified immediately after the second training session. A temporal window is revealed, within which the first odor-shock association is sensitive to disruption by the reversal learning. Strikingly, the disrupted memory appears to correspond quantitatively and temporally to that missing in *amn* mutants. Moreover, the asymptotic level of reversal learning–resistant memory is quantitatively similar to that of ARM. These observations suggest that reversal training disrupts a memory component in wild-type flies that is already genetically impaired in *amn* mutants, thereby eliminating any difference between wild-type and *amn* reversal retention curves. In fact, the reversal retention curves of wild-type and *amn* flies are indistinguishable. These data suggest that MTM may be a genetically distinct component of memory.

Study on a temperature-sensitive *DC0* mutant (*DC0^{ts}*) also supports the existence of a genetically distinct MTM phase (Li et al., 1996). *DC0^{ts}* has a learning defect, but shifting from permissive to restrictive temperature further disrupts memory retention. The temperature-shift-specific effect is indistinguishable from the amnesiac memory retention curve. Thus, disruption in *DC0* function in adult flies appears to disrupt MTM.

4. STM

Far less is known about STM. Memory decay within 30 min of training is faster in *dnc* and *rut* than in *amn* mutants or wild-type flies, suggesting that the former mutants disrupt a memory component that temporally precedes MTM (i.e., STM). Memory decay after the first 30 min slows considerably in *dnc* and *rut*, but retention levels in these mutants are clearly lower than those in *amn*. This observation raises the possibility that MTM is downstream of, and dependent on, STM.

5. LRN

In contrast to *dnc* and *rut*, *lat*, *lio*, *PKA-RI*, and *fasII* all show reduced Pavlovian olfactory learning, but the rate of memory decay thereafter appears normal (Dubnau and Tully, 1998; Cheng et al., 2001). Hence, these genes may be involved exclusively in the initial acquisition of the odor and shock association.

Learning defects traditionally have been suspected to result more often from mutations that disrupt development than from adult biochemistries. This is clearly not the case for *fasII* mutants, however, since induced expression of *fasII* transgenes in adults fully rescues the *fasII* learning defect. Mutants with normal (or near-normal) learning but defective memory decay, in contrast, have been thought to result more likely in biochemical defects. This generalization also is invalid. If memory retention requires distinct anatomical structures, for instance, then abnormal development might yield memory-specific behavioral defects. Thus, eliminating maldevelopment as a possible explanation for learning and memory defects in adults resides solely in the use of conditional mutations or inducible transgenes.

V. SUMMARY

The power of genetic dissection is limited only by the breadth of gene discovery. The ongoing discovery of genes involved with learning and memory stands to revolutionize our understanding of behavioral plasticity, in two ways. First, systematic analyses of the genetic, biochemical, and cell biological requirements of each gene will yield a vertical integration of information across biological levels of organization. Second, the evolutionary conservation of core mechanisms of cell signaling has enabled isolation of vertebrate homologs and transfer of some of the logic underlying invertebrate learning and memory to vertebrate systems, providing a horizontal integration of information across species and model systems. To that extent, the mouse, with its own versatile genetic toolboxes, is beginning to play a more and more important role in learning and memory research.

ACKNOWLEDGMENTS

Y.W. and J.D. contributed equally to the chapter. We would like to thank the NIH and DART Neuroscience, Inc., for funding to J.D., T.T., and Y.Z.

REFRENCES

Aceves-Pina EO, Quinn WG, Smith KD, Steinberger E, and Rodriguez-Rigau LJ (1979) Learning in Normal and Mutant *Drosophila* Larvae. *Science* 206:93–96.
Ballester R, Marchuk D, Boguski M, Saulino A, Letcher R, Wigler M, and Collins F (1990) The NF1 locus encodes a protein functionally related to mammalian GAP and yeast IRA proteins. *Cell* 63:851–859.
Bear MF (1996) Progress in understanding NMDA-receptor-dependent synaptic plasticity in the visual cortex. *Journal of Physiology — Paris* 90:223–227.

Boynton S, and Tully T (1992) Latheo, a new gene involved in associative learning and memory in *Drosophila melanogaster*, identified from P-element mutagenesis. *Genetics* 131:655–672.

Brand AH, Manoukian AS, and Perrimon N (1994) Ectopic expression in *Drosophila*. *Methods Cell Biol* 44:635–654.

Brand AH, and Perrimon N (1993) Targeted gene expression as a means of altering cell fates and generating dominant phenotypes. *Development* 118:401–415.

Broadie K, Rushton E, Skoulakis EM, and Davis RL (1997) Leonardo, a *Drosophila* 14-3-3 protein involved in learning, regulates presynaptic function. *Neuron* 19:391–402.

Buchberg AM, Cleveland LS, Jenkins NA, and Copeland NG (1990) Sequence homology shared by neurofibromatosis type-1 gene and IRA-1 and IRA-2 negative regulators of the RAS cyclic AMP pathway. *Nature* 347:291–294.

Budnik V, Zhong Y, and Wu CF (1990) Morphological plasticity of motor axons in *Drosophila* mutants with altered excitability. *J Neurosci* 10:3754–3768.

Cawthon RM, Weiss R, Xu GF, Viskochil D, Culver M, Stevens J, Robertson M, Dunn D, Gesteland R, O'Connell P, et al. (1990) A major segment of the neurofibromatosis type 1 gene: cDNA sequence, genomic structure, and point mutations. *Cell* 62:193–201.

Chang KT, Shi YJ, and Min KT (2003) The *Drosophila* homolog of Down's syndrome critical region 1 gene regulates learning: Implications for mental retardation. *Proc Natl Acad Sci USA* 100:15794–15799.

Cheng Y, Endo K, Wu K, Rodan AR, Heberlein U, and Davis RL (2001) *Drosophila fasciclinII* is required for the formation of odor memories and for normal sensitivity to alcohol. *Cell* 105:757–768.

Choi KW, Smith RF, Buratowski RM, and Quinn WG (1991) Deficient protein kinase C activity in *turnip*, a *Drosophila* learning mutant. *J Biol Chem* 266:15999–15606.

Comas D, Petit F, and Preat T (2004) *Drosophila* long-term memory formation involves regulation of cathepsin activity. *Nature* 430:460–463.

Connolly JB, Roberts IJ, Armstrong JD, Kaiser K, Forte M, Tully T, and O'Kane CJ (1996) Associative learning disrupted by impaired G, signaling in *Drosophila* mushroom bodies. *Science* 274:2104–2107.

Cooley L, Kelley R, and Spradling A (1988) Insertional mutagenesis of the *Drosophila* genome with single P elements. *Science* 239:1121–1128.

Crittenden JR, Skoulakis EM, Han KA, Kalderon D, and Davis RL (1998) Tripartite mushroom body architecture revealed by antigenic markers. *Learn Mem* 5:38–51.

Dauwalder B, and Davis RL (1995) Conditional rescue of the dunce learning/memory and female fertility defects with *Drosophila* or rat transgenes. *J Neurosci* 15:3490–3499.

Davis GW, Schuster CM, and Goodman CS (1996) Genetic dissection of structural and functional components of synaptic plasticity. III. CREB is necessary for presynaptic functional plasticity. *Neuron* 17:669–679.

Davis RL (1996) Physiology and biochemistry of *Drosophila* learning mutants. *Physiol Rev* 76:299–317.

de Belle JS, and Heisenberg M (1994) Associative odor learning in *Drosophila* abolished by chemical ablation of mushroom bodies. *Science* 263:692–695.

de Belle JS, and Heisenberg M (1996) Expression of *Drosophila* mushroom body mutations in alternative genetic backgrounds: A case study of the mushroom body miniature gene (*mbm*). *Proc Natl Acad Sci USA* 93:9875–9880.

DeZazzo J, Sandstrom D, de Belle S, Velinzon K, Smith P, Grady L, DelVecchio M, Ramaswami M, and Tully T (2000) *nalyot*, a mutation of the *Drosophila* myb-related *Adf1* transcription factor, disrupts synapse formation and olfactory memory. *Neuron* 27:145–158.

DeZazzo J, Xia S, Christensen J, Velinzon K, and Tully T (1999) Developmental expression of an *amn*⁺ transgene rescues the mutant memory defect of *amnesiac* adults. *J Neurosci* 19:8740–8746.

Drain P, Folkers E, and Quinn WG (1991) cAMP-dependent protein kinase and the disruption of learning in transgenic flies. *Neuron* 6:71–82.

Drier EA, Tello MK, Cowan M, Wu P, Blace N, Sacktor TC, and Yin JC (2002) Memory enhancement and formation by atypical PKM activity in *Drosophila melanogaster. Nat Neurosci* 5:316–324.

Dubnau J, Chiang AS, and Tully T (2003a) Neural substrates of memory: From synapse to system. *J Neurobiol* 54:238–253.

Dubnau J, Chiang AS, Grady L, Barditch J, Gossweiler S, McNeil J, Smith P, Buldoc F, Scott R, Certa U, Broger C, and Tully T (2003b) The staufen/pumilio pathway is involved in *Drosophila* long-term memory. *Curr Biol* 13:286–296.

Dubnau J, Grady L, Kitamoto T, and Tully T (2001) Disruption of neurotransmission in *Drosophila* mushroom body blocks retrieval but not acquisition of memory. *Nature* 411:476–480.

Dubnau J, and Tully T (1998) Gene discovery in *Drosophila*: New insights for learning and memory. *Annu Rev Neurosci* 21:407–444.

Dudai Y, Jan YN, Byers D, Quinn WG, and Benzer S (1976) Dunce, a mutant of *Drosophila* deficient in learning. *Proc Natl Acad Sci USA* 73:1684–1688.

Dura JM, Preat T, and Tully T (1993) Identification of *linotte*, a new gene affecting learning and memory in *Drosophila melanogaster. J Neurogenet* 9:1–14.

Folkers E, Drain P, and Quinn WG (1993) *Radish*, a *Drosophila* mutant deficient in consolidated memory. *Proc Natl Acad Sci USA* 90:8123–8127.

Folkers E, Waddell S, Quinn WG (2006) The *Drosophila radish* gene encodes a protein required for anesthesia-resistant memory. *PNAS* 103:17496–17500.

Ge X, Hannan F, Xie Z, Feng C, Tully T, Zhou H, and Zhong Y (2004) Notch signaling in *Drosophila* long-term memory formation. *Proc Natl Acad Sci USA* 101:10172–10176.

Griffith LC, Verselis LM, Aitken KM, Kyriacou CP, Danho W, and Greenspan RJ (1993) Inhibition of calcium/calmodulin-dependent protein kinase in *Drosophila* disrupts behavioral plasticity. *Neuron* 10:501–509.

Grotewiel MS, Beck CD, Wu KH, Zhu XR, and Davis RL (1998) Integrin-mediated short-term memory in *Drosophila. Nature* 391:455–460.

Grunwald DJ, and Eisen JS (2002) Headwaters of the zebrafish — emergence of a new model vertebrate. *Nat Rev Genetics* 3:717–724.

Guo HF, The I, Hannan F, Bernards A, and Zhong Y (1997) Requirement of *Drosophila* NF1 for activation of adenylyl cyclase by PACAP38-like neuropeptides. *Science* 276:795–798.

Guo HF, Tong J, Hannan F, Luo L, and Zhong Y (2000) A neurofibromatosis-1-regulated pathway is required for learning in *Drosophila. Nature* 403:895–898.

Han PL, Meller V, and Davis RL (1996) The *Drosophila* brain revisited by enhancer detection. *J Neurobiol* 31:88–102.

Hannan F, Ho I, Tong JJ, Zhu Y, Nurnberg P, and Zhong Y (2006) Effect of neurofibromatosis type I mutations on a novel pathway for adenylyl cyclase activation requiring neurofibromin and *Ras. Hum Mol Genet* 15:1087–1098.

Heisenberg M (2003) Mushroom body memoir: From maps to models. *Nat Rev Neurosci* 4:266–275.

Heisenberg M, Borst A, Wagner S, and Byers D (1985) *Drosophila* mushroom body mutants are deficient in olfactory learning. *J Neurogenet* 2:1–30.

Isabel G, Pascual A, and Preat T (2004) Exclusive consolidated memory phases in *Drosophila. Science* 304:1024–1027.

Johnston DS, and Nusslein-Volhard C (1992) The origin of pattern and polarity in the *Drosophila* embryo. *Cell* 68:201–219.

Kandel E, and Abel T (1995) Neuropeptides, adenylyl cyclase, and memory storage. *Science* 268:825–826.

Lam G, and Thummel CS (2000) Inducible expression of double-stranded RNA directs specific genetic interference in *Drosophila*. *Curr Biol* 10:957–963.

Lee T, and Luo L (1999) Mosaic analysis with a repressible cell marker for studies of gene function in neuronal morphogenesis. *Neuron* 22:451–461.

Li W, Tully T, and Kalderon D (1996) Effects of a conditional *Drosophila* PKA mutant on olfactory learning and memory. *Learn Mem* 2:320–333.

Liu G, Seiler H, Wen A, Zars T, Ito K, Wolf R, Heisenberg M, and Liu L (2006) Distinct memory traces for two visual features in the *Drosophila* brain. *Nature* 439:551–556.

Livingstone MS, Sziber PP, and Quinn WG (1984) Loss of calcium/calmodulin responsiveness in adenylate cyclase of rutabaga, a *Drosophila* learning mutant. *Cell* 37:205–215.

Margulies C, Tully T, and Dubnau J (2005) Deconstructing memory in *Drosophila*. *Curr Biol* 15: R700–713.

Martin GA, Viskochil D, Bollag G, McCabe PC, Crosier WJ, Haubruck H, Conroy L, Clark R, O'Connell P, Cawthon RM, et al. (1990) The GAP-related domain of the neurofibromatosis type 1 gene product interacts with ras p21. *Cell* 63:843–849.

McGuire SE, Le PT, and Davis RL (2001) The role of *Drosophila* mushroom body signaling in olfactory memory. *Science* 293:1330–1333.

McGuire SE, Le PT, Osborn AJ, Matsumoto K, and Davis RL (2003) Spatiotemporal rescue of memory dysfunction in *Drosophila*. *Science* 302:1765–1768.

McGuire TR, and Hirsch J (1977) Behavior-genetic analysis of *Phormia regina*: Conditioning, reliable individual differences, and selection. *Proc Natl Acad Sci USA* 74:5193–5197.

Nusslein-Volhard C, and Wieschaus E (1980) Mutations affecting segment number and polarity in *Drosophila*. *Nature* 287:795–801.

O'Kane CJ, and Gehring WJ (1987) Detection in situ of genomic regulatory elements in Drosophila. *Proc Natl Acad Sci USA* 84:9123–9127.

Osterwalder T, Yoon KS, White BH, and Keshishian H (2001) A conditional tissue-specific transgene expression system using inducible GAL4. *Proc Natl Acad Sci USA* 98:12596–12601.

Pinto S, Quintana DG, Smith P, Mihalek RM, Hou ZH, Boynton S, Jones CJ, Hendricks M, Velinzon K, Wohlschlegel JA, Austin RJ, Lane WS, Tully T, and Dutta A (1999) *latheo* encodes a subunit of the origin recognition complex and disrupts neuronal proliferation and adult olfactory memory when mutant. *Neuron* 23:45–54.

Presente A, Boyles RS, Serway CN, de Belle JS, and Andres AJ (2004) *Notch* is required for long-term memory in *Drosophila*. *Proc Natl Acad Sci USA* 101:1764–1768.

Quinn WG, Sziber PP, and Booker R (1979) The *Drosophila* memory mutant *amnesiac*. *Nature* 277:212–214.

Riemensperger T, Voller T, Stock P, Buchner E, and Fiala A (2005) Punishment prediction by dopaminergic neurons in *Drosophila*. *Curr Biol* 15:1953–1960.

Rohrbough J, Grotewiel MS, Davis RL, and Broadie K (2000) Integrin-mediated regulation of synaptic morphology, transmission, and plasticity. *J Neurosci* 20:6868–6878.

Rohrbough J, Pinto S, Mihalek RM, Tully T, and Broadie K (1999) *latheo*, a *Drosophila* gene involved in learning, regulates functional synaptic plasticity. *Neuron* 23:55–70.

Roman G, Endo K, Zong L, and Davis RL (2001) P[Switch], a system for spatial and temporal control of gene expression in *Drosophila melanogaster*. *Proc Natl Acad Sci USA* 98:12602–12607.

Rong YS, and Golic KG (2000) Gene targeting by homologous recombination in *Drosophila*. *Science* 288:2013–2018.

Rong YS, Titen SW, Xie HB, Golic MM, Bastiani M, Bandyopadhyay P, Olivera BM, Brodsky M, Rubin GM, and Golic KG (2002) Targeted mutagenesis by homologous recombination in *D. melanogaster*. *Genes Dev* 16:1568–1581.

Rubin GM, and Spradling AC (1982) Genetic transformation of *Drosophila* with transposable element vectors. *Science* 218:348–353.

Sanyal S, Sandstrom DJ, Hoeffer CA, and Ramaswami M (2002) AP-1 functions upstream of CREB to control synaptic plasticity in *Drosophila*. *Nature* 416:870–874.

Schuster CM, Davis GW, Fetter RD, and Goodman CS (1996) Genetic dissection of structural and functional components of synaptic plasticity. II. Fasciclin II controls presynaptic structural plasticity. *Neuron* 17:655–667.

Schwaerzel M, Monastirioti M, Scholz H, Friggi-Grelin F, Birman S, and Heisenberg M (2003) Dopamine and octopamine differentiate between aversive and appetitive olfactory memories in *Drosophila*. *J Neurosci* 23:10495–10502.

Skoulakis EM, and Davis RL (1996) Olfactory learning deficits in mutants for *leonardo*, a *Drosophila* gene encoding a 14-3-3 protein. *Neuron* 17:931–944.

Skoulakis EM, Kalderon D, and Davis RL (1993) Preferential expression in mushroom bodies of the catalytic subunit of protein kinase A and its role in learning and memory. *Neuron* 11:197–208.

Stebbins MJ, and Yin JC (2001) Adaptable doxycycline-regulated gene expression systems for *Drosophila*. *Gene* 270:103–111.

Stemple DL (2004) Tilling — a high-throughput harvest for functional genomics. *Nat Rev Genetics* 5:145–150.

Stocker RF, Lienhard MC, Borst A, and Fischbach KF (1990) Neuronal architecture of the antennal lobe in *Drosophila melanogaster*. *Cell Tissue Res* 262:9–34.

Strauss R (2002) The central complex and the genetic dissection of locomotor behaviour. *Curr Opin Neurobiol* 12:633–638.

The I, Hannigan GE, Cowley GS, Reginald S, Zhong Y, Gusella JF, Hariharan IK, and Bernards A (1997) Rescue of a *Drosophila* NF1 mutant phenotype by protein kinase A. *Science* 276:791–794.

Tong J, Hannan F, Zhu Y, Bernards A, and Zhong Y (2002) Neurofibromin regulates G protein-stimulated adenylyl cyclase activity. *Nat Neurosci* 5:95–96.

Tryon RC (1940) Genetic differences in maze learning in rats. *Yk Natl Soc Stud Educ* 39:111–119.

Tully T (1996) Discovery of genes involved with learning and memory: An experimental synthesis of Hirschian and Benzerian perspectives. *PNAS* 93:13460–13467.

Tully T, Bolwig G, Christensen J, Connolly J, DelVecchio M, DeZazzo J, Dubnau J, Jones C, Pinto S, Regulski M, Svedberg B, and Velinzon K (1996) A return to genetic dissection of memory in *Drosophila*. *Cold Spring Harb Symp Quant Biol* 61:207–218.

Tully T, Boynton S, Brandes C, Dura JM, Mihalek R, Preat T, and Villella A (1990) Genetic dissection of memory formation in *Drosophila melanogaster*. *Cold Spring Harb Symp Quant Biol* 55:203–211.

Tully T, and Quinn WG (1985) Classical conditioning and retention in normal and mutant *Drosophila melanogaster*. *J Comp Physiol* [A] 157:263–277.

Tully T, Preat T, Boynton SC, and Del Vecchio M (1994) Genetic dissection of consolidated memory in *Drosophila*. *Cell* 79:35–47.

Viskochil D, Buchberg AM, Xu G, Cawthon RM, Stevens J, Wolff RK, Culver M, Carey JC, Copeland NG, Jenkins NA, et al. (1990) Deletions and a translocation interrupt a cloned gene at the neurofibromatosis type 1 locus. *Cell* 62:187–192.

Waddell S, Armstrong JD, Kitamoto T, Kaiser K, and Quinn WG (2000) The *amnesiac* gene product is expressed in two neurons in the *Drosophila* brain that are critical for memory. *Cell* 103:805–813.

Waddell S, and Quinn WG (2001) Flies, genes, and learning. *Annu Rev Neurosci* 24:1283–1309.

Wallace MR, Marchuk DA, Andersen LB, Letcher R, Odeh HM, Saulino AM, Fountain JW, Brereton A, Nicholson J, Mitchell AL, et al. (1990) Type 1 neurofibromatosis gene: Identification of a large transcript disrupted in three NF1 patients. *Science* 249:181–186.

Wieschaus E (1996) Embryonic transcription and the control of developmental pathways. *Genetics* 142:5–10.

Winkler S, Schwabedissen A, Backasch D, Bokel C, Seidel C, Bonisch S, Furthauer M, Kuhrs A, Cobreros L, Brand M, and Gonzalez-Gaitan M (2005) Target-selected mutant screen by *Tilling* in *Drosophila*. *Genome Res* 15:718–723.

Xia S, Miyashita T, Fu TF, Lin WY, Wu CL, Pyzocha L, Lin IR, Saitoe M, Tully T, and Chiang AS (2005) NMDA receptors mediate olfactory learning and memory in *Drosophila*. *Curr Biol* 15:603–615.

Xu GF, Lin B, Tanaka K, Dunn D, Wood D, Gesteland R, White R, Weiss R, and Tamanoi F (1990a) The catalytic domain of the neurofibromatosis type 1 gene product stimulates ras GTPase and complements ira mutants of *S. cerevisiae*. *Cell* 63:835–841.

Xu GF, O'Connell P, Viskochil D, Cawthon R, Robertson M, Culver M, Dunn D, Stevens J, Gesteland R, White R, et al. (1990b) The neurofibromatosis type 1 gene encodes a protein related to GAP. *Cell* 62:599–608.

Yin JC, Del Vecchio M, Zhou H, and Tully T (1995) CREB as a memory modulator: Induced expression of a dCREB2 activator isoform enhances long-term memory in *Drosophila*. *Cell* 81:107–115.

Yin JC, Wallach JS, Del Vecchio M, Wilder EL, Zhou H, Quinn WG, and Tully T (1994) Induction of a dominant negative CREB transgene specifically blocks long-term memory in *Drosophila*. *Cell* 79:49–58.

Zamore PD, Tuschl T, Sharp PA, and Bartel DP (2000) RNAi: Double-stranded RNA directs the ATP-dependent cleavage of mRNA at 21 to 23 nucleotide intervals. *Cell* 101:25–33.

Zars T, Fischer M, Schulz R, and Heisenberg M (2000) Localization of a short-term memory in *Drosophila*. *Science* 288:672–675.

Zawistowski SL, and Hirsch J (1984) Conditioned discrimination in the behavior-genetic analysis of the blowfly *Phormia regina*: Controls and bidirectional selection. *Anim Learn Behav* 12: 402–408.

Zhong Y (1995) Mediation of PACAP-like neuropeptide transmission by coactivation of Ras/Raf and cAMP signal transduction pathways in *Drosophila*. *Nature* 375:588–592.

Zhong Y, Budnik V, and Wu CF (1992) Synaptic plasticity in *Drosophila* memory and hyper-excitable mutants: Role of cAMP cascade. *J Neurosci* 12:644–651.

Zhong Y, and Pena LA (1995) A novel synaptic transmission mediated by a PACAP-like neuropeptide in *Drosophila*. *Neuron* 14:527–536.

Zhong Y, and Wu CF (1991) Altered synaptic plasticity in *Drosophila* memory mutants with a defective cyclic AMP cascade. *Science* 251:198–201.

CHAPTER 4

Gene Expression in Learning and Memory

Joe L. Martinez, Jr.

Biology Department, University of Texas, San Antonio, TX 78249

Kenira J. Thompson

Department of Physiology, Ponce School of Medicine, Ponce, PR 00732

Angela M. Sikorski

Biology Department, University of Texas, San Antonio, TX 78249

I. INTRODUCTION

Memories can last a long time, even a lifetime. Science has shown that memories reside in the brain. Since the principal cell of the brain is the neuron, it follows that memories exist in neurons. A memory could be in one neuron that recognizes a particular face (Rolls, 1992), or it could be in a distributed network of neurons (Martinez and Derrick, 1996). Regardless, a memory is a physical entity. As we've said before, "memory is a thing in a place in a brain (Martinez and Derrick, 1996, p. 173)," although the place may vary with time (Kim et al., 1995).

Neurons are eukaryotic cells and all share many properties. Each neuron contains the entire DNA library for a person. However, only a portion of the genes coded in the neuron is active at anytime. Hence we have the environment acting on the genome. In the case of learning, the environment acts in a way that is described as "experience-dependent plasticity." For the most part, cells make permanent changes by expressing genes and translating proteins that make the change. One common example of permanent change is the transformation of children to adolescents, directed by hormone-induced gene

Neurobiology of Learning and Memory, Second Edition

129

expression (Rossmanith et al., 1994). A similar process occurs in your brain when you create a memory (Martinez and Derrick, 1996). Neurons change the shape of their synaptic connections to each other and even create new connections. Interestingly, there is no *a priori* reason to expect the brain to engage is this metabolically costly process. Altering the strength of existing connections could form memories, and, indeed, some nonpermanent memories may be formed in this way (Huang and Kandel, 1994). It appears that the brain seldom engages in the simplest processes to achieve its ends in spite of Occam's razor.

The knowledge that the brain engages in gene expression to form memories is recent. "Genes" were hypothetical constructs as late as 1957, when the structure of DNA was deduced (Watson and Crick, 1953). Soon thereafter the genetic code was discovered (Nirenberg, 2004), and the sequencing of the human, mouse, and *c. elegans* genomes propelled us forward at a tremendous rate. Other discoveries — such as the polymerase chain reaction (PCR), which allows the amplification of RNA, gene knock-in and knockout animals, where genes are added or deleted, and recent RNA silencing techniques — allow the manipulation of genes involved in memory. The development of gene array technology allows the assessment of the state of an entire genome on a few chips. All of these technologies have converged to provide us with a new, though far from complete, understanding of how genes function to form memories. What we know is the subject of this chapter.

Historically there has been much interest in RNA and memory. Early researchers thought a memory could be encoded in an RNA sequence and that the resulting peptide would encode the memory. Thus *scotophobin* a peptide that encodes fear of the dark, was isolated by Ungar (1970). The infamous "worm runners" thought memories encoded in RNA sequences could be transferred from animal to animal (McConnell, 1966). All of this early work kept scientists thinking about RNA and DNA and memory. It turns out there is no "memory molecule," just as there are no unique "memory genes." Memories are represented in cells as long-term changes in the function of proteins translated from mRNA transcribed from genes. If there are 30,000 genes in a mouse genome, then how many of these genes could be involved in memory formation and maintenance? The answer is more than one and less that 30,000. The studies reviewed in this chapter suggest that the number is uncomfortably high, suggesting many parallel processes, but not so high that they cannot be studied.

A memory takes time to create, because it involves gene expression. Hence a synapse that is activated and that becomes part of a memory network has to send a message to the nucleus that its state is now changed, and the nucleus has to send a message back, acknowledging and maintaining that memory state. Thus, there is a time-dependent cascade of events inside a cell that represents the memory. As you might imagine, some of the gene changes have to do with excitability, which is a salient property of neurons; some have to do with transcription to recruit new genes products into the process; some have to do with responding to the bodies reaction to the learning situation in the form of hormones; some are involved in altering the structure of neurons, especially at the synapse where neurons communicate. We do not yet have a clear description of a memory in terms of the cascade of cellular function. However, such an understanding is not too far in the distant future.

II. GENE EXPRESSION AND LEARNING AND MEMORY

A. The Hippocampus and Learning and Memory

1. Types of Learning and Memory

Memory is not a unitary construct. Rather, the memories one has are as diverse as the experiences that produce them. According to Squire and Zola-Morgan, memory is comprised of two broad categories, declarative memory and procedural memory. *Declarative memory* encompasses one's explicit recollection for facts and events and working memory processes. *Procedural memory* includes one's implicit knowledge for skills, priming, and simple classical conditioning (Squire and Zola-Morgan, 1991). Research has shown the anatomical substrates that underlie declarative and procedural memory to be mutually exclusive. For example, the hippocampus is critically required for declarative memory but not for procedural memory, and other cortical and subcortical structures mediate procedural memory and not declarative memory (Eichenbaum and Cohen, 2001).

In humans, memory is relatively easy to examine because we are capable of verbalizing what we know. Because nonhuman animals are nonverbal beings, however, scientists developed innovative strategies by which declarative knowledge could be assessed. Spatial tests of learning and memory are excellent ways in which declarative memory can be assessed in nonhuman animals, because they can behaviorally demonstrate to the experimenter their knowledge of previous experiences. In the Morris water maze (Morris, 1984), for example, animals rely on extramaze environmental cues to remember where the hidden platform is located, and performance is based on measures such as the latency to reach the platform and the percent of time spent in the quadrant containing the platform.

Classical conditioning is sensitive to hippocampal disruption, but only when there is a trace in time between the conditioned stimulus (CS) offset and the unconditioned stimulus (US) onset. It is believed that to learn the US-CS association, the hippocampus must temporarily "hold" the US representation in a working memory store (Clark and Squire, 1998). Damage to the hippocampus disrupts contextual fear conditioning in rodents (McEchron et al., 1998; Rogers et al., 2006) and humans (McGlinchy-Berroth et al., 1997). Evidence to demonstrate the hippocampus' role in trace conditioning comes from studies using the neurotoxin methylazoxymethanol-acetate (MAM) to inhibit neurogenesis. The data show that compared to saline-treated controls, MAM-treated animals exhibit impairments in trace eye-blink conditioning, suggesting that basal levels of hippocampal neurogenesis are required for the normal acquisition of this task (Shors et al., 2002).

B. Learning and Memory Genes

1. Using Invertebrate Model Systems to Examine the Neurobiology of Learning and Memory

A great deal of what we know today about the molecular bases of mammalian learning and memory is based on experiments on the sea snail, *Aplysia* (see Kandel, 2001, for review). Although the *Aplysia* is an invertebrate organism with a relatively simple neural network, it is an ideal model system to examine the neurobiology of mammalian learning and memory. The small quantity of very large neurons, many of which are visible to the naked eye and morphologically distinguishable, make it relatively easy to identify the precise neurons involved in learning and memory-related processes. More importantly, the *Aplysia* demonstrates implicit learning and memory that is similar to mammalian implicit learning and memory. In the *Aplysia*, implicit learning and memory is measured behaviorally in the form of habituation and sensitization of the gill-withdrawal response. In habituation, the repeated administration of an electrical stimulus results in a progressive attenuation in its gill withdrawal, until eventually the animal fails to respond to the stimulus. During sensitization, a weak stimulus is applied to the siphon, and precedes a noxious stimulus to the tail. The temporal pairing of the weak and aversive stimuli results in an augmented gill-withdrawal response when subsequent mild stimulation to the siphon is administered. These manipulations led to the identification of genes involved in both invertebrate and vertebrate memory mechanisms.

2. The Cyclic-AMP Response Element Binding Protein (CREB)

The first protein identified as being associated with the neurobiology of learning and memory was cyclic AMP (cAMP). In the *Aplysia*, cAMP facilitates synaptic transmission between neurons involved in the gill-withdrawal response (Brunelli et al., 1976), and today cAMP-mediated gene expression is known to play a role in a variety of physiological processes (Montminy and Bilezikjian, 1987; Hoeffler et al., 1988; Sassone-Corsi et al., 1988; Montminy et al., 1990),

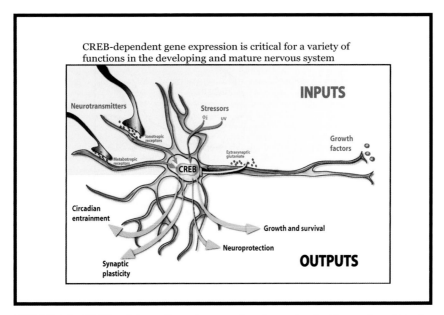

FIGURE 4-1   CREB activity mediates the expression of genes involved in a variety of physiological processes. [From Samuel Feldman (www.cns.nyu.edu/~sam/old/1030_Lect17.ppt).]

including learning and memory (Dash et al., 1990). cAMP affects memory
processes by initiating a signaling cascade that leads to the phosphorylation of
cyclic AMP responsive element binding protein (CREB) (Dash et al., 1990).
CREB is a nuclear transcription factor that mediates the expression of genes
involved in a number of physiological processes (see Fig. 4-1) in organisms
such as amphibians (Lutz et al., 1999), fish (Yoshida and Mishina, 2005), and
mammals (Montminy and Bilezikjian, 1987). In humans, CREB is estimated
to mediate the expression of over 4,000 genes, most of which are involved in
transcription processes (Zhang et al., 2005), by binding the cyclic AMP
response element (CRE).

CREB-mediated gene transcription occurs via a number of signaling pathways, each of which is initiated at the cell surface (Fig. 4-2). Briefly, neurotransmitter, neuromodulator, and growth factor ligands bind to receptors
that, when activated, lead to the expression of protein kinases such as protein
kinase C (PKC), protein kinase A (PKA), mitogen-activated protein kinase
(MAPK), and calcium/calmodulin-dependent protein kinase II (CamKII)
(Kim, Lu, and Quinn, 2000; Josslyn and Nguyen, 2005). Protein kinases
phosphorylate CREB at several residue sites (Giebler et al., 2000; Fimia et al.,
1998). However, phosphorylation at the serine 133 residue (ser133) site is

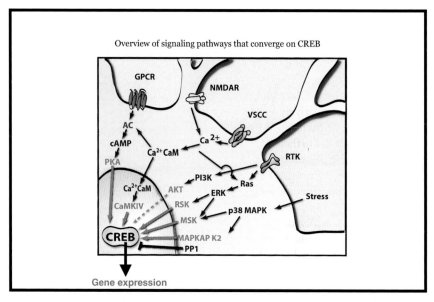

FIGURE 4-2   CREB-mediated gene expression occurs via several protein-kinase pathways. GPCR: G-protein coupled receptors; VSCC: voltage-sensitive calcium channels; RTK: receptor tyrosine kinase (binds growth factors). [From: Samuel Feldman (www.cns.nyu.edu/~sam/old/1030_Lect17.ppt).]

necessary for CREB-activated gene transcription to occur (Gonzales and Montminy, 1989; Shaywitz and Greenberg, 1999).

In the vertebrate, learning and memory enhances CREB expression in a region-specific manner. Animals trained on tasks dependent on the hippocampus, such as passive avoidance and contextual fear conditioning, exhibit significant increases in hippocampal CREB expression (Impey et al., 1998). When treated with the *N*-methyl-D-aspartate (NMDA) antagonist, APV, animals fail to exhibit long-term memory for conditioned fear (Athos et al., 2002) and they show no increase in CREB expression in the hippocampus when compared to vehicle-treated rats (S.M. Rodriguez et al., 2004). In contrast to contextual fear conditioning, auditory cue fear conditioning is not dependent on the hippocampus (Kim and Fanselow, 1992). Rather, learning this memory requires an intact amygdala (Maren et al., 1994), and following auditory cue fear conditioning CREB expression is upregulated in the amygdala and not the hippocampus (Impey et al., 1998).

The importance of CREB in learning and memory processes is strengthened further by the observation that disruption of hippocampal CREB impairs an animal's performance on spatial tasks such as the Morris water maze and contextual fear conditioning (Pittenger et al., 2002; Bourtchuladze et al.,

1994). Animals with mutations to the CREB protein exhibit learning and memory deficits. The degree to which CREB mutations affect learning and memory is dependent on how extensive the CREB mutations are. When compared to control animals, animals with mutations to the CREB α and Δ allele (CREB$_{\alpha\Delta}$) perform less well on the Morris water maze. However, when compared to animals in which all CREB alleles are mutated (CREB$_{comp}$), CREB$_{\alpha\Delta}$ perform relatively well on the Morris water maze because they take significantly less time than CREB$_{comp}$ animals to locate the submerged platform (Gass et al., 1998). Further support for this observation includes experiments using antisense oligonucleotides (ODN) that inhibit the expression of most of CREB's known isoforms. Animals that receive CREB ODN have greater escape latencies and longer swim paths than animals treated with scrambled ODN (Guzowski and McGaugh, 1997).

CREB is also involved in memories that are dependent on other structures besides the hippocampus. For example, in the amygdala–dependent conditioned taste aversion (CTA) task, animals are presented with either a novel taste that produces illness or a novel taste that does not produce illness. In normal animals, a novel taste that produces illness will be avoided when presented on subsequent trials. In contrast, animals with damage to the amygdala continue to sample the aversive taste despite its negative effect on the animal. To investigate whether CTA is mediated by CREB activity, animals with CREB deletions were presented with either flavored water (no malaise) or flavored water with added lithium chloride (LiCl) (malaise). Twenty-four hours later, presenting animals with both flavored waters assessed memory. Results revealed that CREB-mutant mice consumed significantly more of the LiCl-flavored water when compared to wild-type controls, suggesting that their memory for the aversion to the water was significantly impaired (Josselyn et al., 2004). Interestingly, the learning and memory impairments exhibited by CREB-mutant animals are only evident when animals are tested at least 24 hours after training, suggesting that CREB is required for long-term, but not short-term, memory (Kaang et al., 1993; Silva and Jasselyn, 2002).

3. Neuronal Growth-Associated Proteins (nGAPs)

Neuronal growth-associated proteins (nGAPs) are a family of gene products involved in the growth and regeneration of the nervous system and include but are not limited to GAP-43 and SCG10. nGAPs are expressed in the brain throughout life; however, during organism development their expression is most robust (Higo et al., 1999). Recent work has shown that in adulthood nGAP expression may be mediated by experience and is involved in synaptic changes that underlie learning and memory processes. GAP-43, which is involved with neurite outgrowth (Karns et al., 1987) and signal transduction (Akers and Routtenberg, 1985), is altered following tasks such as the Morris water maze (Pascale et al., 2004) and contextual fear conditioning (Young et al., 2000) as well as during drug abuse (Park et al., 2002; Vukosavic et al., 2001) and aging (Sugaya et al., 1998). Routtenberg and colleagues recently examined the extent to which GAP-43 is involved in cognitive processes. By using a transgenic mouse line that overexpressed GAP-43, they found that compared to controls, the GAP-43–enhanced animals performed significantly better on delayed-matching-to-sample tasks and delayed-nonmatching-to-sample tasks (Routtenberg et al., 2000). The role of GAP-43 in hippocampal-dependent memory processes is also illustrated in studies using GAP-43 knockout mice. Rekart and colleagues (2005) trained heterozygous GAP-43 knockout animals ($GAP^{+/-}$) on a contextual fear conditioning task in which a tone was paired with a foot shock. Twenty-four hours later animals received a single test to assess memory retention for the context in which they received the foot shock. When compared to wild-type animals, $GAP^{+/-}$ animals exhibited significantly less time freezing, thus indicating substantial memory impairment. As a control, both wild-type and $GAP^{+/-}$ animals were trained and tested for cue fear conditioning, which is not dependent on the hippocampus. Using this strategy, no behavioral differences were observed between the two groups, suggesting that GAP-43 may be involved specifically in hippocampal-dependent learning and memory.

nGAPs are also involved in learning and memory at the cellular level. GAP-enhanced animals exhibit more robust LTP when compared to controls (Routtenberg et al., 2000); and in both normal and GAP-enhanced animals, LTP induction results in a significant reduction in hippocampal GAP-43 expression (Meberg et al., 1993; Routtenberg et al., 2000). The down-regulation of GAP-43 following LTP induction is presumed to reflect the synaptic stabilization of connections formed during LTP and, in this view, long-term memory maintenance (Meberg et al., 1993).

The nGAP SCG10 (superior cervical ganglia clone 10) is expressed abundantly in the developing brain. The localization of SCG10 to the growth cones and axons of neurons during development suggests that it is involved in synaptogenesis (Stein et al., 1998). Although its expression is significantly lower in the brains of adult animals (Stein et al., 1998), SCG10 is transiently expressed during adulthood in an experience-dependent fashion. Peng et al. (2003) found that LTP induction at the Schaeffer collateral–CA1 pathway in rats resulted in regional changes in SCG10 expression in the hippocampus. Specifically, SCG10 expression was most robust in CA3 when compared to either CA1 or the dentate gyrus (Peng et al., 2003). The LTP-mediated expression of SCG10 in the hippocampus occurs rapidly and is short-lived. Within three hours of LTP induction, SCG10 expression is nearly doubled, and by 24 hours it is back to basal levels (Peng et al., 2004). Like GAP-43, the down-regulation of SCG10 in the hippocampus one day following LTP may reflect its role in synaptic stabilization, and therefore memory.

4. Immediate Early Genes

Immediate early genes (IEGs) are named for their rapid response to cell stimuli. To date, as many as 40 IEGs are known to exist (Lanahan and Worley, 1998), some of which are involved in learning and memory processes. The IEGs *Arc*, *zif/268*, and *c-fos* were examined in the dorsal hippocampus either immediately following or two hours after training in a Morris water maze. While each IEG was significantly up-regulated immediately after training, by two hours they had returned to basal levels (Guzowski et al., 2001). In a more recent study, the expression of *Arc* mRNA in the hippocampus was performed in a region-specific manner. Specifically, *Arc* expression was examined in both the dorsal and ventral regions of CA1, CA3, and the dentate gyrus as well as in the dorsal and ventral regions of the subiculum. *Arc* mRNA was enhanced throughout the hippocampus 24 hours following Morris water maze training and that at 1 month following training all regions except CA1 continued to exhibit enhanced *Arc* mRNA, suggesting that *Arc* may contribute to both memory formation and memory maintenance (Gusev et al., 2005).

Martinez and colleagues used Affymetrix oligonucleotide microarrays and a subtractive hybridization technique to examine regulatory processes underlying spatial learning and LTP induction in the MF-CA3 pathway (Thompson et al., 2003). The results revealed significant changes in hippocampal IEG expression one hour following Morris water maze training or LTP induction. Pathway analysis (Ingenuity Systems, Inc.) was then performed to examine further the array results for MF-CA3 LTP and water maze–trained animals based on their biological functions. The pathway analysis following MF-CA3 LTP included an attractive array of genes, one of which is the IEG *myc* (Fig. 4-3A). In addition to its involvement in MF-CA3 LTP induction, *myc* is known to provoke sustained cell proliferation. Interestingly, stimulation of the granule cell mossy fibers sufficient to induce MF-CA3 LTP increases the number of newly formed granule cells in the dentate gyrus. This indicates that granule cell neurogenesis may be regulated by the induction of LTP at the MF-CA3 pathway (Derrick et al., 2000), a process that may also be associated with certain hippocampal-dependent learning tasks (Shors et al., 2001, 2002).

With regard to water maze training, pathway analysis identified important connections between genes such as *krox24* (EGR1), *Ania-3* (*Homer1* splice variant), and *jun-d*, all of which are linked to memory processes (Fig. 4-3B). *Arc*, an activity-dependent cytoskeleton-associated protein, is also significantly up-regulated following MF-CA3 LTP induction. *Arc* as well as the IEG *Homer* have also been shown to modify dendritic connections in order to strengthen synaptic connectivity (Vazdarjanova et al., 2002). *Homer1* in particular is critically involved in activity-dependent changes of synaptic function (Ammon et al., 2003). As mentioned earlier, a similar expression pattern was observed

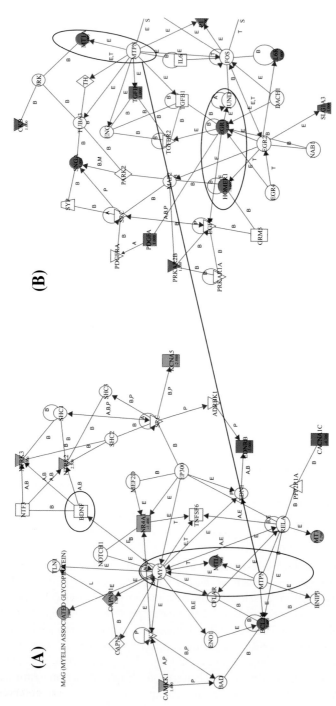

FIGURE 4-3 **(A)** Path analysis (Ingenuity Systems, Inc.) following MF-CA3 LTP indicates important pathways centered around the IEG *myc* (enclosed in black oval). **(B)** Path analysis following spatial learning on a Morris water maze indicates important pathways centered around *knox24*, *Homer-1*, and *jun-d* (enclosed in black oval). MTPN and MT1A are also activated following spatial learning. Red and green color represents increased or decreased expression, respectively. Labels include: (B)inding, (E)xpression, (A)ctivation, (T)ranscription, (P)hosphorylation, Tall rectangle (G-protein coupled receptor), long rectangle (nuclear receptor), circle (other), downward triangle (kinase), square (growth factor), oval (transcription factor) and diamond (enzyme).

for MTPN and MT1A between the MF-CA3 LTP animals and those animals trained on the water maze. This indicates that the MTPN and MT1A-related pathways are potentially altered due to overall plasticity (which occurs in the presence of both LTP and learning), whereas the *krox24, Homer1, jun-d* pathways are directly related to spatial learning.

Pathway analyses also demonstrated a bidirectional pattern of expression between *myc* and genes such as MTPN (myotrophin), metallothionein 1A (MT1A), and BDNF. MTPN attracts integral membrane proteins to cytoskeletal elements, whereas MT1A is protein that is transcriptionally regulated by both heavy metals and glucocorticoids. Both of these genes are also an important component of the network analysis following spatial training in a water maze (Fig. 4-3), and thus the MF-CA3 LTP and spatial learning pathways are connected, as one would expect if LTP were a substrate of memory.

Drugs of abuse also increase several IEGs, including *jun-b, zif/268*, and a family of *fos* proteins (for review, see NIDA research monograph #125), suggesting their involvement in the neurobiology of addiction, which, as described at length later, is also a form of learning. The expression of IEGs following drug administration depends on whether drug delivery is acute or chronic. During acute drug administration, IEGs are rapidly and transiently up-regulated (Hope et al., 1992; Burchett et al., 1999; Parelkar and Wang, 2004); whereas following chronic drug administration, IEG expression is comparable to that of control animals. The pattern of IEG expression during prolonged drug administration suggests that IEGs may develop drug tolerance (Hope et al., 1992).

5. Genes Involved in Drug Addiction and Abuse

Drug addiction is a form of learning. Animals and humans display sensitization and tolerance to a drug's effect, both of which are forms of nonassociative learning. In addition, they exhibit place preference to drugs of abuse, which is a form of associative learning. Amphetamine (AMPH) self-administration affects gene expression in the *nucleus accumbens* (*NAc*), a brain region associated with reward-mediated behavior. Intracranial self-administration (ICSA) up-regulates the expression of the G protein beta 1 subunit (rGB1) as well as genes associated with neuroplasticity, such as C-CAM4, *k-cadherin*, and *vimentin* (J.S. Rodriguez et al., 2002). Animals trained to lever press for methamphetamine (METH) exhibit a robust augmentation in *NAc* oxytocin (OXT) expression; and when pretreated with the OXT antagonist vasotosin, drug-seeking behavior is abolished (J.S. Rodriguez et al., 2002). Drug self-administration also increases CREB in a variety of mesolimbic structures, including the *NAc* (J.S. Rodriguez et al., 2002), and striatum (Konradi et al., 1994). In these brain regions, CREB activates the cocaine- and amphetamine-related transcript (CART), thus leading to changes in dopamine (DA) release (Kuhar et al., 2005). People who have abused cocaine exhibit altered CART mRNA levels in the VTA (Albertson et al., 2004). Interestingly, CART gene mutations are also associated with alcohol abuse (Flatscher-Bader et al., 2005). Together these data suggest that CREB-mediated CART activation in mesolimbic structures plays a role in the neurobiology of drug addiction.

III. LTP AND GENE EXPRESSION

Let us assume that the persistence or repetition of a reverberatory activity (or "trace") tends to induce lasting cellular changes that add to its stability. (Hebb, 1949)

Gene expression has been extensively examined in the hippocampus, a brain structure that is essential for learning and memory. Long-term potentiation (LTP) refers to a persistent increase in synaptic strength that is produced by brief high-frequency stimulation at excitatory afferents. Since its discovery, LTP has been the most accepted model for studying the neural mechanisms that underlie learning and memory in the hippocampus (Bliss and Lomo, 1973). In fact, LTP and its sister process, long-term depression (LTD), are considered cellular correlates of learning and memory. Although LTP is generally associated with the activation of NMDA receptors, other receptor systems are also implicated (see Fig. 4-4). NMDA-dependent LTP is characteristic of the Schaffer collateral–CA1 pathway, the medial perforant path projection to the dentate gyrus, and the commissural–CA3 pathway (Hebb, 1949;

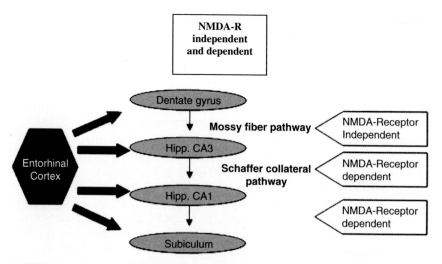

FIGURE 4-4   Characteristic hippocampal pathways. Small arrows indicate whether LTP at this pathway is dependent on or independent of NMDA receptor (NMDA-R) activation.

Harris et al., 1984; Harris and Cotman, 1986), and most agree that this form of LTP requires a postsynaptic increase in Ca^{2+} (Bliss and Collingridge, 1993; Collingridge and Bliss, 1995). Subsequent changes in protein kinases and other cellular cascades, in turn, modulate the increase in synaptic strength.

The lateral perforant path to dentate gyrus and the mossy fiber projection to area CA3 of the hippocampus display a form of LTP that is not as well known as the typical NMDA receptor–dependent forms (Bliss and Lomo, 1973; M.R. Martin, 1983; Harris and Cotman, 1986; Ishihara et al., 1990; Jaffe and Johnston, 1990; Bramham et al., 1991a, 1991b; Derrick et al., 1991; Breindl et al., 1994; Urban and Barrionuevo, 1996). LTP induction at the mossy fiber CA3 synapse depends on the activation of μ-opioid receptors (Derrick et al., 1992) and metabotropic glutamate receptors (MGluRs) (Thompson et al., 2005) and on repetitive mossy fiber activity (Jaffe and Johnston, 1990; Zalutsky and Nicoll, 1990; Zalutsky and Nicoll, 1992; Derrick and Martinez, 1994a, 1994b). The time course for LTP at the mossy fiber–CA3 pathway also differs from NMDA receptor–dependent LTP. Whereas NMDA-dependent LTP reaches its maximum almost immediately and then begins to decay, mossy fiber–CA3 LTP induced *in vivo* takes about 1 hr to reach its maximum and does not decay (Derrick et al., 1991; Derrick et al., 1992; Breindl et al., 1994; Derrick and Martinez, 1994a, 1994b). These differences are important in understanding the mechanisms underlying each form of LTP.

Similar to memory storage, LTP at most hippocampal synapses has distinct temporal phases (Nguyen et al., 1994). In the early phase of LTP, which lasts 1–3 hr, preexisting proteins are covalently modified, whereas the late phase of LTP, typically induced by repeated stimulation, is dependent on new RNA transcription and protein translation and lasts for at least several hours. Although these defined time frames vary across hippocampal synapses, altered gene expression is important across synapses and at all stages of hippocampal LTP.

As mentioned in the introduction to this chapter, recent advances in PCR and oligonucleotide microarray technology have allowed us to monitor the expression patterns of thousands of genes simultaneously following LTP (Thompson et al., 2003; Lee et al., 2005). Genes associated with LTP are temporally classified as immediate early genes (IEGs) or late-responder genes (Abraham et al., 1991; Dragunow, 1996; Peng et al., 2003). In the initial hours after LTP induction, cell surface receptors activate second messenger systems that result in IEG transcription (Walton et al., 1999). These, in turn, encode transcription factors, which, once translated, reenter the nucleus and regulate the expression of late-responder genes. Although many important IEGs and late-responder genes have been identified, further understanding of genes involved in LTP induction and maintenance may help unveil the molecular mechanisms that underlie information storage in the brain. In this section, we outline some of the genes that are implicated in hippocampal LTP.

A. Transcription Factors and Hippocampal LTP

As discussed earlier, the cAMP responsive element binding protein (CREB) is possibly one of the best-studied transcription factors implicated in hippocampal learning and memory. CREB activity is regulated by both cAMP and calcium influx (Brindle and Montminy, 1992), and it is critical for long-term memory (Josselyn and Nguyen, 2005). CREB modulates transcription of genes containing cAMP-responsive elements (CRE sites) in their promoters and is commonly activated by calcium/calmodulin-dependent protein kinase IV (CaMKIV) (Marie et al., 2005). LTP induction and maintenance display a delayed onset of CREB phosphorylation (Bito et al., 1996; Bito, 1998), whereas alphaCaMKII and MAPK2 display an enhanced phosphorylation state throughout the induction, early-, and late-LTP. Interestingly, only the late enhancement of pCREB is clearly dependent on protein synthesis (Ahmed and Frey, 2005). Another group (Balschun et al., 2003) found that conditional knockout strains with a marked reduction or complete deletion of all CREB isoforms in the hippocampus showed no deficits in lasting forms of hippocampal LTP and LTD. Thus, in the adult mouse brain, CREB deletions spare LTP and LTD in paradigms that are sensitive enough to detect deficits in other mutants. This suggests a species-specific or regionally restricted role of CREB in the brain.

The IEG *c-fos* is another transcription factor that is implicated in learning and memory. Mice that lack *c-fos* in the brain show a reduced LTP at CA3-to-CA1 synapses (Gass et al., 2004). Interestingly, LTP-induced levels of *c-fos* mRNA are significantly higher in aged animals, suggesting that age-dependent hippocampal dysfunction may be associated with a selective change in the dynamic activity of signaling pathways upstream of *c-fos* (Lanahan et al., 1997). Similar rapid increases in the IEGs *c-jun* and *jun-B* RNA are associated with dentate gyrus LTP (Abraham et al., 1991), whereas *jun-D* mRNA and protein display a more delayed and persistent increase (Demmer et al., 1993).

Other studies show that induction of LTP at the perforant path to the dentate gyrus synapse results in increases in the expression of the IEGs *zif268* (also termed NGFI-A, *Egr-1*, or *Krox 24*), activity-regulated cytoskeletal associated protein (*Arc*, also termed *Arg 3.1*), and *Homer* (Matsuo et al., 2000; French et al., 2001; Jones et al., 2001). LTP induction in the dentate gyrus also results in robust NMDA-R-dependent transcription of *zif268* (Cole et al., 1989). Further work using *zif286*-mutant mice shows that, although the early phase of dentate gyrus LTP is normal in these mice, the later phases are not present, and the ability of the mice to maintain learned information over a 24-hr period is deficient (Bozon et al., 2002). Recent work has focused on the expression of *Arc* mRNA, which is delivered to dendrites and translated within minutes after tetanic (high-frequency) stimulation. *Arc* protein binds to actin, possibly to regulate cytoskeletal restructuring after synaptic activation (Lynch, 2004). *Arc* disruption using antisense oligonucleotides inhibits LTP maintenance (Guzowski et al., 2000) but not induction. Similar increases in *Arc* and *Homer* were observed following mossy fiber-CA3 LTP induction (Thompson et al., 2005). Others have also shown colocalization of *Arc* and *Homer 1a*, indicating that they may function together to modify dendrites in order to increase synaptic efficacy (Vazdarjanova et al., 2002). However, no changes in *Arc, Homer*, or *zif268* were found following LTP at the commissural projection to CA1 pyramidal cells in vivo (French et al., 2001), indicating that the activation of these IEGs is pathway specific.

Tissue plasminogen activator (tPA) mRNA is significantly increased following mossy fiber LTP (Thompson et al., 2003). Previous work has shown tPA as a serine protease that plays an important role in tissue remodeling and LTP. tPA serves as an immediate early gene and is induced in the hippocampus during seizures, kindling, and LTP (Qian et al., 1993). tPA knockout mice show a decrease in late-phase LTP (Carmeliet et al., 1994) and show deficits in two-way avoidance tasks (Frey et al., 1996). Additionally, overexpression of tPA results in enhanced CA1 LTP and learning (Madani et al., 1999). To date, the role of tPA on hippocampal function is not clear. One possibility is that tPA converts plasminogen, which is the enzyme's main substrate and known to be found in the hippocampus, to the protease plasmin, which in turn can cleave many other extracellular substrates (laminin, for example) to result in alterations of hippocampal structure and function (Chen and Strickland, 1997). Other studies found that binding of tPA to the low-density lipoprotein (LDL) receptor-related protein (LRP) in hippocampal neurons enhances the activity of cAMP-dependent protein kinase, a key molecule in LTP (Zhuo et al., 2000). Overall, the findings suggest an important role for tPA in LTP.

Although CREB, *c-fos, zif268, Arc, Homer*, and tPA expression appear to be important for LTP, their specific role in memory storage is still unclear.

B. Protein Synthesis and LTP

At most NMDA-dependent pathways, the transition from early- to late-phase LTP requires gene expression and protein synthesis. Protein synthesis inhibitors disrupt late-phase LTP at multiple hippocampal synapses (Krug et al., 1984; Otani and Abraham, 1989; Otani et al., 1989; Barea-Rodriguez et al., 2000; Calixto et al., 2003). The expression of early-phase mossy fiber-CA3 LTP is also disrupted by the protein synthesis inhibitor anisomycin in vivo and in vitro (Barea-Rodriguez et al., 2000; Calixto et al., 2003), indicating that early-phase mossy fiber LTP is dependent on protein synthesis. This finding supports the previously described differences between NMDA-dependent and NMDA-independent LTP.

One theory is that LTP-induced protein synthesis induces the morphological changes that occur following LTP and that are essential for synaptic restructuring. For example, LTP increases spine number, spine area, as well as the distribution of synaptic vesicles (Applegate et al., 1987; Desmond and Levy, 1988, 1990; Meshul and Hopkins, 1990). These ultrastructural changes have been observed both pre- and postsynaptically following LTP (Lisman and Harris, 1993; Edwards, 1995; Lynch, 2004), indicating that gene expression (protein) changes are necessary on both sides of the activated synapse.

Many proteins have been linked to hippocampal LTP. For example, proenkephalin was found to be up-regulated following mossy fiber–CA3 LTP (Thompson et al., 2003). This corresponds with previous literature showing that enkephalin peptides released from hippocampal mossy fibers lower the threshold for induction of LTP at mossy fiber synapses (Roberts et al., 1997). Neuropeptide Y (NPY), which was also up-regulated following mossy fiber–CA3 LTP (Thompson et al., 2003), has been previously linked to inhibition of glutamate release and LTP in the dentate gyrus (Whittaker et al., 1999).

The neurotrophin brain-derived neurotrophic factor (BDNF) has been widely implicated in NMDA-R-dependent LTP (Barco et al., 2005) and, more recently, NMDA-R-independent LTP (Thompson et al., 2003). BDNF is thought to trigger long-lasting synaptic strengthening through MEK/ERK pathways (Messaoudi et al., 2002; Ying et al., 2002). Interestingly, BDNF-induced synaptic strengthening in cultured hippocampal neurons increases the expression of the IEGs *c-fos*, early growth response gene 1 (EGR1), and *Arc*, all of which increase following LTP (Alder et al., 2003). Other growth factors, such as vascular growth factor (VGF) and nerve growth factor (NGF), are altered following both NMDA-receptor-dependent and -independent LTP and learning (Sugaya et al., 1998; Alder et al., 2003; Thompson et al., 2003), suggesting a possible role of growth factors in synaptic modification.

C. Synaptic Tagging and LTP

Neural networks allow single synapses or groups of synapses to be activated simultaneously during LTP (and learning). This activation requires both transcription and translation. However, the mechanisms that underlie the movement of gene products to the activated synapse are still not known (K.C. Martin and Kosik, 2002). The synaptic-tagging hypothesis (Frey and Morris, 1997) proposes that a short-lasting LTP tags the activated synapse, allowing it to seize proteins synthesized by the nucleus (Frey and Morris, 1997). On the contrary, weak stimuli do not induce long-lasting changes (L-LTP), because they do not stimulate transcription of mRNAs that encode proteins essential for strengthening the synapse. In this model, learning or LTP-induced protein synthesis targets newly synthesized proteins to the activated synapses in order to make the change permanent (Kandel, 2001). In general, the synaptic-tagging model has been supported, yet there is still some disagreement surrounding the issue of "new" protein synthesis. As part of the controversy, one group proposed that posttranslational modification (PTM) of proteins already located at the synapse is the crucial mechanism underlying LTP and long-term memory (Routtenberg and Rekart, 2005).

Despite the disagreement, multiple mechanisms seem to be involved in LTP induction and maintenance. As a result, multiple genes (synaptic tags) can be activated, depending on the pathway, the temporal distribution of synaptic activity, and many other factors. It appears that subfields of the hippocampus display different transcriptional responses that may contribute to their regionally specific involvement in learning and memory. Although many LTP-related genes have been identified using microarray analysis and cDNA subtractive hybridization, these technologies are still emerging. One of the biggest pitfalls of microarray technology is the precise statistical analysis required to avoid false positives when dealing with thousands of genes simultaneously. Advances in these areas will contribute greatly toward the identification of the gene expression changes that underlie hippocampal LTP and, ultimately, learning and memory.

Learning and memory has a significant genetic influence, and evidence shows that many genes are critically required for both memory formation and long-term memory storage. Knowing which genes are involved in cognitive processes is an important step in the search for treatments for learning and memory disorders, such as Alzheimer's disease. Given how rapidly technology has advanced over the past several years, it is very possible that future therapeutic treatments for cognitive disorders may include the use of pharmacological agents to manipulate gene expression.

III. SUMMARY

1. Memory is a thing in a place in a brain.

2. There are two types of memory. Explicit memories are declarative in nature, whereas implicit memories are not. Behavioral tests such as the Morris water maze and classical conditioning permit scientists to examine memory in nonhuman animals.

3. Memories are represented in cells as long-term changes in the function of proteins translated from mRNA transcribed from genes.

4. The cyclic-AMP response element binding protein (CREB) is the most widely studied transcription factor and is involved in a variety of physiological processes, including learning and memory.

5. Behavioral tests of learning and memory, such as the Morris water maze, trace fear conditioning, and conditioned taste avoidance, lead to significant changes in CREB expression in the brain structures that mediate the type of learning being assessed, and animals with CREB mutations exhibit learning and memory deficits.

6. Neuronal growth-associated proteins (nGAPs), such as SCG10 and GAP-43, as well as immediate early genes (IEGs), such as *cFOS, arc,* and *zif/268,* are also mediated by learning and memory tasks.

7. Experience-dependent increases in synaptic strength underlie memory formation in networks of neurons and are known as Hebb's postulate and long-term potentiation (LTP).

8. There are two fundamental types of LTP, one that is NMDA receptor dependent and one that is NMDA receptor independent, here called opioid-receptor-dependent LTP.

9. Knockout mutant mice revealed the importance of several gene products in LTP induction and learning of a hippocampally dependent learning task, including tissue plasminogen activator (tPA), cFOS, and CREB.

10. The mRNA transcription that encodes proteins essential for strengthening the synapse occurs during late-phase LTP (l-LTP) but not during early-phase LTP (e-LTP), suggesting that synapse strengthening during l-LTP is protein synthesis dependent.

REFERENCES

Abraham, W.C., M. Dragunow, et al. (1991). The role of immediate early genes in the stabilization of long-term potentiation. *Mol Neurobiol* **5**(2–4):297–314.

Ahmed, T., and J.U. Frey (2005). Plasticity-specific phosphorylation of CaMKII, MAP-kinases and CREB during late-LTP in rat hippocampal slices in vitro. *Neuropharmacology* **49**(4): 477–492.

Akers, R.F., and A. Routtenberg (1985). Protein kinase C phosphorylates a 47 Mr protein (F1) directly related to synaptic plasticity. *Brain Res* **13**:147–151.

Albertson, D.N., B. Pruetz, C.J. Schmidt, D.M. Kuhn, G. Kapatos, and M.J. Bannon (2004). Gene expression profile of the nucleus accumbens of human cocaine abusers:Evidence for dysregulation of myelin. *Journal of Neurochemistry* **88**:1211–1219.

Alder, J., S. Thakker-Varia, et al. (2003). Brain-derived neurotrophic factor-induced gene expression reveals novel actions of VGF in hippocampal synaptic plasticity. *J Neurosci* **23**(34):10800–10808.

Ammon, S., P. Mayer, U. Riechert, H. Tischmeyer, and V. Hollt (2003). Microarray analysis of genes expressed in the frontal cortex of rats chronically treated with morphine and after naloxone precipitated withdrawal. *Brain Res Mol Brain Res* **112**:113–125.

Applegate, M.D., D.S. Kerr, et al. (1987). Redistribution of synaptic vesicles during long-term potentiation in the hippocampus. *Brain Res* **401**(2):401–406.

Athos, J., S. Impey, V.V. Pineda, X. Chen, and D.R. Storm (2002). Hippocampal CRE-mediated gene expression is required for contextual memory formation. *Nat Neurosci* **5**(11):1119–1120.

Balschun, D., D.P. Wolfer, et al. (2003). Does cAMP response element-binding protein have a pivotal role in hippocampal synaptic plasticity and hippocampus-dependent memory? *J Neurosci* **23**(15):6304–6314.

Barco, A., S. Patterson, et al. (2005). Gene expression profiling of facilitated L-LTP in VP16-CREB mice reveals that BDNF is critical for the maintenance of LTP and its synaptic capture. *Neuron* **48**(1):123–137.

Barea-Rodriguez, E.J., D.T. Rivera, et al. (2000). Protein synthesis inhibition blocks the induction of mossy fiber long-term potentiation in vivo. *J Neurosci* **20**(22):8528–8532.

Bito, H. (1998). [A potential mechanism for long-term memory: CREB signaling between the synapse and the nucleus]. *Seikagaku* **70**(6):466–471.

Bito, H., K. Deisseroth, et al. (1996). CREB phosphorylation and dephosphorylation: A Ca^{2+}- and stimulus duration-dependent switch for hippocampal gene expression. *Cell* **87**(7):1203–1214.

Bliss, T.V., and G.L. Collingridge (1993). A synaptic model of memory: Long-term potentiation in the hippocampus. *Nature* **361**(6407):31–39.

Bliss, T.V., and T. Lomo (1973). Long-lasting potentiation of synaptic transmission in the dentate area of the anaesthetized rabbit following stimulation of the perforant path. *J Physiol* **232**(2):331–356.

Bourtchuladze, R., B. Frenguelli, J. Blendy, D. Cioffi, G. Schutz, and A.J. Silva (1994). Deficient long-term memory in mice with a targeted mutation of the cAMP-responsive element-binding protein. *Cell* **78**:59–68.

Bozon, B., S. Davis, et al. (2002). Regulated transcription of the immediate-early gene Zif268: Mechanisms and gene dosage-dependent function in synaptic plasticity and memory formation. *Hippocampus* **12**(5):570–577.

Bramham, C.R., N.W. Milgram, et al. (1991a). Activation of AP5-sensitive NMDA receptors is not required to induce LTP of synaptic transmission in the lateral perforant path. *Eur J Neurosci* **3**(12):1300–1308.

Bramham, C.R., N.W. Milgram, et al. (1991b). Delta opioid receptor activation is required to induce LTP of synaptic transmission in the lateral perforant path in vivo. *Brain Res* **567**(1):42–50.

Breindl, A., B.E. Derrick, et al. (1994). Opioid receptor-dependent long-term potentiation at the lateral perforant path-CA3 synapse in rat hippocampus. *Brain Res Bull* **33**(1):17–24.

Brindle, P.K., and M.R. Montminy (1992). The CREB family of transcription activators. *Curr Opin Genet Dev* **2**(2):199–204.

Brunelli, M., V. Castellucci, and E.R. Kandel (1976). Synaptic facilitation and behavioral sensitization in *Aplysia*: Possible role of serotonin and cyclic AMP. *Science* **194**(4270): 1178–1181.

Burchett, S.A., M.J. Bannon, and J.G. Granneman (1999). RGS mRNA expression in rat striatum: Modulation by dopamine receptors and effects of repeated amphetamine administration. *J Neurochem* **72**:1529–1533.

Calixto, E., E. Thiels, et al. (2003). Early maintenance of hippocampal mossy fiber — long-term potentiation depends on protein and RNA synthesis and presynaptic granule cell integrity. *J Neurosci* **23**(12):4842–4849.

Carmeliet, P., L. Schoonjans, et al. (1994). Physiological consequences of loss of plasminogen activator gene function in mice. *Nature* **368**(6470):419–424.

Chen, Z.L., and S. Strickland (1997). Neuronal death in the hippocampus is promoted by plasmin-catalyzed degradation of laminin. *Cell* **91**(7):917–925.

Clark, R.E., and L.R. Squire (1998). Classical conditioning and brain systems: The role of awareness. *Science* **280**:77–81.

Cole, A.J., D.W. Saffen, et al. (1989). Rapid increase of an immediate early gene messenger RNA in hippocampal neurons by synaptic NMDA receptor activation. *Nature* **340**(6233): 474–476.

Collingridge, G.L., and T.V. Bliss (1995). Memories of NMDA receptors and LTP. *Trends Neurosci* **18**(2):54–56.

Dash, P.K., B. Hochner, and E.R. Kandel (1990). Injection of the cAMP-responsive element into the nucleus accumbens of *Aplysia* sensory neurons blocks long-term facilitation. *Nature* **345**:718–721.

Demmer, J., M. Dragunow, et al. (1993). Differential expression of immediate early genes after hippocampal long-term potentiation in awake rats. *Brain Res Mol Brain Res* **17**(3–4):279–286.

Derrick, B.E., and J.L. Martinez, Jr. (1994a). Frequency-dependent associative long-term potentiation at the hippocampal mossy fiber-CA3 synapse. *Proc Natl Acad Sci USA* **91**(22): 10290–10294.

Derrick, B.E., and J.L. Martinez, Jr. (1994b). Opioid receptor activation is one factor underlying the frequency dependence of mossy fiber LTP induction. *J Neurosci* **14**(7):4359–4367.

Derrick, B.E., S.B. Rodriguez, et al. (1992). Mu opioid receptors are associated with the induction of hippocampal mossy fiber long-term potentiation. *J Pharmacol Exp Ther* **263**(2): 725–733.

Derrick, B.E., S.B. Weinberger, et al. (1991). Opioid receptors are involved in an NMDA receptor-independent mechanism of LTP induction at hippocampal mossy fiber-CA3 synapses. *Brain Res Bull* **27**(2):219–223.

Derrick, B.E., A.D. York, and J.L. Martinez (2000). Increased granule cell neurogenesis in the adult dentate gyrus following mossy fiber stimulation sufficient to induce long-term potentiation. *Brain Res* **857**:300–307.

Desmond, N.L., and W.B. Levy (1988). Synaptic interface surface area increases with long-term potentiation in the hippocampal dentate gyrus. *Brain Res* **453**(1–2):308–314.

Desmond, N.L., and W.B. Levy (1990). Morphological correlates of long-term potentiation imply the modification of existing synapses, not synaptogenesis, in the hippocampal dentate gyrus. *Synapse* **5**(2):139–143.

Dragunow, M. (1996). A role for immediate-early transcription factors in learning and memory. *Behav Genet* **26**(3):293–299.

Edwards, F.A. (1995). LTP — a structural model to explain the inconsistencies. *Trends Neurosci* **18**(6):250–255.

Eichenbaum, H., and Cohen, N.J. (2001). *From Conditioning to Conscious Recollection: Memory Systems of the Brain.* New York: Oxford University Press.

Fimia, G.M., D. De Cesare, and P. Sassone-Corsi (1998). Mechanisms of activation by CREB and CREM: Phosphorylation, CBP, and a novel coactivator, ACT. *Cold Spring Harb Symp Quant Biol.* **63**:631–642.

Flatscher-Bader, T., M. van der Brug, J.W. Hwang, P.A. Gochee, I. Matsumoto, S. Niwa, and P.A. Wilce (2005). Alcohol-responsive genes in the frontal cortex and nucleus accumbens of human alcoholics. *J Neurochem* **93**:359–370.

French, P.J., V. O'Connor, et al. (2001). Subfield-specific immediate early gene expression associated with hippocampal long-term potentiation in vivo. *Eur J Neurosci* **13**(5):968–976.

Frey, U., and R.G. Morris (1997). Synaptic tagging and long-term potentiation. *Nature* **385**(6616):533–536.

Frey, U., M. Muller, et al. (1996). A different form of long-lasting potentiation revealed in tissue plasminogen activator mutant mice. *J Neurosci* **16**(6):2057–2063.

Gass, P., A. Fleischmann, et al. (2004). Mice with a fra-1 knock-in into the c-fos locus show impaired spatial but regular contextual learning and normal LTP. *Brain Res Mol Brain Res* **130**(1–2):16–22.

Gass, P., D.P. Wolfer, D. Balschun, D. Rudolph, W. Frey, H. Lipp, and G. Schutz (1998). Deficits in memory tasks of mice with CREB mutations depend on gene dosage. *Learning Memory* **5**:274–288.

Giebler, H.A., I. Lemasson, and J.K. Nyborg (2000). p53 recruitment of CREB binding protein mediated through phosphorylated CREB: A novel pathway of tumor suppressor regulation. *Mol Cell Biol* **20**:49–58.

Gonzalez, G.A., and M.R. Montminy (1989). Cyclic AMP stimulates somatostatin gene transcription by phosphorylation of CREB at serine 133. *Cell* **59**:675–680.

Gusev, P.A., C. Cui, D.L. Alkon, and A.N. Gubin (2005). Topography of Arc/Arg3.1 mRNA expression in the dorsal and ventral hippocampus induced by recent and remote spatial memory recall: Dissociation of CA3 and CA1 activation. *J Neurosci* **25**:9384–9397.

Guzowski, J.F., G.L. Lyford, et al. (2000). Inhibition of activity-dependent arc protein expression in the rat hippocampus impairs the maintenance of long-term potentiation and the consolidation of long-term memory. *J Neurosci* **20**(11):3993–4001.

Guzowski, J.F., and J.L. McGaugh (1997). Antisense oligodeoxynucleotide-mediated disruption of hippocampal cAMP response element binding protein levels impairs consolidation of memory for water maze training. *Proc Natl Acad Sci USA* **94**:2693–2698.

Guzowski, J.F., B. Setlow, E.K. Wagner, and J.L. McGaugh (2001). Experience-dependent gene expression in the rat hippocampus after spatial learning: A comparison of the immediate-early genes Arc, c-fos, and zif268. *J Neurosci* **21**(14):5089–5098.

Harris, E.W., and C.W. Cotman (1986). Long-term potentiation of guinea pig mossy fiber responses is not blocked by N-methyl-D-aspartate antagonists. *Neurosci Lett* **70**(1):132–137.

Harris, E.W., A.H. Ganong, et al. (1984). Long-term potentiation in the hippocampus involves activation of N-methyl-D-aspartate receptors. *Brain Res* **323**(1):132–137.

Hebb, D. (1949). *Organization of Behavior: A Neuropsychological Theory*. New York: Wiley.

Higo, N., T. Oishi, A. Yamashita, K. Matsuda, and M. Hayashi (1999). Quantitative nonradioactive in situ hybridization study of GAP-43 and SCG10 mRNAs in the cerebral cortex of adult and infant macaque monkeys. *Cerebral Cortex* **9**:317–331.

Hoeffler, J.P., T.E. Meyer, Y. Yun, J.L. Jameson, and J.F. Habener (1988). Cyclic AMP-Responsive DNA-binding protein: Structure based on a cloned placental cDNA. *Science* **242**:1430–1433.

Hope, B., B. Kosofsky, S.E. Hyman, and E.J. Nestler (1992). Regulation of immediate early gene expression and AP-1 binding in the rat nucleus accumbens by chronic cocaine. *Proc Nat Acad Sci USA* **89**:5764–5768.

Huang, Y.Y., and E.R. Kandel (1994). Recruitment of long-lasting and protein kinase A-dependent long-term potentiation in the CA1 region of hippocampus requires repeated tetanization. *Learning Memory* **1**:74–82.

Impey, S., D.M. Smith, K. Obrietan, R. Donahue, C. Wade, and D.R. Storm (1998). Stimulation of cAMP response element (CRE)-mediated transcription during contexual learning. *Nature Neurosci* **1**(7):595–601.

Ishihara, K., H. Katsuki, et al. (1990). Different drug susceptibilities of long-term potentiation in three input systems to the CA3 region of the guinea pig hippocampus in vitro. *Neuropharmacology* **29**(5):487–492.

Jaffe, D., and D. Johnston (1990). Induction of long-term potentiation at hippocampal mossy-fiber synapses follows a Hebbian rule. *J Neurophysiol* **64**(3):948–960.

Jones, M.W., M.L. Errington, et al. (2001). A requirement for the immediate early gene Zif268 in the expression of late LTP and long-term memories. *Nat Neurosci* **4**(3):289–296.

Josselyn, S.A., S. Kida, and A.J. Silva (2004). Inducible repression of CREB function disrupts amygdala-dependent memory. *Neurobiol Learning Memory* **82**:159–163.

Josselyn, S.A., and P.V. Nguyen (2005). CREB, synapses and memory disorders: Past progress and future challenges. *Curr Drug Targets CNS Neurol Disord* **4**(5):481–497.

Kaang, B.K., E.R. Kandel, and S.G. Grant (1993). Activation of cAMP-responsive genes by stimuli that produce long-term facilitation in Aplysia sensory neurons. *Neuron* **10**:427–435.

Kandel, E.R. (2001). The molecular biology of memory storage: A dialogue between genes and synapses. *Science* **294**(5544):1030–1038.

Karns, L.R., S.C. Ng, J.A. Freeman, and M.C. Fishman (1987). Cloning of complementary DNA for GAP-43, a neuronal growth-related protein. *Science* **236**:597–600.

Kim, J., J. Lu, and P.G., Quinn (2000). Distinct cAMP response element binding protein (CREB) domains stimulate different steps in a concerted mechanism of transcription activation. Proceedings of the National Academy of Science USA, Oct 10; **97**(12):11292–11296.

Kim, J.J., R.E. Clark, and R.F. Thompson (1995). Hippocampectomy impairs the memory of recently, but not remotely, acquired trace eyeblink conditioned responses. *Behavioral Neuroscience* **109**:195–203.

Kim, J.J., and M.S. Fanselow (1992). Modality-specific retrograde amnesia of fear. *Science* **256**:675–678.

Konradi, C., R.L. Cole, S. Heckers, and S.E. Hyman (1994). Amphetamine regulates gene expression in rat striatum via transcription factor CREB. *Journal of Neuroscience* **14**(9): 5623–5634.

Krug, M., B. Lossner, et al. (1984). Anisomycin blocks the late phase of long-term potentiation in the dentate gyrus of freely moving rats. *Brain Res Bull* **13**(1):39–42.

Kuhar, M.J., J.N. Jaworski, G.W. Hubert, K.B. Philpot, and G. Dominguez (2005). Cocaine- and amphetamine-regulated transcript peptides play a role in drug abuse and are potential therapeutic targets. *AAPS J* **7**:259–265.

Lanahan, A., G. Lyford, et al. (1997). Selective alteration of long-term potentiation-induced transcriptional response in hippocampus of aged, memory-impaired rats. *J Neurosci* **17**(8): 2876–2885.

Lanahan, A., and P. Worley (1998). Immediate-early genes and synaptic function. *Neurobiol Learn Mem* **70**:37–43.

Lee, P.R., J.E. Cohen, et al. (2005). Gene expression in the conversion of early-phase to late-phase long-term potentiation. *Ann NY Acad Sci* **1048**:259–271.

Lisman, J.E., and K.M. Harris (1993). Quantal analysis and synaptic anatomy — integrating two views of hippocampal plasticity. *Trends Neurosci* **16**(4):141–147.

Lutz, B., W. Schmid, C. Niehrs, and G. Schutz (1999). Essential role of CREB family proteins during Xenopus embryogenesis. *Mech Dev* **88**:55–66.

Lynch, M.A. (2004). Long-term potentiation and memory. *Physiol Rev* **84**(1):87–136.

Madani, R., S. Hulo, et al. (1999). Enhanced hippocampal long-term potentiation and learning by increased neuronal expression of tissue-type plasminogen activator in transgenic mice. *Embo J* **18**(11):3007–3012.

Maren, S., G. Aharonov, and M.S. Fanselow (1996). Retrograde abolition of conditional fear after excitotoxic lesions in the basolateral amygdala of rats: Absence of a temporal gradient. *Behav Neursci* **110**:718–726.

Marie, H., W. Morishita, et al. (2005). Generation of silent synapses by acute in vivo expression of CaMKIV and CREB. *Neuron* **45**(5):741–752.

Martin, K.C., and K.S. Kosik (2002). Synaptic tagging — who's it? *Nat Rev Neurosci* **3**(10): 813–820.

Martin, M.R. (1983). Naloxone and long-term potentiation of hippocampal CA3 field potentials in vitro. *Neuropeptides* **4**(1):45–50.

Martinez, J.L., Jr., and B.E. Derrick (1996). Long-term potentiation and learning. *Annu Rev Psychol* **47**:173–203.

Matsuo, R., A. Murayama, et al. (2000). Identification and cataloging of genes induced by long-lasting long-term potentiation in awake rats. *J Neurochem* **74**(6):2239–2249.

McConnell, J.V. (1966). Comparative physiology: Learning in invertebrates. *Annu Rev Physiol* **28**:107–136.

McEchron, M.D., H. Bouwmeester, W. Tseng, C. Weiss, and J.F. Disterhoft (1998). Hippocampectomy disrupts auditory trace fear conditioning and contextual fear conditioning in the rat. *Hippocampus* **6**:638–646.

McGlinchey-Berroth, R., M.C. Carrillo, J.D. Gabrieli, C.M. Brawn, and J.F. Disterhoft (1997). Impaired trace eyeblink conditioning in bilateral, medial-temporal lobe amnesia. *Behav Neurosci* **111**:873–882.

Meberg, P.J., C.A. Barnes, B.L. McNaughton, and A. Routtenberg (1993). Protein kinase C and F1/GAP-43 gene expression in hippocampus inversely related to synaptic enhancement lasting 3 days. *Proc Nat Acad Sci USA* **90**:12050–12054.

Meshul, C.K., and W.F. Hopkins (1990). Presynaptic ultrastructural correlates of long-term potentiation in the CA1 subfield of the hippocampus. *Brain Res* **514**(2):310–319.

Messaoudi, E., S.W. Ying, et al. (2002). Brain-derived neurotrophic factor triggers transcription-dependent, late phase long-term potentiation in vivo. *J Neurosci* **22**(17):7453–7461.

Montminy, M.R., and Bilezikjian, L.M. (1987). Binding of a nuclear protein to the cyclic-AMP response element of the somatostatin gene. *Nature* **328**:175–178.

Montminy, M.R., G.A. Gonzalea, and K.K. Yamamoto (1990). Regulation of c-AMP inducible genes by CREB. *Trends in Neuroscience* **13**:184–188.

Morris, R.G.M. (1984). Development of a water-maze procedure for studying spatial learning in the rat. *J Neurosci Methods* **11**:47–60.

Nirenberg, M. (2004). Historical review: Deciphering the genetic code — a personal account. *Trends Biochem Sci* **29**(1):46–54.

Nguyen, P.V., T. Abel, et al. (1994). Requirement of a critical period of transcription for induction of a late phase of LTP. *Science* **265**(5175):1104–1107.

Otani, S., and W.C. Abraham (1989). Inhibition of protein synthesis in the dentate gyrus, but not the entorhinal cortex, blocks maintenance of long-term potentiation in rats. *Neurosci Lett* **106**(1–2):175–180.

Otani, S., C.J. Marshall, et al. (1989). Maintenance of long-term potentiation in rat dentate gyrus requires protein synthesis but not messenger RNA synthesis immediately post-tetanization. *Neuroscience* **28**(3):519–526.

Parelkar, N.K., and J.Q. Wang (2004). mGluR5-dependent increases in immediate early gene expression in the rat striatum following acute administration of amphetamine. *Brain Res Mol Brain Res* **122**:151–157.

Park, Y.H., L. Kantor, K.K. Wang, and M.E. Gnegy (2002). Repeated, intermittent treatment with amphetamine induces neurite outgrowth in rat pheochromocytoma cells (PC12 cells). *Brain Res* **951**:43–52.

Pascale, A., P.A. Gusev, M. Amadio, T. Dottorini, S. Govoni, D.L. Alkon, and A. Quattrone (2004). Increase of the RNA-binding protein HuD and posttranscriptional up-regulation of the GAP-43 gene during spatial memory. *Proc Natl Acad Sci USA* **101**:1217–1222.

Peng, H., B.E. Derrick, et al. (2003). Identification of upregulated SCG10 mRNA expression associated with late-phase long-term potentiation in the rat hippocampal Schaffer-CA1 pathway in vivo. *J Neurosci* **23**(16):6617–6626.

Peng H., B.E. Derrick, and J.L. Martinez, Jr. (2004). Time-course study of SCG10 mRNA levels associated with LTP induction and maintenance in the rat Schaffer-CA1 pathway in vivo. *Brain Res Mol Brain Res* **120**:182–187.

Pittenger, C., Y.Y. Huang, R.F. Paletzki, R. Bourtchouladze, H. Scanlin, S. Vronskaya, and E.R. Kandel (2002). Reversible inhibition of CREB/ATF transcription factors in region CA1 of the dorsal hippocampus disrupts hippocampus-dependent spatial memory. *Neuron* **34**:447–462.

Qian, Z., M.E. Gilbert, et al. (1993). Tissue-plasminogen activator is induced as an immediate-early gene during seizure, kindling and long-term potentiation. *Nature* **361**(6411):453–457.

Rekart, J.L., K. Meiri, and A. Routtenberg (2005). Hippocampal-dependent memory is impaired in heterozygous GAP-43 knockout mice. *Hippocampus* **15**:1–7.

Roberts, L.A., C.H. Large, et al. (1997). Long-term potentiation in perforant path/granule cell synapses is associated with a post-synaptic induction of proenkephalin gene expression. *Neurosci Lett* **227**(3):205–208.

Rodriguez, J.S., C.F. Phelix, and J.L. Martinez, Jr. (2002). Microarray analysis of altered gene expression associated with D-amphetamine self-administered into the nucleus accumbens via microdialysis. *Soc Neurosci Abstr* **808**:19.

Rodriguez, S.M., G.E. Schafe, and J.E. LeDoux (2004). Molecular mechanisms underlying emotional learning and memory in the lateral amygdala. *Neuron* **44**:75–91.

Rogers, J.L., M.R. Hunsaker, and R.P. Kesner (2006). Effects of ventral and dorsal CA1 subregional lesions on trace fear conditioning. *Neurobiol Learning Mem.*

Rolls, E.T. (1992). Neurophysiological mechanisms underlying face processing within and beyond the temporal cortical visual areas. *Philos Trans R Soc Lond B Biol Sci.* **335**:11–20.

Rossmanith, W.G., D.L. Marks, D.K. Clifton, and R.A. Steiner (1994). Induction of galanin gene expression in gonadotrophin-releasing hormone neurons with puberty in the rat. *Endocrinology*, Oct; **135**(4):1401–1408.

Routtenberg, A., I. Cantallops, S. Zaffuto, P. Serrano, and U. Namgung (2000). Enhanced learning after genetic overexpression of a brain growth protein. *Proc Natl Acad Sci USA* **97**(13):7657–7662.

Routtenberg, A., and J.L. Rekart (2005). Posttranslational protein modification as the substrate for long-lasting memory. *Trends Neurosci* **28**(1):12–19.

Sassone-Corsi, P., J. Visvader, L. Ferland, P.L. Mellon, and I.M. Verma (1988). Induction of proto-oncogene fos transcription through the adenylate cyclase pathway: Characterization of a cAMP-responsive element. *Genes Dev* **2**:1529–1538.

Shaywitz, A.J., and M.E. Greenberg (1999). CREB: A stimulus-induced transcription factor activated by a diverse array of extracellular signals. *Annu Rev Biochem* **68**:821–861.

Shors, T.J., G. Miesegaes, A. Beylin, M. Zhao, T. Rydel, and E. Gould (2001). Neurogenesis in the adult is involved in the formation of trace memories. *Nature* **410**:372–376.

Shors, T.J., D.A. Townsend, M. Zhao, Y. Kozorovitskiy, and E. Gould (2002). Neurogenesis may relate to some but not all types of hippocampal-dependent learning. *Hippocampus* **12**:578–584.

Silva, A., and S.A. Josselyn (2002). The molecules of forgetfulness. *Nature* **418**:929–930.

Squire, L.R., and S. Zola-Morgan (1991). The medial temporal lobe memory system. *Science* **253**:1380–1386.

Stein, R., N. Mori, K. Matthews, L.C. Lo, and D.J. Anderson (1998). The NGF-inducible SCG10 mRNA encodes a novel membrane-bound protein present in growth cones and abundant in developing neurons. *Neuron* **1**:463–476.

Sugaya, K., R. Greene, et al. (1998). Septo-hippocampal cholinergic and neurotrophin markers in age-induced cognitive decline. *Neurobiol Aging* **19**(4):351–361.

Thompson, K.J., M.L. Mata, et al. (2005). Metabotropic glutamate receptor antagonist AIDA blocks induction of mossy fiber-CA3 LTP in vivo. *J Neurophysiol* **93**(5):2668–2673.

Thompson, K.J., W.J. Meilandt, P. Lingala, J. Orfila, H. Peng, and J.L. Martinez (2005). *Arc and Homer-1 Are Differentially Expressed Following* in vivo *Induction of Early-Phase LTP at the Hippocampal Mossy Fiber-CA3 Pathway.* Washington, DC: Society for Neuroscience.

Thompson, K.J., J.E. Orfila, et al. (2003). Gene expression associated with in vivo induction of early phase-long-term potentiation (LTP) in the hippocampal mossy fiber-Cornus Ammonis (CA)3 pathway. *Cell Mol Biol (Noisy-le-grand)* **49**(8):1281–1287.

Ungar, G. (1970). Chemical transfer of learned behavior. *Inflammation Research* **1**(4), 155–163.

Urban, N.N., and G. Barrionuevo (1996). Induction of Hebbian and non-Hebbian mossy fiber long-term potentiation by distinct patterns of high-frequency stimulation. *J Neurosci* **16**(13):4293–4299.

Vazdarjanova, A., B.L. McNaughton, et al. (2002). Experience-dependent coincident expression of the effector immediate-early genes arc and Homer 1a in hippocampal and neocortical neuronal networks. *J Neurosci* **22**(23):10067–10071.

Vukosavic, S., S. Ruzdijic, R. Veskov, L. Rakic, and S. Kanazir (2001). Differential effects of amphetamine and phencyclidine on the expression of growth-associated protein GAP-43. *Neurosci Res* **40**:133–140.

Walton, M., C. Henderson, et al. (1999). Immediate early gene transcription and synaptic modulation. *J Neurosci Res* **58**(1):96–106.

Watson, J.D., and F.H. Crick (1953). Molecular structure of nucleic acids; a structure for deoxyribose nucleic acid. *Nature* **171**(4356):737–738.

Whittaker, E., E. Vereker, et al. (1999). Neuropeptide Y inhibits glutamate release and long-term potentiation in rat dentate gyrus. *Brain Res* **827**(1–2):229–233.

Ying, S.W., M. Futter, et al. (2002). Brain-derived neurotrophic factor induces long-term potentiation in intact adult hippocampus: Requirement for ERK activation coupled to CREB and upregulation of Arc synthesis. *J Neurosci* **22**(5):1532–1540.

Yoshida, T., and M. Mishina (2005). Distinct roles of calcineurin-nuclear factor of activated T-cells and protein kinase A-cAMP response element-binding protein signaling in presynaptic differentiation. *J Neurosci.* **25**:3067–3079.

Young, E.A., E.H. Owen, K.F. Meiri, and J.M. Wehner (2000). Alterations in hippocampal GAP-43 phosphorylation and protein level following contextual fear conditioning. *Brain Res* **860**:95–103.

Zalutsky, R.A., and R.A. Nicoll (1990). Comparison of two forms of long-term potentiation in single hippocampal neurons. *Science* **248**(4963):1619–1624.

Zalutsky, R.A., and R.A. Nicoll (1992). Mossy fiber long-term potentiation shows specificity but no apparent cooperativity. *Neurosci Lett* **138**(1):193–197.

Zhang, X., D.T. Odom, S. Koo, M.D. Conkright, G. Canettieri, J. Best, H. Chen, R. Jenner, E. Herbolsheimer, E. Jocobsen, S. Kadam, J.R. Ecker, B. Emerson, J.B. Hogenesch, T. Unterman, R.A. Young, and M. Montminy (2005). Genome-wide analysis of cAMP-response element binding protein occupancy, phosphorylation, and target gene activation in human tissues. *Proc Natl Acad Sci USA* **102**(12):4459–4464.

Zhuo, M., D.M. Holtzman, et al. (2000). Role of tissue plasminogen activator receptor LRP in hippocampal long-term potentiation. *J Neurosci* **20**(2):542–549.

CHAPTER 5

Mnemonic Contributions of Hippocampal Place Cells

Sherri J. Y. Mizumori, D. M. Smith, and C. B. Puryear

Department of Psychology, University of Washington, Seattle, WA 98195

I. INTRODUCTION

The role of the hippocampus in learning and memory processes has been the subject of intense scrutiny since the 1957 report (Scoville and Milner, 1957) of the remarkably severe amnesia in patient H.M., which resulted from the surgical removal of the hippocampal formation and a large amount of the adjacent tissue in the medial temporal lobes. Since then we have made great strides in documenting the hippocampal contribution to learning and memory, and we have generated a massive literature on the subject. As a result, there is virtually universal consensus that the hippocampus is involved in learning and memory. Nevertheless, many questions remain about the specific kinds of memory subserved by the hippocampus, the specific processes enabled by hippocampal neurocomputation, and the kinds of information encoded by hippocampal neurons.

Accounts of the potentially unique contribution by hippocampus to learning include its role in spatial processing (e.g., O'Keefe and Nadel, 1978; Poucet, 1993; Long and Kesner, 1996; Goodrich-Hunsaker et al., 2005), working memory (Olton et al., 1979), relational learning (Eichenbaum and Cohen,

Neurobiology of Learning and Memory, Second Edition
Copyright © 2007 by Elsevier Inc. All rights reserved.

2001), episodic memory (e.g., Tulving, 2002), context processing (e.g., Hirsh, 1974), declarative memory (Squire et al., 2004), and the encoding of experiences in general (Moscovitch et al., 2005). Attempts to reconcile these disparate accounts will require converging evidence from different levels of analysis, from single-cell analysis to the study of the functions of entire brain structures to studies of the broader network within which the involved brain areas operate. A circuit-level approach involving recording neuronal activity from multiple anatomically connected brain regions has shown that the hippocampus is part of an extended circuitry in which different regions make unique contributions to memory and navigation (Gabriel, 1993; Redish and Touretzky, 1997; Mizumori et al., 2000b; Smith et al., 2004). Large-scale recording of neuronal ensemble activity patterns (e.g., Wilson and McNaughton, 1993) and biologically plausible computational models (e.g., Blum and Abbott, 1996; Wallenstein and Hasselmo, 1997) can begin to suggest principles of network dynamics that might mediate hippocampal mnemonic functions. Studies of synaptic plasticity and investigations of the response characteristics of individual neurons can identify neurobiological mechanisms underlying the circuit dynamics that ultimately result in the learning and memory functions of the hippocampus.

The focus here is to evaluate the potential relationship between neural representation by individual hippocampal neurons and learning and memory. The most commonly reported behavioral correlate of hippocampal output neurons (pyramidal cells) is location-selective firing, referred to as *place fields* (see examples in Fig. 5-1; O'Keefe and Dostrovsky, 1971). The seminal discovery that hippocampal pyramidal neurons exhibit remarkably distinct and reliable firing when rats visit particular regions of the environment led to the highly influential idea that the hippocampus was the neural substrate of the cognitive map (O'Keefe and Nadel, 1978). Over 35 years of study has yielded a vast literature on hippocampal place cells (for reviews, see O'Keefe, 1976; O'Mara, 1995; McNaughton et al., 1996; Muller et al., 1996; Wiener, 1996; Mizumori et al., 1999b). In spite of our remarkably detailed knowledge about factors that influence the response properties of place fields and an extensive literature on the critical role of the hippocampus in certain forms of memory, including apparently nonspatial memory, the fundamental issue of how place fields are related to the broad role of the hippocampus in learning and memory remains unclear. Improved technology for recording large numbers of single neurons (McNaughton et al., 1983b; Gray et al., 1995) and for careful monitoring of behavior during learning and performance in well-designed behavioral tasks has improved our ability to answer questions about the mnemonic significance of place cells (Fig. 5-1). In the following, evidence is provided that demonstrates the sensory and movement responsiveness of place fields. Then we discuss past work that has attempted to address more directly the role of place fields in learning and memory. A context discrimination

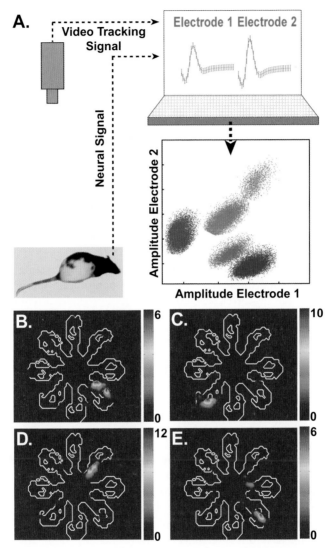

FIGURE 5-1   **A.** Schematic illustration of recordings from freely behaving rats. Output from a video monitor is synchronized in time to incoming neural signals. Multiunit records can be decomposed into signals from individual cells by using a cluster analysis that compares values for spike parameters on two separate electrodes of a tetrode. Shown are clusters that were generated when the amplitude of the signal from one electrode was compared to the amplitude of the signal recorded from an adjacent electrode. **B.–E.** Color density plots illustrated location-selective firing by hippocampal place cells. White lines outline visited locations as a rat performed a spatial working memory task on an eight-arm radial maze. Values to the right of each plot illustrate the maximum firing rate for the cell being illustrated.

hypothesis is then presented to account for the role that place fields may play in hippocampal-dependent mnemonic functions.

II. PLACE FIELDS: SENSORY AND MOVEMENT CORRELATES

Some of the earliest studies showed quite convincingly that many place fields are sensitive to changes in the visual environment (e.g., Ranck, 1973; O'Keefe, 1976; Olton et al., 1978; Muller and Kubie, 1987). More recent studies demonstrate specific aspects of the visual environment that are particularly relevant to place cell firing, such as geometric features of the test environment (e.g., Kubie and Ranck, 1983; Gothard et al., 1996; O'Keefe and Burgess, 1996b; Wiener, 1996, 2002). Also, local cues can have different effects on place fields when compared to distal cues (e.g., O'Keefe and Speakman, 1987; Wiener et al., 1995; Tanila et al., 1997; Truiller et al., 1999). Although it is clear that visual cues exert a strong influence over place fields in navigating rats, it has also been shown that the location-specific and nonlocation-specific firing of pyramidal cells can be correlated with other types of external sensory stimuli, such as olfactory cues (Eichenbaum et al., 1999; Save et al., 2000), auditory cues (O'Keefe and Conway, 1978; Sakurai, 1994; McEchron and Disterhoft, 1999), and somatosensory cues (Young et al., 1994; Tanila et al., 1997). Thus, it is generally accepted that hippocampal pyramidal cells are capable of processing multimodal (external) sensory cue information. Such results have been taken as evidence that the hippocampus represents the external sensory features of a context (e.g., Nadel and Wilner, 1980; Kubie and Ranck, 1983; Nadel and Payne, 2002).

Place fields, however, reflect more than details of the external sensory surround, since they are observed when external cues are essentially absent (O'Keefe and Conway, 1978; Quirk et al., 1990; McNaughton et al., 1996). In the absence of such cues, temporal or internal sensory cue information can shape the characteristics of place fields. As an example, the elapsed time since leaving a goal box was often a better predictor of place fields than the features in the environment (Gothard et al., 1996; Redish et al., 2000). Also, there are many demonstrations that internal (idiothetic) cues that are related to one's own movement impact place fields. For example, the velocity of an animal's movement through a place field has been shown to be correlated with cell firing rates (e.g., McNaughton et al., 1983a). Also, the orientation of an animal, or the direction in which it traverses a field, may importantly affect the firing of place cells (e.g., McNaughton et al., 1983a; Markus et al., 1994). Vestibular (or inertial) information may further inform place cell responses (e.g., Hill and Best, 1981; Knierim et al., 1995; Wiener et al., 1995; Gavrilov et al., 1998). Interestingly, the extent to which place cells are sensitive to idiothetic cues is reported to decline systematically from the septal pole to the temporal pole of the hippocampus (Maurer et al., 2005). Coincident with this trend is the finding that place fields become increasingly larger for place cells recorded along the dorsal–ventral axis (e.g., Jung et al., 1994). Thus, place field selectivity appears related to the degree of sensitivity to idiothetic cues. One factor that may regulate the influence of movement-related information on place fields is the degree to which animals are free to move about in an environment (Foster et al., 1989; Gavrilov et al., 1998; Song et al., 2005). Compared to passive movement conditions, in which rats are made to go through a place field either by being held by the experimenter or by being placed on a movable robotic device, active and unrestrained movement corresponds to the observation of more selective and reliable place fields.

A common interpretation of the sensitivity of place fields to orientation and idiothetic information is that place fields represent a combination of sensory and response cues to update place (visual) representations in familiar environments (e.g., Knierim et al., 1995; McNaughton et al., 1996; Gavrilov et al., 1998; Maurer et al., 2005). In this way, place cells contribute to both egocentric and allocentric information processing, perhaps for the purpose of facilitating a path integration process (e.g., McNaughton et al., 1996; Whishaw, 1998) that allows rats to navigate environments in which external cues become unreliable. The precise relationship between such path integration functions and more general learning and memory processes remains enigmatic. What is clear, however, is that one needs to consider carefully both environmental and idiothetic variables when evaluating place field functions.

III. PLACE FIELDS: RELATIONSHIP TO LEARNING AND MEMORY

Seven approaches can be identified that test more directly the relationship between learning, memory, and place representation by hippocampus (Table 5-1). *One approach* has been to determine whether experimental manipulations have comparable consequences for place fields and learning. For example, the effects of NMDA receptor antagonists have been shown to impair hippocampal-dependent learning and to reduce the stability of place fields (Shapiro and Eichenbaum, 1999). Similarly, mice lacking functional NMDA receptors have poor spatial learning abilities (Tsien et al., 1996) and less specific place fields (McHugh et al., 1996). While this approach has revealed that place field stability/specificity and learning have common underlying neurobiological mechanisms (e.g., NMDA receptor activation, Kentros et al., 1998, 2004; Nakazawa et al., 2004), it offers little in terms of helping us to understand how place representations contribute to the mnemonic functions of hippocampus. A related *second approach* to establishing a connection between place fields and

TABLE 5-1   Approaches Used to Assess the Contribution of Place Cells to Learning and Memory

---

1. Test whether experimental manipulations similarly affect learning and place fields
2. Examine the relationship between synaptic models of learning (e.g., LTP) and place field plasticity
3. Evaluate place field responses during new learning
4. Evaluate place field responses to exposure to novel environment
5. Compare place fields from subjects that differ in learning capability
6. Examine place field responses to changes in familiar information
7. Use tasks with common test environments and behavioral response to evaluate place field responses to changes in cognitive demand

---

learning/memory is to evaluate the relationship between synaptic models of neuroplasticity and place field plasticity. As an example, it was shown that the induction of long-term potentiation (LTP), a leading candidate model of learning-induced synaptic changes (Martin et al., 2000), alters place fields (Dragoi et al., 2003). Also, experience-dependent changes in immediate early gene expression are correlated with experience-dependent changes in place field responses (Guzowski et al., 2005).

There are only a handful of studies that have attempted to test directly how place fields contribute to new learning (a *third approach*). Although the sample size was small, place fields became more localized and/or stable as learning progressed (e.g., Mizumori and Kalyani, 1997). Relatedly, place fields may move toward new goal locations when tested in a familiar room (e.g., Hollup et al., 2001; Lenck-Santini et al., 2001, 2002). Before one can conclude that such changes in place fields are related to new spatial learning, however, it is necessary to show that changes in the place fields were not due to the changes in behavior that accompany learning. A *fourth approach* has been to correlate changes in place field properties (e.g., field specificity and/or reliability, and firing rates) with timed exposure to novel environments (e.g., Wilson and McNaughton, 1993; Frank et al., 2004; Leutgeb et al., 2005). While provocative effects have been described, a difficulty with this approach is that hippocampal-dependent learning is not assessed. A *fifth approach* is to compare place field properties across groups of animals that differ in their learning capabilities. Place fields recorded from young, adult, and aged rats were compared (Mizumori et al., 1996; Tanila et al., 1997; Barnes et al., 1997), but different conclusions were drawn regarding the effects of age on the stability of place fields, and hippocampal-dependent learning was not assessed (Mizumori and Leutgeb, 1999).

A *sixth approach* to understanding the relationship between place fields and learning and memory is to demonstrate that place fields change in response to encounters with specific combinations of cues or to a change in familiar cues.

Numerous studies describe place cells that fire relative to (presumably) learned combinations of cues or spatial relationships (e.g., Young et al., 1994; Cressant et al., 1997; Shapiro et al., 1997; Tanila et al., 1997). While these studies often establish a correlation, the reasons for the correlations are not certain, due to the fact that different behaviors were required to demonstrate learning and/or the learning was not hippocampal dependent. Wiener et al. (1989) showed that a given pyramidal cell may exhibit a place field during performance of a spatial task in one portion of a test compartment and then switch modes of firing such that it fires relative to a nonspatial aspect of the task (i.e., task phase or a particular odor) when rats perform an olfactory discrimination task in a different part of the test chamber. While this study demonstrates that hippocampal pyramidal neurons can shift the nature of their representation (from spatial to nonspatial) depending on prior learning, the differential neural responses may reflect different behaviors exhibited during different test phases. Thus, the significance of the place field remained somewhat ambiguous. Place fields have been reported to change after fear conditioning (Moita et al., 2004), but behavioral and motivational differences across different test phases cloud strong conclusions about the relationship between memory and place representation. This is especially the case because hippocampus appears to process motivational information (Kennedy and Shapiro, 2004). Auditory responses of hippocampal cells have been shown to vary, depending on the learned significance of the auditory stimulus during conditioning (Sakurai, 1994; Freeman et al., 1996; McEchron and Disterhoft, 1999). Since spatial correlates were not evaluated, however, it is not known how such learning correlates are related to place representation. Many studies have tested place cell responses to a change in task demands (e.g., changes in reward location, task phase, or required behavior; Wible et al., 1986; Foster et al., 1987; Markus et al., 1994; Gothard et al., 1996). These experiments showed clear neural responses to changes in task demands. But more often than not, either the animals were not performing a hippocampal-dependent task or a direct evaluation of the relationship to learning was not provided.

To circumvent some of the concerns of past studies, recent investigations of the relationship between place fields and learning or memory employ new behavioral designs to test the responsiveness of place cells to changes in cognitive demand (a *seventh approach*). Ideally, for this kind of test, sensory and motivational influences should be held constant across test conditions. Moreover, behavioral responses should be "clamped." That is, identical behaviors should be required for the phases of the task that are being compared. The latter condition is critical so that differential behavioral expression of learned information cannot be used to explain discriminating responses of place cells. By holding sensory, response, and motivational factors constant, the relationship between learning or memory operations and place field plasticity can be revealed. Based on this approach, it has been suggested that the exhibition of

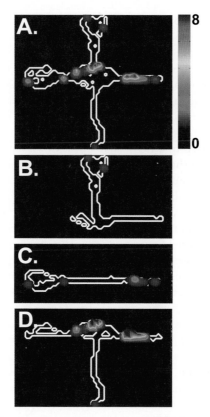

FIGURE 5-2   Illustration that place fields are sensitive to behaviors exhibited prior to entering the field location. A rat was trained to obtain food reward on the east maze arm when started from either the north, west, or south arm of the maze. **A.** When considering the firing pattern for an entire training session, place fields are found on the east arm and the north sector of the central platform of the maze. However, when the data were divided according to the starting arm location, it is clear that the place fields were not exhibited when rats arrived from the north **(B)** or west **(C)** maze arms. **D.** Rather, the place fields were observed only when entering the goal arm from the south maze arm.

place fields is conditional on the recent (exemplifying retrospective coding) or upcoming (exemplifying prospective coding) behavioral sequences or response trajectories (Frank et al., 2000; Wood et al., 2000). Although the authors' conclusion that hippocampus encodes retrospective and prospective informa-tion is appealing, it should be noted that the tasks used were not hippocampal dependent (e.g., Ainge and Wood, 2003). However, Ferbinteanu and Shapiro (2003) used a hippocampal-dependent plus-maze task to replicate the Wood et al. (2000) finding of trajectory-dependent place codes and to more clearly show that place cells are affected by some kind of prospective and retrospective information (see Fig. 5-2 for an example of retrospective trajectory coding from our laboratory). That is, the expectation of a future goal location appeared to influence some place fields, while the exhibition of place fields by other cells depended on recent trajectories taken by a rat on the maze. The trajectory-dependent neural codes degraded during error trials. These findings demonstrate a temporal modulation of spatial information in hippocampus. It was argued that temporally regulated neural codes are likely to be an essential component of the hippocampal contribution to episodic learning and memory.

The features of place fields that have been described so far are certainly consistent with the idea that they make a contribution to the mnemonic functions of hippocampus. However, understanding the specific nature of the contribution by place fields remains a significant challenge. Different reasons may account for the noticeably large gap between findings related to place field characteristics and studies that implicate a specific role for the hippocampus in learning. One reason may be that studies of place cells and studies of the role of the hippocampus in learning and memory involve different levels of analyses. Most place cell studies involve investigation of only one type of hippocampal neuron, the pyramidal cell. Lesion and imaging studies of the hippocampus' role in learning and memory, on the other hand, involve either removing or imaging all cell types within the entire (or circumscribed) hippocampal area(s). A second issue of concern is that place cell studies typically test rats on tasks that are not hippocampal dependent. Of particular relevance to this point is the finding that the extent to which performance of the task depends on hippocampal function can change the degree to which place fields are controlled by the available visual cues (Zinuyk et al., 2000). Finally, a third, but related, issue is that few studies provide unambiguous tests of the relationship between place field characteristics and the hippocampal-dependent acquisition of new information (but see Smith and Mizumori, in press, discussed later).

In the following, we attempt to bridge the gap between place field results and hippocampal-dependent learning by focusing on what appears to be a critical role for the hippocampus in context processing. Extensive animal research has implicated the hippocampus in processing contextual information (for reviews see Myers and Gluck, 1994; Anagnostaras et al., 2001; Maren, 2001). For example, subjects with hippocampal damage do not exhibit conditioned fear responses to contextual stimuli, even though responses to discrete conditional stimuli remain intact (Kim and Fanselow, 1992; Phillips and Ledoux, 1992). Also, subjects with hippocampal damage are insensitive to changes in the context. Intact subjects exhibit decrements in conditioned responding when the context is altered. This does not happen in subjects with lesions of the hippocampus (Penick and Solomon, 1991) or the entorhinal cortex (Freeman et al., 1996). Recently, it was shown that fornix-lesioned subjects trained in different contexts showed severely impaired learning on two different auditory discrimination tasks (Smith et al., 2004). In the same subjects, context-specific firing patterns were degraded in structures that normally receive hippocampal input via the fornix (anterior thalamus and cingulate cortex). Finally, it appears that manipulations that impact hippocampal plasticity (e.g., LTP) also affect context learning (e.g., Shors and Matzel, 1997). These findings converge on a hypothesis that hippocampal processing results in a unique context code, one that presumably modulates processing in downstream structures in a context-dependent manner.

A context-processing interpretation of hippocampal function is not necessarily inconsistent with other views that the hippocampus is necessary for the flexible use of conjunctive, sequential, or relational and spatial processing (e.g., O'Keefe and Nadel, 1978; Foster et al., 1987; Wood et al., 2000; Eichenbaum, 2001; Eichenbaum and Cohen, 2001; O'Reilly and Rudy, 2001). In fact, these specific forms of processing are likely to be required in order for the hippocampus to make context discriminations. Determining the extent to which a learned (or expected) context has changed requires the flexible consideration of sequences of sensory and motor information and/or the significance of (or relationships among) this type of information. Thus a context-processing interpretation of hippocampal function incorporates flexible, spatial, and relational processes. Less clear, however, is how a context-processing theory of hippocampus accounts for the numerous findings regarding hippocampal place cells in freely moving rats. In the following, a context discrimination hypothesis (CDH) is presented not only to account for the wealth of place cell data, but to provide a theoretical framework that may serve to connect the growing body of work on neuroplasticity of place representation and current ideas about the specific role of the hippocampus in especially episodic learning and memory, i.e., learning about and remembering specific meaningful events in our lives (Tulving, 2002).

A. Context Discrimination Hypothesis Account of Hippocampal Neural Representation

Hippocampal place cells may represent spatial contexts (Nadel and Wilner, 1980; Nadel and Payne, 2002) such that the extent to which familiar contexts change can be determined, perhaps by performing a match–mismatch comparison (e.g., Vinogradova, 1995; Mizumori et al., 2000b, 1999b; Anderson and Jeffery, 2003; Jeffery et al., 2004). Computational models provide suggestions for how such error analyses might be implemented within hippocampal circuitry and modulated by neurotransmitters such as dopamine and acetylcholine (e.g., Lisman, 1999; Hasselmo and McGaughty, 2004; Treves, 2004; Hasselmo, 2005a, 2005b). The results of match–mismatch comparisons can be used to distinguish different contexts, a function that is necessary to define significant events or episodes. In support of this view, prior work has shown that fornix lesions impair context discrimination (Smith et al., 2004). A context discrimination hypothesis (CDH), then, postulates that the hippocampus contributes to complex forms of learning, such as those involved in episodic learning and memory, by discriminating meaningful contexts.

According to the CDH, context discrimination is the computational result of context analysis by the entire network of hippocampus neurons, including both hippocampal pyramidal cells *and* interneurons. The combined activity profile of these cells represents an experience-dependent context code that is unique for a particular test situation. For individual cells of both types, a context code might reflect simultaneously a combination of different context-defining features, such as spatial information (i.e., location and heading direction), consequential information (i.e., reward), movement-related feedback (i.e., velocity and acceleration — determinants of response trajectory), external (nonspatial) sensory information, the nature of the currently active memory (defined operationally in terms of task strategy and/or task phase), and the current motivational state. The CDH, then, puts forth a broader definition of context than is typical for place cell studies — a definition that emphasizes the integration of sensory, motivational, response, and memorial input for all hippocampal cell types. The relative strengths of these different types of inputs may vary, depending on task demands, such that a given cell may show, for example, a place correlate during the performance of one task and a nonspatial correlate during the performance of a different task (e.g., Wiener et al., 1989). Also, movement correlates observed in one task may not be observed when the mnemonic component of the context, and not behavior, changes (e.g., Yeshenko et al., 2004; Eschenko and Mizumori, 2007).

The more inclusive and broader definition of context by the CDH is consistent with findings that place fields are sensitive to a variety of types of sensory and response information when rats are tested on different tasks. It should be noted that the CDH is also consistent with descriptions of pyramidal neuron responses during nonspatial task performance, such as olfactory (e.g., Wiener et al., 1989) or auditory discrimination (Sakurai, 1994; Freeman et al., 1996). The CDH also accounts for findings that a variety of correlate types is observed for hippocampal pyramidal cells during the performance of a single task (e.g., Smith and Mizumori, 2006; Gothard et al., 1996). Figure 5-3 illustrates the findings of movement-, reward-, and task phase–related firing by presumed pyramidal neurons recorded during spatial maze training.

A neural representation may demonstrate context sensitivity with changes in firing rates and/or altered temporal patterns of unit activity. Indeed, there are many reports of place fields that rapidly reorganize (i.e., change field location and/or firing rate) when the environment changes, while other place fields persist despite changes in contextual features (see Fig. 5-4 and 5-5 for examples). The responsive place fields may reflect the current contextual features, while the persistent fields may reflect expected contextual features. If

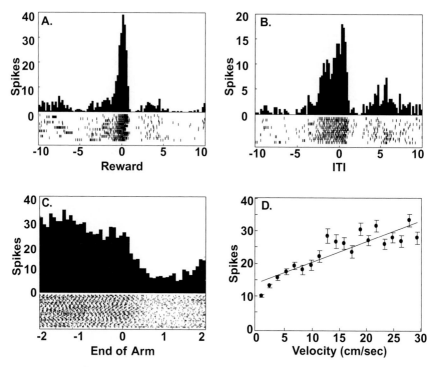

FIGURE 5-3   Illustration of nonspatial correlates of hippocampal neurons when rats solve a spatial maze task. For panels A–C, the *x*-axis shows time (sec) relative to task events. **A.** Elevated firing of a pyramidal neuron is observed when the rat received food reward, regardless of the reward location. **B.** This pyramidal cell showed increased firing just prior to the beginning of the intertrial interval (ITI). **C.** Interneurons typically show elevated firing during ambulation on the maze and reduced firing when the rat reaches the ends of maze arms (i.e., the location of the reward). These cells are often correlated with specific features of an animal's movement, such as movement velocity **(D)**.

the current context is determined to be different from the expected context (i.e., the two contexts are discriminated), then an appropriately changed message may be sent to update cortical memory circuits, which in turn will ultimately update the most recent hippocampal expectation for a context. The latter process should result in a subsequent reorganization of neural activity patterns in the hippocampus. If a context is defined by a unique array of inputs, then, in theory, a change in any one or combination of features should produce an "error" signal that reflects a mismatch (Mizumori et al., 2000b). Indeed, changes in a variety of factors have been shown to have dramatic effects on place fields. Changed hippocampal output (which reflects place field reorganization) could be used to update the selection of ongoing behaviors in context-

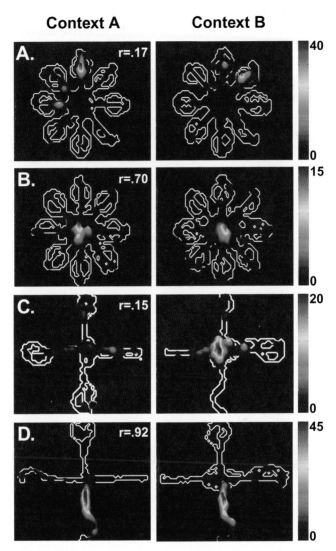

FIGURE 5-4 Some place fields respond, and other place fields do not respond, to the same manipulation of the context regardless of the task. Place fields recorded during baseline trials are shown in the left column. The right column shows the place field for the same cells but after a context manipulation. **A.–B.** A well-trained rat performed the radial maze task before and after the lights were turned off. The cell in panel A showed a field that changed location after the lights went off, whereas the simultaneously recorded cell in panel B showed a place field that did not respond to the imposed darkness. The *r*-values represent a spatial correlation score that compares the spatial distribution of firing across the two test phases. **C.–D.** A rat was trained to find reward on the east maze arm of a plus maze during the baseline trials. The reward location switched to the west maze arm during the manipulation phase. The cell shown in panel C did not show a place field during the baseline condition, but a clear field appeared during the manipulation phase. A simultaneously recorded place cell showed no response to the reward switch (panel **D**).

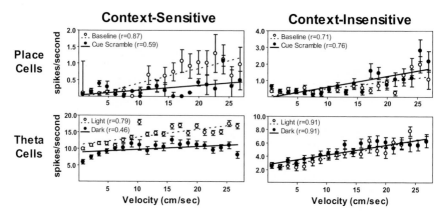

FIGURE 5-5   For each panel, the relationship between firing rate and velocity of an animal's movement on a radial maze is shown for baseline trials and for trials performed after a context manipulation. **(Left)** It can be seen that some cells of each type showed a significant change in velocity correlate after a context manipulation, even though the animal's behavioral performance was unchanged. **(Right)** Other cells of both types showed remarkably stable velocity correlates after a context shift. It appears that some of the movement correlates may reflect context-dependent responses, while others may represent the current movement state.

appropriate ways. If it is determined that the context has not changed (i.e., there is no place field reorganization), a consistent hippocampal output will result in the persistence of currently active neural activity patterns. As a result, the most recently engaged memory system and behavioral expression patterns will be maintained.

Given that situations or events have temporal limits, it seems reasonable to assume that in addition to changes in rate codes, neural firing relative to the occurrence of significant events can be useful for identifying contexts (or events). The ability to retain information "online" in the form of retrospective and prospective codes (e.g., Ferbinteanu and Shapiro, 2003) may provide the hippocampus with a platform on which to compare recent, current, and anticipated contextual information. The outcome of the comparison analysis (i.e., information about the identification of distinct events or episodes) can then be forwarded within an anatomically extended memory system.

B. Relevance of the Context Discrimination Hypothesis for Learning

In our view, the CDH can not only account for a vast number of the reported characteristics of hippocampal place fields, but the CDH also provides a conceptual framework for linking place cell data with theories regarding a hippocampal contribution to learning and memory. During new learning, one's perception of the relationship between environmental stimuli, responses, and consequences is continually changing. Presumably mismatches between expected and experienced contexts are frequently detected, resulting in the continual shaping of long-term memory representations (McClelland et al., 1995). As memory representations become more precise, so will the feedback to hippocampal cells encoding the expected contextual features. Thus, it is predicted that place fields should become more specific/reliable with continued training as one gradually learns about associations relevant to the test environment. While many studies have shown that place fields become more specific and/or reliable with short-term exposure to novel environments (e.g., Muller and Kubie, 1987; Wilson and McNaughton, 1993, Markus et al., 1995; O'Keefe and Burgess, 1996b; Hetherington and Shapiro, 1997; Frank et al., 2004), it is less clear how place fields change during the course of hippocampal-dependent learning.

Learning can be considered to be complete when mismatches no longer occur and consistent memory (and associated neural activity patterns) representations are maintained during behavior. However, to ensure adaptive behavioral functions, the hippocampus must continue to engage in context comparisons in the event that the context shifts. Similarly, the hippocampus should process contextual information even for tasks that do not explicitly require contextual knowledge, in case contextual information becomes relevant. Indeed, we have shown that specific neural codes in the hippocampus remain responsive to changes in context even though contextual learning was not necessary to solve the (response) task (Yeshenko et al., 2004). Thus, processing contextual information by the hippocampus appears automatic and continuous (Morris and Frey, 1997).

If the hippocampus continually processes contextual information, then why do hippocampal lesions disrupt only certain forms of learning and not others? We suggested that lesion effects are observed only when the intrinsic processing by the structure of interest is unique and essential for learning to take place (Mizumori et al., 2004). If other neural circuits can compensate for the lost function after a lesion (for example, by adopting different learning algorithms), then no behavioral impairment will be observed. Thus, stimulus–response learning is not impaired following hippocampal lesions because striatal computations are sufficient to support such learning. This does not mean that the hippocampus does not normally play a role in stimulus–response performance. Hippocampus may contribute by defining the context for the learning, which in turn may serve to make the learned information more adaptive to new situations in the future.

C. Physiological Mechanisms Underlying Hippocampal Context Analysis

1. Cellular Properties

Consideration of the nature of rate and temporal codes of place (pyramidal) cell and movement cell (interneuron) firing reveals that the hippocampus may be ideally suited to engage in context discrimination. The relative influence of context-defining input on spike discharge rates may vary not only according to the strength of each type of afferent input, but also with the intrinsic (membrane) properties of a cell. Place cells exhibit characteristic short-lasting, high-frequency bursts of action potentials when rats pass through a cell's place field (Ranck, 1973). This type of phasic, burst firing pattern may be associated with increased synaptic plasticity using LTP-like mechanisms, possibly resulting in the encoding of discrete features of a situation that do not change very rapidly or often (e.g., significant locations, reward expectations, task phase). Interneurons, on the other hand, discharge signals continuously and at high rates, a pattern that is well suited to encode *rapidly* and *continuously* changing features (e.g., changes in movement and orientation during task performance). Since details of these movement correlates are context sensitive, interneurons may represent to hippocampal networks context-appropriate response information (Yeshenko et al., 2004). Thus, the specific contribution of pyramidal cells and interneurons to context discrimination may differ in large part because of their different intrinsic cellular properties. The complex combination of features encoded by both pyramidal cells and interneurons, then, provides hippocampal computational networks with a rich array of information with which to identify and distinguish unique situations or contexts.

There are likely to be multiple stages to the complex process of comparing expected and current contexts (Mizumori et al., 1999b). Initially, relevant stimuli need to be selected for comparison. The dentate gyrus is thought to engage in pattern separation functions that might serve this purpose by distinguishing potentially important input (O'Reilly and McClelland, 1994; Rolls, 1996; Gilbert et al., 2001). In contrast, CA3 and CA1 may be the primary players when it comes to determining the stability of contextual information.

2. Comparisons Between CA3 and CA1 Place Fields and Learning

Information regarding both expected and current contexts need to be represented, and this information needs to be held "online," in order for a comparison to be made. Both of these conditions appear to be true for the hippocampus. There is ample evidence that the hippocampus represents expected and current contextual information. Place fields may reorganize following context manipulations, indicating that some place cells are sensitive to the status of the current definition of context. Other place fields may not reorganize following the same context manipulation, and these place fields may be driven more by an expectation of the current context based on past experience. The recurrent networks of the CA3 region may provide the short-term buffer necessary for comparisons to be made, perhaps via a type of pattern-completion process (Guzowski et al., 2004; Treves, 2004; Gold and Kesner, 2005).

Interestingly, the differential responses by CA3 and CA1 place cells suggest that these populations of cells make different contributions to the context-comparison process. When rats perform at asymptotic levels on a hippocampal-dependent, spatial-working-memory task, CA3 place fields are smaller (i.e., more spatially selective) than CA1 place fields (Mizumori et al., 1989, 1996; Barnes et al., 1990) and more easily disrupted following cue manipulations (Mizumori et al., 1999b). Also, CA3 place fields are more sensitive than CA1 place fields to disruption following reversible inactivation of the medial septum (Mizumori et al., 1989). The greater sensitivity of CA3 fields to context manipulation seems to occur regardless of the type of task (Lee et al., 2004; Leutgeb et al., 2004), a finding consistent with the view that the CA3 region is continually engaged in the analysis of contexts. It is important to note that despite the greater overall sensitivity of CA3 place fields to changes in contextual information, about 40% of CA3 place fields continued to persist when faced with contextual changes. The combined evidence, then, suggests that CA3 place cells could play a primary role in determining whether specific features of the current context match expected contextual features.

CA1 also has the potential for representing current and expected contextual information. But, relative to CA3, a greater proportion of cells (about 60%) show persistent place fields despite context change (e.g. Mizumori et al., 1989, 1999b; Lee et al., 2004; Leutgeb et al., 2004). One interpretation of the differences observed for CA1 and CA3 place field stability in the face of context changes is that CA1 neurons are preferentially driven by memory representations of the context, perhaps via entorhinal afferents that bypass CA3 (Witter et al., 2000). For this reason, it may be predicted that the behavioral effects of context manipulation will be more directly related to the responses of CA3 place cells than CA1 place cells. Although a direct test of this prediction has yet to be carried out, it is worth noting that several laboratories have begun to report a lack of correlation between CA1 place field reorganization and behavioral responses (e.g., Cooper and Mizumori, 2001; Jeffery et al., 2003).

A question, then, is what is the significance of the CA1 code if CA3 is primarily responsible for the comparison of contextual information? Temporal organization or sequencing of information by the hippocampus had been suggested previously by many investigators (e.g., Olton et al., 1979; Rawlins, 1985; Kesner, 1991), including the possibility of a special role by CA1 (Hampson et al., 1993; Wiener et al., 1995; Gilbert et al., 2001; Kesner et al., 2004; Treves, 2004). Therefore, one possibility is that a primary function of CA1 place cells is to temporally organize, or group, CA3 output to allow for meaningful sequences to be passed on to efferent targets, such as the prefrontal cortex (Jay et al., 1989) and subiculum. Direct and selective entorhinal input to CA1 (Witter et al., 2000) may provide a route whereby neocortical-based memory representations predispose CA1 to temporally organize CA3-based information in experience-dependent ways (Mizumori et al., 1999b). Although the precise nature of the temporal organization remains to be determined, one possibility is that CA1 is more tightly coupled to the rhythmic oscillations of hippocampal EEG, which is thought to contribute to the temporal organization of hippocampal output (Buzsaki, 2005).

3. Temporal Encoding of Contextual Information

It is becoming clearer that important context information is embedded within hippocampal networks in the form of the temporal relationship between the activities of locally and distally located neurons. Many years ago, it was shown that movement through place fields is associated with dynamic changes in spike timing relative to the ongoing theta oscillations in the EEG (O'Keefe and Recce, 1993). That is, on a single pass through a field, the first spike of successive bursts occurs at progressively earlier phases of the theta cycle. The discovery of this so-called *phase precession* effect is considered significant, for it was the first clear evidence that place cells are part of a temporal code that could contribute to the mnemonic processes of hippocampus. Such temporally dynamic changes in spike timing may be a key mechanism by which place fields can provide a link between temporally extended behaviors of an animal and the comparatively rapid synaptic plasticity mechanisms (e.g., LTP) that are thought to subserve learning (e.g., Skaggs et al., 1996). Theoretical models have been generated to explain in more detail how phase precession could explain the link between behavior and neural plasticity mechanisms (e.g., Buzsaki, 2005; Zugaro et al., 2005).

Another form of temporal-based neuroplasticity involves a change in the timing of spike discharge by one cell relative to those of other cells. For example, it has been shown that the temporal coherence of the discharges of place cells is greater in mice with an intact hippocampus compared to mice with deficient NMDA systems (McHugh et al., 1996). Greater synchronization could offer a stronger output signal to efferent structures. Relatedly, experience-dependent temporal codes may be found in terms of the temporal relationships between the firing of cells with adjacent place fields. With continued exposure to a new environment, place fields begin to expand asymmetrically, in that the peak firing rate is achieved with shorter latency upon entrance into the field (Mehta et al., 1997, 2000). It was postulated that with continued exposure to a spatial task, place cells are repeatedly activated in a particular sequence. As a result, synaptic connections between cells with adjacent fields become stronger, such that entry into one place field begins to activate the cell with the adjacent place field at shorter and shorter latency. The backward expansion of place fields may result from asymmetric Hebbian mechanisms, such as LTP, since NMDA antagonism blocks the LTP and the expansion effect (Eckstrom et al., 2001). The asymmetric backward expansion of place fields is thought to provide a neural mechanism for learning directional sequences. Moreover, it has been suggested that the backward-expansion phenomenon may contribute to the transformation of a rate code to a temporal code such as that illustrated in phase precession (Mehta et al., 2002). Perhaps the backward-expansion phenomenon could help to explain other place field phenomenon, such as the tendency for place cells to fire in anticipation of entering a field within a familiar environment (Muller and Kubie, 1989). While the dynamic changes in place field shape are intriguing, it remains to be determined whether the asymmetric expansion is related directly to spatial learning.

D. Context Processing within a Larger Neural System

Hippocampal context processing is likely only one component of a broadly distributed neural system responsible for a complex form of memory such as episodic memory. To fully understand the neurobiology of episodic memory, it is essential that we understand better the anatomical and functional basis for interactions between the hippocampus and connected brain regions. Thus far, recent studies have examined these interactions in different ways. The impact of neocortical function on hippocampal place fields was tested following lesions or reversible inactivations of prefrontal cortex or posterior cortical regions: Place field reliability or specificity became significantly disrupted (Cooper and Mizumori, 2001; Muir and Bilkey, 2001; Kyd and Bilkey, 2005). Also, it has been recently shown that the temporal relationship between place cell firing and the ongoing (local) theta rhythm (i.e., phase precession, O'Keefe and Reece, 1993) is linked to phase precession in prefrontal cortex (Siapas et al., 2005) in a task-dependent manner (Jones and Wilson, 2005). These findings suggest that hippocampal place fields are regulated in part from input from the neocortex and that hippocampal processing may in turn impact neocortical representational organization. The precise mechanism of these interactions is not known.

A common assumption is that long-term memories are represented within neocortical circuitry. Therefore, memories may bias the firing patterns of hippocampal neurons such that they come to represent the expectation of encountering significant locations (in the case of place cells) or experience-dependent

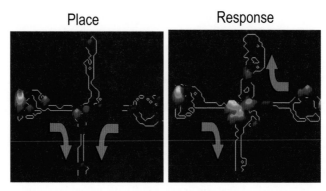

FIGURE 5-6   Rats were trained to shift strategies within a single session of plus-maze training. **(Left)** For place training, the rat started on either the east or west maze arm and found food on only the south maze arm. A place field was observed only on the west maze arm. **(Right)** The rat then switched to the use of a response strategy. Again it started on either the west or east maze arm, but now the reward location changed depending on the starting location. Regardless of the starting location, the rat learned to turn right to obtain reward. It can be seen that when this rat switched strategies, a new place field emerged on the central platform area. Thus, cognitive factors may importantly influence the place code in the hippocampus. In this case, retrieval of different memories resulted in a different spatial organization of the place field.

behavioral responses (in the case of interneurons). Indeed, several studies show that lesion or reversible inactivation of different areas of the neocortex has significant impact on hippocampal place fields (Cooper and Mizumori, 2001; Muir and Bilkey, 2001; Kyd and Bilkey, 2005). We recently tested this hypothesis further by evaluating how retrieving different memories (of the correct behavioral strategy) affected hippocampal neural responses. Specifically, we studied the effects of switching between two learned task strategies on the behavioral correlates of pyramidal cells and interneurons (Yeshenko et al., 2001; Eschenko and Mizumori, 2007). Rats were trained to perform 10 trials on a plus maze according to place strategy (i.e., approach a particular spatial location for reward) and then 10 trials according to a response strategy (i.e., make a particular response, such as turn right, for reward). Other rats were trained to use first a response strategy and then a place strategy. The order of strategy use did not impact the neural responses. Figure 5-6 illustrates that both place cells and movement-sensitive interneurons dramatically changed the nature of their neural representation when rats switched cognitive strategies. Different cells lost, gained, or significantly changed the nature of the original place or movement correlate after a strategy switch. Interestingly, changing which memory is active resulted in a similar pattern of changed neural responses as was shown following context manipulations when rats performed different responses using the same strategy (Yeshenko et al., 2004). Thus, memory influences (or the retrieval of different memories) may comprise a component of the hippocampal context code that is as important as the processing of sensory or response information.

Recent evidence suggests another issue for future consideration regarding the role of hippocampal place cells in complex forms of learning and memory. Location-selective firing has now been observed in many areas of the brain in addition to the hippocampus proper. A question is whether place cells from different brain areas make the same or different functional contribution to learning. Insights may be gained by comparing the properties and response patterns of the place fields of different regions. Place fields are now described for cells in brain regions closely connected to the hippocampus, such as subiculum (Sharp and Green, 1994; Sharp, 1999) and entorhinal cortex (Quirk et al., 1992; Mizumori et al., 1992; Frank et al., 2000; Hafting et al., 2005). One striking difference between place fields within the hippocampus and those in subiculum or entorhinal cortex became evident following subtle changes in the external environment (e.g., enlarging a test environment). It was shown that such changes in a familiar environment could produce drastic and unpredictable hippocampal place field reorganization, while subicular or entorhinal place fields maintained their spatial relationship to external cues. These findings, together with the most recent discovery of entorhinal cells that maintain multiple, regularly spaced place fields regardless of cue manipulation or experience (i.e., grid cells; Hafting et al., 2005), suggest a function for subicular and entorhinal place fields that differs from hippocampal place fields. Sharp (1999) suggested that subiculum contained a cognitive map, and Hafting et al. (2005) contend that entorhinal cortex provides an equivalent to a spatial reference frame within which hippocampal contextual information may become embedded. It appears that the functional contribution of place fields, then, may vary depending on brain structure.

Perhaps more surprising than the finding of place fields in subiculum or entorhinal cortex was the finding of place fields in a very different neural system, the striatum (Mizumori et al., 1999a, 2000a; Ragozzino et al., 2001; Yeshenko et al., 2004). The finding of striatal place fields presented a challenge to current views of the brain organization of memory systems since the striatum was thought not to be important for spatial processing. Rather, striatum was considered to be a specialized network that guides slower forms of learning, such as stimulus–response, habit, and stimulus–stimulus learning (e.g., Packard and White, 1991; Knowlton et al., 1996; Packard and Knowlton, 2002; Cook and Kesner, 1988). Striatal place fields tend to be larger than hippocampal place fields but similar in terms of the reliability with which the place fields are observed. Also, in contrast to hippocampal place fields, almost all of the striatal place fields that have been tested rapidly respond to cue manipulations (Mizumori et al., 2000a; Yeshenko et al., 2004) and strategy shifts (Yeshenko et al., 2001; Eschenko and Mizumori, 2007). It was rare to find striatal place fields that persist after a context manipulation. Thus, when considered as a group, striatal fields are more sensitive to context changes, even more so than CA3 place fields. The characteristics of the striatal place fields, combined with the known behavioral effects of striatal lesions, led to the suggestion (Mizumori et al., 1999a) that striatum uses contextual information to inform basal ganglia computations that are involved in determining the context-appropriate, anticipated reinforcing consequences of behavioral responses (Schultz, 1997, 1998; Schultz and Dickinson, 2000). Continual expression and modification of adaptive responses likely requires continual evaluation of the consequences of behavior. Therefore, it may be more important that striatum has access to the current definition of context than access to the remembered context of earlier training. Perhaps this is why we observe such a high percentage of striatal place fields that are sensitive to contextual manipulations. The source of striatal context information is uncertain. However, the posterior neocortex is a likely candidate since hippocampal output is strong to that area of the cortex, and since the posterior cortex projects to striatum (McGeorge and Faull, 1989).

It appears, then, that place fields found in different areas may or may not represent the same type of information. Thus, to facilitate a neural systems perspective of hippocampal function, it would be prudent to test the different populations of place cells in comparable behavioral and environmental situations so that direct conclusions could be drawn.

IV. FUTURE ISSUES TO CONSIDER

Understanding the relationship between place fields and learning or memory remains a significant challenge. While many impressive findings have been reported and provocative ideas proposed, several important issues need to be resolved before there is a clear answer to the question regarding the contribution of place cells to hippocampal-dependent learning, especially context learning.

One issue concerns the definition of context when referring to hippocampal processing. The evidence provided here challenges the traditional definitions of context to include information in addition to that provided by the background sensory cues. In apparent contrast to this view, the more selective term *spatial context* is often applied to interpretations of place fields with regard to experience-dependent functions (e.g., Nadel and Wilner, 1980; Mizumori et al., 1999b; Jeffery et al., 2004). This term is used to acknowledge the fact that, in freely behaving animals, the defining features of events or situations are almost always tied in some way to the spatial organization of environmental stimuli. Thus, it is argued, a complex form of spatial perceptions likely play a prominent role in hippocampal context analysis.

The call for an expanded definition of the term *context* and the use of the term *spatial context* are not necessarily contradictory. Rather, the hippocampus may be biased to evaluate changes in context according to a spatial reference any time that animals are asked to move through space to demonstrate learning (which happens to be the case for most learning paradigms used to record place cells). Where does this bias come from? Changes in the spatial environment are initially registered by sensory input systems, and these in turn produce continuous and systematic changes in representation at multiple, subsequent levels of neural processing (e.g., Merzenich and deCharms, 1996). Since higher-level sensory representations have been shown to be importantly affected by changes in attention or expectations (e.g., in parietal cortex; Colby and Duhamel, 1996; Colby and Olson, 1999), established memories also appear to influence our spatial perceptions. Parietal cortex information arrives in the hippocampus via entorhinal cortex. Therefore, parietal cortex representations of our spatial perceptions may be more organized within the entorhinal cortex "grid" structure (Hafting et al., 2005), which reflects the spatially extended environment of the animal. As a result, in navigating animals performing any task, the hippocampus is strongly predisposed to carry out a significant part of its context analysis within a spatial reference system (Redish and Touretsky, 1997; Mizumori et al., 1999b, 2000b).

The fact that place fields are such a robust and reliable correlate of hippocampal pyramidal cells in navigating rats (*regardless of the specific task demands or task strategy*) is certainly consistent with the claim that the hippocampus is biased to organize contextual information within a spatial reference frame. In this way, fluctuations of either spatial or nonspatial inputs will be reflected by the reorganization of place fields (a finding common to many studies). It is noteworthy, however, that hippocampal place cells have been described to change their responses from spatial to nonspatial ones in tasks where spatial information is made explicitly not important (e.g., Wiener et al., 1989). Similarly, many have reported that various context manipulations result in a loss of place fields for at least a portion of the place cells being recorded. These results, together with the fact that nonspatial behavioral correlates have been identified for pyramidal neurons (e.g., Eichenbaum, 2001; Smith and Mizumori, in press), clearly demonstrate that the spatial component of hippocampal neural codes can become dissociated from other components. These are important observations because they show that spatial representation is not absolute or obligatory for the hippocampus (Eichenbaum, 2001; Fortin et al., 2002; Kesner et al., 2002).

One way to reconcile the spatial context and nonspatial interpretations of hippocampal function is to argue that both are correct but that the type of correlate observed is in part a reflection of how the rats have been tested. Location-selective firing is a predominant correlate in situations where animals move through space to perform a task, even if the task is not hippocampal dependent. For example, a high percentage of pyramidal cells exhibit context-sensitive place fields during the performance of hippocampal-independent tasks, such as egocentric response training (Yeshenko et al., 2004) and random foraging tasks (Muller and Kubie, 1987). When locomotion is discouraged or made less relevant to accurate performance (e.g., an auditory discrimination task in an operant chamber; Sakurai, 1994), task-related correlates (e.g., tone onset) are observed. Perhaps, movement is fundamentally important for determining the mode of processing within the hippocampus. By definition, a movement-sensitive code reflects indicators of movement through space, such as velocity, acceleration, and direction of movement. Indeed place cells (and interneurons) have been shown to be sensitive to these movement variables (e.g., McNaughton et al., 1983a; Markus et al., 1994; Czurko et al., 1999). Moreover, place fields are generally more specific and reliable when animals are actively engaged in voluntary movement rather than when they are passively transported (Foster et al., 1989; Song et al., 2005). Theta rhythm appears in the EEG record during active and passive movement (Song et al., 2005). Therefore the mere presence or absence of theta cannot account for the movement effects. The view that movement plays a key role in the hippocampal contribution to learning and memory resonates well with previous theories (e.g., McNaughton et al., 1996; Whishaw, 1998; Etienne and Jeffery, 2004). However, rather than interpreting the movement code in terms of its role in idiothetic-based navigation, it is suggested here that the presence of movement codes engages the spatial reference system, which in turn leads to spatial context analysis by the hippocampus.

An implication of the hypothesis that movement changes the mode of hippocampal processing is that movement is required for normal spatial learning to occur. It could be argued that this is not the case since passive observation of a test room (latent learning) facilitates subsequent learning in a water maze task (Keith and McVety, 1988). However, in that experiment, rats were trained to perform the task in a different room prior to the latent learning test. Therefore, upon exposure to the new environment, the spatial reference frame relevant to a similar prior test situation may have been recalled, thereby biasing rats to pay attention to potentially significant stimuli. In this sense, this experiment does not directly test the hypothesis that movement is necessary for new learning.

It might also be argued that since place fields are observed on the first pass through a field by a naive rat, a reference frame must have already existed without prior movement. To demonstrate the existence of a place field, rats have to move through space. Thus, a spatial reference frame could have been called up at the start of the test session prior to entry into the field. It is worth noting that in this situation, it has been shown that although place fields appear quickly in a novel room, they continue to change, becoming more reliable as additional passes are made through the field (Frank et al., 2004). Perhaps it is this experience-dependent change in place fields that is related to new learning. Indeed, Smith and Mizumori (2006) recently reported that hippocampal place fields discriminate contexts only when the significance of the contexts changes.

The demonstrated modifiability of place fields in nonspatial tasks or the existence of nonspatial correlates of pyramidal neurons does not negate the claim of a strong bias by the hippocampus to code contextual information within a spatial reference frame. On the other hand, the existence of place fields does not prove that the hippocampus processes only spatial information. Rather, it is suggested that when discussing place cell results, it remains appropriate to refer to hippocampal processing in terms of its analysis of the spatial context. We must be ready, however, to address the issue of how this spatial context analysis reflects more generally on our understanding of the role of the hippocampus in processing context information.

A second challenge is to understand the relationship between hippocampal context processing, and more specifically place field properties, and the behavioral expression of learning. As described earlier, many past attempts to provide a functional interpretation of place fields have been conducted with tasks that do not require hippocampal processing. A significant advance toward making the connection between place fields and context discrimination is the finding that place fields develop context-specific characteristics only when animals are explicitly trained to distinguish contexts, and not when animals are trained on the same maze but not required to distinguish contexts (Smith and Mizumori, in press).

Although it is expected that place fields will be found regardless of the task, details of place field characteristics may vary as a function of the task demand. These details included variations in the rate code (e.g., place field specificity, in-field firing rates, and place field reliability) and/or changes in the temporal dimensions of place fields (e.g., temporal coherence, correlations with different phases of the ongoing EEG, and spike-timing-dependent plasticity). Given the hypothesis that the outcome of context discrimination by the hippocampus informs a larger neural network responsible for episodic learning and memory, future research needs to focus on the consequences of changes in place field characteristics for processing by extended, yet connected, neural circuitry. Using the hippocampal–prefrontal circuit as a model, recent new work has demonstrated that indeed changes in place field properties impacts the relationship between the hippocampus and connected efferent structures (Siapas et al., 2005).

A third issue that requires attention relates to the attempts to establish a direct link between episodic learning and memory (as defined most recently by Tulving, 2002) and the behavioral performance of rats during place cell recordings. Tulving refers to episodic processing in terms of the learning of significant events that is revealed by the conscious recollection of relevant information. Place fields are referred to in terms of their representation of sensory and motor associations or in terms of an analysis of contextual features.

At first glance, there seems to be a fundamental difference in terms of the nature and level of processing being compared. Therefore, one may question whether it is even possible to understand the neurobiology of episodic learning and memory by studying hippocampal, in particular place cell, functions in rats and mice. Investigators have attempted to establish a potential link between place fields and episodic functions by operationalizing episodic processing into the components of the "when," "what," and "where" information that defines events (e.g., Ferbinteanu and Shapiro, 2003; Ergorul and Eichenbaum, 2004). It is likely that there will continue to be much discussion about the validity of this approach. Part of the discussion should address whether hippocampal context codes reflect or define events.

A final issue here is the need to consider more fully the functional role of hippocampal interneurons in the overall mission of the hippocampus to evaluate changes in context. Interneuron codes may contribute movement-related information that acts to engage intrahippocampal neural networks that are involved in spatial context analyses. In this way, movement through space serves as a signal to evaluate context information within a spatial reference frame. It may also be the case that interneuron movement codes represent knowledge about context-specific behavioral responses, since the movement correlates of many (but not all) interneurons have been shown to change as the context (but not behavior) changes (Yeshenko et al., 2004). Clearly, understanding the interneuron contribution will help us to better understand the information network within which place cells operate.

V. CONCLUSIONS

The initial trend in place field research was to define this most striking behavioral correlate of hippocampal neurons in terms of its sensory and movement-related elements. This work gave us a rich and detailed view of the dynamic neural code that place fields represent. Based on this work, it has become clear that place fields do not reflect the mere detection of sensory stimuli and behavioral responses. Rather, they represent specific neural computations of spatial (geometric) relationships among environmental cues (e.g., O'Keefe and Burgess, 1996a; Hetherington and Shapiro, 1997), velocity and acceleration (McNaughton et al., 1983a), and environment-specific directional heading (e.g., Markus et al., 1994; Leutgeb et al., 2000). A number of more recent studies have attempted to determine how these spatial representations contribute to the vital role that the hippocampus is known to play in learning and memory. An emerging view is that place cells represent contextual information: For a given situation, some place cells encode information about the current context, while other place cells may code the memory-driven expectation of relevant contextual features. In this way, the hippocampus could engage in a match–mismatch comparison to determine the extent to which a given context has changed. The output of such a computation could be used to define events for a larger neural system responsible for episodic learning and memory.

ACKNOWLEDGMENTS

This work has enjoyed, most recently, support from the NIMH (grants MH58755 and MH067399) and, in the past, support from the NSF, HFSP, the University of Washington, and the University of Utah. Also, many talented graduate and undergraduate students have contributed to discussions and data collection that led to the ideas expressed in this chapter.

REFERENCES

Ainge, J.A., and Wood, E.R. (2003). Excitotoxic lesions of the hippocampus impair performance on a continuous alternation t-maze task with short delays but not with no delay. *Soc. Neurosci. 2003 Abstr Viewer/Itin Planner* [91.1].

Anagnostaras, S.G., Gale, G.D., and Fanselow, M.S. (2001). Hippocampus and contextual fear conditioning: Recent controversies and advances. *Hippocampus, 11*, 8–17.

Anderson, M.I., and Jeffery, K.J. (2003). Heterogeneous modulation of place cell firing by changes in context. *J. Neurosci., 23*, 8827–8835.

Barnes, C.A., McNaughton, B.L., Mizumori, S.J.Y., Leonard, B.W., and Lin, L.-H. (1990). Comparison of spatial and temporal characteristics of neuronal activity in sequential stages of hippocampal processing. *Prog. Brain Res., 83*, 287–300.

Barnes, C.A., Suster, M.S., Shen, J., and McNaughton, B.L. (1997). Multistability of cognitive maps in the hippocampus of old rats. *Nature, 388*, 272–275.

Blum, K.I., and Abbott, L.F. (1996). A model of spatial map formation in the hippocampus of the rat. *Neural Comput., 8*, 85–93.

Buzsáki, G. (2005). Theta rhythm of navigation: Link between path integration and landmark navigation, episodic and semantic memory. *Hippocampus, 15*, 827–840.

Colby, C.L., and Duhamel, J.R. (1996). Spatial represention for action in parietal cortex. *Cogn. Brain Res., 5*, 105–115.

Colby, C.L., and Olson, C.R. (1999). Spatial cognition. In *Fundamental Neuroscience*, M.J. Zigmond, F.E. Bloom, S.C. Landis, J.L. Roberts, and L.R. Squire (eds.). New York: Academic Press, pp. 1363–1383.

Cook, D., and Kesner, R.P. (1988). Caudate nucleus and memory for egocentric localization. *Behav. Neural Biol., 49*, 332–343.

Cooper, B.G., and Mizumori, S.J.Y. (2001). Temporary inactivation of retrosplenial cortex causes a transient reorganization of spatial coding in hippocampus. *J. Neurosci., 21*, 3986–4001.

Cressant, A., Muller, R.U., and Poucet, B. (1997). Failure of centrally placed objects to control the firing fields of hippocampal place cells. *J. Neurosci., 17*, 2531–2542.

Czurkó, A., Hirase, H., Csicsvari, J., and Buzsáki, G. (1999). Sustained activation of hippocampal pyramidal cells by "space clamping" in a running wheel. *Eur. J. Neurosci., 11*, 344–352.

Dragoi, G., Harris, K.D., and Buzsáki, G. (2003). Place representation within hippocampal networks is modified by long-term potentiation. *Neuron, 39*, 843–853.

Eckstrom, A.D., Meltzer, J., McNaughton, B.L., and Barnes, C.A. (2001). NMDA receptor antagonism blocks experience-dependent expansion of hippocampal "place fields." *Neuron, 31*, 631–638.

Eichenbaum, H. (2001). The hippocampus and declarative memory: Cognitive mechanisms and neural codes. *Behav. Brain Res., 127,* 199–207.

Eichenbaum, H., and Cohen, N.J. (2001). *From Conditioning to Conscious Recollection: Memory Systems of the Brain.* New York: Oxford University Press.

Eichenbaum, H., Dudchenko, P., Wood, E., Shapiro, M., and Tanila, H. (1999). The hippocampus, memory, and place cells: Is it spatial memory or a memory space? *Neuron, 425,* 184–188.

Ergorul, C., and Eichenbaum, H. (2004). The hippocampus and memory for "what," "where," and "when." *Learn. Mem., 11,* 397–405.

Eschenko, O., and Mizumori, S.J.Y. (2007). Memory influences on hippocampal and striatal codes: Effects of a shift between task rules. *Neurobiol. Learn. Mem.* (in press).

Etienne, A.S., and Jeffery, K.J. (2004). Path integration in mammals. *Hippocampus, 14,* 180–192.

Ferbinteanu, J., and Shapiro, M.L. (2003). Prospective and retrospective memory coding in the hippocampus. *Neuron, 40,* 1227–1239.

Fortin, N.J., Agster, K.L., and Eichenbaum, H.B. (2002). Critical role of the hippocampus in memory for sequences of events. *Nat. Neurosci., 5,* 458–462.

Foster, T.C., Castro, C.A., and McNaughton, B.L. (1989). Spatial selectivity of rat hippocampal neurons: Dependence on preparedness for movement. *Science, 244,* 1580–1582.

Foster, T.C., Christian, E.P., Hampson, R.E., Campbell, K.A., and Deadwyler, S.A. (1987). Sequential dependencies regulate sensory evoked responses of single units in the rat hippocampus. *Brain Res., 408,* 86–96.

Frank, L.M., Brown, E.M., and Wilson, M.A. (2000). Trajectory encoding in the hippocampus and entorhinal cortex. *Neuron, 27,* 169–178.

Frank, L.M., Stanley, G.B., and Brown, E.N. (2004). Hippocampal plasticity across multiple days of exposure to novel environments. *J. Neurosci., 24,* 7681–7689.

Freeman, J.H., Jr., Cuppernell, C., Flannery, K., and Gabriel, M. (1996). Context-specific multi-site cingulate cortical, limbic thalamic, and hippocampal neuronal activity during concurrent discriminative approach and avoidance training in rabbits. *J. Neurosci., 1,* 1538–1549.

Gabriel, M. (1993). Discriminative avoidance learning: A model system. In *Neurobiology of Cingulate Cortex and Limbic Thalamus,* B.A. Vogt, and M. Gabriel (eds.). Boston: Birkhauser, pp. 478–523.

Gavrilov, V.V., Wiener, S.I., and Berthoz, A. (1998). Discharge correlates of hippocampal complex spike neurons in behaving rats passively displaced on a mobile robot. *Hippocampus, 8,* 475–490.

Gilbert, P.E., Kesner, R.P., and Lee, I. (2001). Dissociating hippocampal subregions: Double dissociation between dentate gyrus and CA1. *Hippocampus, 11,* 626–636.

Gold, E., and Kesner, R.P. (2005). The role of the CA3 subregion of the dorsal hippocampus in spatial pattern completion in the rat. *Hippocampus, 15,* 808–814.

Goodrich-Hunsaker, N.J., Hunsaker, M.R., and Kesner, R.P. (2005). Dissociating the role of the parietal cortex and dorsal hippocampus for spatial information processing. *Behav. Neurosci., 119,* 1307–1315.

Gothard, K.M., Skaggs, W.E., Moore, K.M., and McNaughton, B.L. (1996). Binding of hippocampal CA1 neural activity to multiple reference frames in a landmark-based navigation task. *J. Neurosci., 16,* 825–835.

Gray, C.M., Maldonado, P.E., Wilson, M., and McNaughton, B.L. (1995). Tetrodes markedly improve the reliability and yield of multiple single-unit isolation from multi-unit recordings in cat striate cortex. *J. Neurosci. Meth., 63,* 43–54.

Guzowski, J.F., Knierim, J.J., and Moser, E.I. (2004). Ensemble dynamics of hippocampal regions CA3 and CA1. *Neuron, 44,* 581–584.

Guzowski, J.F., Timlin, J.A., Roysam, B., McNaughton, B.L., Worley, P.F., and Barnes, C.A. (2005). Mapping behaviorally relevant neural circuits with immediate-early gene expression. *Curr. Opin. Neurobiol., 15*, 599–606.

Hafting, T., Fyhn, M., Molden, S., Moser, M.B., and Moser, E.I. (2005). Microstructure of a spatial map in the entorhinal cortex. *Nature, 436*, 801–806.

Hampson, R.E., Heyser, C.J., and Deadwyler, S.A. (1993). Hippocampal cell-firing correlates of delayed-match-to-sample performance in the rat. *Behav. Neurosci., 107*, 715–739.

Hasselmo, M.E. (2005a). The role of hippocampal regions CA3 and CA1 in matching entorhinal input with retrieval of associations between objects and context: Theoretical comment on Lee et al. (2005). *Behav. Neurosci. 119*, 342–345.

Hasselmo, M.E. (2005b). What is the function of hippocampal theta rhythm? Linking behavioral data to phasic properties of field potential and unit recording data. *Hippocampus, 15*, 936–949.

Hasselmo, M.E., and McGaughy, J. (2004). High acetylcholine levels set circuit dynamics for attention and encoding and low acetylcholine levels set dynamics for consolidation. *Prog Brain Res., 145*, 207–231.

Hetherington, P.A., and Shapiro, M.L. (1997). Hippocampal place fields are altered by the removal of single visual cues in a distance-dependent manner. *Behav. Neurosci., 11*, 20–34.

Hill, A.J., and Best, P.J. (1981). Effects of deafness and blindness on the spatial correlates of hippocampal unit activity in the rat. *Exp. Neurol., 74*, 204–217.

Hirsh, R. (1974). The hippocampus and contextual retrieval of information from memory: A theory. *Behav. Biol., 12*, 421–444.

Hollup, S.A., Molden, S., Donnett, J.G., Moser, M.B., and Moser, E.I. (2001). Accumulation of hippocampal place fields at the goal location in an annular watermaze task. *J. Neurosci., 21*, 1635–1644.

Jay, T.M., Glowinski, J., and Thierry, A.M. (1989). Selectivity of the hippocampal projection to the prelimbic area of the prefrontal cortex in the rat. *Brain Res., 505*, 337–340.

Jeffery, K.J., Anderson, M.I., Hayman, R., and Chakraborty S. (2004). A proposed architecture for the neural representation of spatial context. *Neurosci. Biobehav. Rev., 28*, 201–218.

Jeffery, K.J., Gilbert, A., Burton, S., and Strudwick, A. (2003). Preserved performance in a hippocampal-dependent spatial task despite complete place cell remapping. *Hippocampus, 13*, 175–189.

Jones, M.W., and Wilson, M.A. (2005). Theta rhythms coordinate hippocampal–prefrontal interactions in a spatial memory task. *PLoS Biol., 3*, e402.

Jung, M.W., Wiener, S.I., and McNaughton, B.L. (1994). Comparison of spatial firing characteristics of units in dorsal and ventral hippocampus of the rat. *J. Neurosci., 14*, 7347–7356.

Keith, J.R., and McVety, K.M. (1988). Latent place learning in a novel environment and the influences of prior training in rats. *Psychobiol., 16*, 146–151.

Kennedy, P.J., and Shapiro, M.L. (2004). Retrieving memories via internal context requires the hippocampus. *J. Neurosci., 24*, 6979–6985.

Kentros, C., Hargreaves, E.L., Hawkins, R.D., Kandel, E.R., Shapiro, M., and Muller, R.U. (1998). Abolition of long-term stability of new hippocampal place cell maps by NMDA receptor blockade. *Science, 280*, 2121–2126.

Kentros, C.G., Agnihotri, N.T., Streater, S., Hawkins, R.D., and Kandel, E.R. (2004). Increased attention to spatial context increases both place field stability and spatial memory. *Neuron, 42*, 283–295.

Kesner, R.P. (1991). Neurobiological views of memory. In *Learning and Memory: A Biological View*, J.L. Martinez, Jr., and R.P. Kesner, (eds.). San Diego, CA: Academic Press, pp. 499–547.

Kesner, R.P., Gilbert, P.E., and Barua, L.A. (2002). The role of the hippocampus in memory for the temporal order of a sequence of odors. *Behav. Neurosci., 116*, 286–290.

Kesner, R.P., Lee, I., and Gilbert, P. (2004). A behavioral assessment of hippocampal function based on a subregional analysis. *Rev. Neurosci., 15*, 333–351.

Kim, J.J., and Fanselow, M.S. (1992). Modality-specific retrograde amnesia of fear. *Science, 256*, 675–677.

Knierim, J.J., Kudrimoti, H., and McNaughton, B.L. (1995). Hippocampal place fields, the internal compass, and the learning of landmark stability. *J. Neurosci., 15*, 1648–1659.

Knowlton, B.J., Mangels, J.A., and Squire, L.R. (1996). A neostriatal habit learning system in humans. *Science, 273*, 1399–1402.

Kubie, J.L., and Ranck, Jr., J. (1983). Sensory-behavioral correlates in individual hippocampus neurons in three situations: Space and context. In *Neurobiology of the Hippocampus*, W. Seifert (ed.). New York: Academic Press, pp. 433–447.

Kubie, J.L., Muller, R.U., and Bostock, E. (1990). Spatial firing properties of hippocampal theta cells. *J. Neurosci., 10*, 1110–1123.

Kyd, R.J., and Bilkey, D.K. (2005). Hippocampal place cells show increased sensitivity to changes in the local environment following prefrontal cortex lesions. *Cereb. Cortex, 15*, 720–731.

Lee, I., Yoganarasimha, D., Rao, G., and Knierim, J.J. (2004). Comparison of population coherence of place cells in hippocampal subfields CA1 and CA3. *Nature, 430*, 456–459.

Lenck-Santini, P.P., Save, E., and Poucet, B. (2001). Evidence for a relationship between place-cell spatial firing and spatial memory performance. *Hippocampus, 11*, 377–390.

Lenck-Santini, P.P., Muller, R.U., Save, E., and Poucet, B. (2002). Relationships between place cell firing fields and navigational decisions by rats. *J. Neurosci., 22*, 9035–9047.

Leutgeb, S., Leutgeb, J.K., Treves, A., Moser, M.B., and Moser, E.I. (2004). Distinct ensemble codes in hippocampal areas CA3 and CA1. *Science, 305*, 1295–1298.

Leutgeb, S., Leutgeb, J.K., Barnes, C.A., Moser, E.I., McNaughton, B.L., and Moser, M.B. (2005). Independent codes for spatial and episodic memory in hippocampal neuronal ensembles. *Science, 309*, 619–623.

Leutgeb, S., Ragozzino, K.E., and Mizumori, S.J.Y. (2000). Convergence of head direction and place information in the CA1 region of HPC. *Neurosci., 100*, 11–19.

Lisman, J.E. (1999). Relating hippocampal circuitry to function: Recall of memory sequences by reciprocal dentate–CA3 interactions. *Neuron, 22*, 233–242.

Long, J.M., and Kesner, R.P. (1996). The effects of dorsal versus ventral hippocampal, total hippocampal, and parietal cortex lesions on memory for allocentric distance in rats. *Behav. Neurosci., 110*, 922–932.

Maren, S. (2001). Neurobiology of Pavlovian fear conditioning. *Annu. Rev. Neurosci., 24*, 897–931.

Markus, E.J., Barnes, C.A., McNaughton, B.L., Gladden, V.L., and Skaggs, W.E. (1994). Spatial information content and reliability of hippocampal CA1 neurons: Effects of visual input. *Hippocampus, 4*, 410–421.

Markus, E.J., Qin, Y.L., Leonard, B., Skaggs, W.E., McNaughton, B.L., and Barnes, C.A. (1995). Interactions between location and task affect the spatial and directional firing of hippocampal neurons. *J. Neurosci., 15*, 7079–7094.

Martin, S.J., Grimwood, P.D., and Morris, R.G. (2000). Synaptic plasticity and memory: An evaluation of the hypothesis. *Annu. Rev. Neurosci., 23*, 649–711.

Maurer, A.P., VanRhoads, S.R., Sutherland, G.R., Lipa, P., and McNaughton, B.L. (2005). Self-motion and the origin of differential spatial scaling along the septo-temporal axis of the hippocampus. *Hippocampus, 15*, 841–852.

McClelland, J.L., McNaughton, B.L., and O'Reilly, R.C. (1995). Why there are complementary learning systems in the hippocampus and neocortex: Insights from the success and failures of connectionist models of learning and memory. *Psychol. Rev., 102*, 419–457.

McEchron, M.D., and Disterhoft, J.G. (1999). Hippocampal encoding of non-spatial trace conditioning. *Hippocampus, 9*, 385–396.

McGeorge, A.J., and Faull, R.L. (1989). The organization of the projection from the cerebral cortex to the striatum in the rat. *Neurosci., 29*, 503–537.

McHugh, T.J., Blum, K.I., Tsien, J.Z., Tonegawa, S., and Wilson, M.A. (1996). Impaired hippocampal representation of space in CA1-specific NMDAR1 knockout mice. *Cell, 87*, 1339–1349.

McNaughton, B.L., Barnes, C.A., Gerrard, J.L., Gothard, K., Jung, M.W., Knierim, J.J., Kudrimoti, H., Qin, Y., Skaggs, W.E., Suster, M., and Weaver, K.L. (1996). Deciphering the hippocampal polyglot: The hippocampus as a path integration system. *J. Exp. Biol., 199*, 173–185.

McNaughton, B.L., Barnes, C.A., and O'Keefe, J. (1983a). The contributions of position, direction, and velocity to single-unit activity in the hippocampus of freely moving rats. *Exp. Brain Res., 52*, 41–49.

McNaughton, B.L., O'Keefe, J., and Barnes, C.A. (1983b). The stereotrode: A new technique for simultaneous isolation of several single units in the central nervous system from multiple unit records. *J. Neurosci. Methods, 8*, 391–397.

Mehta, M.R., Barnes, C.A., and McNaughton, B.L. (1997). Experience-dependent, asymmetric expansion of hippocampal place fields. *Proc. Natl. Acad. Sci. U.S.A., 94*, 8918–8921.

Mehta, M.R., Lee, A.K., and Wilson, M.A. (2002). Role of experience and oscillations in transforming a rate code into a temporal code. *Nature, 417*, 741–746.

Mehta, M.R., Quirk, M.C., and Wilson, M.A. (2000). Experience-dependent asymmetric shape of hippocampal receptive fields. *Neuron, 25*, 707–715.

Merzenich, M.M., and deCharms, R.C. (1996). Neural representation, experience, and change. In *Mind-Brain Continuum: Sensory Processes*, R. Llinas and P.S. Churchland (eds.). Cambridge, MA: MIT Press, pp. 61–81.

Mizumori, S.J.Y., Cooper, B.G., Leutgeb, S., and Pratt, W.E. (2000b). A neural systems analysis of adaptive navigation. *Molec. Neurobiol., 21*(1/2), 57–82.

Mizumori, S.J.Y., and Kalyani, A. (1997). Age and experience-dependent representational reorganization during spatial learning. *Neurobiol. Aging, 18*, 651–659.

Mizumori, S.J.Y., Lavoie, A.M., and Kalyani, A. (1996). Redistribution of spatial representation in the hippocampus of aged rats performing a spatial memory task. *Behav. Neurosci., 110*, 1006–1016.

Mizumori, S.J.Y., and Leutgeb, S. (1999). Interpreting neural representations of aged animals. *Hippocampus, 9*, 607–608.

Mizumori, S.J.Y., McNaughton, B.L., Barnes, C.A., and Fox, K.B. (1989). Preserved spatial coding in hippocampal CA1 pyramidal cells during reversible suppression of CA3 output: Evidence for pattern completion in HPC. *J. Neurosci., 9*, 3915–3928.

Mizumori, S.J.Y., Pratt, W.E., and Ragozzino, K.E. (1999a). Function of the nucleus accumbens within the context of the larger striatal system. *Psychobiol., 27*, 214–224.

Mizumori, S.J.Y., Ragozzino, K.E., and Cooper, B.G. (2000a). Location and head direction representation in the dorsal striatum of rats. *Psychobiol., 28*, 441–462.

Mizumori, S.J.Y., Ragozzino, K.E., Cooper, B.G., and Leutgeb, S. (1999b). Hippocampal representational organization and spatial context. *Hippocampus, 9*, 444–451.

Mizumori, S.J.Y., Ward, K.E., and Lavoie, A.M. (1992). Medial septal modulation of entorhinal single unit activity in anesthetized and freely moving rats. *Brain Res., 570*, 188–197.

Mizumori, S.J.Y., Yeshenko, O., Gill, K., and Davis, D. (2004). Parallel processing across neural systems: Implications for a multiple memory systems hypothesis. *Neurobiol. Learn. Mem., 82*, 278–298.

Moita, M.A., Rosis, S., Zhou, Y., LeDoux, J.E., and Blair, H.T. (2004). Putting fear in its place: Remapping of hippocampal place cells during fear conditioning. *J. Neurosci., 24*, 7015–7023.

Morris, R.G., and Frey, U. (1997). Hippocampal synaptic plasticity: Role in spatial learning or the automatic recording of attended experience? *Phil. Trans. R. Soc. Lond. B Biol. Sci., 352*, 489–503.

Moscovitch, M., Rosenbaum, R.S., Gilboa, A., Addis, D.R., Westmacott, R., Grady, C., McAndrews, M.P., Levine, B., Black, S., Winocur, G., and Nadel, L. (2005). Functional neuroanatomy of remote episodic, semantic and spatial memory: A unified account based on multiple trace theory. *J. Anat., 207*, 35–66.

Muir, G.M., and Bilkey, D.K. (2001). Instability in the place field location of hippocampal place cells after lesions centered on the perirhinal cortex. *J. Neurosci., 21*, 4016–4025.

Muller, R.U., and Kubie, J.L. (1987). The effects of changes in the environment on the spatial firing of hippocampal complex-spike cells. *J. Neurosci., 7*, 1951–1968.

Muller, R.U., and Kubie, J.L. (1989). The firing of hippocampal place cells predicts the future position of freely moving rats. *J. Neurosci., 9*, 4101–4110.

Muller, R.U., Stead, M., and Pach, J. (1996). The hippocampus as a cognitive graph. *J. Gen. Physiol., 107*, 663–694.

Myers, C.E., and Gluck, M. (1994). Context, conditioning, and hippocampal representation in animal learning. *Behav. Neurosci., 108*, 835–847.

Nadel, L., and Payne, J.D. (2002). The hippocampus, wayfinding and episodic memory. In *The Neural Basis of Navigation: Evidence from Single Cell Recording*, P.E. Sharp (ed.). Norwell, MA: Kluwer Academic, pp. 235–248.

Nadel, L., and Wilner, J. (1980). Context and conditioning: A place for space. *Physiol. Psychol., 8*, 218–228.

Nakazawa, K., McHugh, T.J., Wilson, M.A., and Tonegawa, S. (2004). NMDA receptors, place cells and hippocampal spatial memory. *Nature Rev. Neurosci., 5*, 361–372.

O'Keefe, J. (1976). Place units in the HPC of the freely moving rat. *Exp. Neurol., 51*, 78–109.

O'Keefe, J., and Burgess, N. (1996a). Geometric determinants of the place fields of hippocampal neurons. *Nature, 381*, 425–428.

O'Keefe, J., and Burgess, N. (1996b). Spatial and temporal determinants of hippocampal place cell activity. In *Perception, Memory and Emotion: Frontiers in Neuroscience*, T. Ono, B.L. McNaughton, S. Molotchnikoff, E.T. Rolls, and H. Nishijo (eds.). New York: Pergamon Press, pp. 359–373.

O'Keefe, J., and Conway, D.H. (1978). Hippocampal place units in the freely moving rat: Why they fire where they fire. *Exp. Brain Res., 31*, 573–590.

O'Keefe, J., and Dostrovsky, J. (1971). The hippocampus as a spatial map. Preliminary evidence from unit activity in the freely moving rat. *Brain Res., 34*, 171–175.

O'Keefe, J., and Nadel, L. (1978). *The Hippocampus as a Cognitive Map.* Oxford, U.K.: Oxford, University Press.

O'Keefe, J., and Recce, M.L. (1993). Phase relationship between hippocampal place units and the EEG theta rhythm. *Hippocampus, 3*, 317–330.

O'Keefe, J., and Speakman, A. (1987). Single-unit activity in the rat hippocampus during a spatial memory task. *Exp. Brain Res., 68*, 1–27.

Olton, D.S., Becker, J.T., and Handelmann, G.E. (1979). Hippocampus, space, and memory. *Brain Behav. Sci., 2*, 313–365.

Olton, D.S., Branch, M., and Best, P.J. (1978). Spatial correlates of hippocampal unit activity. *Exp. Neurol., 58*, 387–409.

O'Mara, S.M. (1995). Spatially selective firing properties of hippocampal formation neurons in rodents and primates. *Prog. Neurobiol., 45*, 253–274.

O'Reilly, R.C., and McClelland, J.L. (1994). Hippocampal conjunctive encoding, storage, and recall: Avoiding a trade-off. *Hippocampus, 4*, 661–682.

O'Reilly, R.C., and Rudy, J.W. (2001). Conjunctive representations in learning and memory: Principles of cortical and hippocampal function. *Psychol. Rev., 108*, 311–345.

Packard, M.G., and Knowlton, B.J. (2002). Learning and memory functions of the basal ganglia. *Annu. Rev. Neurosci., 25*, 563–593.

Packard, M.G., and White, N.M. (1991). Dissociation of HPC and caudate nucleus memory systems by post-training intracerebral injection of dopamine agonists. *Behav. Neurosci., 105,* 295–306.

Penick, S., and Solomon, P.R. (1991). Hippocampus, context, and conditioning. *Behav. Neurosci., 105,* 611–617.

Phillips, R.G., and LeDoux, J.E. (1992). Differential contribution of amygdala and hippocampus to cued and contextual fear conditioning. *Behav. Neurosci., 106,* 274–285.

Poucet, B. (1993). Spatial cognitive maps in animals: new hypotheses on their structure and neural mechanisms. *Psychol. Rev., 100,* 163–182.

Quirk, G.J., Muller, R.U., and Kubie, J.L. (1990). The firing of hippocampal place cells in the dark depends on the rat's recent experience. *J. Neurosci., 10,* 2008–2017.

Quirk, G.J., Muller, R.U., Kubie, J.L., and Ranck, Jr., J.B. (1992). The positional firing properties of medial entorhinal neurons: Description and comparison with hippocampal place cells. *J. Neurosci., 12,* 1945–1963.

Ragozzino, K.E., Leutgeb, S., and Mizumori, S.J.Y. (2001). Conditional coupling of dorsal striatal head direction and hippocampal place representations during spatial navigation. *Exp. Brain Res., 139,* 372–376.

Ranck, Jr., J.R. (1973). Studies on single neurons in dorsal hippocampus formation and septum in unrestrained rats. Part I. Behavioral correlates and firing repertoires. *Exp. Neurol., 41,* 461–535.

Rawlins, J.N.P. (1985). Associations across time: The hippocampus as a temporary memory store. *Brain Behav. Sci., 8,* 479–496.

Redish, A.D., Rosesnzweig, E.S., Bohanick, J.D., McNaughton, B.L., and Barnes, C.A. (2000). Dynamics of hippocampal ensemble activity realignment: Time versus space. *J. Neurosci., 20,* 9298–9309.

Redish, A.D., and Touretzky, D.S. (1997). Cognitive maps beyond the hippocampus. *Hippocampus, 7,* 15–35.

Rolls, E. (1996). A theory of hippocampal function in memory. *Hippocampus, 6,* 601–620.

Sakurai, Y. (1994). Involvement of auditory cortical and hippocampal neurons in auditory working memory and reference memory in the rat. *J. Neurosci., 14,* 2606–2623.

Save, E., Nerad, L., and Poucet, B. (2000). Contribution of multiple sensory information to place field stability in hippocampal place cells. *Hippocampus, 10,* 64–76.

Schultz, W. (1997). Dopamine neurons and their role in reward mechanisms. *Curr. Opin. Neurobiol., 7,* 191–197.

Schultz, W. (1998). Predictive reward signal of dopamine neurons. *J. Neurophysiol., 80,* 1–27.

Schultz, W., and Dickinson, A. (2000). Neuronal coding of prediction errors. *Annu. Rev. Neurosci., 23,* 473–500.

Scoville, W.B., and Milner, B. (1957). Loss of recent memory after bilateral hippocampal lesions. *J. Neurol. Neurosurg. Psychiat., 20,* 11–21.

Shapiro, M.L., and Eichenbaum, H. (1999). Hippocampus as a memory map: Synaptic plasticity and memory encoding by hippocampal neurons. *Hippocampus, 9,* 365–384.

Shapiro, M.L., Tanila, H., and Eichenbaum, H. (1997). Cues that hippocampal place cells encode: Dynamic and hierarchical representation of local and distal stimuli. *Hippocampus, 7,* 624–642.

Sharp, P.E. (1999). Complementary roles for hippocampal versus subicular/entorhinal place cells in coding place, context, and events. *Hippocampus, 9,* 432–443.

Sharp, P.E., and Green, C. (1994). Spatial correlates of firing patterns of single cells in the subiculum of the freely moving rat. *J. Neurosci., 14,* 2339–2356.

Shors, T.J., and Matzel, L.D. (1997). Long-term potentiation: What's learning got to do with it? *Behav. Brain Sci., 20,* 597–614.

Siapas, A.G., Lubenov, E.V., and Wilson, M.A. (2005). Prefrontal phase locking to hippocampal theta oscillations. *Neuron, 46*, 141–151.

Skaggs, W.E., McNaughton, B.L., Wilson, M.A., and Barnes, C.A. (1996). Theta phase precession in hippocampal neuronal populations and the compression of temporal sequences. *Hippocampus, 6*, 149–172.

Smith, D.M., and Mizumori, S.J.Y. (2006). Learning-related development of context-specific neuronal responses to places and events: The hippocampal role in context processing. *J. Neurosci, 26*, 3154–3163.

Smith, D.M., Wakeman, D., Patel, J., and Gabriel, M. (2004). Fornix lesions impair context-related cingulothalamic neuronal patterns and concurrent discrimination learning. *Behav. Neurosci., 118*, 1225–1239.

Song, E.Y., Kim, Y.B., Kim, Y.H., and Jung, M.W. (2005). Role of active movement in place-specific firing of hippocampal neurons. *Hippocampus, 15*, 8–17.

Squire, L.R., Stark, C.E., and Clark, R.E. (2004). The medial temporal lobe. *Annu. Rev. Neurosci., 27*, 279–306.

Tanila, H., Shapiro, M.L., and Eichenbaum, H. (1997). Discordance of spatial representation in ensembles of hippocampal place cells. *Hippocampus, 7*, 613–623.

Thierry, A., Gioanni, Y., Degenetais, E., and Glowinski, J. (2000). Hippocampo–prefrontal cortex pathway: Anatomical and electrophysiological characteristics. *Hippocampus, 10*, 411–419.

Treves, A. (2004). Computational constraints between retrieving the past and predicting the future, and the CA3-CA1 differentiation. *Hippocampus, 14*, 539–556.

Trullier, O., Shibata, R., Mulder, A.B., and Wiener, S.I. (1999). Hippocampal neuronal position selectivity remains fixed to room cues in rats alternating between place navigation and beacon approach tasks. *Eur. J. Neurosci., 11*, 4381–4388.

Tsien, J.Z., Huerta, P.T., and Tonegawa, S. (1996). The essential role of hippocampal CA1 NMDA receptor–dependent synaptic plasticity in spatial memory. *Cell, 87*, 1327–1338.

Tulving, E. (2002). Episodic memory: From mind to brain. *Annu. Rev. Psychol. 53*: 1–25.

Vanderwolf, C. (1969). Hippocampal electrical activity and voluntary movement in the rat. *Electroencephalogr. Clin., Neurophysiol., 26*, 407–418.

Vinogradova, O.S. (1995). Expression, control, and probably functional significance of the neuronal theta-rhythm. *Prog. Neurobiol., 45*, 523–583.

Wallenstein, G.V., and Hasselmo, M.E. (1997). GABAergic modulation of hippocampal population activity: Sequence learning, place field development, and the phase precession effect. *J. Neurophysiol., 78*, 393–408.

Whishaw, I.Q. (1998). Place learning in hippocampal rats and the path integration hypothesis. *Neurosci. Biobehav. Rev., 22*, 209–220.

Wible, C.G., Findling, R.L., Shapiro, M., Lang, E.J., Crane, S., and Olton, D.S. (1986). Mnemonic correlates of unit activity in the hippocampus. *Brain Res., 399*, 97–110.

Wiener, S.I. (1996). Spatial, behavioral and sensory correlates of hippocampal CA1 complex spike cell activity: Implications for information processing functions. *Prog. Neurobiol., 49*, 335–361.

Wiener, S.I., Berthoz, A., and Zugaro, M.B. (2002). Multisensory processing in the elaboration of place and head direction responses by limbic system neurons. *Brain Res. Cogn. Brain Res., 14*, 75–90.

Wiener, S.I., Korshunov, V.A., Garcia, R., and Berthoz, A. (1995). Inertial, substratal and landmark cue control of hippocampal CA1 place cell activity. *Eur. J. Neurosci., 7*, 2206–2219.

Wiener, S.I., Paul, C.A., and Eichenbaum, H. (1989). Spatial and behavioral correlates of hippocampal neuronal activity. *J. Neurosci., 9*, 2737–2763.

Wilson, M.A., and McNaughton, B.L. (1993). Dynamics of the hippocampal ensemble code for space. *Science, 261*, 1055–1058.

Witter, M.P., Naber, P.A., van Haeften, T., Machielsen, W.C., Rombouts, S.A., Barkhof, F., Scheltens, P., and Lopes da Silva, F.H. (2000). Cortico-hippocampal communication by way of parallel parahippocampal–subicular pathways. *Hippocampus, 10,* 398–410.

Wood, E.R., Dudchenko, P.A., Tobitsek, R.J., and Eichenbaum, H. (2000). Hippocampal neurons encode information about different types of memory episodes occurring in the same location. *Neuron, 27,* 623–633.

Yeshenko, O., Guazzelli, A., and Mizumori, S.J.Y. (2001). Differential hippocampal neuronal activity during place or response performance on a T-maze. *Soc. Neurosci. 2001 Abstr Viewer/ Itin Planner.*

Yeshenko, O., Guazzelli, A., and Mizumori, S.J.Y. (2004). Context-dependent reorganization of spatial and movement representations by simultaneously recorded hippocampal and striatal neurons during performance of allocentric and egocentric tasks. *Behav. Neurosci., 118,* 751–769.

Young, B.J., Fox, G.D., and Eichenbaum, H. (1994). Correlates of hippocampal complex-spike cell activity in rats performing a nonspatial radial maze task. *J. Neurosci., 14,* 6553–6563.

Zinyuk, L., Kubik, S., Kaminsky, Y., Fenton, A.A., and Bures, J. (2000). Understanding hippocampal activity by using purposeful behavior: Place navigation induces place cell discharge in both task-relevant and task-irrelevant spatial reference frames. *Proc. Natl. Acad. Sci. USA, 97,* 3771–3776.

Zugaro, M.B., Monconduit, L., and Buzsaki, G. (2005). Spike phase precession persists after transient intrahippocampal perturbation. *Nat. Neurosci., 8,* 67–71.

CHAPTER 6

Computations in Memory Systems in the Brain

Edmund T. Rolls

Department of Experimental Psychology, University of Oxford, Oxford OX1 3UD, United Kingdom; www.cns.ox.ac.uk

I. INTRODUCTION

This chapter describes memory systems in the brain based on closely linked neurobiological and computational approaches. The neurobiological approaches include evidence from brain lesions, which show the type of memory for which each of the brain systems considered is necessary, and analysis of neuronal activity in each of these systems to show what information is represented in them and the changes that take place during learning. Much of the neurobiology considered is from nonhuman primates as well as humans, because the operation of some of the brain systems involved in memory and the systems connected to them have undergone great development in primates. Some such brain systems include those in the temporal lobe, which develops massively in primates for vision and which sends inputs to the hippocampus via highly developed parahippocampal regions, and the prefrontal cortex. Many memory systems in primates receive outputs from the primate inferior temporal visual cortex, and understanding the perceptual representations in this of objects and how they are appropriate as inputs to different memory systems helps to provide a coherent way to understand the different memory systems in the brain (see Rolls and Deco, 2002, which provides a more extensive treatment of the brain architectures used for perception and memory). The computational

Neurobiology of Learning and Memory, Second Edition
191

approaches are essential in order to understand how the circuitry could retrieve as well as store memories, the capacity of each memory system in the brain, the interactions between memory and perceptual systems, and the speed of operation of the memory systems in the brain.

The architecture, principles of operation, and properties of the main types of network referred to here, autoassociation or attractor networks, pattern-association networks, and competitive networks, are described by Rolls and Treves (1998) and Rolls and Deco (2002).

II. FUNCTIONS OF THE HIPPOCAMPUS IN LONG-TERM MEMORY

The inferior temporal visual cortex projects via the perirhinal cortex and entorhinal cortex to the hippocampus (see Fig. 6-1), which is implicated in long-term memory of, for example, where objects are located in spatial scenes, which can be thought of as an example of episodic memory. The architecture shown in Figure 6-1 indicates that the hippocampus provides a region where visual outputs from the inferior temporal visual cortex can, via the perirhinal cortex and entorhinal cortex, be brought together with outputs from the ends of other cortical processing streams. In this section, we consider how the visual input about objects is in the correct form for the types of memory implemented by the perirhinal and hippocampal systems; how the primate hippocampus contains a representation of the visual space being viewed; how this may be similar computationally to the apparently very different representation of places that is present in the rat hippocampus; how these spatial representations are in a form that could be implemented by a continuous attractor network that could be updated in the dark by idiothetic inputs; and how a unified attractor theory of hippocampal function can be formulated using the concept of mixed attractors. The visual output from the inferior temporal visual cortex may be used to provide the perirhinal and hippocampal systems with information about objects that is useful in visual-recognition memory, in episodic memory of where objects are seen, and for building spatial representations of visual scenes. Before summarizing the computational approaches to these issues, we first summarize some of the empirical evidence that needs to be accounted for in computational models.

A. Effects of Damage to the Hippocampus and Connected Structures on Object-Place and Episodic Memory

Partly because of the evidence that in humans with bilateral damage to the hippocampus and nearby parts of the temporal lobe, anterograde amnesia is

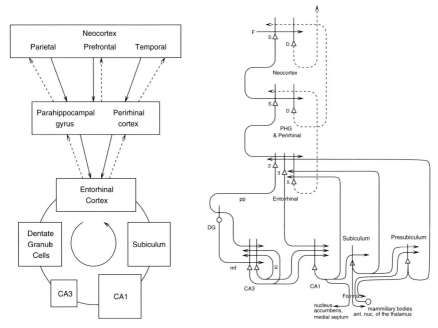

FIGURE 6-1   Forward connections (solid lines) from areas of cerebral association neocortex via the parahippocampal gyrus and perirhinal cortex and entorhinal cortex to the hippocampus; and back-projections (dashed lines) via the hippocampal CA1 pyramidal cells, subiculum, and parahippocampal gyrus to the neocortex. There is great convergence in the forward connections down to the single network implemented in the CA3 pyramidal cells and great divergence again in the back-projections. **Left:** Block diagram. **Right:** More detailed representation of some of the principal excitatory neurons in the pathways. **Abbreviations:** D, deep pyramidal cells; DG, dentate granule cells; F, forward inputs to areas of the association cortex from preceding cortical areas in the hierarchy; mf, mossy fibers; PHG, parahippocampal gyrus and perirhinal cortex; pp, perforant path; rc, recurrent collaterals of the CA3 hippocampal pyramidal cells; S, superficial pyramidal cells; 2, pyramidal cells in layer 2 of the entorhinal cortex; 3, pyramidal cells in layer 3 of the entorhinal cortex; 5, 6, pyramidal cells in the deep layers of the entorhinal cortex. The thick lines above the cell bodies represent the dendrites.

produced (Squire and Knowlton, 2000), there is continuing great interest in how the hippocampus and connected structures operate in memory. The effects of damage to the hippocampus indicate that the very long-term storage of at least some types of information is not in the hippocampus, at least in humans. On the other hand, the hippocampus does appear to be necessary to learn certain types of information, characterized as *declarative*, or "knowing that," as contrasted with *procedural*, or "knowing how," which is spared in amnesia. Declarative memory includes what can be declared or brought to mind as a proposition or an image. Declarative memory includes *episodic* memory (memory for particular episodes), and *semantic* memory (memory for facts) (Squire and Knowlton, 2000).

Lesion studies have shown that damage to the hippocampus or to some of its connections, such as the fornix in monkeys, produces deficits in learning about the places of objects and about the places where responses should be made (Buckley and Gaffan, 2000). For example, macaques and humans with damage to the hippocampal system or fornix are impaired in object-place memory tasks in which not only the objects seen, but where they were seen, must be remembered (Burgess, Maguire, and O'Keefe, 2002; Crane and Milner, 2005; Gaffan, 1994; Gaffan and Saunders, 1985; Parkinson, Murray, and Mishkin, 1988; Smith and Milner, 1981). Posterior parahippocampal lesions in macaques impair even a simple type of object-place learning in which the memory load is just one pair of trial-unique stimuli (Malkova and Mishkin, 2003). (It is further predicted that a more difficult object-place learning task with nontrial-unique stimuli and with many object-place pairs would be impaired by neurotoxic hippocampal lesions.) Further, neurotoxic lesions that selectively damage the primate hippocampus impair spatial scene memory (Murray, Baxter, and Gaffan, 1998). Also, fornix lesions impair conditional left–right discrimination learning, in which the visual appearance of an object specifies whether a response is to be made to the left or the right (Rupniak and Gaffan, 1987). A comparable deficit is found in humans (Petrides, 1985). Fornix-sectioned monkeys are also impaired in learning, on the basis of a spatial cue about which object to choose (e.g., if two objects are on the left, choose object A, but if the two objects are on the right, choose object B) (Gaffan and Harrison, 1989a). Monkeys with fornix damage are also impaired in using information about their place in an environment. For example, Gaffan and Harrison (1989b) found learning impairments when which of two or more objects the monkey had to choose depended on the position of the monkey in the room. Rats with hippocampal lesions are impaired in using environmental spatial cues to remember particular places (Cassaday and Rawlins, 1997; Jarrard, 1993; Kesner, Lee, and Gilbert, 2004; Martin, Grimwood, and Morris, 2000; O'Keefe and Nadel, 1978), to utilize spatial cues or bridge delays (Kesner, 1998; Kesner, Lee, and Gilbert, 2004; Kesner and Rolls, 2001; Rawlins, 1985) or to perform relational operations on remembered material (Eichenbaum, 1997).

One way of relating the impairment of spatial processing to other aspects of hippocampal function (including the memory of recent events or episodes in humans) is to note that this spatial processing involves a snapshot type of memory, in which one whole scene with its often unique set of parts or elements must be remembered. This memory may then be a special case of episodic memory, which involves an arbitrary association of a set of spatial and/or nonspatial events that describe a past episode. Further, the deficit in paired associate learning in humans (Squire and Knowlton, 2000) may be especially evident when this involves arbitrary associations between words, for example, window–lake. The right– (spatial) –left (word) dissociation in the human hippocampus (Burgess, Maguire, and O'Keefe, 2002; Crane and Milner, 2005) could be related to the fact that arbitrary associations between words and places are not required and that the hippocampal commissural system in humans is not well developed.

The perirhinal cortex is involved in recognition memory, in that damage to the perirhinal cortex produces impairments in recognition memory tasks in which several items intervene between the sample presentation of a stimulus and its presentation again as a match stimulus (Malkova, Bachevalier, Mishkin, and Saunders, 2001; Zola-Morgan, Squire, Amaral, and Suzuki, 1989; Zola-Morgan, Squire, and Ramus, 1994). Indeed, damage to the perirhinal cortex rather than to the hippocampus is believed to underlie the impairment in recognition memory found in amnesia in humans associated with medial temporal lobe damage (Buckley and Gaffan, 2000). The functions of the perirhinal cortex in recognition memory and how it could implement long-term familiarity memory are discussed elsewhere (Rolls and Kesner, 2006).

B. Neurophysiology of the Hippocampus and Connected Areas

In the rat, many hippocampal pyramidal cells fire when the rat is in a particular place, as defined, for example, by the visual spatial cues in an environment such as a room (Kubie and Muller, 1991; O'Keefe, 1990, 1991). There is information from the responses of many such cells about the place where the rat is in the environment. When a rat enters a new environment, B, connected to a known environment, A, there is a period of approximately 10 minutes in which, as the new environment is learned, some of the cells that formerly had place fields in A develop instead place fields in B. It is as if the hippocampus sets up a new spatial representation that can map both A and B, keeping the proportion of cells active at any one time approximately constant (Wilson and McNaughton, 1993). Some rat hippocampal neurons are found to be more task related, responding, for example, to olfactory stimuli to which particular behavioral responses must be made (Eichenbaum, 1997), and some of these neurons may in different experiments show place-related responses.

It has been discovered that in the primate hippocampus, many spatial cells have responses not related to the place where the monkey is, but instead related to the place where the monkey is looking (Rolls, 1999b, 1999c; Rolls, Robertson, and Georges-François, 1997). These are called *spatial view cells*, an example of which is shown in Figure 6-2. These cells encode information in allocentric (world-based, as contrasted with egocentric, body-related) coordinates (Georges-François, Rolls, and Robertson, 1999; Rolls, Treves, Robertson, Georges-François, and Panzeri, 1998). They can in some cases

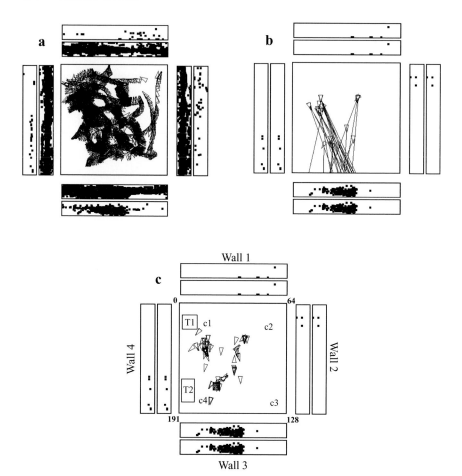

FIGURE 6-2 Examples of the firing of a hippocampal spatial view cell when the monkey was walking around the laboratory. **a.** The firing of the cell is indicated by the spots in the outer set of four rectangles, each of which represents one of the walls of the room. There is one spot on the outer rectangle for each action potential. The base of the wall is toward the center of each rectangle. The positions on the walls fixated during the recording sessions are indicated by points in the inner set of four rectangles, each of which also represents a wall of the room. The central square is a plan view of the room, with a triangle printed every 250 msec to indicate the position of the monkey, thus showing that many different places were visited during the recording sessions. **b.** A similar representation of the same three recording sessions as in (a), but modified to indicate some of the range of monkey positions and horizontal gaze directions when the cell fired at more than 12 spikes/sec. **c.** A similar representation of the same three recording sessions as in (b), but modified to indicate more fully the range of places when the cell fired. The triangle indicates the current position of the monkey, and the line projected from it shows which part of the wall is being viewed at any one time while the monkey is walking. One spot is shown for each action potential. [After Georges-François, Rolls, and Robertson (1999).]

respond to remembered spatial views, in that they respond when the view details are obscured, and use idiothetic (self-motion) cues, including eye position and head direction, to trigger this memory recall operation (Robertson, Rolls, and Georges-François, 1998). Another idiothetic input that drives some primate hippocampal neurons is linear and axial whole-body motion (O'Mara, Rolls, Berthoz, and Kesner, 1994), and, in addition, the primate presubiculum has been shown to contain head direction cells (Robertson, Rolls, Georges-François, and Panzeri, 1999).

Part of the interest of spatial view cells is that they could provide the spatial representation required to enable primates to perform object-place memory, for example, remembering where they saw a person or object, an example of an episodic memory, and indeed similar neurons in the hippocampus respond in object-place memory tasks (Rolls et al., 1989; Rolls, Xiang, and Franco, 2005). Associating such a spatial representation with a representation of a person or object could be implemented by an autoassociation network implemented by the recurrent collateral connections of the CA3 hippocampal pyramidal cells (Rolls, 1989, 1996; Rolls and Treves, 1998). Some other primate hippocampal neurons respond in the object-place memory task to a combination of spatial information and information about the object seen (Rolls et al., 1989; Rolls, Xiang, and Franco, 2005). Further evidence for this convergence of spatial and object information in the hippocampus is that in another memory task for which the hippocampus is needed, learning where to make spatial responses conditional on which picture is shown, some primate hippocampal neurons respond to a combination of which picture is shown and where the response must be made (Cahusac, Rolls, Miyashita, and Niki, 1993; Miyashita, Rolls, Cahusac, Niki, and Feigenbaum, 1989). Another important type of convergence found neurophysiologically in the primate hippocampus is between particular locations and particular rewards (Rolls and Xiang, 2005), and this reward-place memory is important in enabling primates to remember where rewards have been found, which is impaired by hippocampal lesions (Murray, Baxter, and Gaffan, 1998).

These primate spatial view cells are thus unlike place cells found in the rat (Kubie and Muller, 1991; O'Keefe, 1979, 1990, 1991; Wilson and McNaughton, 1993). Primates, with their highly developed visual and eye-movement control systems, can explore and remember information about what is present at places in the environment without having to visit those places. Such spatial view cells in primates would thus be useful as part of a memory system, in that they would provide a representation of a part of space that would not depend on exactly where the monkey or human was and that could be associated with items that might be present in those spatial locations. An example of the utility of such a representation in humans would be remembering where a particular person had been seen. The primate spatial representations would also be useful in remembering trajectories through environments, of use, for example, in short-range spatial navigation (O'Mara, Rolls, Berthoz, and Kesner, 1994; Rolls, 1999b).

The representation of space in the rat hippocampus, which is of the place where the rat is, may be related to the fact that with a much less developed visual system than the primate, the rat's representation of space may be defined more by the olfactory and tactile as well as distant visual cues present and may thus tend to reflect the place where the rat is. An interesting hypothesis on how this difference could arise from essentially the same computational process in rats and monkeys is as follows (de Araujo, Rolls, and Stringer, 2001; Rolls, 1999b). The starting assumption is that in both the rat and the primate, the dentate granule cells and the CA3 and CA1 pyramidal cells respond to combinations of the inputs received. In the case of the primate, a combination of visual features in the environment will over a typical viewing angle of perhaps 10°–20° result in the formation of a spatial view cell, the effective trigger for which will thus be a combination of visual features within a relatively small part of space. In contrast, in the rat, given the very extensive visual field, which may extend over 180°–270°, a combination of visual features formed over such a wide visual angle would effectively define a position in space, that is, a place. The actual processes by which the hippocampal formation cells would come to respond to feature combinations could be similar in rats and monkeys, involving, for example, competitive learning in the dentate granule cells, autoassociation learning in CA3 pyramidal cells, and competitive learning in CA1 pyramidal cells (Rolls, 1989, 1996; Rolls and Kesner, 2006; Rolls and Treves, 1998; Treves and Rolls, 1994). Thus spatial view cells in primates and place cells in rats might arise by the same computational process but be different by virtue of the fact that primates are foveate and view a small part of the visual field at any one time, whereas the rat has a very wide visual field. Although the representation of space in rats therefore may be in some ways analogous to the representation of space in the primate hippocampus, the difference does have implications for theories, and modeling, of hippocampal function.

In rats, the presence of place cells has led to theories that the rat hippocampus is a spatial cognitive map and can perform spatial computations to implement navigation through spatial environments (Burgess and O'Keefe, 1996; Burgess, Recce, and O'Keefe, 1994; O'Keefe, 1991; O'Keefe and Nadel, 1978). The details of such navigational theories could not apply in any direct way to what is found in the primate hippocampus. Instead, what is applicable to both the primate and rat hippocampal recordings is that hippocampal neurons contain a representation of space (for the rat primarily where the rat is, and for the primate primarily of positions "out there" in space), which is a suitable representation for an episodic memory system. In primates, this would enable one to remember, for example, where an object was seen. In rats, it might enable memories to be formed of where particular objects (for example, those defined by olfactory, tactile, and taste inputs) were found. Thus, at least in primates and possibly also in rats, the neuronal representation of space in the hippocampus may be appropriate for forming memories of events (which in these animals usually have a spatial component). Such memories would be useful for spatial navigation, for which, according to the present hypothesis, the hippocampus would implement the memory component but not the spatial computation component. Evidence that what neuronal recordings have shown is represented in the nonhuman primate hippocampal system may also be present in humans is that regions of the hippocampal formation (including the parahippocampal gyrus) can be activated when humans look at spatial views (Burgess, Maguire, and O'Keefe, 2002; Epstein and Kanwisher, 1998; O'Keefe, Burgess, Donnett, Jeffery, and Maguire, 1998; Spiridon, Fischl, and Kanwisher, 2005).

C. Hippocampal Models

These neuropsychological and neurophysiological analyses are complemented by neuronal network models of how the hippocampus could operate to store and retrieve large numbers of memories (Rolls, 1987, 1989, 1996; Rolls and Kesner, 2006; Rolls and Treves, 1998; Treves and Rolls, 1992, 1994).

1. CA3

One key hypothesis (also adopted by McClelland, McNaughton, and O'Reilly, 1995) is that the hippocampal CA3 recurrent collateral connections that spread throughout the CA3 region provide a *single autoassociation network* that enables the firing of *any* set of CA3 neurons representing one part of a memory to be associated with the firing of any other set of CA3 neurons representing another part of the same memory (cf. Marr, 1971). The generic architecture of an attractor network is shown in Figure 6-3. Associatively modifiable synapses in the recurrent collateral synapses allow memories to be stored and then later retrieved from only a part, as described by Hopfield (1982), Amit (1989), Hertz, Krogh, and Palmer (1991), Rolls and Treves (1998), and Rolls and Deco (2002). The number of patterns p each representing a different memory that could be stored in the CA3 system operating as an autoassociation network would be as shown in Equation (1) (see Rolls and Treves, 1998, and Rolls and Deco, 2002, which describe extensions to the analysis developed by Hopfield, 1982).

$$p = \frac{C^{RC}}{a \ln\left(\frac{1}{a}\right)} k \qquad (1)$$

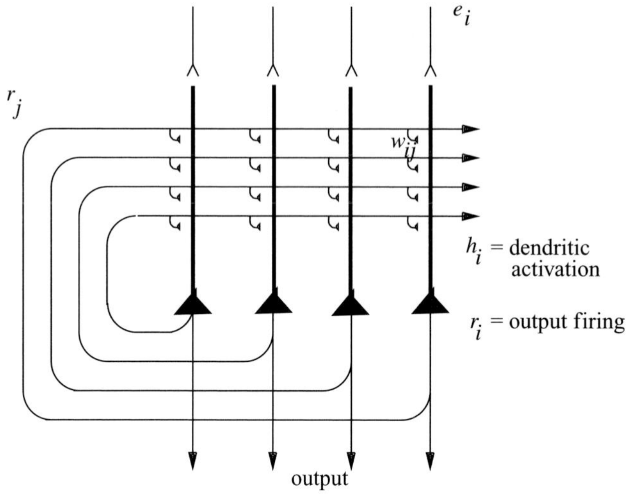

external input

FIGURE 6-3   The architecture of an attractor neural network.

where C^{RC} is the number of synapses on the dendrites of each neuron devoted to the recurrent collaterals from other CA3 neurons in the network, a is the sparseness of the representation, and k is a factor that depends weakly on the detailed structure of the rate distribution, on the connectivity pattern, etc., but is roughly in the order of 0.2–0.3. Given that C^{RC} is approximately 12,000 in the rat, the resulting storage capacity would be greater than 12,000 memories and perhaps up to 36,000 memories if the sparseness a of the representation was as low as 0.02 (Treves and Rolls, 1992, 1994).

The theory of the CA3 as a single autoassociation network holds that parts of a single memory (e.g., an object and place association) are stored in CA3, and later the whole memory can be retrieved from one of its parts (e.g., the place when the object is given as a recall cue). Empirical tests of recall and of pattern completion are consistent with this.

For example, in a one-trial object-place recall task, some macaque CA3 neurons respond to the place being recalled after the object is presented (Rolls and Xiang, 2006). Day, Langston, and Morris (2003) found that injections of AP5, an NMDA receptor blocker, to the rat dorsal hippocampus impaired the learning of a flavor-place recall task. In a hippocampus subregion test of the CA3 hypothesis, rats in a study phase were shown one object in one location and then a second object in another location. (There were 50 possible objects and 48 locations). In the test phase, the rat was shown one object in the start box and then after a 10-sec delay had to go to the correct location (choosing between two marked locations). After rats were trained in the task, CA3 lesions produced chance performance on this one-trial object-place recall task (Warthen and Kesner, 2005). A control fixed visual conditional to place task with the same delay was not impaired, showing that it is recall after one-trial (or rapid) learning that is impaired.

As noted earlier, the CA3 system is predicted to be important in the retrieval of hippocampus-dependent information when there is an incomplete retrieval cue. Support for the pattern-completion process in CA3 can be found in the following studies. Rats were tested on a cheese board with a black curtain with four extramaze cues surrounding the apparatus. (The cheese board is like a dry-land water maze with 177 holes on a 119-cm-diameter board.) Rats were trained to move a sample phase object covering a food well that could appear in one of five possible spatial locations. During the test phase of the task, following a 30-sec delay, the animal needs to find the same food well in order to receive reinforcement with the object now removed. After reaching stable performance in terms of accuracy to find the correct location, rats received lesions in CA3. During postsurgery testing, four extramaze cues were always available during the sample phase. However, during the test phase, zero, one, two, or three cues were removed in different combinations. The results indicated that controls performed well on the task regardless of the availability of one, two, three, or all cues, suggesting intact spatial pattern completion. Following the CA3 lesion, however, there was an impairment in accuracy, compared to the controls, especially when only one or two cues were available, suggesting impairment in spatial pattern completion in CA3-lesioned rats (Gold and Kesner, 2005). A useful aspect of this task is that the test for the ability to remember a spatial location learned in one presentation can be tested with varying number of available cues and many times in which the locations vary, to allow for accurate measurement of pattern-completion ability when the information stored on the single presentation must be recalled. In another study Nakazawa et al. (2002) trained CA3 NMDA receptor-knockout mice in an analogous task, using the water maze. When the animals were required to perform the task in an environment where some of the familiar cues were removed, they were impaired in performing the task. The result suggests that the NMDA receptor-dependent synaptic plasticity mechanisms in CA3 are critical to perform the pattern-completion process in the hippocampus.

2. Dentate Granule Cells

Based on the anatomy of the DG, its input and output pathways, and the development of a computational model, Rolls (1989, 1996) has suggested that the DG can act as a competitive learning network with Hebb-like modifiability to remove redundancy from the inputs, producing a more orthogonal, sparse, and categorized set of outputs. One example of this in the theory is that, to the extent that some entorhinal cortex neurons represent space as a grid (Hafting, Fyhn, Molden, Moser, and Moser, 2005), the correlations in this type of encoding are removed to produce a representation of place by using competitive learning (Rolls and Kesner, 2006; Rolls, Stringer, and Elliot, 2006), with each place encoded differently from other places (Jung and McNaughton, 1993). To the extent, then, that DG acts to produce separate representations of different places, it is predicted that the DG will be especially important when memories must be formed about similar places. I note that to form spatial representations, learned conjunctions of sensory inputs, including vestibular, olfactory, visual, auditory, and somatosensory, may be involved. DG may help to form orthogonal representations based on all these inputs and could thus help to form orthogonal nonspatial as well as orthogonal spatial representations for use in CA3. In any case, the model predicts that for spatial information, the DG should play an important role in hippocampal memory functions when the spatial information is very similar, for example, when the places are close together.

To examine the contribution of the DG to spatial pattern separation, Gilbert, Kesner, and Lee (2001) tested rats with DG lesions using a paradigm that measured one-trial short-term memory for spatial location information as a function of spatial similarity between two spatial locations (Gilbert, Kesner, and DeCoteau, 1998).

Rats were trained to displace an object that was randomly positioned to cover a baited food well in 1 of 15 locations along a row of food wells. Following a short delay, the rats were required to choose between two objects identical to the sample phase object. One object was in the same location as the sample phase object, and the second object was in a different location along the row of food wells. A rat was rewarded for displacing the object in the same position as the sample phase object (correct choice) but received no reward for displacing the foil object (incorrect choice). Five spatial separations, from 15 cm to 105 cm, were used to separate the correct object from the foil object on the choice phase. The results showed that rats with DG lesions were significantly impaired at short spatial separations; however, the performance of the DG-lesioned rats increased as a function of increased spatial separation between the correct object and the foil on the choice phases. The performance of rats with DG lesioned matched controls at the largest spatial separation. The graded nature of the impairment and the significant linear increase in performance as a function of increased separation illustrate the deficit in pattern separation produced by DG lesions. Based on these results, it can be concluded that lesions of the DG decrease efficiency in spatial pattern separation, which resulted in impairments on trials with increased spatial proximity and hence increased spatial similarity among working memory representations. In the same study it was found that CA1 lesions do not produce a deficit in this task.

Additional evidence comes from a study (Goodrich-Hunsaker, Hunsaker, and Kesner, 2005) using a modified version of an exploratory paradigm developed by Poucet (1989), in which rats with DG lesions and controls were tested on tasks involving a metric spatial manipulation. In this task, a rat was allowed to explore two different visual objects that were separated by a specific distance on a cheeseboard maze. On the initial presentation of the objects, the rat explored each object. However, across subsequent presentations of the objects in their respective locations, the rat habituated and eventually spent less time exploring the objects. Once the rat had habituated to the objects and their locations, the metric spatial distance between the two objects was manipulated so that the two objects were either closer together or further apart. The time the rat spent exploring each object was recorded. The results showed that DG lesions impaired detection of the metric distance change, in that rats with DG lesions spent significantly less time exploring the two objects that were displaced.

The results of both experiments provide empirical validation of the role of DG in spatial pattern separation and support the prediction from the computational model presented in this chapter. There are neurophysiological results that are also consistent, in that McNaughton, Barnes, Meltzer, and Sutherland (1989) found that following colchicine-induced lesions of the DG, there is a significant decrease in the reliability of CA3 place-related firing. Therefore, if CA3 cells display less reliability following DG lesions, the cells may not form accurate representations of space, due to decreased efficiency in pattern separation. Also consistent are the findings that the place fields of DG cells (Mizumori, Perez, Alvarado, Barnes, and McNaughton, 1990) and specifically granular cells (Jung and McNaughton, 1993) are small and highly reliable, and this may reflect the role of DG in pattern separation.

3. Mossy Fiber Inputs to CA3 Cells from the Dentate Granule Cells

Another part of the hypothesis is that the very sparse (see Fig. 6-4) but powerful connectivity of the mossy fiber inputs to the CA3 cells from the dentate granule cells is important during learning (but not recall) to force a new, arbitrary, set of firing onto the CA3 cells that dominates the activity of the recurrent collaterals, thereby enabling a new memory represented by the firing of the CA3 cells to be stored (Rolls, 1987, 1989; Treves and Rolls, 1992).

The perforant path input to the CA3 cells, which is numerically much larger than the mossy fiber input but is at the apical end of the dendrites, would be

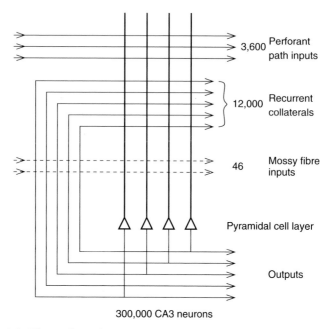

FIGURE 6-4   The numbers of connections from three different sources onto each CA3 cell from three different sources in the rat. [After Treves and Rolls (1992) and Rolls and Treves (1998).]

used to initiate recall from an incomplete pattern (Rolls and Treves, 1998; Treves and Rolls, 1992). The prediction of the theory about the necessity of the mossy fiber inputs to the CA3 cells during learning but not recall has now been confirmed (Lassalle, Bataille, and Halley, 2000). A way to enhance the efficacy of the mossy fiber system relative to the CA3 recurrent collateral connections during learning may be to increase the level of acetyl choline by increasing the firing of the septal cholinergic cells (Hasselmo, Schnell, and Barkai, 1995).

4. Back-Projections to the Neocortex and Recall

Another key part of the quantitative theory is that not only can retrieval of a memory by an incomplete cue be performed by the operation of the associatively modified CA3 recurrent collateral connections, but also that recall of that information to the neocortex can be performed via CA1 and the hippocampo-cortical and cortico-cortical back-projections (Rolls, 1996, 2000a; Rolls and Treves, 1998; Treves and Rolls, 1994) shown in Figure 6-1. In this case, the number of memory patterns p^{BP} that can be retrieved by the back-projection system is

$$p^{BP} \approx \frac{C^{BP}}{a^{BP} \ln\left(\frac{1}{a^{BP}}\right)} k^{BP} \tag{2}$$

where C^{BP} is the number of synapses on the dendrites of each neuron devoted to back-projections from the preceding stage (dashed lines in Fig. 6-1), a^{BP} is the sparseness of the representation in the back-projection pathways, and k^{BP} is a factor that depends weakly on the detailed structure of the rate distribution, on the connectivity pattern, etc., but is roughly in the order of 0.2–0.3. The insight into this quantitative analysis came from treating each layer of the back-projection hierarchy as being quantitatively equivalent to another iteration in a single recurrent attractor network (Treves and Rolls, 1991, 1994). The need for this number of connections to implement recall and more generally to constrain satisfaction in connected networks (see Rolls and Deco, 2002) provides a fundamental and quantitative reason for why there are approximately as many back-projections as forward connections between the adjacent connected cortical areas in a cortical hierarchy. This and other computational approaches to hippocampal function are included in special issues of the journals *Hippocampus* [1996: 6(6)], and *Neural Networks* (2005: 18).

Another aspect of the theory is that the operation of the CA3 system to implement recall and of the back-projections to retrieve the information would be sufficiently fast, given the fast recall in associative networks built of neurons with continuous dynamics (see Rolls and Deco, 2002).

There is also evidence that the hippocampus is involved in memories of spatial sequences, and part of the basis for this may be association between successively presented items within the CA3 network using a temporally asymmetric learning rule (Abbott and Blum, 1996; Abbott and Nelson, 2000; Levy, Wu, and Baxter, 1995; Markram, Pikus, Gupta, and Tsodyks, 1998; Minai and Levy, 1993; Rolls and Kesner, 2006; Wu, Baxter, and Levy, 1996). There is also evidence implicating the hippocampus in mediating associations across time (Kesner, 1998; Rawlins, 1985). Hippocampal subregion analyses show that CA1 lesions impair this (Rolls and Kesner, 2005). Computationally, a strong hypothesis would be that the CA3 system could provide the working memory necessary for hippocampus-dependent associations across time and that the CA3 then influences the CA1 for this function to be implemented. The actual learning could involve holding one item active in CA3 by continuing firing in an attractor state until the next item in the sequence arrives, when it could be associated with the preceding item by temporally asymmetric synaptic associativity. The computational suggestion thus is that associations across time could be implemented in the hippocampus by using the same functionality that may be used for sequence memory.

A further elaboration of this theory, and of empirical tests of it, is provided by Rolls and Kesner (2006).

D. Continuous Spatial Representations, Path Integration, and the Use of Idiothetic Inputs

1. Introduction

The fact that spatial patterns, which imply continuous representations of space, are represented in the hippocampus has led to the application of continuous attractor models to help understand hippocampal function. Such models have been developed by Samsonovich and McNaughton (1997), Battaglia and Treves (1998a), Stringer, Trappenberg, Rolls, and Araujo (2002), Stringer, Rolls, Trappenberg, and Araujo (2002), Stringer, Rolls, and Trappenberg (2005), and Stringer and Rolls (2002) (see Rolls and Deco, 2002). Indeed, we have shown how a continuous attractor network could enable the head-direction cell firing of presubicular cells to be maintained in the dark, and updated by idiothetic (self-motion) head-rotation cell inputs (Robertson, Rolls, Georges-François, and Panzeri, 1999; Stringer, Trappenberg, Rolls, and Araujo, 2002). The continuous attractor model has been developed to understand how place cell firing in rats can be maintained and updated by idiothetic inputs in the dark (Stringer, Rolls, Trappenberg, and Araujo, 2002). The continuous attractor model has also been developed to understand how spatial view cell firing in primates can be maintained and updated by idiothetic eye-movement and head-direction inputs in the dark (Robertson, Rolls, and Georges-François, 1998; Stringer, Rolls, and Trappenberg, 2005).

The way in which path integration could be implemented in the hippocampus or related systems is described next. Single-cell recording studies have shown that some neurons represent the current position along a continuous physical dimension or space even when no inputs are available, for example, in darkness. Examples include neurons that represent the positions of the eyes (i.e., eye direction with respect to the head), the place where the animal is looking in space, head direction, and the place where the animal is located. In particular, examples of such classes of cells include head-direction cells in rats (Muller, Ranck, and Taube, 1996; Ranck, 1985; Taube, Goodridge, Golob, Dudchenko, and Stackman, 1996; Taube, Muller, and Ranck, 1990) and primates (Robertson, Rolls, Georges-François, and Panzeri, 1999), which respond maximally when the animal's head is facing in a particular preferred direction; place cells in rats (Markus et al., 1995; McNaughton, Barnes, and O'Keefe, 1983; Muller, Kubie, Bostock, Taube, and Quirk, 1991; O'Keefe, 1984; O'Keefe and Dostrovsky, 1971), which fire maximally when the animal is in a particular location; and spatial view cells in primates, which respond when the monkey is looking toward a particular location in space (Georges-François, Rolls, and Robertson, 1999; Robertson, Rolls, and Georges-François, 1998; Rolls, Robertson, and Georges-François, 1997). In the parietal cortex there are many spatial representations, in several different coordinate frames (see Andersen, Batista, Snyder, Buneo, and Cohen, 2000; Rolls and Deco, 2002), and they have some capability to remain active during memory periods when the stimulus is no longer present. Even more than this, the dorsolateral pre-frontal cortex networks to which the parietal networks project have the capability to maintain spatial representations active for many seconds or minutes during short-term memory tasks, when the stimulus is no longer present (see later).

A class of network that can maintain the firing of its neurons to represent any location along a continuous physical dimension, such as spatial position, head direction, etc., is a "continuous attractor" neural network (CANN). It uses excitatory recurrent collateral connections between the neurons to reflect the distance between the neurons in the state space of the animal (e.g., head-direction space). These networks can maintain the bubble of neural activity constant for long periods wherever it is started to represent the current state of the animal (head direction, position, etc.) and are likely to be involved in many aspects of spatial processing and memory, including spatial vision. Global inhibition is used to keep the number of neurons in a bubble or packet of actively firing neurons relatively constant and to help to ensure that there is only one activity packet. Continuous attractor networks can be thought of as very similar to autoassociation or discrete attractor networks (see Rolls and Deco, 2002), and they have the same architecture, as illustrated in Figure 6-3. The main difference is that the patterns stored in a CANN are continuous patterns, with each neuron having broadly tuned firing that decreases with, for example, a Gaussian function as the distance from the optimal firing location of the cell is varied and with different neurons having tuning that overlaps throughout the space. Such tuning is illustrated in Figure 6-5. For comparison, autoassociation networks normally have discrete (separate) pat-terns (each pattern implemented by the firing of a particular subset of the neurons), with no continuous distribution of the patterns throughout the space (see Fig. 6-5). A consequent difference is that the CANN can maintain its firing at any location in the trained continuous space, whereas a discrete attrac-tor or autoassociation network moves its population of active neurons toward one of the previously learned attractor states and thus implements the recall of a particular previously learned pattern from an incomplete or noisy (distorted) version of one of the previously learned patterns. The energy landscape of a discrete attractor network (see Rolls and Deco, 2002) has separate energy minima, each one of which corresponds to a learned pattern, whereas the energy landscape of a continuous attractor network is flat, so the activity packet remains stable with continuous firing wherever it is started in the state space.

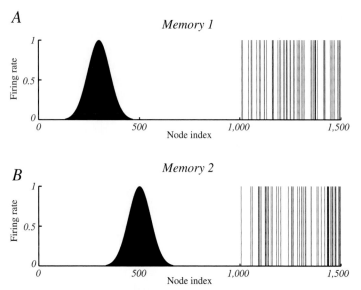

FIGURE 6-5   The types of firing patterns stored in continuous attractor networks are illustrated for the patterns present on neurons 1–1,000 for Memory 1 (when the firing is that produced when the spatial state represented is that for location 300) and for Memory 2 (when the firing is that produced when the spatial state represented is that for location 500). The continuous nature of the spatial representation results from the fact that each neuron has a Gaussian firing rate that peaks at its optimal location. This particular mixed network also contains discrete representations that consist of discrete subsets of active binary firing-rate neurons in the range 1,001–1,500. The firing of these latter neurons can be thought of as representing the discrete events that occur at the location. Continuous attractor networks by definition contain only continuous representations, but this particular network can store mixed continuous and discrete representations and is illustrated to show the difference of the firing patterns normally stored in separate continuous attractor and discrete attractor networks. For this particular mixed network, during learning, Memory 1 is stored in the synaptic weights, then Memory 2, etc., and each memory contains part that is continuously distributed to represent physical space and part that represents a discrete event or object.

(The *state space* refers to set of possible spatial states of the animal in its environment, e.g., the set of possible head directions).

I next describe the operation and properties of continuous attractor networks, which have been studied, for example, by Amari (1977), Zhang (1996), and Taylor (1999). Then, following Stringer, Trappenberg, Rolls, and Araujo (2002), I address four key issues about the biological application of continuous attractor network models.

One key issue in such continuous attractor neural networks is how the synaptic strengths between the neurons in the continuous attractor network could be learned in biological systems (upcoming Section IID3).

A second key issue in such continuous attractor neural networks is how the bubble of neuronal firing representing one location in the continuous state space should be updated based on nonvisual cues to represent a new location in state space. This is essentially the problem of path integration: how a system that represents a memory of where the agent is in physical space could be updated based on idiothetic (self-motion) cues such as vestibular cues (which might represent a head-velocity signal) or proprioceptive cues (which might update a representation of place based on movements being made in the space, for example, during walking in the dark).

A third key issue is how stability in the bubble of activity representing the current location can be maintained without much drift in darkness, when it is operating as a memory system (see Rolls and Deco, 2002; Stringer, Trappenberg, Rolls, and Araujo, 2002).

A fourth key issue is considered later, when I describe networks that store both continuous patterns and discrete patterns (see Fig. 6-5), which can be used, for example, to store the location in (continuous, physical) space where an object (a discrete item) is present.

2. Generic Model of a Continuous Attractor Network

The generic model of a continuous attractor is as follows. [The model is described in the context of head-direction cells, which represent the head direction of rats (Muller, Ranck, and Taube, 1996; Taube, Goodridge, Golob, Dudchenko, and Stackman, 1996) and macaques (Robertson, Rolls, Georges-François, and Panzeri, 1999) and can be reset by visual inputs after gradual drift in darkness.] The model is a recurrent attractor network with global inhibition. It is different from a Hopfield attractor network (Hopfield, 1982) primarily in that there are no discrete attractors formed by associative learning of discrete patterns. Instead there is a set of neurons connected to each other by synaptic weights w_{ij} that are a simple function, for example, Gaussian, of the distance between the states of the agent in the physical world (e.g., head directions) represented by the neurons. Neurons that represent similar states (locations in the state space) of the agent in the physical world have strong synaptic connections, which can be set up by an associative-learning rule, as described in Section IID3. The network updates its firing rates by the following "leaky-integrator" dynamical equations. The continuously changing activation h_i^{HD} of each head-direction cell i is governed by the equation

$$\tau \frac{dh_i^{HD}(t)}{dt} = -h_i^{HD}(t) + \frac{\phi_0}{C^{HD}} \sum_j (w_{ij} - w^{inh}) r_j^{HD}(t) + I_i^v, \qquad (3)$$

where r_j^{HD} is the firing rate of head-direction cell j, w_{ij} is the excitatory (positive) synaptic weight from head-direction cell j to cell i, w^{inh} is a global constant describing the effect of inhibitory interneurons, and τ is the time constant of the system.[1] The term $-h_i^{HD}$ (t) indicates the amount by which the activation decays (in the leaky integrator neuron) at time t. (The network is updated in a typical simulation at much smaller time steps than the time constant of the system, τ.) The next term in Equation 3 is the input from other neurons in the network r_j^{HD} weighted by the recurrent collateral synaptic connections w_{ij} (scaled by a constant ϕ_0 and C^{HD}, which is the number of synaptic connections received by each head-direction cell from other head-direction cells in the continuous attractor). The term I_i^V represents a visual input to head-direction cell i. Each term I_i^V is set to have a Gaussian response profile in most continuous attractor networks, and this sets the firing of the cells in the continuous attractor to have Gaussian response profiles as a function of where the agent is located in the state space (see, e.g., Fig. 6-5), but the Gaussian assumption is not crucial. [It is known that the firing rates of head-direction cells in both rats (Muller, Ranck, and Taube, 1996; Taube, Goodridge, Golob, Dudchenko, and Stackman, 1996) and macaques (Robertson, Rolls, Georges-François, and Panzeri, 1999) are approximately Gaussian.] When the agent is operating without visual input in memory mode, then the term I_i^V is set to zero. The firing rate r_i^{HD} of cell i is determined from the activation r_i^{HD} and the sigmoid function

$$r_i^{HD}(t) = \frac{1}{1 + e^{-2\beta(h_i^{HD}(1)-\alpha)}} \tag{4}$$

where α and β are the sigmoid threshold and slope, respectively.

3. Learning the Synaptic Strengths Between the Neurons That Implement a Continuous Attractor Network

So far we have said that the neurons in the continuous attractor network are connected to each other by synaptic weights w_{ij}, which are a simple function, for example, Gaussian, of the distance between the states of the agent in the physical world (e.g., head directions, spatial views) represented by the neurons. In many simulations, the weights are set by formula to have these appropriate Gaussian values. However, Stringer, Trappenberg, Rolls, and Araujo (2002) showed how the appropriate weights could be set up by learning. They started with the fact that since the neurons have broad tuning that may be Gaussian in shape, nearby neurons in the state space will have overlapping spatial fields and will thus be coactive to a degree that depends on the distance between

---

[1]Note that here I use r rather than y to refer to the firing rates of the neurons in the network, remembering that, because this is a recurrently connected network (see Fig. 6-3), the output from a neuron y_i might be the input x_j to another neuron.

them. They postulated that therefore the synaptic weights could be set up by associative learning based on the coactivity of the neurons produced by external stimuli as the animal moved in the state space. For example, head-direction cells are forced to fire during learning by visual cues in the environment that produce Gaussian firing as a function of head direction from an optimal head direction for each cell. The learning rule is simply that the weights w_{ij} from head-direction cell j with firing rate r_j^{HD} to head-direction cell i with firing rate r_i^{HD} are updated according to an associative (Hebb) rule,

$$\delta w_{ij} = k r_i^{HD} r_j^{HD} \tag{5}$$

where δw_{ij} is the change of synaptic weight and k is the learning rate constant. During the learning phase, the firing rate r_i^{HD} of each head-direction cell i might be the following Gaussian function of the displacement of the head from the optimal firing direction of the cell:

$$r_i^{HD} = e^{-s_{HD}^2 / 2\sigma_{HD}^2} \tag{6}$$

where s_{HD} is the difference between the actual head direction x (in degrees) of the agent and the optimal head direction x_i for head-direction cell i and σ_{HD} is the standard deviation.

Stringer, Trappenberg, Rolls, and Araujo (2002) showed that after training at all head directions, the synaptic connections develop strengths that are an almost Gaussian function of the distance between the cells in head-direction space, as shown in Figure 6-6 (left). Interestingly if a nonlinearity is introduced into the learning rule that mimics the properties of NMDA receptors by allowing the synapses to modify only after strong postsynaptic firing is present, then the synaptic strengths are still close to a Gaussian function of the distance between the connected cells in head-direction space (see Fig. 6-6, left). They showed that after training, the continuous attractor network can support stable activity packets in the absence of visual inputs (see Fig. 6-6, right) provided that global inhibition is used to prevent all the neurons from becoming activated. (The exact stability conditions for such networks have been analyzed by Amari, 1977.) Thus Stringer, Trappenberg, Rolls, and Araujo (2002) demonstrated biologically plausible mechanisms for training the synaptic weights in a continuous attractor using a biologically plausible local learning rule.

4. Idiothetic Update of a Continuous Attractor Network to Implement Path Integration

So far, we have considered how spatial representations could be stored in continuous attractor networks and how the activity can be maintained at any location in the state space in a form of short-term memory when the external (e.g., visual) input is removed. However, many networks with spatial repre-

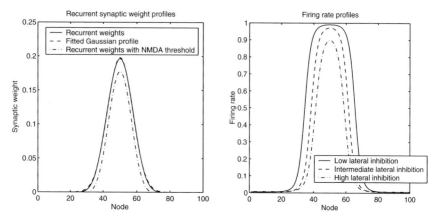

FIGURE 6-6   Training the weights in a continuous attractor network with an associative rule (Equation 5). **Left:** The trained recurrent synaptic weights from head-direction cell 50 to the other head-direction cells in the network arranged in head-direction space (solid curve). The dashed line shows a Gaussian curve fitted to the weights shown in the solid curve. The dash-dot curve shows the recurrent synaptic weights trained with rule equation (5), but with a nonlinearity introduced that mimics the properties of NMDA receptors by allowing the synapses to modify only after strong postsynaptic firing is present. **Right:** The stable firing-rate profiles forming an activity packet in the continuous attractor network during the testing phase when the training (visual) inputs are no longer present. The firing rates are shown after the network has initially been stimulated by visual input to initialize an activity packet and then allowed to settle to a stable activity profile without visual input. The three graphs show the firing rates for low, inter-mediate, and high values of the lateral inhibition parameter w^{inh}. For both left and right plots, the 100 head-direction cells are arranged according to where they fire maximally in the head-direction space of the agent when visual cues are available. [After Stringer, Trappenberg, Rolls, and Araujo (2002).]

sentations in the brain can be updated by internal, self-motion (i.e., idiothetic), cues even when there is no external (e.g., visual) input. Examples are head-direction cells in the presubiculum of rats and macaques, place cells in the rat hippocampus, and spatial view cells in the primate hippocampus. The major question arises about how such idiothetic inputs could drive the activity packet in a continuous attractor network and, in particular, how such a system could be set up biologically by self-organizing learning.

One approach to simulating the movement of an activity packet produced by idiothetic cues (which is a form of path integration whereby the current location is calculated from recent movements) is to employ a lookup table that stores (taking head-direction cells as an example), for every possible head direction and head rotational velocity input generated by the vestibular system, the corresponding new head direction (Samsonovich and McNaughton, 1997). Another approach involves modulating the strengths of the recurrent synaptic weights in the continuous attractor on one but not the other side of a currently represented position so that the stable position of the packet of activity, which

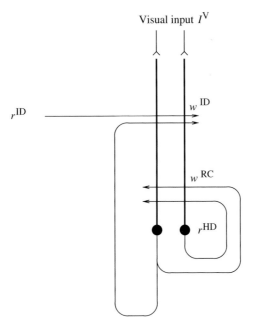

Visual input I^V

r^{ID}

w^{ID}

w^{RC}

r^{HD}

FIGURE 6-7   General network architecture for a one-dimensional continuous attractor model of head-direction cells that can be updated by idiothetic inputs produced by head-rotation cell firing r^{ID}. The head-direction cell firing is r^{HD}, the continuous attractor synaptic weights are w^{RC}, the idiothetic synaptic weights are w^{ID}, and the external visual input is I^V.

requires symmetric connections in different directions from each node, is lost, and the packet moves in the direction of the temporarily increased weights, although no possible biological implementation was proposed of how the appropriate dynamic synaptic-weight changes might be achieved (Zhang, 1996). Another mechanism (for head-direction cells) (Skaggs, Knierim, Kudrimoti, and McNaughton, 1995) relies on a set of cells, termed (head) rotation cells, which are coactivated by head-direction cells and vestibular cells and drive the activity of the attractor network by anatomically distinct connections for clockwise- and counterclockwise-rotation cells, in what is effectively a lookup table. However, no proposal was made about how this could be achieved by a biologically plausible learning process, and this has been the case until recently for most approaches to path integration in continuous attractor networks, which rely heavily on rather artificial preset synaptic connectivities.

Stringer, Trappenberg, Rolls, and Araujo (2002) introduced a proposal with more biological plausibility about how the synaptic connections from idiothetic inputs to a continuous attractor network can be learned by a self-organizing learning process. The essence of the hypothesis is described with Figure 6-7.

The continuous attractor synaptic weights w^{RC} are set up under the influence of the external visual inputs I^V, as described in Section IID3. At the same time, the idiothetic synaptic weights w^{ID} (in which the ID refers to the fact that they are in this case produced by idiothetic inputs, produced by cells that fire to represent the velocity of clockwise and counterclockwise head rotation) are set up by associating the change of head-direction cell firing that has just occurred (detected by a trace memory mechanism described in detail by Stringer, Trappenberg, Rolls, and Araujo, 2002) with the current firing of the head rotation cells r^{ID}. For example, when the trace memory mechanism incorporated into the idiothetic synapses w^{ID} detects that the head-direction cell firing is at a given location (indicated by the firing r^{ID}) and is moving clockwise (produced by the altering visual inputs I^V), and there is simultaneous clockwise head-rotation cell firing, the synapses w^{ID} learn the association, so when that rotation cell firing occurs later without visual input, it takes the current head-direction firing in the continuous attractor into account and moves the location of the head-direction attractor in the appropriate direction.

Simulations demonstrating the operation of this self-organizing learning to produce movement of the location being represented in a continuous attractor network were described by Stringer, Trappenberg, Rolls, and Araujo (2002), and one example of the operation is shown in Figure 6-8. They also showed that, after training with just one value of the head-rotation cell firing, the network showed the desirable property of moving the head direction being represented in the continuous attractor by an amount that was proportional to the value of the head-rotation cell firing. Stringer, Trappenberg, Rolls, and Araujo (2002) also describe a related model of the idiothetic cell update of the location represented in a continuous attractor, in which the rotation cell firing directly modulates in a multiplicative way the strength of the recurrent connections in the continuous attractor in such a way that clockwise rotation cells modulate the strength of the synaptic connections in the clockwise direction in the continuous attractor, and vice versa.

It should be emphasized that although the cells are organized in Figure 6-8 according to the spatial position being represented, there is no need for cells in continuous attractor networks that represent nearby locations in the state space to be close together, because the distance in the state space between any two neurons is represented by the strength of the connection between them, not by where the neurons are physically located. This enables continuous attractor networks to represent spaces with arbitrary topologies, for the topology is represented in the connection strengths (Rolls and Stringer, 2005; Stringer and Rolls, 2002; Stringer, Rolls, and Trappenberg, 2005; Stringer, Rolls, Trappenberg, and Araujo, 2002; Stringer, Trappenberg, Rolls, and Araujo, 2002). Indeed, it is this that enables many different charts, each with its own topology, to be represented in a single continuous attractor network (Battaglia and Treves, 1998a).

Firing rates during rotation

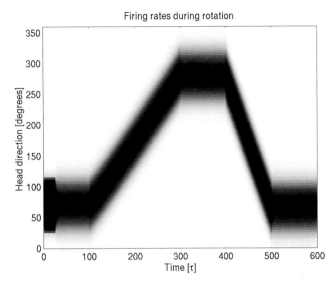

FIGURE 6-8   Idiothetic update of the location represented in a continuous attractor network. The firing rate of the cells with optima at different head directions (organized according to head direction on the ordinate) is shown by the blackness of the plot, as a function of time. The activity packet was initialized to a head direction of 75°, and the packet was allowed to settle without visual input. For $t = 0$ to $t = 100$ there was no rotation cell input, and the activity packet in the continuous attractor remained stable at 75°. For $t = 100$ to $t = 300$ the clockwise-rotation cells were active, with a firing rate of 0.15 to represent a moderate angular velocity, and the activity packet moved clockwise. For $t = 300$ to $t = 400$ there was no rotation-cell firing, and the activity packet immediately stopped and remained still. For $t = 400$ to $t = 500$, the counter-clockwise-rotation cells had a high firing rate of 0.3 to represent a high velocity, and the activity packet moved counterclockwise with a greater velocity. For $t = 500$ to $t = 600$ there was no rotation cell firing, and the activity packet immediately stopped.

5. Continuous Attractor Networks in Two or More Dimensions

Some types of spatial representation used by the brain are of spaces that exist in two or more dimensions. Examples are the two- (or three-) dimensional space representing where one is looking in a spatial scene. Another is the two- (or three-) dimensional space representing where one is located. It is possible to extend continuous attractor networks to operate in higher-dimensional spaces than the one-dimensional spaces considered so far (Stringer, Rolls, Trappenberg, and Araujo, 2002; Taylor, 1999). Indeed, it is also possible to extend the analyses of how idiothetic inputs could be used to update two-dimensional state spaces, such as the locations represented by place cells in rats (Stringer, Rolls, Trappenberg, and Araujo, 2002) and the location at which one is looking represented by primate spatial view cells (Stringer and Rolls, 2002; Stringer, Rolls, and Trappenberg, 2005). Interestingly, the number of terms in the synapses implementing idiothetic update do not need to increase beyond three (as in sigma–pi synapses) even when higher-dimensional state spaces are being considered (Stringer, Rolls, Trappenberg, and Araujo, 2002). Also interestingly, a continuous attractor network can in fact represent the properties of very high-dimensional spaces, because the properties of the spaces are captured by the connections between the neurons of the continuous attractor, and these connections are of course, as in the world of discrete attractor networks, capable of representing high-dimensional spaces (Stringer, Rolls, Trappenberg, and Araujo, 2002). With these approaches, continuous attractor networks have been developed of the two-dimensional representation of rat hippocampal place cells with idiothetic update by movements in the environment (Stringer, Rolls, Trappenberg, and Araujo, 2002) and of primate hippocampal spatial view cells with idiothetic update by eye and head movements (Rolls and Stringer, 2005; Stringer and Rolls, 2002; Stringer, Rolls, and Trappenberg, 2005). It has also been shown that a single attractor network can maintain two packets of activity simultaneously if they are in separate spaces and that the two packets of activity can cross each other stably during idiothetic update (Stringer, Rolls, and Trappenberg, 2004).

E. A Unified Theory of Hippocampal Memory: Mixed Continuous and Discrete Attractor Networks

If the hippocampus is to store and retrieve episodic memories, it may need to associate together patterns that have continuous spatial attributes, and other patterns that represent objects, that are discrete. To address this issue, we have now shown that attractor networks can store both continuous patterns and discrete patterns and can thus be used to store, for example, the location in (continuous, physical) space where an object (a discrete item) is present (see Fig. 6-5, and Rolls, Stringer, and Trappenberg, 2002). In this network, when events are stored that have both discrete (object) and continuous (spatial) aspects, then the whole place can be retrieved later by the object, and the object can be retrieved by using the place as a retrieval cue. Such networks are likely to be present in parts of the brain that receive and combine inputs both from systems that contain representations of continuous (physical) space and from brain systems that contain representations of discrete objects, such as the inferior temporal visual cortex. One such brain system is the hippocampus, which appears to combine and store such representations in a mixed attractor network in the CA3 region, which thus is able to implement episodic memories that typically have a spatial component, for example, where an item such as a key is located.

This network thus shows that in brain regions where the spatial- and object-processing streams are brought together, a single network can represent and learn associations between both types of input. Indeed, in brain regions such as the hippocampal system, it is essential that the spatial- and object-processing streams be brought together in a single network, for it is only when both types of information are in the same network that spatial information can be retrieved from object information, and vice versa, which is a fundamental property of episodic memory. It may also be the case that in the prefrontal cortex, attractor networks can store both spatial and discrete (e.g., object-based) types of information in short-term memory (see Section III).

F. Speed of Operation of Memory Networks: The Integrate-and-Fire Approach

Consider, for example, a real network whose operation has been described by an autoassociative formal model that acquires, with learning, a given attractor structure. How does the state of the network approach, in real time during a retrieval operation, one of those attractors? How long does it take? How does the amount of information that can be read off the network's activity evolve with time? Also, which of the potential steady states is indeed a stable state that can be reached asymptotically by the net? How is the stability of different states modulated by external agents? These are examples of dynamical properties, which to be studied require the use of models endowed with some dynamics. An appropriate such model is one that incorporates integrate-and-fire neurons.

The concept that attractor (autoassociation) networks can operate very rapidly if implemented with neurons that operate dynamically in continuous time is described by Rolls and Treves (1998) and Rolls and Deco (2002). The result described was that the principal factor affecting the speed of retrieval is the time constant of the synapses between the neurons that form the attractor (Battaglia and Treves, 1998b; Panzeri, Rolls, Battaglia, and Lavis, 2001; Rolls and Treves, 1998; Treves, 1993). This was shown analytically by Treves (1993) and described by Rolls and Treves (1998, Appendix 5). If the (inactivation) time constant of AMPA synapses is taken as 10 msec, then the settling time for a single attractor network is approximately 15–17 msec (Battaglia and Treves, 1998b; Panzeri, Rolls, Battaglia, and Lavis, 2001; Rolls and Treves, 1998). A connected series of four such networks (representing, for example, four connected cortical areas), each involving recurrent (feedback) processing implemented by the recurrent collateral synaptic connections, takes approximately 4 × 17 msec to propagate from start to finish, retrieving information from each layer as the propagation proceeds (Panzeri, Rolls, Battaglia, and Lavis, 2001; Rolls and Deco, 2002). This speed of operation is sufficiently rapid that such attractor networks are biologically plausible (Rolls and Deco, 2002; Rolls and Treves, 1998).

The way in which networks with continuous dynamics (such as networks made of real neurons in the brain and networks modeled with integrate-and-fire neurons) can be conceptualized as settling so fast into their attractor states is that spontaneous activity in the network ensures that some neurons are close to their firing threshold when the retrieval cue is presented, so the firing of these neurons is influenced within 1–2 msec by the retrieval cue. These neurons then influence other neurons within milliseconds (given the point that some other neurons will be close to threshold) through the modified recurrent collateral synapses that store the information. In this way, the neurons in networks with continuous dynamics can influence each other within a fraction of the synaptic time constant, and retrieval can be very rapid (Rolls and Deco, 2002; Rolls and Treves, 1998).

III. SHORT-TERM MEMORY SYSTEMS

A. Prefrontal Cortex Short-Term Memory Networks and Their Relation to Temporal and Parietal Perceptual Networks

A common way the brain uses to implement a short-term memory is to maintain the firing of neurons during a short memory period after the end of a stimulus (see Fuster, 2000; Rolls and Deco, 2002; Rolls and Treves, 1998). In the inferior temporal cortex this firing may be maintained for a few hundred milliseconds even when the monkey is not performing a memory task (Desimone, 1996; Rolls and Tovee, 1994; Rolls, Tovee, and Panzeri, 1999; Rolls, Tovee, Purcell, Stewart, and Azzopardi, 1994). In more ventral temporal cortical areas, such as the entorhinal cortex, the firing may be maintained for longer periods in delayed match-to-sample tasks (Suzuki, Miller, and Desimone, 1997) and, in the prefrontal cortex, for even tens of seconds (Fuster, 1997, 2000). In the dorsolateral and inferior convexity prefrontal cortex the firing of the neurons may be related to the memory of spatial responses or objects (Goldman-Rakic, 1996; Wilson, O'Scalaidhe, and Goldman-Rakic, 1993) or both (Rao, Rainer, and Miller, 1997) and, in the principal sulcus/arcuate sulcus region, to the memory of places for eye movements (Funahashi, Bruce, and Goldman-Rakic, 1989) (see Rolls and Deco, 2002). The firing may be maintained by the operation of associatively modified recurrent collateral connections between nearby pyramidal cells, producing attractor states in autoassociative networks (see Rolls and Deco, 2002).

For the short-term memory to be maintained during periods in which new stimuli are to be perceived, there must be separate networks for the perceptual and short-term memory functions, and indeed two coupled networks, one in the inferior temporal visual cortex for perceptual functions and another in the prefrontal cortex for maintaining the short-term memory during intervening

Inferior temporal cortex (IT)   Prefrontal cortex (PF)

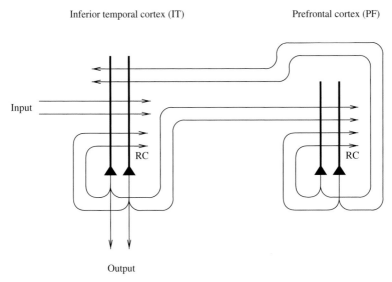

Output

FIGURE 6-9   A short-term memory autoassociation network in the prefrontal cortex could hold active a working memory representation by maintaining its firing in an attractor state. The prefrontal module would be loaded with the to-be-remembered stimulus by the posterior module (in the temporal or parietal cortex) in which the incoming stimuli are represented. Back-projections from the prefrontal short-term memory module to the posterior module would enable the working memory to be unloaded, for example, to influence ongoing perception (see text). RC — recurrent collateral connections.

stimuli, provide a precise model of the interaction of perceptual and short-term memory systems (Renart, Moreno, de la Rocha, Parga, and Rolls, 2001; Renart, Parga, and Rolls, 2000) (see Fig. 6-9). In particular, this model shows how a prefrontal cortex attractor (autoassociation) network could be triggered by a sample visual stimulus represented in the inferior temporal visual cortex in a delayed match-to-sample task and could keep this attractor active during a memory interval in which intervening stimuli are shown. Then when the sample stimulus reappears in the task as a match stimulus, the inferior temporal cortex module showed a large response to the match stimulus, because it is activated both by the visual incoming match stimulus and by the consistent back-projected memory of the sample stimulus still being represented in the prefrontal cortex memory module (see Fig. 6-9). This computational model makes it clear that in order for ongoing perception to occur unhindered, implemented by posterior cortex (parietal and temporal lobe) networks, there must be a separate set of modules capable of maintaining a representation over intervening stimuli. This is the fundamental understanding offered for the evolution and functions of the dorsolateral prefrontal cortex, and it is this ability to provide multiple separate short-term attractor memories that provides, we suggest, the basis for its functions in planning (Deco and Rolls, 2003; Rolls and Deco, 2002).

Renart, Parga, and Rolls (2000) and Renart, Moreno, de al Rocha, Parga, and Rolls (2001) performed analyses and simulations that showed that for working memory to be implemented in this way, the connections between the perceptual and the short-term memory modules (see Fig. 6-9) must be relatively weak. As a starting point, they used the neurophysiological data showing that in delayed match-to-sample tasks with intervening stimuli, the neuronal activity in the inferior temporal visual cortex (IT) is driven by each new incoming visual stimulus (Miller and Desimone, 1994; Miller, Li, and Desimone, 1993), whereas in the prefrontal cortex, neurons start to fire when the sample stimulus is shown and continue the firing that represents the sample stimulus even when the potential match stimuli are being shown (Miller, Erickson, and Desimone, 1996). The architecture studied by Renart, Parga, and Rolls (2000) was as shown in Figure 6-9, with both the intramodular (recurrent collateral) and the intermodular (forward IT to PF, and backward PF to IT) connections trained on the set of patterns with an associative synaptic modification rule. A crucial parameter is the strength of the intermodular connections, g, which indicates the relative strength of the intermodular to the intramodular connections. (This parameter measures effectively the relative strengths of the currents injected into the neurons by the intermodular relative to the intramodular connections, and the importance of setting this parameter to relatively weak values for useful interactions between coupled attractor networks was highlighted by Renart, Parga, and Rolls (1999b, 1999a) (see Rolls and Deco, 2002). The patterns themselves were sets of random numbers, and the simulation utilized a dynamical approach with neurons with continuous (hyperbolic tangent) activation functions (Amit and Tsodyks, 1991; Kuhn, 1990; Kuhn, Bos, and van Hemmen, 1991; Rolls and Deco, 2002; Shiino and Fukai, 1990). The external current injected into IT by the incoming visual stimuli was sufficiently strong to trigger the IT module into a state representing the incoming stimulus. When the sample was shown, the initially silent PF module was triggered into activity by the weak ($g > 0.002$) intermodular connections. The PF module remained firing to the sample stimulus even when IT was responding to potential match stimuli later in the trial, provided that g was less than 0.024, because then the intramodular recurrent connections could dominate the firing (see Fig. 6-10). If g was higher than this, then the PF module was pushed out of the attractor state produced by the sample stimulus. The IT module responded to each incoming potentially matching stimulus provided that g was not greater than approximately 0.024. Moreover, this value of g was sufficiently large that a larger response of the IT module was found when the stimulus matched the sample stimulus (the match-enhancement effect found neurophysiologically, and a mechanism by which the

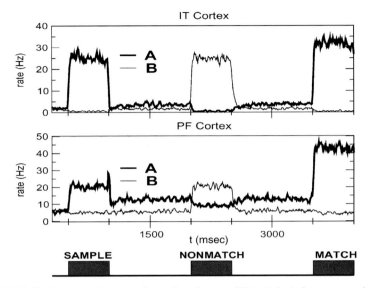

FIGURE 6-10   Interaction between the prefrontal cortex (PF) and the inferior temporal cortex (IT) in a delayed match-to-sample task with intervening stimuli with the architecture illustrated in Figure 6-9. **Above:** Activity in the IT attractor module. **Below:** Activity in the PF attractor module. The thick lines show the firing rates of the set of neurons with activity selective for the sample stimulus (which is also shown as the match stimulus and labeled A), and the thin lines show the activity of the neurons with activity selective for the nonmatch stimulus, which is shown as an intervening stimulus between the sample and match stimuli and labeled B. A trial is illustrated in which A is the sample (and match) stimulus. The prefrontal cortex module is pushed into an attractor state for the sample stimulus by the IT activity induced by the sample stimulus. Because of the weak coupling to the PF module from the IT module, the PF module remains in this sample-related attractor state during the delay periods and even while the IT module is responding to the nonmatch stimulus. The PF module remains in its sample-related state even during the nonmatch stimulus because once a module is in an attractor state, it is relatively stable. When the sample stimulus reappears as the match stimulus, the PF module shows higher sample stimulus–related firing, because the incoming input from IT is now adding to the activity in the PF attractor network. This in turn also produces a match enhancement effect in the IT neurons with sample stimulus-related selectivity, because the back-projected activity from the PF module matches the incoming activity to the IT module. [After Renart, Parga, and Rolls (2000) and Renart, Moreno, de la Rocha, Parga, and Rolls (2001).]

matching stimulus can be identified). This simple model thus shows that the operation of the prefrontal cortex in short-term memory tasks, such as delayed match-to-sample with intervening stimuli, and its relation to posterior perceptual networks can be understood as the interaction of two weakly coupled attractor networks, as shown in Figures 6-9 and 6-10. The paper by Renart, Moreno, de al Rocha, Parga, and Rolls (2001) extended the earlier findings of Renart, Parga, and Rolls (2000) to integrate-and-fire neurons, and it is results from the integrate-and-fire simulations that are shown in Figure 6-10.

B. Computational Necessity for a Separate, Prefrontal Cortex, Short-Term Memory System

The approach just described (in Section A) emphasizes that in order to provide a good brain lesion test of prefrontal cortex short-term memory functions, the task set should require a short-term memory for stimuli over an interval in which other stimuli are being processed, because otherwise the posterior cortex perceptual modules could implement the short-term memory function by their own recurrent collateral connections. This approach also emphasizes that there are many at least partially independent modules for short-term memory functions in the prefrontal cortex [e.g., several modules for delayed saccades; one or more for delayed spatial (body) responses in the dorsolateral prefrontal cortex; one or more for remembering visual stimuli in the more ventral prefrontal cortex; and at least one in the left prefrontal cortex used for remembering the words produced in a verbal fluency task — see Section 10.3 of Rolls and Treves (1998)].

This computational approach thus provides a clear understanding of why a separate (prefrontal) mechanism is needed for working memory functions, as elaborated in Section A. It may also be commented that if a prefrontal cortex module is to control behavior in a working memory task, then it must be capable of assuming some type of executive control. There may be no need to have a single central executive additional to the control that must be capable of being exerted by every short-term memory module. This is in contrast to what has traditionally been assumed for the prefrontal cortex (Shallice and Burgess, 1996).

C. Role of Prefrontal Cortex Short-Term Memory Systems in Visual Search and Attention

The same model shown in Figure 6-9 can also be used to help understand the implementation of visual search tasks in the brain (Renart, Parga, and Rolls, 2000). In such a visual search task, the target stimulus is made known beforehand, and inferior temporal cortex neurons then respond more when the search target (as compared to a different stimulus) appears in the receptive field of the IT neuron (Chelazzi, Duncan, Miller, and Desimone, 1998; Chelazzi, Miller, Duncan, and Desimone, 1993). The model shows that this could be implemented by the same system of weakly coupled attractor networks in PF and IT shown in Figure 6-9 as follows. When the target stimulus is shown, it is loaded into the PF module from the IT module as described for the delayed match-to-sample task. Later, when the display appears with two or more stimuli present, there is an enhanced response to the target stimulus in the receptive field, because of the back-projected activity from PF to IT, which adds to the firing being produced by the target stimulus itself (Renart, Moreno, de la Rocha, Parga, and Rolls, 2001; Renart, Parga, and Rolls, 2000).

The interacting spatial and object networks described by Rolls and Deco (2002) and Deco and Rolls (2004, 2005a) (see Fig. 6-11) take this analysis one stage further and show that once the PF–IT interaction has set up a greater response to the search target in IT, this enhanced response can in turn, by back-projections to topologically mapped earlier cortical visual areas, move the "attentional spotlight" in the parietal cortex to the place where the search target is located. Correspondingly, if the location to which attention should be paid is highlighted in the parietal cortex as a result of a top-down influence from the prefrontal cortex, which holds the to-be-attended location in a short-term memory, then the parietal cortex exerts an influence by its back-projections on topologically mapped earlier areas, such as V2. Then any features in the input image at the highlighted locations in V2 receive extra activation, which can act nonlinearly (Deco and Rolls, 2005b), and those features then help the object represented by them to win the competition in the inferior temporal visual cortex. This produced correct identification of the object at the attended location. In this model of attention, a short-term memory attractor (in the prefrontal cortex) plays a key role by keeping active what has to be paid attention to by continuing firing, and this short-term memory produces top-down effects to bias competition in cortical perceptual areas such as the temporal and parietal cortices.

These models have been extended to decision making (Deco and Rolls, 2003). In this model, a short-term memory holds active the rule by which stimuli can be mapped to behavioral responses in a hierarchical network of modules, from stimulus through intermediate neurons to response neurons (see Fig. 6-12). The rule can be to map the object shown to a response or the position of the object to a response. The short-term memory rule attractor network exerts a top-down effect to bias competition in a population of intermediate neurons that respond to combinations of either the object and the response required or the location of the object and the response required, which are found within the prefrontal cortex. By reversing the rule module, the mapping can be rapidly altered, thus altering how decisions are made (Deco and Rolls, 2005c). Part of the interest of this approach is that the model is implemented at the integrate–and–fire neuronal level, allowing realistic dynamics, direct comparisons between the neurons recorded in the prefrontal cortex and neurons in the model, and the effects of neurotransmitters and pharmacological agents to be simulated and predicted, because different classes of receptors and the dynamics of their synapses are part of the model (Deco and Rolls, 2003).

In a further analysis of decision making, it is possible to simulate the probabilistic nature of decision making by utilizing a single attractor short-term memory network with each decision state represented by a separate memory

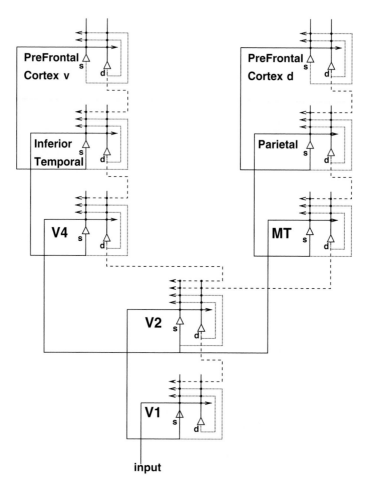

input

FIGURE 6-11   The overall architecture of a model of object and spatial processing and attention, including the prefrontal cortical areas that provide the short-term memory required to hold
active object or spatial target of attention. Forward projections between areas are shown as solid
lines, and back-projections as dashed lines. The triangles represent pyramidal cell bodies, with
the thick vertical line above them the dendritic trees. The cortical layers in which the cells are
concentrated are indicated by *s* (superficial, layers 2 and 3) and *d* (deep, layers 5 and 6). The
prefrontal cortical areas most strongly reciprocally connected to the inferior temporal cortex
"what" processing stream are labeled *v* to indicate that they are in the more ventral part of the
lateral prefrontal cortex, area 46, close to the inferior convexity in macaques. The prefrontal
cortical areas most strongly reciprocally connected to the parietal visual cortical "where" processing stream are labeled *d* to indicate that they are in the more dorsal part of the lateral prefrontal
cortex, area 46, in and close to the banks of the principal sulcus in macaques (see text). V1 is
the primary visual cortex, and V2, V4, and MT are other visual cortical areas (see Rolls and
Deco, 2002, for further details).

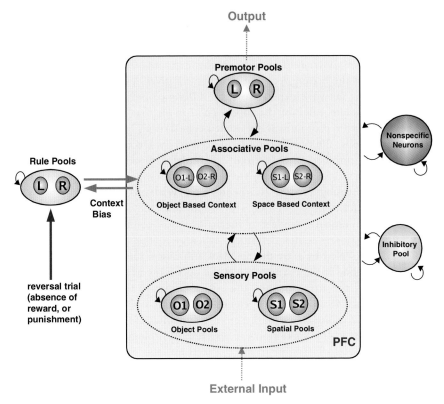

FIGURE 6-12 Network architecture of the prefrontal cortex unified model of attention, working memory, and decision making. There are sensory neuronal populations or pools for object type (O1 and O2) and spatial position (S1 and S2). These connect hierarchically (with stronger forward than backward connections) to the intermediate, or "associative," pools, in which neurons may respond to combinations of the inputs received from the sensory pools for some types of mapping, such as reversal, as described by Deco and Rolls (2003). For the simulation of the data of Asaad, Rainer, and Miller (2000) these intermediate pools respond to O1-L, O2-R, S1-L, or S2-R. These intermediate pools receive an attentional bias, which, in the case of this particular simulation, biases either the O pools or the S pools. The intermediate pools are connected hierarchically to the premotor pools, which in this case code for a Left or Right response. Each of the pools is an attractor network in which there are stronger associatively modified synaptic weights between the neurons that represent the same state (e.g., object type for a sensory pool or response for a premotor pool) than between neurons in the other pools or populations. However, all the neurons in the network are associatively connected by at least weak synaptic weights. The attractor properties, the competition implemented by the inhibitory inter-neurons, and the biasing inputs result in the same network implementing both short-term memory and biased competition; and the stronger feed-forward than feedback connections between the sensory, intermediate, and premotor pools result in the hierarchical property by which sensory inputs can be mapped to motor outputs in a way that depends on the biasing contextual or rule input.

in the network (Deco and Rolls, 2006; Wang, 2002). The evidence relevant to each decision is used to bias the relevant memory in the attractor network, and one of the attractor states, which represents the decision, rises probabilistically from the spontaneous activity in the network. One of the fascinating findings is that the biasing required to achieve a given level of correct decision making in this network depends on the difference between the two biasing inputs/the absolute value of the inputs ($\delta I/I = k$); that is, the way in which the network settles probabilistically implements Weber's law of decision making (Deco and Rolls, 2006). The model described by Deco and Rolls (2006) is different in a number of ways from accumulator or counter models, which may include a noise term and which undergo a random walk in real time, which is a diffusion process (Carpenter and Williams, 1995; Ratcliff, Van Zandt, and McKoon, 1999) (see also Usher and McClelland, 2001; Wang, 2002). In accumulator models, a mechanism for computing the difference between the stimuli is not described, whereas in the current model this is achieved, and scaled by f, by the feedback inhibition included in the attractor network. Second, in the current model the decision corresponds to high firing rates in one of the attractors, and there is no arbitrary threshold that must be reached. Third, the noise in the current model is not arbitrary, but is accounted for by finite-size noise effects of the spiking dynamics of the individual neurons with their Poisson-like spike trains in a system of limited size. Fourth, because the attractor network has recurrent connections, the way in which it settles into a final attractor state (and thus the decision process) can naturally take place over quite a long time, as information gradually and stochastically builds up due to the positive feedback in the recurrent network, the weights in the network, and the biasing inputs. Thus by considering decision making within the context of attractor memory networks, a natural and biologically realistic account of decision making emerges (Deco and Rolls, 2006).

D. Synaptic Modification Is Needed to Set up But Not to Reuse Short-Term Memory Systems

To set up a new short-term memory attractor, synaptic modification is needed to form the new, stable attractor. Once the attractor is set up, it may be used repeatedly, when triggered by an appropriate cue, to hold the short-term memory state active by continued neuronal firing, even without any further synaptic modification (see Kesner and Rolls, 2001; Rolls and Deco, 2002). Thus manipulations that impair the long-term potentiation (LTP) of synapses may impair the formation of new short-term memory states but not the use of previously learned short-term memory states. Kesner and Rolls (2001) analyzed many studies of the effects of blockade of LTP in the hippocampus on spatial working memory tasks and found evidence consistent with this prediction. Interestingly, it was found that if there was a large change in the delay interval over which the spatial information had to be remembered, then the task became susceptible, during the transition to the new delay interval, to the effects of blockade of LTP. The implication is that some new learning is required when the rat must learn the strategy of retaining information for longer periods when the retention interval is changed.

IV. INVARIANT VISUAL-OBJECT RECOGNITION

Rolls (1992) proposed a feature hierarchical model of ventral stream visual-object processing from the primary visual cortex (V1), via V2 and V4, to the inferior temporal visual cortex, which could learn to represent objects invariantly with respect to position on the retina, scale, rotation, and view. The theory uses a short-term ("trace") memory term in an associative-learning rule to help capture the fact that the natural statistics of the visual world reflect the fact that the same object is likely to be present over short time periods, for example, over 1 or 2 seconds, during which an object is seen from different views. A model of the operation of the system has been implemented in a four-layer network, corresponding to brain areas V1, V2, V4, and inferior temporal visual cortex (IT), with convergence to each part of a layer from a small region of the preceding layer and with local competition between the neurons within a layer implemented by local lateral inhibition (Elliffe, Rolls, and Stringer, 2002; Rolls and Deco, 2002; Rolls and Milward, 2000; Wallis and Rolls, 1997) (see Fig. 6-13). During a learning phase, each object is

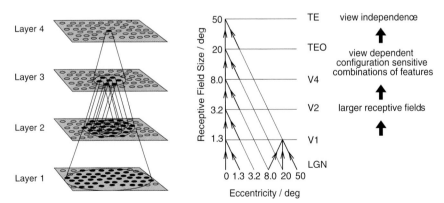

FIGURE 6-13   Convergence in the visual system. **Right:** As it occurs in the brain. V1: visual cortex area V1; TEO: posterior inferior temporal cortex; TE: inferior temporal cortex (IT). **Left:** As implemented in VisNet. Convergence through the network is designed to provide fourth-layer neurons with information from across the entire input retina.

learned. This is done by training the connections between modules using a trace learning rule with the general form

$$\delta w_{ij} = \alpha \overline{y}_i^\tau x_j^\tau \tag{7}$$

where x_j^τ is the jth input to the neuron at time step τ, y_i is the output of the ith neuron, and w_{ij} is the jth weight on the ith neuron.

The trace $\overline{y}_i^\tau$ is updated according to

$$\overline{y}_i^\tau = (1 - \eta) y_i^\tau + \eta \overline{y}_i^{\tau-1} \tag{8}$$

The parameter $\eta \in [0,1]$ controls the relative contributions to the trace $\overline{y}_i^\tau$ from the instantaneous firing rate y_i^τ at time step τ and the trace at the previous time step $\eta \overline{y}_i^{\tau-t}$.

V. VISUAL STIMULUS–REWARD ASSOCIATION, EMOTION, AND MOTIVATION

Learning about which visual and other stimuli in the environment are rewarding, punishing, or neutral is crucial for survival. For example, it takes just one trial to learn if a seen object is hot when we touch it, and associating that visual stimulus with the pain may help us to avoid serious injury in the future. Similarly, if we are given a new food that has an excellent taste, we can learn in one trial to associate the sight of it with its taste so that we can select it in future. In these examples, the previously neutral visual stimuli become conditioned reinforcers by their association with a primary (unlearned) reinforcer such as taste or pain. Our examples show that learning about which stimuli are rewards and punishments is very important in the control of motivational behavior, such as feeding and drinking, and in emotional behavior, such as fear and pleasure (Rolls, 2005). The type of learning involved is pattern association, between the conditioned and the unconditioned stimulus (Rolls, 2005). This type of learning presents a major example of how the visual representations provided by the inferior temporal visual cortex are used by the other parts of the brain (Rolls, 1999a, 2000b, 2005; Rolls and Deco, 2002). In this section we consider where in sensory processing this stimulus-reinforcement association learning occurs, which brain structures are involved in this type of learning, how the neuronal networks for pattern-association learning may actually be implemented in these regions, and how the distributed representation about objects provided by the inferior temporal cortex output is suitable for this pattern-association learning.

The crux of the answer to the last issue is that the inferior temporal cortex representation is ideal for this pattern-association learning because it is a transform-invariant representation of objects and because the code can be read by a neuronal system that performs dot products using neuronal ensembles as

Brain Mechanisms of Emotion

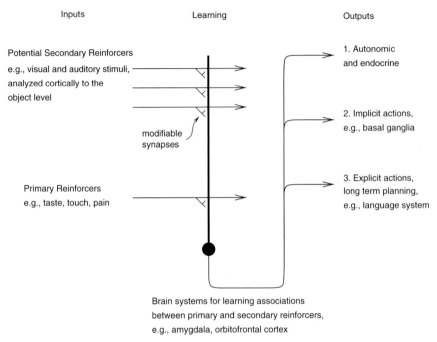

Brain systems for learning associations
between primary and secondary reinforcers,
e.g., amygdala, orbitofrontal cortex

FIGURE 6-14   Schematic diagram showing the organization of brain networks involved in learning reinforcement associations of visual and auditory stimuli. The learning is implemented by pattern-association networks in the amygdala and orbitofrontal cortex. The visual representation provided by the inferior temporal cortex is in an appropriate form for this pattern-association learning, in that information about objects can be read from a population of IT neurons by dot-product neuronal operations.

inputs. This is precisely what pattern associators in the brain need, because they are implemented by neurons that perform, as their generic computation, a dot product of their inputs with their synaptic weight vectors (see Rolls and Deco, 2002; Rolls and Treves, 1998).

A schematic diagram summarizing some of the conclusions reached (Rolls, 1999a, 2005; Rolls and Deco, 2002; Rolls and Treves, 1998) is shown in Figure 6-14. The pathways are shown in more detail in Figure 6-15. The primate inferior temporal visual cortex provides a representation that is independent of reward or punishment and is about objects. The utility of this is that the output of the inferior temporal visual cortex can be used for many memory and related functions (including episodic memory, short-term memory, and reward/punishment memory), independent of whether the visual stimulus is currently rewarding or not. Thus we can learn about objects and place them

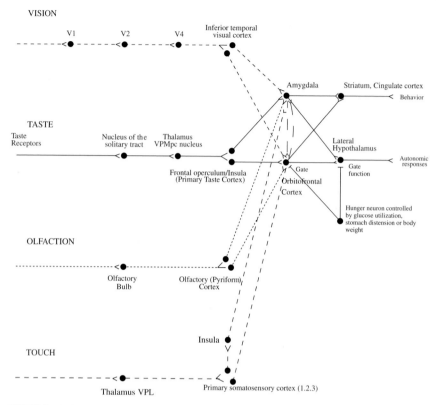

FIGURE 6-15   Diagrammatic representation of some of the connections described in this chapter. V1, striate visual cortex. V2 and V4, cortical visual areas. In primates, sensory analysis proceeds in the visual system as far as the inferior temporal cortex and the primary gustatory cortex; beyond these areas, in, for example, the amygdala and orbitofrontal cortex, the hedonic value of the stimuli and whether they are reinforcing or are associated with reinforcement is represented (see text).

in short-term memory, independent of whether they are currently wanted or not. This is a key feature of brain design. The inferior temporal cortex then projects into two structures, the amygdala and the orbitofrontal cortex, which contain representations of primary (unlearned) reinforcers, such as taste and pain. These two brain regions then learn associations between visual and other previously neutral stimuli and primary reinforcers (Rolls, 1999a, 2005), using what is highly likely to be a pattern-association network, as illustrated in Figure 6-14. A difference between the primate amygdala and orbitofrontal cortex may be that the orbitofrontal cortex is set up to perform reversal of these associations very rapidly, in as little as one trial. Because the amygdala and orbitofrontal cortex represent primary reinforcers and learn associations between these and neutral stimuli, they are key brain regions in emotions (which can be understood as states elicited by reinforcers, that is, rewards and punishers) and in motivational states, such as feeding and drinking (Rolls, 1999a, 2005).

Although pattern association (see Fig. 6-14) is the main mechanism for stimulus–reinforcer association learning in the orbitofrontal cortex and amygdala, there is another mechanism that must be available in the primate orbitofrontal cortex to implement rapid (one-trial) stimulus–reinforcer association reversal, which is important in emotion, for it allows rapid adjustment of emotional responses and behavior based on the reinforcing feedback being received. The need for this is illustrated by the fact that in a Go/NoGo visual discrimination task, if the currently rewarded stimulus is punished on a reversal trial or the very next trial, the primate will choose to make a Go response to the previously punished stimulus, which, it should be noted, has not previously been associated with reward (Deco and Rolls, 2005c; Rolls, 2005; Thorpe, Rolls, and Maddison, 1983). In this reversal learning set, an associative process thus cannot control the behavior. Instead, a rule must be involved, which may specify that one stimulus is currently rewarded and the other is punished. We have built a computational model of this in which a rule attractor biases neurons found in the orbitofrontal cortex (Rolls, 2005; Thorpe, Rolls, and Maddison, 1983) that respond to combinations of the stimulus and whether it is currently associated with reward (see Fig. 6-16). The rule attractor in the model (Deco and Rolls, 2005c) is reversed by using the error neurons also present in the orbitofrontal cortex, which fire only on reversal trials (Rolls, 2005; Thorpe, Rolls, and Maddison, 1983) to quench the rule attractor (i.e., stop it from firing, by acting, for example, on inhibitory interneurons). Then, as a result of a small amount of adaptation in the synapses or neurons in the rule attractor, when activity emerges again from the spontaneous activity in the rule attractor, it is the set of neurons representing the opposite rule that emerges from the noise, and thus the mapping from sensory stimuli to reward neurons is reversed via the effects of the rule attractor on the conditional reward neurons in the intermediate layer of the hierarchical attractor network (Deco and Rolls, 2005c) (see Fig. 6-16).

VI. EFFECTS OF MOOD ON MEMORY AND VISUAL PROCESSING

The current mood state can affect the cognitive evaluation of events or memories (see Blaney, 1986; Rolls and Stringer, 2001). An example is that when people are in a depressed mood, they tend to recall memories that were stored when they were depressed. The recall of depressing memories when depressed can have the effect of perpetuating the depression, and this may be a factor

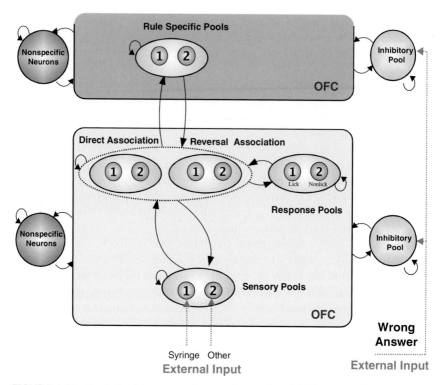

FIGURE 6-16   Cortical architecture or the reward reversal model. There is a rule module **(top)** and a sensory–intermediate neuron–reward module **(below)**. Neurons within each module are fully connected and form attractor states. The sensory–intermediate neuron–reward module consists of three hierarchically organized levels of attractor network, with stronger synaptic connections in the forward- than the back-projection direction. The intermediate level of the sensory–intermediate neuron–reward module contains neurons that respond to combinations of an object and its association with reward or punishment, e.g., object 1–reward (O1R, in the direct association set of pools), and object 1–punishment (O1P in the reversal association set of pools). The rule module acts as a biasing input to bias the competition between the object–reward combination neurons at the intermediate level of the sensory–intermediate neuron–reward module. [After Deco and Rolls (2005c).]

with relevance to the etiology and treatment of depression. A normal function of the effects of mood state on memory recall might be to facilitate continuity in the interpretation of the reinforcing value of events in the environment or in the interpretation of an individual's behavior by others or simply to keep behavior motivated toward a particular goal. Another possibility is that the effects of mood on memory do not have adaptive value but are a consequence of having a general cortical architecture with back-projections. According to the latter hypothesis, the selection pressure is great for leaving the general

Inferior temporal cortex (IT)   Amygdala

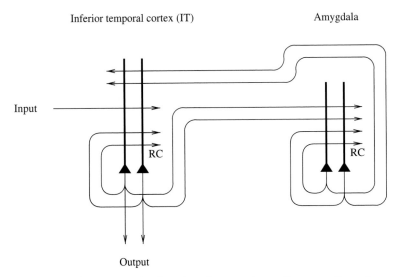

Input

Output

FIGURE 6-17   Architecture used to investigate how mood can affect perception and memory. The IT module represents brain areas, such as the inferior temporal cortex, involved in perception and hippocampus-related cortical areas that have forward connections to regions, such as the amygdala and orbitofrontal cortex, involved in mood. [After Rolls and Stringer (2001).]

architecture operational, rather than trying to find a genetic way to switch off back-projections just for the projections of mood systems back to perceptual systems (cf. Rolls and Stringer, 2000).

Rolls and Stringer (2001) (see also Rolls, 1989, 1999a) have developed a theory of how the effects of mood on memory and perception could be implemented in the brain. The architecture, shown in Figure 6-17, uses the massive back-projections from parts of the brain where mood is represented, such as the orbitofrontal cortex and amygdala, to the cortical areas such as the inferior temporal visual cortex, and hippocampus-related areas (labeled IT in Fig. 6-17) that project into these mood-representing areas (Amaral and Price, 1984; Amaral, Price, Pitkanen, and Carmichael, 1992). The model uses an attractor in the mood module (labeled amygdala in Fig. 6-17), which helps the mood to be an enduring state, and also an attractor in IT. The system is treated as a system of coupled attractors (see Rolls and Deco, 2002), but with an odd twist: Many different perceptual states are associated with any one mood state. Overall, there is a large number of perceptual/memory states and only a few mood states, so there is a many-to-one relation between perceptual/memory states and the associated mood states. The network displays the properties that one would expect (provided the coupling parameters g between the attractors are weak). These include the ability of a perceptual input to trigger a mood state in the "amygdala" module if there is not an existing mood, but greater difficulty to induce a new mood if there is already a strong mood attractor present; and the ability of the mood to affect, via the back-projections, which memories are triggered.

An interesting property that was revealed by the model is that because of the many-to-few mapping of perceptual to mood states, an effect of a mood was that it tended to make all the perceptual or memory states associated with a particular mood more similar then they would otherwise have been. The implication is that the coupling parameter g for the back-projections must be quite weak, for otherwise interference increases in the perceptual/memory module (IT in Fig. 6-17).

VII. CONCLUSION

This chapter has shown how evidence from neurophysiology, the effects of lesions, and computational neuroscience can be combined to generate an understanding of how some memory systems in the brain may operate. Evidence from single-neuron (and multiple single-neuron) recordings is an important part of the approach, for it is at the neuronal level that information is being communicated between the computing elements of the brain, the neurons. This is the case, in that any one cortical neuron typically communicates with on the order of 10,000 other neurons, and this is achieved by the all-or-none spikes traveling relatively long distances along axons to other neurons that are not immediately adjacent to the sending neuron. Thus by recording the spiking activity of neurons, we know all the details of what is represented by neurons in an area and thus what the inputs to the computational system in any one area are and what the outputs from that area are. The computational model then simulates the activity of the different neurons in an area, utilizing evidence on the connectivity of the neurons, the synaptic modification rules, and a theory of what is being computed and how it is being computed, and can be tested by direct comparison with the neuronal activity recorded in a brain area. The computational model is essential, for it is only at the network level that many of the fundamental properties of the system are produced, including generalization, completion, and graceful degradation. Moreover, it is only at the computational neuroscience level that we can understand properties such as the number of memories that can be stored in a brain system, how brain systems interact nonlinearly to produce cognitive effects such as top-down attentional effects, and how enormously complicated computations such as forming view-invariant representations of objects could be achieved by the brain. The functions and properties of the simple networks that are some of the building blocks of neural computation are described by Rolls and Deco (2002) (Chapter 7), by Rolls and Treves (1998), and by Rolls (2007). Approaches to the ways in which different memory systems operate computationally are described by Rolls and Treves (1998) and by Rolls (2007). Ways in which visual computations are performed are described by Rolls and Deco (2002). Ways in which attention operates are described by Rolls and Deco (2002) and Deco and Rolls (2005a). Ways in which networks are involved in decision making are described by Rolls (2005), Deco and Rolls (2005a, 2006), and Rolls (2007).

ACKNOWLEDGMENTS

This research was supported by Medical Research Council Programme Grant PG9826105, by the MRC Interdisciplinary Research Centre for Cognitive Neuroscience, and by the Human Frontier Science Program.

REFERENCES

Abbott, L.F. and Blum, K.I. (1996) Functional significance of long-term potentiation for sequence learning and prediction. *Cerebral Cortex* 6, 406–416.

Abbott, L.F. and Nelson, S.B. (2000) Synaptic plasticity: Taming the beast. *Nature Neuroscience* 3, 1178–1183.

Amaral, D.G. and Price, J.L. (1984) Amygdalo-cortical projections in the monkey (*Macaca fascicularis*). *Journal of Comparative Neurology* 230, 465–496.

Amaral, D.G., Price, J.L., Pitkanen, A., and Carmichael, S.T. (1992) Anatomical organization of the primate amygdaloid complex. In: *The Amygdala* (pp. 1–66), J. P. Aggleton (ed.). New York: Wiley-Liss.

Amari, S. (1977) Dynamics of pattern formation in lateral-inhibition-type neural fields. *Biological Cybernetics* 27, 77–87.

Amit, D.J. (1989) *Modeling Brain Function*. Cambridge, U.K.: Cambridge University Press.

Amit, D.J. and Tsodyks, M.V. (1991) Quantitative study of attractor neural network retrieving at low spike rates. 1: Substrate — spikes, rates and neuronal gain. *Network* 2, 259–273.

Andersen, R.A., Batista, A.P., Snyder, L.H., Buneo, C.A., and Cohen, Y.E. (2000) Programming to look and reach in the posterior parietal cortex. In: *The New Cognitive Neurosciences* (pp. 515–524), M. S. Gazzaniga (ed.). Cambridge, MA: MIT Press.

Asaad, W.F., Rainer, G., and Miller, E.K. (2000) Task-specific neural activity in the primate prefrontal cortex. *Journal of Neurophysiology* 84, 451–459.

Battaglia, F.P. and Treves, A. (1998a) Attractor neural networks storing multiple space representations: A model for hippocampal place fields. *Physical Review E* 58, 7738–7753.

Battaglia, F.P. and Treves, A. (1998b) Stable and rapid recurrent processing in realistic auto-associative memories. *Neural Computation* 10, 431–450.

Blaney, P.H. (1986) Affect and memory: A review. *Psychological Bulletin* 99, 229–246.

Buckley, M.J. and Gaffan, D. (2000) The hippocampus, perirhinal cortex, and memory in the monkey. In: *Brain, Perception, and Memory: Advances in Cognitive Neuroscience* (pp. 279–298); J. J. Bolhuis (ed.). Oxford, U.K.: Oxford University Press.

Burgess, N., Maguire, E.A., and O'Keefe, J. (2002) The human hippocampus and spatial and episodic memory. *Neuron* 35, 625–641.

Burgess, N. and O'Keefe, J. (1996) Neuronal computations underlying the firing of place cells and their role in navigation. *Hippocampus* 6, 749–762.

Burgess, N., Recce, M., and O'Keefe, J. (1994) A model of hippocampal function. *Neural Networks* 7, 1065–1081.

Cahusac, P.M.B., Rolls, E.T., Miyashita, Y., and Niki, H. (1993) Modification of the responses of hippocampal neurons in the monkey during the learning of a conditional spatial response task. *Hippocampus* 3, 29–42.

Carpenter, R.H.S. and Williams, M.L. (1995) Neural computation of log likelihood in control of saccadic eye movements. *Nature* 377, 59–62.

Cassaday, H.J. and Rawlins, J.N. (1997) The hippocampus, objects, and their contexts. *Behavioral Neuroscience* 111, 1228–1244.

Chelazzi, L., Duncan, J., Miller, E.K., and Desimone, R. (1998) Responses of neurons in inferior temporal cortex during memory-guided visual search. *Journal of Neurophysiology* 80, 2918–2940.

Chelazzi, L., Miller, E.K., Duncan, J., and Desimone, R.E. (1993) A neural basis for visual search in inferior temporal cortex. *Nature* 363, 345–347.

Crane, J. and Milner, B. (2005) What went where? Impaired object-location learning in patients with right hippocampal lesions. *Hippocampus* 15, 216–231.

Day, M., Langston, R., and Morris, R.G. (2003) Glutamate-receptor-mediated encoding and retrieval of paired-associate learning. *Nature* 424, 205–209.

de Araujo, I.E.T., Rolls, E.T., and Stringer, S.M. (2001) A view model which accounts for the spatial fields of hippocampal primate spatial view cells and rat place cells. *Hippocampus* 11, 699–706.

Deco, G. and Rolls, E.T. (2003) Attention and working memory: A dynamical model of neuronal activity in the prefrontal cortex. *European Journal of Neuroscience* 18, 2374–2390.

Deco, G. and Rolls, E.T. (2004) A neurodynamical cortical model of visual attention and invariant object recognition. *Vision Research* 44, 621–644.

Deco, G. and Rolls, E.T. (2005a) Attention, short-term memory, and action selection: A unifying theory. *Progress in Neurobiology* 76, 236–256.

Deco, G. and Rolls, E.T. (2005b) Neurodynamics of biased competition and cooperation for attention: A model with spiking neurons. *Journal of Neurophysiology* 94, 295–313.

Deco, G. and Rolls, E.T. (2005c) Synaptic and spiking dynamics underlying reward reversal in orbitofrontal cortex. *Cerebral Cortex* 15, 15–30.

Deco, G. and Rolls, E.T. (2006) A neurophysiological model of decision-making and Weber's law. *European Journal of Neuroscience* 24, 901–916.

Desimone, R. (1996) Neural mechanisms for visual memory and their role in attention. *Proceedings of the National Academy of Science, USA* 93, 13494–13499.

Eichenbaum, H. (1997) Declarative memory: Insights from cognitive neurobiology. *Annual Review of Psychology* 48, 547–572.

Elliffe, M.C.M., Rolls, E.T., and Stringer, S.M. (2002) Invariant recognition of feature combinations in the visual system. *Biological Cybernetics* 86, 59–71.

Epstein, R. and Kanwisher, N. (1998) A cortical representation of the local visual environment. *Nature* 392, 598–601.

Funahashi, S., Bruce, C.J., and Goldman-Rakic, P.S. (1989) Mnemonic coding of visual space in monkey dorsolateral prefrontal cortex. *Journal of Neurophysiology* 61, 331–349.

Fuster, J.M. (1997) Network memory. *Trends in Neurosciences* 20, 451–459.

Fuster, J.M. (2000) *Memory Systems in the Brain*. New York: Raven Press.

Gaffan, D. (1994) Scene-specific memory for objects: A model of episodic memory impairment in monkeys with fornix transection. *Journal of Cognitive Neuroscience* 6, 305–320.

Gaffan, D. and Harrison, S. (1989a) A comparison of the effects of fornix section and sulcus principalis ablation upon spatial learning by monkeys. *Behavioral Brain Research* 31, 207–220.

Gaffan, D. and Harrison, S. (1989b) Place memory and scene memory: Effects of fornix transection in the monkey. *Experimental Brain Research* 74, 202–212.

Gaffan, D. and Saunders, R.C. (1985) Running recognition of configural stimuli by fornix transected monkeys. *Quarterly Journal of Experimental Psychology* 37B, 61–71.

Georges-François, P., Rolls, E.T., and Robertson, R.G. (1999) Spatial view cells in the primate hippocampus: Allocentric view not head direction or eye position or place. *Cerebral Cortex* 9, 197–212.

Gilbert, P.E., Kesner, R.P., and DeCoteau, W.E. (1998) Memory for spatial location: Role of the hippocampus in mediating spatial pattern separation. *Journal of Neuroscience* 18, 804–810.

Gilbert, P.E., Kesner, R.P., and Lee, I. (2001) Dissociating hippocampal subregions: Double dissociation between dentate gyrus and CA1. *Hippocampus* 11, 626–636.

Gold, A.E. and Kesner, R.P. (2005) The role of the CA3 subregion of the dorsal hippocampus in spatial pattern completion in the rat. *Hippocampus* 15, 808–814.

Goldman-Rakic, P.S. (1996) The prefrontal landscape: Implications of functional architecture for understanding human mentation and the central executive. *Philosophical Transactions of the Royal Society B* 351, 1445–1453.

Goodrich-Hunsaker, N.J., Hunsaker, M.R., and Kesner, R.P. (2005) Effects of hippocampus subregional lesions for metric and topological spatial information processing. *Society for Neuroscience Abstracts,* in press.

Hafting, T., Fyhn, M., Molden, S., Moser, M.B., and Moser, E.I. (2005) Microstructure of a spatial map in the entorhinal cortex. *Nature* 436, 801–806.

Hasselmo, M.E., Schnell, E., and Barkai, E. (1995) Dynamics of learning and recall at excitatory recurrent synapses and cholinergic modulation in rat hippocampal region CA3. *Journal of Neuroscience* 15, 5249–5262.

Hertz, J., Krogh, A., and Palmer, R.G. (1991) *An Introduction to the Theory of Neural Computation.* Wokingham, U.K.: Addison-Wesley.

Hopfield, J.J. (1982) Neural networks and physical systems with emergent collective computational abilities. *Proceedings of the National Academy of Science USA* 79, 2554–2558.

Jarrard, E.L. (1993) On the role of the hippocampus in learning and memory in the rat. *Behavioral and Neural Biology* 60, 9–26.

Jung, M.W. and McNaughton, B.L. (1993) Spatial selectivity of unit activity in the hippocampal granular layer. *Hippocampus* 3, 165–182.

Kesner, R.P. (1998) Neural mediation of memory for time: Role of hippocampus and medial prefrontal cortex. *Psychological Bulletin Reviews* 5, 585–596.

Kesner, R.P., Lee, I., and Gilbert, P. (2004) A behavioral assessment of hippocampal function based on a subregional analysis. *Review of Neuroscience* 15, 333–351.

Kesner, R.P. and Rolls, E.T. (2001) Role of long-term synaptic modification in short-term memory. *Hippocampus* 11, 240–250.

Kubie, J.L. and Muller, R.U. (1991) Multiple representations in the hippocampus. *Hippocampus* 1, 240–242.

Kuhn, R. (1990) Statistical methods of neural networks near saturation. In: *Statistical Mechanics of Neural Networks*, L. Garrido (ed.). Berlin: Springer-Verlag.

Kuhn, R., Bos, S., and van Hemmen, J.L. (1991) Statistical mechanics for networks of graded response neurons. *Physical Review A* 243, 2084–2087.

Lassalle, J.M., Bataille, T., and Halley, H. (2000) Reversible inactivation of the hippocampal mossy fiber synapses in mice impairs spatial learning, but neither consolidation nor memory retrieval, in the Morris navigation task. *Neurobiology of Learning and Memory* 73, 243–257.

Levy, W.B., Wu, X., and Baxter, R.A. (1995) Unification of hippocampal function via computational/encoding considerations. *International Journal of Neural Systems* 6, 71–80.

Malkova, L., Bachevalier, J., Mishkin, M., and Saunders, R.C. (2001) Neurotoxic lesions of perirhinal cortex impair visual recognition memory in rhesus monkeys. *Neuroreport* 12, 1913–1917.

Malkova, L. and Mishkin, M. (2003) One-trial memory for object-place associations after separate lesions of hippocampus and posterior parahippocampal region in the monkey. *Journal of Neuroscience* 23, 1956–1965.

Markram, H., Pikus, D., Gupta, A., and Tsodyks, M.V. (1998) Information processing with frequency-dependent synaptic connections. *Neuropharmacology* 37, 489–500.

Markus, E.J., Qin, Y.L., Leonard, B., Skaggs, W., McNaughton, B.L., and Barnes, C.A. (1995) Interactions between location and task affect the spatial and directional firing of hippocampal neurons. *Journal of Neuroscience* 15, 7079–7094.

Marr, D. (1971) Simple memory: A theory for archicortex. *Philosophical Transactions of the Royal Society of London B* 262, 23–81.

Martin, S.J., Grimwood, P.D., and Morris, R.G. (2000) Synaptic plasticity and memory: An evaluation of the hypothesis. *Annual Review of Neuroscience* 23, 649–711.

McClelland, J.L., McNaughton, B.L., and O'Reilly, R.C. (1995) Why there are complementary learning systems in the hippocampus and neocortex: Insights from the successes and failures of connectionist models of learning and memory. *Psychological Review* 102, 419–457.

McNaughton, B.L., Barnes, C.A., Meltzer, J., and Sutherland, R.J. (1989) Hippocampal granule cells are necessary for normal spatial learning but not for spatially selective pyramidal cell discharge. *Experimental Brain Research* 76, 485–496.

McNaughton, B.L., Barnes, C.A., and O'Keefe, J. (1983) The contributions of position, direction, and velocity to single unit activity in the hippocampus of freely-moving rats. *Experimental Brain Research* 52, 41–49.

Miller, E.K. and Desimone, R. (1994) Parallel neuronal mechanisms for short-term memory. *Science* 263, 520–522.

Miller, E.K., Erickson, C., and Desimone, R. (1996) Neural mechanisms of visual working memory in prefrontal cortex of the macaque. *Journal of Neuroscience* 16, 5154–5167.

Miller, E.K., Li, L., and Desimone, R. (1993) Activity of neurons in anterior inferior temporal cortex during a short-term memory task. *Journal of Neuroscience* 13, 1460–1478.

Minai, A.A. and Levy, W.B. (1993) Sequence learning in a single trial. *International Neural Network Society World Congress of Neural Networks* 2, 505–508.

Miyashita, Y., Rolls, E.T., Cahusac, P.M., Niki, H., and Feigenbaum, J.D. (1989) Activity of hippocampal formation neurons in the monkey related to a conditional spatial response task. *Journal of Neurophysiology* 61, 669–678.

Mizumori, S.J., Perez, G.M., Alvarado, M.C., Barnes, C.A., and McNaughton, B.L. (1990) Reversible inactivation of the medial septum differentially affects two forms of learning in rats. *Brain Research* 528, 12–20.

Muller, R.U., Kubie, J.L., Bostock, E.M., Taube, J.S., and Quirk, G.J. (1991) Spatial firing correlates of neurons in the hippocampal formation of freely moving rats. In: *Brain and Space* (pp. 296–333), J. Paillard (ed.). Oxford, U.K.: Oxford University Press.

Muller, R.U., Ranck, J.B., Jr., and Taube, J.S. (1996) Head direction cells: Properties and functional significance. *Current Opinion in Neurobiology* 6, 196–206.

Murray, E.A., Baxter, M.G., and Gaffan, D. (1998) Monkeys with rhinal cortex damage or neurotoxic hippocampal lesions are impaired on spatial scene learning and object reversals. *Behavioral Neuroscience* 112, 1291–1303.

Nakazawa, K., Quirk, M.C., Chitwood, R.A., Watanabe, M., Yeckel, M.F., Sun, L.D., Kato, A., Carr, C.A., Johnston, D., Wilson, M.A., and Tonegawa, S. (2002) Requirement for hippocampal CA3 NMDA receptors in associative memory recall. *Science* 297, 211–218.

O'Keefe, J. (1979) A review of the hippocampal place cells. *Progress in Neurobiology* 13, 419–439.

O'Keefe, J. (1984) Spatial memory within and without the hippocampal system. In: *Neurobiology of the Hippocampus* (pp. 375–403), W. Seifert (ed.). London: Academic Press.

O'Keefe, J. (1990) A computational theory of the hippocampal cognitive map. *Progress in Brain Research* 83, 301–312.

O'Keefe, J. (1991) The hippocampal cognitive map and navigational strategies. In: *Brain and Space* (pp. 273–295), J. Paillard (ed.). Oxford, U.K.: Oxford University Press.

O'Keefe, J., Burgess, N., Donnett, J.G., Jeffery, K.J., and Maguire, E.A. (1998) Place cells, navigational accuracy, and the human hippocampus. *Philosophical Transactions of the Royal Society B* 353, 1333–1340.

O'Keefe, J. and Dostrovsky, J. (1971) The hippocampus as a spatial map: Preliminary evidence from unit activity in the freely moving rat. *Brain Research* 34, 171–175.

O'Keefe, J. and Nadel, L. (1978) *The Hippocampus as a Cognitive Map*. Oxford, U.K.: Clarendon Press.

O'Mara, S.M., Rolls, E.T., Berthoz, A., and Kesner, R.P. (1994) Neurons responding to whole-body motion in the primate hippocampus. *Journal of Neuroscience* 14, 6511–6523.

Panzeri, S., Rolls, E.T., Battaglia, F., and Lavis, R. (2001) Speed of information retrieval in multilayer networks of integrate-and-fire neurons. *Network: Computation in Neural Systems* 12, 423–440.

Parkinson, J.K., Murray, E.A., and Mishkin, M. (1988) A selective mnemonic role for the hippocampus in monkeys: Memory for the location of objects. *Journal of Neuroscience* 8, 4059–4167.

Petrides, M. (1985) Deficits on conditional associative-learning tasks after frontal- and temporal-lobe lesions in man. *Neuropsychologia* 23, 601–614.

Poucet, B. (1989) Object exploration, habituation, and response to a spatial change in rats following septal or medial frontal cortical damage. *Behavioral Neuroscience* 103, 1009–1016.

Ranck, J.B.J. (1985) Head direction cells in the deep cell layer of dorsolateral presubiculum in freely moving rats. In: *Electrical Activity of the Archicortex*, G. Buzsaki and C.H. Vanderwolf (eds.). Budapest: Akademiai Kiado.

Rao, S.C., Rainer, G., and Miller, E.K. (1997) Integration of what and where in the primate prefrontal cortex. *Science* 276, 821–824.

Ratcliff, R., Van Zandt, T., and McKoon, G. (1999) Connectionist and diffusion models of reaction time. *Psychological Review* 106, 261–300.

Rawlins, J.N.P. (1985) Associations across time: The hippocampus as a temporary memory store. *Behavioral and Brain Sciences* 8, 479–496.

Renart, A., Moreno, R., de la Rocha, J., Parga, N., and Rolls, E.T. (2001) A model of the IT-PF network in object working memory which includes balanced persistent activity and tuned inhibition. *Neurocomputing* 38–40, 1525–1531.

Renart, A., Parga, N., and Rolls, E.T. (1999a) Associative memory properties of multiple cortical modules. *Network* 10, 237–255.

Renart, A., Parga, N., and Rolls, E.T. (1999b) Backprojections in the cerebral cortex: Implications for memory storage. *Neural Computation* 11, 1349–1388.

Renart, A., Parga, N., and Rolls, E.T. (2000) A recurrent model of the interaction between the prefrontal cortex and inferior temporal cortex in delay memory tasks. In: *Advances in Neural Information Processing Systems,* Vol. 12 (pp. 171–177), S.A. Solla, T.K. Leen, and K.-R. Mueller (eds.). Cambridge, MA: MIT Press.

Robertson, R.G., Rolls, E.T., and Georges-François, P. (1998) Spatial view cells in the primate hippocampus: Effects of removal of view details. *Journal of Neurophysiology* 79, 1145–1156.

Robertson, R.G., Rolls, E.T., Georges-François, P., and Panzeri, S. (1999) Head direction cells in the primate pre-subiculum. *Hippocampus* 9, 206–219.

Rolls, E.T. (1987) Information representation, processing and storage in the brain: Analysis at the single neuron level. In: *The Neural and Molecular Bases of Learning* (pp. 503–540), J.-P. Changeux and M. Konishi (eds.). Chichester, U.K.: Wiley.

Rolls, E.T. (1989) Functions of neuronal networks in the hippocampus and neocortex in memory. In: *Neural Models of Plasticity: Experimental and Theoretical Approaches* (pp. 240–265), J.H. Byrne and W.O. Berry (eds.). San Diego: Academic Press.

Rolls, E.T. (1992) Neurophysiological mechanisms underlying face processing within and beyond the temporal cortical visual areas. *Philosophical Transactions of the Royal Society of London B* 335, 11–21.

Rolls, E.T. (1996) A theory of hippocampal function in memory. *Hippocampus* 6, 601–620.

Rolls, E.T. (1999a) *The Brain and Emotion*. Oxford, U.K.: Oxford University Press.

Rolls, E.T. (1999b) The representation of space in the primate hippocampus, and its role in memory. In: *The Hippocampal and Parietal Foundations of Spatial Cognition* (pp. 320–344), N. Burgess, K.J. Jeffrey, and J. O'Keefe (eds.). Oxford, U.K.: Oxford Univerisity Press.

Rolls, E.T. (1999c) Spatial view cells and the representation of place in the primate hippocampus. *Hippocampus* 9, 467–480.

Rolls, E.T. (2000a) Hippocampo-cortical and cortico-cortical backprojections. *Hippocampus* 10, 380–388.

Rolls, E.T. (2000b) Memory systems in the brain. *Annual Review of Psychology* 51, 599–630.

Rolls, E.T. (2005) *Emotion Explained*. Oxford, U.K.: Oxford University Press.

Rolls, E.T. (2007) Memory, Attention and decision-Making. Oxford University Press: Oxford.

Rolls, E.T. and Deco, G. (2002) *Computational Neuroscience of Vision*. Oxford, U.K.: Oxford University Press.

Rolls, E.T. and Kesner, R.P. (2006) A computational theory of hippocampal function, and empirical tests of the theory. *Progress in Neurobiology* 79, 1–48.

Rolls, E.T. and Milward, T. (2000) A model of invariant object recognition in the visual system: Learning rules, activation functions, lateral inhibition, and information-based performance measures. *Neural Computation* 12, 2547–2572.

Rolls, E.T., Miyashita, Y., Cahusac, P.M.B., Kesner, R.P., Niki, H., Feigenbaum, J., and Bach, L. (1989) Hippocampal neurons in the monkey with activity related to the place in which a stimulus is shown. *Journal of Neuroscience* 9, 1835–1845.

Rolls, E.T., Robertson, R.G., and Georges-François, P. (1997) Spatial view cells in the primate hippocampus. *European Journal of Neuroscience* 9, 1789–1794.

Rolls, E.T. and Stringer, S.M. (2000) On the design of neural networks in the brain by genetic evolution. *Progress in Neurobiology* 61, 557–579.

Rolls, E.T. and Stringer, S.M. (2001) A model of the interaction between mood and memory. *Network: Computation in Neural Systems* 12, 111–129.

Rolls, E.T. and Stringer, S.M. (2005) Spatial view cells in the hippocampus, and their idiothetic update based on place and head direction. *Neural Networks* 18, 1229–1241.

Rolls, E.T., Stringer, S.M., and Elliot, T. (2006) Entorhinal cortex grid cells can map to hippocampal place cells by competitive learning. *Network: Computation in Neural Systems* 17, 447–465.

Rolls, E.T., Stringer, S.M., and Trappenberg, T.P. (2002) A unified model of spatial and episodic memory. *Proceedings of the Royal Society of London B* 269, 1087–1093.

Rolls, E.T. and Tovee, M.J. (1994) Processing speed in the cerebral cortex and the neurophysiology of visual masking. *Proceedings of the Royal Society of London B* 257, 9–15.

Rolls, E.T., Tovee, M.J., and Panzeri, S. (1999) The neurophysiology of backward visual masking: Information analysis. *Journal of Cognitive Neuroscience* 11, 335–346.

Rolls, E.T., Tovee, M.J., Purcell, D.G., Stewart, A.L., and Azzopardi, P. (1994) The responses of neurons in the temporal cortex of primates, and face identification and detection. *Experimental Brain Research* 101, 473–484.

Rolls, E.T. and Treves, A. (1998) *Neural Networks and Brain Function*. Oxford, U.K.: Oxford University Press.

Rolls, E.T., Treves, A., Robertson, R.G., Georges-François, P., and Panzeri, S. (1998) Information about spatial view in an ensemble of primate hippocampal cells. *Journal of Neurophysiology* 79, 1797–1813.

Rolls, E.T. and Xiang, J.-Z. (2005) Reward-spatial view representations and learning in the hippocampus. *Journal of Neuroscience* 25, 6167– 6174.

Rolls, E.T. and Xiang, J-Z. (2006) Spatial view cells in the primate hippocampus, and memory recall. *Reviews in the Neurosciences* 17, 175–200.

Rolls, E.T., Xiang, J.-Z., and Franco, L. (2005) Object, space and object-space representations in the primate hippocampus. *Journal of Neurophysiology* 94, 833–844.

Rupniak, N.M.J. and Gaffan, D. (1987) Monkey hippocampus and learning about spatially directed movements. *Journal of Neuroscience* 7, 2331–2337.

Samsonovich, A. and McNaughton, B.L. (1997) Path integration and cognitive mapping in a continuous attractor neural network model. *Journal of Neuroscience* 17, 5900–5920.

Shallice, T. and Burgess, P. (1996) The domain of supervisory processes and temporal organization of behaviour. *Philosophical Transactions of the Royal Society B* 351, 1405–1411.

Shiino, M. and Fukai, T. (1990) Replica-symmetric theory of the non-linear analogue neural networks. *Journal of Physics A* 23, 1009–1017.

Skaggs, W.E., Knierim, J.J., Kudrimoti, H.S., and McNaughton, B.L. (1995) A model of the neural basis of the rat's sense of direction. In: *Advances in Neural Information Processing Systems,* Vol. 7 (pp. 173–180), G. Tesauro, D.S. Touretzky, and T.K. Leen (eds.). Cambridge, MA: MIT Press.

Smith, M.L. and Milner, B. (1981) The role of the right hippocampus in the recall of spatial location. *Neuropsychologia* 19, 781–793.

Spiridon, M., Fischl, B., and Kanwisher, N. (2005) Location and spatial profile of category-specific regions in human extrastriate cortex. *Human Brain Mapping,* in press.

Squire, L.R. and Knowlton, B.J. (2000) The medial temporal lobe, the hippocampus, and the memory systems of the brain. In: *The New Cognitive Neurosciences,* Vol. X, Ch. 53 (pp. 765–779), M.S. Gazzaniga (ed.). Cambridge, MA: MIT Press.

Stringer, S.M. and Rolls, E.T. (2002) Invariant object recognition in the visual system with novel views of 3D objects. *Neural Computation* 14, 2585–2596.

Stringer, S.M., Rolls, E.T., and Trappenberg, T.P. (2004) Self-organizing continuous attractor networks with multiple activity packets, and the representation of space. *Neural Networks* 17, 5–27.

Stringer, S.M., Rolls, E.T., and Trappenberg, T.P. (2005) Self-organizing continuous attractor network models of hippocampal spatial view cells. *Neurobiology of Learning and Memory* 83, 79–92.

Stringer, S.M., Rolls, E.T., Trappenberg, T.P., and Araujo, I.E.T. (2002) Self-organizing continuous attractor networks and path integration. Two-dimensional models of place cells. *Network: Computation in Neural Systems* 13, 429–446.

Stringer, S.M., Trappenberg, T.P., Rolls, E.T., and Araujo, I.E.T. (2002) Self-organizing continuous attractor networks and path integration: One-dimensional models of head direction cells. *Network: Computation in Neural Systems* 13, 217–242.

Suzuki, W.A., Miller, E.K., and Desimone, R. (1997) Object and place memory in the macaque entorhinal cortex. *Journal of Neurophysiology* 78, 1062–1081.

Taube, J.S., Goodridge, J.P., Golob, E.J., Dudchenko, P.A., and Stackman, R.W. (1996) Processing the head direction signal: A review and commentary. *Brain Research Bulletin* 40, 477–486.

Taube, J.S., Muller, R.U., and Ranck, J.B.J. (1990) Head-direction cells recorded from the postsubiculum in freely moving rats 1: Description and quantitative analysis. *Journal of Neuroscience* 10, 420–435.

Taylor, J.G. (1999) Neural "bubble" dynamics in two dimensions: Foundations. *Biological Cybernetics* 80, 393–409.

Thorpe, S.J., Rolls, E.T., and Maddison, S. (1983) Neuronal activity in the orbitofrontal cortex of the behaving monkey. *Experimental Brain Research* 49, 93–115.

Treves, A. (1993) Mean-field analysis of neuronal spike dynamics. *Network* 4, 259–284.

Treves, A. and Rolls, E.T. (1991) What determines the capacity of autoassociative memories in the brain? *Network* 2, 371–397.

Treves, A. and Rolls, E.T. (1992) Computational constraints suggest the need for two distinct input systems to the hippocampal CA3 network. *Hippocampus* 2, 189–199.

Treves, A. and Rolls, E.T. (1994) A computational analysis of the role of the hippocampus in memory. *Hippocampus* 4, 374–391.

Usher, M. and McClelland, J.L. (2001) The time course of perceptual choice: The leaky, competing accumulator model. *Psychological Review* 108, 550–592.

Wallis, G. and Rolls, E.T. (1997) Invariant face and object recognition in the visual system. *Progress in Neurobiology* 51, 167–194.

Wang, X.J. (2002) Probabilistic decision making by slow reverberation in cortical circuits. *Neuron* 36, 955–968.

Warthen, M.W. and Kesner, R.P. (2005) CA3 involvement in one-trial short-term memory for cued recall object-place associations, in preparation.

Wilson, F.A.W., O'Scalaidhe, S.P.O., and Goldman-Rakic, P.S. (1993) Dissociation of object and spatial processing domains in primate prefrontal cortex. *Science* 260, 1955–1958.

Wilson, M.A. and McNaughton, B.L. (1993) Dynamics of the hippocampal ensemble code for space. *Science* 261, 1055–1058.

Wu, X., Baxter, R.A., and Levy, W.B. (1996) Context codes and the effect of noisy learning on a simplified hippocampal CA3 model. *Biological Cybernetics* 74, 159–165.

Zhang, K. (1996) Representation of spatial orientation by the intrinsic dynamics of the head-direction cell ensemble: A theory. *Journal of Neuroscience* 16, 2112–2126.

Zola-Morgan, S., Squire, L.R., Amaral, D.G., and Suzuki, W.A. (1989) Lesions of perirhinal and parahippocampal cortex that spare the amygdala and hippocampal formation produce severe memory impairment. *Journal of Neuroscience* 9, 4355–4370.

Zola-Morgan, S., Squire, L.R., and Ramus, S.J. (1994) Severity of memory impairment in monkeys as a function of locus and extent of damage within the medial temporal lobe memory system. *Hippocampus* 4, 483–494.

CHAPTER 7

Modulation of Learning and Memory by Adrenal and Ovarian Hormones

Donna L. Korol

Department of Psychology, Neuroscience Program, and Institute for Genomic Biology, University of Illinois, Champaign, IL 61820

Paul E. Gold

Departments of Psychology and Psychiatry, Neuroscience Program, and Institute for Genomic Biology, University of Illinois, Champaign, IL 61820

I. INTRODUCTION

Hormones have a host of effects on physiological functions, many of which include behavioral actions. Stress hormones not only shift glucose utilization from smooth to striate muscle but also dispose an organism toward classic fight-or-flight responses. Reproductive hormones not only prepare sex organs for reproduction but also dispose an organism toward mating behaviors. Gut hormones not only regulate digestive processes but also dispose an organism toward or away from feeding behaviors. With hormonal regulation of multiple behaviors, it is perhaps not surprising to note that hormones also have effects on learning and memory. Interestingly, the classes of learning and memory effects influenced by a particular hormone include cognitive actions associated with other behavioral functions of the hormone, e.g., stress, reproduction, and digestion from the foregoing list, but also include a wide range of less intuitive effects. For example, stress hormones such as epinephrine and corticosterone influence learning and memory for avoidance tasks, well within the genre of stress, but these hormones also have important effects on learning and memory in appetitive and other nonstressful tasks (for reviews see de Kloet, 2000;

Neurobiology of Learning and Memory, Second Edition

Roozendaal, 2002; Gold and McCarty, 1995). Reproductive hormones such as estrogen also have potent influences on learning in avoidance and appetitive tasks that, on their face, have little direct role in reproduction (Korol, 2004; Dohanich, 2002).

This chapter focuses on the effects on learning and memory of the stress-related adrenal hormones, epinephrine and corticosterone, and the reproduction-related ovarian hormones, estrogen and progesterone. There is much known about each of these hormones in terms of their actions on and via different memory systems in the brain and also in terms of the neurobiological mechanisms by which the hormones influence learning, memory, and neural plasticity.

Before describing experimental findings linking the actions of these hormones to memory, we present some general ideas that guide the later sections of this chapter. Hormones can have both organizational and activational influences on the brain. Generally, organizational effects occur early and act on neural development, leaving lasting changes in brain structure and function and behavior that alter subsequent responses to stimuli, including the hormones themselves that produce the change. In contrast, activational effects of hormones refer mainly to action resulting from the presence of the hormone. In classical terms, organizational effects occur during developmental stages and activational effects occur during adulthood. However, current findings that exposure to hormones in adults can alter neuronal morphology (for review: McEwen, 2001, 2002), among other effects, certainly obscure this distinction between organizational and activational effects.

Thus, while organizational influences on the brain are traditionally studied in the context of development, there are now many demonstrations of hormonal organizational effects, i.e., long-lasting changes in brain structure and function, in adult organisms. Findings that hormones can have both short- and long-lasting effects on learning and memory fit two different views of what the role of hormones is in regulating these processes. One interpretation of demonstrations that hormones administered near the time of learning can enhance later memory for the experience is that the hormones modulate memory formation (Gold and McGaugh, 1975; McGaugh et al., 2002; Gold, 1995, 2005). Modulation of memory involves processes acting on the mechanisms of memory formation or consolidation to enhance or to impair those mechanisms. Results like this suggest that hormonal modulation of memory augments neural plasticity to enhance memory formation.

While both corticosterone and estrogen modulate memory in this sense, these steroids also have additional slower effects on memory that are somewhat different. These hormones can apparently bias the use of one strategy over another to solve a task, e.g., under one condition favoring the use of place strategies sensitive to hippocampal disruption, and under another condition favoring the use of response strategies sensitive to striatal disruption (Kim et al., 2001; Korol and Kolo, 2002; Chapa et al., 2005). These changes in strategy are reflected in differences in the specific responses expressed during tests of learning. Hormonal control over strategy selection can be viewed as a form of metamodulation, in which modulation of neural plasticity can be differentially up- and down-regulated in different neural systems, thereby changing the cognitive structure of what is learned (Korol, 2004).

These general comments are illustrated within the discussion of the stress-related hormones, epinephrine and corticosterone, and the sex-related hormones, estrogen and progesterone, considered in this chapter.

II. STRESS HORMONES AND MEMORY

A. Epinephrine

Epinephrine is perhaps the hormone best studied as a modulator of memory. Epinephrine is released from the adrenal medulla into blood in response to stress (cf. Wortsman, 2002; Wurtman, 2002) and, importantly, is also released in rats in response to handling or exposure to a novel compartment. The hormone is released into blood in a graded manner, from low levels in response to the mild arousal of gentle handling, to placement in a novel environment, to foot shock increasing in a monotonic manner by shock intensity, to the extremely high levels released during immersion in water, as in the swim task (cf. Mabry et al., 1995).

The design and results of an early experiment illustrate modulation of memory by epinephrine (Gold and van Buskirk, 1975). The task used was a one-trial inhibitory (passive) avoidance task in which rats were trained to avoid crossing from a well-lit start compartment into the end of a dark compartment where shock was previously administered. In this type of task, it is not surprising that a mild shock results in smaller increases in latency on the test trial than does a more intense shock. An increase in the intensity of the shock results in better memory, certainly in part because of the increased aversiveness. The key question regarding hormonal modulation of memory is whether the increase in epinephrine release that accompanies increased shock intensity can enhance memory processing by mechanisms separate from changes in aversiveness of the shock. To address this question, rats were trained with a mild shock and were then removed from the training apparatus and given an injection of epinephrine, at one of several doses and times after training. Memory was tested 24 hr later, well after circulating epinephrine levels had returned to baseline. A single injection of intermediate doses of epinephrine administered immediately after training enhanced 24-hr memory (Fig. 7-1, left panel).

As seen for many hormonal and drug treatments that enhance memory (Koob, 1991; Gold, 2006), the dose–response curve had an inverted-U shape

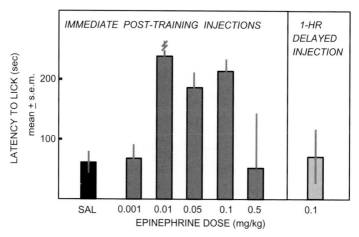

FIGURE 7-1   Effects of epinephrine on memory. Rats received a single injection of epinephrine after training and were tested for memory 24 hr later. Note that epinephrine enhanced 24-hr memory in an inverted-U dose-dependent manner. The effects of epinephrine on memory decreased as the time between training and injection increased. [From Gold and van Buskirk (1975).]

in which low and high doses did not enhance memory; in other experiments, high doses of epinephrine impair memory, i.e., produce retrograde amnesia. One implication of the inverted-U dose–response curve is that the same dose of epinephrine might enhance memory for training under low-foot-shock conditions but impair memory under high-foot-shock conditions. This pattern of results is evident for epinephrine and many other treatments that modulate the formation of new memories (Gold, 2006). Also, the epinephrine doses that enhance memory result in elevations in circulating epinephrine levels that match quite well those seen after training with a higher-intensity foot shock (McCarty and Gold, 1996). The effects of epinephrine on later memory are time dependent, diminishing as a function of time after training (Fig. 7-1, right panel), suggesting that the enhancement of memory reflects actions on the processes of memory formation.

Epinephrine enhances learning and memory in a wide range of tasks, including avoidance, extinction of avoidance, and spatial memory tasks (Gold, 1995). Considerable effort has gone into understanding the mechanisms by which the pervasive effects of epinephrine on memory are manifested. Because of an effective blood–brain barrier mechanism, circulating epinephrine does not appear to have significant direct access to the brain (Axelrod et al., 1959). Therefore, the initial epinephrine signal transduction mechanism relevant to enhancement of memory is likely to be outside the brain.

Two main lines of research have explored peripheral mechanisms that mediate epinephrine effects on memory formation. One has examined the role of vagal afferents to the brain as a mediator of these effects of epinephrine. The vagus nerve contains β-adrenergic receptors, opening the possibility that the nerve conveys to the brain information that circulating epinephrine levels are high. Consistent with this view, drugs that block or activate the brain region that receives vagal input, the nucleus tractus solitarius, block or mimic epinephrine effects on memory and on release of norepinephrine in the amygdala (Williams et al., 1998; Miyashita and Williams, 2004). Vagotomy blocks the effects on memory of some treatments, including cholecystokinin-A (Lemaire et al., 1994; Flood et al., 1987), substance P (Tomaz and Nogueira, 1997), and bombesin and gastrin-related peptide (Flood and Morley, 1988), though to our knowledge there are no direct tests of vagotomy effects on epinephrine enhancement of memory. The potential to do so is suggested, however, by findings that vagotomy blocks epinephrine suppression of brain seizures (Krahl et al., 2000).

A second set of experiments has demonstrated that epinephrine effects on memory are mediated, at least in part, by increases in blood glucose levels subsequent to epinephrine release or administration (for reviews see: McNay and Gold, 2002; Messier, 2004; Gold, 2005). A classic physiological response to increases in circulating epinephrine levels is the liberation of glucose from hepatic stores, resulting in increases in blood glucose levels. Like epinephrine, systemic administration of glucose enhances memory in rodents for a wide range of tasks. The dose–response function has an inverted-W shape, with one peak of memory enhancement seen at 100–300 mg/kg and another at 1–3 g/kg. The underlying biology appears to be different for the two peaks. The higher peak dose is not effective in vagotomized rats (Fig. 7-2;

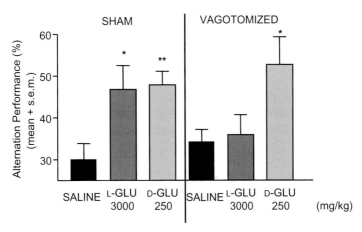

FIGURE 7-2   Vagotomy effects on memory enhancement with D- and L-glucose. Vagotomy blocked the enhancement of memory by L-glucose but not by D-glucose. [From Talley et al. (2002).]

Talley et al., 2002) or in rats after coeliac ganglion removal (White, 1991). However, the lower peak dose remains effective in vagotomized rats (Talley et al., 2002); efficacy of the lower dose after coeliac ganglion removal has not been tested. Of interest, slowly metabolized fructose and nonmetabolized sugars, e.g., 2-deoxyglucose, 3-O-methylglucose, and L-glucose, can also enhance memory (cf. Messier, 2004). The enhancement of memory by one of these sugars, L-glucose, appears at high (3 g/kg) but not low (300 mg/kg) doses, and the efficacy of L-glucose in enhancing memory at high doses is blocked in vagotomized rats (Talley et al., 2002).

Thus, there appear to be two mechanisms that might contribute to the enhancement of memory by sugars, a high-dose range, at which several sugars are effective and for which memory enhancement is blocked by vagotomy, and a low-dose range, at which glucose and not nonmetabolized sugar is effective and for which enhancement of memory processing is not susceptible to vagotomy. When the evidence linking epinephrine release to increases in blood glucose levels is viewed in terms of the underlying biology and in terms of the blood glucose levels optimal for epinephrine enhancement of memory, it appears that the lower peak glucose dose, and not the higher dose, contributes importantly to epinephrine enhancement of memory processes. As has been done in studies of glucose effects on memory, it will be important to assess systematically the effects of vagotomy on memory-enhancing actions of epinephrine.

Thus, glucose may serve as an intermediate step between the release of epinephrine, excluded from the brain, and enhancement of memory. In contrast to epinephrine, glucose is actively transported into the brain from blood. Although the mechanism may be indirect, glucose augments the release of acetylcholine in the hippocampus, while rats perform spatial memory tasks and may interact with potassium-ATP channels to regulate release of other neurotransmitters and modulators (cf. McNay and Gold, 2002; Gold, 2004).

The results of a series of experiments indicate that microinjections of glucose into specific brain regions can modulate memory processes, findings consistent with the view that glucose increases in blood after epinephrine injection or release can act directly on neural systems to promote memory functions. For example, glucose injections into the hippocampus, striatum, amygdala, and medial septum enhance learning and memory, generally on tasks for which learning would be impaired by damage to these brain regions (Canal et al., 2005; Erickson et al., 2006; Schroeder and Packard, 2003; cf. Gold, 2003). Interestingly, up-regulation of learning and memory processing in the striatum by injections of glucose interferes with acquisition of a spatial task sensitive to hippocampal lesions (Pych et al., 2006), suggesting that augmented striatal functions may impair spatial learning by competition across memory systems (White and McDonald, 2002; Gold, 2004).

Glucose is also a potent enhancer of memory in humans, with demonstrations that glucose can enhance memory in healthy young adult and aged individuals as well as in people with Alzheimer's disease (see Korol, 2002; Messier, 2004). Epinephrine has similarly been reported to enhance memory in young adults (Cahill and Alkire, 2003).

The differing findings regarding the mechanisms mediating epinephrine effects on brain functions, i.e., via increases in blood and brain glucose levels or via vagal afferents, may reflect multiple neurobiological actions acting in concert to enhance memory. As reviewed elsewhere (McIntyre et al., 2003; McGaugh et al., 2002), considerable evidence suggests that the convergence of these factors underlying modulation of memory may include actions at the amygdala. Epinephrine may act through vagal afferents to influence modulation of memory by the amygdala and, in parallel, increase blood glucose levels to act directly on the amygdala and other neural systems to enhance memory formation. Controls of brain neurotransmitter release, particularly acetylcholine and norepinephrine, appear to contribute to epinephrine regulation of memory processing (McIntyre et al., 2002; Gold, 2003).

B. Corticosterone

Some of the earliest research examining hormones and memory included demonstrations that ACTH could modulate learning and memory (for review, see de Wied, 1990). Among the findings were results indicating that ACTH fragments, including some that did not release adrenal steroids, influenced memory when administered either systemically or directly into specific brain areas (e.g., van Rijzingen et al., 1996).

Although ACTH therefore appears to have effects on learning and memory independent of adrenal steroids, a significant body of recent work indicates that adrenal steroids, particularly corticosterone, have potent effects on learning and memory (Roozendaal, 2002; McEwen, 2001; Sandi, 2004; Wolf, 2003; Luine, 2002). Like injections of epinephrine, injections of corticosterone enhance memory for a wide range of tasks, e.g., inhibitory avoidance, fear conditioning, and object recognition (Okuda et al., 2004; see left panel, Fig. 7-3). As in Figure 7-3, the dose–response curve for the effects of corticosterone on memory follows an inverted-U dose–response function (e.g., Okuda et al., 2004), as noted earlier for both epinephrine and glucose. These results are consistent with findings that circulating corticosterone levels have an inverted-U relationship with the hippocampal primed-burst form of long-term potentiation (Diamond et al., 1992). Another feature of corticosterone enhancement of memory shared with epinephrine is that the effects of posttraining injections of corticosterone on memory are time dependent: Posttraining injections enhance memory at short but not long intervals between training and corticosterone treatment (Flood et al., 1978; Sandi and Rose, 1994). Thus, the effects of corticosterone on memory likely reflect actions on memory processes.

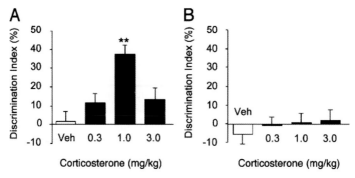

FIGURE 7-3   Effects of corticosterone on 24-hr memory for an object recognition task. **A.** In the absence of prior habituation, corticosterone enhanced memory in an inverted-U dose-dependent manner. **B.** The enhancement of memory was not evident in rats that had been previously habituated to the experimental context. [From Okuda et al. (2004).]

Not surprisingly, as with epinephrine and glucose, the dose–response curve for corticosterone effects on memory interacts with stress at the time of training. For example, corticosterone enhances memory for the spatial version of the swim task when rats are trained in relatively warm but not cold water (Akirav et al., 2004; Sandi et al., 1997). To enhance object-recognition memory, prior habituation of rats to an experimental context, i.e., a novel environment, blocked enhancement of memory by corticosterone, showing that at least modest training-related arousal is important for demonstrations of corticosterone-induced enhancement of memory (Okuda et al., 2004; see Fig. 7-3, right panel).

The amygdala, in particular the basolateral nucleus, appears to play an important role in mediating the effects of corticosterone on memory, as was discussed earlier for epinephrine. Lesions of the amygdala or the stria terminalis or injections of β-adrenergic antagonists into the amygdala block the effects of corticosteroids on memory (Roozendaal, 2002).

Findings also suggest that corticosteroids injected at the time of memory testing impair retrieval of memory (Roozendaal, 2003). These effects do not appear to be the result of direct action of glucocorticoids on the amygdala, since direct injections into the basolateral nucleus of the amygdala do not impair memory retrieval for a swim task. However, lesions of the basolateral nucleus or injections of β-adrenergic antagonists into the amygdala block the effects of intrahippocampal injections of glucocorticoid agonists (Roozendaal et al., 2004). Thus, the basolateral amygdala does not appear to be the primary target of glucocorticoid regulation of retrieval processes but instead is permissive of such effects.

III. GONADAL STEROIDS AND COGNITION

A. Estrogen

Estrogen has potent effects on the structure and function of the adult brain (Brinton, 2001; McEwen, 2002) that, in addition to involvement in reproductive behaviors, may modulate cognition. The direction of estrogen action on learning and memory depends on several variables, including stress levels, type and duration of hormone regimen, and specific task and memory demands (for reviews: Dohanich, 2002; Korol, 2004). Emerging from a growing literature is the idea that estradiol, perhaps the most potent naturally found estrogen in mammals, modulates memory formation and maintenance and biases the learning strategy used to solve a task by altering the relative participation of different neural systems during task performance.

There are numerous reports supporting the finding that specific regimens of estrogen enhance working memory in a variety of tasks (cf. Dohanich, 2002; Korol, 2004), including delayed T-maze, radial maze, and spatial swim tasks. The greatest enhancements on working memory are observed when task difficulty is increased by extending the intertrial interval. Importantly, in some tasks that dissociate types of memory, reference memory, thought to be insensitive to hippocampal manipulations, appears to remain unaffected by estrogen treatment (Dohanich, 2002; Korol, 2004; but also see Heikkinen et al., 2002). It is possible that reference memory components in past studies have not been sufficiently difficult to observe estrogen effects, though assessments of task-difficulty-dependent effects of estrogen on reference memory have not yet been made.

Direct tests of estrogen modulation of memory formation come from studies in which animals are given posttraining treatments and tested for later retention. Systemic treatments of a rapidly metabolized estrogen, estradiol-hydroxypropyl β-cyclodextrin inclusion complex, to ovariectomized rats given immediately but not two hours after training in the swim task enhanced savings during a test 24 hours later (Fig. 7-4; Packard and Teather, 1997b). In addition, memory for inhibitory avoidance training was facilitated by posttraining estrogen injections (Rhodes and Frye, 2006). Object- and place-recognition memory tested with a 4-hr retention interval were enhanced by both natural and synthetic estrogenic compounds that have varying affinities for the two classical estrogen receptor (ER) subtypes, ERα and ERβ (see later; Luine et al., 2003). Again, memory facilitation was seen only with immediate posttraining treatments and not with delayed injections, pointing to actions of estrogen on memory processes per se. Posttraining intrahippocampal infusions of estrogen also enhanced memory in spatial-swim (Packard and Teather, 1997a) and active-avoidance (Farr et al., 2000) paradigms, suggesting that estrogen may engage neurobiological mechanisms involved in different types

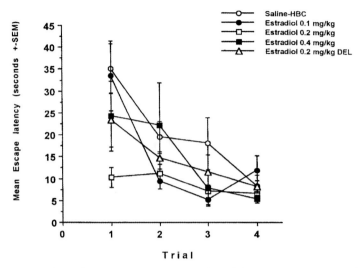

FIGURE 7-4   Effects of posttraining i.p. injections of estradiol on retention-test escape latencies in the spatial swim task. Immediate posttraining estradiol enhanced memory in an inverted-U dose-dependent manner. Memory enhancement was not evident when delayed injections were made 2 hr posttraining (DEL). SEM indicates standard error of the mean. [From Packard and Teather (1997b).]

of memory formation and that certain neural systems may mediate the modulating effects of estrogen.

1. Estrogen and Learning Strategy

One property of estrogen's effects on cognition is that tasks that tap hippocampus function appear to be sensitive to modulation by estrogen, a quality that is not surprising, given the robust effects of estrogen on hippocampus structure and function (for review: McEwen, 2002). Estrogen promotes good performance in hippocampus-sensitive tasks, such as acquisition of a standard eight-arm radial maze test (Daniel et al., 1997) and, as mentioned earlier, tests of working memory in various land- and water-based working memory tasks (Korol, 2004).

The differential involvement of multiple neural systems in memory can be dissociated based on task attributes (White and McDonald, 2002; White, 2004; Kesner and Rogers, 2004). When estrogen effects on memory are taken together and placed within a memory system framework, the findings suggest that estrogen promotes the use of hippocampus-sensitive solutions and may prohibit or impair the use of other, nonhippocampal strategies. Direct tests of this theory show that acute (Fig. 7-5; Korol and Kolo, 2002) and chronic

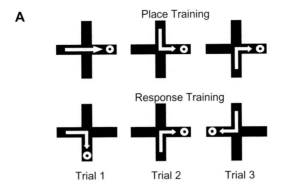

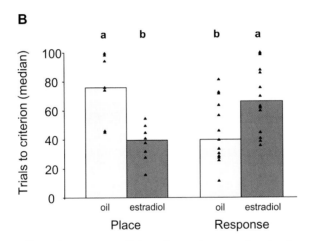

FIGURE 7-5 Effects of systemic estradiol on place and response learning in ovariectomized rats. **A.** Rats were trained to find food in a plus-shaped maze using either a "go there" strategy in place learning or a "turn this way" strategy in response learning. **B.** Estradiol-treated rats learned significantly more quickly than did oil-treated rats in the place-learning task, while the converse was true for response learning. Triangles indicate number of trials for individual rats. Groups marked with different letters are significantly different from each other. [From Korol and Kolo (2002).]

(Davis et al., 2005) estradiol treatments to ovariectomized female rats enhance place learning (hippocampus sensitive) but impair response learning (striatum sensitive) in food-motivated tasks. Conversely, estradiol deprivation promotes response learning but impairs place learning (Fig. 7-5). These results are supported by work using a dual-solution version of the swim task in which chronic estradiol exposure impaired performance in the cued, nonspatial, version, while hormone deprivation impaired performance in a spatial probe test with no cue (Daniel and Lee, 2004). Strategy choice in a land-based dual-solution task in the T-maze, in which either place or response strategies are effective (Tolman et al., 1947; Restle, 1957), was biased toward place learning in cycling rats with endogenously high profiles of estrogen and toward response learning in rats with endogenously low profiles of estrogen (Korol et al., 2004; McElroy and Korol, 2005). Thus, it is becoming clear that estrogen changes not only *how much* is learned but also *what* is learned.

Relative to our knowledge of epinephrine and corticosterone dose–response functions (described eariler), very little is known about effective dose–response functions for estradiol and cognition. Similar to that shown for stress hormones, Packard and colleagues demonstrated a robust inverted-U dose–response function for systemic posttraining estradiol injections on retention in the spatial swim task (Fig. 7-4; Packard and Teather, 1997b). Furthermore, in a series of experiments, different doses of estradiol produced varying effects on cognition, depending on the attributes required — and perhaps the memory systems engaged — by the task of interest. Specifically, in a working-memory task that is thought to depend on an intact prefrontal cortex, low doses of estrogen tended to enhance performance (decrease errors) under low-working-memory load, whereas higher doses impaired memory under higher memory loads (Wide et al., 2004). However, in radial maze working-memory tasks that may tap hippocampal function, physiologically low doses of estradiol enhanced while higher doses impaired working memory (Holmes et al., 2002). In contrast, supraphysiological but not lower doses of estradiol *enhanced* acquisition of a conditioned response in a trace conditioning eye-blink paradigm (Leuner et al., 2004), also a task thought to require hippocampal function. Undoubtedly, to reconcile these differences, more systematic examinations of dose-dependent effects of estrogen across various tasks are needed.

2. Estrogen Modulates Distinct Neural Systems

Framed by the idea of multiple memory systems, the effects of estrogen on cognition can be dissociated by the neural system engaged during learning. To produce such opposing effects on cognition, it is possible that estrogen up-regulates hippocampal involvement and down-regulates striatal involvement during cognition and thus modulates the relative contribution of different neural systems to learning. If so, direct application of estrogen to specific brain areas would mimic the effects of systemic elevations in estrogen for the cognate task. Hippocampal actions of estrogen on spatial learning have been implicated by studies showing that central treatments enhance spatial memory (Packard and Teather, 1997a) and systemic treatments can reverse spatial deficits resulting from hippocampal cholinergic blockade (Packard and Teather, 1997b; Fader et al., 1998; Gibbs, 1999). While these data address the possibility that estrogen acts at the hippocampus, they fail to address whether actions are site and task specific. Recent findings show that direct infusions of estradiol into the hippocampus enhance place learning specifically, whereas infusions into the dorsolateral striatum selectively impair response learning (Zurkovsky et al., 2007). Furthermore, intrahippocampal blockade of estrogen receptors attenuates facilitation of place learning by systemic estrogen (Zurkovsky et al., 2004), and intrastriatal blockade of estrogen receptors prevents impairment by systemic estradiol (Kent et al., 2005). Thus, estrogen likely modulates learning through activation of estrogen receptors in specific neural systems.

How estrogen acts to alter the balance of different neural systems during learning is largely unknown. However, there is a growing literature reporting that estrogen influences the neurobiology of the hippocampus and striatum (Davis et al., 2005; for review: Korol, 2004) as well as other brain areas, such as the prefrontal cortex and the amygdala (Kritzer and Kohama, 1999; J. Wang et al., 2004; Tinkler et al., 2004; Womble et al., 2002). Naturally occurring fluctuations and treatments of estrogen produce dramatic changes in dendritic and spine morphology, in the synthesis, release, and kinetics of neurotransmitter systems such as acetylcholine, GABA, glutamate, and dopamine, and in neural transmission and synaptic plasticity (for reviews: Becker, 1999; Cyr et al., 2001; Dohanich, 2002; Korol, 2004; McEwen, 2002; Woolley, 1998). While many neuronal cell types in hippocampus contain identified estrogen receptors, until recently, when estrogen receptors were localized to the striatum (C. Wang et al., 2005), it was thought that estrogen acts in the striatum through novel membrane sites only.

3. Short-Term Versus Durable Changes: Requirements for Cognition?

It is now known that estrogen has both rapid and slow effects at target tissues, including the brain (McEwen, 2002; Moss et al., 1997; Toran-Allerand et al., 2004). These different actions have been defined as nongenomic and genomic events, respectively, with the slow (>1–2 hr), or "genomic," actions of estrogen thought to be mediated by intranuclear receptor-induced changes in gene expression and the rapid (<1 hr), "nongenomic," actions through several different intracellular signaling cascades (Belchner and Zsarnovsky, 2001; Barnea and Gorski, 1970; Kelly and Wagner, 1999). However, recent data suggest that even rapid effects may act through classical receptors (e.g., ER β; Abraham et al., 2004) and that nongenomic events, while rapid in onset and reversible, produce long-lasting changes through cell signaling pathways (Orchinik and McEwen, 1993), blurring the distinction of genomic and nongenomic actions, but again emphasizing important cellular differences based on duration and timing of estrogen exposure.

Numerous reports document rapid effects of estrogen in the hippocampus that may or may not require the presence of estrogen receptors. Estrogen consistently increases excitability of hippocampal neurons, perhaps leading to increased likelihood of synaptic plasticity (cf. Woolley, 1998), a more durable change. An exciting new finding that 30 min of systemic treatment with estradiol produced robust increases in spine synapse densities in CA1 pyramidal neurons (MacLusky et al., 2005) suggests that some aspects of estrogen-induced restructuring may not require long exposures. Rapid estrogen effects have been well-documented in the striatum (Becker, 1999; Dluzen and Horstink, 2003). For example, estradiol decreases Ca^{2+} currents through membrane-related events within seconds of treatment to dissociated striatal cells (Mermelstein et al., 1996). Activation of c-jun can be measured in striatum as early as 15 min following estrogen administration (Zhou and Dorsa, 1994). Interestingly, estradiol may rapidly modulate dopaminergic G-protein-coupled receptors, the direction of which has been shown in other brain regions to vary by estrogen receptor subtype (Kelly et al., 2003), pointing to the role of estrogen as a metamodulator.

In concert with these processes, estradiol can trigger within seconds to minutes dramatic intraneuronal changes in levels and activity of adenylate cyclase, PKA, PKC, and MAPK pathways (Bi et al., 2000; Singh et al., 2000; Driggers and Segars, 2002). One downstream consequence of PKA activity is phosphorylation of CREB, enhancement of which is observed within 15 min of estrogen treatment in hypothalamus (Abraham et al., 2004) and hippocampus (McEwen et al., 2001). Importantly, estradiol effects on CREB induction are duration sensitive and differ by brain region (Carlstrom et al., 2001).

Consistent with these neurobiological effects, rapid actions of estrogen on cognition have been reported. Estrogen injections given 30 minutes prior to training enhanced recognition memory for objects and place (Luine et al., 2003). Similarly, findings from studies using immediate posttraining injections of estrogen (e.g., Luine et al., 2003; Packard and Teather, 1997b) suggest rapid modulation of cognition by estrogen, because delayed injections are ineffective.

4. Distribution of Estrogen Receptor Subtypes in the Brain

The estrogen receptor belongs to a superfamily of nuclear receptors that regulate gene transcriptional activity. Advances have led to the discovery of at least two different estrogen receptor subtypes, ERα and ERβ (and even more isoforms of ERβ; Kuiper et al., 1996). When activated, the different receptors show unique profiles of gene expression in estrogen-responsive tissues (Pettersson and Gustafsson, 2001) that may translate into different functional outcomes. Both receptor subtypes bind estradiol with high affinity, have a nearly identical DNA-binding domain, and activate transcription through binding to the same estrogen receptor response elements (Cowley et al., 1997).

In brain tissue, the two estrogen receptor subtypes are differentially distributed across anatomically and neurochemically distinct neural systems (Shughrue et al., 1997). Within a given brain area, cells may express one receptor subtype (ERα or ERβ), both subtypes, or neither subtype (Shughrue et al., 1998). Initial studies demonstrating differential distribution of ERα and ERβ together with the behavioral results of studies using ERα and ERβ knockout mice (ERαKOs and ERβKOs) suggested that ERα is involved primarily in reproductive behaviors, whereas ERβ is involved in nonreproductive functions, such as memory and mood (Osterlund et al., 2000; Shughrue and Merchenthaler, 2001; but see Fugger et al., 2000). However, more recently, regions involved in cognition, including the hippocampus, have been found to contain ERα. Interestingly, extranuclear ERα receptors are found both pre- and postsynaptically in the hippocampus, suggesting that estrogen may regulate neurotransmitter release or sensitivity (Mitra et al., 2003). Until recently, it has been thought that the striatum lacks both estrogen receptor subtypes (Shughrue et al., 1997), but new findings highlight the presence of extranuclear ERα in striatal neurons (C. Wang et al., 2005). Thus, the robust effects of estradiol on striatal function and striatum-sensitive learning (Korol, 2004) may act through these receptors, through a novel, uncharacterized cytosolic receptor or through membrane receptor–mediated actions (Kelly and Wagner, 1999; Mermelstein et al., 1996; Toran-Allerand, 2004).

It is clear that an understanding of the estrogen receptor subtype that mediates estrogen's effects on cognition in hippocampus and striatum will direct future work on the cellular mechanisms underlying these effects. In addition, some of the different cellular responses to ERα and ERβ are likely to map against fast and slow effects of estrogen, such that the rapid versus slow effects of estrogen may dissociate by estrogen receptor subtype or brain region. In certain murine brain areas, estrogen-induced CREB phosphorylation required the presence of either ERα or ERβ; which receptor subtype depended on the type of receptor normally expressed in the respective brain area (Abraham et al., 2004). Particularly striking was a lack of rapid estrogen induction of CREB phosphorylation in the striatum. Thus, rapid effects of estradiol can be mediated by both ERs, with specificity of effects likely to be defined by the particular brain area and effector pathway.

Cognitive effects may also be based on receptor subtype, though very few studies have examined receptor specificity in mediating the effects of estrogen on learning and memory. One recent study found that an ERα-selective agonist was ineffective compared to estradiol in enhancing place memory but was effective in enhancing object-recognition memory (Luine et al., 2003). Estrogen receptors are differentially distributed across brain areas, which, in turn, may be differentially engaged in tasks with specific attributes. Thus, these findings suggest again that actions of estrogen at its receptors may underlie different memory effects across structures. The independent effects of ERα and ERβ on cognition have also been demonstrated in a pioneering set of studies by Rissman and colleagues using ERα and ERβ knockout mice

(ERKOs). In the spatial version of the swim task, estradiol treatment to ovariectomized ERαKO mice failed to produce impairing effects seen in wild-type mice. In ERβKOs, however, estradiol effectively impaired performance, pointing to a key role of ERα in estrogen-induced spatial learning deficits reported previously (Rissman et al., 2002). Interestingly, in a different task, i.e., inhibitory avoidance, estradiol acts through an ERα-independent mechanism to enhance memory (Fugger et al., 2000). Recent findings using selective estrogen receptor modulators, termed SERMS, demonstrated that posttraining administration of ERβ-selective compounds tended to enhance spatial-escape memory in the swim task, whereas compounds acting at both ERα and ERβ sites were most effective at enhancing memory for inhibitory avoidance tasks (Rhodes and Frye, 2006). Together, these findings suggest that estrogen may have opposing actions on different forms of memory and that the actions may be through different estrogen receptor subtypes or brain structures.

B. Progesterone

A review of ovarian hormone actions on cognition would be incomplete without a discussion of the effects of progesterone on learning and memory. Progesterone is an ovarian steroid produced by the corpus luteum that, among other things, can act as an anticonvulsant, prohypnotic, and anxiolytic agent (for review: Reddy, 2003; Zinder and Dar, 1999). Examinations of the effects of progesterone on learning and memory are complicated by the fact that progesterone metabolism produces many neuroactive metabolites that may have diverse neurobiological actions. Despite this complexity, there are consistent findings showing that progesterone has dramatic effects on brain function and behavior, particularly related to anxiety and fear.

Endogenous or exogenous elevations in circulating progesterone decrease anxiety on a variety of tasks including the elevated plus maze and open field behavior (Frye and Walf, 2002, 2004; for review: Dubrovsky, 2005). The anxiolytic actions of progesterone on plus-maze and open-field behaviors can be observed with direct amygdala infusions as well (Frye and Walf, 2004). However, progesterone may actually facilitate certain fear responses, e.g., startle (Hiroi and Neumaier, 2006), thereby producing mixed or null effects in tasks that are aversive or stressful by nature.

The effects of progesterone on cognition are mixed and likely depend on factors such as the timing of treatments and testing paradigms. Progesterone treatment was shown to be ineffective in enhancing spatial memory when given prior to training in a radial maze task (Tanabe et al., 2004; Sato et al., 2004) or when given posttraining to rats tested in an inhibitory avoidance task (Rhodes and Frye, 2004), though in these same examples estrogen treatments facilitated memory. In addition, when examined across the menstrual cycle in young adult women (Maki et al., 2002) or rhesus monkeys (Lacreuse et al., 2001), circulating levels of progesterone did not correlate with performance on a subset of cognitive tests, while estradiol levels did. Interestingly, when given to estrogen-primed rats, progesterone can oppose the cognitive actions of estrogen or progesterone alone (Diaz Veliz et al., 1994; Chesler and Juraska, 2000; Sandstrom and Williams, 2001). Interestingly, progesterone administered to old ovariectomized rats reversed the enhancing effects of the ovariectomy on delay-based spatial-memory tasks, suggesting that the higher levels of progesterone in old rats may contribute to some of the age-related memory deficits (Bimonte-Nelson et al., 2004). In contrast, however, impairing effects of progesterone in old rats were not found when weekly treatments were given to estrogen-primed ovariectomized rats and initiated within 3 months of ovariectomy (Gibbs, 2000). Thus, the effects of estrogen and progesterone appear to have complexly interacting characteristics based on such variables as hormone timing, age, and dose.

Actions of progesterone may be mediated by its actions on gene transcription via binding of intracellular progesterone receptors or through modulation of neurotransmitter function, particularly by progesterone metabolites that are produced in the periphery (corpus luteum, adrenal cortex) and the central nervous system (for review: Reddy, 2003). Progesterone's neuroactive 5-alpha reduced metabolite, allopregnanolone ($3\alpha5\alpha$-THP), is a potent positive allosteric $GABA_A$ receptor modulator that produces dramatic and rapid suppression of neural excitability via increases in the frequency or duration of opening of chloride channels (Rupprecht and Holsboer, 1999). It is thought that $3\alpha5\alpha$-THP acts through distinct, neuroactive steroid-binding sites on the $GABA_A$ receptor. However, like benzodiazepines, which also modulate $GABA_A$ receptor function, acute increases in $3\alpha5\alpha$-THP through treatment or across the reproductive cycle are anxiolytic (for review: Reddy, 2003), decreasing anxiety, measured as increased entries into open arms of the plus maze and time spent toward the center of an open-field arena, suggesting that progesterone's antianxiety properties emerge from $GABA_A$ receptor-, and not progesterone receptor-, mediated events (Reddy et al., 2005). Not all treatments of progesterone produce positive regulation of $GABA_A$ receptor; the direction of GABA receptor modulation depends on the duration and timing of progesterone exposure, in addition to the metabolite through which progesterone acts. For example, longer treatments of progesterone and its metabolite $3\alpha5\alpha$-THP (48 hr) have been shown to be anxiogenic, perhaps through alterations in specific $GABA_A$ receptor subunits (Gulinello and Smith, 2003).

Transiently elevated levels of $3\alpha5\alpha$-THP not only influence anxiety but also impair learning and memory in a variety of tasks. Systemic and direct brain injections of $3\alpha5\alpha$-THP impaired learning and memory in the spatial version of the swim task (Johansson et al., 2002; Matthews et al., 2002; Turkmen et al., 2006) but not on a nonspatial, cued task (Matthews et al., 2002). Progesterone and its metabolites undoubtedly produce cognitive effects that depend on the duration and timing of exposure to these neuroactive steroids.

IV. MAJOR POINTS

This chapter has described evidence that both stress hormones and reproductive hormones have robust and reliable effects on learning and memory. There are several features of these findings that are evident across different hormones and experiments.

- *Broad cognitive effects.* Particularly in tests of peripheral hormone fluctuations and manipulations, the hormones modulate learning and memory for wide ranges of tasks, beyond the domain, e.g., stress or reproduction, traditionally associated with each hormone.
- *Timing matters.* Both pre- and posttraining injections of most hormones influence memory. These effects appear to reflect many actions: on both memory formation processing and on learning strategy selection. The steroids, corticosterone, estrogen, and progesterone, have both rapid and slow effects that may participate differently in learning and memory processes.
- *Peripheral and central effects.* Steroidal effects on learning and memory may reflect direct actions on the brain. In contrast, the effects of epinephrine appear to be mediated peripherally by increases in circulating glucose levels and by actions on vagal afferents to the brain. The brain actions are mediated, in part, by local changes in glucose availability and by release of neurotransmitters, such as norepinephrine and acetylcholine, which modulate memory formation.
- *Molecular and cellular bases.* The neural bases of hormone action include genomic effects through classical intracellular receptors and through initiation of signal transduction mechanisms (corticosterone, estrogen, progesterone) and also include secondary effects (metamodulation) (epinephrine, corticosterone, estrogen, progesterone) through modulation of neurotransmitter systems.

ACKNOWLEDGMENTS

Supported by research grants from NSF (IOB 0520876) (DLK) and from NIA (AG 07648), and NIDA (DA 016951) (PEG), and by the University of Illinois Initiative on Aging and the Office of the Vice-Chancellor for Research (DLK, PEG).

REFERENCES

Abraham, I.M., Todman, M.G., Korach, K.S., and Herbison, A.E. (2004). Critical in vivo roles for classical estrogen receptors in rapid estrogen actions on intracellular signaling in mouse brain. *Endocrinology*, **145**, 3055–3061.

Akirav, I., Kozenicky, M., Tal, D., Sandi, C., Venero, C., and Richter-Levin, G. (2004). A facilitative role for corticosterone in the acquisition of a spatial task under moderate stress. *Learning and Memory*, **11**, 188–195.

Axelrod, J., Weil-Malherbe, H., and Tomchick, R. (1959). The physiological disposition of H3-epinephrine and its metabolite metanephrine. *Journal of Pharmacology and Experimental Therapeutics*, **127**, 251–256.

Barnea, A., and Gorski, J. (1970). Estrogen-induced protein. Time course of synthesis. *Biochemistry*, **9**, 1899–1904.

Becker, J.B. (1999). Gender differences in dopaminergic function in striatum and nucleus accumbens. *Pharmacology, Biochemistry and Behavior*, **64**, 803–812.

Belchner, S.M., and Zsarnovsky, A. (2001). Estrogenic actions in the brain: Estrogen, phytoestrogens and rapid intracellular signaling mechanisms. *Perspectives in Pharmacology*, **299**, 408–414.

Bi, R., Broutman, G., Foy, M.R., Thompson, R.F., and Baudry M. (2000). The tyrosine kinase and mitogen-activated protein kinase pathways mediate multiple effects of estrogen in hippocampus. *Proceedings of the National Academy of Sciences*, **97**, 3602–3607.

Bimonte-Nelson, H.A., Singleton, R.S., Williams, B.J., and Granholm, A.C. (2004). Ovarian hormones and cognition in the aged female rat: II. Progesterone supplementation reverses the cognitive enhancing effects of ovariectomy. *Behavioral Neuroscience*, **118**, 707–714.

Cahill, L., and Alkire, M.T. (2003). Epinephrine enhancement of human memory consolidation: Interaction with arousal at encoding. *Neurobiology of Learning and Memory*, **79**, 194–198.

Canal, C., Stutz, S.J., and Gold, P.E. (2005). Glucose injections into the hippocampus or striatum of rats prior to T-maze training: Modulation of learning rates and strategy selection. *Learning and Memory*, **12**, 367–374.

Carlstrom, L., Ke, Z.J., Unnerstall, J.R., Cohen, R.S., and Pandey, S.C. (2001). Estrogen modulation of the cyclic AMP response element-binding protein pathway. Effects of long-term and acute treatments. *Neuroendocrinology*, **74**, 227–243.

Chapa, G.R., Wieczorek, L.A., and Gold, P.E. (2005). Effects of chronic and acute stress on response and place learning in rats. Society for Neuroscience, 35th Annual Meeting, Washington, DC.

Chesler, E.J., and Juraska, J.M. (2000). Acute administration of estrogen and progesterone impairs the acquisition of the spatial morris water maze in ovariectomized rats. *Hormones and Behavior*, **38**, 234–242.

Cowley, S.M., Hoare, S., Mosselman, S., and Parker, M.G. (1997). Estrogen receptors α and β form heterodimers on DNA. *Journal of Biological Chemistry*, **272**, 19858–19862.

Cyr, M., Ghribi, O., Thibault, C., Morissette, M., Landry, M., and Di Paolo, T. (2001). Ovarian steroids and selective estrogen receptor modulators activity on rat brain NMDA and AMPA receptors. *Brain Research Reviews*, **37**, 153–161

Daniel, J.M., Fader, A.J., Spencer, A.L., and Dohanich, G.P. (1997). Estrogen enhances performance of female rats during acquisition of a radial arm maze. *Hormones and Behavior*, **32**, 217–225.

Daniel, J.M., and Lee, C.D. (2004). Estrogen replacement in ovariectomized rats affects strategy selection in the Morris water maze. *Neurobiology of Learning and Memory*, **82**, 142–149.

Davis, D.M., Jacobson, T., Aliakbari, S., and Mizumori, S.J.Y. (2005). Differential effects of estrogen on hippocampal- and striatal-dependent learning. *Neurobiology of Learning and Memory*, **84**, 132–137.

de Kloet, E.R. (2000). Stress in the brain. *European Journal of Pharmacology*, **405**, 187–198.

de Wied, D. (1990). Neurotrophic effects of ACTH/MSH neuropeptides. *Acta Neurobiologiae Experimentalis*, **50**, 353–366.

Diamond, D.M., Bennett, M.C., Fleshner, M., and Rose, G.M. (1992). Inverted-U relationship between the level of peripheral corticosterone and the magnitude of hippocampal primed burst potentiation. *Hippocampus*, **2**, 421–430.

Diaz-Veliz, G., Urresta, F., Dussaubat, N., and Mora, S. (1994). Progesterone effects on the acquisition of conditioned avoidance responses and other motoric behaviors in intact and ovariectomized rats. *Psychoneuroendocrinology*, **19**, 387–394.

Dluzen, D.E., and Horstink, M. (2003). Estrogen as neuroprotectant of nigrostriatal dopaminergic system: Laboratory and clinical studies. *Endocrine*, **21**, 67–75.

Dubrovsky, B.O. (2005). Steroids, neuroactive steroids, and neurosteroids in psychopathology. *Progress in Neuropsychopharmacology and Biological Psychiatry*, **29**, 169–192.

Dohanich, G.P. (2002). Gonadal steroids, learning and memory. In: *Hormones, Brain and Behavior*, D.W. Pfaff, A.P. Arnold, A.M. Etgen, S.E. Fahrbach, and R.T. Rubin (eds.), Academic Press, San Diego, pp. 265–327.

Driggers, P.H., and Segars, J.H. (2002). Estrogen action and cytoplasmic signaling pathways. Part II: The role of growth factors and phosphorylation in estrogen signaling. *Trends in Endocrinology and Metabolism*, **13**, 422–427.

Erickson, E.J., Watts, K.D., and Parent, M.B. (2006). Septal co-infusions of glucose with a GABAB agonist impair memory. *Neurobiology of Learning and Memory*, **85**, 66–70.

Fader, A.J., Hendricson, A.W., and Dohanich, G.P. (1998). Estrogen improves performance of reinforced T-maze alternation and prevents the amnestic effects of scopolamine administered systemically or intrahippocampally. *Neurobiology of Learning and Memory*, **69**, 225–240.

Farr, S.A., Banks, W.A., and Morley, J.E. (2000). Estradiol potentiates acetylcholine and glutamate-mediated post-trial memory processing in the hippocampus. *Brain Research*, **864**, 263–269.

Flood, J.F., and Morley, J.E. (1988). Effects of bombesin and gastrin-releasing peptide on memory processing. *Brain Research*, **460**, 314–322.

Flood, J.F., Smith, G.E., and Morley, J.E. (1987). Modulation of memory processing by cholecystokinin: Dependence on the vagus nerve. *Science*, **236**, 832–834.

Flood, J.F., Vidal, D., Bennett, E.L., Orme, A.E., Vasquez, S., and Jarvik, M.E. (1978). Memory facilitating and anti-amnesic effects of corticosteroids. *Pharmacology, Biochemistry & Behavior*, **8**, 81–87.

Frye, C.A., and Walf, A.A. (2002). Changes in progesterone metabolites in the hippocampus can modulate open field and forced swim test behavior of proestrous rats. *Hormones and Behavior*, **41**, 306–315.

Frye, C.A., and Walf, A.A. (2004). Estrogen and/or progesterone administered systemically or to the amygdale can have anxiety-, fear-, and pain-reducing effects in ovariectomized rats. *Behavioral Neuroscience*, **118**, 306–313.

Fugger, H.N., Foster, T.C., Gustafsson, J.-Å., and Rissman, E.F. (2000). Novel effects of estradiol and estrogen receptor α and β on cognitive function. *Brain Research*, **883**, 258–264.

Gibbs, R.B. (1999). Estrogen replacement enhances acquisition of a spatial memory task and reduces deficits associated with hippocampal muscarinic receptor inhibition. *Hormones and Behavior*, **36**, 222–233.

Gibbs, R.B. (2000). Long-term treatment with estrogen and progesterone enhances acquisition of a spatial memory task by ovariectomized aged rats. *Neurobiology of Aging*, **21**, 107–116.

Gold, P.E. (1995). Modulation of emotional and non-emotional memories: Same pharmacological systems, different neuroanatomical systems. In: J.L. McGaugh, N.M. Weinberger, and G.S. Lynch (Eds.), *Brain and Memory: Modulation and Mediation of Neural Plasticity*, Oxford Press, NY, pp. 41–74.

Gold, P.E. (2003). Acetylcholine modulation of neural systems involved in learning and memory. *Neurobiology of Learning and Memory*, **80**, 194–210.

Gold, P.E. (2004). Coordination of multiple memory systems. *Neurobiology of Learning and Memory*, **82**, 230–242.

Gold, P.E. (2005). Glucose and age-related changes in memory. *Neurobiology of Aging*, **26S**, S60–S64.

Gold, P.E. (2006). The many faces of amnesia. *Learning and Memory*, in press.

Gold, P.E., Hankins, L., Edwards, R.M., Chester, J., and McGaugh, J.L. (1975). Memory interference and facilitation with posttrial amygdala stimulation: Effect on memory varies with footshock level. *Brain Research*, **86**, 509–513.

Gold, P.E., and McCarty, R. (1995). Stress regulation of memory processes: Role of peripheral catecholamines and glucose. In: *Neurobiological and Clinical Consequences of Stress: From Normal Adaptation to PTSD*, M.J. Friedman, D.S. Charney, and A.Y. Deutch (eds.), Lippincott-Raven, Philadelphia, pp. 151–162.

Gold, P.E., and McGaugh, J.L. (1975). A single-trace, two-process view of memory storage processes. In: *Short-Term Memory*, D. Deutsch and J.A. Deutsch (eds.), Academic Press, New York, pp. 355–390.

Gold, P.E., and van Buskirk, R.B. (1975). Facilitation of time-dependent memory processes with posttrial epinephrine injections. *Behavioral Biology*, **13**, 145–153.

Gold, P.E., and van Buskirk, R.B. (1978). Effects of α- and β-adrenergic receptor antagonists on post-trial epinephrine modulation of memory: Relationship to post-training brain norepinephrine concentrations. *Behavioral Biology*, **24**, 168–184.

Gulinello, M., and Smith, S.S. (2003). Anxiogenic effects of neurosteroid exposure: Sex differences and altered $GABA_A$ receptor pharmacology in adult rats. *Journal of Pharmacology and Experimental Therapeutics*, **306**, 541–548.

Heikkinen, T., Puolivali, J., Liu, L., Rissanen, A., and Tanila, H. (2002). Effects of ovariectomy and estrogen treatment on learning and hippocampal neurotransmitters in mice. *Hormones and Behavior*, **41**, 22–32.

Hiroi, R., and Neumaier, J.F. (2006). Differential effects of ovarian steroids on anxiety versus fear as measured by open-field test and fear-potentiated startle. *Behavioral Brain Research*, **166**, 93–100.

Holmes, M.M., Wide, J.K., and Galea, L.A.M. (2002). Low levels of estradiol facilitate, whereas high levels of estradiol impair, working memory performance on the radial arm maze. *Behavioral Neuroscience*, **116**, 928–934.

Hui, G.K., Figueroa, I.R., Poytress, B.S., Roozendaal, B., McGaugh, J.L., and Weinberger, N.M. (2004). Memory enhancement of classical fear conditioning by post-training injections of corticosterone in rats. *Neurobiology of Learning and Memory*, **81**, 67–74.

Johansson, I.-M., Birzniece, V., Lindblad, C., Olsson, T., and Backstrom, T. (2002). Allopregnanolone inhibits learning in the Morris water maze. *Brain Research*, **934**, 125–131.

Kelly, M.J., Qiu, J., and Ronnekleiv, O.K. (2003). Estrogen modulation of G-protein-coupled receptor activation of potassium channels in the central nervous system. *Annals of the New York Academy of Science*, **1007**, 6–16.

Kelly, M.J., and Wagner, E.J. (1999). Estrogen modulation of G-protein-coupled receptors. *Trends in Endocrinology and Metabolism*, **10**, 369–374.

Kent, M.H., Zurkovsky, L., Fornelli, D.C., Fell, J.A., and Korol, D.L. (2005). Intra-striatal antiestrogen ICI 182,780 attenuates the impairing effects of peripheral estradiol treatment on response learning in young adult ovariectomized rats. 35th Annual meeting for the Society for Neuroscience, 883.3.

Kesner, R.P., and Rogers, J. (2004). An analysis of independence and interactions of brain substrates that subserve multiple attributes, memory systems and underlying processes. *Neurobiology of Learning and Memory*, **82**, 199–215.

Kim, J.J., Lee, H.J., Han, J.S., and Packard, M.G. (2001). Amygdala is critical for stress-induced modulation of hippocampal long-term potentiation and learning. *Journal of Neuroscience*, **21**, 5222–5228.

Koob, G.F. (1991). Arousal, stress and inverted-U shaped curves: Implications for cognitive function. In *Perspectives on Cognitive Neuroscience*, R.G. Lister and H.J. Weingartner (eds.), pp. 300–313. Oxford University Press, London.

Korol, D.L. (2002). Enhancing cognitive function across the life span. *Annals of the New York Academy of Sciences*, **959**, 167–179.

Korol, D.L. (2004). Role of estrogen in balancing contributions from multiple memory systems. *Neurobiology of Learning and Memory*, **82**, 309–323.

Korol, D.L., and Kolo, L.L. (2002). Estrogen-induced changes in place and response learning in young adult female rats. *Behavioral Neuroscience*, **116**, 411–420.

Korol, D.L., Malin, E.L., Borden, K.A., Busby, R.A., and Couper-Leo, J. (2004). Shifts in preferred learning strategy across the estrous cycle in female rats. *Hormones and Behavior*, **45**, 330–338.

Krahl, S.E., Senanayake, S.S., and Handforth, A. (2000). Seizure suppression by systemic epinephrine is mediated by the vagus nerve. *Epilepsy Research*, **38**, 171–175.

Kritzer, M.F., and Kohama, S.G. (1999). Ovarian hormones differentially influence immunoreactivity for dopamine beta-hydroxylase, choline acetyltransferas, and serotonin in the dorsolateral prefrontal cortex of adult rhesus monkeys. *Journal of Comparative Neurology*, **409**, 438–451.

Kuiper, G.G.J.M., Enmark, E., Pelto-Huikko, M., Nilsson, S., and Gustafsson, J.-Å. (1996). Cloning of a novel estrogen receptor expression in rat prostate and ovary. *Proceedings of the National Academy of Sciences, USA*, **93**, 5925–5930.

Lacreuse, A., Verreault, M., and Herndon, J.G. (2001). Fluctuations in spatial recognition memory across the menstrual cycle in female rhesus monkeys. *Psychoneuroendocrinology*, **26**, 623–639.

Lemaire, M., Barneoud, P., Bohme, G.A., Piot, O., Haun, F., Roques, B.P., and Blanchard, J.C. (1994). CCK-A and CCK-B receptors enhance olfactory recognition via distinct neuronal pathways. *Learning and Memory*, **1**, 153–164.

Leuner, B., Mendolia-Loffredo, S., and Shors, T.J. (2004). High levels of estrogen enhance associative memory formation in ovariectomized females. *Psychoneuroendocrinology*, **29**, 883–890.

Luine, V. (2002). Sex differences in chronic stress effects on memory in rats. *Stress*, **5**, 205–216.

Luine, V.N., Jacome, L.F., and MacLusky, N.J. (2003). Rapid enhancement of visual and place memory by estrogens in rats. *Endocrinology*, **144**, 2836–2844.

Mabry, T.R., Gold, P.E., and McCarty, R. (1995). Age-related changes in plasma catecholamine responses to acute swim stress. *Neurobiology of Learning and Memory*, **63**, 260–268.

MacLusky, NIJ., Luine, V.N., Hajszan, T., and Leranth, C. (2005). The 17α and 17β isomers of estradiol both induce rapid spine synapse formation in the CA1 hippocampal subfield of ovariectomized female rats. *Endocrinology*, **146**, 287–293.

Maki, P.M., Rich, J.B., and Rosenbaum, R.S. (2002). Implicit memory varies across the menstrual cycle: Estrogen effects in young women. *Neuropsychologia*, **40**, 518–520.

Matthews, D.B., Morrow, A.L., Tokunaga, S., and McDaniel, J.R. (2002). Acute ethanol administration and acute allopregnanolone administration impair spatial memory in the morris water task. *Alcoholism: Clinical and Experimental Research*, **26**, 1747–1751.

McCarty, R., and Gold, P.E. (1996). Catecholamines, stress and disease: A psychobiological perspective. *Psychosomatic Medicine*, **58**, 590–597.

McElroy, M.W., and Korol, D.L. (2005). Intrahippocampal administration of muscimol shifts learning strategy in cycling female rats. *Learning and Memory*, **12**, 150–153.

McEwen, B.S. (2001). Plasticity of the hippocampus: Adaptation to chronic stress and allostatic load. *Annals of the New York Academy of Sciences*, **933**, 265–277.

McEwen, B.S. (2002). Estrogen actions throughout the brain. *Recent Progress in Hormone Research*, **57**, 357–84.

McEwen, B.S., Akama, K., Alves, S., Brake, W.G., Bulloch, K., Lee, S., Li, C., Yuen, G., and Milner, T.A. (2001). Tracking the estrogen receptor in neurons: Implications for estrogen-induced synapse formation. *Proceedings of the National Academy of Sciences*, **98**, 7093–7100.

McGaugh, J.L., McIntyre, C.K., and Power, A.E. (2002). Amygdala modulation of memory consolidation: Interaction with other brain systems. *Neurobiology of Learning and Memory*, **78**, 539–552.

McIntyre, C.K., Hatfield, T., and McGaugh, J.L. (2002). Amygdala norepinephrine levels after training predict inhibitory avoidance retention performance in rats. *European Journal of Neuroscience*, **16**, 1223–1226.

McIntyre, C.K., Power, A.E., Roozendaal, B., and McGaugh, J.L. (2003). Role of the basolateral amygdala in memory consolidation. *Annals of the New York Academy of Sciences*, **985**, 273–293.

McNay, E.C., and Gold, P.E. (2002). Food for thought: Fluctuations in brain extracellular glucose provide insight into the mechanisms of memory modulation. *Behavioral and Cognitive Neuroscience Reviews*, **1**, 264–280.

Mermelstein, P.G., Becker, J.B., and Surmeier, D.J. (1996). Estradiol reduces calcium currents in rat neostriatal neurons via a membrane receptor. *Journal of Neuroscience*, **16**, 595–604.

Messier, C. (2004). Glucose improvement of memory: A review. *European Journal of Pharmacology*, **490**, 33–57.

Mitra, S.W., Hoskin, E., Yudkovitz, J., Pear, L., Wilkinson, H.A., Hayashi, S., Pfaff, D.W., Ogawa, S., Rohrer, S.P., Schaeffer, J.M., McEwen, B.S., and Alves, S.E. (2003). Immunolocalization of estrogen receptor β in the mouse brain: Comparison with estrogen receptor α. *Endocrinology*, **144**, 2055–2067.

Miyashita, T., and Williams, C.L. (2004). Peripheral arousal-related hormones modulate norepinephrine release in the hippocampus via influences on brainstem nuclei. *Behavioral Brain Research*, **153**, 87–95.

Miyashita, T., and Williams, C.L. (2006). Epinephrine administration increases neural impulses propagated along the vagus nerve: Role of peripheral beta-adrenergic receptors. *Neurobiology of Learning and Memory*, **85**, 116–124.

Moss, R.L., Gu, Q., and Wong, M. (1997). Estrogen: Nontranscriptional signaling pathway. *Recent Progress in Hormone Research*, **52**, 33–68.

Okuda, S., Roozendaal, B., and McGaugh, J.L. (2004). Glucocorticoid effects on object recognition memory require training-associated emotional arousal. *Proceedings of the National Academy of Science*, **101**, 853–858.

Orchinik, M., and McEwen, B.S. (1993). Novel and Classical actions of neuroactive steroids. *Neurotransmissions*, **9**, 1–6.

Osterlund, M.K., Gustafsson, J.A., Keller, E., and Hurd, Y.L. (2000). Estrogen receptor beta (ERβ) messenger ribonucleic acid (mRNA) expression within the human forebrain: Distinct distribution pattern to ERα mRNA. *Journal of Clinical Endocrinology and Metabolism*, **85**, 3840–3846.

Packard, M.G., and McGaugh, J.L. (1996). Inactivation of the hippocampus or caudate nucleus with lidocaine differentially affects expression of place and response learning. *Neurobiology of Learning and Memory*, **65**, 65–72.

Packard, M.G., and Teather, L.A. (1997a). Intra-hippocampal estradiol infusion enhances memory in ovariectomized rats. *Neuroreport*, **8**, 3009–3013.

Packard, M.G., and Teather, L.A. (1997b). Posttraining estradiol injections enhance memory in ovariectomized rats: Cholinergic blockade and synergism. *Neurobiology of Learning and Memory*, **68**, 172–188.

Pettersson, K., and Gustafsson, J.A. (2001). Role of estrogen recepteor beta in estrogen action. *Annual Review of Physiology*, **63**, 165–192.

Pugh, C.R., Tremblay, D., Fleshner, M., and Rudy, JW. (1997). A selective role for corticosterone in contextual-fear conditioning. *Behavioral Neuroscience*, **111**, 503–511.

Pych, J.C., Kim, M., and Gold, P.E. (2006). Effects of injections of glucose into the dorsal striatum on learning of place and response mazes. *Behavioral Brain Research*, **167**, 373–378.

Reddy, D.S. (2003). Pharmacology of endogenous neuroactive steroids. *Critical Reviews in Neurobiology*, **15**, 197–234.

Reddy, D.S., O'Malley, B.W., and Rogawski, M.A. (2005). Anxiolytic activity of progesterone in progesterone receptor knockout mice. *Neuropharmacology*, **48**, 14–24.

Restle, F. (1957). Discrimination of cues in mazes: A resolution of the place vs. response controversy. *Psychological Review*, **64**, 217–228.

Rhodes, M.E., and Frye, C.A. (2004). Estrogen has mnemonic-enhancing effects in the inhibitory avoidance task. *Pharmacology, Biochemistry and Behavior*, **78**, 551–558.

Rhodes, M.E., and Frye, C.A. (2006). ERbeta-selective SERMs produce mnemonic-enhancing effects in the inhibitory avoidance and water maze tasks. *Neurobiology of Learning and Memory*, **85**, 183–191.

Rissman, E.F., Heck, A.L., Leonard, J.E., Shupnik, M.A., and Gustafsson, J.-Å. (2002). Disruption of estrogen receptor B gene impairs spatial learning in female mice. *Proceedings of the National Academy of Sciences*, **99**, 3996–4001.

Roozendaal, B. (2002). Stress and memory: Opposing effects of glucocorticoids on memory consolidation and memory retrieval. *Neurobiology of Learning and Memory*, **78**, 578–595.

Roozendaal, B. (2003). Systems mediating acute glucocorticoid effects on memory consolidation and retrieval. *Progress in Neuro-Psychopharmacology and Biological Psychiatry*, **27**, 1213–1223.

Roozendaal, B., Hahn, E.L., Nathan, S.V., de Quervain, D.J., and McGaugh J.L. (2004). Glucocorticoid effects on memory retrieval require concurrent noradrenergic activity in the hippocampus and basolateral amygdala. *Journal of Neuroscience*, **24**, 8161–8169.

Rupprecht, R., and Holsboer, F. (1999). Neuroactive steroids: Mechanisms of action and neuropsychopharmacological perspectives. *Trends in Neurosciences*, **22**, 410–416.

Sandi, C. (2004). Stress, cognitive impairment and cell adhesion molecules. *Nature Reviews Neuroscience*, **5**, 917–930.

Sandi, C., Loscertales, M., and Guaza, C. (1997). Experience-dependent facilitating effect of corticosterone on spatial memory formation in the water maze. *European Journal of Neuroscience*, **9**, 637–642.

Sandi, C., and Rose, S.P. (1994). Corticosterone enhances long-term retention in one-day-old chicks trained in a weak passive avoidance learning paradigm. *Brain Research*, **647**, 106–112.

Sandstrom, N.J., and Williams, C.L. (2001). Memory retention is modulated by acute estradiol and progesterone replacement. *Behavioral Neuroscience*, **115**, 384–393.

Sato, T., Tanaka, K., Ohnishi, Y., Teramoto, T., Irifune, M., and Nishikawa, T. (2004). Effects of estradiol and progesterone on radial maze performance in middle-aged female rats fed a low-calcium diet. *Behavioral Brain Research*, **150**, 33–42.

Schroeder, J.P., and Packard, M.G. (2003). Systemic or intra-amygdala injections of glucose facilitate memory consolidation for extinction of drug-induced conditioned reward. *European Journal of Neuroscience*, **17**, 1482–1488.

Shughrue, P.J., Lane, M.V., and Merchenthaler, I. (1997). Comparative distribution of estrogen receptor-α and -β mRNA in the rat central nervous system. *Journal of Comparative Neurology*, **388**, 507–525.

Shughrue, P.J., and Merchenthaler, I. (2001). Distribution of estrogen receptor β immunoreactivity in the rat central nervous system. *Journal of Comparative Neurology*, **436**, 64–81.

Shughrue, P.J., Scrimo, P.J., and Merchenthaler, I. (1998). Evidence for the colocalization of the estrogen receptor-β mRNA and estrogfen receptor-α immunoreactivity in neurons of the rat forebrain. *Endocrinology*, **139**, 5267–5270.

Singh, M., Sétáló Jr., G., Guan, X., Frail, D.F., and Toran-Allerand, C.D. (2000). Estrogen-induced activation of the MAP kinase cascade in the cerebral cortex of estrogen receptor-alpha knock-out (ERKO) mice. *Journal of Neuroscience*, **20**, 1694–1700.

Talley, C.P., Clayborn, H., Jewel, E., McCarty, R., and Gold, P.E. (2002). Vagotomy attenuates effects of L-glucose but not D-glucose on spontaneous alternation performance. *Physiology and Behavior*, **77**, 243–249.

Tanabe, F., Miyasaka, N., Kubota, T., and Aso, T. (2004). Estrogen and progesterone improve scopolamine-induced impairment of spatial memory. *Journal of Medical and Dental Sciences*, **51**, 89–98.

Tinkler, G.P., Tobin, J.R., and Voytko, M.L. (2004). Effects of two years of estrogen loss or replacement on the nucleus basalis cholinergic neurons and cholinergic fibers to the dorsolateral prefrontal and inferior parietal cortex of monkeys. *Journal of Comparative Neurology*, **469**, 507–521.

Tolman, E.C., Ritchie, B.F., and Kalish, D. (1947). Studies in spatial learning. V. Response learning vs. place learning by the non-correction method. *Journal of Experimental Psychology*, **37**, 285–292.

Tomaz, C., and Nogueira, P.J. (1997). Facilitation of memory by peripheral administration of substance P. *Behavioral Brain Research*, **83**, 143–145.

Toran-Allerand, C.D. (2004). Estrogen and the brain: Beyond ER alpha and ER beta. *Experimental Gerontology*, **39**, 1579–1586.

Turkmen, S., Lofgren, M., Birzniece, V., Backstrom, T., and Johansson, I.-M. (2006). Tolerance development to Morris water maze test impairments induced by acute allopregnanolone. *Neuroscience*, **139**, 651–659.

van Rijzingen, I.M., Gispen, W.H., Dam, R., and Spruijt, B.M. (1996). Chronic and intra-amygdala administrations of the ACTH(4–9) analog ORG 2766 modulate behavioral changes after manipulation of NMDA-receptor activity. *Brain Research*, **717**, 200–203.

Wang, C., Kim, J., Yang, H., Jenkins, W.J., Musatov, S., Hu, M., Kaplitt, M.I., and Becker, J.B. (2005). Estrogen receptor α immunoreactivity in striatum of female rats. 561.14, 2005 Abstract Viewer/Itinerary Planner. Washington, DC: Society for Neurosocience.

Wang, J., Cheng, C.M., Zhou, J., Smith, A., Weickert, C.S., Perlman, W.R., Becker, K.G., Powell, D., and Bondy, C.A. (2004). Estradiol alters transcription factor gene expression in primate prefrontal cortex. *Journal of Neuroscience Research*, **76**, 306–314.

White, N.M. (1991). Peripheral and central memory-enhancing actions of glucose. In: *Peripheral Signaling of the Brain: Role in Neural-Immune Interactions, Learning and Memory*, R.C.A. Frederickson, J.L. McGaugh, and D.L. Felton (eds.). Hogrefe & Huber, Toronto, pp. 421–441.

White, N.M. (2004). The role of stimulus ambiguity and movement in spatial navigation: A multiple memory systems analysis of location discrimination. *Neurobiology of Learning and Memory*, **82**, 216–229.

White, N.M., and McDonald, R.J. (2002). Multiple parallel memory systems in the brain of the rat. *Neurobiology of Learning and Memory*, **77**, 125–184.

Wide, J.K., Hanratty, K., Ting, J., and Galea L.A. (2004). High-level estradiol impairs and low-level estradiol facilitates non-spatial working memory. *Behavioral Brain Research*, **155**, 45–53.

Williams, C.L., Men, D., Clayton, E.C., and Gold, P.E. (1998). Norepinephrine release in the amygdala following systemic injection of epinephrine or escapable footshock: Contribution of the nucleus of the solitary tract. *Behavioral Neuroscience*, **112**, 1414–1422.

Wolf, O.T. (2003). HPA axis and memory. *Best Practice and Research in Clinical Endocrinology and Metabolism*, **17**, 287–299.

CHAPTER 8

Neurobiological Views of Memory

Raymond P. Kesner

Psychology Department, University of Utah, Salt Lake City, UT 84112

I. INTRODUCTION

The structure and utilization of memory is central to one's knowledge of the past, interpretation of the present, and prediction of the future. Therefore, the understanding of the structural and process components of memory systems at the psychological and neurobiological levels is of paramount importance. There have been a number of attempts to divide learning and memory into multiple memory systems. Schacter and Tulving (1994) have suggested that one needs to define memory systems in terms of the kind of information to be represented, the processes associated with the operation of each system, and the neurobiological substrates, including neural structures and mechanisms that subserve each system. Furthermore, it is likely that within each system there are multiple forms or subsystems associated with each memory system, and there are likely to be multiple processes that define the operation of each system. Finally, there are probably multiple neural structures that form the overall substrate of a memory system.

Currently, the most established models of memory can be characterized as dual memory system models with an emphasis on the hippocampus or medial

Neurobiology of Learning and Memory, Second Edition

temporal lobe for one component of the model and a composite of other brain structures as the other component. For example, Squire (1994, 2004) has proposed that memory can be divided into a medial temporal lobe–dependent declarative memory, which provides for conscious recollection of facts and events, and a nonhippocampal-dependent nondeclarative memory, which provides for memory without conscious access for skills, habits, priming, simple classical conditioning, and nonassociative learning. Others have used different terms to reflect the same type of distinction, including a hippocampal-dependent explicit memory versus a nonhippocampal-dependent implicit memory (Schacter, 1987) and a hippocampal-dependent declarative memory based on the representation of relationships among stimuli versus a nonhippocampal-dependent procedural memory based on the representation of a single stimulus or configuration of stimuli (Cohen and Eichenbaum, 1993; Eichenbaum, 2004). Olton (1983) has suggested a different dual memory system, in which memory can be divided into a hippocampal-dependent working memory, defined as memory for the specific, personal, and temporal context of a situation, and a nonhippocampal-dependent reference memory, defined as memory for rules and procedures (general knowledge) of specific situations. Different terms have been used to reflect the same distinction, including episodic versus semantic memory (Tulving, 1983).

However, memory is more complex and involves many neural systems in addition to the hippocampus. To remedy this situation, Kesner (2002) has proposed a tripartite attribute-based theoretical model of memory that is organized into event-based, knowledge-based, and rule-based memory systems. Each system is composed of the same set of multiple attributes or forms of memory, characterized by a set of process-oriented operating characteristics and mapped onto multiple neural regions and interconnected neural circuits. For more detail see Kesner (1998, 2002).

On a psychological level (see Figs. 8-1, 8-2, and 8-3), the event-based memory system provides for temporary representations of incoming data concerning the present, with an emphasis on data and events that are usually personal or egocentric and that occur within specific external and internal contexts. The emphasis is on the processing of new and current information. During initial learning, great emphasis is placed on the event-based memory system, which will continue to be of importance, even after initial learning, in situations where unique or novel trial information needs to be remembered. This system is akin to episodic memory (Tulving, 1983) and some aspects of declarative memory (Squire, 1994, 2004).

The knowledge-based memory system provides for more permanent representations of previously stored information in long-term memory and can be thought of as one's general knowledge of the world. The knowledge-based memory system would tend to be of greater importance after a task has been learned, given that the situation is invariant and familiar. The organization of

Event-based Memory

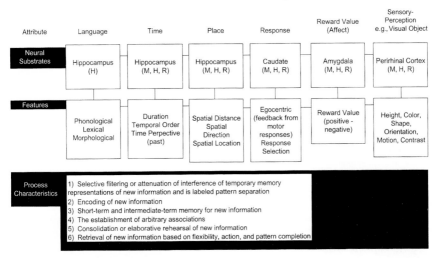

Attribute	Language	Time	Place	Response	Reward Value (Affect)	Sensory-Perception e.g., Visual Object
Neural Substrates	Hippocampus (H)	Hippocampus (M, H, R)	Hippocampus (M, H, R)	Caudate (M, H, R)	Amygdala (M, H, R)	Perirhinal Cortex (M, H, R)
Features	Phonological Lexical Morphological	Duration Temporal Order Time Perpective (past)	Spatial Distance Spatial Direction Spatial Location	Egocentric (feedback from motor responses) Response Selection	Reward Value (positive - negative)	Height, Color, Shape, Orientation, Motion, Contrast

Process Characteristics	1) Selective flitering or attenuation of interference of temporary memory representations of new information and is labeled pattern separation 2) Encoding of new information 3) Short-term and intermediate-term memory for new information 4) The establishment of arbitrary associations 5) Consolidation or elaborative rehearsal of new information 6) Retrieval of new information based on flexibility, action, and pattern completion

Key: M = Monkey, H = Humans, R = Rats

FIGURE 8-1   Representation of the neural substrates, features, and process characteristics associated with the event-based memory system for the language, time, place, response, reward value (affect), and sensory-perception attributes.

these attributes within the knowledge-based memory system can take many forms and are organized as a set of attribute-dependent cognitive maps and their interactions, which are unique for each memory. This system is akin to semantic memory (Tulving, 1983).

The rule-based memory system receives information from the event-based and knowledge-based systems and integrates the information by applying rules and strategies for subsequent action. In most situations, however, one would expect a contribution from all three systems, with a varying proportion of involvement of one relative to the other.

The three memory systems are composed of the same forms, domains, or attributes of memory. Even though there could be many attributes, the most important attributes include *space, time, response, sensory-perception, and reward value (affect)*. In humans a *language* attribute is also added. A spatial (space) attribute within this framework involves memory representations of places or relationships between places. It is exemplified by the ability to encode and remember spatial maps and to localize stimuli in external space. Memory representations of the spatial attribute can be further subdivided into specific spatial features, including allocentric spatial distance, egocentric spatial distance, allocentric direction, egocentric direction, and spatial location. A temporal (time) attribute within this framework involves memory representations

Knowledge-based Memory

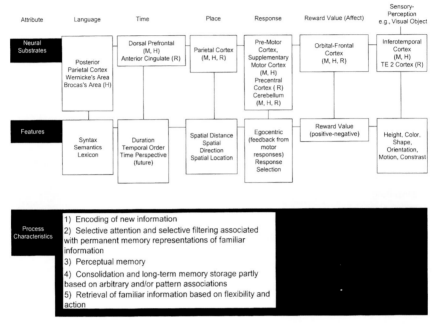

Key: M = Monkey, H = Humans, R = Rats

FIGURE 8-2   Representation of the neural substrates, features, and process characteristics associated with the knowledge-based memory system for the language, time, place, response, reward value (affect), and sensory-perception attributes.

of the duration of a stimulus and the succession or temporal order of temporally separated events or stimuli and, from a time perspective, the memory representation of the past. A response attribute within this framework involves memory representations based on feedback from motor responses (often based on proprioceptive and vestibular cues) that occur in specific situations as well as memory representations of stimulus–response associations. A reward value (affect) attribute within this framework involves memory representations of reward value, positive or negative emotional experiences, and the associations between stimuli and rewards. A sensory-perceptual attribute within this framework involves memory representations of a set of sensory stimuli that are organized in the form of cues as part of a specific experience. Each sensory modality (olfaction, auditory, vision, somatosensory, or taste) can be considered part of the sensory-perceptual attribute component of memory. A language attribute within this framework involves memory representations of phonological, lexical, morphological, syntactical, and semantic information. The

Rule-based Memory

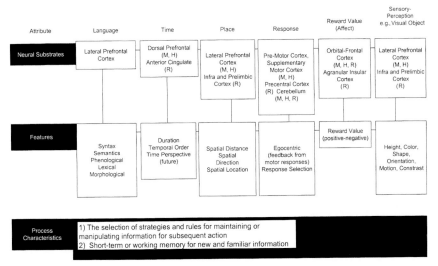

FIGURE 8-3   Representation of the neural substrates, features, and process characteristics associated with the rule-based memory system for the language, time, place, response, reward value (affect), and sensory-perception attributes.

attributes within each memory system can be organized in many different ways and are likely to interact extensively with each other, even though it can be demonstrated that these attributes do in many cases operate independent of each other. The organization of these attributes within the event-based memory system can take many forms and are probably organized hierarchically and in parallel. The organization of these attributes within the knowledge-based memory system can take many forms and are assumed to be organized as a set of cognitive maps or neural nets and their interactions, which are unique for each memory. It is assumed that long-term representations within cognitive maps are more abstract and less dependent on specific features. The organization of these attributes within the rule-based memory system can also take many forms, and they are assumed to be organized to provide flexibility in executive function in developing rules and development of goals and affecting decision processes.

Within each system, attribute information is processed in different ways, based on different operational characteristics. For the event-based memory system, specific processes involve (a) selective filtering or attenuation of interference of temporary memory representations of new information, labeled *pattern separation*, (b) encoding of new information, (c) short-term and intermediate-term memory for new information, (d) the establishment of arbitrary associations, (e) consolidation or elaborative rehearsal of new information, and (f) retrieval of new information based on flexibility, action, and pattern completion.

For the knowledge-based memory system, specific processes include (a) encoding of new information, (b) selective attention and selective filtering associated with permanent memory representations of familiar information, (c) perceptual memory, (d) consolidation and long-term memory storage, partly based on arbitrary and/or pattern associations, and (e) retrieval of familiar information based on flexibility and action.

For the rule-based memory system, it is assumed that information is processed through the integration of information from the event-based and knowledge-based memory systems for the use of major processes, which include the selection of strategies and rules for maintaining or manipulating information for subsequent action as well as short-term or working memory for new and familiar information.

On a neurobiological level (see Figs. 8-1, 8-2, and 8-3) each attribute maps onto a set of neural regions and their interconnected neural circuits. For example, within the event-based memory system it has been demonstrated that in animals and humans (a) the hippocampus supports memory for spatial, temporal, and language attribute information, (b) the caudate mediates memory for response attribute information, (c) the amygdala subserves memory for reward value (affect) attribute information, and (d) the perirhinal and extrastriate visual cortex support memory for visual object attribute information, as an example of a sensory-perceptual attribute. (For more detail see Kesner, 1998, 2002).

Within the knowledge-based memory system, it has been demonstrated that in animals and humans (a) the posterior parietal cortex supports memory for spatial attributes, (b) the dorsal and dorsolateral prefrontal cortex and/or anterior cingulate support memory for temporal attributes, (c) the premotor, supplementary motor, and cerebellum in monkeys and humans and precentral cortex and cerebellum in rats support memory for response attributes, (d) the orbital prefrontal cortex supports memory for reward value (affect) attributes, (e) the inferotemporal cortex in monkeys and humans and TE2 cortex in rats subserves memory for sensory-perceptual attributes, e.g., visual objects, and (f) the parietal cortex and Broca and Wernicke's areas subserve memory for the language attribute. (For more detail see Kesner, 1998, 2002).

Within the rule-based memory system it can be shown that different subdivisions of the prefrontal cortex support different attributes. For example, (a) the dorsolateral and ventrolateral prefrontal cortex in monkeys and humans and the infralimbic and prelimbic cortex in rats support spatial, visual object, and language attributes, (b) the premotor and supplementary motor cortex in monkeys and humans and precentral cortex in rats support response attributes,

(c) the dorsal, dorsolateral, and middorsolateral prefrontal cortex in monkeys and humans and anterior cingulate in rats mediate primarily temporal attributes, and (d) the orbital prefrontal cortex in monkeys and humans and agranular insular cortex in rats support affect attributes. (For more detail see Kesner, 2000a, 2002).

Given the complexity of the nature of memory representations and the multitude of processes involved with learning and memory associated with any specific task, it is clear that prior to analyzing the neural circuits that support mnemonic processing, one must determine which attributes and which systems and associated underlying processes are essential for memory analysis of the proposed task. One example will suffice: If one assumes that the hippocampus supports the processing of the spatial attribute within the event-based memory system, then any task that minimizes the importance of the spatial attribute and emphasizes the importance of reward value, response, and sensory-perceptual attributes are not likely to involve the hippocampus. Because of space limitations, I concentrate in this chapter primarily on the spatial attribute, as revealed by a variety of processes assumed to be mediated by the event-, knowledge-, and rule-based memory systems.

II. SPATIAL ATTRIBUTE: EVENT-BASED MEMORY

Based on a series of experiments, it can be shown that within the event-based memory system, the hippocampus plays an important role in supporting the operations associated with the spatial attribute. The most extensive data set is based on the use of paradigms that measure the short-term or working memory process, such as matching- or nonmatching-to-sample, delayed-conditional-discrimination, or continuous-recognition memory of single or lists of items and paradigms that measure new learning requiring a consolidation process, such as learning an inhibitory avoidance response, taste aversion learning, and water maze spatial navigation and retrieval of previously learned information.

A. Spatial Pattern Separation

It can clearly be demonstrated that single cells within the hippocampus are activated by most sensory inputs, including vestibular, olfactory, visual, auditory, and somatosensory as well as higher-order integration of sensory stimuli (Cohen and Eichenbaum, 1993). The question of importance is whether these sensory inputs have a memory representation within the hippocampus. One possible role for the hippocampus in processing all sensory information might be to provide for sensory markers to demarcate a spatial location so that the hippocampus can more efficiently mediate spatial information. It is, thus, possible that one of the main process functions of the hippocampus is to encode and separate spatial events from each other. This would ensure that new, highly processed sensory information is organized within the hippocampus and enhances the possibility of remembering and temporarily storing in one place as separate from another place. It is assumed that this is accomplished via pattern separation of event information so that spatial events can be separated from each other and spatial interference is reduced. This process is akin to the idea that the hippocampus is involved in orthogonalization of sensory input information (Rolls, 1996; Rolls and Treves, 1998) and indirectly in the utilization of relationships (Cohen and Eichenbaum, 1993).

To assess this function, rats were trained in a *spatial*-pattern-separation task. In this task, rats were required to remember a spatial location dependent on the distance between the study phase object and an object used as a foil. More specifically, during the study phase an object that covers a baited food well was randomly positioned in one of 15 possible spatial locations on a cheese board. During the ensuing test phase, rats were allowed to choose between two objects that were identical to the study-phase object. One object was baited and positioned in the previous study-phase location (correct choice); the other (foil) was unbaited and placed in a different location (incorrect choice). Five distances (minimum = 15 cm, maximum = 105 cm) were randomly used to separate the foil from the correct object. Following the establishment of a criterion of 75% correct averaged across all separation distances, rats were given either large (dorsal and ventral) hippocampal or cortical control lesions dorsal to the dorsal hippocampus. Following recovery from surgery, the rats were retested. The results indicate that whereas control rats matched their presurgery performance for all spatial distances, hippocampal-lesioned rats displayed impairments for short (15–37.5 cm) and medium (60 cm) spatial separations, but performed as well as controls when the spatial separation was long (82.5–105 cm). The fact that the hippocampal-lesioned group was able to perform the task well at large separations indicates that the deficits observed at the shorter separations were not the result of an inability to remember the rule. The results suggest that the hippocampus may serve to separate incoming spatial information into patterns or categories by temporarily storing one place as separate from another place. It can be shown that the ability to remember the long distances was not based on an egocentric response strategy, because if the study phase was presented on one side of the cheese board and the test originated on the opposite side, the hippocampal-lesioned rats still performed the long distances without difficulty. Furthermore, the hippocampal-lesioned group had no difficulty discriminating between two short distances. It is clear that in this task it is necessary to separate one spatial location from another spatial location. Hippocampal-lesioned rats cannot separate these spatial locations very well, so they can perform the task only when the spatial locations are far apart (Gilbert, Kesner, and DeCoteau, 1998). Additional experiments demonstrated that only the dentate gyrus (DG), but not the CA3 or CA1, subregion of the hippocampus can mimic the hippocampal-lesion effect (Gilbert, Kesner, and Lee, 2001). Similar deficits have been observed for new geographical information in patients with hippocampal damage due to a hypoxic episode (Hopkins and Kesner, 1993).

Does spatial pattern separation play a role in novelty detection of changes in spatial distance based on metric changes? Using a modified version of an exploratory paradigm described by Poucet (1993), we tested rats with dorsal hippocampus, dorsal DG, dorsal CA3, and dorsal CA1 lesions and controls on tasks involving either metric spatial or topological spatial manipulations. In the metric manipulation condition, a rat was allowed to explore two different visual objects that were separated by a specific distance on a cheeseboard maze. After habituation to the objects and their locations, the metric spatial distance between the two objects was manipulated that so that the two objects were either closer together or further apart. The time the rat spent exploring each object was recorded. In the topological manipulation condition, rats were allowed to explore four different visual objects that were positioned in a square on the cheeseboard maze. After habituation, the locations of two of the objects were switched and the time the rat spent exploring each object was recorded. The results showed that dorsal hippocampus and dorsal DG lesions impaired detection of the metric manipulation but not of the topological manipulation. In contrast, dorsal CA3 and dorsal CA1 lesions did not impair the metric or the topological manipulation. The results suggest that neurons in the DG may be involved in processing spatial information on a metric scale but may not be necessary for representing topological space (Goodrich-Hunsaker, Hunasaker, and Kesner, 2005a; Goodrich, Hunsaker, and Kesner, 2005b).

Does spatial pattern separation based on spatial interference play a role in the acquisition (consolidation) of a variety of hippocampal-dependent tasks? A few examples will suffice. Because rats are started in different locations in the standard water maze task, there is a great potential for interference among similar and overlapping spatial patterns. Thus, the observation that hippocampal-lesioned rats are impaired in learning and subsequent consolidation of important spatial information in this task could be due to difficulty in separating spatial patterns, resulting in enhanced spatial interference. Support for this idea comes from the observation of Eichenbaum, Stewart, and Morris (1990), who demonstrated that when fimbria-fornix-lesioned rats are trained on the water maze task from only a single starting position (less spatial interference) there are hardly any learning deficits, whereas training from many different starting points resulted in learning difficulties. In a somewhat similar study it was shown that total hippocampal-lesioned rats learned or consolidated rather readily that only one spatial location was correct on an 8-arm maze (Hunt, Kesner, and Evans, 1994). Thus, spatial pattern separation can play a role in the acquisition of new spatial information. It should be noted that the hippocampus also plays an important role in temporal pattern separation (Gilbert, Kesner, and Lee, 2001), but the hippocampus is not involved in pattern separation for visual objects, motor responses, or reward value (Gilbert and Kesner, 2002b, 2003b; Kesner and Gilbert, in press). Furthermore, additional data have demonstrated that the perirhinal cortex, caudate nucleus, and amygdala support pattern separation for visual objects, motor responses, and reward value, respectively (Gilbert and Kesner, 2002b, 2003b; Kesner and Gilbert, in press). Therefore, the role of the hippocampus may be limited to spatial and temporal pattern separation.

B. Spatial Arbitrary Associations

In addition to pattern separation, it has been suggested that the hippocampus and its subregions support the formation of arbitrary associations, including paired-associate learning (Eichenbaum and Cohen, 2001). Rolls (1996) suggested that the hippocampus and specifically a CA3 autoassociative network are responsible for the formation and storage of arbitrary associations. For example, information from parietal cortex regarding the location of an object may be associated with information from temporal cortex regarding the identity of an object. These two kinds of information may be projected to the CA3 region of the hippocampus to enable the organism to remember a particular object and its location.

Behavioral studies have examined the effects of hippocampal lesions on the formation of arbitrary associations using paired-associate learning. Nonhuman primates and rats with hippocampal lesions display deficits in object–place paired-associate learning (Gaffan, 1994; Gaffan and Harrison, 1989; Sziklas, Lebel, and Petrides, 1998). Our lab has designed a series of experiments to test directly the involvement of the hippocampus in spatial paired-associate learning (Gilbert and Kesner, 2002a). Rats were trained on a successive discrimination go/no-go task to examine object–place paired-associate learning. In this task, rats with hippocampal lesions, compared to controls, were severely impaired in learning object–place paired associations (Gilbert and Kesner, 2002a).

In a second task, rats were trained on a successive discrimination go/no-go task to examine odor–place paired-associate learning. In this task, the same procedure was used, except that the rat needs to learn that when an odor is presented in its paired location, the rat should dig in sand mixed with the odor to receive a reward. Rats with hippocampal lesions were severely impaired relative to controls in learning odor–place paired associations (Gilbert and Kesner, 2002a). Data from our laboratory using the aforementioned paradigms indicate that rats with CA3 lesions are severely impaired in object–place and odor–place paired-associate learning. However, animals with DG or CA1 lesions learn the object–place and odor–place tasks as well as controls (Gilbert and Kesner, 2003a). These data support the hypothesis that CA3, but not DG or CA1, supports paired-associate learning when a stimulus is associated with a spatial location.

In a third task, rats were trained on a successive discrimination go/no-go task to examine odor–object paired-associate learning. In this task, the same procedure was used, except that the rat needs to learn that when an odor is presented in front of its paired object, the rat should dig in sand mixed with the odor to receive a reward. The results indicate that rats with hippocampal lesions acquire the odor–object task as quickly as controls (Gilbert and Kesner, 2002a). These data suggest that the hippocampus is clearly involved in paired-associate learning when a stimulus must be associated with a spatial location, but the hippocampus does not appear to be important when a spatial location is not a component of the paired-associate task. Support for this idea comes from a number of studies demonstrating that the hippocampus is not involved in arbitrary associations that involve odor–odor (Bunsey and Eichenbaum, 1993; Li, Matsumoto, and Watanabe, 1999), odor–reward (Wood, Agster, and Eichenbaum, 2004), or object–object association (Cho, Kesner, and Brodale, 1995; Murray, Gaffan, and Mishkin, 1993). Given that the acquisition of an object–odor association is not dependent on the hippocampus, would adding a temporal component to an object–odor association task recruit the hippocampus? To test this idea, rats were given area CA1, CA3, or control lesions prior to learning an object–trace-odor task. The task was run in a 115-cm linear box in which the rat was presented with an object for 10 sec, after which it was removed, followed by a 10-sec trace period and by the presentation of an odor 50 cm away. If the odor and the object were paired, the rat was to dig in the odor cup for a reward. If unpaired, the rat was to refrain from digging. Animals that had CA1 lesions were unable to make the association and never performed above chance, whereas animals that had CA3 lesions performed as well as controls (Kesner, Hunsaker, and Gilbert, 2005). These results support the idea that the hippocampus is involved in forming arbitrary associations that do not necessarily involve spatial information, as long as a temporal component is involved. Furthermore, the CA1 appears to be critical for mediating an association with a temporal component, whereas the CA3 is important for an association that involves a spatial component.

Memory for object locations can be thought of as paired-associate learning for object and location. Smith and Milner (1981, 1989) suggested that object-location memory was mediated by the right hippocampus in humans. Smith and Milner (1981, 1989) presented amnesic subjects with temporal lobe excisions with object-location tests. The subjects with large right hippocampal lesions were impaired relative to controls for the recall of objects and their locations when tested after a delay of 2–4 minutes. In another study (Owen,

Sahakian, Semple, Polkey, and Robbins, 1995), patients with amygdala-hippocampus resection were required to remember up to eight pattern–location associations. The patients with amygdala-hippocampal resections had a marked impairment relative to controls. In a similar experiment, hypoxic subjects with hippocampal damage were shown 16 object–location and 16 abstract pattern–location pairs and asked to learn the 16 pairs. The results indicated that the hypoxic subjects were severely impaired in learning the task and in addition that they were impaired in retention for the pairs, (e.g., object–location and abstract design–location) (Hopkins, Pasker, and Kesner, 1995). Thus, it appears that the hippocampus in both rats and humans is critically involved in processing information associated with learning and subsequent retention of arbitrary associations involving a spatial attribute as one critical component.

C. Spatial Pattern Completion

It has been shown in rats that spatial cells recorded in CA1 or hilar/CA3 subregions of the hippocampus continue to fire in the dark when visual cues are not available. This observation has been interpreted to reflect a spatial pattern completion process. It should be noted that more cells in the CA1 than in the hilar/CA3 region were active for pattern completion (Mizumori, Ragozzino, Cooper, and Leutgeb, 1999). Similar results were reported in monkeys with spatial view cells recorded in CA1 or CA3 subregions of the hippocampus. Similarly when the visual details were obscured, the spatial view cells continued to fire when the monkey looked toward where the view was initiated, with more cells in the CA1 than in the CA3 region fired for pattern completion (Rolls, 1999).

Based on computational models of the hippocampus, it has been suggested that a possible mechanism for memory retrieval may be pattern completion, wherein an autoassociative network recalls previous patterns of activity given noisy or degraded cues. To study pattern completion using a short-term memory paradigm, it is important that only partial or reduced information (relative to the study-phase information) be presented. We measured short-term memory for spatial location as a function of how many components present during the study phase were removed during the test phase. Rats were tested using a cheeseboard maze apparatus on a delayed match-to-sample for spatial location task as described earlier (Gilbert, Kesner, and DeCouteau, 1998). The study phase was identical to that used in the spatial pattern separation experiment, but in this experiment, following a 5-sec delay, the animals were required to find the same location, even though the object had been removed. An animal was rewarded for choosing the same spatial location as the sample-phase object (correct choice) but received no reward for choosing a different location (incorrect choice). In additional manipulations, the object was removed and curtains were lowered to eliminate extramaze cues (spatial condition), the object was removed and the animal was rotated seven times (vestibular condition), or the object was removed, the curtains were lowered, and the animal was rotated (spatial and vestibular conditions). Normal rats readily learn this task, and they perform well in terms of accuracy on the test phase when the object is removed. After preoperative training, rats received cortical control or complete hippocampal lesions. Control rats were able to perform the task and demonstrate pattern completion when visual extramaze or vestibular cues were reliable, but not when the cues were manipulated. Rats with hippocampal lesions were impaired in the baseline condition as well as during all manipulations. These results support the hypothesis that the hippocampus supports spatial pattern completion (Kirwan, Gilbert, and Kesner, 2005).

In a subsequent study (Gold and Kesner, 2005), the number of available visual cues were manipulated using the same delayed matching-to-sample for spatial location task. A black curtain with four extramaze cues surrounded the apparatus. During the study phase of the task, rats were trained to move a small black block covering a food well that could appear in one of five possible spatial locations. During the test phase of the task, following a 30-sec delay, rats were required to find the same food well with the block removed in order to receive reinforcement. After reaching stable performance (i.e., accuracy to find the correct location), the rats received neurotoxic injections into the CA3 subregion of the hippocampus. The control group received vehicle injections into the CA3. After surgery, four cues were always available during the sample phase, but during the test they performed well on the task regardless of the availability of one, two, three, or all cues, suggesting intact spatial pattern completion. The CA3-lesioned rats were impaired compared to the controls, especially when only one or two cues were available, suggesting impaired spatial pattern completion. The data are consistent with the findings of Nakazawa et al. (2002), who reported that CA3 NMDA knockout mice fail to show visual cue pattern completion in a reference memory task, and with computational models (O'Reilly and McClelland, 1994; Rolls and Treves, 1998; Shapiro and Olton, 1994). The computational models predict a deficit following CA3 lesions and suggest that an autoassociative CA3 network may be responsible for the completion of patterns based on incomplete input.

There are few studies that have investigated the role of the hippocampus in spatial pattern completion in humans with hippocampal damage. In a series of studies, Ryan, Althoff, Whitlow, and Cohen (2000) and Ryan and Cohen (2004) have shown that when eye movements are monitored to assess memory of a scene, there is a decrease in sampling of the previously viewed scene as compared to a novel scene, suggesting impaired memory for the previously viewed scene. Furthermore, whenever the relationships between the elements of the scene were changed after long delays, normal subjects spent increased time looking for an element that was not in that particular location. These studies are examples of spatial pattern completion. Subjects with hippocampal damage did not show increased time looking for an element that was not in the previously viewed location, suggesting a problem in the spatial pattern completion process. Subjects with hippocampal damage appear to perform similarly to controls when very short time delays were used. Similar deficits in spatial pattern completion can be observed following hippocampal lesions in rats and humans, even though very different methodologies were used in the rat and human studies.

D. Short- and Intermediate-Term Memory

From a temporal dynamic processing point of view, memory can be divided into three critical time periods: short-term memory, with a duration of seconds, intermediate-term memory, with a duration from minutes to a few hours, and long-term memory, with a duration from hours to days to years. It should be recognized that the boundaries between short-term, intermediate-term, and long-term memory are fuzzy and can vary from task to task.

The mnemonic processes of pattern separation, pattern association, and pattern completion operate primarily within this dynamic temporal framework involving short-term, intermediate-term, and long-term memory representations. However, in most of the aforementioned studies no systematic attempt was made to determine whether the hippocampus supports short-term or intermediate-term memory or both. We now report a series of studies where attempts are made to determine whether the hippocampus does indeed support short-term or intermediate-term memory.

It is thought that the hippocampus supports both short-term and intermediate-term memory but not long-term memory (Lee and Kesner, 2002, 2003a, 2003b). Alternatively the hippocampus supports intermediate-term or perhaps long-term memory but not short-term memory (Alvarez, Zola-Morgan, and Squire, 1994). The most extensive data set aimed at addressing these questions in both rats and humans are based on the use of paradigms that measure the short-term and intermediate-term or long-term memory process, such as matching- or nonmatching-to-sample, delayed-conditional-discrimination, and continuous-recognition memory for single item or lists of items. Based on these tasks it has been suggested that following hippocampal damage, the observation of a complete deficit across all delays implies that the hippocampus is involved in processing both short-term and intermediate-term memory. When there is intact performance at short delays followed by impairments at longer delays, the results are interpreted to reflect hippocampal mediation of intermediate-term but not short-term memory.

To examine this issue in the context of processing spatial information, rats were trained in a recognition memory task for spatial location using a delayed spatial matching-to-sample procedure within an 8-arm radial maze. During the study phase of a trial, each rat was trained to enter a randomly selected arm, in order to obtain reinforcement. Immediately after finding the food and returning to the center (1–4 sec), a linoleum piece was wrapped around the central chamber and the correct arm was baited. The rat was then given a choice between the arm that was previously entered and a new arm (test phase). Correct performance during the test phase of a trial required the rat to return to the previously reinforced arm (i.e., the animal had to use a "win-stay" rule in order to receive an additional reinforcement). After reaching criterion performance (75% correct or better on 16 consecutive trials), the rats received large (dorsal and ventral) hippocampus or cortical control lesions. Following recovery from surgery, the rats were retested daily with four trials a day until they rereached criterion performance. The rats were then tested at longer delays (30 sec). Hippocampal-lesioned rats had a complete deficit (chance performance) at all delays (Kesner, Bolland, and Dakis, 1993). Jackson-Smith, Kesner, and Chiba (1993) tested rats with large hippocampal lesions on a spatial continuous-recognition memory task in a 12-arm maze and found that the rats were impaired for all of the distances associated with spatial performance.

In a different task, rats were trained to remember the distance of 2 or 7 cm between two visual cues on a delayed matching-to-sample task with a very short (a few seconds) delay between the study and test phases. Large hippocampal lesions produced a complete disruption of short-term memory for allocentric distance information (Long and Kesner, 1996).

All of the foregoing tasks used a single item; however, one study used a list of five items, which allowed for serial-position curve to be assessed, with memory for the first items reflecting a "primacy," or intermediate-term memory, effect and memory for the last item reflecting a "recency," or short-term memory, effect. In this study (DiMattia and Kesner, 1984) showed that unless animals are overtrained, the results reflect a serial-position curve. After extensive training, each rat was allowed to visit a sequence of five arms on each trial (one per day), which was randomly selected (study phase). Immediately after the animal had received reinforcement from the last of the five arms, the test phase began. Only one test was given for each trial and consisted of opening of two doors simultaneously, with one door representing an arm previously visited and the other door representing a novel arm. The rule to be learned leading to an additional reinforcement was to choose the arm that had been previously visited during the study phase (win-stay rule). After training, animals received small dorsal hippocampal lesions followed by test trials for each serial position. The results indicate that small hippocampal lesions disrupted performance for the early serial positions (impaired primacy effect) but produced no deficit for the last serial position (intact recency effect). These data suggest that the hippocampus mediates intermediate-term but not short-term memory. However, large hippocampal lesions produced a deficit for all serial positions, suggesting that the hippocampus can also mediate short-term memory. Thus, the hippocampus appears to support both short-term and intermediate-term memory, but with smaller lesions there is some residual short-term memory capacity.

Ever since the observation of relatively normal short-term memory with an impairment of long-term memory in patient H.M., it has been suggested that amnesic patients with presumed hippocampal damage have normal short-term memory for all types of information. However, amnesic subjects have difficulty in transferring this information (consolidation) into intermediate-term or long-term memory.

To examine the temporal dynamic of hippocampal involvement in short-term and intermediate-term memory in the context of processing spatial information, Holdstock, Shaw, and Aggleton (1995) tested patients with hippocampal damage in a delayed matching-to-sample paradigm analogous to tasks used for rats. In this task a single stimulus was presented in a specific location, and, following delays of 3–40 sec, the patients had to remember that location in comparison with a location not previously seen. The results indicated that there were no memory deficits for delays up to 20 sec, followed by a deficit at the 40-sec delay. However, Cave and Squire (1992) found no deficits for short-term memory for a dot on a line or memory for an angle. In a different experiment, hypoxic subjects with bilateral hippocampal damage were tested on a short-term memory test to determine whether the hippocampus supports short-term or intermediate-term memory for a spatial relationship based on distance information. Control subjects and hypoxic subjects with bilateral hippocampal damage were tested for memory for spatial distance information for delays of 1, 4, 8, 12, or 16 seconds. The hypoxic subjects were impaired for memory for distance information at the long but not the short delays, compared to normal controls (Kesner and Hopkins, 2001).

For humans, normal subjects often show excellent memory for the first (primacy effect) and the last (recency effect) items in memory tests of list of words. The recency effect is attributed to information represented within short-term memory, whereas the primacy effect reflects transfer of information into intermediate memory. Tests for item-recognition memory were developed specifically to assess serial-position curves that are analogous to the test of memory for spatial locations previously administered in rats (Hopkins, Kesner, and Goldstein, 1995). In order to test item recognition for spatial locations, six Xs appeared on a grid of 16 possible locations on a computer screen, one at a time. In each trial, a subset of six stimulus locations (Ss) were randomly selected and presented in a sequential manner for 5 seconds. During the test phase, two items were presented simultaneously on the screen, one that occurred in the study phase and a new location. The subject was asked to select the item that occurred in the presented list. Hypoxic subjects with hippocampal atrophy were, relative to normal controls, impaired for item-recognition memory for a list of spatial locations. The hypoxic subjects had a significant deficit for the primacy effect but an intact recency effect for spatial locations (Hopkins, Kesner, and Goldstein, 1995).

In general, it appears that in humans with hippocampal damage there are deficits for spatial location information, primarily for longer delays, with savings at short delays, suggesting that the hippocampus is involved in intermediate-term but not short-term memory. In rats, however, there are a number of examples for spatial information of deficits following complete hippocampal lesions across all delays or for both primacy and recency effects based on serial-position analyses. This could be due to the observation that in many of the rat studies lesions of the hippocampus are complete. Since it is unlikely that there is complete bilateral hippocampal damage in humans, it is possible that the remaining hippocampus is sufficient to serve short-term memory functions. Without the hippocampus, however, the hippocampus supports both short-term and intermediate-term memory.

Thus, in summary for the spatial attribute information, it can be shown that with the use of the aforementioned paradigms to measure short-term or intermediate memory for spatial information, there are severe impairments for rats, monkeys, and humans with right hippocampal damage or bilateral hippocampal damage (Hopkins, Kesner, and Goldstein, 1995; Kesner, 1990; Olton, 1983; Parkinson, Murray, and Mishkin, 1988; Pigott and Milner, 1993; Smith and Milner, 1981).

With the use of short-term memory paradigms to measure short-term memory, it has been shown that for rats and humans with hippocampal damage, there are, in contrast to deficits for spatial and temporal information, no impairments for remembering response attribute, affect attribute, and sensory-perceptual attribute information. This is based on the following observations: (a) Rats with hippocampal lesions are not impaired in short-term or working memory for a right or left turn response (response attribute), visual object (sensory-perceptual attribute), or magnitude of reinforcement (affect attribute) information, but they are impaired for short-term or working memory for spatial location information (Kesner, Bolland, and Dakis, 1993; Kesner and Williams, 1995). (b) Rats with hippocampal lesions display impaired performance in a spatial continuous-recognition memory task (spatial attribute), but they perform without any difficulty in an object continuous-recognition memory task (sensory-perceptual attribute) (Jackson Smith et al., 1993). (c) Right temporal lobe–resected patients, which includes the hippocampus, are impaired in remembering that the location of an object was changed within a scene (spatial attribute) but have no difficulty in remembering that an object was changed for a specific location within a scene (sensory-perceptual attribute), and they are impaired in short-term memory for the spatial location of an array of objects but have no difficulty in free recall of the same objects (sensory-perceptual attribute) (Pigott and Milner, 1993; Smith and Milner, 1981). (d) Patients with right or left temporal lobe resection are not impaired in short-term or working memory for the distance of a motor movement response (response attribute), implicit sensory perceptual priming of various stimuli, or in displaying a liking response based on the mere exposure effect (affect attribute) (Chiba, Kesner, Matsuo, and Heilbrun, 1993; Leonard and Milner, 1991; Shimamura, 1986).

It should be noted that the hippocampus is not the only neural region that mediates short-term memory for spatial information. Using a continuous spatial short-term recognition task, it has been shown that lesions of the pre- and parasubiculum produce profound deficits similar to what has been described for hippocampal lesions, suggesting that other neural regions contribute to the spatial attribute within the event-based memory system (Kesner and Giles, 1998). The exact contribution of each of these areas needs to be investigated, but it should be noted that place cells have also been recorded from the parasubiculum (Taube, Goodridge, Golob, Dudchenko, and Stackman, 1996) and that head-direction cells have been recorded from the lateral dorsal nucleus of the thalamus (Mizumori and Williams, 1993).

E. Consolidation

The hippocampus also plays a role in the encoding and acquisition or learning of new spatial information requiring the consolidation of spatial attributes. This is readily observable in the acquisition of spatial navigation tasks in a water maze, dry-land version of the water maze, and inhibitory avoidance tasks requiring an association of a painful stimulus with a specific spatial location, in that rats with hippocampal lesions are markedly impaired in these tasks (Kesner, 1990; Morris, Garrud, Rawlins, and O'Keefe, 1982; O'Keefe and Nadel, 1978). Furthermore, it has been shown that posttrial disruption of normal hippocampal function with, for example, electrical brain stimulation results in time-dependent memory impairments (Kesner and Wilburn, 1974). These effects reveal that the hippocampus is involved in short-term consolidation processes, because the gradients are usually short, within minutes to a few hours. Long-term temporally graded functions have also been observed for previously learned spatial discriminations prior to surgery in rats and mice, but these long-term gradients (2–4 weeks) are observed primarily following entorhinal cortex lesions rather than hippocampal lesions (Cho and Kesner, 1995, 1996). A long-term gradient following hippocampal lesions has been reported following contextual fear conditioning (Kim and Fanselow, 1992), but other results have questioned whether these gradients can be reliably measured (Maren, Aharonov, and Fanselow, 1996; Weisend, Astur, and Sutherland, 1996). Thus, it is possible that short-term consolidation gradients derive from hippocampal dysfunction, whereas entorhinal cortex dysfunction is necessary to produce long-term retrograde amnesia consolidation gradients. Whether the hippocampus promotes the transfer of spatial information to the knowledge-based system or whether the hippocampus promotes the consolidation of information already processed in the knowledge-based system still needs to be resolved.

Other brain areas, such as the caudate nucleus perirhinal cortex and amygdala, are also involved in the acquisition and consolidation of new information, but in this case the tasks involve the establishment of stimulus–response associations, motor skills (mirror reading or pursuit rotor), probability classification, and conditioned autonomic responses to visual or auditory stimuli (Packard and Knowlton, 2002). Similarly, the perirhinal cortex is involved in the establishment of object–object associations (Buckley and Gaffan, 1998; Higuchi and Miyashita, 1996) and the amygdala in stimulus–reward associations (Fanselow and Poulos, 2005).

III. SPATIAL ATTRIBUTE: KNOWLEDGE-BASED MEMORY

The organization of the spatial attribute within the knowledge-based memory system is assumed to involve the parietal cortex and retrosplenial cortex, parahippocampal gyrus, and entorhinal cortex. The most extensive data set is based on the use of paradigms that measure perceptual memory (e.g., repetition priming), the acquisition of new information, discrimination performance, and the operation of a variety of long-term memory programs.

With respect to spatial attribute information it is assumed that the parietal cortex (PC) processes a subset of the spatial features associated with the spatial attribute and plays an important role in perceptual memory as well as long-term memory encoding, consolidation, and retrieval of spatial information within the knowledge-based memory system.

A. Nature of Spatial Information

In terms of the qualitative representation of spatial information, it has been suggested that the PC processes the spatial location of objects in an egocentric framework based on the relationship between the organism and appropriate object or objects in the external environment (Andersen, 1999; Chen, Lin, Barnes, and McNaughton, 1994; Colby, 1999). In contrast, the hippocampus processes the spatial location of objects in an allocentric framework based on the relative position of objects in the external environment (Morris et al., 1982; O'Keefe, 1979; Rolls, 1999). However, both the PC and the hippocampus mediate learning and retention of tasks that have been suggested to emphasize primarily allocentric information, such as the water maze spatial navigation, the reaction to a spatial change, and the object–place paired-associate tasks (DiMattia and Kesner, 1988b; Gilbert and Kesner, 2002a, Kesner, Farnsworth, and Kametani, 1992; Long, Mellem, and Kesner, 1998; Save, Poucet, Foreman, and Thinus Blanc, 1992). These results suggest the possibility that there might be an interaction between the PC and the hippocampus. Support for this possible interaction comes from a disconnection study showing that crossed lesions of the PC and hippocampus produce profound deficits in the object–place paired-associate learning task and the dry-land version of the water maze, whereas lesions of the PC and the hippocampus on one side does not produce any deficits in these two tasks (Rogers and Kesner, unpublished observations). The hippocampus and PC are also both involved in processing of ideothetic information (one form of egocentric information) based on self-motion and vestibular inputs (Berthoz, 2000; Bures, Fenton, Kaminsky, Rossier, Sacchetti, and Zinyuk, 1999; Chen and Nakamura, 1998; Whishaw, McKenna, and Maaswinkel, 1997). Thus, it appears that both the hippocampus and the PC process egocentric and allocentric information, but it is possible that the hippocampus places a proportionally greater emphasis on allocentric and the PC places a greater emphasis on egocentric processing of spatial information. To test this idea further, two versions of a modified Hebb–Williams maze were used to examine the role of the hippocampus and the PC in processing allocentric and egocentric space during acquisition and retention. Bilateral lesions were made to either the hippocampus or the PC before maze testing (acquisition) or after maze testing (retention). The results indicate that lesions of the hippocampus impair allocentric maze acquisition, whereas lesions of the PC impair egocentric maze acquisition. During retention, lesions of the PC produced a significant impairment on both maze versions, whereas lesions of the hippocampus produced short-lived, transient impairments on both maze versions (Rogers and Kesner, 2006). These results suggest that during acquisition, the hippocampus and the PC process spatial information in parallel; however, long-term retention of spatial information requires the PC, with the hippocampus necessary for retrieval and/or access but not necessarily storage. This suggests the idea that the PC becomes critical when there is an increase in complexity requiring the combination of different spatial features and associations with spatial information. This view would be consistent with Treisman's feature integration theory (Treisman and Gelade, 1980), which suggests that intact parietal function is essential for binding object features with spatial information, such as a "master map of locations" (Robertson, Treisman, Friedman-Hill, and Grabowecky 1997), or a "cognitive map." Integration of egocentric and allocentric spatial information occurring in the PC explains the deficits seen during retention of both maze versions with lesions of the PC. Further support for this idea comes from the observation that in rats neurons have been found within the parietal cortex that encode spatial location and head-direction information and that many of these cells are sensitive to multiple cues, including visual, proprioceptive, sensorimotor, and vestibular cue information (Chen et al., 1994; McNaughton, Chen, and Marcus, 1991). Furthermore, in less complex tasks involving the discrimination or short-term memory for single spatial features, including spatial location, allocentric egocentric spatial distance (Long and Kesner, 1996, 1998), there are no impairments.

Consistent with the finding that PC lesions disrupt topological but not metric information, it appears that the PC is essential to processing topological information (Goodrich-Hunsaker et al., 2005a). The results are consistent with patient RM, who had a bilateral PC lesion. This patient demonstrated impairment for learning topological relationships. RM was asked to determine if a large dot was outside or inside a circle. RM was unable to learn this task, averaging 49% correct (18 out of 37 trials) (Friedman-Hill, Grabowecky, Robertson, and Treisman, 1997).

The emphasis on egocentric and allocentric processing of spatial information would suggest the possibility that short-term spatial information is first encoded in the PC in an egocentric framework to guide action and then processed in long-term memory in the hippocampus in an allocentric framework (Burgess, Jeffery, and O'Keefe, 1999; Save and Poucet, 2000a, 2000b). This model suggests that long-term memory for spatial information resides in the hippocampus, a view also championed by Nadel and Moscovitch (1997). Some support for this view comes from the finding of long-term retrograde amnesia gradients (for a review see Nadel and Moscovitch, 1997).

B. Perceptual Memory

Furthermore, in patients with parietal lesions and spatial neglect there is a deficit in spatial repetition priming without a loss in short-term or working memory for spatial information (Ellis, Sala, and Logie, 1996). One might expect hippocampal lesions to impair spatial short-term and intermediate-term memory or episodic memory (a component of event-based memory; see earlier), whereas PC lesions may impair perceptual memory (a component of knowledge-based memory). Indeed, a double dissociation exists between the hippocampus and the PC with regard to perceptual memory versus short-term memory (STM) or working memory (WM) for spatial information. Keane, Gabrieli, Mapstone, Johnston, and Corkin (1995) reported that a patient with occipital lobe damage (extending into the PC) showed a deficit in perceptual priming but demonstrated no effect on recognition memory, whereas a patient with bilateral medial temporal lobe damage (including hippocampus) had a loss of recognition memory but no loss of perceptual memory. Recently, Chiba, Kesner, and Jackson (2002) tested hippocampal- and PC-lesioned rats on two versions of a spatial continuous-recognition task. These authors demonstrated that rats with lesions of the PC had disrupted perceptual memory (continuous-reinforcement condition), whereas lesions of the hippocampus disrupted spatial working memory (differential-reinforcement condition). Furthermore, Kesner (2000b) reported that lesions of the PC, but not the hippocampus, disrupted positive priming (a component of perceptual memory) for spatial locations, whereas the same lesions had no effect on priming for visual objects. Thus, the hippocampus and PC can operate independent of each other, suggesting the possibility of parallel processing.

C. Encoding, Consolidation, and Retrieval in Long-Term Memory

Support for the idea that the PC is involved in encoding, consolidation, and retrieval can be found in an analysis of human patients with parietal cortex damage. In addition to problems with attention, sensation, and motor control, there is often a deficit associated with spatial aspects of the patients' environment. These include an inability to draw maps or diagrams of familiar spatial locations, to use information to guide them in novel or familiar routes, to discriminate near from far objects, and to solve complex mazes. There is a general loss of "topographic sense," which may involve loss of long-term geographical knowledge as well as an inability to form cognitive maps of new environments. Using PET scan and functional MRI data, it can be shown that complex spatial information results in activation of the parietal cortex (Ungerleider, 1995). Thus, memory for complex spatial information appears to be impaired (Benton, 1969; De Renzi, 1982).

Additional support comes from studies with parietal-lesioned monkeys. These animals demonstrate deficits in place reversal, landmark reversal, distance discrimination, bent wire route-finding, pattern string-finding, and maze-learning tasks (Milner, Ockleford, and DeWar, 1977; Petrides and Iversen, 1979). Furthermore, rats with parietal cortex lesions display deficits in both the acquisition and retention of spatial navigation tasks that are presumed to measure the operation of a spatial cognitive map within a complex environment (DiMattia and Kesner, 1988b; Kesner, Farnsworth, and Kametani, 1992). They also display deficits in the acquisition and retention of spatial recognition memory for a list of five spatial locations (DiMattia and Kesner, 1988a). When the task is more complex, involving the association of objects and places (components of a spatial cognitive map), then the parietal cortex plays an important role. Support for this comes from the finding that rats with parietal lesions are impaired in the acquisition and retention of a spatial location plus object discrimination (paired-associate task) but show no deficits for only spatial or object discriminations (Long and Kesner, 1996; Long, Mellem, and Kesner, 1998).

Finally, there is some support to suggest that the parietal cortex may be a site for long-term representation of complex spatial information. Cho and

Kesner (1996) have shown that rats with parietal cortex lesions have a non-graded retrograde amnesia for four, but not two, spatial discriminations learned prior to surgery, suggesting that the deficit cannot be due to a performance or anterograde amnesia problem, but rather appears to be a function of the amount or complexity of the spatial information to be stored and remembered.

In general, it is suggested that the hippocampus and PC operate independently with regard to intrinsic processes (such as short-term and intermediate-term memory versus perceptual memory), but they are more likely to interact during learning and retention of especially spatial information.

The parietal cortex is probably not the only neural region that mediates long-term memory for spatial information. For example, topographical amnesia has also been reported for patients with parahippocampal lesions, and spatial navigation deficits have also been found following retrosplenial and entorhinal cortex lesions (Habib and Sirigu, 1987; Sutherland and McDonald, 1990). Thus, other neural regions (e.g., parahippocampal cortex, entorhinal cortex, and retrosplenial cortex) may also contribute to the long-term representation of a spatial cognitive map. The medial entorhinal cortex appears to have *grid cells*. These are cells that show multiple peaks of firing distributed across the surface of the rat's total environment (Hafting, Fyhn, Molden, Moser, and Moser, 2006). It is possible that these grid cells reflect the representation of a cognitive map in long-term memory.

IV. SPATIAL ATTRIBUTE: RULE–BASED MEMORY

Evidence supportive of the aforementioned mapping of attributes onto specific brain regions is based in part on the use of paradigms that measure the use of rules within short-term memory based on performance within matching- or nonmatching-to-sample, delayed-conditional-discrimination, or continuous-recognition memory of a single item or lists of items, temporal ordering of information, and sequential learning as well as paradigms that measure the use of rules in cross-modal switching, reversal learning, paired-associate, prospective coding, and problem-solving tasks. For a detailed presentation of the research to support these statements, see Kesner, (2000b).

A. Short-Term or Working Memory

There is good evidence that the prefrontal cortex plays an important role in short-term memory, especially in a delayed matching- or nonmatching-to-sample tasks, where a correct choice response for a stimulus (e.g., spatial location) is required after a delay period (for a review see Kesner, 2000b). Strong electrophysiological correlates can be identified during a short-term delay period within the prefrontal cortex in nonhuman primates in spatial delayed-choice tasks (Constantinidis, Franowicz, and Goldman-Rakic, 2001; Rainer, Asaad, and Miller, 1998). The rodent medial prefrontal cortex also plays an equivalent role in tasks where a delayed-choice response to stimuli is required and lesions of the pre- and infralimbic cortex impair performance in delayed-choice tasks (Delatour and Gisquet-Verrier, 1996; Shaw and Aggleton, 1993). Thus, based on the previously mentioned studies it has been suggested that the prefrontal cortex, as part of the rule-based memory system, is involved in mediating short-term memory for spatial attribute information. The observations with lesions of the PL-IL cortex are similar to what has been reported for monkeys and humans with dorsolateral or ventrolateral prefrontal cortex lesions. In monkeys, for example, lesions of the dorsolateral and ventrolateral regions disrupt performance on delayed-response, delayed-alternation, delayed-occulomotor, spatial search, and visual object-recognition tasks (for a review, see Fuster, 1997; Kesner, 2000b). Furthermore, for short-term or working memory there are delay-specific cells in the dorsolateral and ventrolateral prefrontal cortex in spatial tasks, such as delayed-response, delayed-alternation, and delayed-occulomotor tasks and in visual object–delay tasks. In humans, D'Esposito et al. (1998) reported a meta-analysis of neuroimaging results based on visual object and spatial location working memory tasks that made a strong case for processing of both visual object and spatial location information in working memory in both the dorsolateral and ventrolateral prefrontal cortex.

To what extent is there an interaction between the prelimbic and infralimbic cortex and the hippocampus in mediating short-term and intermediate-term memory? To answer this question, Lee and Kesner (2003b) examined the dynamic interactions between the prefrontal cortex and the hippocampus by training and testing rats on a delayed nonmatching-to-place task on a radial 8-arm maze requiring memory for a single spatial location following short-term (i.e., 10-sec or 5-min) delays. The results showed that inactivating both regions at the same time resulted in a severe impairment of short-term and intermediate-term memory for spatial information, suggesting that one of the structures needs to function properly for intact processing of short-term or intermediate-term spatial memory. Thus, the two regions interact with each other to ensure the processing of spatial information across a dynamic temporal range including both short-term and intermediate-term memory. The current results provide compelling evidence that a mnemonic time window is a critical factor in dissociating the function of the hippocampal system from that of the medial prefrontal cortex in a delayed-choice task. That is, the dorsal hippocampus and medial prefrontal cortex appear to process spatial memory in parallel within a short-term range, whereas the dorsal hippocampal function becomes more essential once the critical time window requires spatial memory for a time period exceeding that range.

B. Cross-Modal Switching

It has been shown that in rats lesions of the prelimbic-infralimbic cortex impair cross-modal switching between place and visual cue or visual cue and place as well as motor response and place and place and motor response (Ragozzino, Detrick, and Kesner, 1999; Ragozzino, Wilcox, Raso, and Kesner, 1999). Similar results were reported by Dias, Robbins, and Roberts (1996), who have shown that in monkeys lateral prefrontal cortex lesions disrupt cross-modal, but not intramodal, switching; and Konishi, Nakajima, Uchida, Kameyama, Nakahara, Sekihara, and Miyashita (1998) have shown, using neuroimaging techniques, that the critical area of activation in humans in a set shifting paradigm is located between the dorsolateral and ventrolateral prefrontal cortex.

C. Goal-Oriented Control

Support for the idea that the prelimbic cortex may be involved in response selection and response readiness comes from the findings that lesions of the prelimbic cortex disrupt goal-directed performance during acquisition, as indicated by an insensitivity to the devaluation of a reward and by disruption of motor readiness to respond by releasing a lever cued by variable duration of a light stimulus (Killcross and Coutureau, 2003; Ostlund and Balleine, 2005).

D. Prospective Coding

It has been suggested that the prefrontal cortex mediates memory for future events, given that memory for the past within either a short or long time frame is not based on temporal order. One might then expect a dissociation between the prefrontal cortex and the hippocampus in spatial tasks that could be solved using both prospective and retrospective codes. In order to test this idea, sham-operated rats or animals with medial prefrontal cortex or hippocampus lesions were tested in a task that provided an opportunity for rats to utilize retrospective and prospective memory codes while remembering items (spatial locations) within short or long lists (Kametani and Kesner, 1989; Kesner, 1989). More specifically, on any one trial a rat was presented with 2, 4, 6, 8, or 10 items (spatial locations) on a 12-arm radial maze, followed 15 minutes later by two win-shift tests comprising a choice between a place previously visited and a novel place. Each animal was given a total of 20 trials, with eight tests for each point of interpolation or each list length (2, 4, 6, 8, or 10). During learning, rats show an increase in errors as the number of places to be remembered is increased from 2 to 6 to 8, reflecting the use of a retrospective memory code (the ability to remember the spatial locations previously visited). These rats also show a decrease in errors as the number of places is increased from 8 to 10, reflecting the use of a prospective memory code (the ability to anticipate the spatial locations that have not yet been visited). Results indicate that sham-operated animals display an increase in errors as a function of point of interpolation or set size (2–8 items), followed by a decrease in errors with a set size of 10 items, suggesting the use of both retrospective and prospective memory codes. In contrast, animals with medial prefrontal cortex lesions made few errors for short lists but a large number of errors for the longer lists, reflecting an inability to shift from a retrospective to a prospective memory code.

V. SUMMARY

Memory is a complex phenomenon, due to a large number of potential interactions that are associated with the organization of memory at the psychological and neural system levels. Most of the neurobiological models of memory postulate an organizational schema involving two or three systems, each supported by different neurobiological substrates and each mediated by different operating characteristics. These systems are labeled event-based, knowledge-based, and rule-based memory, locale versus taxon, working versus reference memory, declarative versus nondeclarative, and declarative versus procedural.

In the Kesner tripartite, multiple-attribute, multiple-process memory model, different forms of memory and its neurobiological underpinnings are represented in terms of the nature, structure, or content of information representation as a set of different attributes, including language, time, place, response, reward value (affect), and visual object as an example of sensory-perception. For each attribute, information is processed in the event-based memory system through operations that involve pattern separation and orthogonalization of specific attribute information, short-term and intermediate-term memory processing, encoding of specific arbitrary associations into long-term memory, consolidation and retrieval of stored information via flexibility and pattern completion. In addition, for each attribute, information is processed in the knowledge-based system through operations of long-term storage, selective attention, perceptual memory, and retrieval of pattern and arbitrary associations. Finally, for each attribute, it is assumed that information is processed in the rule-based memory system through the integration of information from the event-based and knowledge-based memory systems for the use of major processes that include the selection of strategies and rules for maintaining or manipulating information for subsequent action as well as short-term or working memory for new and familiar information. The neural systems that subserve specific attributes within a system can operate independent of each other, even though there are also many possibilities for interactions among the

Spatial Attribute Neural Circuit

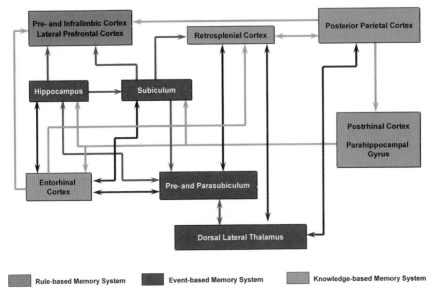

FIGURE 8-4   A representation of the spatial attribute neural circuit incorporating neural regions that mediate event-based, knowledge-based, and rule-based memory.

attributes. Even though the event-based and knowledge-based memory systems are supported by neural substrates and different operating characteristics, suggesting that the two systems can operate independent of each other, there are also important interactions between the two systems, especially during the consolidation of new information and retrieval of previously stored information. Finally, because it is assumed that the rule-based system is influenced by the integration of event-based and knowledge-based memory information, there should be important interactions between the event-based and knowledge-based memory systems and the rule-based memory system. Thus, for each attribute there is a neural circuit that encompasses all three memory systems in representing specific attribute information. Since I have concentrated primarily on the spatial attribute, I present in Figure 8-4 only a depiction of the neural substrates and their interconnections associated with the spatial (place) attribute across all three memory systems. Note that the dorsal lateral thalamus, pre- and parasubiculum, hippocampus, and subiculum represent neural substrates that support the event-based memory system; the entorhinal cortex, parahippocampal gyrus or postrhinal cortex, posterior parietal cortex, and retrosplenial cortex support the knowledge-based memory system; and the lateral prefrontal cortex or pre- and infralimbic cortex support the rule-based memory system. This circuit provides anatomical support for a possible independence in the operation of the hippocampus as part of the event-based memory system and posterior parietal cortex as part of the knowledge-based memory system, in that spatial information that is processed via the dorsal lateral thalamus can activate both the hippocampus and the posterior parietal cortex in parallel. Also, information can reach the lateral prefrontal cortex or pre- and infralimbic cortex as part of the rule-based memory system via direct connections from the posterior parietal cortex part of the knowledge-based memory system and hippocampus as part of the event-based memory system. Finally, spatial information can interact with other specific attributes via a series of direct connections, including, for example, an interaction with reward value attribute information via a hippocampal–amygdala connection or lateral prefrontal cortex–orbital frontal cortex connections or an interaction with response attribute information via hippocampal–caudate or lateral prefrontal–premotor or supplementary motor connections. In general, the tripartite attribute memory model represents the most comprehensive memory model capable of integrating the extant knowledge concerning the neural system representation of memory.

REFERENCES

Alvarez, P., Zola-Morgan, S., and Squire, L.R. (1994). The animal model of human amnesia: Long-term memory impaired and short-term memory intact. *Proceedings of the National Academy of Sciences of the United States of America, 91*, 5637–5641.

Anderson, R.A. (1999). Multimodal integration for the representation of space in the posterior parietal cortex. In N. Burgess, K.J. Jeffery, J. O'Keefe (eds.), *The Hippocampal and Parietal Foundations of Spatial Cognition* (pp. 90–103). Oxford, U.K.: Oxford University Press.

Benton, A.L. (1969). Disorder of spatial orientation. In P.J. Vinken and G.W. Bruyn (eds.), *Handbook of Clinical Neurology (Vol. 3)*. Amsterdam: North Holland.

Berthoz, A. (2000). *The Brain's Sense of Movement*. Cambridge, MA: Harvard University Press.

Buckley, M.J., and Gaffan, D. (1998). Perirhinal cortex ablation impairs configural learning and paired-associate learning equally. *Neuropsychologia, 36*, 535–546.

Bunsey, M., and Eichenbaum, H. (1993). Critical role of the parahippocampal region for paired-associate learning in rats. *Behavioral Neuroscience, 107*, 740–774.

Bures, J., Fenton, A.A., Kaminsky, Y., Rossier, J., Sacchetti, B., and Zinyuk, L. (1999). Dissociation of exteroceptive and ideothetic orientation cues: Effect on hippocampal place cells and place navigation. In N. Burgess, K.J. Jeffery, and J. O'Keefe (eds.), *The Hippocampal and Parietal Foundations of Spatial Cognition* (pp. 167–185). Oxford, U.K.: Oxford University Press.

Burgess, N., Jeffery, K.J., and O'Keefe, J. (1999). Integrating hippocampal and parietal functions: A spatial point of view. In N. Burgess, K.J. Jeffery, and J. O'Keefe (eds.), *The Hippocampal and Parietal Foundations of Spatial Cognition* (pp. 3–32). Oxford, U.K.: Oxford University Press.

Cave, C.B., and Squire, L.R. (1992). Intact verbal and nonverbal short-term memory following.

Chen, L.L., Lin, L., Barnes, C.A., and McNaughton, B.L. (1994). Head-direction cells in the rat posterior cortex II: Contributions of visual and ideothetic information to the directional firing. *Experimental Brain Research, 101,* 24–34.

Chen L.L., and Nakamura, K. (1998). Head-centered representation and spatial memory in rat posterior parietal cortex. *Psychobiology, 26,* 119–127.

Chiba, A.A., Kesner, R.P., and Jackson, P. (2002). Two forms of spatial memory: A double dissociation between the parietal cortex and the hippocampus in the rat. *Behavioral Neuroscience, 116,* 874–883.

Chiba, A.A., Kesner, R.P., Matsuo, F., and Heilbrun, M.P. (1993). A dissociation between affect and recognition following unilateral temporal lobectomy including the amygdala. *Society for Neuroscience Abstracts, 19,* 792.

Cho, Y.H., and Kesner, R.P. (1995). Relational object association learning in rats with hippocampal lesions. *Behavioural Brain Research, 67,* 91–98.

Cho, Y.H., and Kesner, R.P. (1996). Involvement of entorhinal cortex or parietal cortex in long-term spatial discrimination memory in rats: Retrograde amnesia. *Behavioral Neuroscience, 110,* 436–442.

Cho, Y.H., Kesner, R.P., and Brodale, S. (1995). Retrograde and anterograde amnesia for spatial discrimination in rats: Role of hippocampus, entorhinal cortex and parietal cortex. *Psychobiology, 23,* 185–194.

Cohen, N.J., and Eichenbaum, H.B. (1993). *Memory, Amnesia, and Hippocampal Function.* Cambridge, MA: MIT Press.

Colby, C.L. (1999). Parietal cortex constructs action-oriented spatial representations. In N. Burgess, K.J. Jeffery, and J. O'Keefe (eds.), *The Hippocampal and Parietal Foundations of Spatial Cognition* (pp. 104–126). New York: Oxford University Press.

Constantinidis, C., Franowicz, M.N., and Goldman-Rakic, P.S. (2001). The sensory nature of mnemonic representation in the primate prefrontal cortex. *Nature Neuroscience, 4,* 311–316.

D'Esposito, M., Aguirre, G.K., Zarahn, E., Ballard, D., Shin, R.K., and Lease, J. (1998). Functional MRI studies of spatial and nonspatial working memory. *Cognitive Brain Research, 7,* 1–13.

De Renzi, E. (1982). *Disorders of Space Exploration and Cognition.* New York: Wiley.

Delatour, B., and Gisquet-Verrier, P. (1996). Prelimbic cortex specific lesions disrupt delayed-variable response tasks in the rat. *Behavioral Neuroscience, 110,* 1282–1298.

Dias, R., Robbins, T.W., and Roberts, A.C. (1996). Dissociation in prefrontal cortex of affective and attentional shifts. *Nature, 380,* 69–72.

DiMattia, B.V., and Kesner, R.P. (1984). Serial position curves in rats: Automatic versus effortful information processing. *Journal of Experimental Psychology. Animal Behavior Processes, 10,* 557–563.

DiMattia, B.V., and Kesner, R.P. (1988a). The role of the posterior parietal association cortex in the processing of spatial event information. *Behavioral Neuroscience, 102,* 397–403.

DiMattia, B.V., and Kesner, R.P. (1988b). Spatial cognitive maps: Differential role of parietal cortex and hippocampal formation. *Behavioral Neuroscience, 102,* 471–480.

Eichenbaum, H. (2004). Hippocampus: Cognitive processes and neural representations that underlie declarative memory. *Neuron, 44,* 109–120.

Eichenbaum, H., and Cohen, N.J. (2001). *From Conditioning to Conscious Recollection: Memory Systems of the Brain.* New York: Oxford University Press.

Eichenbaum, H., Stewart, C., and Morris, R.G.M. (1990). Hippocampal representation in spatial learning. *Journal of Neuroscience, 10,* 331–339.

Ellis, A.X., Sala, S.D., and Logie, R.H. (1996). The Bailiwick of visuo-spatial working memory: Evidence from unilateral spatial neglect. *Cognitive Brain Research, 3,* 71–78.

Fanselow, M.S., and Poulos, A.M. (2005). The neuroscience of mammalian associative learning. *Annual Review of Psychology, 56,* 207–234.

Friedman-Hill, S., Grabowecky, M., Robertson, L., and Treisman, A. (1997). The interaction of spatial and object pathways: Evidence from Balint's syndrome. *Journal of Cognitive Neuroscience, 9(3)*, 295–317.

Fuster, J.M. (1997). *The Prefrontal Cortex: Anatomy, Physiology, and Neuropsychology of the Frontal Lobe* (3rd ed.). New York: Lippincott-Raven.

Gaffan, D. (1994). Dissociated effects of perirhinal cortex ablation, fornix transection and amygdalectomy: Evidence for multiple memory systems in the primate temporal lobe. *Experimental Brain Research, 99*, 411–422.

Gaffan, D., and Harrison, S. (1989). Place memory and scene memory: Effects of fornix transection. *Experimental Brain Research, 74*, 202–212.

Gilbert, P.E., and Kesner, R.P. (2002a). Role of the rodent hippocampus in paired-associate learning involving associations between a stimulus and a spatial location. *Behavioral Neuroscience, 116*, 63–71.

Gilbert, P.E., and Kesner, R.P. (2002b). The amygdala but not the hippocampus is involved in pattern separation based on reward value. *Neurobiology of Learning and Memory, 77*, 338–353.

Gilbert, P.E., and Kesner, R.P. (2003a). Localization of function within the dorsal hippocampus: The role of the CA3 subregion in paired-associate learning. *Behavioral Neuroscience, 117*, 1385–1394.

Gilbert, P.E., and Kesner, R.P. (2003b). Recognition memory for complex visual discrimination is influenced by stimulus interference in rodents with perirhinal cortex damage. *Learning and Memory, 10*, 525–530.

Gilbert, P.E., Kesner, R.P., and DeCoteau, W. (1998). Memory for spatial location: Role of the hippocampus in mediating spatial pattern separation. *Journal of Neuroscience, 18*, 804–810.

Gilbert, P.E., Kesner, R.P., and Lee, I. (2001). Dissociating hippocampal subregions: A double dissociation between dentate gyrus and CA1. *Hippocampus, 11*, 626–636.

Gold, E., and Kesner, R.P. (2005). The role of the CA3 subregion of the dorsal hippocampus in spatial pattern completion in the rat. *Hippocampus, 15*, 808–814.

Goodrich, N.J., Hunsaker, M.R., and Kesner, R.P. (2005b). Effects of hippocampus subregional lesions for metric and topological spatial information processing. *Society for Neuroscience 35*th *Annual Meeting*, Washington, DC.

Goodrich-Hunsaker, N.J., Hunsaker, M.R., and Kesner, R.P. (2005a). Dissociating the role of the parietal cortex and dorsal hippocampus for spatial information processing. *Behavioral Neuroscience, 119*, 1307–1315.

Habib, M., and Sirigu, A. (1987). Pure topographical disorientation: A definition and anatomical basis. *Cortex, 23*, 73–85.

Hafting, T., Fyhn, M, Molden, S. Moser M.B., and Moser, E.I. (2006). Microstructure of a spatial map in the entorhinal cortex. *Nature, 436*, 801–806.

Higuchi, S., and Miyashita, Y. (1996). Formation of mnemonic neuronal responses to visual paired associates in inferotemporal cortex is impaired by perirhinal and entorhinal lesions. *Proceedings of the National Academy of Sciences, 93*, 739–743.

Holdstock, J.S., Shaw, C., and Aggleton, J.P. (1995). The performance of amnesic subjects on tests of delayed matching-to-sample and delayed matching-to-position. *Neuropsychologia, 33*, 1583–1596.

Hopkins, R.O., and Kesner, R.P. (1993). Memory for temporal and spatial distances for new and previously learned geographical information in hypoxic subjects. *Society for Neuroscience Abstracts, 19*, 1284, Washington, DC.

Hopkins, R.O., Kesner, R.P., and Goldstein, M. (1995). Item and order recognition memory for words, pictures, astract pictures, spatial locations, and motor responses in subjects with hypoxic brain injury. *Brain and Cognition, 27*, 180–201.

Hopkins, R.O., Pasker, M., and Kesner, R.P. (1995). Impaired spatial location — abstract object paired-associate learning in hypoxic subjects. *Society for Neuroscience Abstracts, 21*, 272.

Hunt, M.E., Kesner, R.P., and Evans, R.B. (1994). Memory for spatial location: Functional dissociation of entorhinal cortex and hippocampus. *Psychobiology, 22,* 186–194.

Jackson-Smith, P., Kesner, R.P., and Chiba, A.A. (1993). Continuous recognition of spatial and nonspatial stimuli in hippocampal lesioned rats. *Behavioral and Neural Biology, 59,* 107–119.

Kametani, H., and Kesner, R.P. (1989). Retrospective and prospective coding of information: Dissociation of parietal cortex and hippocampal formation. *Behavioral Neuroscience, 103,* 84–89.

Keane, M.M., Gabrieli, J.D.E., Mapstone, H.C., Johnson, K.A., and Corkin, S. (1995). Double dissociation of memory capacities after bilateral occipital-lobe or medial temporal-lobe lesions. *Brain, 118,* 1129–1148.

Kesner, R.P. (1989). Retrospective and prospective coding of information: Role of the medial prefrontal cortex. *Journal of Experimental Brain Research, 74,* 163–167.

Kesner, R.P. (1990). Learning and memory in rats with an emphasis on the role of the hippocampal formation. In R.P. Kesner and D.S. Olton (eds.), *Neurobiology of Comparative Cognition* (pp. 179–204). Hillsdale, NJ: Erlbaum.

Kesner, R.P. (1998). Neurobiological views of memory. In J.L. Martinez and R.P. Kesner (eds.), *Neurobiology of Learning and Memory* (pp. 361–416). San Diego: Academic Press.

Kesner, R.P. (2000a). Subregional analysis of mnemonic functions of the prefrontal cortex in the rat. *Psychobiology, 28,* 219–228.

Kesner, R.P. (2000b). Behavioral analysis of the contribution of the hippocampus and parietal cortex to the processing of information: Interactions and dissociations. *Hippocampus, 10,* 483–490.

Kesner, R.P. (2002). Memory neurobiology. In V.S. Ramachadran (ed), *Encyclopedia of the Human Brain, Vol. 2* (pp. 783–796). San Diego: Academic Press.

Kesner, R.P., Bolland, B.L., and Dakis, M. (1993). Memory for spatial locations, motor response, and objects: Triple dissociation among the hippocampus, caudate nucleus, and extrastriate visual cortex. *Experimental Brain Research, 93,* 462–470.

Kesner, R.P., Farnsworth, G., and Kametani, H. (1992). Role of parietal cortex and hippocampus in representing spatial information. *Cerebral Cortex, 1,* 367–373.

Kesner, R.P., and Gilbert, P.E. (2006). The role of the medial caudate nucleus, but not the hippocampus, in a matching—to sample task for a motor response. *European Journal of Neuroscience, 23,* 1888–1894.

Kesner, R.P., and Giles, R. (1998). Neural circuit analysis of spatial working memory: Role of pre- and parasubiculum, medial and lateral entorhinal cortex. *Hippocampus, 8,* 416–423.

Kesner, R., and Hopkins, R. (2001). Short-term memory for duration and distance in humans: Role of the hippocampus. *Neuropsychology, 15,* 58–68.

Kesner, R.P., Hunsaker, M.R., and Gilbert P.E. (2005). The role of CA1 in the acquisition of an object-trace-odor paired-associate task. *Behavioral Neuroscience, 119,* 781–786.

Kesner, R.P., and Wilburn, M.W. (1974). A review of electrical stimulation of the brain in context of learning and retention. *Behavioral Biology, 10,* 259–293.

Kesner, R.P., and Williams, J.M. (1995). Memory for magnitude of reinforcement: Dissociation between the amygdala and hippocampus. *Neurobiology of Learning and Memory, 64,* 237–244.

Killcross, S., and Coutureau, E. (2003). Coordination of actions and habits in the medial prefrontal cortex of rats. *Cerebral Cortex, 13,* 400–408.

Kim, J.J., and Fanselow, M.S. (1992). Modality-specific retrograde amnesia of fear. *Science, 256,* 675–677.

Kirwan, C.B., Gilbert, P.E., and Kesner, R.P. (2005). The role of the hippocampus in the retrieval of a spatial location. *Neurobiology of Learning and Memory, 83,* 65–71.

Konishi, S., Nakajima, K., Uchida, I., Kameyama, M., Nakahara, S. Sekihara, K., and Miyashita, Y. (1998). Transient activation of inferior prefrontal cortex during cognitive set shifting. *Nature Neuroscience, 1,* 80–84.

Lee, I., and Kesner, R.P. (2002). Differential contribution of NMDA receptors in hippocampal subregions to spatial working memory. *Nature Neuroscience, 5,* 162–168.

Lee, I., and Kesner, R.P. (2003a). Differential roles of dorsal hippocampal subregions in spatial working memory with short versus intermediate delay. *Behavioral Neuroscience, 117,* 1044–1053.

Lee, I., and Kesner, R.P. (2003b). Time-dependent relationship between the dorsal hippocampus and the prefrontal cortex in spatial memory. *Journal of Neuroscience, 23,* 1517–1523.

Leonard, G., and Milner, B. (1991). Contribution of the right frontal lobe to the encoding and recall of kinesthetic distance information. *Neuropsychologia, 29,* 47–58.

Li, H., Matsumoto, K., and Watanabe, H. (1999). Different effects of unilateral and bilateral hippocampal lesions in rats on the performance of radial maze and odor paired-associate tasks. *Brain Research Bulletin, 48,* 113–119.

Long, J., Mellem, J., and Kesner, R. (1998). The effects of parietal cortex lesions on an object/ spatial location paired-associate task in rats. *Psychobiology, 26,* 128–133.

Long, J.M., and Kesner, R.P. (1996). The effects of dorsal vs. ventral hippocampal, total hippocampal, and parietal cortex lesions on memory for allocentric distance in rats. *Behavioral Neuroscience, 110,* 922–932.

Long, J.M., and Kesner, R.P. (1998). Effects of hippocampal and parietal cortex lesions on memory for egocentric distance and spatial location information in rats. *Behavioral Neuroscience, 112,* 480–495.

Maren, S., Aharonov, G., and Fanselow, M.S. (1996). Retrograde abolition of conditional fear after excitotoxic lesions in the basolateral amygdala of rats: Absence of a temporal gradient. *Behavioral Neuroscience, 110,* 718–726.

McNaughton, B.L., Chen, L.L., and Marcus, E.J. (1991). "Dead Reckoning," landmark learning, and the sense of direction: A neurophysiological and computational hypothesis. *Journal of Cognitive Neuroscience, 3,* 190–202.

Milner, A.D., Ockleford, E.M., and DeWar, W. (1977). Visuo-spatial performance following posterior parietal and lateral frontal lesions in stumptail macaques. *Cortex, 13,* 170–183.

Mizumori, S.J., Ragozzino, K.E., Cooper, B.G., and Leutgeb, S. (1999). Hippocampal representational organization and spatial context. *Hippocampus, 9,* 444–451.

Mizumori, S.J.Y., and Williams, J.D. (1993). Directionally selective mnemonic properties of neurons in the lateral dorsal nucleus of the thalamus of rats. *Journal of Neuroscience, 13,* 4015–4028.

Morris, R.G.M., Garrud, J.N.P., Rawlins, J.N.P., and O'Keefe, J. (1982). Place navigation impaired in rats with hippocampal lesions. *Nature, 297,* 681–683.

Murray, E.A., Gaffan, D., and Mishkin, M. (1993). Neural substrates of visual stimulus–stimulus association in rhesus monkeys. *Journal of Neuroscience, 13,* 4549–4561.

Nadel, L., and Moscovitch, M. (1997). Consolidation, retrograde amnesia and the hippocampal formation. *Current Opinion in Neurobiology, 7,* 217–227.

Nakazawa, K., Quirk, M.C., Chitwood, R.A., Watanabe, M., Yeckel, M.F., Sun, L.D., Kato, A., Carr, C.A., Johnston, D., Wilson, M.A., and Tonegawa, S. (2002). Requirement for hippocampal CA3 NMDA receptors in associative memory recall. *Science, 297,* 211–218.

O'Keefe, J. (1979). A review of the hippocampal place cells. *Progressive Neurobiology, 13,* 419–439.

O'Keefe, J., and Nadel, L. (1978). *The hippocampus as a cognitive map.* Oxford, U.K.: Clarendon Press.

Olton, D.S. (1983). Memory functions and the hippocampus. In W. Seifert (ed.), *Neurobiology of the Hippocampus.* New York: Academic Press.

O'Reilly, R.C., and McClelland, J.L. (1994). Hippocampal conjunctive encoding storage, and recall: Avoiding a trade-off. *Hippocampus, 4,* 661–682.

Ostlund, S.B., and Balleine, B.W. (2005). Lesions of medial prefrontal cortex disrupt the acquisition but not the expression of goal-directed learning. *Journal of Neuroscience, 25*, 7763–7770.

Owen, A.M., Sahakian, B.J., Semple, J., Polkey, E., and Robbins, T.W. (1995). Visuo-spatial short-term recognition memory and learning after temporal lobe excisions, frontal lobe excisions or amygdalo-hippocampectomy in man. *Neuropsychologia, 33*, 1–24.

Packard, M.G., and Knowlton, B.J. (2002). Learning and memory functions of the basal ganglia. *Annual Review in Neuroscience, 25*, 563–593.

Parkinson, J.K., Murray, E.A., and Mishkin, M. (1988). A selective mnemonic role for the hippocampus in monkeys: Memory for the location of objects. *Journal of Neuroscience, 8*, 4159–4167.

Petrides, M., and Iversen, S.D. (1979). Restricted posterior parietal lesions in the rhesus monkey and performance on visuo-spatial tasks. *Brain Research, 161*, 63–77.

Pigott, S., and Milner, B. (1993). Memory for different aspects of complex visual scenes after unilateral temporal- or frontal-lobe resection. *Neuropsychologia, 31*, 1–15.

Poucet, B. (1993). Spatial cognitive maps in animals: New hypotheses on their structure and neural mechanisms. *Psychological Review, 100(2)*, 163–182.

Ragozzino, M.E., Detrick, S., and Kesner, R.P. (1999). Involvement of the prelimbic-infralimbic areas of the rodent prefrontal cortex in behavioral flexibility for place and response learning. *Journal of Neuroscience, 19*, 4585–4594.

Ragozzino, M.E., Wilcox, C., Raso, M., and Kesner, R.P. (1999). Involvement of the medial prefrontal cortex subregions in strategy switching. *Behavioral Neuroscience, 113*, 32–41.

Rainer, G., Asaad, W.F., and Miller, E.K. (1998). Selective representation of relevant information by neurons in the primate prefrontal cortex. *Nature, 393*, 577–579.

Robertson, L., Treisman, A., Friedman-Hill, S., and Grabowecky, M. (1997). The interaction of spatial and object pathways: Evidence from Baliant's syndrome. *Journal of Cognitive Neuroscience, 9*, 295–317.

Rogers, J.L., and Kesner, R.P. (2006). Lesions of the dorsal hippocampus or parietal cortex differentially affect spatial information processing. *Behavioral Neuroscience, 120*, 852–860.

Rolls, E.T. (1996). A theory of hippocampal function in memory. *Hippocampus, 6*, 610–620.

Rolls, E.T. (1999). Spatial view cells and the representation of place in the primate hippocampus. *Hippocampus, 9*, 467–480.

Rolls, E.T., and Treves, A. (1998). *Neural Networks and Brain Function.* Oxford, U.K.: Oxford University Press.

Ryan, J.D., Alhoff, R.R., Whitlow, S., and Cohen, N.J. (2000). Amnesia is a deficit in relational memory. *Psychological Science, 11*, 454–461.

Ryan, J.D., and Cohen, N.J. (2004). Processing and short-term retention of relational information in amnesia. *Neuropsychologia, 42*, 497–511.

Save, E., and Poucet, B. (2000a). Hippocampal–parietal cortical interactions in spatial cognition. *Hippocampus, 10*, 491–499.

Save, E., and Poucet, B. (2000b). Involvement of the hippocampus and associative parietal cortex in the use of proximal and distal landmarks for navigation. *Behavioral Brain Research, 109*, 195–206.

Save, E., Poucet, B., Foreman, N., and Thinus Blanc, C. (1992). Object exploration and reactions to spatial and nonspatial changes in hooded rats following damage to parietal cortex or hippocampal formation. *Behavioral Neuroscience, 106*, 447–456.

Schacter, D.L. (1987). Implicit memory: History and current status. *Journal of Experimental Psychology: Learning, Memory, and Cognition, 13*, 501–518.

Schacter, D.L., and Tulving, E. (1994). *Memory Systems 1994.* Cambridge, MA: MIT Press.

Shapiro, M.L., and Olton, D.S. (1994). Hippocampal function and interference. In D.L. Schacter, E. Tulving (eds.), *Memory Systems 1994.* (pp. 141–146). Cambridge, MA: MIT press.

Shaw, C., and Aggleton, J.P. (1993). The effects of fornix and medial prefrontal lesions on delayed non-matching-to-sample by rats. *Behavioural Brain Research, 54*, 91–102.

Shimamura, A.P. (1986). Priming effects in amnesia: Evidence for a dissociable memory function. *Quarterly Journal of Experimental Psychology, 38A*, 619–644.

Smith, M.L., and Milner, B. (1989). Right hippocampal impaiment in the recall of spatial location: Encoding deficit or rapid forgetting? *Neuropsychologia, 27*, 71–81.

Smith, M.L., and Milner, B. (1981). The role of the right hippocampus in the recall of spatial location. *Neuropsychologia, 19*, 781–793.

Squire, L.R. (1994). Declarative and nondeclarative memory: Multiple brain systems supporting learning and memory. In D.L. Schacter and E. Tulving (eds.), *Memory Systems 1994* (pp. 203–231). Cambridge, MA: MIT Press.

Squire, L.R. (2004). Memory systems of the brain: A brief history and current perspective. *Neurobiology of Learning and Memory, 82*, 171–177.

Sutherland, R.J., and McDonald, R.J. (1990). Hippocampus, amygdala, and memory deficits in rats. *Behavioural Brain Research, 37*, 57–79.

Sziklas, V., Lebel, S., and Petrides, M. (1998). Conditional associative learning and the hippocamal system. *Hippocampus, 8*, 131–137.

Taube, J.S., Goodridge, J.P., Golob, E.J., Dudchenko, P.A., and Stackman, R.W. (1996). Processing the head direction cell signal: A review and commentary. *Brain Research Bulletin, 40*, 447–486.

Treisman, A., and Gelade, G. (1980). A feature integration theory of attention. *Cognitive Psychology, 12*, 97–136.

Tulving, E. (1983). *Elements of Episodic Memory.* Oxford, U.K.: Clarendon Press.

Ungerleider, L.G. (1995). Functional brain imaging studies of cortical mechanisms of memory. *Science, 270*, 769–775.

Weisend, M.P., Astur, R.S., and Sutherland, R.J. (1996). The specificity and temporal characteristics of retrograde amnesia after hippocampal lesions. *Society for Neuroscience Abstracts, 22*, 1118.

Whishaw, I.Q., McKenna, J.E., and Maaswinkel, H. (1997). Hippocampal lesions and path integration. *Currrent Opinion in Neurobiology, 7*, 228–234.

Wood, E.R., Agster, K.M., and Eichenbaum, H. (2004). One-trial odor–reward association: A form of event memory not dependent on hippocampal functions. *Behavioral Neuroscience, 118*, 526–539.

CHAPTER 9

The Medial Temporal Lobe and Memory

Alison R. Preston and Anthony D. Wagner

Department of Psychology and Neurosciences Program, Stanford University, Stanford, CA 94305

I. INTRODUCTION

Memory is a central component of our cognitive lives, and it is accomplished by multiple neurobiological memory systems that encode, retain, and retrieve different forms of information about past experience. Memory for the past can be used to build predictive models in the present and thus can guide current thoughts, decisions, and actions. For example, memory enables us to discriminate novel people, places, and things from stimuli that are familiar due to prior encounters. At other times, memory allows us to recollect specific details from a past experience, providing relevant information that may inform current decisions. Given the centrality of memory, there has been an intense focus on delineating the neural mechanisms through which knowledge and experience are learned, represented, and later accessed.

In this chapter, we focus on the role of medial temporal lobe (MTL) structures — the hippocampus and surrounding parahippocampal, perirhinal, and entorhinal cortices — in memory processing and representation. Extensive evidence indicates that there are multiple forms of long-term memory, with only some depending on the function of the MTL. In particular, bilateral

Neurobiology of Learning and Memory, Second Edition

damage to the MTL gives rise to impairments in *declarative memory* — long-term memory for general facts (semantic memory) and specific events (episodic memory) (Cohen and Squire, 1980; Eichenbaum, 2000; Eichenbaum and Cohen, 2001; Tulving, 1983). In the face of profound impairments in declarative memory following bilateral MTL damage (Milner, 1972; Scoville and Milner, 1957), there is remarkable preservation of immediate forms of memory (working memory) as well as of nondeclarative forms of long-term memory, in which knowledge is expressed as experience-induced changes in performance unrelated to awareness (Gabrieli, 1998; Squire, 1992; Squire et al., 2004). This review focuses on the role of the MTL in episodic memory, first considering relevant aspects of MTL anatomy and then considering neuropsychological and neuroimaging data in humans that bear on understanding how the MTL supports memory for events.

II. MTL ANATOMY

A. Cortical Projections to and within the MTL

The MTL memory circuit comprises multiple structures, including the hippocampal formation and the surrounding entorhinal, perirhinal, and parahippocampal cortices (Figs. 9-1a and b). The hippocampal formation is further composed of the dentate gyrus (DG), CA fields, and the subiculum. Connections between hippocampus and entorhinal, perirhinal, and parahippocampal cortices are hierarchically organized (Lavenex and Amaral, 2000; Fig. 9-1c). Perirhinal cortex and parahippocampal cortex receive input from unimodal and polymodal association cortices in the lateral temporal, frontal, and parietal lobes via distinct pathways (Jones and Powell, 1970; Suzuki and Amaral, 1994a; Tranel et al., 1988; Van Hoesen and Pandya, 1975; Van Hoesen et al., 1975). In infrahuman primates, the predominant inputs to perirhinal cortex come from unimodal visual association areas in the adjacent inferior temporal cortex, a region important for visual-object processing (Suzuki and Amaral, 1994a). In contrast, parahippocampal cortex receives its predominant input from posterior visual association areas and posterior parietal cortex, whose functions are more visuospatial in nature. Parahippocampal cortex also receives inputs from unimodal auditory association cortex in the superior temporal gyrus (Suzuki and Amaral, 1994a; Tranel et al., 1988).

Perirhinal and parahippocampal cortices provide the major inputs to the second level of the MTL hierarchy — entorhinal cortex — which also receives limited information from polymodal association areas (Insausti et al., 1987). The topographic organization of connections between these regions and entorhinal cortex is distinct. Perirhinal cortex projects primarily to the anterior two-thirds of entorhinal cortex, while parahippocampal cortical projections

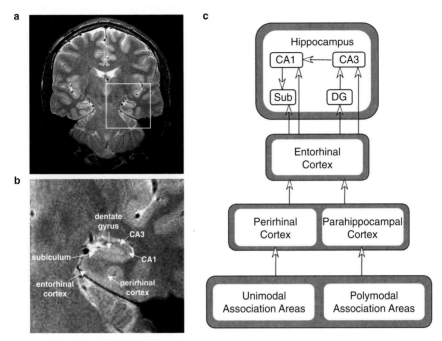

FIGURE 9-1   Anatomy of the medial temporal lobes. **a.** A coronal slice from a T-2 weighted structural MR image through the medial temporal lobe (MTL) region, as indicated by the box. **b.** MTL subregions displayed on the magnified structural image. **c.** Simplified circuit diagram of the neuroanatomical connectivity of the MTL region reflecting the dominant pathways.

terminate primarily in the posterior third (Suzuki and Amaral, 1994b). Parahippocampal connectivity with entorhinal cortex has a further topographic dimension: Medial regions of parahippocampal cortex project to medial, posterior entorhinal cortex, and lateral parahippocampal regions project to lateral, posterior entorhinal cortex. Collectively, this pattern of connectivity suggests that the segregation of different neocortical inputs to perirhinal and parahippocampal cortices might be preserved within entorhinal cortex, resulting in functional differences across subregions of entorhinal cortex (Hargreaves et al., 2005).

B. Hippocampal Circuitry

The entorhinal cortex provides the major inputs to hippocampus at the apex of the MTL hierarchy (Amaral et al., 1987; Suzuki and Amaral, 1990; Van Hoesen and Pandya, 1975; Witter and Amaral, 1991; Witter et al., 1989).

Entorhinal cortex projects its output to the granule cells of the dentate gyrus via the polysynaptic (trisynaptic) pathway (Amaral and Insausti, 1990; Lavenex and Amaral, 2000). These dentate gyrus granule cells project to the CA3 region of hippocampus via the mossy fiber pathway. In turn, projections from the CA3 pyramidal cells include collaterals to other CA3 pyramidal cells comprising an extensive system of associational connections within the region. These collateral connections are thought to be central to the role of hippocampus in the formation of memory conjunctions that capture the cooccurrence of multiple, disparate neocortical activation patterns during an event. Projections from CA3 also include the Schaffer collaterals, which constitute the major projection to the CA1 pyramidal cells. CA1 receives additional input from entorhinal cortex as part of the direct (monosynaptic) pathway (Amaral and Insausti, 1990; Duvernoy, 1998). CA1 then projects to both the subiculum and entorhinal cortex. Unlike the CA3 field, there are few associational connections within CA1 (Lavenex and Amaral, 2000).

It is noteworthy that, as with cortical projections to entorhinal cortex, projections from entorhinal cortex to hippocampus are topographically organized (Duvernoy, 1998; Witter and Amaral, 1991; Witter et al., 1989). In the infrahuman primate, lateral entorhinal cortex projects preferentially to posterior dentate gyrus, while medial entorhinal cortex projects to anterior dentate gyrus. Different anterior–posterior levels of entorhinal cortex also project to different proximal and distal regions of CA1 and subiculum. Given the topographical organization of inputs to entorhinal cortex from perirhinal and parahippocampal cortices, processing in different regions of dentate gyrus, CA fields, and subiculum may be influenced differentially by information from perirhinal and parahippocampal cortices (Small, 2002; Small et al., 2001; Suzuki and Amaral, 1994b). Parahippocampal cortex, which projects more medially than perirhinal cortex (Suzuki and Amaral, 1994b), may influence more anterior regions of dentate gyrus. In addition, perirhinal and parahippocampal cortices terminate at different anterior–posterior levels of entorhinal cortex, suggesting a possible difference in the distribution of input from these regions to CA1 and subiculum (Suzuki and Amaral, 1994b).

III. MEMORY IMPAIRMENTS RESULTING FROM MTL DAMAGE

A. Anterograde Amnesia

Damage to MTL structures may result from a variety of causes, including resection, herpes simplex encephalitis, anoxia, infarction, and sclerosis (Gabrieli, 1998). Lesions to the MTL region produce a profound anterograde amnesia resulting from an inability to learn new information after the onset of the injury. Critically, such impairments in new learning are limited to the acquisition of declarative memory for facts and events (Cohen and Squire, 1980; Milner et al., 1998; Scoville and Milner, 1957; Squire, 1992). Unilateral MTL lesions result in material-specific declarative memory deficits, with left lateralized lesions producing greater declarative memory deficits for verbal information and right lateralized lesions leading to greater deficits for nonverbal information (Milner, 1971). Bilateral MTL lesions produce a global antero-grade amnesia that extends to multiple domains, regardless of the sensory modality or verbal content of information (Gabrieli et al., 1988; Levy et al., 2003; Milner, 1972; Squire et al., 2001b). Strikingly, selective lesions limited to the CA1 field of the hippocampus are sufficient to produce anterograde amnesia (Rempel-Clower et al., 1996; Zola-Morgan et al., 1986).

A central question raised by the neuropsychological study of amnesic patients with MTL damage is whether the declarative memory deficits observed in anterograde amnesia are the consequence of impairments in encoding new information or of failures to retrieve newly acquired memories or both. Neuroimaging studies in healthy humans provide evidence that MTL structures are involved in both the encoding and retrieval of new episodic information. MTL activation has been associated with episodic encoding of novel information (Gabrieli et al., 1997; Schacter and Wagner, 1999; Stern et al., 1996), with MTL encoding activation being predictive of later remembering or forgetting (Brewer et al., 1998; Henson et al., 1999; Kirchhoff et al., 2000; Paller and Wagner, 2002; Wagner et al., 1998; Fig. 9-2). At retrieval, activation in MTL regions has been associated with the successful retrieval of declarative memory for prior events (Lepage et al., 1998; Schacter and Wagner, 1999), including recognizing and recalling previously encountered stimuli (Gabrieli et al., 1997; Henson, 2005; Rugg and Yonelinas, 2003; Schacter et al., 1996; Small et al., 2001) and remembering specific contextual details about a past encounter (Dobbins et al., 2003; Eldridge et al., 2000).

B. Memory and Awareness

An equally important question regarding MTL mnemonic function concerns the forms of memory that are supported by this neural circuit. A leading theory focuses on conscious awareness as a diagnostic feature of MTL involvement in memory. Extensive evidence indicates that MTL damage results in profound impairment to declarative memory (Cohen and Squire, 1980), as measured by explicit memory tasks that require conscious awareness of the retrieved episodic or semantic knowledge (Graf and Schacter, 1985). Performance is preserved, however, on implicit tests of nondeclarative (or procedural) memory, such as repetition priming and skill learning, in which knowledge is expressed as experience-induced performance changes unrelated to awareness (Gabrieli,

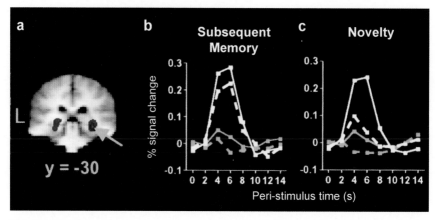

FIGURE 9-2   Activation in medial temporal lobe cortex during encoding. MTL regions that predicted memory outcome also demonstrated sensitivity to stimulus novelty. **a.** Parahippocampal cortex demonstrated greater activation for items subsequently remembered versus forgotten items. **b.** Time courses demonstrating subsequent memory effects. Remembered pictures, solid white line; forgotten pictures, dashed white line; remembered words, solid yellow line; forgotten words, dashed yellow line. **c.** Time courses from the same regions demonstrating encoding activation related to stimulus novelty. Novel pictures, solid white line; repeated pictures, dashed white line; novel words, solid yellow line; repeated words, dashed yellow line. [Adapted from Kirchhoff et al. (2000).]

1998; Squire, 1992). In contrast to this declarative/nondeclarative perspective, relational memory theory proposes that MTL function is independent of conscious awareness, with MTL processes being essential for the binding of distinct elements of an event into a memory representation that captures the relationships between the elements (Eichenbaum and Cohen, 2001). Such relational representations are thought to underlie mnemonic flexibility that allows for novel use of stored knowledge about the relations between elements, expressed as either conscious remembering or nonconscious changes in behavior.

The declarative and relational theories of MTL function both account for why patients show deficits on conscious recall and recognition tests following MTL injury while simultaneously expressing intact repetition priming and skill learning. However, the two views differ on the consequence of MTL damage for relational memory acquired or expressed in the absence of awareness. Unfortunately, there are few tasks that index relational knowledge in the absence of awareness in humans, and awareness is difficult to measure in nonhuman animals. Thus, it has been difficult to determine whether MTL structures contribute to memory only when critical knowledge is accessible to awareness, or whether the hippocampus mediates relational processing irrespective of awareness.

Of the limited evidence available, important data have come from classical conditioning paradigms that permit a test of the declarative hypothesis that

MTL function is critically related to conscious awareness. In delay conditioning, a conditioned stimulus (CS) onsets prior to the onset of an unconditioned stimulus (US), with the two stimuli then overlapping temporally. In contrast, trace-conditioning paradigms impose a brief interval between the offset of the CS and the onset of the US. Animals and humans with lesions to the MTL demonstrate intact delay conditioning but impaired trace conditioning (Clark and Squire, 1998; Gabrieli et al., 1995; McGlinchey-Berroth et al., 1997). Importantly for present purposes, in healthy humans, trace conditioning requires awareness of the CS–US contingency, whereas delay conditioning is unrelated to awareness (Clark et al., 2001; Clark and Squire, 1998; Manns et al., 2000a, 2000b). Thus, the form of conditioning (trace) that is impaired following MTL injury is also dependent on conscious awareness in the intact brain. Taken together, these data suggest that, as with other declarative memory tasks, trace conditioning may critically rely on the MTL because conscious knowledge must be acquired and expressed (Clark and Squire, 1998).

Other recent data, however, suggest that the MTL is critical for performance on tasks that require knowledge of the relations between stimuli, even when that knowledge is not consciously accessible. For example, using eye movements as an index of memory for previously viewed scenes, amnesic patients with MTL damage failed to demonstrate memory for the relations between elements of a scene — expressed as a failure to orient their eye movements to regions of the scene where manipulations of the relations between elements had occurred (J.D. Ryan et al., 2000; J.D. Ryan and Cohen, 2004). In contrast, healthy control subjects demonstrated an increase in viewing time to such regions where relational manipulations had occurred, and these expressions of sensitivity to the relational changes occurred even in the absence of conscious awareness. Thus, MTL injury resulted in a memory deficit for relational knowledge that was acquired and expressed outside of awareness in healthy subjects.

Studies using the contextual cueing paradigm also suggest a critical role for the MTL in relational memory that occurs independent of awareness. The contextual cueing task requires subjects to perform a visual search of an array of stimuli to locate a target object embedded within the context of a field of distractors. Reaction times for target identification in both novel and repeated displays decrease with task training, which reflects *procedural learning* (Chun and Jiang, 1998). Moreover, an additional context-dependent learning effect is observed for targets repeated within the same context. That is, there is a greater decrease in reaction time to identify target stimuli repeated within the same context of distractors relative to target stimuli repeated within a novel context (the *contextual cueing effect*), which suggests that acquiring knowledge about the spatial relations between the target and the distractors across repeated trials serves to facilitate visual search (Chun and Jiang, 1998). Importantly, the contextual cueing effect occurs even when healthy subjects cannot explicitly recognize the repeated contexts, suggesting that these changes can occur without conscious awareness of the target–context relations (Chun and Jiang, 2003). Amnesic patients with large MTL lesions demonstrate intact procedural learning in this task but fail to demonstrate contextual cueing and explicit recognition (Chun and Phelps, 1999; Manns and Squire, 2001). Collectively, these findings are consistent with the relational theory of MTL function, wherein MTL structures mediate the acquisition and retrieval of knowledge of the relations between event elements, independent of conscious awareness of the relations.

C. Retrograde Amnesia

While damage to the MTL region consistently results in anterograde amnesia, the loss of memories encoded before the onset of amnesia — retrograde amnesia — varies in severity and extent. Retrograde amnesia is typically temporally graded, such that memory losses are most severe for the time periods just prior to neural injury and are less severe or absent for more remote memories (Gabrieli, 1998; Squire et al., 2004). Retrograde memory deficits suggest that while the MTL is critical for encoding and retrieving new episodic memories, such memories are not permanently stored in the MTL. Rather, the temporal gradient seen in retrograde amnesia provides evidence for memory consolidation processes that contribute to the permanent storage of memories outside the MTL (McClelland et al., 1995; McGaugh, 2000; Squire and Alvarez, 1995).

According to standard models of memory consolidation, MTL structures initially work in concert with neocortical processing regions to encode new declarative memory traces. Within neocortex, representations of the elements that comprise an episode are distributed across multiple processing regions according to their content. Structures in the MTL rapidly bind this distributed cortical information into a single episodic memory trace, thus acting as an index to representations in neocortical structures. As time passes, consolidation mechanisms allow for a gradual strengthening of direct connections between neocortical regions that eventually enable the cortical memory to be accessed independently of the MTL representation (McClelland et al., 1995; McGaugh, 2000; Squire and Alvarez, 1995).

The temporal profile of retrograde amnesia and the role of MTL structures in declarative memory consolidation have emerged as recent points of controversy (Nadel and Moscovitch, 1997; Spiers et al., 2001; Squire et al., 2001a). Challenging the standard consolidation theory, multiple-trace models of memory suggest that the degree of retrograde memory loss caused by damage to the MTL depends on the type of memory being assessed (Moscovitch et al., 2005; Nadel and Moscovitch, 1997; Nadel et al., 2000). Like standard consolidation models, multiple-trace theory proposes that MTL structures rapidly bind cortical information representing an episode into a single memory trace, such that the elemental information in neocortical circuits that comprise an episode is bound by the MTL representation. According to multiple-trace theory, accessing an MTL-neocortical memory trace in the service of memory retrieval leads to re-encoding, yielding an additional trace sharing some of the information contained in the original memory trace. The proliferation of multiple traces is argued to facilitate the extraction of factual information that is common across episodes, whereby such semantic information is separated from the original context and stored independent of it. Accordingly, retrieval of semantic information can eventually occur independent of the MTL memory circuit through the gradual encoding of semantic knowledge by neocortex. In contrast to standard consolidation theory, which proposes that episodic memory also gradually becomes independent of MTL structures through the implementation of consolidation mechanisms over time (Squire and Alvarez, 1995), multiple-trace theory posits that the unique MTL representation constituting an episodic (autobiographical) memory will always depend on the continued involvement of MTL structures (Nadel and Moscovitch, 1997). From this latter perspective, retrograde amnesia for episodic memory should take a flat rather than temporally graded form.

Numerous studies of autobiographical memory after damage to the MTL demonstrate a temporal gradient of retrograde amnesia, with impairments for more recent autobiographical memories (Reed and Squire, 1998; Rempel-Clower et al., 1996; Zola-Morgan et al., 1986) and spared performance for more remote autobiographical knowledge (Henke et al., 1999; Kapur and Brooks, 1999; Mayes et al., 1997; Reed and Squire, 1998). Temporally graded retrograde amnesia occurs even when damage is limited to the hippocampal region (Kapur and Brooks, 1999), with a sparing of both remote semantic memory (Manns et al., 2003b) and rote episodic memory for autobiographical information (Bayley et al., 2003, 2005). Larger MTL lesions lead to retrograde memory loss that affects longer periods of time (Reed and Squire, 1998), raising the possibility that MTL cortical regions play some role in the storage of episodic information that is complementary to that played by the hippocampus.

In contrast to such observations of temporal gradients in retrograde amnesia, other studies have documented patients with extensive retrograde impairments, wherein autobiographical memory loss encompasses all periods of a patient's lifetime (Bayley et al., 2005; Cipolotti et al., 2001; Hirano and Noguchi, 1998). Such findings may suggest a continued role for the MTL in retrieval of remote autobiographical information. To resolve these apparent inconsistencies, proponents of the multiple-trace theory argue that graded deficits are the result of incomplete MTL lesions, with any preserved autobiographical memory capacity being subserved by the remaining MTL structures (Moscovitch et al., 2005; Nadel and Moscovitch, 1997; Nadel et al., 2000). Standard consolidation theorists argue that the extensive and temporally flat retrograde deficits observed in some patients are due to additional damage to structures outside of the

MTL, including frontal and lateral temporal regions that may be important for strategic retrieval and storage of long-term memories, respectively (Bayley et al., 2003, 2005; Squire et al., 2004).

Several neuroimaging studies have sought to address how MTL structures are involved in the retrieval of recent and remote autobiographical memories in the intact brain. While some studies have demonstrated hippocampal (Haist et al., 2001) and parahippocampal activation (Niki and Luo, 2002) that varied as a function of a memory's age, other studies have observed similar patterns of activation in MTL regions for both recent and remote memories (Maguire et al., 2001; L. Ryan et al., 2001). For example, two studies point to potential functional differences across MTL regions in the retrieval of remote autobiographical memories. In one study, hippocampal involvement was only observed during retrieval of autobiographical information acquired within the last few years, while entorhinal cortex was associated with retrieval of both recent and remote autobiographical memories acquired up to 20 years in the past (Haist et al., 2001; Fig. 9-3). A second study demonstrated lateralized patterns of activation within the hippocampus: Right hippocampus demonstrated a temporal gradient, decreasing in activity the more remote the autobiographical memories, whereas left hippocampus demonstrated no such gradient, suggesting a continued role in remembering autobiographical events throughout the

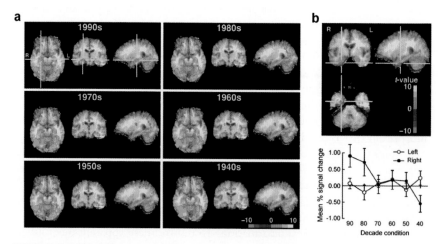

FIGURE 9-3 Medial temporal lobe activation during retrieval of famous faces from across different decades. **a.** A region in the right hippocampus (indicated by crosshairs) demonstrated activation during retrieval of recently acquired information, famous faces from the 1990s, but not during retrieval of more remotely acquired information. **b.** Activation in the right entorhinal cortex was associated with a linear decrease from recent to remote decades, demonstrating significant activation during retrieval of both recent and remotely acquired information acquired up to 20 years in the past. [Adapted from Haist et al. (2001).]

life span (Maguire and Frith, 2003). Thus, some (but not all) neuroimaging data suggest that MTL activation can vary according to the age of the episodic memory being retrieved, a pattern that would appear most compatible with standard consolidation theory. Together with neuropsychological data suggesting an influence of lesion size on the temporal extent of retrograde amnesia, these neuroimaging observations also highlight the importance of understanding how component regions of the MTL circuit differentially contribute to the acquisition, consolidation, and retrieval of episodic memories.

IV. FUNCTIONAL SEGREGATION WITHIN MTL

Hypotheses regarding the functions of MTL substructures draw heavily on knowledge of the anatomical connectivity of the region with structures in neocortex as well as information about the intrinsic connectivity within MTL. The anatomical organization of MTL suggests that component regions of MTL may differentially mediate the acquisition and retrieval of specific classes of stimuli and of different types of declarative memory representations.

The extrinsic connectivity of MTL subregions — notably perirhinal and parahippocampal cortices — with distinct neocortical processing regions suggests constraints on the classes of stimuli or items for which these MTL subregions may contribute to memory. Perirhinal and parahippocampal cortices may encode and retrieve information regarding different types of stimuli, as visuo-object and conceptual information appears to project differentially to perirhinal cortex and spatial information to parahippocampal cortex. Consistent with this possibility, functional neuroimaging studies have demonstrated that the encoding of novel visual-object information is associated with selective activation in perirhinal cortex (Pihlajamaki et al., 2003, 2004), while parahippocampal cortex demonstrates activation related to the encoding of visuospatial information, such as the encoding of scenes, houses, and known landmarks (Epstein et al., 1999; Epstein and Kanwisher, 1998; Maguire et al., 1997; Pihlajamaki et al., 2004; Sommer et al., 2005) and encoding the locations of objects in complex virtual environments (Burgess et al., 2002; Maguire et al., 1998). Paralleling these imaging findings, damage inclusive of human perirhinal cortex results in greater recognition memory deficits for objects compared to damage restricted to the hippocampal formation (Buffalo et al., 1998), whereas damage to human parahippocampal cortex results in impaired performance on spatial memory tasks (Bohbot et al., 1998). Accordingly, the representational capacity of perirhinal and parahippocampal cortical structures may be highly related to incoming sensory and conceptual information (for data on conceptual encoding, see O'Kane et al., 2005), suggesting that these cortices may mediate item-based memory processes linked to sensory and conceptual aspects of stimuli.

The hierarchical organization of the MTL circuit further highlights potential functional segregation between the hippocampus and the surrounding MTL cortices. One hypothesis suggests that (a) the hippocampus, at the apex of the MTL hierarchy, is differentially necessary for performance on tasks that require the formation and retrieval of *conjunctions* between multiple stimuli and between stimuli and context (e.g., associative recognition; contextual or source recollection; recall), whereas (b) MTL cortices mediate performance on tasks that rely on memory for individual stimuli or items (e.g., item recognition) (Aggleton and Brown, 1999; Brown and Aggleton, 2001; Eichenbaum, 2000). Conjunctive representations separately code the elements of an event, maintaining the compositionality of the elemental representations and organizing them in terms of their relations to one another (Eichenbaum et al., 1996). The elemental nature of such conjunctions allows for reactivation of the extended representation from partial input, a process termed *pattern completion* (O'Reilly and Rudy, 2001).

From this theoretical perspective, the hippocampus has the unique capacity to form conjunctive representations rapidly in one trial, while MTL cortex requires multiple learning trials to abstract gradually the statistical co-occurrences of elements (Norman and O'Reilly, 2001). The convergence of widespread cortical information in the hippocampus suggests that hippocampal processing can combine and extend representations built in the MTL cortices. The architecture of the intrahippocampal subfields further suggests that CA3 mechanisms are central to the formation of conjunctive memories that link event elements and are critical for subsequent pattern completion that constitutes episodic retrieval (Kesner and Hopkins, in press; Marr, 1971; McClelland et al., 1995; O'Reilly and Rudy, 2001). Specifically, the encoding of conjunctive representations may critically depend on the widespread collateral connections within CA3, which comprise a powerful associative learning mechanism that allows for the rapid binding of co-occurring event inputs distributed to multiple CA3 neurons. In contrast to the hippocampus, MTL cortical regions may be limited in their ability to form conjunctive representations, partially because perirhinal and parahippocampal cortices each have access to only specific classes of information arriving from distinct neocortical regions.

In contrast to the hypothesis that hippocampus and MTL cortices differentially support conjunctive versus item memory, respectively (Norman and O'Reilly, 2003; O'Reilly and Rudy, 2001), an alternate hypothesis suggests that, while functional distinctions exist within the MTL memory circuit, these differences do not correspond to the conjunctive–item dimension. Rather, MTL cortex and hippocampus are posited to similarly support conjunctive and item processing (Squire et al., 2004; Stark and Squire, 2003; Stark et al., 2002). Accordingly, this perspective proposes that all declarative memory tasks, including those that require memory for single items, depend on the function of both the hippocampus and the surrounding MTL cortex (Stark and Squire, 2003).

In the following sections, we review evidence from neuropsychological and neuroimaging data in humans that bear on putative functional distinctions within the MTL circuit, specifically addressing whether memory for items and memory for conjunctions depend on processing in the same or different subregions of the MTL.

V. CONSEQUENCES OF SELECTIVE HIPPOCAMPAL LESIONS

Selective damage to the hippocampus is sufficient to produce significant memory impairments. Patients with damage limited to the CA1 region of the hippocampus demonstrate moderately severe anterograde amnesia, including deficits on both recall and recognition memory tests; and damage extending beyond the CA1 region but still limited to the hippocampus results in more severe anterograde memory impairments (Rempel-Clower et al., 1996; Zola-Morgan et al., 1986; Fig. 9-4). Thus, severity of memory impairments may depend on the locus and the extent of damage within the MTL and, further, within the hippocampus. However, inconsistent patterns of memory deficits have emerged across studies of patients with damage thought to be limited to the hippocampal formation, resulting in controversy regarding the particular forms of declarative memory that depend on the hippocampus and on the surrounding MTL cortices.

Much of the debate has centered on the nature of the recognition memory deficits observed following MTL damage. The two opposing theories of MTL functional organization introduced earlier make distinct predictions about the importance of the hippocampus in item recognition and in the recollection of item–context or item–item conjunctions. Whereas some posit that the hippocampus and surrounding MTL cortex are essential for both item recognition and recollection (Squire et al., 2004), others posit that the hippocampus is necessary for only the recollective aspects of recognition, with structures in MTL cortex supporting item recognition (Yonelinas et al., 1998, 2002). This second view builds on dual-process models of recognition memory that propose that recognition decisions can be based on two forms of memory (Mandler, 1980), (a) assessments of stimulus *familiarity* and/or (b) *recollection* of specific details surrounding an item's prior encounter, such as recollecting contextual or source information about the study event (Yonelinas, 2002). Recollection may depend on the encoding of conjunctive representations and subsequent pattern completion mechanisms that result in retrieval of extended memory

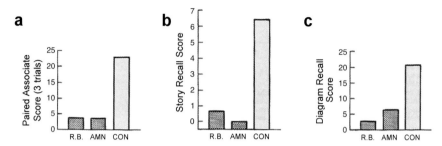

FIGURE 9-4   Memory impairment with selective damage to CA1 field of the hippocampus. Performance on three tests of new learning for R.B., amnesic patients (AMN) with more extensive MTL damage, and healthy controls (CON). **a.** Paired-associate learning on 10 word pairs after three consecutive exposures of each pair. **b.** Recall of a short prose passage after a 10 to 12-min delay. **c.** Recall of diagrams after a 10 to 20-min delay. [Adapted from Zola-Morgan et al. (1986).]

details from partial input (Norman and O'Reilly, 2003; O'Reilly and Rudy, 2001), representations, and processes that are also thought to support performance on tests of recall (Quamme et al., 2004).

The two competing hypotheses of MTL functional organization make different predictions regarding the effects of selective hippocampal lesions on recognition memory. From the distributed-function perspective, damage limited to the hippocampus should result in similar impairments in recall and recognition, and, within recognition, such damage should result in similar impairments in item recognition, associative recognition, and recollection-based recognition judgments. Consistent with these predictions, several studies have reported equivalent impairments in recall and recognition following lesions that are thought to be selective to hippocampus (Manns et al., 2003a; Manns and Squire, 1999; Reed et al., 1997). These patients also appear similarly impaired on recognition memory tests for single items and for associations or conjunctions between items (Stark and Squire, 2003; Stark et al., 2002), and they demonstrate impairments in familiarity-based recognition judgments when indexed using the introspective judgments of "remember" and "know" (Manns et al., 2003a; Fig. 9-5a). Recent investigation using receiver operating characteristics (ROC) to assess the status of distinct aspects of recognition performance in these patients further suggests that recollection and familiarity are similarly impaired in this population of patients with hippocampal damage (Wais et al., 2006).

In contrast to such findings of similar impairments in recall and recognition and in recollective- and familiarity-based recognition, studies of other patients with apparently selective hippocampal damage have documented spared item

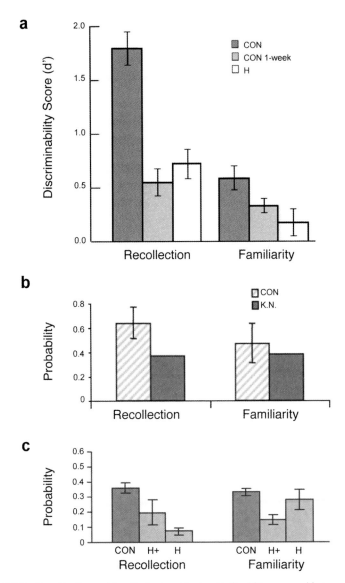

FIGURE 9-5 Recollection and familiarity in patients with focal hippocampal lesions. Estimates of recollection and familiarity during recognition based on the remember/know procedure. **a.** Recollection- and familiarity-based recognition judgments for eight tests of verbal and nonverbal memory for patients with hippocampal damage (H), healthy controls (CON), and healthy controls tested after a one-week retention interval (CON 1-week) (adapted from Manns et al., 2003a). Performance of the hippocampal patients in this study was similarly impaired on recollection and familiarity. **b.** Performance of a patient with focal hippocampal lesions (K.N.) relative to controls (CON) on word recognition (adapted from Aggleton et al., 2005). **c.** Performance on word recognition for healthy controls (CON), patients with focal hippocampal damage (H), and patients with more extensive medial temporal lobe lesions (H+) (adapted from Yonelinas et al., 2002). In both of these latter studies, damage limited to the hippocampus had greater effects on recollection than on familiarity.

recognition relative to both recall (Baddeley et al., 2001; Mayes et al., 2002; Quamme et al., 2004; Vargha-Khadem et al., 1997) and associative recognition (Holdstock et al., 2005; Mayes et al., 2004). This pattern of preserved item recognition following selective hippocampal damage has been interpreted as a sparing of item familiarity processes that are sufficient to support recognition memory (Holdstock et al., 2002), raising the possibility that item recognition supported by familiarity is relatively preserved following hippocampal damage because it relies on mechanisms in MTL cortices (Brown and Aggleton, 2001; Yonelinas et al., 1998). Using both remember/know and ROC procedures to measure recollective- and familiarity-based recognition in patients with selective hippocampal damage, several studies have demonstrated impairments in recollective aspects of recognition, with familiarity-based recognition remaining intact (Aggleton et al., 2005; Yonelinas et al., 2002, 2004; Figs. 9-5b and c). Other data indicate that more extensive MTL damage that encompasses regions in both hippocampus and MTL cortex can lead to impairments in both recollective- and familiarity-based recognition (Yonelinas et al., 2002). Thus, for declarative memory that depends on one-trial learning, these data are consistent with the conjunction–item hypothesis of functional differentiation between hippocampus and MTL cortical structures.

At present, it is unclear how to resolve the inconsistencies between these sets of neuropsychological data. One possible relevant distinction is that between patients with early onset of selective hippocampal damage (Baddeley et al., 2001; Vargha-Khadem et al., 1997) and those who suffer MTL insult later in life (Squire et al., 2004; Squire and Zola, 1998). The functional organization within MTL in the former group may be distinct from that which typically emerges in the healthy brain due to compensatory plasticity that may be unique to early developmental periods. However, age of onset accounts for only some of the conflicts in extant data, and there are a number of instances of late-onset amnesia following putatively selective hippocampal lesion that are associated with selective impairment of conjunctive memory (Aggleton et al., 2005; Holdstock et al., 2005; Mayes et al., 2002, 2004; Quamme et al., 2004; Yonelinas et al., 2002, 2004) and others that result in comparable impairment of conjunctive and item memory (Manns et al., 2003a; Manns and Squire, 1999; Reed et al., 1997; Stark and Squire, 2003; Stark et al., 2002; Wais et al., 2006). These latter inconsistencies may stem from differences in the precise locus of the lesions, for it is possible that patients with "selective" hippocampal lesion differ in the degree to which the lesion may extend modestly into the surrounding MTL cortical structures and the degree to which any such modest damage has functional relevance. A further challenge for understanding the conflicts within the patient literature is the fact that different approaches have been used to index item and conjunctive memory status across studies, raising the possibility that some of the inconsistencies might be resolved if identical procedures were adopted.

VI. NEUROIMAGING OF ITEM AND CONJUNCTIVE MEMORY

A. MTL Encoding Processes

While not capable of addressing the necessity of specific MTL subregions for specific forms of declarative memory, neuroimaging data can complement the neuropsychological literature by revealing how activation within the MTL circuit relates to item and conjunctive memory in the healthy human brain. To this end, recent neuroimaging studies using functional MRI (fMRI) have examined the degree to which subregions of MTL may differentially support declarative memory, focusing on how hippocampus and MTL cortical regions support the acquisition and retrieval of conjunctive and item memory representations (Cansino et al., 2002; Davachi et al., 2003; Davachi and Wagner, 2002; Eldridge et al., 2000; Henson et al., 2003; Ranganath et al., 2003).

In contrast to the neuropsychological literature, there has been strong convergence across fMRI studies of episodic encoding. In particular, multiple fMRI studies using the subsequent-memory paradigm have now documented that greater hippocampal activation during an experience is predictive of a higher probability of subsequent successful recollection (Davachi et al., 2003; Davachi and Wagner, 2002; Ranganath et al., 2003) and of subsequent associative recognition (Jackson and Schacter, 2004; Kirwan and Stark, 2004; Sperling et al., 2003) at retrieval, but it is not predictive of subsequent item recognition (Davachi et al., 2003; Davachi and Wagner, 2002; Kirwan and Stark, 2004; Ranganath et al., 2003). In stark contrast, greater encoding activation in anterior MTL cortex, at or near perirhinal cortex, is predictive of subsequent item recognition but not subsequent recollection (Davachi et al., 2003; Davachi and Wagner, 2002; Ranganath et al., 2003) or subsequent associative recognition (Kirwan and Stark, 2004; c.f., Jackson and Schacter, 2004; Fig. 9-6). These functional dissociations between hippocampus and anterior MTL cortex are consistent with the conjunctive–item hypothesis of MTL functional differentiation, though it should be noted that these studies also consistently indicate that encoding activation in parahippocampal cortex parallels that in hippocampus rather than that in anterior MTL cortex. One speculative possibility is that this functional distinction along the anterior–posterior extent of parahippocampal gyrus is a reflection of the different neocortical inputs to perirhinal and parahippocampal cortex, with the effective encoding of the spatial inputs to parahippocampal cortex (via interactions between parahippocampal cortex, entorhinal cortex, and hippocampus) being critical for successful subsequent recollection.

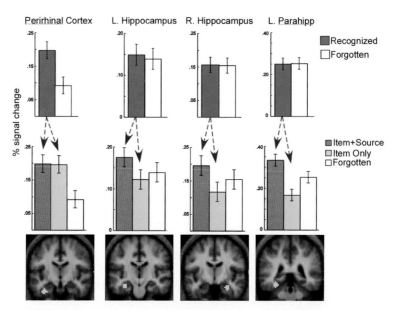

a

Perirhinal Cortex L. Hippocampus R. Hippocampus L. Parahipp

% signal change

■ Recognized
□ Forgotten

■ Item+Source
□ Item Only
□ Forgotten

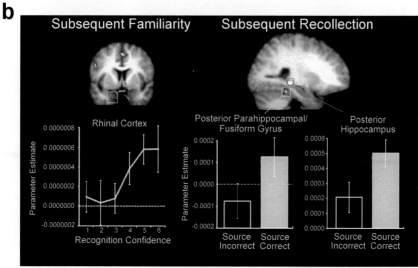

b

Subsequent Familiarity   Subsequent Recollection

Rhinal Cortex

Parameter Estimate

Recognition Confidence

Posterior Parahippocampal/
Fusiform Gyrus

Parameter Estimate

Source Incorrect   Source Correct

Posterior Hippocampus

Source Incorrect   Source Correct

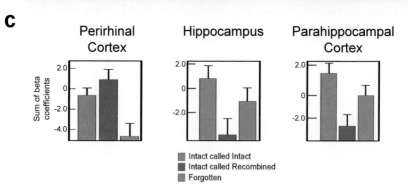

c

Perirhinal
Cortex

Hippocampus

Parahippocampal
Cortex

Sum of beta coefficients

■ Intact called Intact
■ Intact called Recombined
■ Forgotten

FIGURE 9-6   Medial temporal lobe activation during episodic encoding. **a.** Encoding activation in MTL subregions. Activation in perirhinal cortex was associated with successful item encoding, differentiating between items later recognized relative to those later forgotten. In contrast, encoding activation in hippocampus and parahippocampal cortex demonstrated greater activation for items later accompanied by contextual recollection relative to items later recognized without contextual information and relative to forgotten items (adapted from Davachi et al., 2003). **b.** Subsequent familiarity effects in left anterior MTL cortex, as indexed by linear increases in activity with increasing recognition confidence, and subsequent recollection effects in hippocampus and parahippocampal cortex, as indexed by memory for contextual source information (adapted from Ranganath et al., 2003). **c.** Encoding activation in perirhinal cortex was associated with later item recognition. In contrast, encoding activation in hippocampus and parahippocampal cortex was associated with later contextual recognition, demonstrating greater activation for studied paired associates correctly identified as intact at retrieval relative to paired associates incorrectly classified as recombined and relative to pairs forgotten (adapted from Kirwan and Stark, 2004).

B. MTL Retrieval Processes

Neuroimaging studies of episodic retrieval have yielded complementary findings, wherein increased hippocampal activation accompanies (a) the conscious recollection (or remembering) of a studied episode (Eldridge et al., 2000; Fig. 9-7), (b) the recollection of contextual details surrounding an item's prior encounter (Cansino et al., 2002; Dobbins et al., 2003), and (c) associative recognition (Giovanello et al., 2004). Hippocampal activation has also been differentially associated with high-confidence recognition decisions (Yonelinas et al., 2005), which some have argued reflect recollection-based recognition (Yonelinas, 1994, 1999; Yonelinas et al., 1996, 1998; c.f., Wixted and Stretch, 2004). Importantly, in some studies, whereas hippocampal activation increased during expressions of conjunctive memory retrieval (e.g., remembering and context recollection), hippocampal activation did not increase during item recognition based on familiarity, suggesting a selective role for hippocampus in reactivating conjunctive representations (Eldridge et al., 2000; Giovanello et al., 2004; Fig. 9-7). Nevertheless, it should be noted that not all fMRI retrieval studies suggest a hippocampal insensitivity to item recognition (Kirwan and Stark, 2004), raising important questions about the boundary conditions for selective hippocampal activation.

Complementary evidence for the role of the hippocampus in the retrieval of conjunctive representations comes from fMRI studies of transitive inference tasks, wherein participants must infer a relationship between items that are indirectly related through a shared association with one or more intervening items. Neuroimaging measures of MTL activation during the critical transitive inference (or mediated retrieval) trials have demonstrated activation in anterior hippocampus uniquely associated with transitive judgments relative to control conditions (Heckers et al., 2004; Preston et al., 2004), complementing animal work

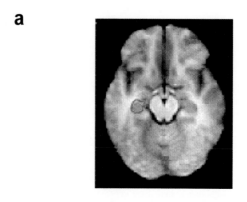

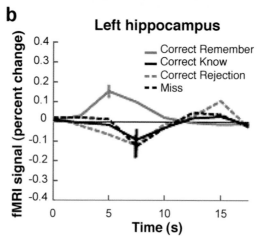

FIGURE 9-7 Hippocampal activation during recollection of studied words. **a.** Axial slice through the MTL, with the left hippocampal region of interest outlined in red. **b.** Averaged event-related responses in the left hippocampus during word recognition. Hippocampal regions demonstrated greater activation when memory for words was accompanied by recollection of studied details (remember responses) relative to recognized words that were familiar but were not accompanied by contextual information (know responses), forgotten words (misses), and new words (correct rejections). [Adapted from Eldridge et al. (2000).]

documenting impaired transitive inference judgments following hippocampal lesions (Bunsey and Eichenbaum, 1996; Dusek and Eichenbaum, 1997; Fig. 9-8).

Recently, neuroimaging correlates of MTL retrieval activation during familiarity-based item recognition have begun to emerge. This work has been motivated by electrophysiological data in nonhuman animals that implicate MTL cortical regions, particularly perirhinal cortex, in recognition memory for items (Brown and Aggleton, 2001; Brown and Xiang, 1998; Xiang and Brown, 1998). In particular, electrophysiological data have demonstrated

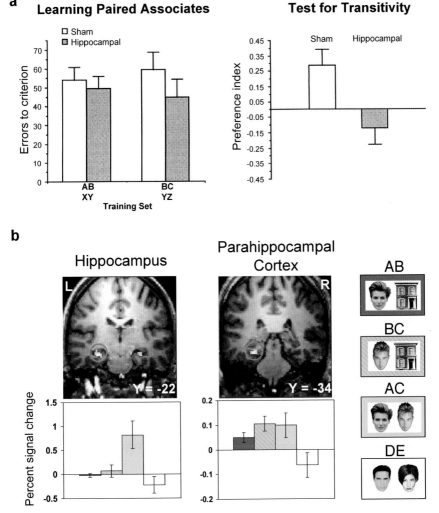

FIGURE 9-8   Hippocampal contribution to the flexible expression of memory. **a.** Performance of intact (sham) and hippocampally lesioned rats during learning of overlapping sets of odor paired associates and subsequent tests for transitivity. Hippocampally lesioned rats were not impaired during learning, but thus demonstrated impaired performance when required to infer a relationship between two studied odors that were associated with a single common odor during learning (adapted from Bunsey and Eichenbaum, 1996). **b.** MTL activation during a transitive inference task in humans. Hippocampal regions were selectively engaged during transitive inference trials that required inferential judgments based on mediated retrieval. In contrast, activation in parahippocampal cortex was associated with retrieval of visuo-spatial information when presented as a retrieval cue and when accessed to mediate judgments on transitive trials (adapted from Preston et al., 2004).

reductions in neuronal firing in perirhinal neurons upon repeated presentation of a stimulus relative to its initial presentation, a phenomenon termed *repetition suppression* (Fig. 9-9a). Complementing these firing-rate reductions in nonhuman perirhinal cortex, emerging fMRI data from humans suggest that MTL cortex demonstrates a particular sensitivity to item familiarity. For example, activation levels in anterior MTL cortex decrease during the processing of previously encountered items as compared to novel items — a putative parallel of the repetition suppression seen in electrophysiological studies — with the magnitude of these activation reductions being unaffected by behavioral expressions of source recollection (Henson et al., 2003, 2005; Fig. 9-9b). Moreover, decreased activation in MTL cortex is observed during correct recognition of old items relative to old items incorrectly classified as new, indicating that these effects relate to memory perception rather than the actual history of a test probe (Weis et al., 2004a, 2004b).

It has been hypothesized that when items are reexperienced, the reduction in MTL cortical activation relative to the activation present during the item's first encounter serves as a basis for discriminating between familiar and novel stimuli (Brown and Aggleton, 2001; Weis et al., 2004a). Behavioral evidence indicates that discrimination between novel and previously encountered stimuli depends, at least partially, on an assessment of item memory strength, which varies in a continuous manner and may underlie the subjective perception of stimulus familiarity (Jacoby, 1991; Mandler, 1980; Wixted and Stretch, 2004; Yonelinas, 2001, 2002). A recent direct test of this possibility, using fMRI and magnetoencephalography (MEG), demonstrated that the magnitude of experience-dependent activation reductions in perirhinal and parahippocampal cortices tracks subjective reports of perceived item familiarity (Fig. 9-9c), with the timing of these effects emerging within the first 200 msec of stimulus processing (Gonsalves et al., 2005; Fig. 9-9d). Importantly, the building body of evidence documenting a relation between MTL cortical deactivation and perceived item familiarity dissociates from the extensive evidence that hippocampal activation increases with markers of conjunctive memory retrieval, including conscious recollection. Thus, fMRI data at retrieval document that distinct subregions within the MTL circuit tend to correlate with distinct forms of declarative memory, with differences in the direction of activation change (increases in hippocampus versus decreases in MTL cortex) further suggesting dissociable function across these subregions.

VII. HIPPOCAMPAL SUBFIELD FUNCTION

A central question regarding MTL function is how the subfields of hippocampus contribute to declarative memory. A leading hypothesis is that CA3 mechanisms are central to the formation of conjunctive representations during

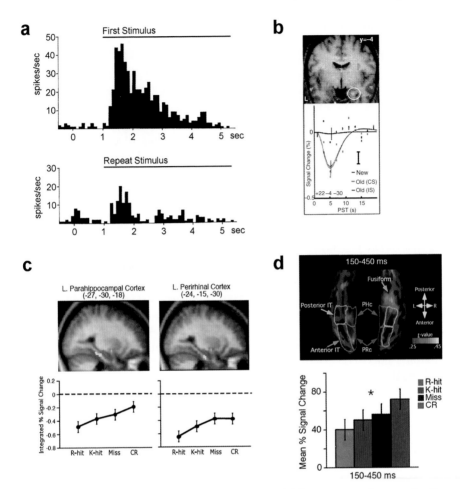

FIGURE 9-9   Repetition suppression in medial temporal lobe cortex. **a.** An example of a repetition-suppression response recorded from a monkey perirhinal neuron to initial (top) and repeated (bottom) presentations of pictures during a serial recognition task (from Brown and Xiang, 1998). **b.** Repetition-suppression response in anterior MTL cortex in an fMRI study during retrieval of item and contextual information. Event–related responses are from the maxima within the circled region (adapted from Henson et al., 2003). **c.** Activation in left parahippocampal and perirhinal cortices whose magnitude of activation significantly declined as perceived memory strength increased, accompanied by a linear regression map of MEG effects along the ventral surface of the inflated temporal lobes. **d.** Mean anatomically constrained MEG current estimates from perirhinal cortex that revealed a graded pattern across varying levels of perceived memory strength (adapted from Gonsalves et al., 2005).

learning, with CA3 pattern-completion mechanisms then serving to retrieve or reinstate such representations during remembering (McClelland et al., 1995; O'Reilly and Rudy, 2001). Recurrent collateral connections within CA3 may critically support encoding of conjunctive memories by enabling binding of the co-occurring inputs that converge on distributed CA3 neurons. In addition, theorists have argued that interactions between CA3 and dentate gyrus may support critical pattern-separation processes during conjunctive encoding that minimize overlap between similar inputs, thus limiting interference between competing MTL memory representations (Gilbert et al., 2001; McClelland and Goddard, 1996; McClelland et al., 1995; O'Reilly and Rudy, 2001).

Evidence supporting the importance of CA3 in conjunctive encoding and subsequent pattern completion has come from the study of CA3-NR1 knockout mice. The strength of recurrent collateral synapses in CA3 can be modified in an NMDA receptor (NR) dependent manner, and the NR1 knockout of CA3 pyramidal cells selectively disrupts NR-dependent long-term potentiation at these recurrent synapses, thus affecting function specific to the CA3 region of hippocampus (Nakazawa et al., 2002). At encoding, CA3-NR1 knockouts demonstrate impaired learning on tasks that require the rapid acquisition of conjunctive information (Nakazawa et al., 2003; Rajji et al., 2006). For example, in one study, knockout mice were impaired during the rapid acquisition of a novel platform location in the Morris water maze but demonstrated preserved performance when tested with familiar platform locations (Nakazawa et al., 2003). CA1 place cell tuning was abnormal in these knockout mice, further suggesting that CA3 learning mechanisms have downstream effects on memory processing and representation in other hippocampal regions. At retrieval, data from knockout mice support the hypothesis that pattern completion critically depends on CA3 and its interaction with CA1 and entorhinal cortex. For example, CA3-NR1 knockout mice demonstrate impaired retrieval when cued by a partial set of inputs, as evidenced by a failure to reactivate encoding patterns in CA1 (Nakazawa et al., 2002). Paralleling these findings, a lesion study of performance during delayed matching-to-sample for spatial location (Gold and Kesner, 2005) revealed that as the number of retrieval cues decreased during the test phase of the task, error rates increased linearly for rats with selective lesions to CA3. This increased probability of retrieval failure with reductions in cue support suggests that retrieval of stored information from partial input critically relies on CA3 pattern-completion processes that recover associated knowledge when cued with limited information.

The acquisition of high-resolution fMRI data from human MTL has recently begun to illustrate complementary functional segregation within human hippocampus. To date, potential functional differences across hippocampal subfields have been noted when comparing memory encoding and retrieval. In particular, two studies have demonstrated activation in CA3/

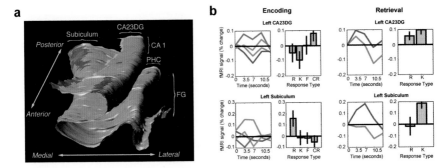

FIGURE 9-10   High-resolution imaging of MTL contributions to encoding and retrieval. **a.** An average three-dimensional rendering of the MTL. Encoding-related activation in CA2/CA3/DG regions shown in red, recall-related activation localized to subiculum in blue, and overlapping encoding and retrieval activation in purple (from Zeineh et al., 2003). **b.** Averaged event-related responses in left CA2/CA3/DG and subiculum sorted according to remember-know performance during encoding (left) and retrieval (right). Remember, red lines; Know, blue lines; Forgot, cyan lines; Correct Rejection, green lines (adapted from Eldridge et al., 2005).

dentate gyrus associated with the formation of episodic memories, while activation in a subicular region was associated with the recollection of learned information (Eldridge et al., 2005; Zeineh et al., 2003; Fig. 9-10). Though a mechanistic understanding of why CA3/dentate gyrus and subiculum are differentially associated with episodic encoding and retrieval is unclear, this high-resolution imaging approach promises to permit greater bridging between animal and human work on hippocampal function.

VIII. SUMMARY

The MTL circuit is central to declarative memory, including the encoding, consolidation, and retrieval of episodic memories. As illustrated in this review, neuropsychological and neuroimaging studies of human MTL have markedly advanced understanding of how the MTL allows us to build memories of our daily experiences and subsequently to retrieve these memories to inform future goal-directed behavior. Important questions remain regarding how the sub-components of the MTL circuit mediate declarative memory, including how interactions within the circuit give rise to various forms of knowledge acquisition and different states of remembering. Continued technological advances, such as in anatomical and functional MRI, will undoubtedly increase the field's ability to integrate data across species and across levels of analysis, empowering researchers to better understand how the MTL accomplishments its amazing mnemonic feats.

REFERENCES

Aggleton, J.P., and Brown, M.W. (1999). Episodic memory, amnesia, and the hippocampal-anterior thalamic axis. *Behav Brain Sci, 22*(3), 425–444; discussion 444–489.

Aggleton, J.P., Vann, S.D., Denby, C., Dix, S., Mayes, A.R., Roberts, N., et al. (2005). Sparing of the familiarity component of recognition memory in a patient with hippocampal pathology. *Neuropsychologia, 43*(12), 1810–1823.

Amaral, D.G., and Insausti, R. (1990). Hippocampal formation. In G. Paxinos (ed.), *The Human Nervous System* (pp. 711–755). San Diego: Academic Press.

Amaral, D.G., Insausti, R., and Cowan, W.M. (1987). The entorhinal cortex of the monkey: I. Cytoarchitectonic organization. *J Comp Neurol, 264*(3), 326–355.

Baddeley, A., Vargha-Khadem, F., and Mishkin, M. (2001). Preserved recognition in a case of developmental amnesia: Implications for the acquisition of semantic memory? *Journal of Cognitive Neuroscience, 13*, 357–369.

Bayley, P.J., Gold, J.J., Hopkins, R.O., and Squire, L.R. (2005). The neuroanatomy of remote memory. *Neuron, 46*(5), 799–810.

Bayley, P.J., Hopkins, R.O., and Squire, L.R. (2003). Successful recollection of remote autobiographical memories by amnesic patients with medial temporal lobe lesions. *Neuron, 38*(1), 135–144.

Bohbot, V.D., Kalina, M., Stepankova, K., Spackova, N., Petrides, M., and Nadel, L. (1998). Spatial memory deficits in patients with lesions to the right hippocampus and to the right parahippocampal cortex. *Neuropsychologia, 36*(11), 1217–1238.

Brewer, J.B., Zhao, Z., Desmond, J.E., Glover, G.H., and Gabrieli, J.D. (1998). Making memories: Brain activity that predicts how well visual experience will be remembered. *Science, 281*, 1185–1187.

Brown, M.W., and Aggleton, J.P. (2001). Recognition memory: What are the roles of the perirhinal cortex and hippocampus? *Nature Review Neuroscience, 2*, 51–61.

Brown, M.W., and Xiang, J.Z. (1998). Recognition memory: Neuronal substrates of the judgment of prior occurrence. *Prog Neurobiol, 55*(2), 149–189.

Buffalo, E.A., Reber, P.J., and Squire, L.R. (1998). The human perirhinal cortex and recognition memory. *Hippocampus, 8*(4), 330–339.

Bunsey, M., and Eichenbaum, H. (1996). Conservation of hippocampal memory function in rats and humans. *Nature, 379*(6562), 255–257.

Burgess, N., Maguire, E.A., and O'Keefe, J. (2002). The human hippocampus and spatial and episodic memory. *Neuron, 35*(4), 625–641.

Cansino, S., Maquet, P., Dolan, R.J., and Rugg, M.D. (2002). Brain activity underlying encoding and retrieval of source memory. *Cerebral Cortex, 12*, 1048–1056.

Chun, M.M., and Jiang, Y. (1998). Contextual cueing: Implicit learning and memory of visual context guides spatial attention. *Cognitive Psychology, 36*, 28–71.

Chun, M.M., and Jiang, Y. (2003). Implicit, long-term spatial contextual memory. *J Exp Psychol Learn Mem Cogn, 29*(2), 224–234.

Chun, M.M., and Phelps, E.A. (1999). Memory deficits for implicit contextual information in amnesic subjects with hippocampal damage. *Nature Neuroscience, 2*, 844–847.

Cipolotti, L., Shallice, T., Chan, D., Fox, N., Scahill, R., Harrison, G., et al. (2001). Long-term retrograde amnesia: The crucial role of the hippocampus. *Neuropsychologia, 39*(2), 151–172.

Clark, R.E., Manns, J.R., and Squire, L.R. (2001). Trace and delay eyeblink conditioning: Contrasting phenomena of declarative and nondeclarative memory. *Psychol Sci, 12*(4), 304–308.

Clark, R.E., and Squire, L.R. (1998). Classical conditioning and brain systems: The role of awareness. *Science, 280*(5360), 77–81.

Cohen, N.J., and Squire, L.R. (1980). Preserved learning and retention of pattern-analyzing skill in amnesia: Dissociation of knowing how and knowing that. *Science, 210*, 207–210.

Davachi, L., Mitchell, J., and Wagner, A.D. (2003). Multiple routes to memory: Distinct medial temporal lobe processes build item and source memories. *Proc Natl Acad Sci USA, 100,* 2157–2162.

Davachi, L., and Wagner, A.D. (2002). Hippocampal contributions to episodic encoding: Insights from relational and item-based learning. *J Neurophysiol, 88*(2), 982–990.

Dobbins, I.G., Rice, H.J., Wagner, A.D., and Schacter, D.L. (2003). Memory orientation and success: Separable neurocognitive components underlying episodic recognition. *Neuropsychologia, 41,* 318–333.

Dusek, J.A., and Eichenbaum, H. (1997). The hippocampus and memory for orderly stimulus relations. *Proceedings of the National Academy of Sciences USA, 94,* 7109–7114.

Duvernoy, H.M. (1998). *The Human Hippocampus.* New York: Springer.

Eichenbaum, H. (2000). A cortical-hippocampal system for declarative memory. *Nature Rev Neurosci, 1,* 41–50.

Eichenbaum, H., and Cohen, N.J. (2001). *From Conditioning to Conscious Recollection: Memory Systems of the Brain.* New York: Oxford University Press.

Eichenbaum, H., Schoenbaum, G., Young, B., and Bunsey, M. (1996). Functional organization of the hippocampal memory system. *Proc Natl Acad Sci USA, 93*(24), 13500–13507.

Eldridge, L.L., Engel, S.A., Zeineh, M.M., Bookheimer, S.Y., and Knowlton, B.J. (2005). A dissociation of encoding and retrieval processes in the human hippocampus. *J Neurosci, 25*(13), 3280–3286.

Eldridge, L.L., Knowlton, B.J., Furmanski, C.S., Bookheimer, S.Y., and Engel, S.A. (2000). Remembering episodes: A selective role for the hippocampus during retrieval. *Nature Neuroscience, 3,* 1149–1152.

Epstein, R., and Kanwisher, N. (1998). A cortical representation of the local visual environment. *Nature, 392,* 598–601.

Epstein, R., Harris, A., Stanley, D., and Kanwisher, N. (1999). The parahippocampal place area: Recognition, navigation, or encoding? *Neuron, 23,* 115–125.

Gabrieli, J.D. (1998). Cognitive neuroscience of human memory. *Annu Rev Psychol, 49,* 87–115.

Gabrieli, J.D., Cohen, N.J., and Corkin, S. (1988). The impaired learning of semantic knowledge following bilateral medial temporal-lobe resection. *Brain Cognition, 7,* 157–177.

Gabrieli, J.D.E., Brewer, J.B., Desmond, J.E., and Glover, G.H. (1997). Separate neural bases of two fundamental memory processes in the human medial temporal lobe. *Science, 276,* 264–266.

Gabrieli, J.D.E., Carrillo, M.C., Cermak, L.S., Mcglinchey-Berroth, R., Gluck, M.A., and Disterhoft, J.F. (1995). Intact delay-eyeblink classical conditioning in amnesia. *Behav Neurosci, 109,* 819–827.

Gilbert, P.E., Kesner, R.P., and Lee, I. (2001). Dissociating hippocampal subregions: Double dissociation between dentate gyrus and CA1. *Hippocampus, 11*(6), 626–636.

Giovanello, K.S., Schnyer, D.M., and Verfaellie, M. (2004). A critical role for the anterior hippocampus in relational memory: Evidence from an fMRI study comparing associative and item recognition. *Hippocampus, 14*(1), 5–8.

Gold, A.E., and Kesner, R.P. (2005). The role of the CA3 subregion of the dorsal hippocampus in spatial pattern completion in the rat. *Hippocampus, 15*(6), 808–814.

Gonsalves, B.D., Kahn, I., Curran, T., Norman, K.A., and Wagner, A.D. (2005). Memory strength and repetition suppression: Multimodal imaging of medial temporal cortical contributions to recognition. *Neuron, 47,* 751–761.

Graf, P., and Schacter, D.L. (1985). Implicit and explicit memory for new associations in normal and amnesic subjects. *J Exper Psychol: Learning Mem Cogn, 11,* 501–518.

Haist, F., Bowden Gore, J., and Mao, H. (2001). Consolidation of human memory over decades revealed by functional magnetic resonance imaging. *Nature Neuroscience, 4,* 1139–1145.

Hargreaves, E.L., Rao, G., Lee, I., and Knierim, J.J. (2005). Major dissociation between medial and lateral entorhinal input to dorsal hippocampus. *Science, 308*(5729), 1792–1794.

Heckers, S., Zalesak, M., Weiss, A.P., Ditman, T., and Titone, D. (2004). Hippocampal activation during transitive inference in humans. *Hippocampus, 14*(2), 153–162.

Henke, K., Kroll, N.E., Behniea, H., Amaral, D.G., Miller, M.B., Rafal, R., et al. (1999). Memory lost and regained following bilateral hippocampal damage. *J Cogn Neurosci, 11*(6), 682–697.

Henson, R.N. (2005). A mini-review of fMRI studies of human medial temporal lobe activity associated with recognition memory. *Q J Exp Psychol B, 58*(3–4), 340–360.

Henson, R.N., Cansino, S., Herron, J.E., Robb, W.G., and Rugg, M.D. (2003). A familiarity signal in human anterior medial temporal cortex? *Hippocampus, 13,* 301–304.

Henson, R.N., Hornberger, M., and Rugg, M.D. (2005). Further dissociating the processes involved in recognition memory: An fMRI study. *J Cogn Neurosci, 17*(7), 1058–1073.

Henson, R.N., Rugg, M.D., Shallice, T., Josephs, O., and Dolan, R.J. (1999). Recollection and familiarity in recognition memory: An event-related functional magnetic resonance imaging study. *J Neurosci, 19,* 3962–3972.

Hirano, M., and Noguchi, K. (1998). Dissociation between specific personal episodes and other aspects of remote memory in a patient with hippocampal amnesia. *Percept Mot Skills, 87*(1), 99–107.

Holdstock, J.S., Mayes, A.R., Gong, Q.Y., Roberts, N., and Kapur, N. (2005). Item recognition is less impaired than recall and associative recognition in a patient with selective hippocampal damage. *Hippocampus, 15*(2), 203–215.

Holdstock, J.S., Mayes, A.R., Roberts, N., Cezayirli, E., Isaac, C.L., O'Reilly, R.C., et al. (2002). Under what conditions is recognition spared relative to recall after selective hippocampal damage in humans? *Hippocampus, 12,* 341–351.

Insausti, R., Amaral, D.G., and Cowan, W.M. (1987). The entorhinal cortex of the monkey: II. Cortical afferents. *J Comp Neurol, 264*(3), 356–395.

Jackson, O., 3rd, and Schacter, D.L. (2004). Encoding activity in anterior medial temporal lobe supports subsequent associative recognition. *Neuroimage, 21*(1), 456–462.

Jacoby, L.L. (1991). A process dissociation framework: Separating automatic from intentional uses of memory. *J Memory Language, 30*(5), 513–541.

Jones, E.G., and Powell, T.P. (1970). An anatomical study of converging sensory pathways within the cerebral cortex of the monkey. *Brain, 93*(4), 793–820.

Kapur, N., and Brooks, D.J. (1999). Temporally specific retrograde amnesia in two cases of discrete bilateral hippocampal pathology. *Hippocampus, 9*(3), 247–254.

Kesner, R.P., and Hopkins, R.O. (in press). Mnemonic functions of the hippocampus: A comparison between animals and humans. *Biol Psychol.*

Kirchhoff, B.A., Wagner, A.D., Maril, A., and Stern, C.E. (2000). Prefrontal-temporal circuitry for novelty encoding and subsequent memory. *J Neurosci, 20,* 6173–6180.

Kirwan, C.B., and Stark, C.E. (2004). Medial temporal lobe activation during encoding and retrieval of novel face-name pairs. *Hippocampus, 14*(7), 919–930.

Lavenex, P., and Amaral, D.G. (2000). Hippocampal–neocortical interaction: A hierarchy of associativity. *Hippocampus, 10,* 420–430.

Lepage, M., Habib, R., and Tulving, E. (1998). Hippocampal pet activations of memory encoding and retrieval: The hiper model. *Hippocampus, 8,* 313–322.

Levy, D.A., Manns, J.R., Hopkins, R.O., Gold, J.J., Broadbent, N.J., and Squire, L.R. (2003). Impaired visual and odor recognition memory span in patients with hippocampal lesions. *Learn Mem, 10*(6), 531–536.

Maguire, E.A., Frackowiak, R.S., and Frith, C.D. (1997). Recalling routes around london: Activation of the right hippocampus in taxi drivers. *J Neurosci, 17*(18), 7103–7110.

Maguire, E.A., and Frith, C.D. (2003). Lateral asymmetry in the hippocampal response to the remoteness of autobiographical memories. *J Neurosci, 23*(12), 5302–5307.

Maguire, E.A., Frith, C.D., Burgess, N., Donnett, J.G., and O'Keefe, J. (1998). Knowing where things are parahippocampal involvement in encoding object locations in virtual large-scale space. *J Cogn Neurosci, 10*(1), 61–76.

Maguire, E.A., Henson, R.N., Mummery, C.J., and Frith, C.D. (2001). Activity in prefrontal cortex, not hippocampus, varies parametrically with the increasing remoteness of memories. *Neuroreport, 12*(3), 441–444.

Mandler, G. (1980). Recognizing: The judgment of previous occurrence. *Psychol Rev, 87*, 252–271.

Manns, J.R., Clark, R.E., and Squire, L.R. (2000a). Awareness predicts the magnitude of single-cue trace eyeblink conditioning. *Hippocampus, 10*(2), 181–186.

Manns, J.R., Clark, R.E., and Squire, L.R. (2000b). Parallel acquisition of awareness and trace eyeblink classical conditioning. *Learn Mem, 7*(5), 267–272.

Manns, J.R., Hopkins, R.O., Reed, J.M., Kitchener, E.G., and Squire, L.R. (2003a). Recognition memory and the human hippocampus. *Neuron, 37*(1), 171–180.

Manns, J.R., Hopkins, R.O., and Squire, L.R. (2003b). Semantic memory and the human hippocampus. *Neuron, 38*(1), 127–133.

Manns, J.R., and Squire, L.R. (1999). Impaired recognition memory on the doors and people test after damage limited to the hippocampal region. *Hippocampus, 9*(5), 495–499.

Manns, J.R., and Squire, L.R. (2001). Perceptual learning, awareness, and the hippocampus. *Hippocampus, 11*(6), 776–782.

Marr, D. (1971). Simple memory: A theory for archicortex. *Philos Trans R Soc Lond B Biol Sci, 262*(841), 23–81.

Mayes, A.R., Daum, I., Markowisch, H.J., and Sauter, B. (1997). The relationship between retrograde and anterograde amnesia in patients with typical global amnesia. *Cortex, 33*(2), 197–217.

Mayes, A.R., Holdstock, J.S., Isaac, C.L., Hunkin, N.M., and Roberts, N. (2002). Relative sparing of item recognition memory in a patient with adult-onset damage limited to the hippocampus. *Hippocampus, 12*(3), 325–340.

Mayes, A.R., Holdstock, J.S., Isaac, C.L., Montaldi, D., Grigor, J., Gummer, A., et al. (2004). Associative recognition in a patient with selective hippocampal lesions and relatively normal item recognition. *Hippocampus, 14*(6), 763–784.

McClelland, J.L., and Goddard, N.H. (1996). Considerations arising from a complementary learning systems perspective on hippocampus and neocortex. *Hippocampus, 6*(6), 654–665.

McClelland, J.L., McNaughton, B.L., and O'Reilly, R.C. (1995). Why there are complementary learning systems in the hippocampus and neocortex: Insights from the successes and failures of connectionist models of learning and memory. *Psycholo Rev, 102*, 419–457.

McGaugh, J.L. (2000). Memory — a century of consolidation. *Science, 287*(5451), 248–251.

McGlinchey-Berroth, R., Carrillo, M.C., Gabrieli, J.D., Brawn, C.M., and Disterhoft, J.F. (1997). Impaired trace eyeblink conditioning in bilateral, medial-temporal lobe amnesia. *Behav Neurosci, 111*(5), 873–882.

Milner, B. (1971). Interhemispheric differences in the localization of psychological processes in man. *Br Med Bull, 27*(3), 272–277.

Milner, B. (1972). Disorders of learning and memory after temporal-lobe lesions in man. *Clin Neurosurg, 19*, 421–446.

Milner, B., Squire, L.R., and Kandel, E.R. (1998). Cognitive neuroscience and the study of memory. *Neuron, 20*(3), 445–468.

Moscovitch, M., Rosenbaum, R.S., Gilboa, A., Addis, D.R., Westmacott, R., Grady, C., et al. (2005). Functional neuroanatomy of remote episodic, semantic and spatial memory: A unified account based on multiple trace theory. *J Anat, 207*(1), 35–66.

Nadel, L., and Moscovitch, M. (1997). Memory consolidation, retrograde amnesia and the hippocampal complex. *Curr Opin Neurobiol, 7*, 217–227.

Nadel, L., Samsonovich, A., Ryan, L., and Moscovitch, M. (2000). Multiple trace theory of human memory: Computational, neuroimaging, and neuropsychological results. *Hippocampus, 10*, 352–368.

Nakazawa, K., Quirk, M.C., Chitwood, R.A., Watanabe, M., Yeckel, M.F., Sun, L.D., et al. (2002). Requirement for hippocampal CA3 NMDA receptors in associative memory recall. *Science, 297*(5579), 211–218.

Nakazawa, K., Sun, L.D., Quirk, M.C., Rondi-Reig, L., Wilson, M.A., and Tonegawa, S. (2003). Hippocampal CA3 NMDA receptors are crucial for memory acquisition of one-time experience. *Neuron, 38*(2), 305–315.

Niki, K., and Luo, J. (2002). An fMRI study on the time-limited role of the medial temporal lobe in long-term topographical autobiographic memory. *J Cogn Neurosci, 14*(3), 500–507.

Norman, K.A., and O'Reilly, R.C. (2003). Modeling hippocampal and neocortical contributions to recognition memory: A complementary-learning-systems approach. *Psychol Rev, 110*(4), 611–646.

O'Kane, G., Insler, R.Z., and Wagner, A.D. (2005). Conceptual and perceptual novelty effects in human medial temporal cortex. *Hippocampus, 15*, 326–332.

O'Reilly, R.C., and Rudy, J.W. (2001). Conjunctive representations in learning and memory: Principles of cortical and hippocampal function. *Psychol Rev, 108*, 311–345.

Paller, K.A., and Wagner, A.D. (2002). Observing the transformation of experience into memory. *Trends Cogn Sci, 6*, 93–102.

Pihlajamaki, M., Tanila, H., Hanninen, T., Kononen, M., Mikkonen, M., Jalkanen, V., et al. (2003). Encoding of novel picture pairs activates the perirhinal cortex: An fMRI study. *Hippocampus, 13*(1), 67–80.

Pihlajamaki, M., Tanila, H., Kononen, M., Hanninen, T., Hamalainen, A., Soininen, H., et al. (2004). Visual presentation of novel objects and new spatial arrangements of objects differentially activates the medial temporal lobe subareas in humans. *Eur J Neurosci, 19*(7), 1939–1949.

Preston, A.R., Shrager, Y., Dudukovic, N.M., and Gabrieli, J.D. (2004). Hippocampal contribution to the novel use of relational information in declarative memory. *Hippocampus, 14*(2), 148–152.

Quamme, J.R., Yonelinas, A.P., Widaman, K.F., Kroll, N.E., and Sauve, M.J. (2004). Recall and recognition in mild hypoxia: Using covariance structural modeling to test competing theories of explicit memory. *Neuropsychologia, 42*(5), 672–691.

Rajji, T., Chapman, D., Eichenbaum, H., and Greene, R. (2006). The role of CA3 hippocampal NMDA receptors in paired associate learning. *Journal of Neuroscience, 26*, 908–915.

Ranganath, C., Yonelinas, A.P., Cohen, M.X., Dy, C.J., Tom, S.M., and D'Esposito, M. (2003). Dissociable correlates of recollection and familiarity within the medial temporal lobes. *Neuropsychologia, 42*(1), 2–13.

Reed, J.M., Hamann, S.B., Stefanacci, L., and Squire, L.R. (1997). When amnesic patients perform well on recognition memory tests. *Behav Neurosci, 111*, 1163–1170.

Reed, J.M., and Squire, L.R. (1998). Retrograde amnesia for facts and events: Findings from four new cases. *J Neurosci, 18*(10), 3943–3954.

Rempel-Clower, N.L., Zola, S.M., Squire, L.R., and Amaral, D.G. (1996). Three cases of enduring memory impairment after bilateral damage limited to the hippocampal formation. *J Neurosci, 16*, 5233–5255.

Rugg, M.D., and Yonelinas, A.P. (2003). Human recognition memory: A cognitive neuroscience perspective. *Trends Cogn Sci, 7*(7), 313–319.

Ryan, J.D., Althoff, R.R., Whitlow, S., and Cohen, N.J. (2000). Amnesia is a deficit in relational memory. *Psychol Sci, 11*(6), 454–461.

Ryan, J.D., and Cohen, N.J. (2004). Processing and short-term retention of relational information in amnesia. *Neuropsychologia, 42*(4), 497–511.

Ryan, L., Nadel, L., Keil, K., Putnam, K., Schnyer, D., Trouard, T., et al. (2001). The hippocampal complex and retrieval of recent and very remote autobiographical memories: Evidence from functional magnetic resonance imaging in neurologically intact people. *Hippocampus, 11*, 707–714.

Schacter, D.L., Alpert, N.M., Savage, C.R., Rauch, S.L., and Albert, M.S. (1996). Conscious recollection and the human hippocampal formation: Evidence from positron emission tomography. *Proc Natl Acad Sci USA, 93*, 321–325.

Schacter, D.L., and Wagner, A.D. (1999). Medial temporal lobe activations in fMRI and PET studies of episodic encoding and retrieval. *Hippocampus, 9*, 7–24.

Scoville, W.B., and Milner, B. (1957). Loss of recent memory after bilateral hippocampal lesions. *J Neurol Neurosurg Psych, 20*, 11–21.

Small, S.A. (2002). The longitudinal axis of the hippocampal formation: Its anatomy, circuitry, and role in cognitive function. *Rev Neurosci, 13*(2), 183–194.

Small, S.A., Nava, A.S., Perera, G.M., DeLaPaz, R., Mayeux. R., and Stern, Y. (2001). Circuit mechanisms underlying memory encoding and retrieval in the long axis of the hippocampal formation. *Nat Neurosci., 4*, 442–449.

Sommer, T., Rose, M., Glascher, J., Wolbers, T., and Buchel, C. (2005). Dissociable contributions within the medial temporal lobe to encoding of object–location associations. *Learn Mem, 12*(3), 343–351.

Sperling, R., Chua, E., Cocchiarella, A., Rand-Giovannetti, E., Poldrack, R., Schacter, D.L., et al. (2003). Putting names to faces: Successful encoding of associative memories activates the anterior hippocampal formation. *Neuroimage, 20*(2), 1400–1410.

Spiers, H.J., Maguire, E.A., and Burgess, N. (2001). Hippocampal amnesia. *Neurocase, 7*(5), 357–382.

Squire, L.R. (1992). Memory and the hippocampus: A synthesis from findings with rats, monkeys, and humans. *Psychol Rev, 99*, 195–231.

Squire, L.R., and Alvarez, P. (1995). Retrograde amnesia and memory consolidation: A neurobiological perspective. *Curr Opin Neurobiol, 5*, 169–177.

Squire, L.R., Clark, R.E., and Knowlton, B.J. (2001a). Retrograde amnesia. *Hippocampus, 11*(1), 50–55.

Squire, L.R., Schmolck, H., and Stark, S.M. (2001b). Impaired auditory recognition memory in amnesic patients with medial temporal lobe lesions. *Learn Mem, 8*(5), 252–256.

Squire, L.R., Stark, C.E., and Clark, R.E. (2004). The medial temporal lobe. *Annu Rev Neurosci, 27*, 279–306.

Squire, L.R., and Zola, S.M. (1998). Episodic memory, semantic memory, and amnesia. *Hippocampus, 8*, 205–211.

Stark, C.E., Bayley, P.J., and Squire, L.R. (2002). Recognition memory for single items and for associations is similarly impaired following damage to the hippocampal region. *Learn Mem, 9*, 238–242.

Stark, C.E., and Squire, L.R. (2003). Hippocampal damage equally impairs memory for single items and memory for conjunctions. *Hippocampus, 13*(2), 281–292.

Stern, C.E., Corkin, S., Gonzalez, R.G., Guimaraes, A.R., Baker, J.R., Jennings, P.J., et al. (1996). The hippocampal formation participates in novel picture encoding: Evidence from functional magnetic resonance imaging. *Proc Natl Acad Sci USA, 93*, 8660–8665.

Suzuki, W.A., and Amaral, D.G. (1990). Cortical inputs to the CA1 field of the monkey hippocampus originate from the perirhinal and parahippocampal cortex but not from area TE. *Neurosci Lett, 115*(1), 43–48.

Suzuki, W.A., and Amaral, D.G. (1994a). Perirhinal and parahippocampal cortices of the macaque monkey: Cortical afferents. *J Comp Neurol, 350*(4), 497–533.

Suzuki, W.A., and Amaral, D.G. (1994b). Topographic organization of the reciprocal connections between the monkey entorhinal cortex and the perirhinal and parahippocampal cortices. *J Neurosci, 14*, 1856–1877.

Tranel, D., Brady, D.R., Van Hoesen, G.W., and Damasio, A.R. (1988). Parahippocampal projections to posterior auditory association cortex (area TPT) in old-world monkeys. *Exp Brain Res, 70*, 406–416.

Tulving, E. (1983). *Elements of Episodic Memory.* Cambridge; UK: Oxford University Press.

Van Hoesen, G., and Pandya, D.N. (1975). Some connections of the entorhinal (area 28) and perirhinal (area 35) cortices of the rhesus monkey. I. Temporal lobe afferents. *Brain Res., 95*, 1–24.

Van Hoesen, G., Pandya, D.N., and Butters, N. (1975). Some connections of the entorhinal (area 28) and perirhinal (area 35) cortices of the rhesus monkey. II. Frontal lobe afferents. *Brain Res, 95*(1), 25–38.

Vargha-Khadem, F., Gadian, D.G., Watkins, K.E., Connelly, A., Van Paesschen, W., and Mishkin, M. (1997). Differential effects of early hippocampal pathology on episodic and semantic memory. *Science, 277*, 376–380.

Wagner, A.D., Schacter, D.L., Rotte, M., Koutstaal, W., Maril, A., Dale, A.M., et al. (1998). Building memories: Remembering and forgetting of verbal experiences as predicted by brain activity. *Science, 281*, 1188–1191.

Wais, P.E., Wixted, J.T., Hopkins, R.O., and Squire, L.R. (2006). The hippocampus supports both the recollection and the familiarity components of recognition memory. *Neuron, 49*(3), 459–466.

Weis, S., Klaver, P., Reul, J., Elger, C.E., and Fernandez, G. (2004a). Temporal and cerebellar brain regions that support both declarative memory formation and retrieval. *Cereb Cortex, 14*(3), 256–267.

Weis, S., Specht, K., Klaver, P., Tendolkar, I., Willmes, K., Ruhlmann, J., et al. (2004b). Process dissociation between contextual retrieval and item recognition. *Neuroreport, 15*(18), 2729–2733.

Witter, M.P., and Amaral, D.G. (1991). Entorhinal cortex of the monkey: V. Projections to the dentate gyrus, hippocampus, and subicular complex. *J Comp Neurol, 307*(3), 437–459.

Witter, M.P., Van Hoesen, G.W., and Amaral, D.G. (1989). Topographical organization of the entorhinal projection to the dentate gyrus of the monkey. *J Neurosci., 9*, 216–228.

Wixted, J.T., and Stretch, V. (2004). In defense of the signal detection interpretation of remember/know judgments. *Psychol Bull Rev, 11*(4), 616–641.

Xiang, J.Z., and Brown, M.W. (1998). Differential neuronal encoding of novelty, familiarity and recency in regions of the anterior temporal lobe. *Neuropharmacology, 37*, 657–676.

Yonelinas, A.P. (1994). Receiver-operating characteristics in recognition memory: Evidence for a dual-process model. *J Exp Psychol Learn Mem Cogn, 20*(6), 1341–1354.

Yonelinas, A.P. (1999). The contribution of recollection and familiarity to recognition and source-memory judgments: A formal dual-process model and an analysis of receiver operating characteristics. *J Exp Psychol Learn Mem Cogn, 25*(6), 1415–1434.

Yonelinas, A.P. (2001). Consciousness, control, and confidence: The 3 Cs of recognition memory. *J Exp Psychol Gen, 130*, 361–379.

Yonelinas, A.P. (2002). The nature of recollection and familiarity: A review of 30 years of research. *J Mem Lang, 46*, 441–517.

Yonelinas, A.P., Dobbins, I., Szymanski, M.D., Dhaliwal, H.S., and King, L. (1996). Signal-detection, threshold, and dual-process models of recognition memory: Rocs and conscious recollection. *Conscious Cogn, 5*(4), 418–441.

Yonelinas, A.P., Kroll, N.E., Dobbins, I., Lazzara, M., and Knight, R.T. (1998). Recollection and familiarity deficits in amnesia: Convergence of remember-know, process dissociation, and receiver operating characteristic data. *Neuropsychology, 12*(3), 323–339.

Yonelinas, A.P., Kroll, N.E., Quamme, J.R., Lazzara, M.M., Sauve, M.J., Widaman, K.F., et al. (2002). Effects of extensive temporal lobe damage or mild hypoxia on recollection and familiarity. *Nature Neuroscience, 5*, 1236–1241.

Yonelinas, A.P., Otten, L.J., Shaw, K.N., and Rugg, M.D. (2005). Separating the brain regions involved in recollection and familiarity in recognition memory. *J Neurosci, 25*(11), 3002–3008.

Yonelinas, A.P., Quamme, J.R., Widaman, K.F., Kroll, N.E., Sauve, M.J., and Knight, R.T. (2004). Mild hypoxia disrupts recollection, not familiarity. *Cogn Affect Behav Neurosci, 4*(3), 393–400; discussion 401–406.

Zeineh, M.M., Engel, S.A., Thompson, P.M., and Bookheimer, S.Y. (2003). Dynamics of the hippocampus during encoding and retrieval of face-name pairs. *Science, 299*(5606), 577–580.

Zola-Morgan, S., Squire, L.R., and Amaral, D.G. (1986). Human amnesia and the medial temporal region: Enduring memory impairment following a bilateral lesion limited to field ca1 of the hippocampus. *J Neurosci, 9*, 2950–2967.

CHAPTER 10

Bootstrapping Your Brain: How Interactions Between the Frontal Cortex and Basal Ganglia May Produce Organized Actions and Lofty Thoughts

Earl K. Miller and Timothy J. Buschman

The Picower Institute for Learning and Memory, RIKEN-MIT Neuroscience Research Center, and Department of Brain and Cognitive Sciences, Massachusetts Institute of Technology, Cambridge, MA 02139

I. INTRODUCTION

Reflexes, habits, and other reactions to the immediate environment are relatively straightforward: A familiar stimulus activates well-established neural pathways that produce stereotyped behaviors. This is so-called bottom-up, stimulus-driven, processing. Goal-directed behavior — acting on, not just reacting to, the environment — relies on a different mode of brain operation. Navigating novel or complex situations to achieve long-planned goals cannot rely on uncoordinated reactions to the environment. Instead, goal-directed thoughts and actions are orchestrated "top-down" from within oneself: With our knowledge of how the world works, we predict what outcomes are desirable and determine what strategies will aid in attaining them.

Learning and memory are the essence of both classes of behavior: automatic habits seem to be established by repeated coactivation of neural pathways, as if "ruts" were being carved into the brain. These habits can simply be triggered, fired off in a "ballistic" fashion with little variation, and hence require little internal oversight. In contrast, with truly sophisticated goal-directed behavior, simply recording and replaying experiences is not sufficient.

Neurobiology of Learning and Memory, Second Edition

Goal-relevant relationships need to be sorted out from spurious coincidences, and, importantly, long-term goals require more than figuring out the world piecemeal, one situation at a time. Smart animals get the "big picture" of the jigsaw puzzle of their experiences, the common structure across a wide range of experiences. This forms the overarching principles and general concepts that are the underpinnings for high-level thought and action and provides the foresight needed for far-off goals.

The goal of this chapter is to review evidence that goal-directed behavior depends on interactions between two different "styles" of learning mechanisms in different frontal lobe systems, specifically that ever-more-complex thought and action can be bootstrapped through recursive interactions between fast (larger-synaptic-weight changes), dopaminergic- (reward-) based plasticity in the basal ganglia and slower (smaller-synaptic-weight changes), more Hebbian- (less reward-) based plasticity in the frontal cortex. By having these two systems interact in recursive processing loops, the brain can learn new things quickly but also take the time to link in more experiences and build up abstract, big-picture thoughts and sophisticated actions.

II. COGNITIVE CONTROL AND THE PFC

The ability to coordinate thought and action and direct them toward a goal is called *cognitive control*. Virtually all goal-directed behaviors are learned and thus depend on a cognitive system that can acquire the rules of the game: what outcomes are possible, what actions might be successful at achieving them, what their costs might be, etc. Take, for example, dining in a restaurant. We are not born knowing that this can be a very rewarding experience nor how to act in the situation. Instead, our experiences arm us with expectations about the important sensory information deserving our attention (the wine list), typical events occurring during the meal, appropriate actions, and expected consequences (paying the bill). These rules orchestrate processing in diverse brain regions along a common, internal theme. It is widely accepted that the prefrontal cortex (PFC) is centrally involved in this process.

The prefrontal cortex is situated at the anterior end of the brain and reaches its greatest elaboration and relative size in the primate, especially human, brain (Fuster, 1995) and is thus presumably responsible for our advanced cognitive capabilities and goal-directed behaviors. Indeed, recent imaging work has suggested the size of the prefrontal cortex is directly correlated with intelligence in adult humans (Haier et al., 2004). The PFC seems anatomically well situated to play a role in cognitive control. It seems to be a microcosm of cortical processing, able to synthesize a wide range of external and internal information and also to exert control over much of the cortex (Fig. 10-1). The PFC receives and sends projections to most of the cerebral cortex (with the exception of primary sensory and motor cortices) as well as all of the major subcortical

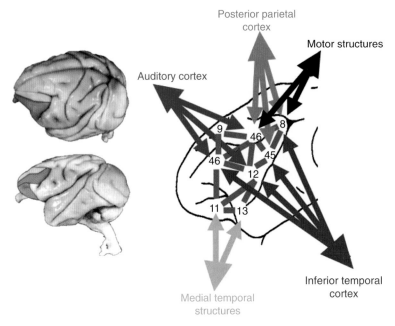

FIGURE 10-1   The monkey prefrontal cortical (PFC) areas with some connections. **Left:** Dorsal is yellow; dorsolateral is red; ventrolateral is green; orbital is blue. **Right:** Some extrinsic and intrinsic connections. Not all connections are depicted. The figure is meant to illustrate the integrative, multimodal nature of prefrontal cortex anatomy.

systems, such as the hippocampus, amygdala, cerebellum, and, most importantly for this chapter, the basal ganglia (Porrino et al., 1981; Amaral and Price, 1984; Amaral, 1986; Selemon and Goldman-Rakic, 1988; Barbas and De Olmos, 1990; Eblen and Graybiel, 1995; Croxson et al., 2005). Different PFC subdivisions have distinct patterns of interconnections with other brain systems (e.g., lateral–sensory and motor cortex, orbital–limbic), but there are prodigious connections both within and between PFC subdivisions (Pandya and Barnes, 1987; Barbas and Pandya, 1989; Pandya and Yeterian, 1990; Barbas et al., 1991; Petrides and Pandya, 1999). The anatomical architecture suggests an infrastructure ideal for learning, one that can act as a large associative network for detecting and storing associations between diverse events, experiences, internal states, etc. After learning, such a network can complete or "recall" an entire pattern, given a subset of its inputs, and can retrieve associated patterns.

Indeed, neurophysiological studies in animals and imaging studies in humans have shown that the PFC is highly multimodal, its neurons responsive to a wide range of information, and that it has other properties useful for cognitive control (Miller, 2000). Its neurons sustain their activity to maintain information across short, multisecond, memory delays (Pribram et al., 1952; Fuster and Alexander, 1971; Fuster, 1973; Funahashi et al., 1989; Miller et al., 1996). This property is crucial for goal-directed behavior, which, unlike "ballistic" reflexes, typically extended over time. After training on a wide range of operant tasks, many PFC neurons (from one-third to one-half of the population) reflect the learned task contingencies: the logic or rules of the task (White and Wise, 1999; Asaad et al., 2000; Wallis et al., 2000; Mansouri et al., 2006). For example, a neuron might be selectively activated by a given cue (e.g., a green light) when it has a specific meaning (like "go"), whereas another neuron might be activated when that cue means something different (like "stop"). Some neurons might activate in anticipation of a forthcoming expected reward or a relevant cue (Watanabe, 1996; Rainer et al., 1999; Wallis and Miller, 2003; Padoa-Schioppa and Assad, 2006), and neurons have even been found to reflect whether a monkey is currently following abstract principles like "same" and "different" (White and Wise, 1999; Wallis et al., 2000). In short, the PFC does indeed act like a brain area that absorbs and reflects the rules needed to guide goal-directed, volitional behavior.

Miller and Cohen (2001) argued that all of this indicates that the cardinal PFC function is to acquire and actively maintain patterns of activity that represent goals and the means to achieve them (rules) *and* the cortical pathways needed to perform the task ("maps," hence "rulemaps"). Under this model, activation of a PFC rulemap sets up bias signals that propagate throughout much of the rest of the cortex, affecting sensory systems as well as systems responsible for response execution, memory retrieval, emotional evaluation, etc. The aggregate effect is to guide the flow of neural activity along pathways that establish the proper mappings between inputs, internal states, and outputs to best perform the task. Establishing the proper mapping is especially important whenever stimuli are ambiguous (i.e., when they activate more than one input representation) or when multiple responses are possible and the task-appropriate response must compete with stronger, more habitual alternatives. In short, task information is acquired by the PFC, which provides support to related information in posterior brain systems, effectively acting as a global attentional controller, or traffic cop, if you will.

Of course, the PFC does not work in isolation. In the rest of this chapter, we present evidence that the PFC works in close collaboration with the basal ganglia in the learning of goal-directed behaviors. We argue that output of the basal ganglia helps train PFC networks, propelling the PFC to build up the high-level representations that underlie abstract thought.

III. THE BASAL GANGLIA

The basal ganglia (BG) comprise a collection of subcortical nuclei: cortical inputs arrive largely via the striatum (which includes both the caudate and the

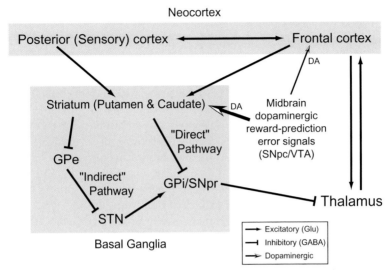

FIGURE 10-2   Simplified circuit diagram for the basal ganglia. See text for explanation. The heavier projection of midbrain dopaminergic neurons to the striatum than to the cortex is depicted by the heavier arrow.

putamen), are processed through the globus pallidus (GP), the subthalamic nucleus (STN), and the substantia nigra (SN), and are then directed back into the cortex via the thalamus (Fig. 10-2).

Like the PFC, the basal ganglia are a major site of cortical convergence. Most of the cortex projects directly onto the striatum (Kemp and Powell, 1970; Kitai et al., 1976) and does so in a slightly divergent manner, in that the cortical afferents make connections with multiple striatal neurons (Flaherty and Graybiel, 1991). The striatum is segregated into *striosomes* and the *matrix* (Graybiel and Ragsdale, 1978), with striosomes receiving inputs preferentially from the entire cerebral cortex and the matrix receiving inputs primarily from the limbic and hippocampal systems and from the prefrontal cortex (Donoghue and Herkenham, 1986; Gerfen, 1992; Eblen and Graybiel, 1995). Anatomical tracing techniques have suggested that functionally similar cortical areas project into the same striosome (Yeterian and Van Hoesen, 1978; Van Hoesen et al., 1981; Flaherty and Graybiel, 1991). For example, both sensory and motor areas relating to the arm seem to innervate the same striosome preferentially. The basal ganglia maintain a degree of topographical separation in different "channels" throughout their nuclei, ensuring that the output via the thalamus is largely to the same cortical areas that gave rise to the initial inputs to the BG (Selemon and Goldman-Rakic, 1985; Alexander et al., 1986; Parthasarathy et al., 1992; Hoover and Strick, 1993; Kelly and Strick, 2004). The frontal cortex receives the largest portion of BG outputs, suggesting some form of close collaboration between these structures (Middleton and Strick, 1994, 2000, 2002).

The majority of neurons found in both the striasomes and matrix are spiny cells (as high as 90%) (Kemp and Powell, 1971). These neurons earned their moniker for their high density of synaptic boutons along their dendritic arbor. Cortical inputs converge onto the dendrites along with an input from midbrain dopaminergic neurons. The latter have been suggested to provide a reward-based "teaching signal" that gates plasticity in the striatum (more on this later). Plus, some evidence suggests that spiny cells may have unique electrical properties — the resting potential of the membrane can exist in either a "down" or an "up" state (Wilson, 1992; Wilson and Kawaguchi, 1996). In the down state, the neuron is not very responsive to inputs, with little spiking response. However, a large enough input can kick the cell into the up state, where it becomes very sensitive to further input, readily producing spikes. All of this has suggested that that striatum has an ideal infrastructure for rapid, supervised learning, i.e., the quick formation of connections between cortical inputs that predict reward.

Projections from the striatum are distributed along two parallel routes: the *direct* and *indirect* pathways (Fig. 10-2) (Mink, 1996; Graybiel, 2000). The direct pathway leads from the striatum into the globus pallidus internal (GPi) and the substantia nigra pars reticulata (SNpr). These regions project directly onto the thalamus. All projections from the striatum are thought to be GABAergic and therefore inhibit downstream neurons in the GPi/SNpr. Neurons in GPi/SNpr inhibit the thalamus, making the direct pathway effectively excitatory — activity in the striatum releases inhibition on the thalamus. The indirect pathway involves striatal projections to the globus pallidus external (GPe), which in turn projects to the subthalamic nucleus (STN), which projects onto the GPi/SNpr. Similar to the other connections in the basal ganglia, GPe inputs into STN are inhibitory, but STN provides glutamatergic, excitatory input into the GPi/SNpr. Due to the added inhibitory synapse, the indirect pathway increases inhibition on the thalamus. These two pathways are believed to exist in an equilibrium that allows for the release of desired patterns while inhibiting unintended ones. Though cortical inputs into the striatum had a divergent nature, connections between the striatum and GPi/SNpr and GPe are believed to be highly convergent (Flaherty and Graybiel, 1993; Parent and Hazrati, 1993; Flaherty and Graybiel, 1994). This convergence of inputs is effectively a reduction in the dimensionality of the patterns and may allow for a certain degree of integration and generalization across specific cortical inputs.

Damage or dysfunction to the basal ganglia, such as in Parkinson's disease or Huntington's disease, causes profound deficits, ranging from motor (such as difficulty initiating volitional movement) to cognitive (Taylor et al., 1986; Cronin-Golomb et al., 1994; Lawrence et al., 1998; Middleton and Strick,

2000). Lesions of the striatum in monkeys and rodents produce impairments in learning new operant behaviors, and, in general, damage to different parts of the striatum causes deficits similar to lesioning the area of cortex that loops with the affected region of the striatum (Divac et al., 1967; Goldman and Rosvold, 1972).

IV. DOPAMINERGIC TEACHING SIGNALS

Goal-directed plasticity requires guidance. Simply strengthening the synapse between two coactivated neurons is not sufficient. Relevant events and predictive relationships need to be sorted out from spurious coincidences, so plasticity needs to be guided by information about which associations are predictive of desirable outcomes.

This guidance appears to come in the form of a *reinforcement signal* and is suggested to be provided by dopaminergic neurons in the midbrain (both the ventral tegmental area, VTA, and the substantia nigra, pars compacta, SNpc) (Schultz et al., 1992, 1997; Schultz, 1998). Neurons in these areas show activity that corresponds directly to the reward-prediction error signals suggested by models of animal learning. They activate and release dopamine (DA) widely throughout the subcortex and cortex (especially in the frontal lobe) whenever animals are unexpectedly rewarded, and they will pause when an expected reward is withheld. Over time the cells will learn to respond to an unexpected event that predicts a reward directly: The event "stands in" for the reward (Schultz et al., 1993). Dopaminergic neurons will now respond to the predictive event when it is unexpected but will no longer respond to the actual reward event, for it is now expected. In short, they seem to correspond to a teaching signal that says, "Something good happened; you did not predict it, so remember what just happened so that you can predict it in the future." If these signals resulted in a net strengthening of connections between neurons in the PFC and BG that were recently active, the result might be to strengthen representation of a network of reward-predicting associations.

Midbrain dopaminergic neurons project to the frontal cortex in a gradient fashion; the projections are heavier anteriorially and then drop off posteriorly (Thierry et al., 1973; Goldman-Rakic et al., 1989). However, the midbrain DA input into the striatum is much heavier than that to the PFC, by as much as an order of magnitude (Lynd-Balta and Haber, 1994). Further, evidence suggests that neither strengthening nor weakening of synapses in the striatum by long-term depression or potentiation can occur without dopaminergic input (Calabresi et al., 1992, 1997; Otani et al., 1998; Kerr and Wickens, 2001). Integrating what we have reviewed about the frontal cortex, we propose that the two regions are specialized for two different modes of learning. This is discussed next.

V. FAST, SUPERVISED BG PLASTICITY VERSUS SLOWER, LESS SUPERVISED CORTICAL PLASTICITY

At first blush, it might seem that the greatest benefit would come from learning that proceeded as quickly as possible. Fast learning has obvious advantages: One can learn to get to resources and avoid obstacles faster and better than competitors. But fast learning comes at a cost; it does not allow the benefits that come from generalizing over multiple experiences, so by necessity it tends to be specific and error prone. Take, for example, an example of one-trial learning: conditioned taste aversion. Many of us have eaten something and then become ill for another reason but developed an aversion to that food anyway, an erroneous attribution. Extending learning across multiple episodes allows organisms to pick up on the regularities of predictive relationships and leave behind spurious (inconsistent) associations and coincidences.

In addition to avoiding errors, slow learning also allows detection of common structure across many different experiences, the regularities and commonalities that form abstractions, general principles, concepts, and symbolisms that are the medium of the sophisticated thought, and the "big pictures" needed for truly long-term goals. Indeed, this is key to proactive thought and action. Generalizing among many past experiences naturally endows the ability to generalize to the future, to imagine possibilities we have not yet experienced but would like to. In addition, abstraction may aid in cognitive flexibility. Generalized representations, by their nature, are concise because they lack details. It is probably easier to quickly change thought and action if the switching is between concise representations rather than only elaborate, detailed ones.

So how does the brain balance the obvious pressure to learn as quickly as possible with the advantages of slower learning? One suggestion comes from O'Reilly's group, who suggested that fast-learning and slow-learning systems interact with one another (McClelland et al., 1995; O'Reilly and Munakata, 2000). O'Reilly and colleagues specifically suggested that the consolidation of long-term memories is a result of the output of fast plasticity mechanisms in the hippocampus "training" slower plasticity cortical networks. Fast versus slow means large versus small changes in synaptic weights. The idea is that the hippocampus is specialized for the rapid acquisition of new information; each learning trial produces large weight changes. The output of the hippocampus will then repeatedly activate cortical networks that have smaller weight changes per episode. Continued hippocampal-mediated reactivation of the cortical representations allows the cortex gradually to connect them with other experiences. That way, the structure shared across experiences can be detected and stored, and the memory can be interleaved with others so that it can be readily accessed.

A similar relationship may exist between the PFC and the BG. A recent experiment by Pasupathy and Miller (2005) provides suggestive evidence. As monkeys learned to associate a visual cue and a directional eye movement over a few tens of trials, neural activity in the striatum showed rapid, almost bistable, changes compared to a much slower trend in the PFC. Interestingly, however, the slower PFC seemed to be the final arbiter of behavior; the monkeys' improvement in selecting the correct response more closely matched PFC changes.

These results may reflect how the BG and the PFC relate to one another as the animal learns: Specific stimulus–response associations are quickly learned in the basal ganglia, slowly training the prefrontal cortex. The strong-weight-changes (fast) plasticity in the striatum is more suited for the rapid formation of "concrete" rules, associations between a specific cue and response. However, as noted earlier, fast learning tends to be error prone, and, indeed, striatal neurons began predicting the forthcoming behavioral response early in learning when that response was often wrong. By contrast, the smaller weight changes in the PFC may have allowed it to accumulate more evidence and more slowly and judiciously arrive at the correct answer.

The faster learning-related changes in the striatum reported by Pasupathy and Miller (2005) are consistent with this notion, as is the observation that there was stronger modulation of activity in the striatum than in the PFC by performance of these specific, concrete rules. But what about abstracted, generalized rules? Our model of fast BG plasticity versus slower PFC plasticity predicts the opposite, namely that abstract rules should have a stronger effect on PFC than BG activity because the slower PFC plasticity is more suited to learn them. A recent experiment by Muhammad et al. (2006) showed just that. Building on the work of Wallis et al. (2001), monkeys were trained to apply the abstract rules *same* and *different* to pairs of pictures. If the *same* rule was in effect, monkeys responded if the pictures were identical; whereas if the *different* rule was in effect, monkeys responded if the pictures were different. The rules were abstract, since the monkeys were able to apply the rules to novel stimuli, stimuli for which there could be no preexisting stimulus–response association. This is the definition of an abstract rule. Muhammad et al. recorded neural activity from the same PFC and striatal regions as Pasupathy and Miller (2005) and found that, in contrast to the specific cue–response associations, the abstract rules were reflected more strongly in PFC activity (more neurons with effects and larger effects) than in BG activity, the opposite of what Pasupathy and Miller (2005) reported for the specific cue–response associations.

In fact, this architecture (fast learning in more primitive, noncortical structures training the slower, more advanced cortex) may be a general brain strategy; in addition to being suggested for the relationship between the hippocampus and cortex, it has also been proposed for the cerebellum and cortex (Houk and Wise, 1995). This makes sense: The first evolutionary pressure on our cortexless ancestors was presumably toward faster learning, while only later did we add on a slower, more judicious, and flexible cortex.

VI. FRONTAL CORTEX–BASAL GANGLIA LOOPS: RECURSIVE PROCESSING AND BOOTSTRAPPING

As already noted, the cortex seems to form closed anatomical loops with the BG. Channels within the BG return outputs, via the thalamus, to the same cortical areas that gave rise to their initial cortical input, forming BG–cortex loops. This structure suggests a recursive system where the results from one iteration can be fed back through the loop for further processing. Such a system is ideal for bootstrapping: Repeated iterations can link in more and more information, building itself into ever-increasing elaboration and sophistication. This may allow the bootstrapping of neural representations to increasing complexity and, with the slower learning in the PFC, greater abstractions.

This model also reflects a hallmark of human intelligence: It is easiest for us to understand new concepts if they can be grounded in familiar ones. We learn to multiply through serial addition, and we understand quantum mechanics by constructing analogies to waves and particles. Interactions between the basal ganglia and the prefrontal cortex may support this type of cognitive bootstrapping. As more complex and generalized representations are learned in the prefrontal cortex, they are passed down through the basal ganglia for further expansion. Recursive processing must surely underlie the ability of the midbrain DA neurons to respond to earlier and earlier events in a predictive chain leading to reward. Both the frontal cortex and striatum send projections to the midbrain DA neurons, but whether or not these inputs' pathways are critical for DA neurons' predictive response is still not yet known.

The frontal cortex–BG loops also suggest an autoassociative type of network, such as that seen in the CA3 of the hippocampus, where it is thought to play a role in recall. The outputs looping back on the inputs allow the network to learn to complete (i.e., recall) previously learned patterns, given a degraded version or a subset of the original inputs (Hopfield, 1982). It may be that BG–cortex loops have similar principles, but, unlike the hippocampus, one dependent on midbrain DA signals and, therefore, more goal oriented.

Another feature of autoassociative networks is that if an activity pattern is delayed from feeding back immediately on the inputs, it can learn sequences of patterns and thus make predictions (Kleinfeld, 1986; Sompolinsky and Kanter, 1986). The lag allows a given pattern to delay its feedback until the next pattern has appeared. Then, if one inputs a given pattern in a learned sequence, the network produces — predicts — the next one. Sequencing and prediction are important faculties for complex, goal-directed behaviors, which typically extend over time. Evidence for a role for the BG in sequencing and prediction comes from observations that striatal neural activity reflects a forthcoming event in a behavioral task (Mushiake and Strick, 1995; Jog et al., 1999) and that lesioning the striatum causes deficits in producing learned sequences (Miyachi et al., 1997; Bailey and Mair, 2006).

One way to add this lag is via a memory buffer. The PFC is well known for this type of property; its neurons can sustain their activity to bridge short-term memory delays. This can act as a bridge for learning contingencies across several seconds, even minutes. The fact that BG loops use inhibitory synapses may also contribute to the unique properties of the recurrent connections. The "direct" pathway has two inhibitory synapses, the result being a net excitatory effect on the cortex via disinhibition of the thalamus. The "indirect" one has three inhibitory synapses, and this makes it net inhibitory; it may be used to countermand current processing in the direct loop. But why evolve a loop out of inhibitory synapses? No doubt there are several answers to this question. It can prevent runaway excitation and thus allows greater control over processing (Wong et al., 1986; Connors et al., 1988; Wells et al., 2000). But another possibility is that the inhibitory synapses slow the circulation of activity through the loops. Many inhibitory synapses are mediated by potassium channels with slow time courses (Couve et al., 2000). The lag may be enough to kick the network into a default mode of sequencing and prediction.

VII. SUMMARY: FRONTAL CORTEX–BG LOOPS FOR GOAL-DIRECTED LEARNING

Here is the general idea: Goal-directed thought and actions are learned via dopaminergic-gated plasticity in frontal cortex–basal ganglia loops. Fast-learning mechanisms in the basal ganglia (specifically the striatum) are more specialized for the detection and storage of specific experiences that can and do lead to reward (i.e., activation of the midbrain DA signals). The output of the BG trains slower-learning mechanisms in the frontal cortex. The slower cortical learning is not only less error prone; it also allows the frontal cortex to build up abstract, generalized representations that reflect the regularities across many different experiences. Recursive iterations of these loops allow bootstrapping of ever-more-complex, and ever-more-predictive, rules and greater abstractions. One result of this plasticity are "rulemaps" in the PFC, the representation of the logic of a likely successful course of thought and action in terms of which cortical pathways are needed. The appropriate rulemap can be activated in a given situation. This sets up top-down signals that feed back to most of the rest of the cortex, dynamically establishing those pathways.

A comparison may be useful. The hippocampus is widely regarded as a system for quickly forming associations between things that are experienced together. In a process called *consolidation*, it manages to etch those pathways in the cortex, resulting in long-term declarative memories of events (Scoville and Milner, 1957; Zola-Morgan and Squire, 1990). The frontal cortex–basal ganglia system may also be a rapid associative learning system, but with some key differences. Its learning is gated by a dopaminergic reward signal that is, in turn, driven by the results of previous learning, making its memories goal oriented rather than episodic like the hippocampus. Also, the results of frontal cortex–BG learning are not the permanent establishment of cortical pathways (memories) like the hippocampus but, rather, the temporary, dynamic establishment of cortical pathways via PFC activity patterns. In this way, the frontal cortex–BG system may produce a PFC activity pattern that can serve to bias and direct the rest of the brain.

In this view, the BG is the "engine" driving goal-directed learning. The frontal cortex, and the PFC more specifically, is an "add on"; it evolved to add greater sophistication to the learning. Both the PFC and the BG can drive behavior (they both send projections to cortical and subcortical motor system structures). However, the PFC may have greater leverage because it has widespread cortical connections and thus can influence cortical processing at many levels. By virtue of its feedback projections to sensory cortex, for example, the PFC can play a direct role in filtering out potentially distracting sensory information (i.e., attention). Thus, the PFC and the BG may work together as different parts of a frontal lobe system for goal-directed learning. But because the PFC has the big picture and can exert a top-down influence on much of the cortex, it may be the executive component of this system.

Finally, it is important to note that this was not meant to be an exhaustive survey of PFC and BG function, nor was it meant to imply that these are the only functions of these structures. For example, work by Kesner and colleagues suggest that the BG may select behavioral responses based, in part, on inputs from the PFC (Kesner and Rogers, 2004). Models by Bullock and Grossberg suggest that the BG has a gating function, allowing or disallowing expression of certain neural representations in the PFC (and other brain areas) (Brown et al., 2004). Graybiel and colleagues emphasize the role of the BG in the formation of habits (Graybiel, 1998). We also point the reader to the work of Robbins, Arnsten, and colleagues on the role of ascending neurotransmitter systems in, and interactions between, the PFC and BG in mediating cognitive flexibility (Arnsten and Robbins, 2002; Robbins, 2005). None of this is inconsistent with what we propose here. This review is only one slice through the multivariate functions of these structures, one that emphasizes their role in acquiring and using the knowledge needed for sophisticated goal-directed behavior.

REFERENCES

Alexander GE, DeLong MR, Strick PL (1986) Parallel organization of functionally segregated circuits linking basal ganglia and cortex. *Annu Rev Neurosci 9*:357–381.

Amaral DG (1986) Amygdalohippocampal and amygdalocortical projections in the primate brain. *Adv Exp Med Biol 203*:3–17.

Amaral DG, Price JL (1984) Amygdalo-cortical projections in the monkey (*Macaca fascicularis*). *J Comp Neurol 230*:465–496.

Arnsten AFT, Robbins TW (2002) Neurochemical modulation of prefrontal cortical function. In: *Principles of Frontal Lobe Function* (Stuss DT, Knight RT, eds). New York: Oxford University Press.

Asaad WF, Rainer G, Miller EK (2000) Task-specific activity in the primate prefrontal cortex. *J Neurophysiol 84*:451–459.

Bailey KR, Mair RG (2006) The role of striatum in initiation and execution of learned action sequences in rats. *J Neurosci 26*:1016–1025.

Barbas H, De Olmos J (1990) Projections from the amygdala to basoventral and mediodorsal prefrontal regions in the rhesus monkey. *J Comp Neurol 300*:549–571.

Barbas H, Henion TH, Dermon CR (1991) Diverse thalamic projections to the prefrontal cortex in the rhesus monkey. *J Comp Neurol 313*:65–94.

Barbas H, Pandya DN (1989) Architecture and intrinsic connections of the prefrontal cortex in the rhesus monkey. *J Comp Neurol 286*:353–375.

Brown JW, Bullock D, Grossberg S (2004) How laminar frontal cortex and basal ganglia circuits interact to control planned and reactive saccades. *Neural Netw 17*:471–510.

Calabresi P, Maj R, Pisani A, Mercuri NB, Bernardi G (1992) Long-term synaptic depression in the striatum: Physiological and pharmacological characterization. *J Neurosci 12*:4224–4233.

Calabresi P, Saiardi A, Pisani A, Baik JH, Centonze D, Mercuri NB, Bernardi G, Borrelli E, Maj R (1997) Abnormal synaptic plasticity in the striatum of mice lacking dopamine D2 receptors. *J Neurosci 17*:4536–4544.

Connors B, Malenka R, Silva L (1988) Two inhibitory postsynaptic potentials, and GABAA and GABAB receptor–mediated responses in neocortex of rat and cat. *J Physiol (Lond) 406*:443–468.

Couve A, Moss SJ, Pangalos MN (2000) GABAB receptors: A new paradigm in G protein signaling. *Mol Cell Neurosci 16*:296–312.

Cronin-Golomb A, Corkin S, Growdon JH (1994) Impaired problem solving in Parkinson's disease: Impact of a set-shifting deficit. *J Neurosci 32*:579–593.

Croxson PL, Johansen-Berg H, Behrens TEJ, Robson MD, Pinsk MA, Gross CG, Richter W, Richter MC, Kastner S, Rushworth MFS (2005) Quantitative investigation of connections of the prefrontal cortex in the human and macaque using probabilistic diffusion tractography. *J Neurosci 25*:8854–8866.

Divac I, Rosvold HE, Szwarcbart MK (1967) Behavioral effects of selective ablation of the caudate nucleus. *J Comp Physiol Psychol 63*:184–190.

Donoghue JP, Herkenham M (1986) Neostriatal projections from individual cortical fields conform to histochemically distinct striatal compartments in the rat. *Brain Res 365*:397–403.

Eblen F, Graybiel A (1995) Highly restricted origin of prefrontal cortical inputs to striosomes in the macaque monkey. *J Neurosci 15*:5999–6013.

Flaherty A, Graybiel A (1991) Corticostriatal transformations in the primate somatosensory system. Projections from physiologically mapped body-part representations. *J Neurophysiol 66*:1249–1263.

Flaherty A, Graybiel A (1993) Output architecture of the primate putamen. *J Neurosci 13*:3222–3237.

Flaherty A, Graybiel A (1994) Input–output organization of the sensorimotor striatum in the squirrel monkey. *J Neurosci* 14:599–610.

Funahashi S, Bruce CJ, Goldman-Rakic PS (1989) Mnemonic coding of visual space in the monkey's dorsolateral prefrontal cortex. *J Neurophysiol* 61:331–349.

Fuster JM (1973) Unit activity in prefrontal cortex during delayed-response performance: Neuronal correlates of transient memory. *J Neurophysiol* 36:61–78.

Fuster JM (1995) *Memory in the Cerebral Cortex*. Cambridge, MA: MIT Press.

Fuster JM, Alexander GE (1971) Neuron activity related to short-term memory. *Science* 173:652–654.

Gerfen CR (1992) The neostriatal mosaic: Multiple levels of compartmental organization. *Trends Neurosci* 15:133–139.

Goldman PS, Rosvold HE (1972) The effects of selective caudate lesions in infant and juvenile rhesus monkeys. *Brain Res* 43:53–66.

Goldman-Rakic PS, Leranth C, Williams SM, Mons N, Geffard M (1989) Dopamine synaptic complex with pyramidal neurons in primate cerebral cortex. *Proc Natl Acad Sci USA* 86:9015–9019.

Graybiel AM (1998) The basal ganglia and chunking of action repertoires. *Neurobiol Learn Mem* 70:119–136.

Graybiel AM (2000) The basal ganglia. *Curr Biol* 10:R509–511.

Graybiel AM, Ragsdale CW, Jr. (1978) Histochemically distinct compartments in the striatum of human, monkeys, and cat demonstrated by acetylthiocholinesterase staining. *Proc Natl Acad Sci USA* 75:5723–5726.

Haier RJ, Jung RE, Yeo RA, Head K, Alkire MT (2004) Structural brain variation and general intelligence. *Neuroimage* 23:425–433.

Hoover JE, Strick PL (1993) Multiple output channels in the basal ganglia. *Science* 259:819–821.

Hopfield JJ (1982) Neural networks and physical systems with emergent collective computational abilities. *Proc Natl Acad Sci* 79:2554–2558.

Houk JC, Wise SP (1995) Distributed modular architectures linking basal ganglia, cerebellum, and cerebral cortex: Their role in planning and controlling action. *Cereb Cortex* 5:95–110.

Jog MS, Kubota Y, Connolly CI, Hillegaart V, Graybiel AM (1999) Building neural representations of habits. *Science* 286:1745–1749.

Kelly RM, Strick PL (2004) Macro-architecture of basal ganglia loops with the cerebral cortex: Use of rabies virus to reveal multisynaptic circuits. *Prog Brain Res* 143:449–459.

Kemp JM, Powell TP (1970) The cortico-striate projection in the monkey. *Brain* 93:525–546.

Kemp JM, Powell TP (1971) The structure of the caudate nucleus of the cat: Light and electron microscopy. *Philos Trans R Soc Lond B Biol Sci* 262:383–401.

Kerr JND, Wickens JR (2001) Dopamine D-1/D-5 receptor activation is required for long-term potentiation in the rat neostriatum in vitro. *J Neurophysiol* 85:117–124.

Kesner RP, Rogers J (2004) An analysis of independence and interactions of brain substrates that subserve multiple attributes, memory systems, and underlying processes. *Neurobiol Learn Mem* 82:199–215.

Kitai ST, Kocsis JD, Preston RJ, Sugimori M (1976) Monosynaptic inputs to caudate neurons identified by intracellular injection of horseradish peroxidase. *Brain Res* 109:601–606.

Kleinfeld D (1986) Sequential state generation by model neural networks. *PNAS* 83:9469–9473.

Lawrence AD, Hodges JR, Rosser AE, Kershaw A, French-Constant C, Rubinsztein DC, Robbins TW, Sahakian BJ (1998) Evidence for specific cognitive deficits in preclinical Huntington's disease. *Brain* 121:1329–1341.

Lynd-Balta E, Haber SN (1994) The organization of midbrain projections to the ventral striatum in the primate. *Neuroscience* 59:609–623.

Mansouri FA, Matsumoto K, Tanaka K (2006) Prefrontal cell activities related to monkeys' success and failure in adapting to rule changes in a Wisconsin card-sorting test analog. *J Neurosci* 26:2745–2756.

McClelland J, McNaughton B, O'Reilly R (1995) Why there are complementary learning systems in the hippocampus and neocortex: Insights from the successes and failurs of connectionist models of learning and memory. *Psychological Rev* 102:419–457.

Middleton FA, Strick PL (1994) Anatomical evidence for cerebellar and basal ganglia involvement in higher cognitive function. *Science* 266:458–461.

Middleton FA, Strick PL (2000) Basal ganglia output and cognition: Evidence from anatomical, behavioral, and clinical studies. *Brain Cogn* 42:183–200.

Middleton FA, Strick PL (2002) Basal-ganglia "projections" to the prefrontal cortex of the primate. *Cereb Cortex* 12:926–935.

Miller EK (2000) The prefrontal cortex and cognitive control. *Nature Rev Neurosci* 1:59–65.

Miller EK, Cohen JD (2001) An integrative theory of prefrontal function. *Annu Rev Neurosci* 24:167–202.

Miller EK, Erickson CA, Desimone R (1996) Neural mechanisms of visual working memory in prefrontal cortex of the macaque. *J Neurosci* 16:5154–5167.

Mink J (1996) The basal ganglia: Focused selection and inhibition of competing motor programs. *Prog Neurobiol* 50:381–425.

Miyachi S, Hikosaka O, Miyashita K, Karadi Z, Rand MK (1997) Differential roles of monkey striatum in learning of sequential hand movement. *Exp Brain Res* 115:1–5.

Muhammad R, Wallis JD, Miller EK (2006) A comparison of abstract rules in the prefrontal cortex, premotor cortex, the inferior temporal cortex and the striatum. *J Cogn Neurosci* 18:974–989.

Mushiake H, Strick PL (1995) Pallidal neuron activity during sequential arm movements. *J Neurophysiol* 74:2754–2758.

O'Reilly RC, Munakata Y (2000) *Computational Explorations in Cognitive Neuroscience: Understanding the Mind.* Cambridge, MA: MIT Press.

Otani S, Blond O, Desce JM, Crepel F (1998) Dopamine facilitates long-term depression of glutamatergic transmission in rat prefrontal cortex. *Neuroscience* 85:669–676.

Padoa-Schioppa C, Assad JA (2006) Neurons in the orbitofrontal cortex encode economic value. Advanced online publication. *Nature* 441:223–226.

Pandya DN, Barnes CL (1987) Architecture and connections of the frontal lobe. In: *The Frontal Lobes Revisited* (Perecman E, ed.), pp 41–72. New York: IRBN Press.

Pandya DN, Yeterian EH (1990) Prefrontal cortex in relation to other cortical areas in rhesus monkey. *Architecture Connections* 85:63–94.

Parent A, Hazrati LN (1993) Anatomical aspects of information processing in primate basal ganglia. *Trends Neurosci* 16:111–116.

Parthasarathy H, Schall J, Graybiel A (1992) Distributed but convergent ordering of corticostriatal projections: Analysis of the frontal eye field and the supplementary eye field in the macaque monkey. *J Neurosci* 12:4468–4488.

Pasupathy A, Miller EK (2005) Different time courses of learning-related activity in the prefrontal cortex and striatum. *Nature* 433:873–876.

Petrides M, Pandya DN (1999) Dorsolateral prefrontal cortex: Comparative cytoarchitectonic analysis in the human and the macaque brain and corticocortical connection patterns. *Eur J Neurosci* 11:1011–1036.

Porrino LJ, Crane AM, Goldman-Rakic PS (1981) Direct and indirect pathways from the amygdala to the frontal lobe in rhesus monkeys. *J Comp Neurol* 198:121–136.

Pribram KH, Mishkin M, Rosvold HE, Kaplan SJ (1952) Effects on delayed-response performance of lesions of dorsolateral and ventromedial frontal cortex of baboons. *J Comp Physiol Psychol* 45:565–575.

Rainer G, Rao SC, Miller EK (1999) Prospective coding for objects in the primate prefrontal cortex. *J Neurosci 19*:5493–5505.

Robbins TW (2005) Chemistry of the mind: Neurochemical modulation of prefrontal cortical function. *J Comp Neurol 493*:140–146.

Schultz W (1998) Predictive reward signal of dopamine neurons. *J Neurophysiol 80*:1–27.

Schultz W, Apicella P, Ljungberg T (1993) Responses of monkey dopamine neurons to reward and conditioned stimuli during successive steps of learning a delayed response task. *J Neurosci 13*:900–913.

Schultz W, Apicella P, Scarnati E, Ljungberg T (1992) Neuronal activity in monkey ventral striatum related to the expectation of reward. *J Neurosci 12*:4595–4610.

Schultz W, Dayan, P., Montague PR (1997) A neural substrate of prediction and reward. *Science 275*:1593–1599.

Scoville WB, Milner B (1957) Loss of recent memory after bilateral hippocampal lesions. *J Neurol Neurosurg Psychiatr 20*:11–12.

Selemon L, Goldman-Rakic P (1988) Common cortical and subcortical targets of the dorsolateral prefrontal and posterior parietal cortices in the rhesus monkey: Evidence for a distributed neural network subserving spatially guided behavior. *J Neurosci 8*:4049–4068.

Selemon LD, Goldman-Rakic (1985) Longitudinal topography and interdigitation of corticostriatal projections in the rhesus monkey. *J Neurosci 5*:776–794.

Sompolinsky H, Kanter II (1986) Temporal association in asymmetric neural networks. *Physical Rev Lett 57*:2861–2864.

Taylor AE, Saint-Cyr JA, Lang AE (1986) Frontal lobe dysfunction in Parkinson's disease. The cortical focus of neostriatal outflow. *Brain 109*:845–883.

Thierry AM, Blanc G, Sobel A, Stinus L, Glowinski J (1973) Dopaminergic terminals in the rat cortex. *Science 182*:499–501.

Van Hoesen GW, Yeterian EH, Lavizzo-Mourey R (1981) Widespread corticostriate projections from temporal cortex of the rhesus monkey. *J Comp Neurol 199*:205–219.

Wallis JD, Anderson KC, Miller EK (2000) Neuronal representation of abstract rules in the orbital and lateral prefrontal cortices (PFC). *Soc Neurosci Abs* in press.

Wallis JD, Anderson KC, Miller EK (2001) Single neurons in the prefrontal cortex encode abstract rules. *Nature 411*:953–956.

Wallis JD, Miller EK (2003) Neuronal activity in the primate dorsolateral and orbital prefrontal cortex during performance of a reward preference task. *Eur J Neurosci 18*:2069–2081.

Watanabe M (1996) Reward expectancy in primate prefrontal neurons. *Nature 382*:629–632.

Wells JE, Porter JT, Agmon A (2000) GABAergic inhibition suppresses paroxysmal network activity in the neonatal rodent hippocampus and neocortex. *J Neurosci 20*:8822–8830.

White IM, Wise SP (1999) Rule-dependent neuronal activity in the prefrontal cortex. *Exp Brain Res 126*:315–335.

Wilson C (1992) Dendritic morphology, inward rectification and the functional properties of neostriatal neurons. In: *Single Neuron Computation* (McKenna T, Davis J, Zornetzer S, eds.), pp 141–171: San Diego: Academic Press.

Wilson C, Kawaguchi Y (1996) The origins of two-state spontaneous membrane potential fluctuations of neostriatal spiny neurons. *J Neurosci 16*:2397–2410.

Wong RK, Traub RD, Miles R (1986) Cellular basis of neuronal synchrony in epilepsy. *Adv Neurol 44*:583–592.

Yeterian EH, Van Hoesen GW (1978) Cortico-striate projections in the rhesus monkey: The organization of certain cortico-caudate connections. *Brain Res 139*:43–63.

Zola-Morgan SM, Squire LR (1990) The primate hippocampal formation: Evidence for a time-limited role in memory storage. *Science 250*:288–290.

CHAPTER 11

Role of the Striatum in Learning and Memory

Michael E. Ragozzino

Department of Psychology, University of Illinois at Chicago, Chicago, IL 60607

I. INTRODUCTION

The striatum is a collection of forebrain structures that include the caudate, putamen, nucleus accumbens, olfactory tubercule, and globus pallidus. These areas are named striatum because of the numerous fibers coursing through these structures that give the areas a striped appearance. Early ideas about the function of the striatum focused on its role in the control of motor behavior (Wilson, 1914). This idea was supported, in part, by neuropathological findings from individuals with Parkinson's or Huntington's disease, who displayed prominent movement abnormalities. Despite early conceptions that the striatum was an area involved primarily in the control of movement, a couple of experiments in the early part of the twentieth century investigated the effects of rat striatal lesions areas in learning (Lashley, 1921; Ghiselli and Brown, 1938). Lashley (1921) initially reported that cautery lesions of the dorsal striatum and overlying cortex did not impair acquisition or retention of a light/dark discrimination. Subsequently, Ghiselli and Brown (1938) demonstrated that lesions that included portions of the basal ganglia (including the striatum) and thalamus impaired discrimination learning between inclines of different slopes that involve the use of proprioceptive and kinesthetic information. However, not until the late 1950s and early 1960s did researchers began to examine in a more specific and systemic fashion the role of the striatum in learning and memory. This developed, in part, because of a study by Rosvold and Delgado (1956) that intended to understand the effects of electrical stimulation or lesioning of the monkey frontal cortex in a delayed-alternation test. Histological analysis indicated that in certain cases the delayed-alternation deficit likely resulted from striatal damage. In particular, the delayed-alternation impairment likely resulted from lesioning the caudate nucleus but not putamen.

There are two important developments from experimentation in the mammalian striatum that followed. First, the empirical findings that arose from investigating the striatum in learning and memory led to initial proposals of what role the striatum plays in mnemonic processing (Battig, Rosvold, and Mishkin, 1960; Chorover and Gross, 1963; Divac, Rosvold, and Sczwarcbart, 1967; Green, Beatty, and Schwartzbaum, 1967; Kirkby, 1969; Mikulas, 1966; Potegal, 1969; R.L. Thompson, 1959). Second, anatomical studies that paralleled these behavioral studies revealed that cortical areas projected to different striatal regions (Carman, Cowan, and Powell, 1963; Domesick, 1969; Nauta, 1964; Webster, 1961). This began investigations to determine whether different areas of the striatum play distinct roles in learning and memory (Divac, Rosvold, and Szwarcbart 1967; Winocur, 1974; Winocur and Mills, 1969). Thus, the viewpoint that the striatum does not serve a unitary role in learning and memory but that separate striatal areas may differentially contribute to mnemonic functioning first emerges at this time.

Subsequent to these investigations, the concept of multiple memory systems emerged that has had a major influence on understanding the neurobiology of learning and memory (Cohen, 1984; Cohen and Squire, 1980; Kesner and DiMattia, 1987; McDonald and White, 1993; Mishkin, Malamut, and Bacheliver, 1984). This idea originated to explain differential effects of hippocampal lesions on learning and memory tests (Eichenbaum, Fagan, Mathews, and Cohen, 1988; Gaffan, 1985; O'Keefe and Nadel, 1978; Olton and Papas, 1979; Squire, 1986; Warrington and Weiskrantz, 1982). In the development of the multiple memory systems concept, research concentrated on whether the hippocampus and striatum, as well as other forebrain structures, make unique contributions to learning and memory (Becker, Walker, and Olton, 1980; Kesner, Bolland, and Dakis, 1993; McDonald and White, 1993; Packard, Hirsh, and White, 1989; Packard and White, 1990; Packard and McGaugh, 1992, 1996). The findings from these experiments led to the idea that the striatum serves a specific role in learning and memory, e.g., learning of arbitrary stimulus–response associations or egocentric response memory (Cook and Kesner, 1988; Kesner, Bolland, and Dakis, 1993; McDonald and White, 1993; Packard, Hirsh, and White, 1989).

Although different theories have emphasized a unitary role for the striatum in learning and memory (Cook and Kesner, 1988; Graybiel, 1998; Knowlton,

Mangels, and Squire, 1996; McDonald and White, 1993; Mishkin, Malamut, and Bachevalier, 1984; Packard and Knowlton, 2002; Potegal, 1969), evidence is accumulating that the striatum supports multiple processes that enable learning and memory. The focus of this chapter is on how separate striatal areas are involved in different aspects of learning and memory. The discussion is based primarily on experiments investigating the caudate-putamen complex in learning and memory. The empirical findings come mainly from studies in rats as well as some studies in nonhuman primates. In primates, the caudate and putamen are separated by the internal capsule. In rodents, there is not a clear differentiation between the caudate and putamen. A more common nomenclature used for rodents is the dorsomedial striatum and dorsolateral striatum. Beyond a discussion of specific striatal circuitry involvement in learning and memory, there has also been an examination of the neurochemical mechanisms in striatal subregions that support learning and memory. The dopaminergic and cholinergic neurotransmitter systems are two commonly investigated. Thus, to further understand the role of the striatum in learning and memory, experiments that investigate dopamine and cholinergic actions in the striatum are also discussed.

II. FEATURES OF STRIATAL ANATOMY

There are a number of anatomical, physiological, and neurochemical characteristics that indicate the complexity of striatal circuitry, which may also suggest that the striatum plays a complex role in learning and memory. There are at least two prominent anatomical features of the striatum that suggest this area is heterogeneous in function. First, the striatum receives input from all major areas of neocortex as well as the hippocampal formation and pyriform cortex, but different striatal regions receive input from different cortical areas (Groenewegen and Berendse, 1994; McGeorge and Faull, 1989). In the rat, for example, somatosensory and motor cortex primarily project to the lateral striatum (McGeorge and Faull, 1989). Within the lateral striatum different parts of the motor and somatosensory cortex project to separate areas of the lateral striatum along the dorsal–ventral plane. In addition to a major input from the somatosensory and motor cortex, there is a small projection from the entorhinal and perirhinal cortex to the lateral sector of the striatum. Moreover, the posterior portion of the rat lateral striatum receives input from the secondary visual cortex and rostral auditory cortex. In contrast, the medial sector of the striatum receives input from a wider range of cortical areas that are different from the cortical input to the lateral striatum. In particular, the medial striatum receives a major input from the medial and lateral prefrontal cortex and cingulate cortices. The medial striatum, particularly in the posterior portion, also receives a significant input from the occipital and temporal cortices. Furthermore, the medial striatum receives input from archicortical areas and the entorhinal cortex. Comparable to the lateral striatum, the origin of cortical inputs to the medial striatum differs along the dorsal–ventral plane. However, the distinct cortical afferents in the lateral and medial portions of the rat striatum indicate these two areas of the striatum process different types of cortical information and may possibly contribute in distinct ways to learning and memory.

A second and related anatomical feature of the striatum that suggests separate functional aspects of striatal subregions is the well-described frontal cortical–basal ganglia–thalamic loops (Groenewegen and Berendse, 1994; Alexander, DeLong, and Strick, 1986). These loops involve different cortical areas projecting to distinct areas of the caudate and putamen and then to the pallidum, which then projects to separate subareas of the mediodorsal thalamic nucleus and finally back to the same cortical area. There are a series of these cortico–basal ganglia–thalamic circuits that run in parallel to each other and remain largely separate at all levels of the loop. These different loops that involve different areas of the striatum have been proposed to be involved in distinct functions (Alexander and Crutcher, 1990).

Besides cortical–striatal connections being important for understanding striatal function, investigation of different neurotransmitter systems in the striatum has revealed how this brain area contributes to learning and memory. The striatum is made up predominantly of medium-size spiny neurons. The medium spiny neurons represent the principle output neurons in the striatum and contain the neurotransmitter GABA (Smith and Bolam, 1990). Despite the predominance of these GABAergic neurons in the striatum, a significant number of studies have examined cholinergic and dopaminergic actions in the striatum. The origin of acetylcholine in the striatum is from interneurons. These neurons have large cell bodies with extensive dendritic branching and axonal fields (Bolam, Ingham, and Smith, 1984). This morphology indicates that cholinergic interneurons can integrate information over a wide range of synaptic input and influence striatal output over an extended area. At the receptor level, muscarinic and nicotinic receptors are both found in the striatum (Clarke, Pert, and Pert, 1984; Hersch, Gutekunst, Rees, Heilman, and Levey, 1994). Behavioral studies investigating the striatum in learning and memory have focused predominantly on muscarinic cholinergic receptors.

The main source of dopamine terminals in the striatum originates from the substantia nigra pars compacta (Fuxe, Hokfelt, and Ungerstedt, 1970; Moore and Bloom, 1978). However, there is a small number of dopamine neurons that are intrinsic to the striatum (Cossette, Lecomte, and Parent, 2005). An early view was that dopamine preferentially influenced the activity of striatal cholinergic interneurons (Hornykiewicz, 1972). More recent evidence indicates that dopamine actions in the striatum can directly affect medium spiny neurons and corticostriatal input (Bamford, Robinson, Palmiter, Joyce, Moore, and Meshul, 2004; Voulalas, Holtzclaw, Wolstenholme, Russell, and Hyman, 2005). With regard to dopamine receptors, all types of receptors, D1–D5, can be found in the striatum (Nicola, Surmeier, and Malenka, 2000; Weiner, Levey, Sunahara, Niznik, O'Dowd, Seeman, and Brann, 1991). Examination of different dopamine receptors in the striatum related to learning and memory have concentrated on dopamine D1 and D2 receptors.

III. INVOLVEMENT OF DORSAL STRIATUM IN LEARNING AND MEMORY

The main idea proposed in this chapter is that separate regions of the dorsal striatum make different contributions to learning and memory. In particular, I propose that the lateral and medial striatum are involved in different processes that enable learning and memory. Before describing the evidence to suggest this idea, there are a couple of points to consider. First, as described earlier, the rat dorsal striatum, or caudate-putamen complex, is not as clearly delineated as the caudate and putamen are in primates. Thus, there is no distinct border between the dorsomedial and dorsolateral striatum. This becomes an important issue when discussing functional differences between the dorsolateral and dorsomedial striatum because in several experiments a large area of the striatum was lesioned that would commonly be considered part of the dorso-lateral and dorsomedial striatum. Because damage covered both the dorsomedial and dorsolateral striatum, unknown from these studies is whether a behavioral impairment resulted from one or both striatal areas. In still other studies, a drug injection or lesion was made not in either the dorsomedial or dorsolateral striatum, but in the area between the lateral and medial striatum. In these experiments, as well, unclear is whether an impairment or lack of impairment is due to the particular location of the treatment (drug injection or lesion). Because of these interpretational issues and the thesis that the medial and lateral striatal areas are involved in different aspects of learning and memory, studies that clearly investigate the medial or lateral portions of the striatum are the focus of our discussion.

Another point to consider when discussing the role of the dorsal striatum is the use of the term itself. The ventral striatum commonly refers to the nucleus accumbens and/or olfactory tubercle. Specifically, the nucleus accumbens is located in the ventromedial striatum. However, there are experiments in which the term *dorsolateral striatum* was used, but clearly it was not the dorsal portion of the lateral striatum that was manipulated but, instead, the ventro-lateral part of the striatum. In other cases, the entire lateral striatum was lesioned. This may be important because, as described in the anatomical section, the dorsolateral and ventrolateral striatum receive somewhat different cortical input. Thus, when appropriate, the dorsolateral or ventrolateral striatum will be specified.

IV. CONTRIBUTIONS OF THE LATERAL STRIATUM TO LEARNING AND MEMORY

One theory that has been prominent in explaining the role of the striatum in learning and memory is the idea that the striatum is critical for acquisition of arbitrary stimulus–response associations (McDonald and White, 1993; Packard and Knowlton, 2002). The idea that the striatum is important for stimulus–response associations has also been referred to as *habit formation* (Mishkin, Malamut, and Bachevalier, 1984). Several studies have investigated involvement of the dorsolateral striatum in the initial learning of tasks thought to require stimulus–response learning. An early experiment by Winocur (1974) performed lesions of the posterior dorsolateral striatum and subsequently trained rats in an active-avoidance task. This test required rats to enter the opposite chamber with the onset of a light. Lesions in the posterior dorsolateral striatum impaired acquisition of this test, suggesting that the dorsolateral striatum is involved in initial learning of a stimulus–response association. More recent studies have also explored whether the dorsolateral striatum is important for initial learning of a stimulus–response association using discrete-cue information. In particular, the effects of lidocaine infusions or neurotoxic lesions of the striatum has been studied in the acquisition of a visual-cue discrimination task in an eight-arm radial maze (Kantak, Green-Jordan, Valencia, Kremin, and Eichenbaum, 2001; McDonald and Hong, 2004). This task involved discriminating between four arms that were lit and four arms that were not lit. Here again, either inactivation or lesioning of the dorsolateral striatum impaired initial learning of the visual cue discrmination. In a comparable fashion, a learning deficit was observed with dorsolateral or ventrolateral striatal lesions in the acquisition of a visual cue or light/tone conditional discrimination in an operant chamber (Featherstone and McDonald, 2004a; Reading, Dunnett, and Robbins, 1991). Moreover, single-unit recordings from the dorsolateral striatum during acquisition of a tone conditional discrimination revealed that neurons developed correlated firing patterns to different procedural aspects of the task (Jog, Kubota, Connolly, Hillegaart, and Graybiel, 1999). Taken together, the results suggest that the dorsolateral and ventrolateral striatum support acquisition of stimulus–response associations.

One study also suggests that dopamine receptor activation in the lateral striatum enhances acquisition of stimulus–response learning (Packard and White, 1991). In this study, rats were trained in either the visual-cue discrimination in the radial-maze or the spatial version in the radial-maze that requires the learning of visuospatial information. Posttraining injections of amphetamine, a dopamine D_1 agonist, or a dopamine D_2 agonist into the ventrolateral striatum facilitated acquisition of the visual-cue discrimination but not the spatial version. Therefore, activation of dopamine receptors in the lateral striatum can facilitate acquisition of stimulus–response learning.

In contrast to the findings just described, lesions aimed at the dorsolateral striatum do not impair initial learning of the cue version in the water maze or a visual-cue discrimination in a cross maze (Devan and White, 1999; Devan, McDonald and White, 1999; McDonald and White, 1994; Pisa and Cyr, 1990). Both the cue version of the water maze and visual-cue discrimination in a cross maze require the learning of an association between a stimulus and a particular response. Thus, if the dorsolateral striatum is critical for stimulus–response learning, a deficit should be observed in the acquisition of these tasks. One possible explanation for the differential effects in these two tests versus tests where a learning deficit is observed is task complexity. Specifically, the visual-cue discrimination in the eight-arm radial maze required discriminating among four lit arms and four unlit arms. Furthermore, each lit arm was to be entered twice in a trial. Therefore, in this task there were multiple relevant stimuli to respond to and multiple irrelevant choice arms not to respond to. In the visual-cue and light/tone conditional discriminations, there were two different stimuli that required two opposing responses. However, in the cue version of the water maze task and the visual discrimination in the cross maze, a rat learned to respond to a single cue. Thus, in these "simpler" versions of acquiring a stimulus–response association, dorsolateral striatal lesions may not be sufficient to impair initial learning.

An alternative explanation for the role of the striatum in learning and memory is that this area is critical for learning and remembering of egocentric-response information (Kesner and DiMattia, 1987; Potegal, 1969). A comparable idea has been put forth that the striatum enables response selection in egocentric space (Brasted, Robbins, and Dunnett, 1999). The egocentric-response hypothesis proposes that the learning between a specific cue in the environment and a response related to cue presentation is not the essential feature of striatal involvement in learning and memory. Instead, the striatum facilitates learning and memory when a situation demands the use of egocentric-response information based on proprioceptive and kinesthetic feedback. Because the dorsolateral striatum receives both somatosensory and motor cortical input, this striatal subregion may play an intimate role in egocentric-response learning. There is evidence to support this idea. A recent study found that lidocaine infusions into the dorsolateral striatum impaired acquisition of a response discrimination in which a rat was required to turn the same direction independent of spatial location (Chang and Gold, 2004). In a comparable manner, NMDA receptor blockade in the dorsolateral striatum impairs the acquisition of a two-choice egocentric-response discrimination (Palencia and Ragozzino, 2005). Other studies do not support the idea that the dorsolateral striatum is involved in learning egocentric-response information. Specifically two studies found that neurotoxic lesions of the dorsolateral striatum do not impair acquisition of a response discrimination (Gabriela, Oliveira, Orlando, Pomarico, and Gugliano, 1997; Pisa and Cyr, 1990). The conflicting results may be explained by differences in the test procedures. A response discrimination deficit occurred only in studies when testing was completed in a single session. In contrast, no deficit occurred in studies that tested rats across several sessions for 5–10 trials per session. In these experiments, testing across multiple sessions allowed for the response information to be consolidated for long-term storage. The procedure allowing for the consolidation of the response information might have recruited a wider and/or different network of brain areas, which made manipulation of the dorsolateral striatum less susceptible to causing a response learning impairment as compared with the procedure in which response information had to be rapidly acquired in a single session. Thus, the dorsolateral striatum may also support acquisition of egocentric-response information, but it may do so when the information must be acquired in a more rapid time frame.

The findings just discussed suggest that the dorsolateral striatum plays an important role in the initial learning of both stimulus–response associations and egocentric-response information. One issue is whether the dorsolateral striatum plays a selective role in acquisition of this type of information or whether the dorsolateral striatum also plays a role in memory storage. Investigation of a brain area in memory storage processes routinely involves training a subject in a task, followed by manipulation of that brain area shortly after training, e.g., drug injection, lesion, or electrical stimulation, with a subsequent retention test some time after the manipulation. A retention deficit in a posttraining manipulation paradigm is commonly interpreted as an impairment in memory storage or memory retrieval.

In general, fewer studies have specifically investigated the role of the dorsolateral striatum in memory employing a posttraining-manipulation paradigm. Several experiments examined whether posttraining electrical stimulation of the striatum impaired retention in an inhibitory-avoidance paradigm (Deadwyler, Montgomery, and Wyers, 1972; Gold and King, 1972; Haycock, Deadwyler, Sideroff, and McGaugh, 1973; Wilburn and Kesner, 1972; Wyers and Deadwyler, 1971). Although it is unclear where in the striatum the electrical stimulation occurred in some studies, in other experiments electrical stimulation of the dorsolateral striatum or ventrolateral striatum occurred in at least a portion of rats. The studies often involved training a rat to navigate to a water tube once it was inserted in a chamber. After the rat learned to locomote rapidly to the water tube when placed into the apparatus, a foot shock was delivered, followed by electrical stimulation. These experiments consistently demonstrated that posttraining electrical stimulation impaired retention in an inhibitory-avoidance task (Deadwyler, Montgomery, and Wyers, 1972;

Haycock, Deadwyler, Sideroff, and McGaugh, 1973; Wyers and Deadwyler, 1971). A more recent study examined the effects of an NMDA receptor antagonist, an AMPA receptor antagonist, and protein synthesis inhibitors into the ventrolateral striatum on step-down inhibitory avoidance (Cammarota, Bevilaqua, Kohler, Medina, and Izquierdo, 2005). Immediate posttraining infusion of these different pharmacological agents did not impair retention following one training trial but did impair retention when administered after two training trials. Based on the pattern of results, this region of the striatum has been proposed to be involved in memory consolidation with additional learning of a response pattern (Cammarota, Bevilaqua, Kohler, Medina, and Izquierdo, 2005). Taken as a whole, interpretation of the inhibitory-avoidance results is open to debate on whether memory was disrupted preferentially for a learned stimulus–response association or an egocentric response or even another type of learning. However, the inhibitory-avoidance findings do suggest that the lateral striatum supports memory storage processes.

Other studies have examined the lateral striatum in memory storage with tasks that use discrete stimulus cues. Specifically, posttraining infusion of an NMDA antagonist into the ventrolateral striatum impaired retention of the visual cue version of the water maze task but not retention in the spatial version (Packard and Teather, 1997). Similarly, dorsolateral striatal lesions following training in a tone conditional discrimination also impairs retention (Adams, Kesner, and Ragozzino, 2001). There is other evidence that activation of dopamine receptors in the lateral striatum facilitates memory storage for stimulus–response learning. Specifically, amphetamine or specific dopamine D1 or D2 agonists infused into the lateral striatum facilitate retention of stimulus–shock pairings (White and Viaud, 1991; White and Salinas, 2003). These tasks are interpreted as assessing stimulus–response learning as the presentation of the stimulus, e.g., tone, becomes associated with a particular withdrawal response. Thus, dopamine receptor activation in the lateral striatum can enhance memory consolidation for stimulus–response associations. Moreover, lateral striatal lesions also impair retention of an egocentric-response learning test (Abraham, Potegal, and Miller, 1983). Taken together, the lateral striatum may support not only initial learning, but also memory consolidation of stimulus–response associations as well as egocentric-response information.

In addition to behavioral paradigms employed to study the role of the striatum in initial learning or memory consolidation processes, other tasks have been used to examine the role of the striatum in the expression of a learned choice pattern or strategy. The place-response paradigm is one such task to examine the expression of a place or response strategy (Packard and McGaugh, 1996; W.G. Thompson, Guilford, and Hicks, 1980). In a modified T-maze, a rat is started from the stem arm and learns to enter the same choice arm for a reinforcement. After several trials of training, a rat is started from the opposite arm. This probe trial exposes the rat to the same two choice arms and determines whether a rat will enter the same arm as in training (place strategy) or turn in the same direction of training and therefore into the choice arm not reinforced in training (response strategy). Early in training, rats preferentially express the use of a place strategy, but with extended training rats preferentially express the use of a response strategy (Packard and McGaugh, 1996). Early in training, dorsolateral striatal inactivation prior to the probe trial did not affect the preferential expression of a place strategy. In contrast, dorsolateral striatal inactivation prior to the probe trial after extensive training shifted a preferential expression of a response strategy to rats preferentially expressing a place strategy (Packard and McGaugh, 1996). This study was followed up by a demonstration that posttraining glutamate infusions into the dorsolateral striatum early in training facilitated the expression of a response strategy (Packard, 1999). In combination, the results suggest that the dorsolateral striatum is important for expression of an egocentric-response strategy.

Another experiment investigated the role of the dorsolateral striatum in expression of a learned response habit using a devaluation procedure. Extensive lever press training for a sucrose reward becomes insensitive to a devaluation manipulation, i.e., when sucrose is paired with lithium chloride (Yin, Knowlton, and Balleine, 2004). The idea is that extensive lever press training leads to the formation of a stimulus–response habit in which a subject becomes less sensitive to the response–reward relationship; thus, when the reward is devalued, the habit of lever pressing is still manifested. If the dorsolateral striatum is critical for the expression of this habitual behavior, then lesions of the area should prevent the insensitivity to the reward devaluation such that a rat exhibits decreased lever pressing following reward devaluation. Indeed, rats that received dorsolateral striatal lesions and extensive lever press training for a sucrose reward decreased lever pressing following the devaluation of the sucrose reward (Yin, Knowlton, and Balleine, 2004). Thus, the dorsolateral striatal lesions prevented the formation of the habitual lever-pressing behavior. Overall, the set of experiments investigating the expression of a learned choice pattern reveal that the dorsolateral striatum plays a critical role in the expression of learned response habits.

To better understand the neurochemical mechanisms that may support the expression of a learned response habit, a pair of studies has focused on the activity of cholinergic interneurons in the dorsolateral striatum (Chang and Gold, 2003; Pych, Chang, Colon-Rivera, and Gold, 2005). Measurement of acetylcholine efflux from the dorsolateral striatum during training in a place-response test revealed that acetylcholine output gradually increases across training and reaches maximal output after an extended number of trials when rats preferentially express a response strategy (Chang and Gold, 2003). This pattern of acetylcholine efflux from the dorsolateral striatum also occurred during a continuously reinforced spontaneous alternation in a Y-maze. In this test rats began to alternate in a fixed response pattern after extended testing. It was during this later test period that rats exhibited not only the expression of a response strategy, but also a maximal increase in dorsolateral striatal acetylcholine output (Pych, Chang, Colon-Rivera, and Gold, 2005). Taken together, these findings indicate that activation of cholinergic interneurons in the dorsolateral striatum may enhance the formation and expression of a response habit.

Our description has focused on what role the lateral striatum plays in learning and memory based mainly on deficits observed following a manipulation to the lateral striatum. A more complete understanding of the role the lateral striatum plays in learning and memory can be appreciated by examining the conditions under which the lateral striatum is not critical for learning and memory. One finding that is consistent across studies is that manipulations of the lateral striatum do not affect the acquisition, retention, or expression of a place discrimination or place strategy (Devan and White, 1999; Devan, McDonald, and White, 1999; Dunnett and Iversen, 1981; Gabriela, Oliveria, Orlando, Pomarico, and Gugliano, 1997; Livesey and Muter, 1976; Packard and McGaugh, 1992, 1996). These negative results have frequently been shown in conjunction with manipulations of the hippocampal formation that often lead to such deficits (Packard and White, 1991; Packard and McGaugh, 1992, 1996). A second condition in which lateral striatal lesions do not affect acquisition or retention is a conditioned cue preference (Everitt, Morris, O'Brien, and Robbins, 1991; Kantak, Green-Jordan, Valencia, Kremin, and Eichenbaum, 2001; McDonald and Hong, 2004). This task is thought to require the learning of a stimulus–reward association in which the amygdala as well as nucleus accumbens are brain areas that support learning of a conditioned cue preference. A third condition in which the lateral striatum is not critical for learning is situations in which there is a change in task contingencies, e.g., reversal learning. For example, manipulations of the lateral striatum do not impair place, response, or visual cue reversal learning (Burk and Mair, 2001; Dunnett and Iversen, 1981; Gabriela, Oliveria, Orlando, Pomarico, and Gugliano, 1997; Palencia and Ragozzino, 2005; Pisa and Cyr, 1990). This may be somewhat surprising, particularly for response reversal learning, because some experiments indicate that the dorsolateral striatum supports initial learning of an egocentric-response discrimination and, further, there is evidence that the dorsolateral striatum enables the expression of a learned response habit. In response reversal learning, a subject must inhibit one learned response pattern and learn a different response choice. Thus, this task simply involves forming a different egocentric-response habit. Yet, lesions or drug infusions into the lateral striatum do not produce response reversal learning deficits (Palencia and Ragozzino, 2005; Pisa and Cyr, 1990). These findings suggest that while the dorsolateral striatum may be critical for the initial learning and expression of an egocentric-response strategy, when the situation requires a shift to a new or different response pattern the dorsolateral striatum is not critical.

In summary, current findings indicate that the rat lateral striatum supports the acquisition, consolidation, and expression of stimulus–response associations as well as egocentric-response information. Furthermore, activation of dopamine receptors in the lateral striatum facilitates the acquisition and memory storage of stimulus–response associations. Activation of cholinergic interneurons in the dorsolateral striatum may enable the formation and expression of an egocentric-response strategy. Unknown at present is how dopamine and acetylcholine may interact in the lateral striatum to enable learning and memory.

V. CONTRIBUTIONS OF THE MEDIAL STRIATUM TO LEARNING AND MEMORY

This section explores the involvement of the dorsomedial striatum in learning and memory. Unlike the previous section, which described studies that examined either the dorsolateral or ventrolateral striatum, the findings described in this section are limited to studies that investigated the dorsomedial striatum and not those that investigated the ventromedial striatum (nucleus accumbens and olfactory tubercle).

Early evidence implicating the dorsomedial striatum in memory emerged from a study in monkeys in which lesions to the caudate were found to impair performance in a delayed-alternation test (Rosvold and Delgado, 1956). Several studies in monkeys followed that replicated this original report of delayed-alternation deficits following caudate lesions (Battig, Rosvold, and Mishkin, 1960; Butters and Rosvold, 1968; Divac, Rosvold, and Szwarcbart, 1967; Rosvold, Mishkin, and Szwarcbart, 1958). These subsequent studies demonstrated that increasing the delay exacerbated the delayed-alternation deficit in monkeys with caudate lesions, but these same lesions did not impair a visual discrimination. The pair of results indicates that the deficit was not due to a general learning and memory impairment, but instead that the caudate is part of a neural system that contributes to working memory. A more recent experiment suggested that different parts of the primate caudate may support different types of working memory based on attribute information. Specifically, 2-deoxyglucose activity was measured in the striatum following performance on a spatial delayed-alternation or object delayed-alternation test (Levy, Friedman, Davachi, and Goldman-Rakic, 1997). The head of the caudate nucleus showed greatest activation following the spatial delayed-alternation test, while the tail of the caudate displayed greatest activation during the object delayed-alternation test. Thus, these different areas of the caudate may enable separate forms of working memory.

Comparable to caudate lesions in monkeys, lesions of the rat dorsomedial striatum produce delayed-alternation deficits (Chorover and Gross, 1963;

Divac, Markowitsch, and Pritzel, 1978; Dunnett, Nathwani, and Brasted, 1999; Furtado and Mazurek, 1996). Furthermore, blockade of NMDA receptors in the dorsomedial striatum also impairs performance in a win-shift version of the eight-arm radial maze (Smith-Roe, Sadeghian, and Kelley, 1999). Kesner, Bolland, and Dakis (1993) investigated the effects of lesions centered in the dorsomedial striatum in different working-memory tests. Lesions were found to impair working memory for egocentric responses but not working memory for spatial locations or visual objects. In a comparable manner, dorsomedial striatal lesions impaired a delayed match-to-sample test for direction information (DeCoteau, Hoang, Huff, Stone, and Kesner, 2004). In this test, rats were trained in the dark to enter a maze arm by either navigating straight or navigating at a 45° direction to the left or right. In the test phase, a rat had a choice between two open arms, one in the same direction as the study phase and one in a different direction. A rat had to navigate in the same direction as on the study phase to make a correct choice. Transfer tests indicated that rats learned the test by using direction information and not spatial information. Again, dorsomedial striatal lesions impaired performance in the delayed match-to-sample for direction information but did not impair learning to discriminate between two different directions. These results suggest that dorsomedial striatum is not critical for the general learning of directional information, but when directional information must be retained in short-term memory and changes trial to trial, the dorsomedial striatum is critical (DeCoteau et al., 2004).

The selective impairments in working memory for egocentric responses and directional information differ, with other findings indicating that the dorsomedial striatum is also important in working memory for spatial or object information. The exact reason for these differences is unclear, although in the study of Kesner, Bolland, and Dakis (1993), the electrolytic lesions were in the posterior portion of the dorsomedial striatum and appeared to damage lateral portions of the dorsal striatum as well. A future issue to address is whether the dorsomedial striatum plays a selective role in working memory or whether the dorsomedial striatum supports multiple forms of working memory in which specific areas of the dorsomedial striatum preferentially contribute to different types of working memory based on attribute information. In addition, it will be important to understand what process or processes the dorsomedial striatum supports to facilitate working memory. For example, does the dorsomedial striatum support working memory related to the maintenance of information and/or response selection? Despite these unanswered questions there is significant evidence indicating that the rat dorsomedial striatum plays an important role in working memory.

The findings from other studies indicate that the effect of dorsomedial striatal lesions in working memory is not the result of a general learning deficit. This is because inactivation or lesions of the dorsomedial striatum generally do not impair the acquisition of a wide range of discrimination tasks (Botreau, El Massioui, Cheruel, and Gisquet-Verrier, 2004; DeCoteau et al., 2004b; Featherstone and McDonald, 2004; Kirkby, 1969; Kolb, 1977; Livesey and Muter, 1976; Palencia and Ragozzino, 2004; Pisa and Cyr, 1990; Ragozzino, Ragozzino, Mizumori, and Kesner, 2002; Ragozzino and Choi, 2004; Ragozzino, Jih, and Tzavos, 2002). An important aspect about this is that dorsomedial striatal lesions do not impair acquisition of some of the same discrimination tests that dorsolateral striatal lesions do impair (Featherstone and McDonald, 2004a, 2004b). For example, while dorsolateral striatal lesions impair the acquisition of a light/tone conditional discrimination, dorsomedial striatal lesions do not produce a deficit in this task (Featherstone and McDonald, 2004a). This pattern of findings suggest that the dorsolateral striatum, but not the dorsomedial striatum, enables the acquisition of arbitrary stimulus–response associations.

In contrast to acquisition of different discrimination tests, inactivation or lesions of the dorsomedial striatum commonly leads to reversal-learning impairments (Kirkby, 1969; Livesey and Muter, 1976; Pisa and Cyr, 1990; Ragozzino, Ragozzino, Mizumori, and Kesner, 2002; Ragozzino, Jih, and Tzavos, 2002; Ragozzino and Choi, 2004). The reversal-learning impairments are observed in a variety of tests, e.g., place, response, visual cue, although an impairment in visual-cue reversal learning following dorsomedial striatal lesions is not always observed (Pisa and Cyr, 1990; Livesey and Muter, 1976). Because dorsomedial striatal manipulations lead to an impairment in reversal learning but not acquisition, this set of findings suggests that the dorsomedial striatum is critically involved in learning when a situation requires a shift in choice patterns. However, the role of the dorsomedial striatum in facilitating learning with a change in task requirements is not limited to reversal-learning tests, but also includes extra-dimensional shifts. For example, dorsomedial striatal inactivation does not impair the acquisition of a visual cue or egocentric-response discrimination but does impair a shift between the use of visual-cue and egocentric-response information (Ragozzino, Ragozzino, Mizumori, and Kesner, 2002). This contrasts with the role of the medial prefrontal cortex in behavioral flexibility in which medial prefrontal cortex inactivation impairs extra-dimensional shifts but not reversal learning (Ragozzino, 2002; Ragozzino, Kim, Hassert, Minniti, and Kiang, 2003; Ragozzino, Detrick, and Kesner, 1999). Thus, although the medial prefrontal cortex projects to the dorsomedial striatum, this striatal region may play a broader role in learning when conditions require a shift in strategies. These findings also indicate that the dorsomedial and dorsolateral striatum differentially contribute to behavioral flexibility, with the dorsomedial striatum, but not dorsolateral striatum, playing a prominent role.

An examination of the error pattern during reversal learning and extra-dimensional shifts has also led to a further understanding of what process the dorsomedial striatum supports to facilitate a shift in strategies. In a series of experiments, we separated the errors into two separate categories: perseverative and regressive errors. *Perseverative* errors are those in which a subject initially chooses the previously correct choice until making a correct choice. For example, in spatial discrimination a rat may first learn to enter spatial location A and avoid entering spatial location B. In reversal learning, a rat now has to learn to enter spatial location B and avoid entering spatial location A. The number of trials a rat initially chooses spatial location A before choosing spatial location B would be counted as perseverative errors. Thus, perseverative errors measure the ability initially to inhibit a previously correct choice pattern. A deficit due to perseveration may occur because of the inability to inhibit a previously relevant strategy and/or the inability to generate a new strategy. After a rat makes a new correct choice, it may occasionally regress to the previously correct choice before reaching criterion in reversal learning. The number of trials in which a rat regresses to the previously correct choice after making a presently correct choice are counted as *regressive* errors. A deficit due to regressive errors may reflect the inability to reliably execute or maintain a new correct choice. The reversal-learning or extra-dimensional shift deficits following dorsomedial striatal inactivation result in a significant increase in regressive errors but not perseverative errors (Ragozzino, Ragozzino, Mizumori, and Kesner, 2002; Ragozzino, Jih, and Tzavos, 2002; Ragozzino and Choi, 2004). The prelimbic and infralimbic areas of the medial prefrontal cortex project to the dorsomedial striatum (Groenewegen and Berendse, 1994). Inactivation or blockade of dopamine D_1 receptors of the prelimbic and infralimbic areas also produces an impairment in an extra-dimensional shift. However, prelimbic-infralimbic inactivation selectively increases perseverative errors but not regressive errors (Ragozzino, 2002; Ragozzino, Kim, Hassert, Minniti, and Kiang, 2003). This dissociation suggests that these two areas play a distinct but complementary role in enabling learning when conditions demand a shift in strategies. More specifically, the medial prefrontal cortex is critical for the initial inhibition of a previously relevant strategy and/or generation of a new strategy, while the dorsomedial striatum supports the reliable execution of a new strategy once generated.

Although manipulations of the dorsomedial striatum have generally not impaired acquisition of discrimination tests, there are studies indicating that dorsomedial striatal lesions do impair initial learning of a discrimination test. These results would appear to contradict the idea that the dorsomedial striatum only supports learning when conditions require inhibition of one strategy and learning of a new or different strategy. For example, a neurotoxic lesion of the dorsomedial striatum was found to impair acquisition of a spatial discrimination (Dunnett and Iversen, 1981). However, in this study rats were required to enter the spatial location that was opposite of their original spatial bias. Because rats had a strong spatial bias and had to enter the location against their spatial bias, acquisition testing may have been more similar to spatial reversal learning.

Acquisition deficits in shock-avoidance tests have also been reported following dorsomedial striatal lesions (Green, Beatty, and Schwartzbaum, 1967;
Sanberg, Lehmann, and Fibiger, 1978). In these tests an active-avoidance procedure was used such that a rat routinely had to switch its behavioral pattern
to avoid shock. Thus, a rat frequently had to execute opposite responses to
exhibit learning. Because this requires a flexible shifting between different
response patterns, the deficits with dorsomedial striatal lesions may result from
an impairment in behavioral flexibility.

Another test in which dorsomedial striatal lesions impair acquisition is in
the visual cue and spatial version of the water maze test (Devan and White,
1999; Devan, McDonald, and White, 1999; Whishaw, Mittleman, Bunch, and
Dunnett, 1987). An analysis of the navigational pattern in rats with dorsomedial striatal lesions reveals that lesioned rats are strongly biased toward circling
near the edge of the pool before navigating to the correct spatial location or
cued location (Devan and White, 1999; Whishaw, Mittleman, Bunch, and
Dunnett, 1987). These findings suggest that dorsomedial striatal lesions impair
the proper execution of an optimal navigational strategy.

A similar pattern of results occurred in which rats were trained on either
a *declarative* or a *procedural* version of the eight-arm radial maze (DeCoteau and
Kesner, 2000). In the declarative version, a sequence of six arms were baited
with cereal reinforcement. To choose an arm in the sequence correctly a rat
had to orient to the door of the arm. When a rat oriented to the proper door,
the door was lowered and the rat could enter the arm. A rat had to do this
for each arm in the sequence. In the procedural version, a sequence of six arms
was again used, but the door for the arm next in sequence was lowered while
a rat was in the previous arm. So in this version the correct arm was always
available without the rat's first having to choose the correct arm. In this
version, the rat's time to complete the sequence decreases across sessions and
may be used as an index of learning. Dorsomedial striatal lesions impaired learning of the procedural sequence but not the declarative sequence
(DeCoteau and Kesner, 2000). Control rats when learning the procedural
sequence develop an egocentric-response strategy, e.g., turn in a particular
direction when exiting an arm. However, rats with dorsomedial striatal lesions
appeared to use more of an allocentric spatial strategy (DeCoteau and Kesner,
2000). Thus, dorsomedial striatal lesions may have prevented the learning or
use of the optimal strategy. Therefore, in many of the studies that indicate an
acquisition deficit with dorsomedial striatal lesions, the deficit may not be due
to a general learning deficit but actually to impairments in reliably shifting or
executing strategies.

There are additional studies that have not used reversal-learning tests that
also suggest that the dorsomedial striatum is critically involved in strategy
switching and the flexible use of response patterns. In the five-choice serial
reaction-time task, used as a test of sustained attention, a rat is required to

respond to a brief presentation of a visual stimulus that can occur in one of five locations. Therefore, a rat must constantly monitor the location of the visual stimulus and be able to switch a particular choice pattern from trial to trial. Dorsomedial striatal lesions not only decrease accuracy in this task but also increase premature responses and reversions to a previously correct location (Rogers, Baunez, Everitt, and Robbins, 2001). This reversion to a previously correct location is comparable to the increased regressive errors in reversal-learning tests following dorsomedial striatal inactivation.

In another study the activity of individual dorsomedial striatal neurons were recorded during a spatial memory test in an eight-arm radial maze (Mizumori, Ragozzino, and Cooper, 2000). In particular, the firing of head-direction cells was examined during a spatial-memory test under light and dark conditions. In these different conditions, a rat must switch what information it uses to complete the test accurately. The head-direction cell activity is dependent on visual information when the spatial-memory test was conducted in the light. However, when the rat performed the task in the dark, head-direction cell activity was influenced by ideothetic information. Thus, the activity of these neurons in the dorsomedial striatum dynamically changes to the environmental conditions in which a rat needs to implement a different navigational strategy to complete the test successfully.

The results just described provide significant evidence that the dorsomedial striatum plays a crucial role in facilitating a shift in strategies with a change in task contingencies. Recent experiments have begun to explore the neuro-chemical mechanisms within the dorsomedial striatum that enable a shift in strategies. Because cholinergic interneurons have extensive dendritic branching and axonal fields, these neurons may be able to integrate a wide array of synaptic input and have widespread influence on striatal output. Based on this morphology of cholinergic interneurons, changes in cholinergic activity in the dorsomedial striatum was hypothesized to underlie behavioral flexibility. To test this hypothesis, acetylcholine output was measured during the acquisition and reversal learning of a place discrimination (Ragozzino and Choi, 2004). Previous studies demonstrated that electrical stimulation or lesions of the dorsomedial striatum impair place reversal learning (Dunnett and Iversen, 1981; Kolb, 1977; Livesey and Muter, 1976). The findings revealed that acetylcholine output did not change during acquisition of the place discrimination. However, acetylcholine output significantly increased in the dorsomedial striatum during place reversal learning. Furthermore, the increase in acetylcholine efflux showed a distinct pattern during reversal learning. Acetylcholine output did not initially change at the beginning of the reversal-learning session when a rat was predominantly choosing the previously correct spatial location. As a rat began to choose the new correct choice there was a significant increase in acetylcholine efflux that returned back to basal levels once the rat learned the new choice pattern. Because acetylcholine selectively increases during reversal learning, the results suggest that activation of cholinergic interneurons in the dorsomedial striatum support a shift in strategies.

Other studies have investigated the cholinergic receptor mechanisms in the dorsomedial striatum that support learning when conditions demand a shift in choice patterns. Blockade of muscarinic cholinergic receptors in the dorsomedial striatum does not impair acquisition of a discrimination test but does impair reversal learning (Ragozzino, Jih, and Tzavos, 2002). In contrast, blockade of nicotinic cholinergic receptors in the dorsomedial striatum does not impair acquisition or reversal learning (Ragozzino, Jih, and Tzavos, 2002). Subsequently, pirenzepine, a relatively selective M_1 muscarinic cholinergic antagonist, was found selectively to impair reversal learning (Tzavos, Jih, and Ragozzino, 2004). Thus, activation of muscarinic cholinergic receptors but not nicotinic cholinergic receptors in the dorsomedial striatum may enable a shift in strategies.

VI. CONCLUSIONS

Accumulating evidence indicates that the lateral and medial striatum make distinct contributions to learning and memory. The lateral striatum is critical for learning, remembering, and the expression of stimulus–response associations. The contribution of the lateral striatum to the learning and memory of stimulus–response associations is not limited to a particular stimulus modality but can occur for several different modalities. Furthermore, the lateral striatum appears to play a particularly critical role in acquiring stimulus–response associations when more "complex" stimulus–response associations must be formed, as in conditional discriminations. Moreover, posttraining injections of dopamine agonists indicate that dopamine actions in the lateral striatum facilitate memory consolidation of learned stimulus–response associations. Besides stimulus–response associations, the lateral striatum is also involved in the learning and remembering of egocentric-response information that involves the use of proprioceptive and kinesthetic information. That is, when a subject must learn and remember particular motor responses, e.g., always make a 90° turn to the right, the lateral striatum contributes to this form of learning. As the use of an egocentric-response strategy is formed, acetylcholine output gradually increases in the lateral striatum. Thus, acetylcholine actions in the dorsolateral striatum may be critical for enabling the formation and expression of an egocentric-response strategy.

One function of the dorsomedial striatum is to enable working memory. Unclear is whether the dorsomedial striatum supports multiple forms of working memory that involve different attribute information or whether the dorsomedial striatum is involved in only a specific type of working memory, e.g., egocentric-response information. Moreover, the dorsomedial striatum is likely part of a larger neural system that underlies working memory. A better understanding is needed to determine how the dorsomedial striatum may contribute to similar or distinct processes as other interconnected brain areas that support working memory, e.g., maintenance of information in short-term memory or response selection. The dorsomedial striatum also facilitates learning when conditions require inhibition of a previously relevant strategy and the learning of a new strategy. This is in contrast to the dorsolateral striatum, which does not seem critical for the flexible shifting of response patterns. The role of the dorsomedial striatum in strategy switching is not related to the initial inhibition or generation of a new strategy, but it appears to facilitate the maintenance of a new correct strategy once generated. In contrast to specific prefrontal cortex subregions, which appear to support specific forms of cognitive flexibility, the dorsomedial striatum facilitates learning under a broad range of conditions that require inhibition of a previously relevant choice pattern and the learning of a different choice pattern. At a neurochemical level, acetylcholine output in the dorsomedial striatum does not increase during initial learning but selectively increases during reversal learning. In particular, acetylcholine output does not change during the initial inhibition of a previous strategy but selectively increases during the learning of a new strategy. The effects of acetylcholine in the dorsomedial striatum on cognitive flexibility are mediated by muscarinic cholinergic receptors but not nicotinic cholinergic receptors. Thus, acetylcholine actions in the dorsomedial striatum appear to enable the learning and reliable execution of a new strategy once selected.

Overall, studies using lesions or pharmacological manipulations support the idea that the dorsomedial and dorsolateral striatum make distinct contributions to learning and memory. Future studies will be important for better understanding the neurochemical mechanisms in these striatal subregions that underlie learning and memory as well as for revealing how these two areas may interact to facilitate learning and memory.

REFERENCES

Abraham L., Potegal M., and Miller S. (1983). Evidence for caudate nucleus involvement in an egocentric spatial task: Return from passive transport. *Physiological Psychology, 11*, 11–17.

Adams S., Kesner R.P., and Ragozzino M.E. (2001). Role of the medial and lateral caudate-putamen in mediating an auditory conditional response association. *Neurobiology of Learning and Memory, 76*, 106–116.

Alexander G.E., and Crutcher M.D. (1990). Functional architecture of basal ganglia circuits: Neural substrates of parallel processing. *Trends in Neuroscience, 13*, 266–271.

Alexander G.E., DeLong M.R., and Strick P.L. (1986). Parallel organization of functionally segregated circuits linking basal ganglia and cortex. *Annual Review of Neuroscience, 9*, 357–381.

Bamford N.S., Robinson S., Palmiter R.D., Joyce J.A., Moore C., and Meshul C.K. (2004). Dopamine modulates release from corticostriatal terminals. *Journal of Neuroscience, 24*, 9541–9552.

Battig K., Rosvold H.E., and Mishkin M. (1960). Comparison of the effects of frontal and caudate lesions on delayed response and alternation in monkeys. *Journal of Comparative and Physiological Psychology, 53*, 400–404.

Becker J.T., Walker J.A., and Olton D.S. (1980). Neuroanatomical bases of spatial memory. *Brain Research, 200*, 307–320.

Bolam J.P., Ingham C.A., and Smith A.D. (1984). The section-Golgi-impregnation procedure. 3. Combination of Golgi impregnation with enzyme histochemistry and electron microscopy to characterize acetylcholinesterase-containing neurons in the rat neostriatum. *Neuroscience, 12*, 687–709.

Botreau F., El Massioui N., Cheruel F., and Gisquet-Verrier P.(2004). Effects of medial prefrontal cortex and dorsal striatum lesions on retrieval processes in rats. *Neuroscience, 129*, 529–553.

Brasted P.J., Robbins T.W., and Dunnett S.B. (1999). Distinct roles for striatal subregions in mediating response processing revealed by focal excitotoxic lesions. *Behavioral Neuroscience, 113*, 253–264.

Burk J.A., and Mair R.G. (2001). Effects of dorsal and ventral striatal lesions on delayed matching trained with retractable levers. *Behavioural Brain Research, 122*, 67–78.

Butters N., and Rosvold H.E. (1968). Effect of caudate and septal nuclei lesions on resistance to extinction and delayed alternation. *Journal of Comparative and Physiological Psychology, 65*, 397–403.

Cammarota M., Bevilaqua L.R.M., Kohler C., Medina J.H., and Izquierdo I. (2005). Learning twice is different from learning once and from learning more. *Neuroscience, 132*, 273–279.

Carman J.B., Cowan W.M., and Powell T.P.S. (1963). The orangization of cortico-striate connections in the rabbit. *Brain, 86*, 525–562.

Chang Q., and Gold P.E. (2003). Switching memory systems during learning: Changes in patterns of brain acetylcholine release in the hippocampus and striatum in rats. *Journal of Neuroscience, 23*, 3001–3005.

Chang Q., and Gold P.E. (2004). Inactivation of dorsolateral striatum impairs acquisition of response learning cue-deficient, but not cue-available, conditions. *Behavioral Neuroscience, 118*, 383–388.

Chorover S.L., and Gross C.G. (1963). Caudate nucleus lesions: Behavioral effects in the rat. *Science, 141*, 826–827.

Clarke P.B.S., Pert C.B., and Pert A. (1984). Autoradiographic distribution of nicotinic receptors in the rat brain. *Brain Research, 323*, 390–395.

Cohen N.J. (1984). Preserved learning capacity in amnesia: Evidence for multiple memory systems. In N. Butters and L.R. Squire (eds.), *The Neuropsychology of Memory* (pp. 83–103). New York: Guilford Press.

Cohen N.J., and Squire L.R. (1980). Reserved learning and retention of pattern analyzing skill in amnesia: Dissociation of knowing how and knowing that. *Science, 210*, 207–209.

Cook D., and Kesner R.P. (1988). Caudate nucleus and memory for egocentric localization. *Behavioral and Neural Biology, 49*, 332–343.

Cossette M., Lecomte F., and Parent A. (2005). Morphology and distribution of dopamine neurons intrinsic to the human striatum. *Journal of Chemical Neuroanatomy, 29*, 1–11.

Deadwyler S.A., Montgomery D., and Wyers E.J. (1972). Passive avoidance and carbachol excitation of the caudate nucleus. *Physiology and Behavior, 8*, 631–635.

DeCoteau W.E., Hoang L., Huff L., Stone A., and Kesner R.P. (2004). Effects of hippocampal and medial caudate nucleus lesions on memory for directional information in rats. *Behavioral Neuroscience, 118*, 540–545.

DeCoteau W.E., and Kesner R.P. (2000). A double dissociation between the rat hippocampus and medial caudoputamen in processing two forms of knowledge. *Behavioral Neuroscience, 114*, 1096–1108.

Devan B.D., McDonald R.J., and White N.M. (1999). Effects of medial and lateral caudate-putamen lesions on place- and cue-guided behaviors in the water maze: Relation to thigmotaxis. *Behavioural Brain Research, 100,* 1–14.

Devan B.D., and White N.M. (1999). Parallel information processing in the dorsal striatum: Relation to hippocampal function. *Journal of Neuroscience, 19,* 2789–2798.

Divac I., Markowitsch H.J., and Pritzel M. (1978). Behavioral and anatomical consequences of small intrastriatal injections of kainic acid in the rat. *Brain Research, 151,* 523–532.

Divac I., Rosvold H.E., and Szwarcbart M.K. (1967). Behavioral effects of selective ablation of the caudate nucleus. *Journal of Comparative Physiological Psychology, 63,* 184–190.

Domesick V.B. (1969). Projections from the cingulate cortex in the rat. *Brain Research, 12,* 296–320.

Dunnett S.B., and Iversen S.D. (1981). Learning impairments following selective kainic acid–induced lesions within the neostriatum of rats. *Behavioural Brain Research, 2,* 189–209.

Dunnett S.B., Nathwani F., and Brasted P.J. (1999). Medial prefrontal and neostriatal lesions disrupt performance in an operant delayed alternation task in rats. *Behavioural Brain Research, 106,* 13–28.

Eichenbaum H., Fagan A., Matthews P., and Cohen N.J. (1988). Hippocampal system dysfunction and odor discrimination learning in rats: Impairment or facilitation depending on representational demands. *Behavioral Neuroscience, 102,* 331–339.

Everitt B.J., Morris K.A., O'Brien A., and Robbins T.W. (1991). The basolateral amygdala–ventral striatal system and conditioned place preference: Further evidence of limbic–striatal interactions underlying reward-related processes. *Neuroscience, 42,* 1–18.

Featherstone R.E., and McDonald R.J. (2004a). Dorsal striatum and stimulus–response learning: Lesions of the dorsolater, but not dorsomedial, striatum impair acquisition of a stimulus–response based instrumental discrimination task, while sparing conditioned place preference learning. *Neuroscience, 124,* 23–31.

Featherstone R.E., and McDonald R.J. (2004b). Dorsal striatum and stimulus–response learning: lesions of the dorsolateral, but not dorsomedial, striatum impair acquisition of a simple discrimination task. *Behavioural Brain Research, 150,*15–23.

Furtado J.C.S., and Mazurek M.F. (1996). Behavioral characterization of quinolinate-induced lesions of the medial striatum: Relevance for Huntington's disease. *Experimental Neurology, 138,* 158–168.

Fuxe K., Hokfelt T., and Ungerstedt U. (1970). Morphological and functional aspects of central monoamine neurons. *International Reviews in Neurobiology, 13,* 93–126.

Gaffan D. (1985). Hippocampus: memory, habit and voluntary movement. *Philosophical Transactions of the Royal Society of Biological Sciences, 308,* 87–99.

Ghiselli E.E., and Brown C.W. (1938). Subcortical mechanisms in learning. V. Inclined plane discrimination. *Journal of Comparative Psychology, 31,* 271–285.

Gold P.E., and King R.A. (1972). Caudate stimulation and retrograde amnesia: Amnesia threshold and gradient. *Behavioral Biology, 7,* 709–715.

Graybiel A.M. (1998). The basal ganglia and chunking of action repertories. *Neurobiology of Learning and Memory, 70,* 119–136.

Green R.H., Beatty W.W., and Schwartzbaum J.S. (1967). Comparative effects of septohippocampal and caudate lesions on avoidance behavior in rats. *Journal of Comparative and Physiological Psychology, 64,* 444–452.

Groenewegen H.J., and Berendse H.W. (1994). Anatomical relationships between the prefrontal cortex and basal ganglia in the rat. In A.M. Thierry, J. Glowinski, P.S. Goldman-Rakic and Y. Christen (eds.), *Motor and Cognitive Functions of the Prefrontal Cortex* (pp. 51–77), Berlin: Springer-Verlag.

Haycock J.W., Deadwyler S.A., Sideroff S.I., and McGaugh J.L. (1973). Retrograde amnesia and cholinergic systems in the caudate-putamen complex and dorsal hippocampus of the rat. *Experimental Neurology, 41,* 201–213.

Hersch S.M., Gutekunst C.A., Rees H.D., Heilman C.J., and Levey A.I. (1994). Distribution of m1–m4 muscarinic receptor proteins in the rat striatum: Light and electron microscopic immunocytochemistry using subtype-specific antibodies. *Journal of Neuroscience, 14*, 3351–3363.

Hornykiewicz O. (1972). Biochemical and pharmacological aspects of akinesia. In J. Siegfried, ed., *Parkinson's Disease*, (pp. 128–149). Berne, Suntzarland: H. Huber.

Jog M., Kubota Y., Connolly C., Hillegaart V., and Graybiel, A. (1999). Building neural representations of habits. *Science, 286*, 1745–1749.

Kantak K.M., Green-Jordan K., Valencia E., Kremin T., and Eichenbaum H.B. (2001). Cognitive task performance after lidocaine-induced inactivation of different sites within the basolateral amygdala and dorsal striatum. *Behavioral Neuroscience, 115*, 589–601.

Kesner R.P., Bolland B.L., and Dakis M. (1993). Memory for spatial locations, motor responses, and objects: Triple dissociation among the hippocampus, caudate nucleus, and extrastriate visual cortex. *Experimental Brain Research, 93*, 462–470.

Kesner R.P., and DiMattia B.V. (1987). Neurobiology of an attribute model of memory. In *Progress in Psychobiology and Physiological Psychology*. New York: Academic Press.

Kirkby R.J. (1969). Caudate nucleus lesions and perseverative behavior. *Physiology and Behavior, 4*, 451–454.

Knowlton B.J., Mangels J.A., and Squire L.R. (1996). A neostriatal habit learning system in humans. *Science, 272*, 1399–1402.

Kolb B. (1977). Studies on the caudate-putamen and the dorsomedial thalamic nucleus of the rat: Implications for mammalian frontal-lobe functions. *Physiology and Behavior, 18*, 237–244.

Lashley K.S. (1921). Studies of cerebral function in learning: III. The motor areas. *Brain, 44*, 255–285.

Levy R., Friedman H.R., Davachi L., and Goldman-Rakic P.S. (1997). Differential activation of the caudate nucleus in primates performing spatial and nonspatial working memory tasks. *Journal of Neuroscience, 17*, 3870–3882.

Livesey P.J., and Muter V. (1976). Functional differentiation within the neostriatum of the rat using electrical (blocking) stimulation during discrimination learning. *Journal of Comparative and Physiological Psychology, 90*, 203–211.

McDonald R.J., and Hong N.S. (2004). A dissociation of dorsolateral striatum and amygdala function on the same stimulus–response habit task. *Neuroscience, 124*, 507–513.

McDonald R.J., and White N.M. (1993). A triple dissociation of memory systems: Hippocampus, amygdala, and dorsal striatum. *Behavioral Neuroscience, 107*, 3–22.

McDonald R.J., and White N.M. (1994). Parallel information processing in the water maze: Evidence for independent memory systems involving dorsal striatum and hippocampus. *Behavioral and Neural Biology, 61*, 260–270.

McGeorge A.J., and Faull R.L. (1989). The organization of the projection from the cerebral cortex to the striatum in the rat. *Neuroscience., 29*, 503–537.

Mikulas W.L. (1966). Effects of lights at the choice point on spatial alternation and position learning by normal rats and rats with bilateral lesions of the caudate nucleus. *Psychonomic Science, 5*, 275–276.

Mishkin, M., Malamut, B., and Bachevalier, J. (1984). Memories and habits: Two neural systems. In J. L. McGaugh, G. Lynch, and N. Weinberger (eds.), *The Neurobiology of Learning and Memory* (pp. 65–77). New York: Guilford Press.

Mizumori S.J.Y., Ragozzino K.E., and Cooper B.G. (2000). Location and head direction representation in the dorsal striatum of rats. *Psychobiology, 28*, 441–462.

Moore R.Y., and Bloom F.E. (1978). Central catecholamine neuron systems: Anatomy and physiology of the dopamine systems. *Annual Review of Neuroscience, 1*, 129–169.

Nauta W.J.H. (1964). Some efferent connections of the prefrontal cortex in the monkey. In J.M. Warren and K. Akert, eds. *The Frontal Granular Cortex and Behavior*, (pp. 397–409). New York: McGraw-Hill.

Nicola S.M., Surmeier D.J., and Malenka R.C. (2000). Dopaminergic modulation of neuronal excitability in the striatum and nucleus accumbens. *Annual Review of Neuroscience, 23,* 185–215.

O'Keefe J.A., and Nadel L. (1978). *The Hippocampus as a Cognitive Map.* London: Oxford University Press.

Oliveira M.G., Bueno O.F., Pomarico A.C., and Gugliano E.B. (1997). Strategies used by hippocampal- and caudate-putamen-lesioned rats in a learning task. *Neurobiology of Learning and Memory, 68,* 32–41.

Olton D.S., and Papas B.C. (1979). Spatial memory and hippocampal function. *Neuropsychologia, 17,* 669–682.

Packard M.G. (1999). Glutamate infused posttraining into the hippocampus or caudate-putamen differentially strengthens place and response learning. *Proceedings of the National Academy of Sciences, 96,* 12881–12886.

Packard M.G., Hirsh R., and White N.M. (1989). Differential effects of fornix and caudate lesions on two radial maze tasks: Evidence for multiple memory systems. *Journal of Neuroscience, 9,* 1465–1472.

Packard M.G., and Knowlton B.J. (2002). Learning and memory functions of the basal ganglia. *Annual Review of Neuroscience, 25,* 563–593.

Packard M.G., and McGaugh J.L. (1992). Double dissociation of fornix and caudate nucleus lesions on acquisition of two water maze tasks: Further evidence for multiple memory systems. *Behavioral Neuroscience, 106,* 439–446.

Packard M.G., and McGaugh J.L. (1996). Inactivation of hippocampus or caudate nucleus with lidocaine differentially affects expression of place and response learning. *Neurobiology of Learning and Memory, 65,* 65–72.

Packard M.G., and Teather L.A. (1997). Double dissociation of hippocampal and dorsal-striatal memory systems by posttraining intracerebral injections of 2-amino-5-phosphonopentanoic acid. *Behavioral Neuroscience, 111,* 543–551.

Packard M.G., and White N.M. (1990). Lesions of the caudate nucleus selectively impair "reference memory" acquisition in the radial maze. *Behavioral and Neural Biology, 53,* 39–50.

Packard M.G., and White N.M. (1991). Dissociation of hippocampus and caudate nucleus memory systems by posttraining intracerebral injection of dopamine agonists. *Behavioral Neuroscience, 105,* 295–306.

Palencia C.A., and Ragozzino M.E. (2004). The influence of NMDA receptors in the dorsomedial striatum on response reversal learning. *Neurobiology of Learning and Memory, 82,* 81–89.

Palencia C.A., and Ragozzino M.E. (2005). The contribution of NMDA receptors in the dorsolateral striatum to egocentric response learning. *Behavioral Neuroscience, 119,* 953–960.

Pisa M., and Cyr J. (1990). Regionally selective roles of the rat's striatum in modality-specific discrimination learning and forelimb reaching. *Behavioural Brain Research, 37,* 281–292.

Potegal M. (1969). Role of the caudate nucleus in spatial orientation of rats. *Journal of Comparative and Physiological Psychology, 69,* 756–764.

Pych J.C., Chang Q., Colon-Rivera C., and Gold P.E .(2005). Acetylcholine release in hippocampus and striatum during a rewarded spontaneous alternation task. *Neurobiology of Learning and Memory, 84,* 93–101.

Ragozzino M.E. (2002). The effects of dopamine D1 receptor blockade in the prelimbic-infralimbic area on behavioral flexibility. *Learning & Memory, 9,* 18–28.

Ragozzino M.E., and Choi D. (2004). Dynamic changes in medial striatal acetylcholine output during place reversal learning. *Learning & Memory, 14,* 70–77.

Ragozzino M.E., Detrick S., and Kesner R.P. (1999). Involvement of the prelimbic-infralimbic areas in shifting between place and response strategies. *Journal of Neuroscience 19,* 4585–4594.

Ragozzino M.E., Jih J., and Tzavos A. (2002). Involvement of the dorsomedial striatum to behavioral flexibility: Role of muscarinic cholinergic receptors. *Brain Research, 953,* 205–214.

Ragozzino M.E., Kim J., Minniti N., Hassert D., and Kiang C. (2003). The contributions of the prelimbic-infralimbic areas to different forms of task switching. *Behavioral Neuroscience, 117*, 1054–1065.

Ragozzino M.E., Ragozzino K.E., Mizumori S.J.Y., and Kesner R.P. (2002). Role of the dorsomedial striatum in behavioral flexibility for response and visual cue discrimination learning. *Behavioral Neuroscience, 116*, 105–115.

Reading P.J., Dunnett S.B., and Robbins T.W. (1991). Dissociable roles of the ventral, medial and lateral striatum on the acquisition and performance of a complex visual stimulus–response habit. *Behavioural Brain Research, 45*, 147–161.

Rogers R.D., Baunez C., Everitt B.J., and Robbins T.W. (2001). Lesions of the medial and lateral striatum in the rat produce differential deficits in attentional performance. *Behavioral Neuroscience, 115*, 799–811.

Rosvold H.E., and Delgado J.M.R. (1956). Effects on delayed-alternation test performance of stimulating or destroying electrically structures within frontal lobes of monkey's brain. *Journal of Comparative and Physiological Psychology, 49*, 365–372.

Rosvold H.E., Mishkin M., and Szwarcbart M.K. (1958). Effects of subcortical lesions in monkeys on visual discrimination and single-alternation performance. *Journal of Comparative and Physiological Psychology, 51*, 437–444.

Sanberg P.R., Lehmann J., and Fibiger H.C. (1978). Impaired learning and memory after kainic acid lesions of the striatum: A behavioral model of Huntington's disease. *Brain Research, 149*, 546–551.

Smith A.D., and Bolam J.P. (1990). The neural network of the basal ganglia as revealed by the study of synaptic connections of identified neurones. *Trends in Neuroscience, 13*, 259–265.

Smith-Roe S.L., Sadeghian K., and Kelley A.E. (1999). Spatial learning and performance in the radial arm maze is impaired after N-methyl-D-aspartate (NMDA) receptor blockade in striatal subregions. *Behavioral Neuroscience, 113*, 703–717.

Squire, L. R. (1986). Mechanisms of memory. *Science, 232*, 1612–1619.

Thompson R.L. (1959). Effects of lesions in the caudate nuclei and dorsofrontal cortex on conditioned avoidance behavior in cats. *Journal of Comparative and Physiological Psychology, 52*, 650–659.

Thompson W.G., Guilford M.O., and Hicks L.H. (1980). Effects of caudate and cortical lesions on place and response learning in rats. *Physiological Psychology, 8*, 473–479.

Tzavos A., Jih J., and Ragozzino M.E. (2004). Differential contribution of muscarinic and nicotinic receptors in the dorsomedial striatum to behavioral flexibility. *Behavioural Brain Research, 154*, 245–253.

Voulalas P.J., Holtzclaw L., Wolstenholme J., Russell J.T., and Hyman S.E. (2005). Metabotropic glutamate receptors and dopamine receptors cooperate to enhance extracellular signal-regulated kinase phosphorylation in striatal neurons. *Journal of Neuroscience, 25*, 3763–3773.

Warrington E.K., and Weiskrantz L. (1982). Amnesia: A disconnection syndrome? *Neuropsychologia, 20*, 233–248.

Webster K.E. (1961). Cortico-striate interrelations in the albino rat. *Journal of Anatomy, 95*, 532–544.

Weiner D.M., Levey A.I., Sunahara R.K., Niznik H.B., O'Dowd B.F., Seeman P., and Brann M.R. (1991). D1 and D2 dopamine receptor mRNA in rat brain. *Proceedings of the National Academy of Sciences, 88*, 1859–1863.

Whishaw I.Q., Mittleman G., Bunch S.T., and Dunnett S.B. (1987). Impairments in the acquisition, retention and selection of spatial navigation strategies after medial caudate-putamen lesions in rats. *Behavioural Brain Research, 24*, 125–138.

White N.M., and Salinas J.A. (2003). Mnemonic functions of dorsal striatum and hippocampus in aversive conditioning. *Behavioural Brain Research, 142*, 99–107.

White N.M., and Viaud M. (1991). Localized intracaudate dopamine D2 receptor activation during the posttraining period improves memory for visual or olfactory conditioned emotional responses in rats. *Behavioral and Neural Biology, 55*, 255–269.

Wilburn M.W., and Kesner R.P. (1972). Differential amnestic effects produced by electrical stimulation of the caudate nucleus and nonspecific thalamic system. *Experimental Neurology, 34*, 45–50.

Wilson S.A.K. (1914). An experimental research into the anatomy and physiology of the corpus striatum. *Brain, 36*, 427–492.

Winocur G. (1974). Functional dissociation within the caudate nucleus of rats. *Journal of Comparative and Physiological Psychology, 86*, 432–439.

Winocur G., and Mills J.A. (1969). Effects of caudate lesions on avoidance behavior in rats. *Journal of Comparative and Physiological Psychology, 68*, 552–557.

Wyers E.J., and Deadwyler S.A. (1971). Duration and nature of retrograde amnesia produced by stimulation of caudate nucleus. *Physiology & Behavior, 6*, 97–103.

Yin H.H., Knowlton B.J., and Balleine B.W. (2004). Lesions of the dorsolateral striatum preserve outcome expectancy but disrupt habit formation in instrumental learning. *European Journal of Neuroscience, 19*, 181–189.

CHAPTER 12

Neural Systems Involved in Fear and Anxiety Based on the Fear-Potentiated Startle Test

Michael Davis

Department of Psychiatry, Emory University School of Medicine, Atlanta, GA 30322

I. CONDITIONED AND UNCONDITIONED FEAR

Since early 1980s, a great deal of progress has been made in delineating the neural pathways and the cellular and molecular mechanisms involved in fear, anxiety, and extinction of fear. Conditioned fear is a hypothetical construct used to explain the cluster of behavioral effects produced when an initially neutral stimulus is consistently paired with an aversive stimulus. For example, when a light that initially has no behavioral effect is paired with an aversive stimulus such as a foot shock, the light alone can now elicit a constellation of behaviors typically used to define a state of fear in animals. To explain these findings, it is generally assumed (cf. McAllister and McAllister, 1971) that during light-shock pairings (training session) the shock activates a central fear state that results in a variety of behaviors that can be used to infer a central state of fear (unconditioned responses — Fig. 12-1). After pairing, the light can now produce the same central fear state and thus the same set of behaviors formerly produced by the shock. Moreover, the behavioral effects that are produced in animals by this formerly neutral stimulus (now called a *conditioned stimulus*, CS) are similar in many respects to the constellation of behaviors used

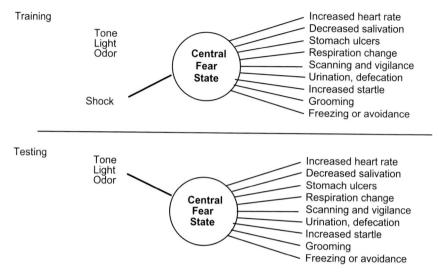

FIGURE 12-1   Hypothetical description of unconditioned and conditioned fear. Prior to train-ing, shock activates a hypothetical central fear state that produces a variety of behaviors that collectively define a state of *unconditioned fear.* Following cue–shock pairings, the cue can now activate the same hypothetical central fear state and hence produce a variety of behaviors that collectively define *conditioned fear.*

to diagnose generalized anxiety in humans. It is important to realize that the responses to the CS do not necessarily mimic those to the shock. For example, when a rat is given foot shock, the immediate reaction is to jump around to try to avoid the shock. However, when the rat is later presented with the CS, it does not jump around but, instead, freezes, quite the opposite of what the rat did during the shock. Thus the conditioned response in the case of fear conditioning often does not mimic the reaction to foot shock but, instead, the emotional effect of the foot shock. Treatments that block fear conditioning do not prevent the rat from jumping around but presumably do prevent the shock from having its usual emotional impact (Blair, Sotres-Bayon, Moita, and Ledoux, 2005; Miserendino, Sananes, Melia, and Davis, 1990).

II. FEAR VERSUS ANXIETY

Fear typically is very stimulus specific. Thus, a rat trained to fear a light by pairing it with a shock will not show a fear reaction to another stimulus, such as a tone or an odor. Fear comes on quickly and dissipates quickly. Thus, we react with fear when a snake crosses our path but get over our fear reaction quickly once the snake is out of sight and we are sure it will not return. Anxiety has many of the same symptoms as fear; in fact, many of the laboratory measures of fear are identical to those used by psychiatrists to diagnose anxiety. However, even though the symptoms of fear and anxiety often are very similar, the stimuli or situations that elicit a state of anxiety often are not very specific. We feel anxious but are not quite sure what it is that bothers us. Anxiety is usually not abrupt in onset, like a state of fear, but instead comes on slowly and often lasts for a long time, in fact more or less continuously in certain patients.

III. ANIMAL MODELS OF FEAR AND ANXIETY

Table 12-1 lists a number of animal models of fear and anxiety. Those in boldface are the most widely used in the field. Although exact boundaries cannot always be established, tests in Section 1 of the table are primarily tests of stimulus-specific fear, typically after pairing a stimulus with a shock and then testing behavior in the presence of that stimulus, whereas those in Sections 2 and 3 are more like anxiety. These latter tests put the animal at "risk," although the stimuli eliciting risk are not as predictable as those where a stimulus has consistently been paired with shock. For example, a 3-sec tone whose offset has consistently been paired with a foot shock elicits freezing, and this is clearly a test of conditioned fear. In contrast, a context that has been associated with shock is more like a test of anxiety because the animal is at "risk" when it reenters that context, although the exact time when a shock might occur is not predictable.

The two most widely used measures of conditioned fear are freezing and fear-potentiated startle (Fendt and Fanselow, 1999) and the systematic study of these behaviors by a host of investigators has rapidly led to a detailed understanding of the neural pathways and the cellular and molecular mechanisms of both the acquisition and expression of conditioned fear. Freezing is a species-typical response of rodents, to stop all behaviors, except for respiration, in the face of an imminent danger, such as the site of a predator or the presentation of a stimulus previously paired with shock, or placement in a context previously paired with shock. The freezing response is reliable, quantifiable, and easy and inexpensive to measure. The other major measure of conditioned fear is the fear-potentiated startle test. Because this is the test I happen to use and because it has certain advantages over the freezing measure, I use data gathered from this test to illustrate many of the basic principles and neural pathways involved in conditioned fear, because conclusions from fear-potentiated startle and freezing have largely, although not entirely, led to similar conclusions. At the end of this chapter I also discuss how different, but highly interconnected, brain structures are involved in fear versus anxiety, and finally I discuss selected

TABLE 12-1   Models of Fear and Anxiety in Rats

Measures of fe ar and anxiety in animals	Possible human analogue
1. Tests measuring classically conditioned fear (response in presence of conditioned stimulus)	
Freezing to CS or context paired with shock	Cessation of ongoing behavior
Fear-potentiated startle (more startle to loud sound in presence of conditioned stimulus — CS)	Increased startle
Conditioned emotional response (less bar pressing in presence of CS)	Cessation of ongoing behavior
Change in autonomic measures in presence of CS (heart rate, respiration, salivation, stomach acidity)	Heart pounding, panting, dry mouth, upset stomach
Tail flick test (rat takes longer time to move tail from hot light in presence of CS)	Stress or fear induced analgesia
Formalin test (rat takes longer time to lick paw in presence of CS)	Stress or fear induced analgesia
Active avoidance (animal makes response lever press, hurdle) to avoid shock in presence of CS	Avoidance of bad things
Inhibitory avoidance (animal does not return to place where shocked — used more as measure of memory)	Avoidance of bad places
Exposure to predator (rat freezes)	Fear of a lion
Electrical stimulation of central grey (rat escapes)	Panic???
Electrical stimulation of locus coeruleus (monkey grimaces, tries to escape, scans)	Scanning, vigilant behavior?
2. Tests requiring prior conditioning or deprivation to produce a non-zero baseline	
Freezing to context paired with shock	Cessation of ongoing behavior
Operant conflict test (shock suppresses bar pressing)	Conflict (sex but then AIDS)
Lick suppression (shock suppresses lick rate)	Conflict
Negative contrast (press less for 4% solution after 32% compared to 4% all along)	Disappointment
Drug discrimination (rat presses pentylenetetrazol lever)	Pentylenetetrazol — induced anxiety
3. Tests not requiring conditioning	
Elevated plus or X maze (rat stays in closed arms)	Fear of heights
Social interaction (rats don't interact)	Social phobias
Light-dark box (rat goes to dark side)	Aversion to bright lights, open spaces
Open field test (rat stays near edge)	Aversion to large open spaces
Open field (rat grooms excessively)	Nail biting, fidgeting
Open field + food (hungry rat takes long time to eat)	Loss of appetite when afraid or anxious
Defensive withdrawal — open field test with "burrow" (tin can) (rat stays in tin can)	Hiding when afraid
Intraventricular infusion of CRF	Stress induced anxiety?
Light-enhanced startle	Aversion to bright lights
Ultrasonic vocalizations in rat pups separated from mother (rat emits 22–25 kHz sounds)	Crying as a result of separation
Urination, defecation following shock or in open field	Frequent urination, diarrhea

studies on extinction of conditioned fear and the clinical relevance of these findings.

IV. THE FEAR-POTENTIATED STARTLE EFFECT

Brown, Kalish, and Farber (1951) demonstrated that the amplitude of the acoustic startle reflex in the rat can be augmented by presenting the eliciting auditory startle stimulus in the presence of a cue (e.g., a light) that has previously been paired with a shock. This phenomenon, termed the *fear-potentiated startle effect*, has been replicated using either an auditory or a visual CS when startle is elicited by either a loud sound or an air puff (Davis, 1986). In this paradigm we typically pair a 3.7-sec light that coterminates with a 0.5-sec 0.4-mA shock. This is called the *training* session (Fig. 12-2). One to two days later, or a month or two if we are looking at remote memory, the rat is placed in a cage specially designed to measure the amplitude of the startle

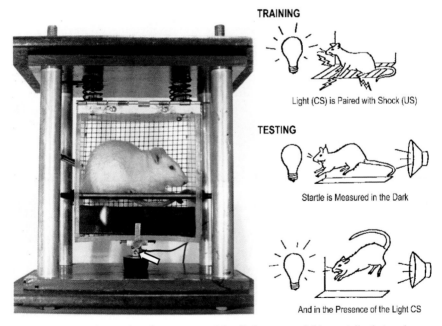

TRAINING

Light (CS) is Paired with Shock (US)

TESTING

Startle is Measured in the Dark

And in the Presence of the Light CS

FIGURE 12-2   The startle reflex is measured by displacement of this specially designed cage. When the rat is startled it moves the cage, which moves an accelerometer (arrow) that puts out a voltage proportionate to cage movement. Fear-potentiated startle involves a training phase, where a cue, such as a light, is paired with foot shock, followed by a test phase, where startle is elicited by a loud sound in the presence or absence of the light. Fear-potentiated startle is defined by greater startle amplitude in the presence versus the absence of the light.

reflex elicited by a burst of noise at the time the shock was presented in training [e.g., 3.2 sec after onset of the light (light–noise test trial) or in darkness (noise-alone)]. Conditioned fear is operationally defined by elevated startle amplitude in the presence of versus the absence of the cue previously paired with a shock (fear-potentiated startle — see Fig. 12-2). Thus, the CS does not elicit startle. Furthermore, the startle-eliciting stimulus is never paired with a shock; instead, the CS is paired with a shock and startle is elicited by another stimulus, in either the presence or the absence of the CS. Facilitation of a simple reflex is used to assay the hypothetical state of fear, which would be expected to facilitate reflexes. Fear-potentiated startle occurs only following paired versus unpaired or "random" presentations of the CS and the shock, which indicates that it is a valid measure of classical conditioning (Davis and Astrachan, 1978). Discriminations between visual and auditory conditioned stimuli (Davis, Hitchcock, and Rosen, 1987) or between auditory cues or visual cues that differ in duration (Davis, Schlesinger, and Sorenson, 1989; Siegel, 1967) have also been demonstrated with potentiated startle. Odors are especially good conditioned stimuli for fear-potentiated startle, in which reliable conditioning can be found after only a single pairing of the an odor with foot shock (Paschall and Davis, 2002). Increased startle in the presence of the CS still occurs very reliably at least one month after the original training, making it appropriate for the study of long-term memory as well (Campeau, Liang, and Davis, 1990). In fact, one of the major reasons that conditioned fear is used to investigate the neural mechanisms of memory is that fear lasts so long. Using freezing as a measure, Gale et al. (2004) reported that fear of a tone was still evident 16 months after original conditioning using rats, which have a life span of about 2–2.5 years. Fear-potentiated startle also can be measured in mice (Falls, Carlson, Turner, and Willott, 1997) and rhesus monkeys (Winslow, Parr, and Davis, 2002).

V. FEAR-POTENTIATED STARTLE IN HUMANS

Fear-potentiated startle can be seen in humans, using several different ways to elicit fear. In humans the eye-blink component of startle is the most easily measured and the most reliable because, although it habituates with repeated presentation of startle stimuli, it typically reaches a nonzero asymptote, so both excitatory and inhibitory effects can be measured. One way to potentiate startle in humans is via conditioning, using procedures that closely parallel those in rats (Grillon and Davis, 1997; Hamm, Greenwald, Bradley, Cuthbert, and Lang, 1991; Hamm, Start, and Vaitl, 1990; Lipp, Sheridan, and Siddle, 1994). For example, Christian Grillon and I (Grillon and Davis, 1997) presented undergraduates with a light consistently paired with a shock (paired group), a light explicitly unpaired with a shock (unpaired group), or a light that served as a signal to push a button as soon as a second light came on (reaction time group). Startle was measured in the presence or absence of these lights prior to and following conditioning and then measured again 1 week later. In the paired group, startle magnitude was greater in the presence versus the absence of the light in both the first and second sessions, indicating retention of conditioning over a 1-week interval. Startle was not elevated when elicited in the presence of the light in either session in the unpaired group or in the reaction time group. This indicates that fear-potentiated startle only occurred following explicit pairing of a cue with a shock and was not simply a function of heightened arousal following shock presentation or instructions to perform in a reaction time experiment, consistent with earlier work (Hamm, Start, and Vaitl, 1990; Lipp, Sheridan, and Siddle, 1994). Interestingly, these earlier studies showed that arousal associated with either shock or with a reaction time experiment increased the galvanic skin response, a measure of activation of the sympathetic nervous system. Thus changes in startle reflected a change in valence (threat versus safe) and arousal, whereas the galvanic skin response did not differentiate between valence and arousal. Finally, in Session 2 there was a pronounced increase in startle amplitude in the beginning of the session in the unpaired group but not in the paired group, indicative of context conditioning, a result predicted from contemporary learning theory (Rescorla and Wagner, 1972). Thus, fear-potentiated startle measured using conditioning procedures in humans closely parallels work done in rodents.

Another way to potentiate startle is simply to tell people that when a certain colored light comes on they might get a shock (Grillon, Ameli, Woods, Merikangas, and Davis, 1991). Thus, even though they have never actually received a shock, just the anticipation of this possibility, which is rated to be very fearful, is enough to increase startle magnitude in humans. Finally, startle elicited in the presence of pictures of scary scenes, such as a snake or dog ready to attack, is potentiated as compared to when it is elicited in the presence of neutral pictures, such as baskets or cans (Lang, Bradley, and Cuthbert, 1990). In contrast, startle is actually inhibited when elicited in the presence of pleasant pictures, such a babies or sexy scenes, whereas the galvanic skin response is increased in the presence of both scary and pleasant scenes. Once again, therefore, startle is sensitive to valence but not simply arousal (Lang, Bradley, and Cuthbert, 1990).

VI. NEURAL PATHWAYS INVOLVED IN FEAR-POTENTIATED STARTLE

One of the major advantages of the fear-potentiated startle test is that the hypothetical state of fear is inferred from an increase in a simple reflex. Moreover, because the acoustic startle reflex has such a short latency (e.g., 8 msec

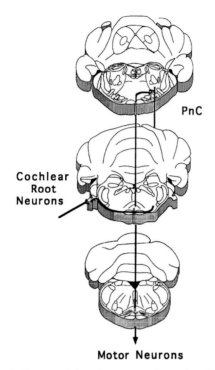

PnC

Cochlear Root Neurons

Motor Neurons

FIGURE 12-3   Schematic diagram of the primary acoustic startle reflex pathway. Axons from sensory receptors in the cochlea synapse onto cochlear root neurons (CRNs) embedded in the auditory nerve. CRNs project to a ventrolateral region of the nucleus reticularis pontis caudalis (PnC). PnC axons form the reticulospinal tract make mono- and polysynaptic connections in the spinal cord onto motoneurons that innervate muscles in the neck, forelimbs, and hind limbs that mediate the acoustic startle reflex.

measured electromyographically in the hind leg, 5 msec in the neck), it must be mediated by a simple neural pathway. We now believe that the primary acoustic startle reflex pathway involves three central synapses: (1) auditory nerves fibers to cochlear root neurons (CRNs) (2), CRN axons to cells in the nucleus reticularis pontis caudalis (PnC), and (3) PnC axons to motor neurons in the facial motor nucleus (pinna reflex) or spinal cord (whole body startle — Fig. 12-3).

A. Cochlear Root Neurons

In rats there is a small group (about 20 on each side) of very large cells (35 μm in diameter) embedded in the cochlear nerve, called *cochlear root neurons*. These neurons receive direct input from the spiral ganglion cells in the cochlea, making them the first acoustic neurons in the central nervous system (Lopez, Merchan, Bajo, and Saldana, 1993). They send exceedingly thick axons (sometimes as wide as 7 μm) through the trapezoid body, at the very base of the brain, to the contralateral side, to an area just medial and ventral to the lateral lemniscus, and continue on up to the deep layers of the superior colliculus. However, they give off thick axon collaterals that terminate directly in the PnC (Lingenhohl and Friauf, 1994; Lopez, Merchan, Bajo, and Saldana, 1993), exactly at the level known to be critical for the acoustic startle reflex (cf. Lee, Lopez, Meloni, and Davis, 1996). Nodal and Lopez (2003) showed direct connections between CRN axons to cells in the PnC that projected to the spinal cord, based on double-labeling techniques. Electron microscopy of the labeled CRNs axons and terminals showed that even the thinnest processes were myelinated, consistent with very rapid transmission. Multiple CRNs synapse onto single reticulospinal neurons in PnC, where most of the connections were axodendritic, with multiple asymmetric synapses, consistent with an excitatory input.

Bilateral chemical lesions of the cochlear root neurons essentially eliminate acoustic startle in rats, and the magnitude of decrease in startle was highly correlated with the number of CRNs destroyed (Lee, Lopez, Meloni, and Davis, 1996). Although damage to the auditory root, where the cochlear root neurons reside, has not been fully ruled out, other tests indicated that these animals could clearly orient to auditory stimuli (e.g., suppression of licking) and had normal compound action potentials recorded from the cochlear nucleus (Lee, Lopez, Meloni, and Davis, 1996).

B. Nucleus Reticularis Pontis Caudalis (PnC)

Very discrete N-methyl-D-aspartate (NMDA)-induced lesions of cell bodies in the PnC completely eliminated startle, whereas NMDA-induced lesions of the ventral nucleus of the lateral lemniscus or the area just ventral and medial to it did not, provided the lesion did not extend to the PnC (Lee, Lopez, Meloni, and Davis, 1996). Local infusion of the NMDA antagonist D-L-2-amino-5-phosphonopentanoic acid (AP5) into the PnC reduced startle by 80–90% (Miserendino and Davis, 1993), at doses 1/60 of those that depressed startle after infusion into the area of the ventral lateral lemniscus (Spiera and Davis, 1988). Moreover, comparably low doses of the non-NMDA antagonist 6-cyano-7-nitroquinoxaline-2,3-dione (CNQX) also depressed startle after local infusion into the PnC (Miserendino and Davis, 1993) but had no effect when infused into the area of the ventral lateral lemniscus, even using much higher doses (Lee and Davis, unpublished). Single-pulse electrical stimulation of the PnC elicited startle responses, with a latency of about 5 msec recorded in the hind leg, compared to about 8 msec when elicited acoustically (Davis, Gendelman, Tischler, and Gendelman, 1982). In humans, startle stimuli lead to an increase in blood flow in the PnC, and this effect habituates with repetition of the startle stimulus (Pissiota, Frans, Fredrikson, Langstrom, and Flaten, 2002).

C. Facial and Spinal Motor Neurons

In rats the pinna component of the startle reflex consists of a rapid backward movement of the pinna, which covers and protects the ear, and the pinna reflex shows many of the features of whole-body startle, including fear-potentiated startle (Cassella and Davis, 1986). The motor neurons that innervate the relevant pinna muscles are located in the dorsolateral division of the facial motor nucleus to which the PnC has direct projections. Startle stimuli elicit action potentials in facial motor nucleus neurons, with a latency of 5 msec (Cassella and Davis, 1987) prior to movement of the pinna muscles, and local infusion of the AMPA/kainite antagonist CNQX into the facial motor nucleus eliminated the click-elicited pinna reflex on the ipsilateral but not the contralateral side (Meloni and Davis, unpublished observations).

Motor neurons in the lumbar spinal cord innervate muscles in the hind leg that provide the major extension/flexion component of startle in rodents (Davis, 1984). When startle is measured electromyographically (EMG) in the hind leg, two distinct components can be measured, a short latency component (~8 msec) and a slightly longer latency component (~15 msec). Infusion in the space between the spinal cord and the membranes that surround the spinal cord (intrathecal infusion) of the AMPA/kainate antagonist CNQX in the vicinity of the lumbar motor neurons eliminated the short latency component but not the longer latency component, whereas infusion of the NMDA antagonist AP5 had just the opposite effect (Boulis, Kehne, Miserendino, and Davis, 1990). Infusion of both compounds together totally eliminated the EMG component of startle in the hind leg. This suggests that the acoustic startle reflex involves motor neurons in the spinal cord that are activated by release of glutamate acting on both non-NMDA and NMDA receptors. Intrathecal administration of cAMP or cAMP analogues markedly facilitate acoustic startle amplitude (Boulis and Davis, 1990; Kehne, Astrachan, Astrachan, Tallman, and Davis, 1986), probably by increasing the release of glutamate from the terminals of neurons in the PnC activated by the startle stimulus. Intrathecal administration of the glycine receptor antagonist strychnine markedly increases startle amplitude (Kehne, Gallager, and Davis, 1981), as do norepinephrine (Davis, Astrachan, Kehne, Commissaris, and Gallager, 1984) and serotonin agonists (Davis, Astrachan, Gendelman, and Gendelman, 1980; Davis, Astrachan, Kehne, Commissaris, and Gallager, 1984), which are known to facilitate the response of motor neurons to glutamate.

VII. FEAR-POTENTIATED STARTLE MEASURED ELECTROMYOGRAPHICALLY

Having delineated what we believe is the primary acoustic startle pathway, we hoped to use this information to deduce where fear ultimately alters neural transmission so as to increase acoustic startle amplitude. Because startle can be measured with a latency of only 8 msec, the light should potentiate this 8-msec response. Typically, however, startle is not measured electromyographically, but instead it is measured as a movement of a cage over a relatively long interval after onset of the startle-eliciting stimulus (e.g., 200 msec). Hence it is possible that the visual CS does not actually alter the very short-latency startle response, but instead it might facilitate transmission in other auditory systems, which could produce cage movements at longer latencies. If so, this might mean that the visual CS would not actually alter transmission along the short-latency pathway outlined in Figure 12-3. However, if the light did increase the very short short-latency startle reflex, we would have to conclude that it alters transmission at some point in the short-latency pathway. In fact, we found that a light previously paired with a foot shock markedly potentiated the short-latency startle response, measured electromyographically in the neck muscles (Fig. 12-4 Cassella, Harty, and Davis, 1986) indicating that the visual CS must ultimately alter neural transmission somewhere along the short-latency pathway outlined in Figure 12-3.

VIII. THE POINT IN THE STARTLE PATHWAY WHERE FEAR MODULATES TRANSMISSION

Having demonstrated that fear facilitates transmission in this very short-latency pathway, the next task was to try to deduce where a light previously paired with a shock ultimately modulates transmission in this short-latency pathway. We had previously shown that startle could be elicited with single electrical pulses at various points along the startle pathway, with progressively shorter latencies as the electrode was moved from the cochlear root axons to the reticulospinal axons connecting the PnC with spinal motor neurons (Davis, Gendelman, Tischler, and Gendelman, 1982), and used this method to deduce where habituation and sensitization occurred within the startle pathway (Davis, Parisi, Gendelman, Tischler, and Kehne, 1982). The logic of this approach to try to determine where fear ultimately might alter transmission in the startle pathway is shown in Figure 12-5. Let us assume that the light, after being paired with a shock, goes from the retina and, in one way or another, ultimately modulates transmission in the PnC to increase acoustic startle amplitude. If one were to elicit startle acoustically or electrically by eliciting startle from points in the pathway upstream from the PnC, then both acoustically

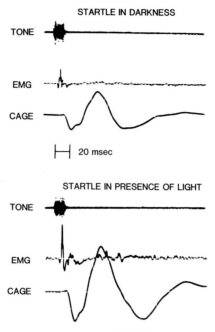

STARTLE IN DARKNESS

TONE

EMG

CAGE

⊢—⊣ 20 msec

STARTLE IN PRESENCE OF LIGHT

TONE

EMG

CAGE

FIGURE 12-4   Startle measured electromyographically in the neck muscles as well as by output of an accelerometer on tone-alone trials (TA) or light-tone trials (LT). **TA:** *Upper trace*: Output of decibel meter on storage oscilloscope showing arrival of a 10-msec burst of white noise (called TA for Tone-Alone), with a 2.5-msec onset and offset duration, used to elicit startle with a sweep time of either 40 msec per division (left) or 10 msec per division (right). *Middle trace*: Muscle activity measured electromyographically in the neck muscles showing that an acoustic startle stimulus elicits a startle reflex with a latency of 5 msec. *Bottom trace*: Voltage output of an accelerometer that measures mainly force on the cage produced by the hind legs, showing a latency of about 15 msec. **LT:** Same data on light-tone trials (LT). Note that the light came on 3.2 sec prior to tone onset, long before the measurements in Figure 12-4 began, and that the very early, 5-msec-latency response is larger in the presence (LT) versus the absence (TA) of the light.

and electrically elicited startle should be facilitated by the light because the startle signal would have to pass through the PnC. On the other hand, if startle were elicited electrically from points downstream from the PnC, it would not be increased in the presence of the light because the startle signal would not pass through the PnC. Although this logic requires a number of assumptions, this is exactly what we found (Berg and Davis, 1985). Thus, startle elicited electrically from CRN axons adjacent to the ventral cochlear nucleus or farther along the base of the brain on route to the PnC was facilitated by the light, whereas startle elicited in the PnC or the reticulospinal tract was not, even though acoustically elicited startle was increased in both cases. Systemic administration of diazepam (Valium), which reduces fear and anxiety in people,

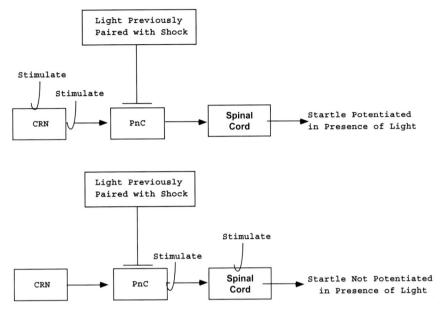

FIGURE 12-5 Logic of using electrically elicited startle to deduce where fear ultimately modulates transmission along the acoustic startle pathway.

selectively decreased fear-potentiated startle elicited electrically from points afferent to the PnC, indicating that eliciting startle in this way could pick up the anxiolytic effect of diazepam (Berg and Davis, 1984).

IX. PROJECTIONS TO THE PNC

Having determined that the PnC was the probable site where fear ultimately altered transmission to increase startle amplitude, we posed the next question: What parts of the brain project to the part of the PnC critical for startle, and are these projections critical for fear-potentiated startle. After several years now, we believe that there are three parallel pathways, each of which may play a part in fear-potentiated startle.

A. Direct Projections from the Central Nucleus of the Amygdala

Local infusion of the retrograde tracer FluoroGold into the part of the PnC critical for startle resulted in labeling of neurons in the medial division of the central nucleus of the amygdala (Rosen, Hitchcock, Sananes, Miserendino, and

Davis, 1991). This was an exciting finding because earlier work in several laboratories (Blanchard and Blanchard, 1972; Gentile, Jarrel, Teich, McCabe, and Schneiderman, 1986; Iwata, LeDoux, Meeley, Arneric, and Reis, 1986; Kapp, Frysinger, Gallagher, and Haselton, 1979) as well as our own (Hitchcock and Davis, 1986, 1987) had implicated the central nucleus of the amygdala in conditioned and unconditioned fear using several different measures. Local infusion into the central nucleus of the amygdala of an anterograde tracer confirmed this connection and was used to delineate the course of the pathway from the central nucleus of the amygdala to the PnC (Fig. 12-6). Electrolytic lesions at various points along this pathway blocked fear-potentiated startle but had no effect on baseline startle amplitude (Hitchcock and Davis, 1991). In contrast, electrolytic lesions of outputs of the central nucleus of the amygdala to the bed nucleus of the stria terminalis had no effect on fear-potentiated startle, consistent with earlier work (LeDoux, Iwata, Cicchetti, and Reis, 1988).

B. Indirect Projections from the Central Nucleus of the Amygdala via the Deep Mesencephalic Reticular Formation

Although these results with electrolytic lesions were consistent with idea that this direct projection mediates fear-potentiated startle, it is still possible that synaptic, rather than direct, projections might also be involved. For example, injection of a retrograde tracer into the PnC showed that several nuclei that lie along this direct pathway contained neurons that also projected directly to the PnC. One of the most prominent of these was in the mesencephalic reticular formation and deep layers of the superior colliculus (deep SC/DpMe). The amygdala sends heavy, broad projections to this part of the rostral midbrain (Rosen, Hitchcock, Sananes, Miserendino, and Davis, 1991), which in turn projects to the PnC (Cameron, Iqbal, Westlund, and Willis, 1995; Meloni and Davis, 1999). Collision tests using electrical brain stimulation suggested that a synapse existed between the amygdala and the midbrain, and electrolytic lesions in the midbrain blocked fear-potentiated startle (Yeomans and Pollard, 1993). Thus, the rostral midbrain was proposed to be a relay between the amygdala and the PnC in fear-potentiated startle (Yeomans and Pollard, 1993).

Consistent with this hypothesis, inactivation of the deep layers of the superior colliculus/the deep mesencephalic nucleus (deep SC/DpMe) with muscimol blocked the expression but not the acquisition of fear-potentiated startle (Meloni and Davis, 1999), suggesting an effect on the output circuitry of the amygdala rather than a blockade of sensory input to the amygdala. Although these results confirmed a critical relay in the midbrain in mediating fear-potentiated startle, the precise part of the midbrain remained unclear. Recently

AMYGDALA
EFFERENTS

LESION
LOCATIONS

LESION EFFECTS ON
POTENTIATED STARTLE

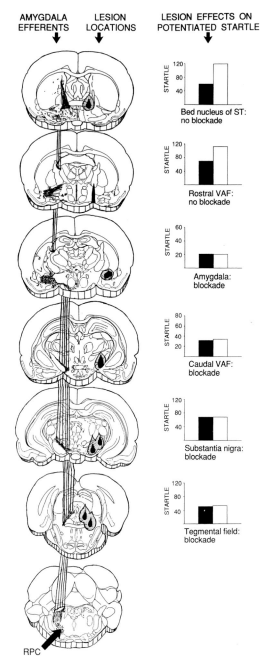

Bed nucleus of ST:
no blockade

Rostral VAF:
no blockade

Amygdala:
blockade

Caudal VAF:
blockade

Substantia nigra:
blockade

Tegmental field:
blockade

RPC

FIGURE 12-6   Schematic diagram of output of the central nucleus of the amygdala, rostrally
to the bed nucleus of the stria terminalis and then caudally down the ventral amygdalofugal tract
(VAF), ultimately terminating in the nucleus reticularis pontis caudalis (PnC).

we found that local infusion of the AMPA/Kainate glutamatergic receptor antagonist NBQX [2,3-dihydroxy-6-nitro-7-sulphamoylbenzo(F)-quinoxaline], which has a very limited area of diffusion (Walker, Paschall, and Davis, 2005), blocked the expression but not the acquisition of fear-potentiated startle if infused into the deep SC/DpMe (Zhao and Davis, 2004). In contrast, infusion of the same doses, either 1 mm lateral into the lateral mesencephalic reticular formation or 1 mm medial into the dorsal/lateral periaqueductal gray or into the superficial layers of the superior colliculus had no effect. The infusions altered not the baseline startle response but only startle facilitated by a conditioned fear stimulus. These data suggest that fear-potentiated startle is mediated by release of glutamate into the deep SC/DpMe and support the idea of an indirect amygdalo-tecto-PnC pathway in which the rostral midbrain serves as a relay between the amygdala and the PnC to mediate fear-potentiated startle (Yeomans and Pollard, 1993).

C. Indirect Projections from the Medial Nucleus of the Amygdala via the Ventromedial Hypothalamus and Ventral Periaqueductal Gray

Despite clear evidence that the medial nucleus of the amygdala is involved in sexual behavior in both males and females, a growing body of literature points to a role of the medial nucleus of the amygdala in stress, especially so-called psychological stress. For example, Dayas et al. (2001) found that psychological stressors such as noise, restraint, and forced swim elicited high levels of *c-fos* expression in the medial nucleus of the amygdala as compared to the central nucleus of the amygdala, whereas physical stressors such as hemorrhage and immune challenge produced the opposite pattern of *c-fos* expression. The medial nucleus of the amygdala and its outputs to the hypothalamus and periaqueductal gray have been implicated in defensive behavior in cats (Adamec, 1994), and we recently found that blockade of AMPA/kainate glutamate receptors in the medial nucleus of the amygdala blocked conditioned fear elicited not only by an odor but also by a light previously paired with foot shock (Walker, Paschall, and Davis, 2005).

In a series of studies we have found that local infusion into the medial nucleus of the amygdala of either morphine (Davis, Yong, Shi, and Zhao, in preparation) or substance P antagonists (Zhao, Yong, and Davis, in preparation) blocked the expression of fear-potentiated startle, without any effect on baseline startle amplitude. The medial nucleus of the amygdala sends heavy projections to the ventral medial hypothalamus, and local infusion of either NBQX, morphine, or substance P antagonists into this region also totally blocked fear-potentiated startle, without any effect on baseline startle amplitude. Although the ventral medial hypothalamus does not project directly to the PnC, it does project to the periaqueductal gray, which in turn projects to the PnC. Recall, however, that local infusion of NBQX into the periaqueductal gray did not block the expression of fear-potentiated startle. Nonetheless, local infusion of either morphine or substance P antagonists into the periaqueductal gray did block fear-potentiated startle. This suggests that the fear-potentiated startle may be mediated or importantly modulated by the release of substance P in the medial nucleus of the amygdala, ventral medial hypothalamus, and the periaqueductal gray. The effect of morphine at each of these areas might be due to its ability to decrease the release of substance P by acting on terminal mu opioid autoreceptors.

Thus far the effects of substance P antagonists into the PnC have not been tested on fear-potentiated startle. However, substance P increases the responsiveness of reticulospinal neurons to acoustic stimuli (Krase, Koch, and Schnitzler, 1994), and local infusion into the PnC of a substance P antagonist completely blocked the normal sensitizing effect of foot shock on startle (Krase, Koch, and Schnitzler, 1994), which has previously been deduced ultimately to modulate transmission at the level of the PnC (Boulis and Davis, 1989). Because substance P is positively coupled to cAMP in some brain areas (Mitsuhashi, Osashi, Shichijo, Christian, Sudduth-Klinger, Harrowe, and Payan, 1992), it is possible that it acts in the PnC via activation of cAMP.

Thus, there appear to be three parallel routes whereby the amygdala can modulate startle during a state of conditioned fear: (1) a direct pathway from the central nucleus of the amygdala to the PnC; (2) an indirect pathway from the central nucleus of the amygdala to the deep SC/Me to the PnC, where glutamate acting on AMPA/kainate receptors seems to be critical; and (3) an indirect pathway from the medial nucleus of the amygdala to the ventral medial hypothalamus to the periaqueductal gray to the PnC, where substance P receptors seem to be critical (Fig. 12-7).

X. ROLE OF THE AMYGDALA IN FEAR

I have just given you a detailed description for how fear modulates a simple reflex in terms of the neural circuitry involved in the reflex, using the acoustic startle reflex as an example, and the way in which the amygdala connects to the reflex pathway. Moreover, this is just one example of many showing that outputs of the central nucleus of the amygdala to the hypothalamus and brainstem are involved in many of the specific signs of fear and anxiety, as illustrated in Figure 12-8. However, this is only the "output" side of the story. One still needs to explain how the sensory stimuli, including foot shocks, activate the amygdala and how pairing sensory stimuli with foot shock can produce a "memory" that can last for a very long time.

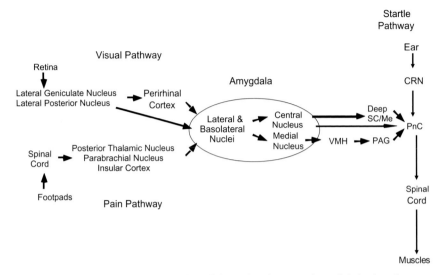

FIGURE 12-7   Schematic diagram of parallel visual pathways and parallel shock pathways to the amygdala; projections from the basolateral amygdala to the central and medial nuclei of the amygdala; and parallel outputs from the central and medial nuclei to the startle pathway. VMH — ventromedial hypothalamus, PAG — periacqueductal gray, deep SC/Me — deep white layers of the superior colliculus/deep mesencephalic reticular nucleus, CRN — cochlear root neurons, PnC — nucleus reticularis pontis caudalis.

A. Anatomy of the Amygdala — Intrinsic Connections

The amygdala complex is a complicated group of interconnected nuclei, each of which comprises several subdivisions. In the rodent the amygdala generally has been divided into the basolateral amygdala, which includes the lateral, basal, and accessory basal nuclei, and several structures surrounding the basolateral amygdala, including the central, medial, and cortical nuclei.

1. Lateral Nucleus

The lateral nucleus comprises three subdivisions: the dorsolateral, ventrolateral, and medial divisions. The dorsolateral nuclei receives both thalamic and cortical input and projects to the medial division, which also receives input from multisensory cortical processing regions, including the prefrontal and perirhinal cortex (Pitkanen, Savander, and LeDoux, 1997). Thus, the medial division of the lateral nucleus receives sensory information via thalamic and cortical inputs, some of which is processed in the dorsolateral division, and sends outputs to other parts of the amygdala, which include the accessory basal nucleus, the basal nucleus, the periamygdaloid cortex, the medial nucleus, the

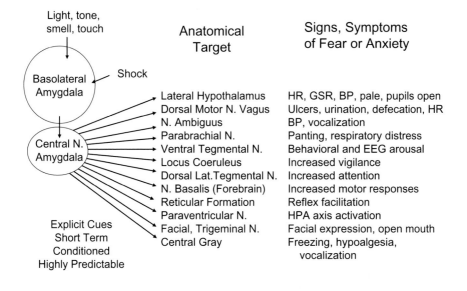

FIGURE 12-8   Schematic diagram of the outputs of the central nucleus to various hypothalamic and brainstem targets and how these targets are involved in specific signs of fear.

posterior cortical nucleus, the capsular division of the central nucleus, and the lateral division of the amygdalohippocampal area (Pitkanen, Stefanacci, Farb, Go, LeDoux, and Amaral, 1995).

2. Basal and Accesory Basal Nuclei

The basal amygdala comprises the magnocellular, intermediate, and parvicellular divisions (Savander, Go, LeDoux, and Pitkanen, 1995). The parvicellular division is the source of most of the projections within the basal nucleus, projecting to both the magnocellular and intermediate divisions. The main projections from the basal nucleus to other parts of the amygdala include the lateral olfactory tract, the anterior amygdaloid area, the medial and capsular divisions of the central nucleus, the anterior cortical nucleus, and the amygdalohippocampal area (Savander, Go, LeDoux, and Pitkanen, 1995). In addition, the magnocellular and intermediate divisions of the basal nucleus send heavy projections to homonymous regions of the amygdala on the contralateral side. The accessory basal nucleus sends projections to the medial and capsular divisions of the central nucleus, the medial division of the amygdalohippocampal area, the medial division of the lateral nucleus, the central division of the medial nucleus, and the posterior cortical nucleus (Savander, Go, Ledoux, and

Pitkanen, 1996). Thus, outputs from the lateral nucleus can affect large parts of the amygdala via relays in the basal and accessory basal nucleus.

3. Central Nucleus of the Amygdala

The central nucleus receives substantial inputs from most of the amygdaloid nuclei, including the lateral, basal, accessory basal, and anterior cortical nuclei, as well as the nucleus of the lateral olfactory tract, the periamygdaloid cortex, and the intercalated nuclei (cf. Jolkkonen and Pitkanen, 1998). In the rat, the central nucleus of the amygdala comprises four major subdivisions: the medial, lateral, lateral capsular, and the intermediate subdivisions (McDonald, 1982). Inputs from the amygdaloid and extra-amygdaloid areas terminate in various divisions of the central nucleus. The capsular division of the central nucleus receives input from the lateral, basal, and accessory basal nuclei, and the medial division gets input from the basal nucleus and accessory basal nuclei. The lateral division of the central nucleus projects to the capsular and medial divisions (Jolkkonen and Pitkanen, 1998). The intermediate division does not seem to project to any of the other divisions of the central nucleus. Interestingly, the lateral division of the central nucleus receives input from other amygdala nuclei but also gets direct input from several cortical areas (Jolkkonen and Pitkanen, 1998).

B. Inputs into the Amygdala Relevant for Fear Conditioning

The amygdala receives input from numerous areas of the brain, many of which are critical for fear conditioning.

1. Pain

Because fear conditioning almost always is produced by pairing some sensory stimulus with pain, the emotional aspect of which becomes the conditioned response, I begin by describing how pathways providing pain information get to the amygdala and the relevance of these pathways for fear conditioning (Fig. 12-7). During fear conditioning, foot-shock information is transmitted to the amygdala via parallel pathways that include the posterior intralaminar nuclei in the thalamus and the parietal insular cortex. Besides receiving acoustic inputs from the inferior colliculus, the posterior intralaminar nuclei also receive somatic pain inputs from the spinal cord and in turn project to the amygdala, particularly the lateral amygdaloid nucleus (cf. Shi and Davis, 1999). Electrical stimulation of this area is an effective unconditioned stimulus for fear conditioning, similar to foot shock (Cruikshank, Edeline, and Weinberger, 1992). Thus, this thalamo-amygdaloid pathway may serve as the unconditioned stimulus pathway during emotional learning. However, pretraining lesions of the posterior intralaminar nuclei alone did not prevent the acquisition of fear conditioning (Campeau and Davis, 1995b; Romanski and LeDoux, 1992b), indicating that additional pathway(s) must contribute foot-shock information to the amygdala.

The caudal part of insular cortex, the so called *parietal insula*, receives convergent inputs from somatosensory cortices, ventroposterior and posterior thalamic nuclei, posterior intralaminar nuclei, and the midbrain parabrachial nucleus (cf. Shi and Davis, 1999). Further, this portion of the insular cortex is probably a primary source in providing cortical somatosensory information to the amygdala. Both the parietal insular cortex and posterior intralaminar nuclei of thalamus in turn project to the lateral, basolateral, basomedial, and central nuclei of the amygdala.

Consistent with this, combined lesions of both parietal insular cortex and posterior intralaminar nuclei of the thalamus were necessary to interrupt the transmission of foot-shock information to the amygdala and thus block the acquisition of fear-potentiated startle (Shi and Davis, 1999). Importantly, however, these lesions did not block the expression of fear-potentiated startle once conditioning had taken place, as one would expect if these pathways were involved in fear acquisition. These combined lesions also reduced the degree to which rats reacted to footshock, which we believe is modulated by the amygdala. Thus, even though the major immediate reactions of a rat to foot shock involve brainstem and spinal cord reflex circuits, these reflexes are most probably modulated by the amygdala, just as the startle reflex is, which also is mediated by a brainstem and spinal cord reflex circuit. Hence lesions that interrupt shock inputs to the amygdala would be expected to influence shock reactivity. This may explain why chemical lesions in some of these thalamic nuclei failed to block fear conditioning, because they also failed to alter shock reactivity (Brunzell and Kim, 2001). We believe the difference was due not to the use of chemical lesions, but, rather, to incomplete lesions, because we have found that complete chemically induced lesions of the posterior intralaminar nuclei block acquisition of fear-potentiated startle and also reduce shock reactivity (Shi and Davis, unpublished observations).

2. Hearing

A great deal of work has been done using auditory cues to study the role of the amygdala in fear conditioning, as exemplified by the elegant work in Dr. Joseph LeDoux's laboratory. Auditory inputs from modality-specific areas of thalamus and cortex exclusively or primarily target the dorsolateral and ventrolateral divisions of the lateral amygdaloid nucleus (cf. Romanski, Clugnet, Bordi, and LeDoux, 1993), and single-unit recording studies find that cells in the dorsolateral division fire with the shortest latencies (12–25 msec) to an auditory stimulus (Bordi and LeDoux, 1992). This division also receives input from somatosensory areas activated by foot shock (Romanski, Clugnet, Bordi, and LeDoux, 1993), and the firing rate of cells activated at short latencies (mean ~25 msec) is increased when these tones are paired with foot shocks (Quirk, Repa, and LeDoux, 1995). Both electrolytic and excitotoxic posttraining lesions of the lateral nucleus of the amygdala, sparing a large number of basolateral neurons, disrupted fear-potentiated startle to both auditory and visual conditioned stimuli (Campeau and Davis, 1995a), consistent with earlier work using auditory cues, pretraining lesions, and freezing as the measure of fear (LeDoux, Cicchetti, Xagoraris, and Romanski, 1990). All auditory inputs to the lateral nucleus of the amygdala ultimately arise from the auditory thalamus, and complete electrolytic or excitotoxic lesions of the entire auditory thalamus specifically disrupted fear-potentiated startle to an auditory but not a visual CS, whether the lesions were made before or after conditioning (Campeau and Davis, 1995b), consistent with earlier work using pretraining lesions and freezing (LeDoux, Sakaguchi, Iwata, and Reis, 1986; LeDoux, Sakaguchi, and Reis, 1984).

Although it has been argued that the direct projection from the thalamus to the amygdala is critical for conditioned fear to an auditory stimulus (Romanski and LeDoux, 1992b), this conclusion is based on results where lesions of a given pathway are made prior to fear conditioning. When lesions are made after fear conditioning, we find that the subcortical pathway probably is not normally used but instead can take over if the thalamo-cortical pathway is disrupted. The cortical pathway I am referring to is not the primary auditory cortex but instead a secondary multisensory cortex called the *perirhinal cortex*. These conclusions are based on the following observations.

Lesions of the ventral and dorsal divisions of the medial geniculate body, giving rise to the main thalamo-cortico-amygdala pathway, significantly disrupted fear-potentiated startle to an auditory but not a visual CS (Campeau and Davis, 1995b). In contrast, animals with posterior thalamic lesions, which project directly to the lateral nucleus of the amygdala, actually had higher levels of fear-potentiated startle, especially to the auditory CS. However, the subcortical pathway can be recruited to mediate fear-potentiated startle to an auditory conditioned stimulus when animals sustaining ventral and dorsal medial geniculate body lesions were retrained. This could explain why lesions of the cortical pathway made prior to fear conditioning, such as those done in the LeDoux lab, would not disrupt conditioned fear, because under these circumstances the subcortical pathway would take over.

Neither pre- nor posttraining auditory cortex ablations, mostly restricted to the primary auditory area, had a reliable effects on fear-potentiated startle (Campeau and Davis, 1995b). In contrast, posttraining lesions to the secondary auditory and perirhinal cortices completely blocked fear-potentiated startle to both auditory and visual CSs, but, importantly, pretraining lesions did not reliably affect fear-potentiated startle to either CS. The posttraining deficits were observed only after the lesions included most of the rostral-caudal extent of the perirhinal area, which also receives visual input. The differences observed between pre- and posttraining lesions did not arise from differences in lesion size, because the extent of the lesions in each study was similar. These results are consistent with the findings of LeDoux's lab (Romanski and LeDoux, 1992a, 1992b) showing that pretraining perirhinal area lesions do not reliably disrupt conditioned fear response to an auditory stimulus as well as with those of our lab (Rosen, Hitchcock, Miserendino, Falls, Campeau, and Davis, 1992) showing that posttraining lesions disrupt fear-potentiated startle to a visual stimulus. Thus, by using posttraining lesions we conclude that the pathway going from the auditory thalamus to the perirhinal cortex to the lateral nucleus of the amygdala is normally used in fear-potentiated startle when an auditory CS is used. The difference between this conclusion and earlier ones reflects the use of pre- versus posttraining lesions and has been confirmed by later work in Ledoux's lab using posttraining lesions of the perirhinal cortex (Corodimas and LeDoux, 1995).

3. Vision

Like the auditory system, the use of posttraining lesions has led us to conclude that subcortical projections from the visual thalamus (e.g., the lateral posterior nucleus) to the amygdala are not normally used in fear-potentiated startle using a visual CS. Instead, we believe that projections from the lateral posterior nucleus of the thalamus to the perirhinal cortex and then into the amygdala are the ones normally used. This is based on the following evidence. Lesions or chemical inactivation of superficial layers of superior colliculus, which receive massive retinal input, do not disrupt the expression of fear-potentiated startle to a visual CS (Meloni and Davis, 1999; Tischler and Davis, 1983). The lateral posterior nucleus of the thalamus also receives direct projections from the retina, and anterograde anatomical tract–tracing studies in our laboratory (Shi and Davis, 2001) showed it sends heavy projections to area TE2 and the dorsal perirhinal cortex and moderate projections to the lateral amygdaloid nucleus. However, posttraining lesions restricted to the lateral posterior nucleus did not block the expression of fear conditioning using a visual CS. Posttraining lesions of neither the dorsal lateral geniculate nucleus (Shi and Davis, 2001), which receives retinal inputs, nor the visual cortex, including V1 and V2, prevented the expression of conditioned fear responses using a visual CS (Falls and Davis, 1994; LeDoux, Romanski, and Xagoraris, 1989; Rosen, Hitchcock, Miserendino, Falls, Campeau, and Davis, 1992; Tischler and Davis, 1983).

However, both Te2 and the perirhinal cortex receive visual inputs from lateral posterior nucleus (Shi and Davis, 2001) and visual cortices and in turn project to the amygdala, and combined lesions of both dorsal lateral geniculate nucleus and lateral posterior nucleus, which would cut off both thalamic and cortical routes to Te2 and perirhinal cortex, totally blocked the expression of conditioned fear using a visual CS. Local infusion of the glutamate antagonist NBQX had the same effect, suggesting that the lesion effect did not result from damage to fibers of passage. Based on these and other results, we conclude that visual input carried by projections from the dorsal lateral geniculate nucleus and lateral posterior nucleus via connections through Te2 and perirhinal cortex to the amygdala normally are involved in conditioned fear using a visual CS.

4. Smell

Olfactory stimuli are unique because olfactory receptors in the nose send axons to the olfactory bulb, which then projects directly to the corticomedial nucleus of the amygdala (i.e., one synapse between the receptor and the amygdala). The olfactory bulb also projects to the piriform cortex, which then projects to the basolateral nucleus of the amygdala. As mentioned earlier, inactivation of the medial nucleus of the amygdala blocks the expression of fear-potentiated startle using either an olfactory or a visual CS. However, as is discussed shortly, NMDA antagonists infused into the basolateral nucleus of the amygdala block the acquisition of conditioned fear, and this is also true for fear-potentiated startle using olfactory cues as conditioned stimuli (Walker, Paschall, and Davis, 2005). However, infusion of NMDA antagonists into the medial nucleus of the amygdala does not block the acquisition of fear-potentiated startle, suggesting that it is more on the output than on the input side in terms of conditioned fear using olfactory cues. Thus, the more indirect pathway from the olfactory bulb to the piriform cortex to the basolateral nucleus of the amygdala is probably the route necessary for conditioned fear, and both pre- and posttraining lesions of the basolateral nucleus of the amygdala blocked fear conditioning to olfactory cues. Finally, like fear conditioning to visual or auditory cues (see later), the perirhinal cortex, to which the piriform cortex projects, may also be important for fear conditioning using olfactory cues (Herzog and Otto, 1997; Schettino and Otto, 2001), although we have not looked at its role or the role of piriform cortex in olfactory mediated fear potentiated startle.

C. Plasticity in the Amygdala During Fear-Potentiated Startle Training

I have now described how the amygdala receives sensory input from several modalities along with shock input known to be critical for fear conditioning. I have also used the fear-potentiated startle effect to illustrate the role of these various sensory pathways in the acquisition and expression of fear-potentiated startle and how parallel outputs from the central and medial nuclei of the amygdala connect to the startle pathway, which, in one way or another, increases acoustic startle amplitude at the level of the PnC in the presence of a conditioned fear stimulus. However, it remains to be determined how sensory stimuli paired with foot shock endow those formerly neutral stimuli with the ability now to potentiate startle for a very long time.

D. Role of Glutamate Receptors in the Amygdala in Fear-Potentiated Startle

Several studies have shown that high-frequency stimulation of amygdala afferents can result in a long-term potentiation (LTP) of neurotransmission at amygdala synapses. The mechanisms that underlie LTP may be similar to those engaged by fear conditioning (e.g., Rogan, Staubli, and LeDoux, 1997). Because the induction but not the expression of LTP most often involves NMDA receptors, we wondered whether amygdala NMDA receptors might also play a special role in fear learning. In fact, we found that local infusion into the basolateral nucleus of the amygdala, which has the highest density of NMDA receptors in the amygdala, of the NMDA antagonist AP5 blocked the acquisition but not the expression of fear-potentiated startle using either a visual (Miserendino, Sananes, Melia, and Davis, 1990), auditory (Campeau, Miserendino, and Davis, 1992), or olfactory (Walker, Paschall, and Davis, 2005) cues as conditioned fear stimuli. Importantly, the same doses did not disrupt the ability of conditioned fear stimuli to potentiate startle when infused prior to testing. Because the amygdala is essential for the expression of fear-potentiated startle (Campeau and Davis, 1995a; Hitchcock and Davis, 1987; Kim, Campeau, Falls, and Davis, 1993; Sananes and Davis, 1992; Walker and Davis, 1997b), these findings indicate that the effects of NMDA receptor blockade on fear learning cannot be attributed to a general disruption of amygdala activity or to a more specific disruption of the ability of rats to process the CS. These findings also indicate that the effects on learning cannot be attributed to anxiolytic influences, insofar as such influences should also disrupt fear-potentiated startle when NMDA receptor antagonists are infused prior to testing.

Although the inability of pretest infusions to disrupt fear-potentiated startle indicates that the effects of pretraining infusions cannot be attributed to a failure to process the CS, it could still be argued that AP5-induced learning impairments are attributable to a disruption of processing of the foot shock. Although these infusions did not alter the degree to which rats reacted to foot shock, as argued earlier, this may simply be a measure of a brainstem, spinal cord reflex, not directly related to fear conditioning. However, Gewirtz and Davis (1997) reported that intra-amygdala AP5 infusions blocked second-order fear conditioning — a procedure in which a previously trained CS substitutes for shock as the aversive reinforcing stimulus. In this study, rats received pairings of an auditory stimulus (i.e., first-order stimulus) and foot shock. On other days, the same rats were given second-order conditioning trials, in which a light (i.e., the second-order stimulus) was paired, not with shock, but with the fear-eliciting first-order auditory stimulus. Prior to these second-order conditioning trials, rats received intra-amygdala infusions of either artificial cerebrospinal fluid (ACSF) or D,L-AP5. When subsequently tested, both groups showed fear-potentiated startle to the auditory stimulus. However, rats that had received AP5 did not show fear-potentiated startle to the light. Because AP5 was only given prior to light–tone pairings, the ability of AP5 to block fear learning could not be attributed to analgesic actions or to a disruption of neural transmission in pathways that convey foot-shock information to the amygdala. Furthermore, in the same rats where AP5 blocked second-order fear conditioning using the noise as the reinforcement, AP5 did not disrupt fear-potentiated startle to the first-order auditory CS. These data strongly suggest that AP5 disrupted the acquisition of fear by preventing the association between light and noise, rather than by preventing amygdala activation by the noise stimulus that was used as the reinforcement in second-order conditioning.

More recently it has been found that local infusion of AP5 antagonists into the amygdala block the expression of several other measures of conditioned fear, including freezing. However, this appears to be due to actions of AP5 on a particular subtype of the NMDA receptor, the NR2A subtype, because infusion of ifenprodil, another NMDA antagonist, which acts at the NR2B and not the NR2A subtype, blocked acquisition of conditioned freezing without having any effect on its expression (Rodrigues, Schafe, and LeDoux, 2001).

E. Involvement of AMPA Receptors in the Basolateral and Central Nucleus of the Amygdala in Fear Learning

Because the basolateral complex is a primary site of sensory convergence within the amygdala (see earlier), we wondered whether this subdivision might play a more prominent role in fear acquisition as compared to the central nucleus of the amygdala. To our surprise, pretraining infusions of NBQX into either area significantly disrupted fear learning, suggesting that both areas play a role in conditioning (Walker and Davis, 2000). Although it is difficult to rule out completely the possibility that infusions into the central nucleus disrupted fear learning by diffusing to the basolateral amygdala, we believe this is unlikely. In an earlier study using the same dose (3 µg/side), infusion volume (0.3 µl), infusion rate (0.1 µl/min), and stereotaxic coordinates, we were able to demonstrate differential effects of infusions into the basolateral versus central nucleus infusions on light-enhanced startle (Walker and Davis, 1997b) — an anxiety paradigm in which sustained exposure to bright light elevates startle amplitude (Walker and Davis, 1997a). In that experiment, NBQX infusions into the basolateral amygdala but not the central nucleus of the amygdala disrupted light-enhanced startle. Also, basolateral but not central nucleus AP5 infusions were able to disrupt fear learning (Fanselow, Kim, Yipp, and De Oca, 1994), presumably because NMDA receptors are more highly concentrated within the basolateral as compared to the central nucleus of the amygdala (Monaghan and Cotman, 1985). Our results and those of others (cf. Samson and Pare, 2005) are consistent with idea that both areas participate in fear learning and recent evidence that long-term potentiation can occur in the central nucleus of the amygdala (Samson and Pare, 2005).

XI. INTRACELLULAR EVENTS INVOLVED IN FEAR-POTENTIATED STARTLE

A. Broad-Based Survey of Gene Changes in the Amygdala Following Fear Conditioning

Many years ago we found an increase in the immediate early gene, *c-fos*, in the amygdala in the presence of a conditioned fear stimulus (Campeau, Hayward, Hope, Rosen, Nestler, and Davis, 1991), an effect that has been replicated many times, looking at both *c-fos* message and c-Fos protein. Interestingly, *c-fos* most consistently is seen in the medial nucleus of the amygdala during fear or stress (cf. Campeau, Falls, Cullinan, Helmreich, Davis, and Watson, 1997), and this may explain why inactivation of this part of the amygdala is so effective in blocking the expression of fear-potentiated startle. Recently, we began a much wider-based, *in situ* hybridization analysis of gene expression in the brain associated with the acquisition of fear-potentiated startle. We examined 21 genes known to be involved in neural plasticity based on their induction with kainic acid–induced seizures (Ressler, Paschall, Zhao, and Davis, 2002). We found a substantial number of these genes were transcriptionally regulated during consolidation of fear conditioning in the amygdala as well as in several other brain areas (Table 12-2). These mRNA changes occurred only when the conditioned and unconditioned stimuli were paired and not when unpaired or when the unconditioned stimulus was presented alone. These results suggest fear memory consolidation occurs within a broad neural circuit *that includes, but is not limited to, the amygdala.* It is associated with early and late changes in gene expression of a variety of transcription factors, cytoskeletal proteins, adhesion molecules, and receptor stabilization molecules, which together may contribute to the neural plasticity underlying long-term memory in mammals.

TABLE 12-2 Gene Changes in the Amygdala and Extra-Amygdala Areas After Fear Conditioning

	Gene	Gene Product	KA-Inducible Areas	Changes in Fear Conditioning	Peak Time of Change
1	*fos*	Transcription factor	hipp, amyg, striatum, ctx	+++	0–60 min
2	zif268/EGR1	Transcription factor	hipp, amyg, striatum, ctx	+++	0–60 min
3	jun	Transcription factor	hipp, amyg, ctx	++	0–60 min
4	Neurofilament-L	Cytoskeletal protein	dentate, amyg, pir ctx	+++	0–60 min
5	Gephyrin	GlyR/GABAR Anchor	*Decrease*-hipp, amyg, pir ctx	++	30–120 min
6	RC3/Neurogranin	2nd msgr modulation	*Decrease*-hipp, amyg, pir ctx	+++	30–240 min
7	Nurr1	Transcription factor	hipp, amyg, pir ctx	+++	1–2 hr
8	16C8	Protease inhibitor	dentate	++	2–4 hr
9	α-actinin	NMDAR/GluR anchor	hipp	++	2–4 hr
10	n-cadherin	ECM/cell adhesion	hipp	+	2–4 hr
11	Ier5/RM5	Transcription factor	dentate, amyg	+/-	1–4 hr
12	Tenascin	ECM/cell adhesion	cortex	+/-	1–4 hr
13	VGF	Neuropeptide	hipp	+/-	1–4 hr
14	EGR2/Krox20	Transcription factor	dentate	-	
15	CREM	Transcription factor	hipp	-	
16	EGR4	Transcription factor	hipp, pir ctx	-	
17	AARP-21	Signal transduction	dentate, pir ctx	-	
18	Rheb2	Signal transduction	dentate, pir ctx	-	
19	PAI-2	Phosphatase inhibitor	dentate	-	
20	ab-crystallin	Chaperone protein	dentate, ctx	-	
21	CRF/CRH	Neuropeptide	amyg, hypoth	-	

1. Gephyrin

Although several genes were up-regulated in the amygdala following fear conditioning, mRNA that codes for a protein called gephyrin, involved in the clustering of GABA and glycine receptors, was down-regulated (Ressler, Paschall, Zhao, and Davis, 2002). We also found this to be true when we measured the protein, as well as seeing a decrease in the surface expression of GABA-A receptors in the basolateral amygdala after fear conditioning, as evidenced by decreased binding of H3-flunitrazepam (Chhatwal, Myers, Ressler, and Davis, 2005). Because a decrease in the level of this clustering protein would be expected to decrease GABA transmission, this suggests that fear conditioning leads to a period of increased excitability in the amygdala for several hours. This is interesting because it is not possible to establish long-term potentiation in amygdala brain slices unless GABA antagonists are added. Although this seems unphysiological, these results with gephyrin suggest that fear conditioning down-regulates GABA-A in the amygdala, perhaps to allow long-term potentiation to take place, which may be important for consolidation of long term memory.

2. Brain-Derived Neurotropic Factor (BDNF)

Brain-derived neurotrophic factor (BDNF) is a member of the neurotrophins, a family of structurally related proteins first described for their role in promoting neuronal survival and differentiation during development. Since its original discovery, a large number of studies have demonstrated that BDNF plays a diverse role in regulating neuronal structure and function in both the developing and adult central nervous system, as well as long-term potentiation and learning and memory in hippocampally dependent tasks (cf. Rattiner, Davis, and Ressler, 2005). We found that BDNF mRNA was elevated in the basolateral amygdala 2 hours following fear conditioning (Rattiner, Davis, French, and Ressler, 2004). Furthermore, levels of TrkB receptor immunostaining appeared to decline in the amygdala, while levels of phosphorylated Trk receptors increased following fear conditioning, suggesting activation and modification of the receptor triggered by BDNF binding. Dominant-negative inhibition of TrkB within the amygdala impaired fear-potentiated startle without disrupting baseline amygdala function. These results strongly suggest a requirement for TrkB signaling in the acquisition and consolidation of fear memory.

Until recently, mechanisms of neural plasticity within the amygdala have focused primarily on NMDA receptor–mediated events (Rodrigues, Schafe, and LeDoux, 2001; Walker and Davis, 2000). There is now increasing evidence that BDNF/TrkB-dependent mechanisms of neural plasticity act in parallel with NMDA-dependent mechanisms in many neuronal cell types. Recent work has shown that PI3 kinase (PI3K) is a critical intracellular mediator required for synaptic plasticity during fear conditioning (Lin, Yeh, Lu, Leu, Chang, and Gean, 2001). It is quite likely that activation of TrkB is a critical step in PI3K-mediated signaling within the amygdala because TrkB has been shown to be a powerful regulator of this kinase. In our studies we have found significant but not complete blockade of fear learning following disruption of TrkB signaling, suggesting that other intracellular cascades likely act in parallel with those activated by BDNF to mediate long-term fear conditioning. Future studies will examine the differential roles of NMDA- versus TrkB-dependent plasticity events within the amygdala as well as the interactions between the different intracellular pathways activated by these receptors.

3. c-AMP Response Element Binding (CREB)

Local infusion of MAP kinase inhibitors into the amygdala blocked acquisition but not expression of conditioned fear using freezing to a tone (Schafe, Atkins, Swank, Bauer, Sweatt, and LeDoux, 2000; Schafe, Nadel, Sullivan, Harris, and LeDoux, 1999), and we found similar results with fear-potentiated startle (Lu, Shi, Tang, and Davis, 2000). Because each of these kinases can phosphorylate the transcription factor, c-AMP response element binding protein (CREB) and because CREB is both downstream and upstream from BDNF, we wondered whether we could establish a role for CREB in the amygdala in fear conditioning. After finding too much toxicity with local infusion of CREB antisense, we used viral vector gene transfer to up-regulate CREB to see if it would facilitate fear conditioning using suboptimal parameters (massed, as opposed to spaced, training trials). We found a dramatic increase in the magnitude of fear-potentiated startle associated with increased CREB protein during training, but not testing, in the basolateral amygdala following local infusion of herpes simplex virus (HSV) CREB in the amygdala (Josselyn, Shi, Carlezon, Jr., Neve, Nestler, and Davis, 2001). This only occurred with massed and not spaced training, consistent with earlier work in *Drosophila* (Yin, Del Vecchio, Zhou, and Tully, 1995) and CREB mutant mice (Kogan, Frankland, Blendy, Coblentz, Marowitz, Schutz, and Silva, 1997). There was no effect when lights and shocks were not paired or when spaced training was given with weak versus strong shocks or on shock reactivity during training and no effect on short-term memory. These data represented the first instance of a gain of function in a mammal following an increase in CREB, localized to a specific brain region, and is considered by some to be the best single piece of evidence implicating CREB in mammalian memory and a role for CREB in the amygdala in fear conditioning (Nguyen, 2001). We have now replicated this using a totally different paradigm, namely social defeat in hamsters (Jasnow, Shi, Israel, Davis, and Huhman, 2005), which is an amygdala-dependent form of long-term fear conditioning in this species.

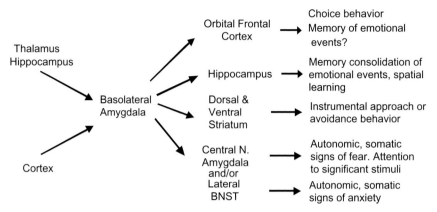

FIGURE 12-9   Schematic diagram of the outputs of the basolateral nucleus of the amygdala to various target structures and possible functions of these connections.

B. Role of the Bed Nucleus of the Stria Terminalis in Anxiety

The BLA projects to a variety of brain areas that are involved in fear and anxiety (Fig. 12-9). Two structures are of particular interest — the central nuclei of the amygdala (CeA) and the bed nucleus of the stria terminalis (BNST). As we have seen, the CeA is critical for the expression and probably the acquisition of conditioned fear. Recall, however, that lesions of the BNST did not block fear-potentiated startle or conditioned freezing. However, the lateral BNST and CeA are anatomically, neurochemically, cytoarchitectonically, and embryologically related (c.f., Alheid, de Olmos, and Beltramino, 1995) and that the BNST has the same downstream projections as the CeA (Fig. 12-10). Hence, we wondered how it might be involved in fear and anxiety.

1. CRH-Enhanced Startle

Infusions of corticotrophin-releasing hormone (CRH) into the lateral cerebral ventricle markedly increase the amplitude of the acoustic startle response in rats, and this effect was blocked by the anxiolytic chlordiazepoxide (Swerdlow, Geyer, Vale, and Koob, 1986). CRH-enhanced startle did not occur with intrathecal infusion and was not disrupted by lesions of the paraventricular nucleus of the hypothalamus (Liang, Melia, Campeau, Falls, Miserendino, and Davis, 1992), indicating mediation by CRH receptors in the brain that did not involve activation of the hypothalamic–pituitary–adrenal axis. Lee and Davis (1997a, 1997b) found that excitotoxic lesions of the BNST but not of the septum, hippocampus, BLA, or CeA completely blocked CRH-enhanced

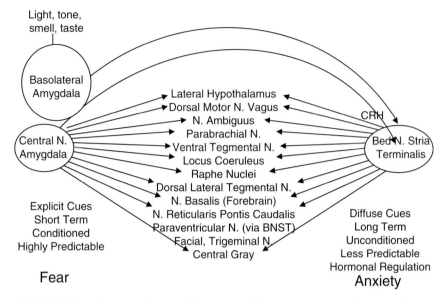

FIGURE 12-10   Schematic diagram of the outputs of the central nucleus or the lateral division of the bed nucleus of the stria terminalis (BNST) to various target structures and possible functions of these connections.

startle, as did intra-BNST infusions of the CRH antagonist αhCRH. Infusions of CRH directly into the BNST increased startle amplitude at doses much lower than those that were required with i.c.v. administration (80 versus 1,000 ng). Neither BNST lesions nor intra-BNST αhCRH infusions disrupted fear-potentiated startle. Moreover, local infusion of CRH into the CeA failed to increase startle amplitude (Liang, Melia, Campeau, Falls, Miserendino, and Davis, 1992), and infusion of αhCRH there failed to block CRH-enhanced startle.

3. Light-Enhanced Startle

Walker and Davis (1997a) described a new animal model of anxiety termed *light-enhanced startle*, in which startle amplitude is increased when rats are exposed to bright light for 20 minutes. Like CRH-enhanced startle, light-enhanced startle was dependent on the BNST and not on the CeA (Walker and Davis, 1997b). Thus, local infusion of the AMPA receptor antagonist NBQX into the BNST but not the CeA blocked light-enhanced startle. The amygdala was still involved, however, because local infusion of NBQX into the BLA did block light-enhanced startle, probably because it is visual information transmitted to the BNST through the BLA. Thus, as with the CRH

experiments described earlier, these experiments demonstrated a double dissociation between the roles of the BNST and the CeA in startle increases
produced by fear-inducing or anxiogenic stimuli.

4. Long-Term Sensitization of the Acoustic Startle Response by Repeated Foot Shock

To examine whether the inability of BNST lesions to block fear-potentiated
startle was related to the strength of conditioning, we used a procedure in
which acquisition of fear-potentiated startle can be measured by giving a few
training and test trials each day (Gewirtz, McNish, and Davis, 1998; Kim and
Davis, 1993). Even at early time points, when fear-potentiated startle was relatively weak, sham- and BNST-lesioned rats showed comparable levels of fear-
potentiated startle. Unexpectedly, BNST lesions did influence one aspect of
performance. In shocked but not nonshocked control rats, baseline startle
amplitude (i.e., startle amplitude to the 10 noise bursts delivered at the beginning of each test session) grew steadily over the course of training. The increase
did not appear to reflect contextual fear conditioning, but seemed instead to
reflect a long-term sensitization to startle stimuli produced by repeated foot-
shock administration and was absent in BNST-lesioned rats.

XII. WHAT DOES THE BNST DO? A PROVISIONAL HYPOTHESIS BASED ON RESULTS FROM FEAR-CONDITIONING AND ACOUSTIC STARTLE STUDIES

Fear-potentiated startle to a specific cue is a highly predictable situation that
involves prior conditioning and uses a rather short cue that reliably predicts an
aversive event. Fear-potentiated startle develops very rapidly, once the light
comes on, and dissipates very quickly, once the light goes off (Davis, Schlesinger,
and Sorenson, 1989). In contrast, light-enhanced startle is a situation where
the animal is exposed to a potentially dangerous situation that is less predictable, does not depend on any obvious conditioning, and involves a long period
of anticipation that something bad might happen. To try to explain why
manipulations of the CeA affect fear-potentiated startle and not light-enhanced
startle and why manipulations of the BNST affect light-enhanced startle and
not fear-potentiated startle, we suggested two alternatives. One hypothesis was
that the CeA mediates conditioned fear responses, whereas the BNST mediates
unconditioned fear responses. Support for this idea came from the finding that
startle that was increased in the presence of the smell of fox feces, presumably
an unconditioned fear stimulus, was blocked by inactivation of the BNST but
not the amygdala (Fendt, Endres, and Apfelbach, 2003). More recently,
however, it was reported that post-training lesions of the BNST blocked the

expression of context conditioning measured with freezing (Sullivan, Apergis, Bush, Johnson, Hou, and Ledoux, 2004), which clearly is a conditioned response to a context previously paired with shock, a result not consistent with the idea that the BNST is only involved in unconditioned fear. Our second hypothesis was that maybe the CeA mediates fear reactions activated by relatively short stimuli in highly predictable situations, whereas the BNST mediates fear responses to relatively long cues under conditions where the perceived danger is not highly predictable and requires a sustained state of defensive preparedness. We now believe this second alternative is the right conclusion.

In the light-enhanced-startle paradigm we found the light had to be on for a least 5 min to see maximal light-enhanced startle. At shorter intervals (i.e., 60 sec) the excitatory effect was weak, and at very brief intervals (i.e., 3.2 sec) the effect of light is often inhibitory (Davis, Schlesinger, and Sorenson, 1989). When the light is turned off, after a 20-min on-time, startle does not abruptly return to baseline but remains elevated for sometime thereafter (Walker and Davis, unpublished observations; de Jongh, Groenink, van der Gugten, and Olivier, 2002). Thus, light-enhanced startle requires a long-duration stimulus, and the effect of this stimulus far outlasts the period when the light is actually on. It also is an inherently unpredictable situation where the animal may feel "at risk" without knowing exactly when something might happen and how bad it might be. In fact, in humans we find that startle is increased in the dark (Grillon, Pellowski, Merikangas, and Davis, 1997) and that this effect is much larger in patients with posttraumatic stress disorder (Grillon, Morgan, Davis, and Southwick, 1998). When we asked these patients how they felt when the light went out, they often reported they felt like they were back in their bunker, anticipating a mortar attack but not knowing when this would happen.

CRH-enhanced startle may be similarly characterized. CRH-enhanced startle appears to be a slow-onset (20 min) and slow-offset effect (several hours), at least with i.c.v. administration (Lee and Davis, 1997a; Liang, Melia, Miserendino, Falls, Campeau, and Davis, 1992). It is not clear whether this protracted time course and slow decay of CRH-enhanced startle reflects response characteristics of the BNST itself, the time required for CRH to occupy and then dissociate from CRH receptors, or emergent properties of the neural circuitry within which the BNST is embedded. For example, Koob (1999) suggested that CRH-responsive neurons in the BNST and elsewhere, once activated by emotional stressors, excite brainstem noradrenergic nuclei, which then feed back to CRH-responsive neurons to stimulate further CRH release.

The effect on startle of repeated foot shock fits the pattern also. In Gewirtz, McNish, and Davis (1998), the effect developed gradually over many days and persisted for at least 24 hours (i.e., the interval between the final shock on the preceding training day and the baseline test on the following day).

TABLE 12-3   Role of the BNST in Anxiety and Stress

- Cue-induced drug craving
- Opiate withdrawal
- Social defeat
- "Learned helplessness"
- Context conditioning
- Predator odors

Overall, then, the data presently available argue for the existence of two phenomenologically and anatomically dissociable response systems, each capable of mediating increases in the amplitude of the acoustic startle response (Fig. 12-10). One, which includes the CeA as an integral component, can be characterized as a rapid response system that mediates short-term responses to specific threat cues (i.e., stimulus-specific fear responses). The other, which includes as an integral component the BNST, can be characterized as a sluggish response system that, once activated, continues to influence behavior long after the initiating stimulus has been terminated. We refer to the first, a stimulus-specific, short-lasting type of response, as "fear" and the second, a more sustained type of response, as "anxiety." Moreover, these two different systems show perfect additivity, consistent with independent, parallel systems that elevate startle (Walker and Davis, 2002). Finally, many other laboratories are finding that the BNST plays a more general role in stress, depression, and anxiety, using many different experimental paradigms, including drug craving and withdrawal (Table 12-3, cf. Walker, Toufexis, and Davis, 2003).

XIII. EXTINCTION OF FEAR-POTENTIATED STARTLE

If, following fear-potentiated startle to a visual stimulus, the light is presented over and over again without shock, there will be a significant decrease in the magnitude of fear-potentiated startle as a direct function of the number of presentations of the light in the absence of shock (Walker, Ressler, Lu, and Davis, 2002). This procedure is known as *extinction training*, and the theoretical process that accounts for this decrease in conditioned fear is known as *extinction*. Behavioral observations indicate that extinction is a form of learning in its own right, rather than an "unlearning" or forgetting of previous learning (for a review see Myers and Davis, 2002). We found that local infusion in the BLA of the NMDA antagonist AP5 completely blocked the development of extinction when animals were tested the next day drug free (Falls, Miserendino, and Davis, 1992). This impairment could not be attributed to an effect on NMDA receptors outside the amygdala, to damage to the amygdala, or to an impairment of sensory transmission during extinction training, and it has been confirmed in several laboratories. Blocking NMDA receptors after extinction training also blocks extinction, suggesting that NMDA receptors are important for the consolidation of extinction (Santini, Muller, and Quirk, 2001).

In light of these findings, the question arose as to whether it would be possible to enhance extinction by enhancing the functioning of the NMDA receptor. It is known that a compound called D-cycloserine (DCS) binds to the NMDA receptor and makes it work better. Thus, we predicted that giving DCS prior to extinction training would enhance extinction. D-Cycloserine given either systemically or directly into the amygdala prior to extinction training dose-dependently enhanced extinction in rats exposed to lights in the absence of shock but not in control rats that did not receive extinction training when testing occurred 24 hr later in the absence of the drug (Walker, Ressler, Lu, and Davis, 2002), an effect now replicated with freezing to a tone (Ledgerwood, Richardson, and Cranney, 2003). This group also found that DCS could still facilitate extinction when given up to about 3 hr after extinction training, a finding consistent with the idea that DCS facilitates consolidation of extinction.

XIV. FROM BENCH TO BEDSIDE

Because treatments for PTSD and other anxiety disorders typically involve a process similar to extinction, we tested whether DCS would enhance exposure-based psychotherapy in people suffering from an inordinate fear of heights in a double-blind placebo controlled study. The exposure therapy used a virtual-reality situation developed by Barbara Rothbaum and colleagues in which patients rode in a virtual glass elevator to progressively higher floors (Ressler, Rothbaum, Tannenbaum, Anderson, Graap, Zimand, Hodges, and Davis, 2004). This situation is very frightening to patients just entering treatment, but it becomes considerably more tolerable with increasing exposure to the virtual environment, typically over six to eight sessions. Thirty patients were rated for their initial fear of heights and divided into three groups that had comparable levels of fear as well as being similar on other variables, such as age and sex, and then they received only two exposure sessions, purposely suboptimal, to detect improvement. Single doses of placebo or D-cycloserine (50 or 500 mg) were taken 2 hrs prior to each of the two sessions of virtual-reality exposure therapy. Exposure therapy combined with D-cycloserine resulted in significantly larger reductions of acrophobia symptoms on all main outcome measures than the same amount of exposure in combination with placebo. Compared to subjects receiving the placebo, subjects receiving DCS had significantly more improvement within the virtual environment both 1 week and 3 months after treatment. They also showed significantly greater decreases in posttreatment skin conductance fluctuations and greater improvement on general measures of real-world acrophobia symptoms and number of self-exposures to real-world heights. Because of these promising results, DCS is now being tested in combination with psychotherapy all over the world for all the major anxiety disorders.

XV. SUMMARY

By studying how a simple reflex can be augmented when elicited in the presence of a fearful stimulus (fear-potentiated startle), we and others have been able to delineate a good deal about the neural circuitry involved in conditioned fear. This involves visual or auditory as well as shock pathways that project via the thalamus and perirhinal or insular cortex to the basolateral nucleus of the amygdala (Bla). The Bla projects to the central (CeA) and medial (MeA) nuclei of the amygdala, which project indirectly to a particular part of the acoustic startle pathway in the brainstem. NMDA and BDNF receptors, as well as various intracellular cascades, in the amygdala are critical for fear learning, which is then mediated by glutamate acting in the CeA and perhaps substance P acting in the MeA. Less predictable stimuli, such as a long-duration bright light and a fearful context, activate the Bla, which projects to the bed nucleus of the stria terminalis (BNST), which projects to the startle pathway, much as the CeA does. The anxiogenic peptide corticotrophin-releasing hormone increases startle by acting directly in the BNST. Based on various differences between CeA- and BNST-mediated behaviors, we have suggested that CeA-mediated behaviors represent stimulus-specific fear, whereas BNST-mediated behaviors are more akin to anxiety. NMDA receptors are also involved in the extinction of conditioned fear, and both extinction in rats and exposure-based psychotherapy in humans are facilitated by an NMDA partial agonist called D-cycloserine.

ACKNOWLEDGMENTS

This work, which began around 1975, has been supported generously by several grants from the National Science Foundation, RO1 grants from the National Institutes of Mental Health (NIMH) and Neurological and Communication Disorders and Stroke, a Research Scientist Award for 25 years, back-to-back MERIT Awards, and several Center and Program Project Grants from NIMH, by the Air Force Office of Research, the Connecticut Mental Health Center and the State of Connecticut, Emory University and the Woodruff Foundation, the Yerkes National Primate Center base grant, and a National Science Foundation Science and Technology Center (the Center for Behavioral Neuroscience under Agreement No. IBN-9876754).

REFERENCES

Adamec, R. (1994). Modelling anxiety disorders following chemical exposures. *Toxicol Ind Health*, *10*(4–5), 391–420.

Alheid, G., deOlmos, J.S., and Beltramino, C.A. (1995). Amygdala and extended amygdala. In G. Paxinos (ed.), *The Rat Nervous System* (pp. 495–578). New York: Academic Press.

Berg, W.K., and Davis, M. (1984). Diazepam blocks fear-enhanced startle elicited electrically from the brainstem. *Physiology and Behavior*, *32*, 333–336.

Berg, W.K., and Davis, M. (1985). Associative learning modifies startle reflexes at the lateral lemniscus. *Behavioral Neuroscience*, *99*, 191–199.

Blair, H.T., Sotres-Bayon, F., Moita, M.A., and Ledoux, J.E. (2005). The lateral amygdala processes the value of conditioned and unconditioned aversive stimuli. *Neuroscience*, *133*(2), 561–569.

Blanchard, D.C., and Blanchard, R.J. (1972). Innate and conditioned reactions to threat in rats with amygdaloid lesions. *Journal of Comparative Physiology and Psychology*, *81*, 281–290.

Bordi, F., and LeDoux, J. (1992). Sensory tuning beyond the sensory system: An initial analysis of auditory response properties of neurons in the lateral amygdaloid nucleus and overlying areas of the striatum. *J Neurosci*, *12*(7), 2493–2503.

Boulis, N., and Davis, M. (1989). Footshock-induced sensitization of electrically elicited startle reflexes. *Behavioral Neuroscience*, *103*, 504–508.

Boulis, N., and Davis, M. (1990). Blockade of the spinal excitatory effect of cAMP on the startle reflex by intrathecal administration of the isoquinoline sulfonamide H-8: Comparison to the protein kinase C inhibitor H-7. *Brain Research*, *525*(2), 198–204.

Boulis, N., Kehne, J.H., Miserendino, M.J.D., and Davis, M. (1990). Differential blockade of early and late components of acoustic startle following intrathecal infusion of 6-cyano-7-nitroquinoxaline-2,3-dione (CNQX) or D,L-2-amino-5-phosphonovaleric acid (AP-5). *Brain Research*, *520*, 240–246.

Brown, J.S., Kalish, H.I., and Farber, I.E. (1951). Conditional fear as revealed by magnitude of startle response to an auditory stimulus. *Journal of Experimental Psychology*, *41*, 317–328.

Brunzell, D.H., and Kim, J.J. (2001). Fear conditioning to tone, but not to context, is attenuated by lesions of the insular cortex and posterior extension of the intralaminar complex in rats. *Behavioral Neuroscience*, *115*(2), 365–375.

Cameron, A.A., Iqbal, A.K., Westlund, K.N., and Willis, W.D. (1995). The efferent projections of the periaqueductal gray in the rat: A *Phaseolus vulgaris*-leucoagglutinin study. II. Descending projections. *Journal of Comparative Neurology*, *351*, 585–601.

Campeau, S., and Davis, M. (1995a). Involvement of the central nucleus and basolateral complex of the amygdala in fear conditioning measured with fear-potentiated startle in rats trained concurrently with auditory and visual conditioned stimuli. *Journal of Neuroscience*, *15*(3 Pt 2), 2301–2311.

Campeau, S., and Davis, M. (1995b). Involvement of subcortical and cortical afferents to the lateral nucleus of the amygdala in fear conditioning measured with fear-potentiated startle in rats trained concurrently with auditory and visual conditioned stimuli. *Jouranl of Neuroscience*, *15*(3 Pt 2), 2312–2327.

Campeau, S., Falls, W.A., Cullinan, W.E., Helmreich, D.L., Davis, M., and Watson, S.J. (1997). Elicitation and reduction of fear: Behavioral and neuroendocrine indices and brain induction of the immediate-early gene *c-fos*. *Neuroscience*, *78*, 1087–1104.

Campeau, S., Hayward, M.D., Hope, B.T., Rosen, J.B., Nestler, E.J., and Davis, M. (1991). Induction of the *c-fos* proto-oncogene in rat amygdala during unconditioned and conditioned fear. *Brain Research*, *565*, 349–352.

Campeau, S., Liang, K.C., and Davis, M. (1990). Long-term retention of fear-potentiated startle following a short training session. *Animal Learning and Behavior, 18*, 462–468.

Campeau, S., Miserendino, M.J.D., and Davis, M. (1992). Intra-amygdala infusion of the N-methyl-D-Aspartate receptor antagonist AP5 blocks acquisition but not expression of fear-potentiated startle to an auditory conditioned stimulus. *Behavioral Neuroscience, 106*, 569–574.

Cassella, J.V., and Davis, M. (1986). Habituation, prepulse inhibition, fear conditioning, and drug modulation of the acoustically elicited pinna reflex in rats. *Behavioral Neuroscience, 100*(1), 39–44.

Cassella, J.V., and Davis, M. (1987). A technique to restrain awake rats for recording single-unit activity with glass micropipettes and conventional microdrives. *Journal of Neuroscience Methods, 19*(2), 105–113.

Cassella, J.V., Harty, P.T., and Davis, M. (1986). Fear conditioning, pre-pulse inhibition and drug modulation of a short latency startle response measure electromyographically from neck muscles in the rat. *Physiology and Behavior, 36*, 1187–1191.

Chhatwal, J.P., Myers, K.M., Ressler, K.J., and Davis, M. (2005). Regulation of gephyrin and GABAA receptor binding within the amygdala after fear acquisition and extinction. *Journal of Neuroscience, 25*(2), 502–506.

Corodimas, K., and LeDoux, J.E. (1995). Disruptive effects of posttraining perirhinal cortex lesions on conditioned fear: Contributions of contextual cues. *Behavioral Neuroscience, 109*, 613–619.

Cruikshank, S.J., Edeline, J.M., and Weinberger, N.M. (1992). Stimulation at a site of auditory-somatosensory convergence in the medial geniculate nucleus is an effective unconditioned stimulus for fear conditioning. *Behavioral Neuroscience, 106*, 471–483.

Davis, M. (1984). The mammalian startle response. In R.C. Eaton (ed.), *Neural Mechanisms of Startle Behavior* (pp. 287–351). New York: Plenum Press.

Davis, M. (1986). Pharmacological and anatomical analysis of fear conditioning using the fear-potentiated startle paradigm. *Behavioral Neuroscience, 100*, 814–824.

Davis, M., and Astrachan, D.I. (1978). Conditioned fear and startle magnitude: Effects of different footshock or backshock intensities used in training. *Journal of Experimental Psychology: Animal Behavior Processes, 4*(2), 95–103.

Davis, M., Astrachan, D.I., Gendelman, P.M., and Gendelman, D.S. (1980). 5-Methoxy-N,N-dimethyltryptamine: Spinal cord and brainstem mediation of excitatory effects on acoustic startle. *Psychopharmacology, 70*(2), 123–130.

Davis, M., Astrachan, D.I., Kehne, J.H., Commissaris, R.L., and Gallager, D.W. (eds.) (1984). *Catecholamine Modulation of Sensorimotor Reactivity Measured with Acoustic Startle*. New York: Alan R. Liss.

Davis, M., Gendelman, D.S., Tischler, M.D., and Gendelman, P.M. (1982). A primary acoustic startle circuit: Lesion and stimulation studies. *Journal of Neuroscience, 6*, 791–805.

Davis, M., Hitchcock, J.M., and Rosen, J.B. (1987). Anxiety and the amygdala: Pharmacological and anatomical analysis of the fear-potentiated startle paradigm. In G.H. Bower (ed.), *The Psychology of Learning and Motivation* (Vol. 21, pp. 263–305). New York: Academic Press.

Davis, M., Parisi, T., Gendelman, D.S., Tischler, M., and Kehne, J.H. (1982). Habituation and sensitization of startle reflexes elicited electrically from the brainstem. *Science, 218*(4573), 688–690.

Davis, M., Schlesinger, L.S., and Sorenson, C.A. (1989). Temporal specificity of fear-conditioning: Effects of different conditioned stimulus-unconditioned stimulus intervals on the fear-potentiated startle effect. *Journal of Experimental Psychology: Animal Behavior Processes, 15*, 295–310.

Dayas, C.V., Buller, K.M., Crane, J.W., Xu, Y., and Day, T.A. (2001). Stressor categorization: acute physical and psychological stressors elicit distinctive recruitment patterns in the amygdala and in medullary noradrenergic cell groups. *European Journal of Neuroscience, 14*(7), 1143–1152.

de Jongh, R., Groenink, L., van Der Gugten, J., and Olivier, B. (2002). The light-enhanced startle paradigm as a putative animal model for anxiety: effects of chlordiazepoxide, flesinoxan and fluvoxamine. *Psychopharmacology (Berl), 159*(2), 176–180.

Falls, W.A., Carlson, S., Turner, J.G., and Willott, J.F. (1997). Fear-potentiated startle in two strains of inbred mice. *Behavioral Neuroscience, 111*(4), 855–861.

Falls, W.A., Miserendino, M.J., and Davis, M. (1992). Extinction of fear-potentiated startle: Blockade by infusion of an NMDA antagonist into the amygdala. *Journal of Neuroscience, 12*(3), 854–863.

Falls, W.A., and Davis, M. (1994). Visual cortex ablations do not prevent extinction of fear-potentiated startle using a visual conditioned stimulus. *Behavioral and Neural Biology, 60,* 259–270.

Fanselow, M.S., Kim, J.J., Yipp, J., and De Oca, B. (1994). Differential effects of the N-methyl-D-aspartate antagonist D-L-2-amino-5-phosphonovalerate on acquisition of fear of auditory and contextual cues. *Behavioral Neuroscience, 108,* 235–240.

Fendt, M., Endres, T., and Apfelbach, R. (2003). Temporary inactivation of the bed nucleus of the stria terminalis but not of the amygdala blocks freezing induced by trimethylthiazoline, a component of fox feces. *Journal of Neuroscience, 23*(1), 23–28.

Fendt, M., and Fanselow, M.S. (1999). The neuroanatomical and neurochemical basis of conditioned fear. *Neuroscience and Biobehavioral Reviews, 23,* 743–760.

Gale, G.D., Anagnostaras, S.G., Godsil, B.P., Mitchell, S., Nozawa, T., Sage, J.R., Wiltgen, B., and Fanselow, M.S. (2004). Role of the basolateral amygdala in the storage of fear memories across the adult lifetime of rats. *Journal of Neuroscience, 24*(15), 3810–3815.

Gentile, C.G., Jarrel, T.W., Teich, A., McCabe, P.M., and Schneiderman, N. (1986). The role of amygdaloid central nucleus in the retention of differential pavlovian conditioning of bradycardia in rabbits. *Behavioral Brain Research, 20,* 263–273.

Gewirtz, J., and Davis, M. (1997). Second-order fear conditioning prevented by blocking NMDA receptors in the amygdala. *Nature, 388,* 471–474.

Gewirtz, J.C., McNish, K.A., and Davis, M. (1998). Lesions of the bed nucleus of the stria terminalis block sensitization of the acoustic startle reflex produced by repeated stress, but not fear-potentiated startle. *Progress in Neuro-Psychopharmacology and Biological Psychiatry, 22,* 625–648.

Grillon, C., Ameli, R., Woods, S.W., Merikangas, K., and Davis, M. (1991). Fear-potentiated startle in humans: Effects of anticipatory anxiety on the acoustic blink reflex. *Psychophysiology, 28,* 588–595.

Grillon, C., and Davis, M. (1997). Fear-potentiated startle conditioning in humans: Effects of explicit and contextual cue conditioning following paired vs. unpaired training. *Psychophysiology, 34,* 451–458.

Grillon, C., Pellowski, M., Merikangas, K.R., and Davis, M. (1997). Darkness facilitates the acoustic startle reflex in humans. *Biol Psychiatry, 42*(6), 453–460.

Grillon, C., Morgan, C.A., 3rd, Davis, M., and Southwick, S.M. (1998).

Hamm, A.O., Greenwald, M.K., Bradley, M.M., Cuthbert, B.N., and Lang, P.J. (1991). The fear-potentiated startle effect. Blink reflex modulation as a result of classical aversive conditioning. *Integr Physiol Behav Sci, 26*(2), 119–126.

Hamm, A.O., Start, and Vaitl, D. (1990). Classical fear conditioning and the startle probe reflex. *Psychophysiology, 27,* S37.

Herzog, C., and Otto, T. (1997). Odor-guided fear conditioning in rats: 2. Lesions of the anterior perirhinal cortex disrupt fear conditioned to the explicit conditioned stimulus but not to the training context. *Behavioral Neuroscience, 111*(6), 1265–1272.

Hitchcock, J.M., and Davis, M. (1986). Lesions of the amygdala, but not of the cerebellum or red nucleus, block conditioned fear as measured with the potentiated startle paradigm. *Behavioral Neuroscience, 100*, 11–22.

Hitchcock, J.M., and Davis, M. (1987). Fear-potentiated startle using an auditory conditioned stimulus: Effect of lesions of the amygdala. *Physiology and Behavior, 39*(3), 403–408.

Hitchcock, J.M., and Davis, M. (1991). The efferent pathway of the amygdala involved in conditioned fear as measured with the fear-potentiated startle paradigm. *Behavioral Neuroscience, 105*, 826–842.

Iwata, J., LeDoux, J.E., Meeley, M.P., Arneric, S., and Reis, D.J. (1986). Intrinsic neurons in the amygdala field projected to by the medial geniculate body mediate emotional responses conditioned to acoustic stimuli. *Brain Research, 383*, 195–214.

Jasnow, A.M., Shi, C., Israel, J.E., Davis, M., and Huhman, K.L. (2005). Memory of social defeat is facilitated by cAMP response element-binding protein overexpression in the amygdala. *Behavioral Neuroscience, 119*(4), 1125–1130.

Jolkkonen, E., and Pitkanen, A. (1998). Intrinsic connections fo the rat amygdaloid complex: Projections originating in the central nucleus. *Journal of Comparative Neurology, 395*, 53–72.

Josselyn, S.A., Shi, C.-J., Carlezon, Jr., W.A., Neve, R., Nestler, E.J., and Davis, M. (2001). Long-term memory is facilitated by cAMP response element-binding protein overexpression in the amygdala. *Journal of Neuroscience, 21*, 2404–2412.

Kapp, B.S., Frysinger, R.C., Gallagher, M., and Haselton, J.R. (1979). Amygdala central nucleus lesions: Effect on heart rate conditioning in the rabbit. *Physiology and Behavior, 23*, 1109–1117.

Kehne, J.H., Astrachan, D.I., Astrachan, E., Tallman, J.F., and Davis, M. (1986). The role of spinal cord cyclic AMP in the acoustic startle response in rats. *Journal of Neuroscience, 6*, 3250–3257.

Kehne, J.H., Gallager, D.W., and Davis, M. (1981). Strychnine: Brainstem and spinal mediation of excitatory effects on acoustic startle. *European Journal of Pharmacology, 76*, 177–186.

Kim, M., Campeau, S., Falls, W.A., and Davis, M. (1993). Infusion of the non-NMDA receptor antagonist CNQX into the amygdala blocks the expression of fear-potentiated startle. *Behavioral and Neural Biology, 59*, 5–8.

Kim, M., and Davis, M. (1993). Electrolytic lesions of the amygdala block acquisition and expression of fear-potentiated startle even with extensive training, but do not prevent re-acquisition. *Behavioral Neuroscience, 107*(4), 580–595.

Kogan, J.H., Frankland, P.W., Blendy, J.A., Coblentz, J., Marowitz, Z., Schutz, G., and Silva, A.J. (1997). Spaced training induces normal long-term memory in CREB mutant mice. *Current Biology, 7*(1), 1–11.

Koob, G.F. (1999). Corticotropin-releasing factor, norepinephrine, and stress. *Biol Psychiatry, 46*(9), 1167–1180. Acoustic startle in Vietnam veterans with PTSD. *Am J Psychiatry, 155*(6), 812–817.

Krase, W., Koch, M., and Schnitzler, H.U. (1994). Substance P is involved in the sensitization of the acoustic startle response by footshock in rats. *Behavioral Brain Research, 63*, 81–88.

Lang, P.J., Bradley, M.M., and Cuthbert, B.N. (1990). Emotion, attention, and the startle reflex. *Psychological Reviews, 97*, 377–395.

Ledgerwood, L., Richardson, R., and Cranney, J. (2003). Effects of D-cycloserine on extinction of conditioned freezing. *Behav Neurosci, 117*(2), 341–349.

LeDoux, J.E., Cicchetti, P., Xagoraris, A., and Romanski, L.M. (1990). The lateral amygdaloid nucleus: Sensory interface of the amygdala in fear conditioning. *Journal of Neuroscience, 10*, 1062–1069.

LeDoux, J.E., Iwata, J., Cicchetti, P., and Reis, D.J. (1988). Different projections of the central amygdaloid nucleus mediate autonomic and behavioral correlates of conditioned fear. *Journal of Neuroscience, 8*(7), 2517–2529.

LeDoux, J.E., Romanski, L., and Xagoraris, A. (1989). Indelibility of subcortical memories. *Journal of Cognitive Neuroscience, 1,* 238–243.

LeDoux, J.E., Sakaguchi, A., Iwata, J., and Reis, D.J. (1986). Interruption of projections from the medial geniculate body to an archi-neostriatal field disrupts the classical conditioning of emotional responses to acoustic stimuli. *Neuroscience, 17,* 615–627.

LeDoux, J.E., Sakaguchi, A., and Reis, D.J. (1984). Subcortical efferent projections of the medial geniculate nucleus mediate emotional responses conditioned to acoustic stimuli. *Journal of Neuroscience, 4,* 683–698.

Lee, Y., and Davis, M. (1997a). Role of the hippocampus, bed nucleus of the stria terminalis and amygdala in the excitatory effect of corticotropin releasing hormone on the acoustic startle reflex. *Journal of Neuroscience, 17,* 6434–6446.

Lee, Y., and Davis, M. (1997b). Role of the septum in the excitatory effect of corticotropin releasing (CRH) hormone on the acoustic startle reflex. *Journal of Neuroscience, 17,* 6424–6433.

Lee, Y., Lopez, D.E., Meloni, E.G., and Davis, M. (1996). A primary acoustic startle circuit: Obligatory role of cochlear root neurons and the nucleus reticularis pontis caudalis. *Journal of Neuroscience, 16*(11), 3775–3789.

Liang, K.C., Melia, K.R., Campeau, S., Falls, W.A., Miserendino, M.J.D., and Davis, M. (1992). Lesions of the central nucleus of the amygdala, but not of the paraventricular nucleus of the hypothalamus, block the excitatory effects of corticotropin releasing factor on the acoustic startle reflex. *Journal of Neuroscience, 12,* 2313–2320.

Liang, K.C., Melia, K.R., Miserendino, M.J.D., Falls, W.A., Campeau, S., and Davis, M. (1992). Corticotropin-releasing factor: Long-lasting facilitation of the acoustic startle reflex. *Journal of Neuroscience, 12,* 2303–2312.

Lin, C.H., Yeh, S.H., Lu, K.T., Leu, T.H., Chang, W.C., and Gean, P.W. (2001). A role for the PI-3 kinase signaling pathway in fear conditioning and synaptic plasticity in the amygdala. *Neuron, 31*(5), 841–851.

Lingenhohl, K., and Friauf, E. (1994). Giant neurons in the rat reticular formation: A sensorimotor interface in the elementary acoustic startle circuit? *Journal of Neuroscience, 14*(3), 1176–1194.

Lipp, O.V., Sheridan, J., and Siddle, D.A. (1994). Human blink startle during aversive and nonaversive Pavlovian conditioning. *Journal of Experimental Psychology: Animal Behavior Processes, 20*(4), 380–389.

Lopez, D.E., Merchan, M.A., Bajo, V.M., and Saldana, E. (1993). The cochlear root neurons in the rat, mouse and gerbil. In M.A. Merchan (ed.), *The Mammalian Cochlear Nuclei: Organization and Function* (pp. 291–301). New York: Plenum Press.

Lu, K.T., Shi, C.-J., Tang, Z.L., and Davis, M. (2000). Pre-training administration of the MAPK inhibitor PD098095 into the amygdala blocks the acquisition of fear-potentiated startle. *Society for Neuroscience Abstracts, 26,* 192.

McAllister, W.R., and McAllister, D.E. (1971). Behavioral measurement of conditioned fear. In F.R. Brush (ed.), *Aversive Conditioning and Learning* (pp. 105–179). New York: Academic Press.

McDonald, A.J. (1982). Cytoarchitecture of the central amygdaloid nucleus of the rat. *Journal of Comparative Neurology, 208*(4), 401–418.

Meloni, E.G., and Davis, M. (1999). Muscimol in the deep layers of the superior colliculus/mesencephalic reticular formation blocks expression but not acquisition of fear-potentiated startle in rats. *Behavioral Neuroscience, 113*(6), 1152–1160.

Miserendino, M.J., and Davis, M. (1993). NMDA and non-NMDA antagonists infused into the nucleus reticularis pontis caudalis depress the acoustic startle reflex. *Brain Research, 623,* 215–222.

Miserendino, M.J., Sananes, C.B., Melia, K.R., and Davis, M. (1990). Blocking of acquisition but not expression of conditioned fear-potentiated startle by NMDA antagonists in the amygdala. *Nature, 345*(6277), 716–718.

Mitsuhashi, M., Osashi, Y., Shichijo, S., Christian, C., Sudduth-Klinger, J., Harrowe, G., and Payan, D.G. (1992). Multiple intracellular signaling pathways of the neuropeptide substance P receptor. *Journal of Neuroscience Research, 32*, 437–443.

Monaghan, D.T., and Cotman, C.W. (1985). Distribution of N-methyl-D-aspartate-sensitive L-[3H]glutamate binding sites in rat brain. *Journal of Neuroscience, 5*, 2909–2919.

Myers, K.M., and Davis, M. (2002). Behavioral and neural analysis of extinction: A review. *Neuron, 36*, 567–584.

Nguyen, P. (2001). CREB and the enhancement of memory. *Trends in Neuroscience, 24*, 314.

Nodal, F.R., and Lopez, D.E. (2003). Direct input from cochlear root neurons to pontine reticulospinal neurons in albino rat. *Journal of Comparative Neurology, 460*(1), 80–93.

Paschall, G.Y., and Davis, M. (2002). Olfactory-mediated fear-potentiated startle. *Behavioral Neuroscience, 116*, 4–12.

Pissiota, A., Frans, O., Fredrikson, M., Langstrom, B., and Flaten, M.A. (2002). The human startle reflex and pons activation: A regional cerebral blood flow study. *European Journal of Neuroscience, 15*(2), 395–398.

Pitkanen, A., Savander, V., and LeDoux, J.E. (1997). Organization of intra-amygdaloid circuitries in the rat: An emerging framework for understanding functions of the amygdala. *Trends in Neuroscience, 20*, 517–523.

Pitkanen, A., Stefanacci, L., Farb, C.R., Go, C.G., LeDoux, J.E., and Amaral, D.G. (1995). Intrinsic connections of the rat amygdaloid complex: Projections originating in the lateral nucleus. *Journal of Comparative Neurology, 356*(2), 288–310.

Quirk, G.J., Repa, C., and LeDoux, J.E. (1995). Fear conditioning enhances short-latency auditory responses of lateral amygdala neurons: Parallel recordings in the freely behaving rat. *Neuron, 15*(5), 1029–1039.

Rattiner, L.M., Davis, M., French, C.T., and Ressler, K.J. (2004). Brain-derived neurotrophic factor and tyrosine kinase receptor B involvement in amygdala-dependent fear conditioning. *Journal of Neuroscience, 24*(20), 4796–4806.

Rattiner, L.M., Davis, M., and Ressler, K.J. (2005). Brain-derived neurotrophic factor in amygdala-dependent learning. *Neuroscientist, 11*(4), 323–333.

Rauch, S.L., Shin, L.M., and Wright, C.I. (2003). Neuroimaging studies of amygdala function in anxiety disorders. *Annals of the New York Academy of Science, 985*, 389–410.

Rescorla, R.A., and Wagner, A.R. (1972). A theory of Pavlovian conditioning: Variations in the effectiveness of reinforcement and nonreinforcement. In A.H. Black and W.F. Prokasy (eds.), *Classical Conditioning II* (current research and theory, pp. 64–99). New York: Appleton-Century-Crofts.

Ressler, K.J., Paschall, G.Y., Zhao, X.L., and Davis, M. (2002). Induction of synaptic plasticity genes in a distributed neural circuit during consolidation of conditioned fear learning. *Journal of Neuroscience, 22*, 7892–7903.

Ressler, K.J., Rothbaum, B.O., Tannenbaum, L., Anderson, P., Graap, K., Zimand, E., Hodges, L., and Davis, M. (2004). Cognitive enhancers as adjuncts to psychotherapy: use of D-cycloserine in phobic individuals to facilitate extinction of fear. *Arch Gen Psychiatry, 61*(11), 1136–1144.

Rodrigues, S.M., Schafe, G.E., and LeDoux, J.E. (2001). Intra-amygdala blockade of the NR2B subunit of the NMDA receptor disrupts the acquisition but not the expression of fear conditioning. *Journal of Neuroscience, 21*(17), 6889–6896.

Rogan, M.T., Staubli, U.V., and LeDoux, J.E. (1997). Fear conditioning induces associative long-term potentiation in the amygdala. *Nature, 390*, 604–607.

Romanski, L.M., Clugnet, M.C., Bordi, F., and LeDoux, J.E. (1993). Somatosensory and auditory convergence in the lateral nucleus of the amygdala. *Behavioral Neuroscience, 107*, 444–450.

Romanski, L.M., and LeDoux, J.E. (1992a). Bilateral destruction of neocortical and perirhinal projection targets of the acoustic thalamus does not disrupt auditory fear conditioning. *Neuroscience Letters, 142,* 228–232.

Romanski, L.M., and LeDoux, J.E. (1992b). Equipotentiality of thalamo-amygdala and thalamo-cortico amygdala circuits in auditory fear conditioning. *Journal of Neuroscience, 12,* 4501–4509.

Rosen, J.B., Hitchcock, J.M., Miserendino, M.J.D., Falls, W.A., Campeau, S., and Davis, M. (1992). Lesions of the perirhinal cortex but not of the frontal, medial prefrontal, visual, or insular cortex block fear-potentiated startle using a visual conditioned stimulus. *Journal of Neuroscience, 12,* 4624–4633.

Rosen, J.B., Hitchcock, J.M., Sananes, C.B., Miserendino, M.J.D., and Davis, M. (1991). A direct projection from the central nucleus of the amygdala to the acoustic startle pathway: Antero-grade and retrograde tracing studies. *Behavioral Neuroscience, 105,* 817–825.

Samson, R.D., and Pare, D. (2005). Activity-dependent synaptic plasticity in the central nucleus of the amygdala. *Journal of Neuroscience, 25*(7), 1847–1855.

Sananes, C.B., and Davis, M. (1992). N-Methyl-D-aspartate lesions of the lateral and basolateral nuclei of the amygdala block fear-potentiated startle and shock sensitization of startle. *Behavioral Neuroscience, 106,* 72–80.

Santini, E., Muller, R.U., and Quirk, G.J. (2001). Consolidation of extinction learning involves transfer from NMDA-independent to NMDA-dependent memory. *Journal of Neuroscience, 21*(22), 9009–9017.

Savander, V., Go, C.G., LeDoux, J.E., and Pitkanen, A. (1995). Intrinsic connections of the rat amygdaloid complex: Projections originating in the basal nucleus. *Journal of Comparative Neurology, 361*(2), 345–368.

Savander, V., Go, C.G., LeDoux, J.E., and Pitkanen, A. (1996). Intrinsic connections of the rat amygdaloid complex: Projections originating in the accessory basal nucleus. *Journal of Comparative Neurology, 374*(2), 291–313.

Schafe, G.E., Atkins, C.M., Swank, M.W., Bauer, E.P., Sweatt, J.D., and LeDoux, J.E. (2000). Activation of ERK/MAP kinase in the amygdala is required for memory consolidation of pavlovian fear conditioning. *Journal of Neuroscience, 20*(21), 8177–8187.

Schafe, G.E., Nadel, N.V., Sullivan, G.M., Harris, A., and LeDoux, J.E. (1999). Memory consolidation for contextual and auditory fear conditioning is dependent on protein synthesis, PKA, and MAP kinase. *Learning and Memory, 6,* 97–110.

Schettino, L.F., and Otto, T. (2001). Patterns of Fos expression in the amygdala and ventral perirhinal cortex induced by training in an olfactory fear conditioning paradigm. *Behavioral Neuroscience, 115*(6), 1257–1272.

Shi, C., and Davis, M. (1999). Pain pathways involved in fear conditioning measured with fear-potentiated startle: Lesion studies. *Journal of Neuroscience, 19*(1), 420–430.

Shi, C.-J., and Davis, M. (2001). Visual pathways involved in fear conditioning measured with fear-potentiated startle: Behavior and anatomic studies. *Journal of Neuroscience, 21,* 9844–9855.

Siegel, A. (1967). Stimulus generalization of a classically conditioned response along a temporal dimension. *Journal of Comparative Physiology and Psychology, 64,* 461–466.

Spiera, R.F., and Davis, M. (1988). Excitatory amino acid antagonists depress acoustic startle after infusion into the ventral nucleus of the lateral lemniscus or paralemniscal zone. *Brain Research, 445,* 130–136.

Sullivan, G.M., Apergis, J., Bush, D.E., Johnson, L.R., Hou, M., and Ledoux, J.E. (2004). Lesions in the bed nucleus of the stria terminalis disrupt corticosterone and freezing responses elicited by a contextual but not by a specific cue-conditioned fear stimulus. *Neuroscience, 128*(1), 7–14.

Swerdlow, N.R., Britton, K.T., and Koob, G.F. (1989). Potentiation of acoustic startle by corticotropin-releasing factor (CRF) and by fear are both reversed by alpha-helical CRF (9–41). *Neuropsychopharmacology, 2,* 285–292.

Swerdlow, N.R., Geyer, M.A., Vale, W.W., and Koob, G.F. (1986). Corticotropin-releasing factor potentiates acoustic startle in rats: Blockade by chlordiazepoxide. *Psychopharmacology*, *88*, 147–152.

Tischler, M.D., and Davis, M. (1983). A visual pathway that mediated fear-conditioned enhancement of acoustic startle. *Brain Research*, *276*, 55–71.

Walker, D.L., and Davis, M. (1997a). Anxiogenic effects of high illumination levels assessed with the acoustic startle paradigm. *Biological Psychiatry*, *42*, 461–471.

Walker, D.L., and Davis, M. (1997b). Double dissociation between the involvement of the bed nucleus of the stria terminalis and the central nucleus of the amygdala in light-enhanced versus fear-potentiated startle. *Journal of Neuroscience*, *17*, 9375–9383.

Walker, D.L., and Davis, M. (2000). Involvement of N-methyl-D-aspartate (NMDA) receptors within the amygdala in short- versus long-term memory for fear conditioning as assessed with fear-potentiated startle. *Behavioral Neuroscience*, *114*, 1019–1033.

Walker, D.L., and Davis, M. (2002). Quantifying fear potentiated startle using absolute versus percent increase scoring methods: Implications for the neurocircuitry of fear and anxiety. *Psychopharmacology*, *164*, 318–328.

Walker, D.L., Paschall, G.Y., and Davis, M. (2005). Glutamate receptor antagonist infusions into the basolateral and medial amygdala reveal differential contributions to olfactory vs. context fear conditioning and expression. *Learning and Memory*, *12*(2), 120–129.

Walker, D.L., Ressler, K.J., Lu, K.-T., and Davis, M. (2002). Facilitation of conditioned fear extinction by systemic administration or intra-amygdala infusions of D-cycloserine as assessed with fear-potentiated startle in rats. *Journal of Neuroscience*, *22*, 2343–2351.

Walker, D.L., Toufexis, D.J., and Davis, M. (2003). Role of the bed nucleus of the stria terminalis versus the amygdala in fear, stress, and anxiety. *European Journal of Pharmacology*, *463*, 199–216.

Winslow, J.T., Parr, L.A., and Davis, M. (2002). Acoustic startle, prepulse inhibition and fear-potentiated startle measured in rhesus monkeys. *Biological Psychiatry*, *51*, 859–866.

Yeomans, J.S., and Pollard, B.A. (1993). Amygdala efferents mediating electrically evoked startle-like responses and fear potentiation of acoustic startle. *Behavioral Neuroscience*, *107*, 596–610.

Yin, J.C., Del Vecchio, M., Zhou, H., and Tully, T. (1995). CREB as a memory modulator: Induced expression of a dCREB2 activator isoform enhances long-term memory in *Drosophila*. *Cell*, *81*(1), 107–115.

Zhao, Z., and Davis, M. (2004). Fear-potentiated startle in rats is mediated by neurons in the deep layers of the superior colliculus/deep mesencephalic nucleus of the rostral midbrain through the glutamate non-NMDA receptors. *Journal of Neuroscience*, *24*(46), 10326–10334.

Cerebellar Learning

Tatsuya Ohyama and Michael D. Mauk

Department of Neurobiology and Anatomy, University of Texas Medical School, Houston, TX 77030

I. INTRODUCTION

Since Marr's groundbreaking theory on the cerebellar cortex (Marr, 1969), the issue of cerebellar learning has been actively if not harmoniously investigated. Marr's theory was based on the equally seminal work of Eccles, Ito, and Szenthàgotai (1967), which laid the foundations for providing a nearly complete circuit diagram of the cerebellum. The nearly 40 years of cerebellar research since has provided not only a wealth of evidence in support of the theory, but additional evidence to include other sites of plasticity outside of the cerebellar cortex, including the deep cerebellar and vestibular nuclei.

Here we review evidence relating to three questions. (1) Does the cerebellum learn? (2) How is this learning accomplished? (3) What is the functional role of cerebellar learning? We first describe evidence of cerebellar involvement in a number of forms of motor learning and adaptation. Then, based on the unique advantages of eyelid conditioning, we review evidence on the mechanisms underlying cerebellar learning. The evidence suggests that cerebellar learning is mediated by plasticity in the cerebellar cortex and in its downstream targets in the deep cerebellar nuclei. Finally, we examine the behavioral

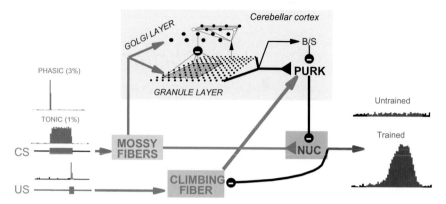

FIGURE 13-1   How Pavlovian eyelid conditioning, an associative form of motor learning, engages the cerebellum. In eyelid conditioning, repeated pairings of a peripheral stimulus such as a tone (the conditioned stimulus, or CS) and mild electrical stimulation around the eye (the unconditioned stimulus, or US) gradually establishes a conditioned eyelid response to the previously neutral CS. Research since the early 1980s has shown that the CS and US, respectively, activate the mossy and climbing fiber afferents to the cerebellum and that the activity of one of the deep cerebellar nuclei drives the expression of the conditioned response. Current evidence suggests that eyelid conditioning is mediated by plasticity at two sites, one at granule cell-to-Purkinje cell (gr-Pkj) synapses in the cerebellar cortex and another at the excitatory mf-nuc synapses of mossy fibers onto deep nucleus cells.

properties of eyelid conditioning to infer the contribution of this learning to cerebellar information processing. We conclude by suggesting that the cerebellum learns temporally specific feed-forward predictions that are used to improve the operation of any neural process occurring on a relatively short time scale.

II. SYNAPTIC ORGANIZATION OF THE CEREBELLUM

As just noted, the architecture of the cerebellum — its cell types, connectivity, and synaptic organization — is known in great detail (Chan-Palay, 1977; Eccles et al., 1967; Ito, 1984; Llinas, 1981). The input to the cerebellum is conveyed by two major afferents, the mossy fibers and climbing fibers (Fig. 13-1). Mossy fibers arise from cells in various nuclei within the brainstem, thereby conveying activity from virtually every area in the brain, including the cerebral cortex. Mossy fibers project profusely into the cerebellar cortex, making excitatory contacts with inhibitory Golgi cells and granule cells and providing excitatory collaterals to cells in the deep cerebellar and vestibular nuclei. Thus, there is a great divergence of mossy fiber input onto granule cells, the most numerous type of cell in the brain. Granule cells give rise to parallel fibers that course through the cerebellar folium and synapse on the inhibitory Purkinje cells, the sole output neurons of the cerebellar cortex projecting to the deep cerebellar and vestibular nuclei. Parallel fibers also contact inhibitory interneurons, which in turn make synapses with Purkinje cells. Purkinje cells receive inputs from up to 200,000 parallel fibers, making them one of the cells with the greatest convergence of inputs in the brain. Mossy fibers influence cerebellar output indirectly via a pathway through the cerebellar cortex and directly via their collateral projections to the deep cerebellar nuclei.

Climbing fibers arise in the inferior olivary nuclei and project directly onto Purkinje cells. Each climbing fiber contacts about 10 Purkinje cells, and each Purkinje cell receives input from only one climbing fiber. The connection between a climbing fiber and a Purkinje cell is one of the most powerful synapses in the brain. Climbing fibers also make far weaker synaptic contacts with cells in the deep cerebellar and vestibular nuclei via collaterals. Therefore, the divergence and convergence ratios for climbing fiber inputs are far less than those for mossy fiber inputs.

III. THE CEREBELLUM LEARNS

An extensive body of evidence from eyelid conditioning and adaptation of eye movements suggests that the cerebellum learns. To varying degrees, the evidence points to the cerebellum as a key site of plasticity for each of these forms of learning.

A. Adaptation of Smooth Pursuit

Smooth-pursuit eye movements are those that we and other primates make when tracking a moving stimulus. When the motion of the stimulus is predictable, an early open-loop portion of the eye movement also adapts to anticipate this stimulus (Barnes and Donelan, 1999; Kettner, Leung, and Peterson, 1996; Leung and Kettner, 1997; Medina, Carey, and Lisberger, 2005; Thier and Ilg, 2005). Currently, evidence for a cerebellar locus for this form of learning comes from lesion studies. Lesions of the cerebellum abolish smooth pursuit (Robinson, Straube, and Fuchs, 1997; Westheimer and Blair, 1974), and lesions specific to the dorsal (oculomotor) vermis prevent pursuit adaptation (Takagi, Zee, and Tamargo, 2000). Computer simulations are also consistent with a cerebellar locus for learning since they demonstrate that network models with plasticity in the cerebellar cortex (see later) can adequately reproduce complex predictive pursuit movements in monkeys (Kettner et al., 1997; Kettner, Suh, Davis, and Leung, 2002).

B. Saccade Adaptation

Saccades are rapid eye movements that redirect the gaze from one location to another. They are commonly studied by having humans or monkeys make an eye movement to a peripherally presented target stimulus. The amplitude of the saccade can be increased or decreased if during the primary saccade the target is moved horizontally in the same or the opposite direction of the initial saccade (forward or backward adaptation, respectively) (Hopp and Fuchs, 2004). Lesion studies suggest that the cerebellum is required for adaptation of saccades induced by artificially weakening extraocular muscles (Optican and Robinson, 1980) and that the oculomotor vermis is necessary for the behaviorally induced adaptation of saccades (Hopp and Fuchs, 2004). Lesions of the caudal fastigial nucleus downstream of the oculomotor vermis also lead to dysmetric saccades and prevent saccade adaptation (Robinson, Fuchs, and Noto, 2002; Robinson, Straube, and Fuchs, 1993). The caudal fastigial nucleus, which projects to the brainstem burst generators that drive saccades, exhibits neural activity that changes with saccade adaptation (Scudder, 2002; Scudder and McGee, 2003) and whose time of termination may be determined by population activity in the oculomotor vermis (Thier, Dicke, Haas, and Barash, 2000; Thier, Dicke, Haas, Thielert, and Catz, 2002). Although these data do not pinpoint the site of plasticity underlying saccade adaptation, they support a cerebellar locus for learning.

C. Adaptation of the Vestibulo-Ocular Reflex (VOR)

The VOR is a reflex that compensates for movements of the head to stabilize visual inputs on the retina (Ito, 1970). A turn of the head induces a compensatory turn of the eyes in the opposite direction, with an amplitude (referred to as *gain*, often taken as the ratio of the velocity of the eyes to that of the head) approximating that of the former (gain ~1). The VOR gain can be modified by manipulations in which visual input is changed so that a gain of 1 does not appropriately compensate for head movement (Gonshor and Jones, 1973).

Based on the proposals by Marr (1969) and Albus (1971), Ito (1972) suggested that VOR adaptation could be mediated by changes in the strengths of granule cell–to–Purkinje cell (*gr-Pkj*) synapses. The VOR is mediated by a three-neuron arc consisting of the semicircular canals, the primary vestibular neurons, and the oculomotor neurons, with the cerebellum attached to this trisynaptic reflex pathway as a side loop (Ito, 1970, 1972). This constrains the possible sites of plasticity to a relatively small number (Miles and Lisberger, 1981). Although an enormous body of evidence now exists for plasticity at *gr-Pkj* synapses (Hansel, Linden, and D'Angelo, 2001; Ito, 2001), the evidence that VOR adaptation is mediated by this plasticity has come largely from studies showing that cerebellar cortex lesions (including genetic modifications in mice that impair plasticity mechanisms) prevent adaptation and from recording studies showing activity patterns consistent with such a mechanism (Ito, 2001). Although there remains disagreement over the essential site(s) of plasticity, it is generally agreed that plasticity at one or more sites within the cerebellar cortico-nuclear microcomplex (i.e., a computational unit involving populations of neurons in the cerebellar cortex, their targets in the deep cerebellar or vestibular nuclei, and inferior olive) mediates VOR adaptation.

D. Eyelid Conditioning

With its long history of use in experimental psychology, eyelid conditioning is one of the most extensively characterized forms of associative learning. Eyelid conditioning involves repeatedly pairing a neutral stimulus such as a tone and either electrical stimulation of the periorbital muscles or a puff of air directed at the eye (the unconditioned stimulus, or US). This training eventually imbues the tone with the power to evoke eyelid closure on its own, rendering it a classical, or Pavlovian, conditioned stimulus (CS).

Several lines of evidence suggest that eyelid conditioning is mediated by learning in the cerebellum. First, lesions of a deep cerebellar nucleus, the anterior interpositus nucleus (AIN) ipsilateral to the trained eye, permanently abolishes learning and retention of eyelid conditioning (McCormick, Clark, Lavond, and Thompson, 1982; McCormick and Thompson, 1984a; Yeo, Hardiman, and Glickstein, 1985a). Since learning to the contralateral side is normal (McCormick et al., 1982; McCormick and Thompson, 1984a), the deficit is not due to perceptual, motivational, or motor impairments. Second, stimulation and recording studies indicate that neural activity in the AIN drives the learned response — stimulating the AIN drives eyelid closure, and activity in the AIN anticipates and models the learned response (McCormick and Thompson, 1984a, 1984b) and drives the premotor and motor neurons via the red nucleus (Chapman, Steinmetz, Sears, and Thompson, 1990). Third, reversible-lesion studies indicate that the underlying plasticity occurs at or upstream of the AIN — inactivation of the AIN but not the red nucleus or superior cerebellar peduncle prevents learning (Krupa, J.K. Thompson, and Thompson, 1993; Krupa and R.F. Thompson, 1995). Finally, stimulation studies show that learning occurs downstream from the pontine nucleus and inferior olive, the sources of cerebellar input from the CS and US, respectively (Fig. 13-1). That is, learning occurs when either a tone or mossy fiber stimulation is paired with stimulation of the inferior olive (Mauk, Steinmetz, and Thompson, 1986; Steinmetz, Lavond, and Thompson, 1989); and in decerebrate ferrets, conditioned responses to a forelimb stimulation CS can be elicited with mossy fiber stimulation even when antidromic activation of their cell bodies in the pons is prevented (Hesslow, Svensson, and Ivarsson, 1999). Collectively, these data suggest strongly that eyelid conditioning is mediated by the cerebellum. They also indicate a remarkably direct mapping of the training stimuli of eyelid conditioning onto cerebellar afferents (CS — mossy fiber, US — climbing fiber) and cerebellar output onto behavioral output (AIN activity — eyelid closure). This implies that the behavioral properties of eyelid conditioning provide a relatively direct window into cerebellar learning and the computation that it supports.

IV. HOW THE CEREBELLUM LEARNS

A. Sites of Plasticity

As already noted, the inputs from the stimuli that induce learning in eyelid conditioning (e.g., the CS and US in eyelid conditioning) converge at the Purkinje cells and cells in the AIN. For this reason, the two sites that have received the greatest attention as candidate sites of plasticity are the *gr-Pkj* synapses in the cerebellar cortex and the synapses of mossy fibers onto the AIN (*mf-nuc* synapses) (Fig. 13-1).

1. Cerebellar Cortex

From the earliest studies in the 1980s, there were indications that plasticity in the cerebellar cortex contributed to eyelid conditioning, although its precise contribution was unclear. Whereas some early studies found that lesions of certain regions of the cerebellar cortex completely abolished previously acquired responses and prevent their reacquisition (Yeo, Hardiman, and Glickstein, 1984, 1985b), other studies showed that learning or relearning was slowed but not completely prevented (Lavond and Steinmetz, 1989; Yeo and Hardiman, 1992). Since 1994, several studies of eyelid conditioning in mice without Purknje cells or with impaired long-term depression (LTD) at *gr-Pkj* synapses (see later) have in general produced the latter pattern of results (Aiba et al., 1994; L. Chen, Bao, Lockard, Kim, and Thompson, 1996; Kishimoto et al., 2001; Koekkoek et al., 2003, 2005; Shibuki et al., 1996) [but see (Miyata et al., 2001)]. The discrepancies between the early rabbit studies prevented definitive statements about the precise contribution of the cerebellar cortex to learning. Similarly, the generally incomplete deficits in mouse studies have proved difficult to interpret due to numerous factors, including potential extracerebellar contributions to learning in mice (Koekkoek et al., 2003, 2005). Nonetheless, these studies are consistent with the hypothesis that the cerebellar cortex is one site of plasticity underlying cerebellar learning.

In part to clarify the disagreement between the early rabbit studies, in one of the first studies in our laboratory we removed large portions of the ipsilateral cerebellar cortex via aspiration after training rabbits to respond to two separate CSs with differently timed responses (Perrett, Ruiz, and Mauk, 1993). Cerebellar cortex lesions did not abolish learned responses but disrupted their timing, revealing responses to both CSs with short, fixed latencies (Fig. 13-2A). This effect was correlated with the rostro-caudal extent of the lesion and whether or not it included the anterior lobe (Fig. 13-2B). Since the timing of conditioned eyelid responses is learned (Hoehler and Leonard, 1976; Kehoe, Graham-Clarke, and Schreurs, 1989; Mauk and Ruiz, 1992; Millenson, Kehoe,

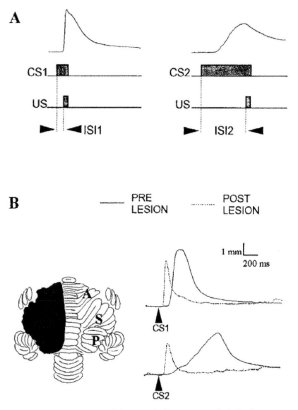

FIGURE 13-2   Posttraining lesions of the cerebellar cortex abolish the timing of the conditioned response. **A.** Conditioned responses (top panels) established using a differential conditioning procedure to one stimulus (CS1) paired with the US at an interstimulus interval (ISI) of 150 ms (left) and a second stimulus (CS2) paired with the US at an ISI of 750 ms (right). Note that the timing of the learned eyelid closure (e.g., upward deflections of the trace) is maximal near the expected time of the US for each CS. **B.** Posttraining lesions of the cerebellar cortex including the anterior lobe (A) unmask responses with short, fixed latencies independent of the ISI. S, ansiform lobule; P, paramedian lobule. [Reprinted with permission from Perret, Ruiz, and Mauk (1993).]

and Gormezano, 1977) (see below), these data, together with recent studies in mice (Koekkoek et al., 2003, 2005), suggest that plasticity in the cerebellar cortex is essential for this component of learning. This lesion effect has now been replicated in a number of studies across different laboratories. The residual short-latency responses (SLRs) have been observed after electrolytic lesions (Garcia, Steele, and Mauk, 1999; Medina, Garcia, Nores, Taylor, and Mauk, 2000) and reversible lesions in which the GABA$_A$ antagonist picrotoxin was infused into the AIN to disconnect it from the cerebellar cortex (Aksenov, Serdyukova, Irwin, and Bracha, 2004; Bao, Chen, Kim, and Thompson, 2002; Garcia and Mauk, 1998; Medina, Garcia, and Mauk, 2001; Ohyama and Mauk, 2001; Ohyama, Nores, and Mauk, 2003; Ohyama, Nores, Medina, Riusech, and Mauk, 2006).

As we describe later, SLRs reflect plasticity at a second site in the AIN. The presence of SLRs after a lesion of the cerebellar cortex is then taken as a functional index that a lesion of the cerebellar cortex is complete, and it provides a way to assess its precise contribution to learning. If the cerebellar cortex is necessary for inducing plasticity in the AIN (Medina and Mauk, 1999; Miles and Lisberger, 1981) (see upcoming Section B, on plasticity rules), then complete lesions should completely prevent further learning. In contrast, if plasticity is induced in the AIN independent of the cerebellar cortex, then complete lesions should only partly prevent new learning. We tested these alternatives by training subjects to one CS, after which we made electrolytic lesions of the cerebellar cortex to unmask SLRs (Fig. 13-3). After recovery, we then trained the subjects to a second CS of a different modality (Garcia, Steele, and Mauk, 1999). Subjects for whom the lesions unmasked SLRs completely failed to learn to the new CS. In contrast, control subjects with lesions outside of the anterior lobe learned to the new CS. These data, together with the results discussed earlier, indicate that the role of plasticity in the cerebellar cortex is two-fold — to mediate adaptive timing of learned responses and to provide a signal for inducing plasticity in the AIN.

2. Deep Cerebellar Nucleus

The SLRs spared by cerebellar cortex lesions are mediated by a form of neural plasticity in the AIN whose induction is associative. First, SLRs are associative, since they are observed neither before training nor after unpaired training but only after paired training (Medina, Garcia, and Mauk, 2001; Ohyama, Nores, and Mauk, 2003; Ohyama et al., 2006). Second, SLRs are highly input specific (Ohyama, Nores, and Mauk, 2003; Perrett and Mauk, 1995), in that they are less likely to be evoked by test stimuli that are less similar to the original CS. These data suggest that SLRs are caused not by a nonassociative process (Attwell, Ivarsson, Millar, and Yeo, 2002) but by an associative form of neural plasticity. Further evidence suggests that the essential site of the plasticity

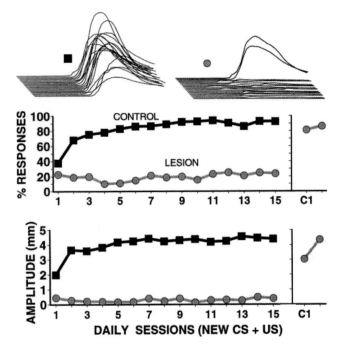

FIGURE 13-3 Lesions of the cerebellar cortex prevent subsequent learning to a new CS. Rabbits were first trained to establish robust responding to either an auditory or a tactile CS, after which electrolytic lesions were made in the cerebellar cortex. Subsequently, they were trained for 15 sessions in which a new CS of the alternative modality was paired with the US. Lesions that unmasked short-latency responses to the previously trained CS completely prevented learning to the new CS (gray circles), whereas new learning was normal for lesions which did not unmask short-latency responses (black circles). This effect is not attributable to nonspecific sensory, motivational, or motor deficits since learning to the new stimulus proceeded normally when training was switched to the opposite eye (CS1). Each trace in the upper panels show a response averaged across CS-alone test trials during a training session for a control (left) and lesioned (right) animal. [Reprinted with permission from Garcia, Steele, and Mauk (1999).]

underlying SLRs is in the AIN (Ohyama et al., 2006). First, SLRs are abolished when glutamatergic synaptic transmission in this nucleus is blocked, suggesting that the expression pathway underlying SLRs includes the AIN. Second, direct stimulation of mossy fibers supports SLRs, suggesting that the site of plasticity is downstream of these fibers. Finally, SLRs can be learned during inactivation of the red nucleus, suggesting that the site of plasticity is upstream of the red nucleus. Since the only site of plasticity downstream of mossy fibers and upstream of the red nucleus is in the AIN, these data collectively indicate that an input-specific form of plasticity in the AIN underlies SLRs (Ohyama et al., 2006).

B. Plasticity Rules

1. LTD at the gr-Pkj Synapse

Inspired by early computational theories (Albus, 1971; Marr, 1969) and their potential to explain VOR adaptation (Ito, 1972), Ito and colleagues discovered that synaptic transmission at the *gr-Pkj* synapse undergoes long-term depression (LTD) after pairing parallel and climbing fiber activation (Ito and Kano, 1982; Ito, Sakurai, and Tongroach, 1982). Climbing fiber–induced LTD (C. Chen and Thompson, 1995; Coesmans, Weber, De Zeeuw, and Hansel, 2004; Ekerot and Kano, 1985, 1989; Hirano, 1990; Lev-Ram, Mehta, Kleinfeld, and Tsien, 2003; Lev-Ram, Wong, Storm, and Tsien, 2002; Linden, Dickinson, Smeyne, and Connor, 1991; Sakurai, 1987; Wang and Linden, 2000) is a candidate mediator of *acquisition* of eyelid conditioning since it meets the criteria of (1) necessity for convergence, (2) sufficiency for induction, (3) capacity for expression, and (4) necessity for learning (Mauk, 1997; Nores, Medina, Steele, and Mauk, 2000).

First, *gr-Pkj* LTD occurs at a site where inputs from the to-be-associated stimuli converge — the tone and US, respectively, are conveyed to Purkinje cells via the mossy and climbing fibers. Second, the rules for inducing *gr-Pkj* LTD are consistent with those for establishing conditioning — both occur only when two stimuli occur close in time. It has been argued that because the fine temporal parameters for induction differ — LTD is induced optimally when parallel and climbing fibers are coactivated (Hirano, 1990; Ito and Kano, 1982; Ito, Sakurai, and Tongroach, 1982; Lev-Ram et al., 2003; Lev-Ram, Wong, Storm, and Tsien, 2002; Sakurai, 1987) (but see C. Chen and Thompson, 1995; Ekerot and Kano, 1985, 1989; Karachot, Kado, and Ito, 1994), whereas eyelid conditioning is optimal when the interval between tone onset and the US is 200–500 msec (Gormezano, Kehoe, and Marshall, 1983; Kehoe and Macrae, 2002) — *gr-Pkj* LTD cannot be the mechanism underlying learning (De Schutter, 1995; Llinas, Lang, and Welsh, 1997; Llinas and Welsh, 1993; Schreurs and Alkon, 1993). However, this argument requires that the tone CS activates granule cells only at its onset (Mauk and Donegan, 1997; Nores et al., 2000). This is highly unlikely, given that mossy fiber inputs are active throughout the extent of a tone (Aitkin and Boyd, 1978). If mossy fiber input is converted into a distributed time-varying granule cell representation (Buonomano and Mauk, 1994; Mauk and Donegan, 1997; Medina, Garcia, et al., 2000; Medina and Mauk, 2000; Nores et al., 2000), then the temporal requirements for eyelid conditioning (Gormezano et al., 1983; Kehoe and Macrae, 2002) can be reproduced, assuming that *gr-Pkj* synapses become eligible for LTD after 100 msec (Kettner et al., 1997; Raymond and Lisberger, 1998; Voicu and Mauk, in press).

Third, LTD of *gr-Pkj* synapses activated by the CS would enable expression of learned eyelid responses by decreasing the normally high rate of Purkinje cell activity during the CS, thereby driving eyelid closure via disinhibition of the AIN (Hesslow, 1994). In anesthetized ferrets in which conditioned eyelid responses were established by pairing a forelimb stimulation CS and a periorbital US, Purkinje cell activity showed *decreases* during the CS that were not evident before training (Hesslow and Ivarsson, 1994). Only Purkinje cells controlling eyelid closure were recorded in this study (selected based on whether stimulating them could evoke a delayed rebound in eyelid EMG activity), so it is likely that the pause in Purkinje cell activity was causing the conditioned responses. In addition, the finding that electrically stimulating Purkinje cells with pulse trains suppresses previously acquired conditioned responses (Hesslow, 1994) is consistent with the sufficiency-for-expression criterion.

Fourth, *gr-Pkj* LTD appears necessary for learning, since lesions of the cerebellar cortex prevent learning (Bao, Chen, Kim, and Thompson, 2002; Garcia, Steele, and Mauk, 1999). In a study using rabbits we lesioned the cerebellar cortex after first establishing learning to a tone or vibratory CS. Using the functional criterion that lesions should spare SLRs (Garcia, Steele, and Mauk, 1999), cerebellar cortex lesions completely prevented subsequent learning to a new CS, suggesting that *gr-Pkj* LTD is necessary for learning. This contrasts with prior studies in rabbits (Lavond and Steinmetz, 1989; Yeo and Hardiman, 1992) and various mouse knockout studies (Aiba et al., 1994; Bao, Chen, Qiao, Knusel, and Thompson, 1998; L. Chen et al., 1996; Shibuki et al., 1996) in which lesions of the cerebellar cortex (or deletion of processes necessary for *gr-Pkj* LTD) partly spared the capacity for learning. A major problem with such studies is that the lesions are often incomplete. Assessment of knockout mice is also complicated by numerous problems, including the potential for (1) compensatory processes, (2) movement artifacts in EMG recording studies, and (3) noncerebellar contributions to eyelid conditioning (Koekkoek et al., 2003, 2005). Future studies should address this issue using treatments that specifically and completely block *gr-Pkj* LTD as well as training protocols that preclude noncerebellar contributions to learning in mice.

2. LTP at the gr-Pkj Synapse

Although *gr-Pkj* LTD meets certain criteria for acquisition, other evidence suggests that it alone is not sufficient. Long-term potentiation (LTP) of *gr-Pkj* synapses is induced when granule cells are activated alone (Coesmans et al., 2004; Hirano, 1990; Lev-Ram et al., 2003, 2002; Sakurai, 1987; Salin, Malenka, and Nicoll, 1996). Although both pre- and postsynaptic forms of LTP have been identified, only the latter reverses climbing fiber–induced LTD (Coesmans et al., 2004; Lev-Ram, Mehta, et al., 2003; Salin et al., 1996) and could thus prevent synapses from saturating (Kenyon, Medina, and Mauk, 1998a, 1998b). Synaptic saturation challenges the LTD hypothesis of learning because spontaneous activity of climbing fibers (1–2 Hz) and granule cells (10–50 Hz) (Eccles et al., 1967) would potentially depress all synapses with sufficient time and abolish any specificity to learning (Llinas, Lang, and Welsh, 1997; Llinas and Welsh, 1993). However, computational analyses suggest that a mutually reversing (but not independent) LTD/LTP plasticity rule combined with inhibitory nucleo-olivary feedback allows simulations of the olivo-cerebellar circuit to both acquire and extinguish conditioned responses without synaptic saturation (Kenyon et al., 1998a, 1998b) (see later). This suggests that postsynaptic gr-Pkj LTP is also necessary for proper learning. There is no direct evidence for this prediction.

LTP of gr-Pkj synapses could underlie *extinction*, or the gradual decline of learned responses with repeated presentation of the CS alone. As will be elaborated later, gr-Pkj LTP occurs at a site where CS-evoked granule cell activity and a suppression of climbing fiber activity due to the inhibitory nucleo-olivary feedback (Medina, Nores, and Mauk, 2002) converge at the Purkinje cell. Like LTD and acquisition, gr-Pkj LTP is induced when gr-Pkj synapses are activated in the absence of climbing fiber activity (Coesmans et al., 2004; Hirano, 1990; Lev-Ram et al., 2003, 2002; Sakurai, 1987; Salin et al., 1996), much as behavioral extinction is induced when the CS is presented in the absence of the US. Induction of gr-Pkj LTP during extinction could increase CS-evoked granule cell activity and thereby restore suppression of AIN cells, consistent with a recording study in which extinction was correlated with a disappearance of the CS-evoked Purkinje cell pause (Hesslow and Ivarsson, 1994). Finally, recent studies are consistent with the hypothesis that gr-Pkj LTP is necessary for extinction, since preventing inhibitory nucleo-olivary feedback prevents extinction (Medina, Garcia, et al., 2002; Nilaweera, Zenitsky, and Bracha, 2005; Ramnani and Yeo, 1996).

3. Plasticity Downstream of the Cerebellar Cortex

A number of plasticity mechanisms at neurons immediately downstream of the Purkinje cells (the deep cerebellar and vestibular nuclei, respectively, for eyelid conditioning and VOR adaptation) could contribute to cerebellar learning. For eyelid conditioning, at least three known forms of plasticity in the AIN could contribute to SLRs spared by cerebellar cortex lesions: the formation of new mf-nuc synapses, LTP at existing mf-nuc synapses, and/or increased intrinsic excitability of AIN neurons. A recent study using electron microscopy to count synapses in the AIN of rats found that whereas the number of inhibitory synapses remained unchanged after either paired or explicitly unpaired training, excitatory synapses increased only after paired training (Kleim et al., 2002). Since excitatory input to the AIN comes largely from mossy fibers, these results raise the possibility that *de novo* formation of *mf-nuc* synapses is required before subsequent activity-dependent plasticity induced at these synapses (Pugh and Raman, 2006; Racine, Wilson, Gingell, and Sunderland, 1986) leads to the gradual emergence of SLRs. A cell-wide increase in the intrinsic excitability of AIN neurons (Aizenman and Linden, 2000) (although by itself this cannot account for the stimulus specificity of SLRs) could contribute to SLRs by modulating the induction of *mf-nuc* plasticity (Pugh and Raman, 2006).

Synaptic plasticity (*de novo* formation and/or LTP) of *mf-nuc* synapses meets the conditions of necessity for convergence, sufficiency for induction, capacity for expression, and necessity for acquisition of SLRs. As with *gr-Pkj* plasticity, CS and US inputs converge at the deep cerebellar nucleus. The conditions for inducing synaptic *mf-nuc* plasticity and learning agree. In intact rats, paired training selectively increases excitatory *mf-nuc* synapses as well as field potentials in the deep cerebellar nucleus driven by white-matter stimulation (Kleim et al., 2002), and *mf-nuc* LTP (Racine et al., 1986) is induced in cerebellar slices stimulated with parameters that mimicking the expected pattern of activity during eyelid conditioning (Pugh and Raman, 2006). Either form of synaptic *mf-nuc* plasticity would suffice to mediate SLRs observed in the absence of the cerebellar cortex. Finally, assuming that induction of SLRs is necessary for learning, that NMDA antagonists or protein synthesis/kinase inhibitors infused into the AIN impair acquisition of eyelid responses without preventing their expression (Bracha et al., 1998; G. Chen and Steinmetz, 2000a, 2000b) is also consistent with the notion that blocking *mf-nuc* plasticity prevents the induction of SLRs. Future studies should address this more directly using SLRs as a dependent measure.

In principle, three postsynaptic signals could control the induction of *mf-nuc* plasticity: (1) climbing fiber activity conveyed via collaterals to the deep nucleus, (2) deep nucleus activity (a Hebbian rule), and (3) Purkinje cell activity (Medina and Mauk, 1999). Several converging lines of evidence point to a Purkinje cell rule. For instance, in contrast to earlier work (Lavond and Steinmetz, 1989; Yeo and Hardiman, 1992), more recent studies show that physical or reversible lesions of the cerebellar cortex completely prevent new learning (Bao, Chen, Kim, and Thompson, 2002; Garcia et al., 1999). In vitro studies have also shown that transiently releasing the AIN neurons from hyperpolarization leads to the rebound excitation (Aizenman and Linden, 1999; Llinas and Muhlethaler, 1988) essential for inducing both increased excitability (Aizenman and Linden, 2000; Zhang, Shin, and Linden, 2004) and *mf-nuc* LTP (Pugh and Raman, 2006). These results suggest that input from the cerebellar cortex, in the form of either learned pause in Purkinje cell activity via LTD at *gr-Pkj* synapses (Hesslow and Ivarsson, 1994; Medina, Garcia et al., 2000) or a climbing fiber–induced pause in simple spike activity (Bell and Grimm, 1969; Eccles et al., 1967; Sato, Miura, Fushiki, and Kawasaki, 1992), is necessary for inducing *mf-nuc* plasticity. Proof of this concept is provided by a recent computational analysis of cerebellar learning, which showed that only a simulated cerebellum with the Purkinje rule could produce a stable set of *gr-Pkj* weights maintained in the presence of background activity in the olivo-cerebellar system (Medina and Mauk, 1999), consistent with earlier analyses of VOR adaptation (Miles and Lisberger, 1981). Studies showing the dynamics of plasticity induction in the cerebellar cortex and AIN (Medina, Garcia, and Mauk, 2001; Ohyama and Mauk, 2001) as well as the differential effects of cerebellar cortex lesions on adapted eye movements depending on the time of the lesion (Broussard and Kassardjian, 2004; Kassardjian et al., 2005; Shutoh, Ohki, Kitazwa, Itohara, and Nagao, 2006) are also consistent with the Purkinje cell rule.

C. Relative Roles of the Cerebellar Cortex and Deep Cerebellar Nuclei

1. The Cerebellar Cortex

Lesions of the cerebellar cortex affect the timing of learned eyelid responses. Since adaptively timed responses are learned even when temporally invariant mossy fiber stimulation is used as the CS (Hesslow, Svensson, and Ivarsson, 1999), precerebellar inputs with varying times to onset (Moore, Desmond, and Berthier, 1989) are not necessary for proper timing. Together with evidence pointing to the cerebellum as the essential site of plasticity, this suggests that the cerebellar cortex generates a temporal code. Simulations of the cerebellum (Buonomano and Mauk, 1994; Medina, Garcia et al., 2000; Medina and Mauk, 2000) again provide proof of this concept. Presented with patterns of mossy and climbing fiber inputs obtained from previous recording studies (Aitkin and Boyd, 1978; Sears and Steinmetz, 1991), the simulations reproduce adaptively timed learning (Medina, Garcia, et al., 2000; Medina and Mauk, 2000). An examination of the simulation revealed a spectrum of granule cells peaking in activity at different times since the onset of the tone — some firing only at the onset of the tone, some only during the early portion of the tone, some only near the end of the tone, and others throughout the tone (Medina, Garcia et al., 2000).

Adaptively timed cerebellar output is achieved by coupling the temporal code with *gr-Pkj* plasticity. Selective LTD at *gr-Pkj* synapses active late in the tone coinciding with the US is necessary, but not sufficient, to produce adaptively timed responses. Consistent with this view, a recent study showed that *gr-Pkj* LTD is necessary for learning-dependent timing (Koekkoek et al., 2003). Transgenic L7-PKCi mutant mice, deficient in *gr-Pkj* LTD due to selective inhibition of protein kinase C in Purkinje cells, were trained with a tone CS and periorbital US. These mice acquired eyelid responses with significantly shorter latencies than wild-type mice late in training; but unlike their wild-type counterparts, they could not adjust the timing of their responses when the ISI was increased. Interestingly, short-latency eyelid responses survived lesions of the AIN, pointing to a noncerebellar contribution to learning in mice (Koekkoek et al., 2003, 2005). Subtracting out the noncerebellar component, the data are consistent with the notion that gr-Pkj LTD is necessary (but not sufficient) for temporally specific learning.

Simulations and experiments indicate that gr-Pkj LTP is also necessary for acquiring adaptively timed responses. In a simulated cerebellum, the learned responses were not appropriately delayed when LTP was inactivated during training trials (Medina, Garcia et al., 2000). Thus, gr-Pkj LTP helps to shape appropriately delayed responses by selectively strengthening synapses active only during the early portion of the CS (in the absence of the US) to suppress the increased cerebellar output due to selective LTD of synapses active later in the CS (in the presence of the US) (Medina, Garcia et al., 2000). To test this hypothesis, we removed a fraction of the Purkinje cells after establishing robust learning to unmask responses with a small short-latency component in addition to a later adaptively timed component (Medina, Garcia et al., 2000). If the hypothesis is correct, then with continued training the short-latency component should eventually disappear due to LTP of the remaining early active gr-Pkj synapses. This prediction was confirmed when partial cerebellar cortex lesions initially unmasked a bimodal response whose early component gradually extinguished with further training (Medina, Garcia et al., 2000).

2. The Deep Cerebellar Nucleus

It has long been known that the AIN is required for the expression of learning, at least in rabbits (McCormick, Clark et al., 1982; McCormick and Thompson, 1984a). Current evidence is consistent with the hypothesis that the induction of plasticity at mf-nuc synapses is necessary for learning. Together with evidence from VOR adaptation, plasticity at mf-nuc synapses may consolidate long-term changes of cerebellar output in responses to stable changes in the pattern of cerebellar inputs. In addition, there is some evidence that these long-term changes mediate savings of learning (see later).

One line of evidence that the induction of mf-nuc plasticity is necessary for learning comes from reversible-lesion studies. A simulated cerebellum (Medina, Garcia, and Mauk, 2001) with bidirectional plasticity rules at gr-Pkj and mf-nuc synapses, respectively controlled by climbing fiber and Purkinje cells, respectively, predicts that learning in the cerebellar cortex should occur first, followed by an increase in cerebellar output as mf-nuc plasticity is induced. This prediction was confirmed when reversible lesions of the cerebellar cortex at different time points during acquisition revealed that learning and SLRs developed in parallel, suggesting that the induction of AIN plasticity (as measured by the presence of SLRs) determines the rate of learning (Medina, Garcia, and Mauk, 2001). A related study tested whether the expression of learning required AIN plasticity by utilizing the temporal specificity of learning in the cerebellar cortex (Ohyama and Mauk, 2001). In the first phase, subjects were trained with a long tone/ISI to subthreshold levels in order to restrict plasticity to the cerebellar cortex. As expected, when the cerebellar cortex was pharmacologically disconnected, no SLRs were observed. In the second phase, subjects were trained robustly with a short tone/ISI to induce AIN plasticity. As expected, disconnecting the cerebellar cortex after this phase unmasked SLRs. In the crucial test, presenting the original long tone evoked conditioned responses with two peaks, each timed to the ISIs in the second and first phases, respectively. Although blocking glutamatergic transmission in the AIN does not significantly affect the expression of normal conditioned eyelid responses (Aksenov, Serdyukova, Bloedel, and Bracha, 2005; Attwell et al., 2002), this may indicate that an increased excitability of AIN neurons is sufficient for expressing the learning in the cerebellar cortex (Ohyama et al., 2006).

Several rabbit studies have shown that infusing agents into the AIN that might block cellular processes necessary for inducing *mf-nuc* plasticity prevents acquisition but not expression of conditioned eyelid responses. This pattern of results has been observed after infusing the NMDA receptor antagonist AP5 (G. Chen and Steinmetz, 2000b), the protein kinase inhibitor H7 (G. Chen and Steinmetz, 2000a), the transcription inhibitor actinomycin D (Gomi et al., 1999), or the protein synthesis inhibitor anisomycin (Bracha et al., 1998) during and after training. Although these studies are subject to the caveat that the drugs may have leaked to the cerebellar cortex, collectively they are consistent with the notion that induction of AIN plasticity is required for acquisition.

The induction of *mf-nuc* plasticity may contribute to savings (Medina, Garcia, and Mauk, 2001), which refers to the observation that relearning or learning to a new CS is faster than original acquisition (Kehoe, 1988). In conditioning of the rabbit's third eyelid, both rapid reacquisition after extinction (Macrae and Kehoe, 1999; Napier, Macrae, and Kehoe, 1992) and cross-modal savings (Holt and Kehoe, 1985; Kehoe and Holt, 1984; Kehoe, Macrae, and Horne, 1995; Kehoe, Morrow, and Holt, 1984; Kehoe and Napier, 1991; Macrae and Kehoe, 1999; Schreurs and Kehoe, 1987) have been observed. Simulation analyses indicate that extinction is governed largely by reversal of *gr-Pkj* plasticity, which slows the reversal of *mf-nuc* plasticity (Medina, Garcia, and Mauk, 2001). This suggests that the degree of residual *mf-nuc* plasticity determines the degree of savings during subsequent acquisition (Medina, Garcia, and Mauk, 2001). This prediction was tested in animals trained for at least five sessions with a tone CS and US and then extinguished for up to 45 days. Querying *mf-nuc* plasticity at various time points during the course of extinction revealed that SLRs could continue to be unmasked by reversible cerebellar cortex lesions despite the complete extinction of normal conditioned responses. The decline of SLRs was much slower than the extinction of normal learned responses, and the percentage of SLRs was correlated with the degree of savings observed during a subsequent reacquisition session.

These results suggest that other forms of savings, such as cross-modal savings (e.g., from tones to lights, and vice versa), might also be mediated by AIN plasticity (Hansel, Linden, and D'Angelo, 2001; Medina, Garcia, and Mauk, 2001). A recent study found that SLRs did not generalize across auditory and visual modalities. This suggests that *mf-nuc* plasticity is not involved in cross-modal savings. Instead, a global increase in intrinsic excitability of AIN neurons may be involved (Hansel, Linden, and D'Angelo, 2001; Ohyama, Nores, and Mauk, 2003). Future studies could address this hypothesis.

D. Bidirectional Learning and Control of Climbing Fiber Activity

As noted earlier, a major challenge for the *gr-Pkj* LTD hypothesis of learning (Albus, 1971; Ito, 1972) is that both granule cells (Eccles et al., 1967) (10–50 Hz) and climbing fibers are spontaneously active (De Schutter, 1995; De Zeeuw et al., 1998; Gilbert, 1975; Glickstein, 1992; Llinas and Welsh, 1993) (1–2 Hz). Hence, if *gr-Pkj* LTD were the only mechanism, all synapses would eventually become weakened. Recent studies suggest that bidirectional plasticity at the *gr-Pkj* synapse and inhibitory nucleo-olivary feedback (Kenyon et al., 1998a, 1998b; Medina, Garcia et al., 2000; Medina and Mauk, 1999) allow the cerebellum to learn and extinguish adaptively timed responses despite this spontaneous activity.

The key to success is the triple-negative feedback control implemented by the olivo-cortico-nuclear loop (Fig. 13-1). Coupled with bidirectional *gr-Pkj* plasticity, this loop is critical for keeping climbing fiber activity within the range of 1–2 Hz. For instance, a treatment that increased (decreased) climbing fiber activity (Llinas and Volkind, 1973) would lead to greater *gr-Pkj* LTD (LTP) and hence decreased (increased) simple-spike activity (i.e., the Purkinje cell spikes due to granule cell activity). The resulting increase (decrease) in AIN activity in turn would decrease (increase) the level of climbing fiber activity back to its spontaneous rate via increased (decreased) nucleo-olivary inhibition. This self-regulation stabilizes synaptic weights at *gr-Pkj* synapses and maintains climbing fiber activity at an equilibrium level (Bloedel and Bracha, 1998; Miall, Keating, Malkmus, and Thach, 1998) (Fig. 13-4A).

In this context, acquisition and extinction of adaptively timed conditioned responses are the olivo-cerebellar system's solutions to restoring the equilibrium level of climbing fiber activity (Kenyon et al., 1998a, 1998b) on

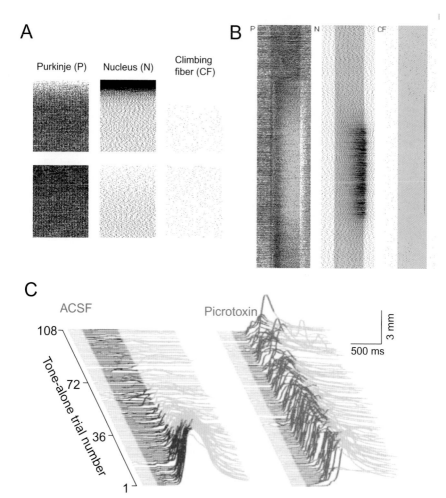

FIGURE 13-4 Inhibitory feedback from the deep cerebellar nucleus to the inferior olive contributes to self-regulation of climbing fiber activity and bidirectional learning. **A.** Simulated raster plots showing self-regulation of climbing fiber activity. The inhibitory feedback from the deep cerebellar nucleus (N) to the inferior olive completes a triple-negative feedback loop that drives climbing fiber (CF) activity back to an equilibrium firing rate (1–2 Hz, right panels) regardless of whether the initial conditions of the simulation promote high (top panels) or low (bottom panels) cerebellar output. **B.** Simulated raster plots showing Purkinje cell (P), deep cerebellar nucleus cell (N), and climbing fiber (CF) activity in response to a CS presented alone before training, during pairings of the CS and US, and extinction (i.e., the CS presented alone). Each row shows 2,000 ms from a single trial [500 ms before, 1,050 ms during (shaded area), and 450 ms after the CS]. During paired trials, the 50-ms US coterminates with the CS. Time progresses from top to bottom. Paired training gradually leads to a suppression of Purkinje cell activity (left) and increase in deep nucleus cell activity (center) during the CS, which in turn decreases the probability of the US activating a climbing fiber (right). Extinction leads to a gradual restoration of high Purkinje cell activity (bottom left) and decreased cerebellar output (bottom center), accompanied by a transient suppression of climbing fiber activity (bottom right) during the CS. **C.** Extinction of conditioned eyelid responses is blocked by preventing inhibition of climbing fiber activity via infusions of the GABA$_A$ antagonist picrotoxin into the inferior olive (red traces). The left panel shows normal extinction during ACSF infusions into the inferior olive (blue traces). [Reprinted with permission from Medina, Nores, and Mauk (2002).]

a shorter time scale. For instance, acquisition disrupts climbing fiber equilibrium by systematically increasing activity at the end of the CS (Fig. 13-4B). Due to temporal coding and plasticity in the cerebellar cortex, Purkinje cells learn to suppress activity later in the CS (Hesslow and Ivarsson, 1994; Medina, Garcia et al., 2000). This allows greater activity in the cerebellar nucleus, which in turn suppresses climbing fiber activity via nucleo-olivary feedback (Hesslow and Ivarsson, 1996; Sears and Steinmetz, 1991), especially near the time of the US. In this way the acquisition of eyelid responses restores climbing fiber activity to its equilibrium level. The reverse occurs in extinction. The omission of the US causes a transient decrease in the probability of climbing fiber activity around its expected time of occurrence due to nucleo-olivary inhibition from the conditioned response. This signal for the absence of the US, which depends on a nonzero level of spontaneous climbing fiber activity (Albus, 1971; Medina, Nores, and Mauk, 2002), induces *gr-Pkj* LTP and decreases cerebellar output back to baseline (Fig. 13-4B).

One prediction of this view is that eliminating the inhibitory nucleo-olivary feedback should block the transient suppression of climbing fiber activity and thereby prevent the extinction of previously learned responses (Fig. 13-4C). This prediction was recently tested in rabbits trained to a tone CS and then given infusions of the GABA$_A$ antagonist picrotoxin into the inferior olive to block nucleo-olivary feedback during extinction (Medina, Nores, and Mauk, 2002). Consistent with the prediction, conditioned responses failed to extinguish when nucleo-olivary inhibition was blocked, while infusing the AMPA antagonist NBQX into the inferior olive in a subsequent training session mimicked extinction despite continued presentation of the US. A similar effect was observed in an experiment in which training to one CS (CS1) normally prevents subsequent learning to a second CS2 paired with CS1 and the US. Consistent with the notion that this so-called blocking effect (Kamin, 1969) is due to suppression of the inferior olive by the learned response to CS1, infusing picrotoxin during the second phase of training rescued learning to CS2 (Kim, Krupa, and Thompson, 1998).

V. CONTRIBUTION OF LEARNING TO CEREBELLAR INFORMATION PROCESSING

Eyelid conditioning displays two temporal properties that provide insights into what the cerebellum computes. First, learning is associative and optimal within a limited range of interstimulus intervals (ISIs: intervals between CS and US onsets) (Gormezano et al., 1983; Kehoe and Macrae, 2002). Learning initially increases as the ISI increases above 100 ms and then decreases gradually as the ISI increases beyond 500 ms (Gormezano et al., 1983; Kehoe and Macrae, 2002; Ohyama, Nores, Murphy, and Mauk, 2003) (Fig. 13-5). This suggests

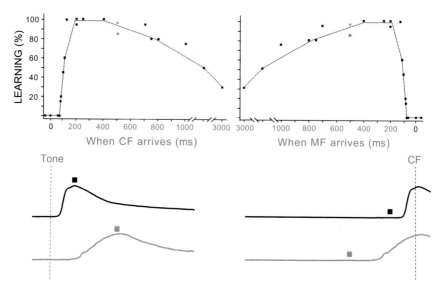

FIGURE 13-5   The contribution of learning to cerebellar computation. In the classical terminology of learning psychology, the ISI, or the interval between the onsets of mossy fiber (i.e.,
tone CS) and climbing fiber (i.e., periorbital US) inputs, determines both the level of learning
achieved (top left) and the timing of the learned response with respect to CS onset (bottom left).
This relationship can be recast in terms of when mossy fiber input arrives relative to the climbing
fiber input if the data are aligned with respect to the US-evoked climbing fiber activity (right).
This reveals that cerebellar output increases only when mossy fiber input reliably precedes climbing fiber input within a relatively short time period (approximately a few seconds) and that
increase in cerebellar output is timed so as to be maximal when the climbing fiber activity arrives.
These two temporal properties of cerebellar learning fulfill the key requirements for the cerebellum to generate temporally specific feed-forward control signals (see text for details). [Adapted
with permission from Ohyama, Nores, Murphy, and Mauk (2003).]

that for cerebllar output to increase, the CS-activated mossy fibers must reliably
precede the US-driven climbing fiber input by at least 100 ms and no more
than a few seconds. Second, learning is adaptively timed — the conditioned
eyelid response is gauged such that maximum closure occurs at or near the
expected time of the US (Frey and Ross, 1968; Schneiderman and Gormezano,
1964; Smith, 1968). Thus, when mossy fiber activity reliably signals a climbing
fiber response, not only does cerebellar output increase, but it does so at the
appropriate time.

   These properties illuminate how the cerebellum contributes to movements.
Sensory input is essential for motor control (Sanes, Mauritz, Evarts, Dalakas,
and Chu, 1984), and one way in which it can be used to control movements
is via feedback. A thermostat uses feedback to control temperature — on
detecting a difference between actual and desired temperature, it activates the

cooler or heater. The problem with this strategy is that it is slow, since adjustments are made only after a difference between desired and actual values is detected. This leads feedback systems to oscillate when driven fast (Kawato and Gomi, 1992; Massaquoi and Topka, 2002; Wolpert and Miall, 1996). In contrast, feed-forward systems anticipate predictable events. If opening a window reliably decreases temperature, a well-adjusted feed-forward thermostat heats the room rapidly whenever the window is subsequently opened. This makes feed-forward systems fast, but at a price — they must first learn to associate changes (temperature drop) with cues (opening a window) to produce an anticipatory response. To be accurate, feed-forward systems must also anticipate *when* change occurs in relation to a particular cue. If opening a small window decreases temperature more slowly than opening a large one, a thermostat would need to know when to start heating in each case. Therefore, an accurate feed-foward control system displays learning that is both associative *and* temporally specific (Ohyama, Nores, Murphy, and Mauk, 2003; Wolpert and Miall, 1996).

This is precisely the kind of learning displayed by eyelid conditioning. The cerebellum thus computes a temporally specific feed-foward prediction that minimizes climbing fiber responses driven by either external (e.g., somatosensory) or internal (e.g., cortical) input. Like previous authors, we suggest that the cerebellum comprises part of a feed-forward controller (Ito, 1970; Kawato and Gomi, 1992; Massaquoi and Topka, 2002), but we additionally emphasize the temporal aspects of its computation. Temporally specific feed-forward predictions probably apply generally across the cerebellum, because the organization of the cerebellar cortex is remarkably uniform (Eccles et al., 1967; Ito, 1984) and strong parallels are observed across learned eye movements that depend on the cerebellum (Ohyama, Nores, Murphy, and Mauk, 2003; Raymond, Lisberger, and Mauk, 1996). For instance, like eyelid conditioning, adaptive gain modification of the VOR (Ito, 1982; Miles and Lisberger, 1981; Robinson, 1976), saccades (Barash et al., 1999; Takagi, Zee, and Tamargo, 1998) and smooth pursuit (Rambold, Churchland, Selig, Jasmin, and Lisberger, 2002; Takagi, Zee, and Tamargo, 2000) all require the cerebellum, and cerebellar output appears crucial for the timing of saccades (Barash et al., 1999; Robinson and Fuchs, 2001; Robinson, Straube, and Fuchs, 1993; Takagi, Zee, and Tamargo, 1998; Thier, Dicke et al., 2000), VOR (Pastor, de la Cruz, and Baker, 1994; Raymond, Lisberger, and Mauk, 1996), and predictive pursuit (Barnes and Donelan, 1999; Kettner et al., 2002; Medina, Carey, and Lisberger, 2005; Vercher and Gauthier, 1988). Studies of single- and multi-joint limb movements also suggest that the cerebellum learns internal models of movement dynamics (Ebner, 1998; Imamizu et al., 2000; Kawato, 1999; Mussa-Ivaldi, 1999; Thach, 1998; Thoroughman and Shadmehr, 2000), i.e., feed-forward control signals that compensate for climbing fiber inputs in an adaptively timed manner. Finally, the temporal constraints of eyelid conditioning suggest that the computation contributes to any process, be it sensory, cognitive (Highstein and Thach, 2002; Schmahmann, 1997), emotive, or motor, that occurs on a relatively short time scale.

VI. CONCLUSION

The cerebellum clearly learns. Recent studies involving detailed computer simulation analyses and permanent and/or reversible lesions suggest that the plasticity underlying this learning is distributed between the cerebellar cortex and its downstream targets, with plasticity at *gr-Pkj* synapses mediating temporally specific learning and plasticity at *mf-nuc* synapses increasing the gain of cerebellar output. Bidirectional plasticity in the cerebellar cortex and deep cerebellar nucleus, respectively, are controlled by climbing fibers and Purkinje cells. The network dynamics of the cerebellar cortex could create a temporal code in the form of granule cells firing with different temporal delays during a CS, which can help shape the timing of cerebellar output via *gr-Pkj* plasticity. Inhibitory nucleo-olivary feedback is crucial for controlling bidirectional plasticity in the cerebellar cortex, a notion that has yet to be tested in other forms of cerebellar learning and adaptation.

The remarkably straightforward way in which eyelid conditioning engages the cerebellum — the CS and US, respectively, activate the mossy and climbing fibers, and the learned behavioral response is driven by the output of a deep cerebellar nucleus — makes the behavioral properties of eyelid conditioning a relatively direct window into what the cerebellum computes. The temporal properties of eyelid conditioning suggest that the cerebellar learning is involved in making temporally specific feed-forward predictions, a computation that is likely across the cerebellum, given its uniform circuitry. Such a computation can encapsulate the many functions previously attributed to the cerebellum. It also offers an explanation for the dysmetric and uncoordinated movements seen in patients with cerebellar pathology, and it provides a foundation for understanding how the cerebellum might contribute to nonmotor functions.

REFERENCES

Aiba, A., Kano, M., Chen, C., Stanton, M.E., Fox, G.D., Herrup, K., et al. (1994). Deficient cerebellar long-term depression and impaired motor learning in mGluR1 mutant mice. *Cell, 79*(2), 377–388.

Aitkin, L.M., and Boyd, J. (1978). Acoustic input to the lateral pontine nuclei. *Hear Res, 1*(1), 67–77.

Aizenman, C.D., and Linden, D.J. (1999). Regulation of the rebound depolarization and spontaneous firing patterns of deep nuclear neurons in slices of rat cerebellum. *J Neurophysiol, 82*(4), 1697–1709.

Aizenman, C.D., and Linden, D.J. (2000). Rapid, synaptically driven increases in the intrinsic excitability of cerebellar deep nuclear neurons. *Nat Neurosci, 3*(2), 109–111.

Aksenov, D., Serdyukova, N., Irwin, K., and Bracha, V. (2004). GABA neurotransmission in the cerebellar interposed nuclei: Involvement in classically conditioned eyeblinks and neuronal activity. *J Neurophysiol, 91*(2), 719–727.

Aksenov, D.P., Serdyukova, N.A., Bloedel, J.R., and Bracha, V. (2005). Glutamate neurotransmission in the cerebellar interposed nuclei: Involvement in classically conditioned eyeblinks and neuronal activity. *J Neurophysiol, 93*(1), 44–52.

Albus, J.S. (1971). A theory of cerebellar function. *Math Biosci, 10*, 25–61.

Attwell, P.J., Ivarsson, M., Millar, L., and Yeo, C.H. (2002). Cerebellar mechanisms in eyeblink conditioning. *Ann NY Acad Sci, 978*, 79–92.

Bao, S., Chen, L., Kim, J.J., and Thompson, R.F. (2002). Cerebellar cortical inhibition and classical eyeblink conditioning. *Proc Natl Acad Sci USA, 99*(3), 1592–1597.

Bao, S., Chen, L., Qiao, X., Knusel, B., and Thompson, R.F. (1998). Impaired eye-blink conditioning in waggler, a mutant mouse with cerebellar BDNF deficiency. *Learn Mem, 5*(4–5), 355–364.

Barash, S., Melikyan, A., Sivakov, A., Zhang, M., Glickstein, M., and Thier, P. (1999). Saccadic dysmetria and adaptation after lesions of the cerebellar cortex. *J Neurosci, 19*(24), 10931–10939.

Barnes, G.R., and Donelan, S.F. (1999). The remembered pursuit task: Evidence for segregation of timing and velocity storage in predictive oculomotor control. *Exp Brain Res, 129*(1), 57–67.

Bell, C.C., and Grimm, R.J. (1969). Discharge properties of Purkinje cells recorded on single and double microelectrodes. *J Neurophysiol, 32*(6), 1044–1055.

Bloedel, J.R., and Bracha, V. (1998). Current concepts of climbing fiber function. *Anat Rec, 253*(4), 118–126.

Bracha, V., Irwin, K.B., Webster, M.L., Wunderlich, D.A., Stachowiak, M.K., and Bloedel, J.R. (1998). Microinjections of anisomycin into the intermediate cerebellum during learning affect the acquisition of classically conditioned responses in the rabbit. *Brain Res, 788*(1–2), 169–178.

Broussard, D.M., and Kassardjian, C.D. (2004). Learning in a simple motor system. *Learn Mem, 11*(2), 127–136.

Buonomano, D.V., and Mauk, M.D. (1994). Neural network model of the cerebellum: Temporal discrimination and the timing of motor responses. *Neural Comp., 6*, 38–55.

Chan-Palay, V. (1977). *Cerebellar Dentate Nucleus: Organization, Cytology and Transmitters*. New York: Springer-Verlag.

Chapman, P.F., Steinmetz, J.E., Sears, L.L., and Thompson, R.F. (1990). Effects of lidocaine injection in the interpositus nucleus and red nucleus on conditioned behavioral and neuronal responses. *Brain Res, 537*(1–2), 149–156.

Chen, C., and Thompson, R.F. (1995). Temporal specificity of long-term depression in parallel fiber–Purkinje synapses in rat cerebellar slice. *Learn Mem, 2*(3–4), 185–198.

Chen, G., and Steinmetz, J.E. (2000a). Microinfusion of protein kinase inhibitor H7 into the cerebellum impairs the acquisition but not the retention of classical eyeblink conditioning in rabbits. *Brain Res, 856*(1–2), 193–201.

Chen, G., and Steinmetz, J.E. (2000b). Intra-cerebellar infusion of NMDA receptor antagonist AP5 disrupts classical eyeblink conditioning in rabbits. *Brain Res, 887*(1), 144–156.

Chen, L., Bao, S., Lockard, J.M., Kim, J.K., and Thompson, R.F. (1996). Impaired classical eyeblink conditioning in cerebellar-lesioned and Purkinje cell degeneration (pcd) mutant mice. *J Neurosci, 16*(8), 2829–2838.

Coesmans, M., Weber, J.T., De Zeeuw, C.I., and Hansel, C. (2004). Bidirectional parallel fiber plasticity in the cerebellum under climbing fiber control. *Neuron, 44*(4), 691–700.

De Schutter, E. (1995). Cerebellar long-term depression might normalize excitation of Purkinje cells: A hypothesis. *Trends Neurosci, 18*(7), 291–295.

De Zeeuw, C.I., Simpson, J.I., Hoogenraad, C.C., Galjart, N., Koekkoek, S.K., and Ruigrok, T.J. (1998). Microcircuitry and function of the inferior olive. *Trends Neurosci, 21*(9), 391–400.

Ebner, T.J. (1998). A role for the cerebellum in the control of limb movement velocity. *Curr Opin Neurobiol, 8*(6), 762–769.

Eccles, J.C., Ito, M., and Szentàgothai, J. (1967). *The Cerebellum as a Neuronal Machine*. Berlin Springer-Verlag.

Ekerot, C.F., and Kano, M. (1985). Long-term depression of parallel fibre synapses following stimulation of climbing fibres. *Brain Res, 342*(2), 357–360.

Ekerot, C.F., and Kano, M. (1989). Stimulation parameters influencing climbing fibre induced long-term depression of parallel fibre synapses. *Neurosci Res, 6*(3), 264–268.

Frey, P.W., and Ross, L.E. (1968). Classical conditioning of the rabbit eyelid response as a function of interstimulus interval. *J Comp Physiol Psychol, 65*(2), 246–250.

Garcia, K.S., and Mauk, M.D. (1998). Pharmacological analysis of cerebellar contributions to the timing and expression of conditioned eyelid responses. *Neuropharmacology, 37*(4–5), 471–480.

Garcia, K.S., Steele, P.M., and Mauk, M.D. (1999). Cerebellar cortex lesions prevent acquisition of conditioned eyelid responses. *J Neurosci, 19*(24), 10940–10947.

Gilbert, P. (1975). How the cerebellum could memorise movements. *Nature, 254*(5502), 688–689.

Glickstein, M. (1992). The cerebellum and motor learning. *Curr Opin Neurobiol, 2*(6), 802–806.

Gomi, H., Sun, W., Finch, C.E., Itohara, S., Yoshimi, K., and Thompson, R.F. (1999). Learning induces a CDC2-related protein kinase, KKIAMRE. *J Neurosci, 19*(21), 9530–9537.

Gonshor, A., and Jones, G.M. (1973). Proceedings: Changes of human vestibulo-ocular response induced by vision-reversal during head rotation. *J Physiol, 234*(2), 102P–103P.

Gormezano, I., Kehoe, E.J., and Marshall, M.S. (1983). Twenty years of classical conditioning research with the rabbit. In J.M. Sprague and A.N. Epstein (eds.), *Progress in Psychobiology and Physiological Psychology* (pp. 197–275). New York: Academic Press.

Hansel, C., Linden, D.J., and D'Angelo, E. (2001). Beyond parallel-fiber LTD: The diversity of synaptic and nonsynaptic plasticity in the cerebellum. *Nat Neurosci, 4*(5), 467–475.

Hesslow, G. (1994). Inhibition of classically conditioned eyeblink responses by stimulation of the cerebellar cortex in the decerebrate cat. *J Physiol, 476*(2), 245–256.

Hesslow, G., and Ivarsson, M. (1994). Suppression of cerebellar Purkinje cells during conditioned responses in ferrets. *Neuroreport, 5*(5), 649–652.

Hesslow, G., and Ivarsson, M. (1996). Inhibition of the inferior olive during conditioned responses in the decerebrate ferret. *Exp Brain Res, 110*(1), 36–46.

Hesslow, G., Svensson, P., and Ivarsson, M. (1999). Learned movements elicited by direct stimulation of cerebellar mossy fiber afferents. *Neuron, 24*(1), 179–185.

Highstein, S.M., and Thach, W.T. (eds.). (2002). *The Cerebellum: Recent Developments in Cerebellar Research* (Vol. 978). New York: New York Academy of Sciences.

Hirano, T. (1990). Depression and potentiation of the synaptic transmission between a granule cell and a Purkinje cell in rat cerebellar culture. *Neurosci Lett, 119*(2), 141–144.

Hoehler, F.K., and Leonard, D.W. (1976). Double responding in classical nictitating membrane conditioning with single-CS dual-ISI training. *Pavlov J Biol Sci, 11*(3), 180–190.

Holt, P.E., and Kehoe, E.J. (1985). Cross-modal transfer as a function of similarities between training tasks in classical conditioning of the rabbit. *Anim Learn Behav, 13*, 51–59.

Hopp, J.J., and Fuchs, A.F. (2004). The characteristics and neuronal substrate of saccadic eye movement plasticity. *Prog Neurobiol, 72*(1), 27–53.

Imamizu, H., Miyauchi, S., Tamada, T., Sasaki, Y., Takino, R., Putz, B., et al. (2000). Human cerebellar activity reflecting an acquired internal model of a new tool. *Nature, 403*(6766), 192–195.

Ito, M. (1970). Neurophysiological aspects of the cerebellar motor control system. *Int J Neurol, 7*(2), 162–176.

Ito, M. (1972). Neural design of the cerebellar motor control system. *Brain Res, 40*(1), 81–84.

Ito, M. (1982). Cerebellar control of the vestibulo-ocular reflex — around the flocculus hypothesis. *Annu Rev Neurosci, 5*, 275–296.

Ito, M. (1984). *The Cerebellum and Neural Control*. New York: Raven Press.

Ito, M. (2001). Cerebellar long-term depression: Characterization, signal transduction, and functional roles. *Physiol Rev, 81*(3), 1143–1195.

Ito, M., and Kano, M. (1982). Long-lasting depression of parallel fiber–Purkinje cell transmission induced by conjunctive stimulation of parallel fibers and climbing fibers in the cerebellar cortex. *Neurosci Lett, 33*(3), 253–258.

Ito, M., Sakurai, M., and Tongroach, P. (1982). Climbing fiber-induced depression of both mossy fiber responsiveness and glutamate sensitivity of cerebellar Purkinje cells. *J Physiol, 324*, 113–134.

Kamin, L. (1969). Selective association and conditioning. In W.K. Honig (ed.), *Fundamental Issues in Associative Learning*. Halifax, Nova Scotia: Dalhousie Press.

Karachot, L., Kado, R.T., and Ito, M. (1994). Stimulus parameters for induction of long-term depression in in vitro rat Purkinje cells. *Neurosci Res, 21*(2), 161–168.

Kassardjian, C.D., Tan, Y.F., Chung, J.Y., Heskin, R., Peterson, M.J., and Broussard, D.M. (2005). The site of a motor memory shifts with consolidation. *J Neurosci, 25*(35), 7979–7985.

Kawato, M. (1999). Internal models for motor control and trajectory planning. *Curr Opin Neurobiol, 9*(6), 718–727.

Kawato, M., and Gomi, H. (1992). The cerebellum and VOR/OKR learning models. *Trends Neurosci, 15*(11), 445–453.

Kehoe, E.J. (1988). A layered network model of associative learning: Learning to learn and configuration. *Psychol Rev, 95*(4), 411–433.

Kehoe, E.J., Graham-Clarke, P., and Schreurs, B.G. (1989). Temporal patterns of the rabbit's nictitating membrane response to compound and component stimuli under mixed CS-US intervals. *Behav Neurosci, 103*(2), 283–295.

Kehoe, E.J., and Holt, P. (1984). Transfer across CS-US intervals and sensory modalities in classical conditioning of the rabbit. *Anim Learn Behav, 12*(2), 122–128.

Kehoe, E.J., and Macrae, M. (2002). Fundamental behavioral methods and findings in classical conditioning. In J.W. Moore (ed.), *A Neuroscientist's Guide to Classical Conditioning* (pp. 171–231). Berlin: Springer.

Kehoe, E.J., Macrae, M., and Horne, A. (1995). Learning to learn — real-time features and a connectionist model. *Adapt Behav, 3*(3), 235–271.

Kehoe, E.J., Morrow, L.D., and Holt, P.E. (1984). General transfer across sensory modalities survives reductions in the original conditioned reflex in the rabbit. *Anim Learn Behav, 12*(2), 129–136.

Kehoe, E.J., and Napier, R.M. (1991). Temporal specificity in cross-modal transfer of the rabbit nictitating membrane response. *J Exp Psychol Anim Behav Process, 17*(1), 26–35.

Kenyon, G.T., Medina, J.F., and Mauk, M.D. (1998a). A mathematical model of the cerebellar-olivary system I: Self-regulating equilibrium of climbing fiber activity. *J Comput Neurosci, 5*(1), 17–33.

Kenyon, G.T., Medina, J.F., and Mauk, M.D. (1998b). A mathematical model of the cerebellar-olivary system II: Motor adaptation through systematic disruption of climbing fiber equilibrium. *J Comput Neurosci, 5*(1), 71–90.

Kettner, R.E., Leung, H.C., and Peterson, B.W. (1996). Predictive smooth pursuit of complex two-dimensional trajectories in monkey: Component interactions. *Exp Brain Res, 108*(2), 221–235.

Kettner, R.E., Mahamud, S., Leung, H.C., Sitkoff, N., Houk, J.C., Peterson, B.W., et al. (1997). Prediction of complex two-dimensional trajectories by a cerebellar model of smooth pursuit eye movement. *J Neurophysiol, 77*(4), 2115–2130.

Kettner, R.E., Suh, M., Davis, D., and Leung, H.C. (2002). Modeling cerebellar flocculus and paraflocculus involvement in complex predictive smooth eye pursuit in monkeys. *Ann NY Acad Sci, 978*, 455–467.

Kim, J.J., Krupa, D.J., and Thompson, R.F. (1998). Inhibitory cerebello-olivary projections and blocking effect in classical conditioning. *Science, 279*(5350), 570–573.

Kishimoto, Y., Kawahara, S., Suzuki, M., Mori, H., Mishina, M., and Kirino, Y. (2001). Classical eyeblink conditioning in glutamate receptor subunit delta 2 mutant mice is impaired in the delay paradigm but not in the trace paradigm. *Eur J Neurosci, 13*(6), 1249–1253.

Kleim, J.A., Freeman, J.H.J., Bruneau, R., Nolan, B.C., Cooper, N.R., Zook, A., et al. (2002). Synapse formation is associated with memory storage in the cerebellum. *Proc Natl Acad Sci USA, 99*, 13228–13231.

Koekkoek, S.K., Hulscher, H.C., Dortland, B.R., Hensbroek, R.A., Elgersma, Y., Ruigrok, T.J., et al. (2003). Cerebellar LTD and learning-dependent timing of conditioned eyelid responses. *Science, 301*(5640), 1736–1739.

Koekkoek, S.K., Yamaguchi, K., Milojkovic, B.A., Dortland, B.R., Ruigrok, T.J., Maex, R., et al. (2005). Deletion of FMR1 in Purkinje cells enhances parallel-fiber LTD, enlarges spines, and attenuates cerebellar eyelid conditioning in fragile X syndrome. *Neuron, 47*(3), 339–352.

Krupa, D.J., Thompson, J.K., and Thompson, R.F. (1993). Localization of a memory trace in the mammalian brain. *Science, 260*(5110), 989–991.

Krupa, D.J., and Thompson, R.F. (1995). Inactivation of the superior cerebellar peduncle blocks expression but not acquisition of the rabbit's classically conditioned eye-blink response. *Proc Natl Acad Sci USA, 92*(11), 5097–5101.

Lavond, D.G., and Steinmetz, J.E. (1989). Acquisition of classical conditioning without cerebellar cortex. *Behav Brain Res, 33*(2), 113–164.

Leung, H.C., and Kettner, R.E. (1997). Predictive smooth pursuit of complex two-dimensional trajectories demonstrated by perturbation responses in monkeys. *Vision Res, 37*(10), 1347–1354.

Lev-Ram, V., Mehta, S.B., Kleinfeld, D., and Tsien, R.Y. (2003). Reversing cerebellar long-term depression. *Proc Natl Acad Sci USA, 100*, 15989–15993.

Lev-Ram, V., Wong, S.T., Storm, D.R., and Tsien, R.Y. (2002). A new form of cerebellar long-term potentiation is postsynaptic and depends on nitric oxide but not cAMP. *Proc Natl Acad Sci USA, 99*, 8389–8393.

Linden, D.J., Dickinson, M.H., Smeyne, M., and Connor, J.A. (1991). A long-term depression of AMPA currents in cultured cerebellar Purkinje neurons. *Neuron, 7*(1), 81–89.

Llinas, R. (1981). Electrophysiology of cerebellar networks. In J.M. Brookhart and V.B. Mountcastle (eds.), *Handbook of Physiology, Section 1: The Nervous System, Volume II. Motor Control* (Vol. 2, pp. 831–876). Bethesda, MD: American Physiological Society.

Llinas, R., Lang, E.J., and Welsh, J.P. (1997). The cerebellum, LTD, and memory: Alternative views. *Learn Mem, 3*(6), 445–455.

Llinas, R., and Muhlethaler, M. (1988). Electrophysiology of guinea-pig cerebellar nuclear cells in the in vitro brain stem-cerebellar preparation. *J Physiol, 404*, 241–258.

Llinas, R., and Volkind, R.A. (1973). The olivo-cerebellar system: Functional properties as revealed by harmaline-induced tremor. *Exp Brain Res, 18*(1), 69–87.

Llinas, R., and Welsh, J.P. (1993). On the cerebellum and motor learning. *Curr Opin Neurobiol, 3*(6), 958–965.

Macrae, M., and Kehoe, E.J. (1999). Savings after extinction in conditioning the rabbit's nictitating membrane response. *Psychobiology, 27,* 85–94.

Marr, D. (1969). A theory of cerebellar cortex. *J Physiol, 202*(2), 437–470.

Massaquoi, S.G., and Topka, H. (2002). Models of cerebellar function. In M.U. Manto and M. Pandolfo (eds.), *The Cerebellum and Its Disorders* (p. 612). Cambridges, U.K.: Cambridge University Press.

Mauk, M.D. (1997). Roles of cerebellar cortex and nuclei in motor learning: Contradictions or clues? *Neuron, 18*(3), 343–346.

Mauk, M.D., and Donegan, N.H. (1997). A model of Pavlovian eyelid conditioning based on the synaptic organization of the cerebellum. *Learn Mem, 4*(1), 130–158.

Mauk, M.D., and Ruiz, B.P. (1992). Learning-dependent timing of Pavlovian eyelid responses: Differential conditioning using multiple interstimulus intervals. *Behav Neurosci, 106*(4), 666–681.

Mauk, M.D., Steinmetz, J.E., and Thompson, R.F. (1986). Classical conditioning using stimulation of the inferior olive as the unconditioned stimulus. *Proc Natl Acad Sci USA, 83,* 5349–5353.

McCormick, D.A., Clark, G.A., Lavond, D.G., and Thompson, R.F. (1982). Initial localization of the memory trace for a basic form of learning. *Proc Natl Acad Sci USA, 79*(8), 2731–2735.

McCormick, D.A., and Thompson, R.F. (1984a). Cerebellum: Essential involvement in the classically conditioned eyelid response. *Science, 223*(4633), 296–299.

McCormick, D.A., and Thompson, R.F. (1984b). Neuronal responses of the rabbit cerebellum during acquisition and performance of a classically conditioned nictitating membrane-eyelid response. *J Neurosci, 4*(11), 2811–2822.

Medina, J.F., Carey, M.R., and Lisberger, S.G. (2005). The representation of time for motor learning. *Neuron, 45*(1), 157–167.

Medina, J.F., Garcia, K.S., and Mauk, M.D. (2001). A mechanism for savings in the cerebellum. *J Neurosci, 21*(11), 4081–4089.

Medina, J.F., Garcia, K.S., Nores, W.L., Taylor, N.M., and Mauk, M.D. (2000). Timing mechanisms in the cerebellum: Testing predictions of a large-scale computer simulation. *J Neurosci, 20*(14), 5516–5525.

Medina, J.F., and Mauk, M.D. (1999). Simulations of cerebellar motor learning: Computational analysis of plasticity at the mossy fiber to deep nucleus synapse. *J Neurosci, 19*(16), 7140–7151.

Medina, J.F., and Mauk, M.D. (2000). Computer simulation of cerebellar information processing. *Nat Neurosci, 3 Suppl,* 1205–1211.

Medina, J.F., Nores, W.L., and Mauk, M.D. (2002). Inhibition of climbing fibres is a signal for the extinction of conditioned eyelid responses. *Nature, 416,* 330–333.

Miall, R.C., Keating, J.G., Malkmus, M., and Thach, W.T. (1998). Simple spike activity predicts occurrence of complex spikes in cerebellar Purkinje cells. *Nat Neurosci, 1*(1), 13–15.

Miles, F.A., and Lisberger, S.G. (1981). Plasticity in the vestibulo-ocular reflex: A new hypothesis. *Annu Rev Neurosci, 4,* 273–299.

Millenson, J.R., Kehoe, E.J., and Gormezano, I. (1977). Classical conditioning of the rabbit's nictitating membrane response under fixed and mixed CS-US intervals. *Learn Motiv, 8*(3), 351–366.

Miyata, M., Kim, H.T., Hashimoto, K., Lee, T.K., Cho, S.Y., Jiang, H., et al. (2001). Deficient long-term synaptic depression in the rostral cerebellum correlated with impaired motor learning in phospholipase C beta4 mutant mice. *Eur J Neurosci, 13*(10), 1945–1954.

Moore, J.W., Desmond, J.E., and Berthier, N.E. (1989). Adaptively timed conditioned responses and the cerebellum: A neural network approach. *Biol Cybern, 62*(1), 17–28.

Mussa-Ivaldi, F.A. (1999). Modular features of motor control and learning. *Curr Opin Neurobiol,* *9*(6), 713–717.

Napier, R.M., Macrae, M., and Kehoe, E.J. (1992). Rapid reaquisition in conditioning of the rabbit's nictitating membrane response. *J Exp Psychol Anim Behav Process, 18*(2), 182–192.

Nilaweera, W.U., Zenitsky, G.D., and Bracha, V. (2005). Inactivation of the brachium conjunctivum prevents extinction of classically conditioned eyeblinks. *Brain Res, 1045*(1–2), 175–184.

Nores, W.L., Medina, J.F., Steele, P.M., and Mauk, M.D. (2000). Relative contributions of the cerebellar cortex and cerebellar nucleus to eyelid conditioning. In D.S. Woodruff and J.E. Steinmetz (eds.), *Eyeblink Classical Conditioning: Animal Models* (Vol. 2). Boston: Kluwer Academic.

Ohyama, T., and Mauk, M. (2001). Latent acquisition of timed responses in cerebellar cortex. *J Neurosci, 21*(2), 682–690.

Ohyama, T., Nores, W.L., and Mauk, M.D. (2003). Stimulus generalization of conditioned eyelid responses produced without cerebellar cortex: Implications for plasticity in the cerebellar nuclei. *Learn Mem, 10*(5), 346–354.

Ohyama, T., Nores, W.L., Murphy, M., and Mauk, M.D. (2003). What the cerebellum computes. *Trends Neurosci, 26*(4), 222–227.

Optican, L.M., and Robinson, D.A. (1980). Cerebellar-dependent adaptive control of primate saccadic system. *J Neurophysiol, 44*(6), 1058–1076.

Pastor, A.M., de la Cruz, R.R., and Baker, R. (1994). Cerebellar role in adaptation of the goldfish vestibuloocular reflex. *J Neurophysiol, 72*(3), 1383–1394.

Perrett, S.P., and Mauk, M.D. (1995). Extinction of conditioned eyelid responses requires the anterior lobe of cerebellar cortex. *J Neurosci, 15*(3 Pt 1), 2074–2080.

Perrett, S.P., Ruiz, B.P., and Mauk, M.D. (1993). Cerebellar cortex lesions disrupt learning-dependent timing of conditioned eyelid responses. *J Neurosci, 13*(4), 1708–1718.

Pugh, J.R., and Raman, I.M. (2006) Potentiation of mossy fiber EPSCs in the cerebellar nuclei by NMDA receptor activation followed by postinhibitory rebound current. Neuron, *51*, 113–123.

Racine, R.J., Wilson, D.A., Gingell, R., and Sunderland, D. (1986). Long-term potentiation in the interpositus and vestibular nuclei in the rat. *Exp Brain Res, 63*(1), 158–162.

Rambold, H., Churchland, A., Selig, Y., Jasmin, L., and Lisberger, S.G. (2002). Partial ablations of the flocculus and ventral paraflocculus in monkeys cause linked deficits in smooth pursuit eye movements and adaptive modification of the VOR. *J Neurophysiol, 87*(2), 912–924.

Ramnani, N., and Yeo, C.H. (1996). Reversible inactivations of the cerebellum prevent the extinction of conditioned nictitating membrane responses in rabbits. *J Physiol, 495 (Pt 1),* 159–168.

Raymond, J.L., and Lisberger, S.G. (1998). Neural learning rules for the vestibulo-ocular reflex. *J Neurosci, 18*(21), 9112–9129.

Raymond, J.L., Lisberger, S.G., and Mauk, M.D. (1996). The cerebellum: A neuronal learning machine? *Science, 272*(5265), 1126–1131.

Robinson, D.A. (1976). Adaptive gain control of vestibuloocular reflex by the cerebellum. *J. Neurophysiol, 39*, 954–969.

Robinson, F.R., and Fuchs, A.F. (2001). The role of the cerebellum in voluntary eye movements. *Annu Rev Neurosci, 24*, 981–1004.

Robinson, F.R., Fuchs, A.F., and Noto, C.T. (2002). Cerebellar influences on saccade plasticity. *Ann NY Acad Sci, 956*, 155–163.

Robinson, F.R., Straube, A., and Fuchs, A.F. (1993). Role of the caudal fastigial nucleus in saccade generation. II. Effects of muscimol inactivation. *J Neurophysiol, 70*(5), 1741–1758.

Robinson, F.R., Straube, A., and Fuchs, A.F. (1997). Participation of caudal fastigial nucleus in smooth pursuit eye movements. II. Effects of muscimol inactivation. *J Neurophysiol, 78*(2), 848–859.

Sakurai, M. (1987). Synaptic modification of parallel fibre-Purkinje cell transmission in in vitro guinea-pig cerebellar slices. *J Physiol, 394,* 463–480.

Salin, P.A., Malenka, R.C., and Nicoll, R.A. (1996). Cyclic AMP mediates a presynaptic form of LTP at cerebellar parallel fiber synapses. *Neuron, 16,* 797–803.

Sanes, J.N., Mauritz, K.H., Evarts, E.V., Dalakas, M.C., and Chu, A. (1984). Motor deficits in patients with large-fiber sensory neuropathy. *Proc Natl Acad Sci USA, 81*(3), 979–982.

Sato, Y., Miura, A., Fushiki, H., and Kawasaki, T. (1992). Short-term modulation of cerebellar Purkinje cell activity after spontaneous climbing fiber input. *J Neurophysiol, 68*(6), 2051–2062.

Schmahmann, J.D. (ed.). (1997). *The Cerebellum and Cognition.* San Diego: Academic Press.

Schneiderman, N., and Gormezano, I. (1964). Conditioning of the nictitating membrane of the rabbit as a function of CS-US interval. *J Comp Physiol Psychol, 57,* 188–195.

Schreurs, B., and Kehoe, E. (1987). Cross-modal transfer as a function of initial training level in classical-conditioning with the rabbit. *Anim Learn Behav, 15*(1), 47–54.

Schreurs, B.G., and Alkon, D.L. (1993). Rabbit cerebellar slice analysis of long-term depression and its role in classical conditioning. *Brain Res, 631*(2), 235–240.

Scudder, C.A. (2002). Role of the fastigial nucleus in controlling horizontal saccades during adaptation. *Ann NY Acad Sci, 978,* 63–78.

Scudder, C.A., and McGee, D.M. (2003). Adaptive modification of saccade size produces correlated changes in the discharges of fastigial nucleus neurons. *J Neurophysiol, 90*(2), 1011–1026.

Sears, L.L., and Steinmetz, J.E. (1991). Dorsal accessory inferior olive activity diminishes during acquisition of the rabbit classically conditioned eyelid response. *Brain Res, 545*(114–122.).

Shibuki, K., Gomi, H., Chen, L., Bao, S., Kim, J.J., Wakatsuki, H., et al. (1996). Deficient cerebellar long-term depression, impaired eyeblink conditioning, and normal motor coordination in GFAP mutant mice. *Neuron, 16*(3), 587–599.

Shutoh, F., Ohki, M., Kitazwa, H., Itohara, S., and Nagao, S. (2006). Memory trace of motor learning shifts transsynaptically from cerebellar cortex to nuclei for consolidation. *Neuroscience, 139*(2), 767–777.

Smith, M.C. (1968). CS-US interval and US intensity in classical conditioning of the rabbit's nictitating membrane response. *J Comp Physiol Psychol, 66*(3), 679–687.

Steinmetz, J.E., Lavond, D.G., and Thompson, R.F. (1989). Classical conditioning in rabbits using pontine nucleus stimulation as a conditioned stimulus and inferior olive stimulation as an unconditioned stimulus. *Synapse, 3*(3), 225–233.

Takagi, M., Zee, D.S., and Tamargo, R.J. (1998). Effects of lesions of the oculomotor vermis on eye movements in primate: Saccades. *J Neurophysiol, 80*(4), 1911–1931.

Takagi, M., Zee, D.S., and Tamargo, R.J. (2000). Effects of lesions of the oculomotor cerebellar vermis on eye movements in primate: Smooth pursuit. *J Neurophysiol, 83*(4), 2047–2062.

Thach, W.T. (1998). A role for the cerebellum in learning movement coordination. *Neurobiol Learn Mem, 70*(1–2), 177–188.

Thier, P., Dicke, P.W., Haas, R., and Barash, S. (2000). Encoding of movement time by populations of cerebellar Purkinje cells. *Nature, 405*(6782), 72–76.

Thier, P., Dicke, P.W., Haas, R., Thielert, C.D., and Catz, N. (2002). The role of the oculomotor vermis in the control of saccadic eye movements. *Ann NY Acad Sci, 978,* 50–62.

Thier, P., and Ilg, U.J. (2005). The neural basis of smooth-pursuit eye movements. *Curr Opin Neurobiol, 15*(6), 645–652.

Thoroughman, K.A., and Shadmehr, R. (2000). Learning of action through adaptive combination of motor primitives. *Nature, 407*(6805), 742–747.

Vercher, J.L., and Gauthier, G.M. (1988). Cerebellar involvement in the coordination control of the oculo-manual tracking system: effects of cerebellar dentate nucleus lesion. *Exp Brain Res, 73*(1), 155–166.

Wang, Y.T., and Linden, D.J. (2000). Expression of cerebellar long-term depression requires postsynaptic clathrin-mediated endocytosis. *Neuron, 25*(3), 635–647.

Westheimer, G., and Blair, S.M. (1974). Function organization of primate oculomotor system revealed by cerebellectomy. *Exp Brain Res, 21*(5), 463–472.

Wolpert, D.M., and Miall, R.C. (1996). Forward models for physiological motor control. *Neural Netw, 9*(8), 1265–1279.

Yeo, C.H., and Hardiman, M.J. (1992). Cerebellar cortex and eyeblink conditioning: A reexamination. *Exp Brain Res, 88*(3), 623–638.

Yeo, C.H., Hardiman, M.J., and Glickstein, M. (1984). Discrete lesions of the cerebellar cortex abolish the classically conditioned nictitating membrane response of the rabbit. *Behav Brain Res, 13*(3), 261–266.

Yeo, C.H., Hardiman, M.J., and Glickstein, M. (1985a). Classical conditioning of the nictitating membrane response of the rabbit. I. Lesions of the cerebellar nuclei. *Exp Brain Res, 60*(1), 87–98.

Yeo, C.H., Hardiman, M.J., and Glickstein, M. (1985b). Classical conditioning of the nictitating membrane response of the rabbit. II. Lesions of the cerebellar cortex. *Exp Brain Res, 60*(1), 99–113.

Zhang, W., Shin, J.H., and Linden, D.J. (2004). Persistent changes in the intrinsic excitability of rat deep cerebellar nuclear neurones induced by EPSP or IPSP bursts. *J Physiol, 561*(Pt 3), 703–719.

Part III

Applications of the Importance of Learning and Memory to Applied Issues

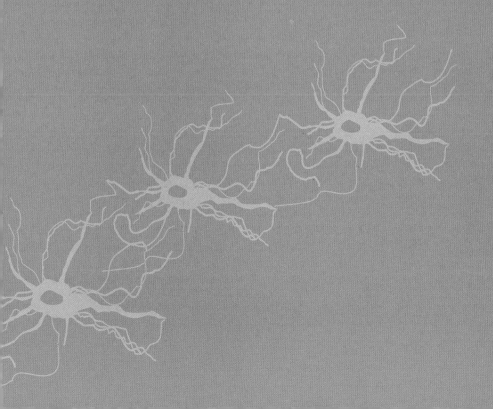

CHAPTER 14

Reward and Drugs of Abuse

Ryan T. LaLumiere

Department of Neurosciences, Medical University of South Carolina, Charleston, SC 29425

Peter W. Kalivas

Department of Neurosciences, Medical University of South Carolina, Charleston, SC 29425

I. INTRODUCTION

Although a number of perspectives on drug addiction exist, this chapter focuses on drug addiction as a type of pathology of learning and memory systems, a viewpoint that has become increasingly common since the mid-1990s. Both natural rewards and drugs of abuse appear to use the same systems within the brain to influence and reinforce behavior, and it is well known that these systems are involved in learning and memory, particularly in connecting motivations and memories with behaviors. The next section of the chapter provides an introduction to the major issues of reward and addiction. The following three sections examine drug addiction in a manner based loosely on the stages of memory — namely, acquisition, consolidation, and storage/retrieval. Although we shall see that drug addiction does not mirror these stages perfectly, this model provides a starting point for understanding how learning and memory, reward, and addiction interact at a neurobiological level.

II. REWARD, ADDICTION, AND LEARNING

A. Learning Systems for Reward

The ability of animals to adapt behaviorally in response to external stimuli and to maintain adaptations over a long time is critical for survival. In particular, animals must be able to make adjustments in their nervous systems in response to environmental stimuli that permit adaptive behavioral responses to future encounters with the stimuli. Thus, after discovering food in a particular location, an animal would want to remember that location and seek it out when hungry and do the opposite for locations where danger may exist. A motivational brain circuit provides a neuroanatomical substrate whereby the internal state of the animal (hunger, thirst, etc.) interacts with memory-processing areas and decision-making areas to influence behavior. When an animal becomes hungry, both the desire for food and information regarding how and where food may be acquired are integrated to influence the behavior of the animal. Thus, the neural circuits involved provide mechanisms by which rewards reinforce behavior.

Learning to perform a behavior in order to receive a reward or avoid a negative consequence is termed *instrumental* or *response learning*. Such learning is critical to survival and for guiding future behaviors. At the same time an animal is engaged in response learning, it is also learning associations of various stimuli in its environment. This associative (Pavlovian) learning occurs continuously and is a basic mechanism for establishing adaptive responses to environmental stimuli. Therefore, in any response-learning paradigm, it is expected that the animal is engaging in associative learning, and, as we shall see, this associative learning is very much involved in drug addiction and relapse.

Figure 14-1 is a basic diagram of the structures and their connections involved in the motivation/reward/addiction circuit; the reader is referred elsewhere for a detailed description (McFarland and Kalivas, 2003). The ventral tegmental area (VTA) lies in the midbrain and provides the dopaminergic input to most of the forebrain, including the nucleus accumbens (NAc), the prefrontal cortex (PFC), and the basolateral amygdala (BLA). The BLA plays a critical role in modulating emotionally influenced memory consolidation (McGaugh 2002; LaLumiere and McGaugh, 2005). The PFC is involved in executive control and decision-making functions and is known to be involved in working memory (Seamans and Yang, 2004). The NAc, like the BLA and PFC, is important for modulating memories (LaLumiere et al., 2005) but is also critical for integrating information from these other structures and influencing motor/behavioral output. It does so via projections to the VTA and ventral pallidum (VP). The VP, in turn, projects to the medial dorsal thalamus (MD), thus completing thalamic–cortico–accumbens–pallido loop. This circuitry provides a pathway by which structures involved in decision making, memory, and

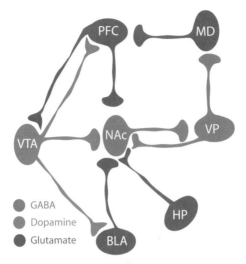

FIGURE 14-1 Schematic diagram of structures and connections involved in reward and drug learning in the motive circuit. PFC: prefrontal cortex; MD: mediodorsal thalamus; HP: hippocampus; BLA: basolateral amygdala; VTA: ventral tegmental area; NAc: nucleus accumbens.

motivation are integrated by the NAc and can then directly affect the behaviors of the animal. Addiction, of course, is characterized by detrimental changes in the interaction of decision making, memory, and motivation.

B. How Drug Addiction and Memory Are Related

In recent years, addiction has been conceptualized as a disorder of learning and memory (Heyne et al., 2000; Kelley, 2004a). To the extent that consumption of drugs of abuse depends on associative learning mechanisms, drug use mimics normal learning. However, drug addiction extends beyond mere reward processing. For someone addicted to a drug, the motivation for consuming the drug exceeds all other motivations, and the person faces a nearly irresistible urge to seek out and take the drug. Thus, the hallmark of drug addiction is the propensity to relapse, even after long periods of abstinence. People who have consumed a drug of abuse over an extended period and then cease consuming the drug remain for years at a significant risk of relapsing to drug use, perhaps even for the remainder of their lives. This risk is not limited to the effects of the withdrawal symptoms that accompany the early stages of drug abstinence. Rather, relapse appears to be a symptom of changes in the memory and motivation circuitry such that certain stimuli trigger a set of uncontrollable behaviors, with the goal of obtaining and consuming the drug.

Such findings suggest that addiction is a problem of learning and memory. That is, in addicted individuals, the storage of information regarding the reward of the drug has led to maladaptive neural changes that prevent people from retaining control over their behavior and cause them to seek out and consume the drug, even when they know that doing so is against their own best interests. It would also appear that the neural changes involved in drug addiction should map onto the same systems responsible for the processing and storage of natural rewards. And, as we shall see, the systems are similar, but some of the long-term changes induced by the different rewards are different.

C. Animal Models of Drug Addiction

One of the critical components in the investigation of behavioral neurobiology is the use of an appropriate animal model for the behavior under study. In addiction research, several models have been developed for investigating the long-term changes induced by repeated drug administration. One major model is drug-induced locomotor sensitization. In rodents, acute administration of many drugs, including psychostimulants and opiates, increases the animal's locomotor activity. Repeated administration of the drug induces neurobiological changes such that later acute administration of the drug produces even greater increases in locomotor activity. This increase in drug-induced locomotion following the final drug administration compared to the first drug administration is called *sensitization*. Another often-used model is conditioned place preference (CPP). In CPP, animals are administered the drug of abuse before being placed into one chamber of a two-chamber apparatus. They are then given a control injection before being placed into the other chamber. Through repeated pairings, the animals learn to associate one chamber with the drug of abuse. Animals are tested by being given free exploration of the entire apparatus, and animals that have learned the drug–chamber association spend more time in the drug-associated chamber. This model has been particularly useful for investigating the contextual learning that occurs during drug consumption.

The "gold standard" for addiction is self-administration. In this paradigm, animals are trained to perform a response (often, pressing a lever) in order to receive a small administration of the drug. This model is also often used for the study of natural rewards as well. The advantage of this model is that, as with humans who must perform certain actions to obtain and consume the drug of abuse, the animal must perform particular behaviors in order to receive the drug, unlike sensitization and CPP, where experimenters noncontingently administer the drug. Although many findings from noncontingent administration and self-administration have been in agreement, evidence suggests neurobiological differences between self-administration of a drug and experimenter administration of the drug (Mark et al., 1999; Stefanski et al., 1999). In addition, this model allows for examination of the most problematic aspect of drug addiction: relapse. After reaching a maintenance level of drug self-administration, animals undergo extinction training, in which further responses do not produce drug administration. After responding has been extinguished, the response behaviors can be "reinstated" by experimenter administration of the drug, by presentation of cues associated with the drug taking, or by administration of a stressor. Each of these reinstatement paradigms models a different aspect of relapse in humans and is addressed later in the chapter.

Clearly, CPP and self-administration are learning-based paradigms. The kinds of learning involved in these tasks (Pavlovian learning in CPP; response learning in self-administration) are also used in normal learning and memory investigations, including the study of aversive learning and memory. Locomotor sensitization, however, does not have any obvious learning component, for the administration of the drug is neither contingent on any behavior nor associated with particular cues. This model would appear to suggest that some neurobiological changes of addiction are due primarily to the pharmacological actions of the drug. However, the fact that sensitization can be affected by manipulating the test versus daily injection environment suggests a role for learning in sensitization (Badiani et al., 1995; Browman et al., 1998).

D. Acquisition, Consolidation, and Retrieval

As noted earlier, this chapter's structure is based on the three basic parts of learning and memory: the acquisition of information from the environment, the consolidation and storage of such information through neuroplasticity mechanisms, and the stimulus–induced retrieval of the information. These have been well studied in the field of learning and memory, and, in fact, the study of drugs of abuse has also examined these issues but through different models. However, it is important to note that studies with drugs of abuse are not easily parsed into these three categories. It is especially difficult to distinguish between acquisition and consolidation because *memory consolidation* refers to the plasticity (short- and long-term) that occurs immediately after the acquisition to store the acquired information. With drug addiction, each consumption of the drug is a learning event, and, thus, consolidation occurs after each event. However, it is the cumulative *consolidation* from chronic use of the drug that produces the pathological changes. Therefore, we consider *acquisition* as the initial reinforcing effects of rewards and drugs of abuse, and the *acquisition* section addresses those issues. The *consolidation* section examines the long(er)-term genetic, molecular, and biochemical changes induced by drug reward. In addition, the consolidation section addresses how chronic consumption of the drug produces neural changes not found with acute consumption and how changes in the neurocircuitry store the information regarding the drug reward. The *retrieval* section explores how the changes induced by the chronic use of addictive drugs produce the irresistible compulsion to seek out and consume the drug.

III. ACQUISITION

A. Neurobiology of Reward and Reward Learning

How is "reward" represented neurobiologically? As noted, Figure 14-1 provides a basic diagram of the structures involved in the interaction of memories and motivational states and their ability to influence behavior. The NAc is considered to be a crucial point of integration of information by receiving emotional, mnemonic, and cognitive inputs and by projecting to motor output regions (Mogenson et al., 1980; Kelley, 2004b). The basolateral amygdala, which projects to the medial PFC (mPFC) and the NAc (Kelley et al., 1982; Robinson and Beart, 1988; Pitkanen, 2000), is involved in processing emotion and modulating memory consolidation for emotionally influenced learning (McGaugh, 2002, 2004a). The mPFC, while involved in many processes, appears to be critically involved in decision making and impulse control (Kelley, 2004a). The hippocampus mediates short-term declarative memories in humans and spatial and contextual memories in rats (Squire et al., 2004). Together, these three structures send glutamatergic projections to the NAc (Kelley et al., 1982; Christie et al., 1987).

The NAc is divided into two regions: the core, which surrounds the anterior commissure, and the shell, which surrounds the core primarily on the medial and ventral sides. These two subregions project to different areas, with the core projecting to the ventral and lateral substantia nigra (SN) and to the dorsal VP, whereas the shell projects to the VTA and the ventromedial VP (Zahm and Heimer, 1990; Heimer et al., 1991, 1995). Based on connectivity as well as functional studies, it has been suggested that the core is similar to the dorsal striatum and more directly connected to motor output systems and, thus, is involved in the instrumental behaviors, whereas the shell is more akin to the extended amygdala and involved in emotional processing (Kelley, 2004b). However, it is important to note that studies have found roles for both structures in instrumental learning and that experiments continue to try to dissociate the functions of these subregions (Di Chiara, 2002).

The NAc, along with the hippocampus, the mPFC, and the BLA, receives dopamine (DA) input from the VTA (Fallon and Loughlin, 1995). Most of the DA neurons that innervate the forebrain are located in the midbrain region, specifically in the VTA and the SN (Fallon and Loughlin, 1995). The SN innervates the dorsal striatum (caudate-putamen), whereas the VTA provides input to the rest of the forebrain, including the ventral striatum (the NAc), the PFC, the amygdala, and the hippocampus. Early theories on drugs of abuse and natural rewards suggested that activation of DA neurons, particularly in the VTA, and release of DA in target structures signaled reward (Ungless, 2004). In fact, plenty of evidence suggests that rewarding or pleasurable stimuli increase DA neuron activation in the VTA and DA release in a variety of structures, especially the NAc (Di Chiara, 2002). However, in recent years it has become increasingly clear that the release of DA does not signal exclusively reward and, in contrast to early theories on DA, does not mediate the pleasurable or hedonic effects of the reward (Di Chiara, 2002). Unpleasant or aversive stimuli increase DA release in a variety of structures, including the NAc, demonstrating a role for DA beyond reward (Inglis and Moghaddam, 1999). But it should be noted that evidence indicates differential responses by DA to aversive versus rewarding stimuli (Di Chiara, 2002; Schultz, 2002). Thus, rather than mediating strictly reward, DA appears to be critical for signaling motivationally relevant stimuli, particularly those that are novel.

Considerable evidence suggests that DA neurons, particularly in the VTA, respond to rewarding stimuli in a phasic manner and, over time, respond to any previously neutral stimuli that are predictive of the reward. However, it appears that the DA responses depend on the predictability of the reward (for review, see Schultz, 2002). Predicted rewards do not cause DA release, whereas unpredicted rewards do; moreover, expected rewards that do *not* occur lead to a decrease in DA responses. Thus, DA may provide a prediction-error signal. This DA signaling then influences the NAc, presumably by modifying learning regarding the predicted reward and its predictive stimuli.

It is believed that the convergence of DA and glutamate at synapses in the NAc is critical for the integration of information. The glutamatergic signals provide sensory, emotional, and motor information, whereas the DA signals indicate the unpredicted rewards and/or salient events (Kelley, 2004a). Findings support critical roles for both glutamate and DA. Blockade of NMDA or D1 receptors in the NAc core impairs acquisition, but not consolidation, of lever pressing for sugar pellets (Hernandez et al., 2005), suggesting that activation of these neurotransmitters' receptors is important for the online processing of reward stimuli. Despite the lack of a role for those neurotransmitters in consolidation in that study, it is clear that both glutamate and DA are important in the NAc for long-term potentiation (Floresco et al., 2001). Moreover, post-training inhibition of protein kinase A, which is activated by DA receptors, or protein synthesis in the NAc impairs consolidation of instrumental memories (Baldwin et al., 2002; Hernandez et al., 2002), demonstrating that consolidation occurs in the NAc for instrumental learning for natural rewards. Through the coincident signaling of DA and glutamate, memories are formed regarding the circumstances under which the stimuli occur and the actions that must be performed in order to obtain the stimuli. However, it is important to note that predicted rewards do not increase DA responses, demonstrating that, under normal conditions, this neural system adapts to the rewards.

B. Drugs of Abuse and Reinforcement

Akin to natural reward, addictive drugs encode and reinforce drug-seeking behaviors by regulating DA and glutamate release in the NAc. The acute administration of all drugs of abuse, from alcohol to cocaine to heroin, increases the release of DA in the NAc that is critical for drugs to reinforce behavior. Thus, rhesus monkeys, squirrel monkeys, and rats will self-administer D1 agonists after learning to self-administer cocaine (Self and Stein, 1992; Weed et al., 1993; Grech et al., 1996; Self et al., 1996). Conversely, in rats trained to self-administer cocaine, infusions of the D1 receptor antagonist SCH 23390 into the NAc significantly increase the rate of lever-pressing for cocaine, suggesting that blockade of the D1 receptors in the NAc decreases the reinforcing properties of the cocaine (Maldonado et al., 1993).

DA receptors are G-protein-coupled receptors that often influence activity in the cAMP-PKA pathway by modulating activity of adenylyl cyclase. In particular, D1 receptors are coupled to G_s proteins, whereas the D2-class receptors are usually coupled to G_i/G_o proteins. Intra-NAc infusions of a PKA inhibitor reduce baseline cocaine self-administration and shift the dose–response curve for administering cocaine to the left, suggesting that the inhibitor increases the rewarding effects of the cocaine administration (Self et al., 1998). Conversely, intra-NAc infusions of a PKA activator increase baseline cocaine self-administration and shift the dose–response curve to the right, suggesting that increased levels of PKA reduce the rewarding effects of cocaine. These findings suggest that tonic up-regulation of the cAMP-PKA pathway may underlie tolerance to cocaine, thereby explaining why PKA activators increase cocaine self-administration, whereas PKA inhibitors decrease cocaine self-administration.

In contrast with effects on cocaine self-administration, intra-NAc infusions of the PKA inhibitor have no effect on food-reinforced lever pressing, indicating a specific role for PKA in the NAc for drug self-administration (Self et al., 1998). The lack of influence by PKA regulation on biological reward probably results from the fact that, unlike cocaine, which increases DA release with every administration, repeated exposure to a biological reward reduces the release of DA to presentation of the primary reward, and DA is more effectively released by stimuli associated with reward rather than the reward itself. Thus, DA release comes under control of associative learning engendered by biological rewards, while DA remains under pharmacological control for drugs of abuse. Therefore, inasmuch as DA is signaling motivational circuitry to learn associations that predict a rewarding stimulus, drugs of abuse continuously engender new associations, while natural rewards will do so only if the reward changes. Indeed, it is the continuous release of DA that may cause the learning associated with drugs of abuse to become pathological.

IV. CONSOLIDATION — LONG-TERM CHANGES FOLLOWING CHRONIC DRUG USE

A. Difficulty in Examining "Consolidation"

In traditional learning and memory paradigms, *consolidation* refers to the time following a learning event in which the memory for the event is susceptible to external influences on its subsequent retention. During this time, a variety of processes occur in the brain that store the memory in a more-or-less permanent form, but these processes require time following the event. For addiction, however, it is difficult to isolate a specific consolidation period because it is not the one-time use of the drug that leads to the addiction but the chronic consumption of the drug that produces the neural changes underlying addiction. As already discussed, the fact that each drug administration releases DA as if it were the first experience with a reward probably produces the pathological changes in learning that lead to addiction. Therefore, this section examines some of the long-term cellular changes produced by chronic use of drugs of abuse.

B. Changes in Gene and Protein Expression

Given the role of PKA signaling in mediating DA receptor stimulation and the lack of tolerance to drug-induced DA release, it is not surprising that chronic cocaine increases the activity of PKA and the PKA-regulated transcriptional regulator CREB (Terwilliger et al., 1991). Thus, CREB activation in the NAc regulates the motivation to take drugs (Carlezon et al., 1998; Barrot et al., 2002). For example, overexpression of CREB reduces an animal's sensitivity to the rewarding aspects of the drug, whereas reduced activity of CREB increases the rewarding aspects of the drug (Nestler, 2004). One gene in the NAc regulated by CREB encodes the opioid peptide dynorphin (Nestler, 2004). Dynorphin activates κ opioid receptors, thereby reducing the rewarding effects of drugs of abuse (Shippenberg and Rea, 1997); in fact, κ opioid receptor antagonists reverse the effects of CREB activation (Carlezon et al., 1998). Activation of κ opioid receptors is believed to reduce the rewarding effects of drugs of abuse by presynaptically inhibiting DA release from the DA neurons that innervate the NAc (Nestler, 2004).

Nestler and colleagues have examined another transcription factor whose induction profile lends itself to the regulation of the long-term changes induced by chronic drug use. Whereas the acute administration of drugs of abuse induces rapid but transient expression of a number of Fos and Jun transcriptional proteins (Graybiel et al., 1990; Nestler, 2004), one member of the Fos family, ΔFosB, accumulates slowly in the NAc over repeated administrations of the drug (Hope et al., 1994; Chen et al., 1995, 1997). Elevated ΔFosB persists for weeks after discontinuing drug administration. Studies with transgenic mice overexpressing ΔFosB indicate that the mice have enhanced sensitivity to cocaine and morphine and show greater motivation to get cocaine (Kelz et al., 1999; Colby et al., 2003; Nestler, 2004). These mice also show enhanced sensitivity to natural rewards (Werme et al., 2002; Nestler, 2004).

Changes in ΔFosB expression regulate other genes and their proteins, and it is assumed that it is through such changes in protein expression that chronic drug use produces its lasting effects. Several targets of ΔFosB, and its downstream target AP-1, have been identified. NAC1 is an AP-1-regulated protein that is up-regulated in the NAc following an acute injection of cocaine as well as following withdrawal from chronic administration of cocaine (Cha et al., 1997). Over-expression of NAC1 in the NAc impairs the development, but not expression, of sensitization (Mackler et al., 2000), whereas NAC1 antisense infusions into the NAc potentiates the motor stimulant effects of cocaine (Kalivas et al., 1999). Another ΔFosB-regulated protein is cyclin-dependent kinase, Cdk5 (Chen et al., 2000), which is involved in neural growth. Chronic cocaine administration increases Cdk5 protein expression (Bibb et al., 2001), and infusions of a Cdk5 inhibitor into the NAc prevents cocaine-induced dendritic growth (Norrholm et al., 2003). Other targets, such as the transcription factor nuclear factor-κB, have also been identified but are still being investigated (Nestler, 2004).

Although considerable work remains in the investigation of the downstream targets of ΔFosB and CREB and how they contribute to addiction, CREB and ΔFosB regulate many genes in opposite directions, consistent with behavioral findings from studies on each protein. It would appear that, during early drug use, increased CREB activity causes compensatory transcriptional events that resist changes induced by the drugs of abuse. However, such compensation increases the amount of the drug taken on subsequent occasions in order to produce the same rewarding effects, perhaps eventually overcoming CREB's effects. Conversely, rather than being involved in compensatory actions, ΔFosB actually mediates the effects of this long-term drug use on gene transcription and protein expression. Presumably, these changes underlie some of the long-term plasticity responsible for addiction (see Fig. 14-2).

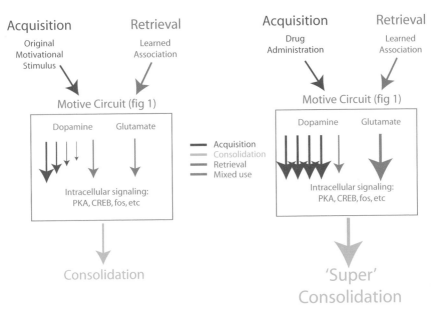

FIGURE 14-2 Schematic diagram of processes underlying learning and retrieval for normal rewards (left) and for drug rewards (right). Across repeated normal rewards, DA responses attenuate, but DA responses do not attenuate and may sensitize across repeated drug use. Thus, the intracellular mechanisms are repeatedly activated at high levels, leading to dysfunctional "super-consolidation." This is manifested through the retrieval-induced increase in glutamate release in the motive circuit and compulsive drug seeking and consumption.

C. Long-Term Changes Involved in Drug Addiction

Although the DA system has been a major focus for investigations into chronic drug-induced long-term changes, recent evidence suggests most changes in DA transmission associated with chronic drug use dissipate during abstinence. In contrast, the enduring pathology of addiction appears to be manifested by cellular dysfunctions in glutamatergic transmission in the NAc core. These changes include a decrease in glutamatergic tone, a potentiated increase in glutamate release following an acute cocaine injection, and changes in proteins that regulate postsynaptic glutamate signaling.

As noted, withdrawal from repeated cocaine administration reduces basal glutamate levels in the NAc core of rats (Baker et al., 2003). Basal levels of glutamate are maintained by a cystine–glutamate exchanger (xc-) and only minimally by synaptically released glutamate (Timmerman and Westerink,

1997; Baker et al., 2002). The xc- exchanges one intracellular glutamate for one extracellular cystine molecule. Although repeated cocaine administration does not change basal levels of extracellular cystine, cocaine-treated rats have decreased cystine–glutamate exchange (Baker et al., 2003). Reverse dialyzing cystine into the NAc or giving a systemic injection of N-acetylcysteine, which elevates brain cysteine levels, restores glutamate levels of cocaine-treated rats to normal control levels and prevents the increase in glutamate following an acute injection that mediates cocaine-induced reinstatement of lever pressing. Together, the findings regarding xc- strongly indicate that chronic cocaine administration changes the mPFC-NAc glutamatergic projections in a manner that underlies cocaine-induced relapse.

How do the decreased basal glutamate levels in the NAc affect the propensity to relapse? One mechanism appears to be through reduced tone of presynaptic metabotropic glutamate receptors (mGluR2/3) (Moran et al., 2005). Normally, mGluR2/3 receptors provide inhibitory feedback on presynaptic glutamate release and systemic administration of mGluR2/3 receptor agonists reduces reinstatement for cocaine or heroin seeking (Baptista et al., 2004; Bossert et al., 2004). xc- regulated glutamate in the NAc regulates glutamate release via providing tone on mGluR2/3 (Moran et al., 2005). Thus, reduced tone on mGluR2/3 on PFC afferents, along with changes to glutamatergic neurons themselves, leads to increased glutamate release when mPFC neurons fire. This, in turn, provides greater glutamatergic activation of NAc neurons and initiates drug-seeking behavior.

Evidence also suggests a critical role for Homer proteins in long-term changes induced by chronic cocaine. The Homer family regulates synaptic proteins and has three members (Homer1–3). Homer is known to play a role in learning-induced neural plasticity, and cocaine reduces levels of Homer proteins (Swanson et al., 2001). A role for reduced Homer in addiction is indicated by the fact that reducing Homer produces behavioral sensitization and facilitates cocaine self-administration (Ghasemzadeh et al., 2003; Szumlinski et al., 2004).

ΔFosB appears to have a role in changes in glutamate signaling. ΔFosB overexpression and cocaine induce expression of GluR2, and overexpression of GluR2 in the NAc increases the sensitivity of animals to acute injections of cocaine (Kelz et al., 1999; Peakman et al., 2003). GluR2 is a subunit of the AMPA receptor, and its presence reduces calcium conductance in AMPA receptors (Thomas et al., 2001). Interestingly, these findings would suggest that long-term drug use leads to less glutamate-induced activation of NAc neurons, in contrast to the previous findings. Future work will have to resolve this issue.

Together, these findings indicate that the glutamatergic system, particularly in the connections from the mPFC to the NAc, undergo long-term, if not permanent, changes following chronic use of a drug. It appears that changes in mPFC neurons and the reduced basal glutamate levels in the NAc lead to increased glutamate release following stimulation of mPFC neurons that subsequently drive NAc-regulated behavior. That these changes persist beyond withdrawal from the drug indicates that they are long-lasting and, as the evidence suggests, underlie the propensity to relapse. Section V provides more evidence suggesting that these long-term changes in the glutamatergic system, rather than the dopaminergic system, are critical for reinstatement in animals.

D. Structural Changes

In addition to changes in protein expression and function, the storage of memories is also believed to be mediated by structural changes in neurons, particularly in neuronal connectivity (Moser et al., 1994). Addiction research has focused on changes in morphology of dendrites and dendritic spines, for these regions of the neuron receive the vast majority of synaptic inputs and appear to be important for experience-induced changes (Kasai et al., 2003). The dendritic spines are of special interest because, in the NAc, excitatory, glutamatergic inputs make contact at the head of the spine and dopaminergic axons synapse on the shaft of the spines. This "triad" of the NAc medium spiny neuron, the glutamatergic input, and the DA input is believed to be critically involved in the plasticity induced by drugs of addiction in the NAc.

Long-term use of cocaine, amphetamine, nicotine, or morphine has been shown to change the dendritic morphology in the NAc. Cocaine and amphetamine, whether self-administered or administered by the experimenter, increase spine density and dendritic branching in the NAc (Robinson and Kolb, 1997, 1999a; Robinson et al., 2001; Li et al., 2003; Norrholm et al., 2003; Crombag et al., 2005; Ferrario et al., 2005). These studies have also found increases in spine density and dendritic branching in the mPFC pyramidal neurons. As already noted, both regions appear to be critically involved in the long-term behavioral changes induced by chronic drug use. Nicotine appears to have morphological effects similar to those of cocaine and amphetamine (Brown and Kolb, 2001), but morphine decreases dendritic branching and spine density in the NAc and the mPFC (Robinson and Kolb, 1999b; Robinson et al., 2002). Researchers have also found a reduction in DA neuron size in the VTA following chronic morphine treatment and withdrawal (Sklair-Tavron et al., 1996; Spiga et al., 2003).

These morphological adaptations persist for over three months (Kolb et al., 2003). Although there is some evidence for differences between self-administered and experimenter-administered drugs on dendritic morphology (Robinson et al., 2002), most studies find similar effects, suggesting that any learning associated with self-administration is unnecessary/unrelated to structural changes. That is, these changes are due to the unconditional effects of the drugs themselves and may be unrelated to learning during drug taking.

V. RETRIEVAL

A. Different Kinds of Retrieval/Reinstatement for Drugs of Addiction

The most insidious problem and, in fact, the hallmark of addiction is the propensity to relapse. For example, approximately 90–95% of cocaine addicts relapse within six months of undergoing treatment for their addiction. Relapse can be evoked in three distinct ways that can be experimentally modeled. First, addicts may relapse if they consume a small amount of the drug. Second, addicts may relapse if they encounter stimuli that they associate with their drug taking. And third, addicts may relapse following a particularly stressful event. Each of these types of relapse has been examined in rats with the self-administration/reinstatement model of addiction. When the animal's lever pressing has been extinguished, the animal undergoes a reinstatement session in which it receives a priming injection of the drug, a cue that was previously paired with the drug training is presented, or a foot shock is administered to the rat. However, lever presses during reinstatement do not themselves produce any drug infusions. These three types of reinstatement (drug priming-induced, cue-induced, and stress-induced) are used as models of the three categories of relapse in humans. From a learning-and-memory perspective, these three reinstatement methods are types of retrieval of the original memory, which lead to the inability of the rat to resist seeking out the drug of abuse.

B. Circuitry Underlying Reinstatement of Drug Seeking

Studies have examined the circuitry involved in reinstatement and identified the critical neural pathways involved. Using GABA receptor agonists to inactivate structures, McFarland and Kalivas (2001) found that the dorsal PFC, the VTA, the NAc core, and the VP (but not the ventral PFC, the NA shell, the SN, the central nucleus of the amygdala, the BLA, or the mediodorsal thalamus) are necessary for cocaine-induced reinstatement. Moreover, of the four structures involved in cocaine-induced reinstatement, only the VP is necessary for reinstatement of food-seeking behavior, suggesting that the dorsal PFC, the VTA, and the NAc core are specific to reinstatement of drug seeking but not to reinstatement of natural reward seeking. Because cocaine increases DA release, cocaine-induced reinstatement must depend on activation of DA receptors within the brain. However, despite the critical role of DA receptors in the NAc core during acquisition of lever pressing for cocaine, these receptors do not appear to be important for reinstatement, because blockade of DA receptors in the core or the VP does not prevent cocaine-induced reinstatement (McFarland and Kalivas, 2001). However, blockade of DA receptors in the dorsal PFC prevents such reinstatement, suggesting that changes in the PFC and the PFC projection to the core are the critical ones involved in relapse. Supporting this, microdialysis experiments demonstrate that cocaine-induced reinstatement is associated with a rise in glutamate levels in the NAc but not with a rise in DA levels (McFarland et al., 2003).

Conditioned cue-induced reinstatement appears to depend on the same structures involved in cocaine-induced reinstatement, including the NAc core (but not the shell) (Fuchs et al., 2004) and the dorsal PFC (McLaughlin and See, 2003). However, in addition, the BLA is selectively involved in conditioned cue-induced reinstatement (Kantak et al., 2002; McLaughlin and See, 2003).

Foot-shock-induced reinstatement appears to require activity in circuitry similar to that with cocaine-induced reinstatement requires (i.e. dorsal PFC, NAc core, and VP) but, in addition, also depends on the central nucleus of the amygdala, the bed nucleus of the stria terminalis, and the NAc shell (McFarland et al., 2004). It appears that foot-shock stress activates limbic circuitry. The central nucleus of the amygdala, which is part of this circuitry, projects to the VTA, and this projection is believed to mediate activation of the DA neurons projecting to the PFC and initiating reinstatement.

C. Connection to Other Systems

One of the most interesting neurobiological problems in the field of drug addiction is how a particular set of behaviors and not others is triggered during relapse. In this chapter, we have presented findings regarding how drugs of abuse are rewarding, how they induce long-term changes, and how, through these long-term changes, reinstatement or relapse to drug seeking and drug taking can be triggered. Based on such findings, it might seem that the basic structure of addiction has been identified — i.e., dysfunction in the prefrontal cortex–nucleus accumbens system allows particular triggers to reinstate drug seeking. However, while the primary infrastructure has been identified, these findings do not address the major issue of how this infrastructure selects drug-related behaviors over more adaptive behaviors. The systems involved in drug addiction did not evolve to subserve the consumption of drugs of abuse and serve to integrate motivational states with memories to control behaviors (see earlier discussion). Although it is clear that dysfunction in decision making and impulse control is important, we do not know how the desire to consume drugs overrides other desires and motivations in drug addicts. It seems logical to assume that the competition between other desires and the desire for drugs must occur in the mPFC. Based on that assumption, there must then be subsets or networks of neurons within the mPFC that maintain "memories" of the repeated drug use, suggesting that there must also be networks of neurons maintaining memories for other motivations.

How does the desire to seek drugs trigger specific drug-seeking behaviors? Although we often refer to the accumbens-pallidal output as simply "behavioral" or "motor," the effects of drug primes, cue primes, or stress do not trigger general changes in behavior. Instead, they trigger very specific behaviors to seek out and consume drugs of abuse. In the self-administration model, animals given a drug prime do not engage in random behaviors or even random lever pressing. Rather, they selectively press the lever that had been paired with the drug during initial training and ignore the lever that had been paired with nothing. Therefore, the output of the accumbens and VP is not just to increase behaviors but to activate a specific set of behaviors. This implies that, just as in the mPFC, there are specific neuronal networks that are selectively changed by repeated drug use and selectively triggered during reinstatement/relapse. Such neuronal networks could then connect with downstream structures and trigger specific drug-seeking behaviors. These networks must be "bound" to each other, because drug effects on the mPFC networks during reinstatement lead to activation of drug-related neuronal networks in the NAc and the VP. However, due to the difficulty in examining such issues in both learning and memory and addiction, research has not provided any answers as to how this information remains bound and how activation in the mPFC leads to a specific set of behaviors.

D. Reconsolidation

In considering the development of treatments for addiction, the largest problem is the enduring nature of addiction. Relapse remains a significant risk for most individuals for a long time, if not permanently, indicating that the changes induced by the chronic use of drugs of abuse are virtually unalterable. A recently revived idea in the field of learning and memory is that, upon retrieval, memories undergo a kind of *reconsolidation* in which the retrieval of the memory induces lability in the memory trace again, rendering it susceptible to external influences again (Nader et al., 2000; Debiec et al., 2002). Such findings have produced great interest on the part of researchers because they raise the possibility of eliminating old, troublesome memories, and addiction memories would appear to be excellent candidates for such elimination (Centonze et al., 2005). In fact, recent work has investigated the susceptibility of "addiction" memories to reconsolidation. Lee et al. (2005) trained rats to self-administer cocaine, infusions of which were paired with a light cue. After the training was completed, rats underwent a single cue exposure session, prior to which they received intra-BLA infusions of Zif268 antisense oligonucleotides. Subsequent testing demonstrated that rats receiving this treatment had impaired memory for the cue–drug association. Infusions of the antisense oligonucleotides alone (with no cue exposure) had no effect, indicating that cue exposure reactivated the memory and rendered it labile again. Using a CPP model, Miller and Marshall (2005) found that pre- or posttest infusions of a MEK inhibitor into the NAc core impaired memory when tested on a subsequent retrieval test. Infusions of the inhibitor alone (with no retrieval test) had no effect on subsequent retrieval tests. Together, these findings indicate the ability to interfere with drug memories through manipulations after retrieval of the memory.

Such results would appear to be very exciting, but at the moment they should be put into the context of a larger debate within the learning and memory field on reconsolidation (McGaugh, 2004b). Evidence suggests that reconsolidation is not merely a recapitulation of consolidation (Debiec and Ledoux, 2004; von Hertzen and Giese, 2005). The processes involved in consolidation and reconsolidation are qualitatively different, and thus the term *reconsolidation* is actually inappropriate. Moreover, a number of groups have found significant caveats to the reconsolidation hypothesis, including the following problems. (1) Reconsolidation does not occur for some learning tasks (Cammarota et al., 2004); (2) reconsolidation effects are (sometimes) temporary (Judge and Quartermain, 1982; Lattal and Abel, 2004); and (3) the time span between original learning and the retrieval test can affect whether reconsolidation occurs (Milekic and Alberini, 2002). Because these problems have yet to be solved in traditional learning and memory experiments and because addiction–reconsolidation experiments have not yet systematically addressed these issues, it is difficult to know whether such investigations hold promise for addiction treatment. In particular, if older memories are less susceptible to reconsolidation (Milekic and Alberini, 2002), addiction may not be amenable to such treatment, for there is typically a significant interval between the beginning of addiction and the seeking of treatment.

VI. CONCLUSIONS

A. Addiction as Disorder of Memory

This chapter has focused on addiction in terms of its relationship with learning and memory. We have presented some of the important findings regarding drug reward and addiction as part of the stages of memory. If addiction is a kind of pathological memory, then it is reasonable to ask where, within these stages, this pathology exists. Certainly, rewarding stimuli and the learning that accompanies such stimuli are not pathological and do not normally induce memories or behaviors considered to be pathological. In addition, many of the initial effects of drugs of abuse are quite similar to those of natural rewards. Therefore, it seems unlikely that the pathology exists from the first consumption of the drug. It is also unlikely that the problem of addiction lies in retrieval processes, for evidence indicates significant changes in neural processing even in the absence of retrieval.

Instead, the genesis of addiction can be conceptualized as pathological memory consolidation. The evidence presented in this chapter suggests that it is not the consolidation of the first experience with a drug that is pathological but, rather, the continued use over a period of time. Although the neurobiological responses to natural rewards and drugs of abuse are similar, these responses diverge when the rewards/drugs are repeatedly administered or encountered. Clearly, the brain has evolved to have adaptive processes that prevent a natural reward from taking exclusive control of the motivational processes. However, the mechanisms for preventing that are circumvented by the consumption of drugs that induce repeated activation of mesocorticolimbic DA release beyond what would occur naturally. Thus, a neural system, highly sensitive to motivational stimuli in order to promote survival, is not prepared for the effects of repeated pharmacological activation. Figure 14-2 presents schematic diagrams illustrating the processes underlying normal reward learning and drug–reward learning.

Through repeated drug use, the memories for these motivational stimuli, the behavioral patterns used to satisfy the desires, and the previously neutral stimuli that become associated with the drugs become *superconsolidated*. When the associated stimuli, such as the sight of the drug or the context in which the drug was consumed, trigger these superconsolidated memories, the desire to take the drug become all-encompassing, and the individual engages in the behaviors that he/she knows will lead to consumption of the drug, regardless of the person's best interest. This idea of a superconsolidated memory has recently become part of the understanding for the development of post-traumatic stress disorder (PTSD) (Schelling, 2002; Schelling et al., 2003, 2004). PTSD develops following a traumatic incident or series of traumatic incidents that appear to be "seared" into their memories. Patients with PTSD usually suffer intrusive memories, and these memories can be triggered by particular stimuli, especially those that remind them of stimuli present during the initial traumatic incidents. Unfortunately, due to the lack of a good animal model for PTSD, it is difficult to assess the neurobiological similarities between PTSD and drug addiction. However, it is interesting that there is a high degree of comorbidity between addiction and PTSD, suggesting overlapping neurobiological vulnerabilities.

B. Summary

Learning about natural rewards and learning about drugs of abuse utilize initially the same pathways in the brain. However, repeated use of drugs produces changes in the brain not normally found with repeated natural rewards, which is most likely due to the ability of drugs to go directly to the brain and have their pharmacological effects. In doing so, the drugs alter the functioning of a number of structures in the brain, of which the NAc and the PFC are particularly prominent due to their apparent roles in addiction. These alterations include changes in gene and protein expression, morphological changes in the neurons, and changes in glutamatergic signaling. In many ways, the process of drug addiction mimics the stages of memory, from acquisition to consolidation and finally retrieval. But in drug addiction, the changes induced by the drugs are detrimental to people in terms of their ability to resist future consumption of the drug. In particular, the pathology of drug addiction appears to arise from the repeated use of a drug that induces a kind of superconsolidation of the memories for the drug. Whether such superconsolidation can be reversed is unknown, but treatments based on counteracting some of the long-term dysfunctional changes are being developed (Kalivas and Volkow, 2005). Nevertheless, because addiction is conceived as pathological learning, the fields of learning and memory and addiction will become more intertwined and will continue to inform one another on the basic functioning of motivational learning.

REFERENCES

Badiani, A., Browman, K.E., and Robinson, T.E. (1995). Influence of novel versus home environments on sensitization to the psychomotor stimulant effects of cocaine and amphetamine. *Brain Res* **674**: 291–298.

Baker, D.A., McFarland, K., Lake, R.W., Shen, H., Tang, X.C., Toda, S., and Kalivas, P.W. (2003). Neuroadaptations in cystine-glutamate exchange underlie cocaine relapse. *Nat. Neurosci* **6**: 743–749.

Baker, D.A., Xi, Z.-X., Shen, H., Swanson, C.J., and Kalivas, P.W. (2002). The primary source and neuronal function of in vivo extracellular glutamate. *J Neurosci* **22**: 9134–9141.

Baldwin, A.E., Sadeghian, K., Holahan, M.R., and Kelley, A.E. (2002). Appetitive instrumental learning is impaired by inhibition of camp-dependent protein kinase within the nucleus accumbens. *Neurobiol Learn Mem* **77**: 44–62.

Baptista, M.A., Martin-Fardon, R., and Weiss, F. (2004). Preferential effects of the metabotropic glutamate 2/3 receptor agonist ly379268 on conditioned reinstatement versus primary reinforcement: Comparison between cocaine and a potent conventional reinforcer. *J Neurosci* **24**: 4723–4727.

Barrot, M., Olivier, J.D., Perrotti, L.I., DiLeone, R.J., Berton, O., Eisch, A.J., Impey, S., Storm, D.R., Neve, R.L., Yin, J.C., Zachariou, V., and Nestler, E.J. (2002). CREB activity in the nucleus accumbens shell controls gating of behavioral responses to emotional stimuli. *Proc Natl Acad Sci USA* **99**: 11435–11440.

Bibb, J.A., Chen, J., Taylor, J.R., Svenningsson, P., Nishi, A., Snyder, G.L., Yan, Z., Sagawa, Z.K., Ouimet, C.C., Nairn, A.C., Nestler, E.J., and Greengard, P. (2001). Effects of chronic exposure to cocaine are regulated by the neuronal protein cdk5. *Nature* **410**: 376–380.

Bossert, J.M., Liu, S.Y., Lu, L., and Shaham, Y. (2004). A role of ventral tegmental area glutamate in contextual cue-induced relapse to heroin seeking. *J Neurosci* **24**: 10726–10730.

Browman, K.E., Badiani, A., and Robinson, T.E. (1998). The influence of environment on the induction of sensitization to the psychomotor activating effects of intravenous cocaine in rats is dose-dependent. *Psychopharmacology (Berl)* **137**: 90–98.

Brown, R.W., and Kolb, B. (2001). Nicotine sensitization increases dendritic length and spine density in the nucleus accumbens and cingulate cortex. *Brain Res* **899**: 94–100.

Cammarota, M., Bevilaqua, L.R., Medina, J.H., and Izquierdo, I. (2004). Retrieval does not induce reconsolidation of inhibitory avoidance memory. *Learn Mem* **11**: 572–578.

Carlezon, W.A., Jr., Thome, J., Olson, V.G., Lane-Ladd, S.B., Brodkin, E.S., Hiroi, N., Duman, R.S., Neve, R.L., and Nestler, E.J. (1998). Regulation of cocaine reward by creb. *Science* **282**: 2272–2275.

Centonze, D., Siracusano, A., Calabresi, P., and Bernardi, G. (2005). Removing pathogenic memories: A neurobiology of psychotherapy. *Mol Neurobiol* **32**: 123–132.

Cha, X.Y., Pierce, R.C., Kalivas, P.W., and Mackler, S.A. (1997). Nac-1, a rat brain mRNA, is increased in the nucleus accumbens three weeks after chronic cocaine self-administration. *J Neurosci* **17**: 6864–6871.

Chen, J., Kelz, M.B., Hope, B.T., Nakabeppu, Y., and Nestler, E.J. (1997). Chronic fos-related antigens: Stable variants of deltafosb induced in brain by chronic treatments. *J Neurosci* **17**: 4933–4941.

Chen, J., Nye, H.E., Kelz, M.B., Hiroi, N., Nakabeppu, Y., Hope, B.T., and Nestler, E.J. (1995). Regulation of delta fosb and fosb-like proteins by electroconvulsive seizure and cocaine treatments. *Mol Pharmacol* **48**: 880–889.

Chen, J., Zhang, Y., Kelz, M.B., Steffen, C., Ang, E.S., Zeng, L., and Nestler, E.J. (2000). Induction of cyclin-dependent kinase 5 in the hippocampus by chronic electroconvulsive seizures: Role of [delta]fosb. *J Neurosci* **20**: 8965–8971.

Christie, M.J., Summers, R.J., Stephenson, J.A., Cook, C.J., and Beart, P.M. (1987). Excitatory amino acid projections to the nucleus accumbens septi in the rat: A retrograde transport study utilizing d[3h]aspartate and [3h]gaba. *Neuroscience* **22**: 425–439.

Colby, C.R., Whisler, K., Steffen, C., Nestler, E.J., and Self, D.W. (2003). Striatal cell type-specific overexpression of deltafosb enhances incentive for cocaine. *J Neurosci* **23**: 2488–2493.

Crombag, H.S., Gorny, G., Li, Y., Kolb, B., and Robinson, T.E. (2005). Opposite effects of amphetamine self-administration experience on dendritic spines in the medial and orbital prefrontal cortex. *Cereb Cortex* **15**: 341–348.

Debiec, J., and Ledoux, J.E. (2004). Disruption of reconsolidation but not consolidation of auditory fear conditioning by noradrenergic blockade in the amygdala. *Neuroscience* **129**: 267–272.

Debiec, J., LeDoux, J.E., and Nader, K. (2002). Cellular and systems reconsolidation in the hippocampus. *Neuron* **36**: 527–538.

Di Chiara, G. (2002). Nucleus accumbens shell and core dopamine: Differential role in behavior and addiction. *Behav Brain Res* **137**: 75–114.

Fallon, J.H. and Loughlin, S.E. (1995). Substantia nigra. In Paxinos, G. (ed.), *The Rat Nervous System*. Academic Press, San Diego.

Ferrario, C.R., Gorny, G., Crombag, H.S., Li, Y., Kolb, B., and Robinson, T.E. (2005). Neural and behavioral plasticity associated with the transition from controlled to escalated cocaine use. *Biol Psychiatry* **58**: 751–759.

Floresco, S.B., Blaha, C.D., Yang, C.R., and Phillips, A.G. (2001). Dopamine d1 and NMDA receptors mediate potentiation of basolateral amygdala-evoked firing of nucleus accumbens neurons. *J Neurosci* **21**: 6370–6376.

Fuchs, R.A., Evans, K.A., Parker, M.C., and See, R.E. (2004). Differential involvement of the core and shell subregions of the nucleus accumbens in conditioned cue-induced reinstatement of cocaine seeking in rats. *Psychopharmacology (Berl)* **176**: 459–465.

Ghasemzadeh, M.B., Permenter, L.K., Lake, R., Worley, P.F., and Kalivas, P.W. (2003). Homer1 proteins and ampa receptors modulate cocaine-induced behavioural plasticity. *Eur J Neurosci* **18**: 1645–1651.

Graybiel, A.M., Moratalla, R., and Robertson, H.A. (1990). Amphetamine and cocaine induce drug-specific activation of the c-fos gene in striosome-matrix compartments and limbic subdivisions of the striatum. *Proc Natl Acad Sci USA* **87**: 6912–6916.

Grech, D.M., Spealman, R.D., and Bergman, J. (1996). Self-administration of d1 receptor agonists by squirrel monkeys. *Psychopharmacology (Berl)* **125**: 97–104.

Heimer, L., Zahm, D.S., and Alheid, G.F. (1995). Basal ganglia. In Paxinos, G. (ed.), *The Rat Nervous System*. Academic Press, San Diego.

Heimer, L., Zahm, D.S., Churchill, L., Kalivas, P.W., and Wohltmann, C. (1991). Specificity in the projection patterns of accumbal core and shell in the rat. *Neuroscience* **41**: 89–125.

Hernandez, P.J., Andrzejewski, M.E., Sadeghian, K., Panksepp, J.B., and Kelley, A.E. (2005). Ampa/kainate, NMDA, and dopamine d1 receptor function in the nucleus accumbens core: A context-limited role in the encoding and consolidation of instrumental memory. *Learn Mem* **12**: 285–295.

Hernandez, P.J., Sadeghian, K., and Kelley, A.E. (2002). Early consolidation of instrumental learning requires protein synthesis in the nucleus accumbens. *Nat Neurosci* **5**: 1327–1331.

Heyne, A., May, T., Goll, P., and Wolffgramm, J. (2000). Persisting consequences of drug intake: Towards a memory of addiction. *J Neural Transm* **107**: 613–638.

Hope, B.T., Nye, H.E., Kelz, M.B., Self, D.W., Iadarola, M.J., Nakabeppu, Y., Duman, R.S., and Nestler, E.J. (1994). Induction of a long-lasting ap-1 complex composed of altered fos-like proteins in brain by chronic cocaine and other chronic treatments. *Neuron* **13**: 1235–1244.

Inglis, F.M., and Moghaddam, B. (1999). Dopaminergic innervation of the amygdala is highly responsive to stress. *J Neurochem* **72**: 1088–1094.

Judge, M.E., and Quartermain, D. (1982). Characteristics of retrograde amnesia following reactivation of memory in mice. *Physiol Behav* **28**: 585–590.

Kalivas, P.W., Duffy, P., and Mackler, S.A. (1999). Interrupted expression of nac-1 augments the behavioral responses to cocaine. *Synapse* **33**: 153–159.

Kalivas, P.W., and Volkow, N.D. (2005). The neural basis of addiction: A pathology of motivation and choice. *Am J Psychiatry* **162**: 1403–1413.

Kantak, K.M., Black, Y., Valencia, E., Green-Jordan, K., and Eichenbaum, H.B. (2002). Dissociable effects of lidocaine inactivation of the rostral and caudal basolateral amygdala on the maintenance and reinstatement of cocaine-seeking behavior in rats. *J Neurosci* **22**: 1126–1136.

Kasai, H., Matsuzaki, M., Noguchi, J., Yasumatsu, N., and Nakahara, H. (2003). Structure–stability–function relationships of dendritic spines. *Trends Neurosci* **26**: 360–368.

Kelley, A.E. (2004a). Memory and addiction: Shared neural circuitry and molecular mechanisms. *Neuron* **44**: 161–179.

Kelley, A.E. (2004b). Ventral striatal control of appetitive motivation: Role in ingestive behavior and reward-related learning. *Neurosci Biobehav Rev* **27**: 765–776.

Kelley, A.E., Domesick, V.B., and Nauta, W.J. (1982). The amygdalostriatal projection in the rat — an anatomical study by anterograde and retrograde tracing methods. *Neuroscience* **7**: 615–630.

Kelz, M.B., Chen, J., Carlezon, W.A., Jr., Whisler, K., Gilden, L., Beckmann, A.M., Steffen, C., Zhang, Y.J., Marotti, L., Self, D.W., Tkatch, T., Baranauskas, G., Surmeier, D.J., Neve, R.L., Duman, R.S., Picciotto, M.R., and Nestler, E.J. (1999). Expression of the transcription factor deltafosb in the brain controls sensitivity to cocaine. *Nature* **401**: 272–276.

Kolb, B., Gorny, G., Li, Y., Samaha, A.N., and Robinson, T.E. (2003). Amphetamine or cocaine limits the ability of later experience to promote structural plasticity in the neocortex and nucleus accumbens. *Proc Natl Acad Sci USA* **100**: 10523–10528.

LaLumiere, R.T., and McGaugh, J.L. (2005). Memory enhancement induced by posttraining intrabasolateral amygdala infusions of beta-adrenergic or muscarinic agonists requires activation of dopamine receptors: Involvement of right, but not left, basolateral amygdala. *Learn Mem* **12**: 527–532.

LaLumiere, R.T., Nawar, E.M., and McGaugh, J.L. (2005). Modulation of memory consolidation by the basolateral amygdala or nucleus accumbens shell requires concurrent dopamine receptor activation in both brain regions. *Learn Mem* **12**: 296–301.

Lattal, K.M., and Abel, T. (2004). Behavioral impairments caused by injections of the protein synthesis inhibitor anisomycin after contextual retrieval reverse with time. *Proc Natl Acad Sci USA* **101**: 4667–4672.

Lee, J.L., Di Ciano, P., Thomas, K.L., and Everitt, B.J. (2005). Disrupting reconsolidation of drug memories reduces cocaine-seeking behavior. *Neuron* **47**: 795–801.

Li, Y., Kolb, B., and Robinson, T.E. (2003). The location of persistent amphetamine-induced changes in the density of dendritic spines on medium spiny neurons in the nucleus accumbens and caudate-putamen. *Neuropsychopharmacology* **28**: 1082–1085.

Mackler, S.A., Korutla, L., Cha, X.Y., Koebbe, M.J., Fournier, K.M., Bowers, M.S., and Kalivas, P.W. (2000). Nac-1 is a brain poz/btb protein that can prevent cocaine-induced sensitization in the rat. *J Neurosci* **20**: 6210–6217.

Maldonado, R., Robledo, P., Chover, A.J., Caine, S.B., and Koob, G.F. (1993). D1 dopamine receptors in the nucleus accumbens modulate cocaine self-administration in the rat. *Pharmacol Biochem Behav* **45**: 239–242.

Mark, G.P., Hajnal, A., Kinney, A.E., and Keys, A.S. (1999). Self-administration of cocaine increases the release of acetylcholine to a greater extent than response-independent cocaine in the nucleus accumbens of rats. *Psychopharmacology (Berl)* **143**: 47–53.

McFarland, K., Davidge, S.B., Lapish, C.C., and Kalivas, P.W. (2004). Limbic and motor circuitry underlying footshock-induced reinstatement of cocaine-seeking behavior. *J Neurosci* **24**: 1551–1560.

McFarland, K., and Kalivas, P.W. (2001). The circuitry mediating cocaine-induced reinstatement of drug-seeking behavior. *J Neurosci* **21**: 8655–8663.

McFarland, K., and Kalivas, P.W. (2003). Motivational systems. In M. Gallgher and R.J. Nelson, (eds.), *Handbook of Psychology*, vol. 3, pp. 379–404. John Wliley & Sons, Hoboken, NJ.

McFarland, K., Lapish, C.C., and Kalivas, P.W. (2003). Prefrontal glutamate release into the core of the nucleus accumbens mediates cocaine-induced reinstatement of drug-seeking behavior. *J Neurosci* **23**: 3531–3537.

McGaugh, J.L. (2002). Memory consolidation and the amygdala: A systems perspective. *Trends Neurosci* **25**: 456.

McGaugh, J.L. (2004a). The amygdala modulates the consolidation of memories of emotionally arousing experiences. *Annu Rev Neurosci* **27**: 1–28.

McGaugh, J.L. (2004b). Memory reconsolidation hypothesis revived but restrained: Theoretical comment on Biedenkapp and Rudy (2004). *Behav Neurosci* **118**: 1140–1142.

McLaughlin, J., and See, R.E. (2003). Selective inactivation of the dorsomedial prefrontal cortex and the basolateral amygdala attenuates conditioned-cued reinstatement of extinguished cocaine-seeking behavior in rats. *Psychopharmacology (Berl)* **168**: 57–65.

Milekic, M.H., and Alberini, C.M. (2002). Temporally graded requirement for protein synthesis following memory reactivation. *Neuron* **36**: 521–525.

Miller, C.A., and Marshall, J.F. (2005). Molecular substrates for retrieval and reconsolidation of cocaine-associated contextual memory. *Neuron* **47**: 873–884.

Mogenson, G.J., Jones, D.L., and Yim, C.Y. (1980). From motivation to action: Functional interface between the limbic system and the motor system. *Prog Neurobiol* **14**: 69–97.

Moran, M.M., McFarland, K., Melendez, R.I., Kalivas, P.W., and Seamans, J.K. (2005). Cystine/glutamate exchange regulates metabotropic glutamate receptor presynaptic inhibition of excitatory transmission and vulnerability to cocaine seeking. *J Neurosci* **25**: 6389–6393.

Moser, M.B., Trommald, M., and Andersen, P. (1994). An increase in dendritic spine density on hippocampal ca1 pyramidal cells following spatial learning in adult rats suggests the formation of new synapses. *Proc Natl Acad Sci USA* **91**: 12673–12675.

Nader, K., Schafe, G.E., and Le Doux, J.E. (2000). Fear memories require protein synthesis in the amygdala for reconsolidation after retrieval. *Nature* **406**: 722–726.

Nestler, E.J. (2004). Molecular mechanisms of drug addiction. *Neuropharmacology* **47 Suppl 1**: 24–32.

Norrholm, S.D., Bibb, J.A., Nestler, E.J., Ouimet, C.C., Taylor, J.R., and Greengard, P. (2003). Cocaine-induced proliferation of dendritic spines in nucleus accumbens is dependent on the activity of cyclin-dependent kinase-5. *Neuroscience* **116**: 19–22.

Peakman, M.C., Colby, C., Perrotti, L.I., Tekumalla, P., Carle, T., Ulery, P., Chao, J., Duman, C., Steffen, C., Monteggia, L., Allen, M.R., Stock, J.L., Duman, R.S., McNeish, J.D., Barrot, M., Self, D.W., Nestler, E.J., and Schaeffer, E. (2003). Inducible, brain region–specific expression of a dominant negative mutant of c-jun in transgenic mice decreases sensitivity to cocaine. *Brain Res* **970**: 73–86.

Pitkanen, A. (2000). Connectivity of the rat amygdaloid complex. In Aggleton, J. (ed.), *The Amygdala*. Oxford University Press, Oxford, UK.

Robinson, T.E., Gorny, G., Mitton, E., and Kolb, B. (2001). Cocaine self-administration alters the morphology of dendrites and dendritic spines in the nucleus accumbens and neocortex. *Synapse* **39**: 257–266.

Robinson, T.E., Gorny, G., Savage, V.R., and Kolb, B. (2002). Widespread but regionally specific effects of experimenter- versus self-administered morphine on dendritic spines in the nucleus accumbens, hippocampus, and neocortex of adult rats. *Synapse* **46**: 271–279.

Robinson, T.E., and Kolb, B. (1997). Persistent structural modifications in nucleus accumbens and prefrontal cortex neurons produced by previous experience with amphetamine. *J Neurosci* **17**: 8491–8497.

Robinson, T.E., and Kolb, B. (1999a). Alterations in the morphology of dendrites and dendritic spines in the nucleus accumbens and prefrontal cortex following repeated treatment with amphetamine or cocaine. *Eur J Neurosci* **11**: 1598–1604.

Robinson, T.E., and Kolb, B. (1999b). Morphine alters the structure of neurons in the nucleus accumbens and neocortex of rats. *Synapse* **33**: 160–162.

Robinson, T.G., and Beart, P.M. (1988). Excitant amino acid projections from rat amygdala and thalamus to nucleus accumbens. *Brain Res Bull* **20**: 467–471.

Schelling, G. (2002). Effects of stress hormones on traumatic memory formation and the development of posttraumatic stress disorder in critically ill patients. *Neurobiol Learn Mem* **78**: 596–609.

Schelling, G., Richter, M., Roozendaal, B., Rothenhausler, H.B., Krauseneck, T., Stoll, C., Nollert, G., Schmidt, M., and Kapfhammer, H.P. (2003). Exposure to high stress in the intensive care unit may have negative effects on health-related quality-of-life outcomes after cardiac surgery. *Crit Care Med* **31**: 1971–1980.

Schelling, G., Roozendaal, B., and De Quervain, D.J. (2004). Can posttraumatic stress disorder be prevented with glucocorticoids? *Ann NY Acad Sci* **1032**: 158–166.

Schultz, W. (2002). Getting formal with dopamine and reward. *Neuron* **36**: 241–263.

Seamans, J.K., and Yang, C.R. (2004). The principal features and mechanisms of dopamine modulation in the prefrontal cortex. *Prog Neurobiol* **74**: 1–58.

Self, D.W., Belluzzi, J.D., Kossuth, S., and Stein, L. (1996). Self-administration of the d1 agonist skf 82958 is mediated by d1, not d2, receptors. *Psychopharmacology (Berl)* **123**: 303–306.

Self, D.W., Genova, L.M., Hope, B.T., Barnhart, W.J., Spencer, J.J., and Nestler, E.J. (1998). Involvement of camp-dependent protein kinase in the nucleus accumbens in cocaine self-administration and relapse of cocaine-seeking behavior. *J Neurosci* **18**: 1848–1859.

Self, D.W., and Stein, L. (1992). The d1 agonists skf 82958 and skf 77434 are self-administered by rats. *Brain Res* **582**: 349–352.

Shippenberg, T.S., and Rea, W. (1997). Sensitization to the behavioral effects of cocaine: Modulation by dynorphin and kappa-opioid receptor agonists. *Pharmacol Biochem Behav* **57**: 449–455.

Sklair-Tavron, L., Shi, W.X., Lane, S.B., Harris, H.W., Bunney, B.S., and Nestler, E.J. (1996). Chronic morphine induces visible changes in the morphology of mesolimbic dopamine neurons. *Proc Natl Acad Sci USA* **93**: 11202–11207.

Spiga, S., Serra, G.P., Puddu, M.C., Foddai, M., and Diana, M. (2003). Morphine withdrawal–induced abnormalities in the VTA: Confocal laser scanning microscopy. *Eur J Neurosci* **17**: 605–612.

Squire, L.R., Stark, C.E., and Clark, R.E. (2004). The medial temporal lobe. *Annu Rev Neurosci* **27**: 279–306.

Stefanski, R., Ladenheim, B., Lee, S.H., Cadet, J.L., and Goldberg, S.R. (1999). Neuroadaptations in the dopaminergic system after active self-administration but not after passive administration of methamphetamine. *Eur J Pharmacol* **371**: 123–135.

Swanson, C.J., Baker, D.A., Carson, D., Worley, P.F., and Kalivas, P.W. (2001). Repeated cocaine administration attenuates group I metabotropic glutamate receptor–mediated glutamate release and behavioral activation: A potential role for homer. *J Neurosci* **21**: 9043–9052.

Szumlinski, K.K., Dehoff, M.H., Kang, S.H., Frys, K.A., Lominac, K.D., Klugmann, M., Rohrer, J., Griffin 3rd, W., Toda, S., Champtiaux, N.P., Berry, T., Tu, J.C., Shealy, S.E., During, M.J., Middaugh, L.D., Worley, P.F., and Kalivas, P.W. (2004). Homer proteins regulate sensitivity to cocaine. *Neuron* **43**: 401–413.

Terwilliger, R.Z., Beitner-Johnson, D., Sevarino, K.A., Crain, S.M., and Nestler, E.J. (1991). A general role for adaptations in g-proteins and the cyclic AMP system in mediating the chronic actions of morphine and cocaine on neuronal function. *Brain Res* **548**: 100–110.

Thomas, M.J., Beurrier, C., Bonci, A., and Malenka, R.C. (2001). Long-term depression in the nucleus accumbens: A neural correlate of behavioral sensitization to cocaine. *Nat Neurosci* **4**: 1217–1223.

Timmerman, W., and Westerink, B.H. (1997). Brain microdialysis of GABA and glutamate: What does it signify? *Synapse* **27**: 242–261.

Ungless, M.A. (2004). Dopamine: The salient issue. *Trends Neurosci* **27**: 702–706.

von Hertzen, L.S., and Giese, K.P. (2005). Memory reconsolidation engages only a subset of immediate-early genes induced during consolidation. *J Neurosci* **25**: 1935–1942.

Weed, M.R., Vanover, K.E., and Woolverton, W.L. (1993). Reinforcing effect of the d1 dopamine agonist skf 81297 in rhesus monkeys. *Psychopharmacology (Berl)* **113**: 51–52.

Werme, M., Messer, C., Olson, L., Gilden, L., Thoren, P., Nestler, E.J., and Brene, S. (2002). Delta fosb regulates wheel running. *J Neurosci* **22**: 8133–8138.

Zahm, D.S., and Heimer, L. (1990). Two transpallidal pathways originating in the rat nucleus accumbens. *J Comp Neurol* **302**: 437–446.

CHAPTER 15

Memory Changes with Age: Neurobiological Correlates

Marsha R. Penner and Carol A. Barnes

Arizona Research Laboratories, Division of Neural Systems, Memory and Aging and Evelyn F. McKnight Brain Institute, University of Arizona, Tucson, AZ 85724

I. INTRODUCTION

> *Aging is a natural process that must be studied intensively, for it remains one of the most agonizing problems in all biology. Not only must gerontologists continue to concentrate on performing the research necessary to understand aging, but they must assume an increasing role in the application of their knowledge for betterment of the status of the aged.*
>
> Lawton, 1965, p. 31.

One way to underscore the urgency for research in the area of aging is to appreciate the dramatic demographic shift in the United States and in other countries of the world. According to the Administration on Aging, only 2% of the United States population was age 65 and older just 100 years ago, whereas today over 12% of the population is in that age category. By 2030 the projected percentage of people over the age of 65 is 20%. Accompanying this demographic shift is an increase in the incidence of disease associated with advancing age, for example, stroke, heart disease, and Alzheimer's disease. Research aimed at understanding the aging process, including how learning and memory processes change as a function of age, is therefore needed to provide the basis for the development of better preventative strategies and treatment strategies to enable an increasingly older population to age successfully.

In the general field of aging research, substantial effort has been focused on pathological aging that may be associated with diseases such as Alzheimer's disease. While this disease (and many others) is certainly devastating in terms of its impact on quality of life and learning and memory function, many of us will not suffer from a dementing condition as we age. Instead, most of us will develop mild memory deficits known as *age-associated memory impairment* (AAMI; Crook et al., 1986; Crook and Ferris, 1992; Ferris and Kluger, 1996). The memory deficits associated with AAMI are relatively subtle in comparison to those associated with a dementing disease, but nevertheless those who experience such changes in their memory function may find this troublesome. Because we will all experience some change in our memory function as we get older, the need for research aimed at identifying normal nonpathological aging is of great importance.

This chapter highlights selected domains of cognition that can be studied across mammalian species and that have known age-related neurobiological underpinnings. Studies conducted on both human and nonhuman species will be considered, although data from rodent studies will make up the bulk of the following discussion.

II. METHODS MATTER

How old would you be if you didn't know how old you was?

Satchel Paige

A large literature supports the idea that memory and other cognitive processes do decline or change with advancing age. Important methodological and other considerations must, however, be taken into account before interpreting such changes as negative consequences of brain aging. At least three primary issues require consideration: (1) What does it mean to be "old"? (2) How can aging be studied? (3) What variable are you studying, and how are you measuring it?

Aging is difficult to define, in either biological or cognitive terms. The simplest solution to the question of "how old" an individual is to assign a number to that person based on the number of years he or she has lived. This assessment of aging, however, has a number of shortcomings (Bourliere, 1970; Ingram, 1983). To take an extreme example, some older individuals, for example, centenarians, may be more cognitively intact than someone 30 years younger than themselves (see Perls, 2004). This highlights a key concept: Individuals appear to age at different rates in chronological time, and therefore biological age does not always match the number of years we have been alive.

In terms of mnemonic function, our cognitive abilities change throughout the course of our development (from birth to death); but for each of us, this change will take its own individualistic course. Although such issues certainly complicate the study of aging, they also provide an opportunity to discover the processes that allow some of us to age successfully, in the absence of debilitating learning and memory decline. Nevertheless, chronological age is still the most widely used predictor of the functional or biological age of the organism being studied (Costa and McCrae, 1980). For humans, "old" is typically considered to begin at 65 years, and for many laboratory animals, particularly rats, ages past the point of a 50% mortality rate for the particular organism or strain being studied are considered "old."

Experimental design is another critical issue to consider in aging research. Many human studies in the aging field employ either a cross-sectional design or a longitudinal design, and both methods of data collection offer important advantages and disadvantages in the study of aging. Typical cross-sectional studies compare a group of young adults in their late teens or early 20s (young adults are typically recruited from university introductory psychology classes) to a group of aged adults, yielding data on age differences. This study design is probably the most efficient and cost-effective means of conducting an aging study. Cross-sectional studies, however, are subject to confounding cohort differences[1] and according to some (e.g., Hofer and Sliwinski, 2001; Salthouse and Nesselroade, 2002) tend to exaggerate age differences.

On the other hand, longitudinal studies,[2] which follow the same individuals over time, are expensive and logistically difficult to conduct, but essentially they eliminate many confounds introduced as a result of cohort effects. These types of studies yield data on age changes. Moreover, longitudinal studies address the issue discussed earlier, that individuals can age at different rates. However, longitudinal studies also have disadvantages, which include underestimation of age differences because of selective subject attrition (i.e., as a result of illness) and practice effects.

Most animal studies employ a cross-sectional strategy, for many of the same reasons as discussed earlier, and therefore many of the same issues apply. Cohort effects in animal studies may be due to differences in housing conditions (i.e., individual or group housing) or to other life history differences of the animals being used. Some of the difficulties inherent in simple cross-sectional and

---

[1]*Cohert differences* are group differences that arise as a result of factors such as socioeconomic status, cultural differences, and educational status in groups of individuals who are born at different time periods.

[2]Currently, there are several large-scale longitudinal studies under way. One example is the Seattle Longitudinal study, which began in 1956. Every seven years, all people who have previously participated are retested. A new group of participants is also added at seven-year intervals. In total, over 6,000 people have participated in this particular longitudinal study (Schaie, 1996).

longitudinal studies can be circumvented using a cross-sequential design, in which groups of individuals of different ages are tested repeatedly over several years. For example, two groups of subjects, one in their 20s and one in their 60s, can be tested to determine age differences in performance on a memory task and then later retested to yield data concerning age changes on the same task.

A third issue to consider is the type of learning or memory under investigation (e.g., long-term memory versus short-term memory) and how performance is assessed (e.g., number of correct responses versus reaction time). Some studies may be focused on long-term memory changes, for example, how vividly autobiographical memories are recalled in an aged population. Other studies may focus on short-term memory changes, for example, working memory. This is a good place to point out that the outlook for the healthy aging individual is not as grim as many suppose. While many age-associated cognitive changes do occur, there are also many things that remain stable as we get older, and thus the aggressively negative stereotype of the elderly is largely unfounded. For example, although working memory, episodic memory, and declarative memory abilities may be affected as one ages, other types of memory, such as autobiographical memory, and semantic knowledge tend to remain stable over the adult life span (for review see Hedden and Gabrieli, 2004). Paradoxically, if vocabulary is tested in isolation, older adults would be found to outperform younger adults (Schaie, 1996; Park et al., 2002); while if tests of "executive function" are given (such as the Wisconsin card-sorting task), then even those individuals in their 40s will likely show age-related decline (for a recent review and meta-analysis see Rhodes, 2004).

One variable that influences the outcome of any study is how performance is measured. For example, if performance on a memory task is based on measures of cued recall or recognition, older individuals often perform as well as their younger counterparts; whereas if the task is one of recall, aged individuals tend to have more difficulty than do younger subjects (e.g., Schonfield and Robertson, 1966; Harwood and Naylor, 1969; Craik and McDowd, 1987). In addition, older individuals tend to be less able to learn new information quickly when compared to younger adults, but this age difference can be reduced by allowing the aged individual to have more practice time or to determine his or her own pace of learning (e.g., Canestrani, 1963; Monge and Hultsch; 1971). Thus, laboratory studies of learning and memory that incorporate timed tasks may not be the best measures of learning and memory abilities in aged individuals. This also holds true for studies using rats (or other animal species) since the aged animals used for any study may be frail or less physically capable as compared to their younger counterparts. If older animals are tested in terms of "how fast" they perform (such as how fast they reach a goal on a maze), large deficits can be observed; but these deficits can be reduced if accuracy measures that are independent of speed are used (such as the distance that is traveled to reach the goal). Designing studies that minimize these issues or take them into account is therefore equally important in studies using human or nonhuman subjects.

III. LEARNING AND MEMORY CHANGES ASSOCIATED WITH AGING

There is a large literature that has examined the question of whether learning and memory function is altered during the aging process. Tasks in which age-related impairments have been found include classical conditioning, such as eyeblink and heart rate conditioning, conditioned taste aversion, fear conditioning, operant tasks such delayed matching-to-sample and delayed nonmatching-to-sample tasks, and instrumental tasks such as active avoidance, passive avoidance, and maze learning tasks (for review see Barnes, 1991; Ingram, 2001; Rosenzweig and Barnes, 2003). The performance of older animals has been shown, in some instances, to be markedly different from that of younger animals, whereas performance on other tasks is unaffected or only marginally affected in aged subjects. Instead of providing a comprehensive review of the literature regarding age-related cognitive changes, the following discussion focuses on one particular behavior that consistently changes with age, that of spatial learning and memory. This form of learning and memory involves the ability of the organism to acquire and retain information that is critical for successful navigation through space.

The rationale for choosing this particular behavior is based on the notion that the study of aging, using nonhuman animal models, is most effective when the behavior under investigation has some analog to human behavior. If the behavior under investigation changes in an age-dependent manner in both the nonhuman animal model and the human case, then the task of making infer-ences from the nonhuman subject is greatly simplified. Moreover, a great deal of accumulated evidence indicates that in all mammals tested, spatial abilities are altered during aging. Spatial learning and memory has been the subject of laboratory studies for decades, does not require language to test, and therefore represents an excellent example of a behavior that can be readily tested in both human and nonhuman subjects. This is the focus of the following section.

A. Spatial Learning and Memory: Rodent Models

Maze-learning tasks are very useful methods for investigating age-related changes in rodents, in part because rodents are naturally excellent foragers and therefore tend to learn these sorts of tasks exceptionally well. In such tasks, the animal begins at a specific start position in the maze and then must navigate to a goal position to receive food or water reward (positive reinforcement) or to avoid foot shock, bright light, or cool water (negative reinforcement). Simple mazes require the animal to learn a limited number of discriminations to reach the goal target, and delays can be imposed to complicate the task. Complex mazes, on the other hand, require the animal to learn and remember a greater number of discriminations in order to reach the target goal (see Fig. 15-1).

Behavioral investigation of age-associated memory changes using mazes dates back at least to the seminal work of Stone (1929a, 1929b), who used a battery of tests to investigate age-associated learning and memory changes in rats up to 24 months of age. Although the results reported by Stone did not reveal significant differences between groups of rats when the data are considered on a per-animal basis, a small proportion of the oldest rats tested by Stone did show deficits. One possible explanation for the weak aging effect reported by Stone is that he used a long-lived strain of rat with an average life span of 36 months. Thus, he may have been collecting data from "biologically younger" animals or rats in late maturity as opposed to old age. In fact, subsequent studies have consistently found age-related deficits using multiple T-mazes (similar to one the Stone himself used) and shorter-lived rat strains (e.g., Goodrick, 1968; Klein and Michel, 1977; Skalicky et al., 1984; Ingram, 1985; Goldman et al., 1987; Lohninger et al., 2001). For example, an early study by Verzar-McDougall (1957) reported that with advancing age, rats have greater difficulty with both the acquisition and the retention of the complex T-maze task (Fig. 15-1A). Moreover, after 20 months of age, individual variability increases significantly, with the oldest rats performing the worst. These studies highlight the importance of considering the issue of whether the subjects tested are truly "aged."

Many other tasks have been used to assess spatial memory in rats of different ages. One such task is the circular platform task (also known as the Barnes maze), which was developed specifically for age comparisons of spatial memory (Fig. 15-1B). This task is well suited for testing aged animals because it does not require food or water deprivation for motivation, and performance is not affected by age-related changes in speed or stamina. For this task, the rat is placed onto a large, open, brightly lit platform and must discover which of 18 holes, located on the perimeter of the platform, leads to a dark escape chamber. Rats prefer dark enclosures to brightly lit open spaces and are therefore motivated to complete the task based on this preference. The escape hole is always placed in the same location of the room with respect to distal cues in the environment. Thus, the most effective strategy for solving this task over multiple trials is to remember the location of the escape box in relation to the distal cues in the room (i.e., spatial strategy). Compared to young rats, aged rats tend to have difficulty learning the location of the escape tunnel and retaining the memory of the tunnel's location from day to day (e.g., Barnes,

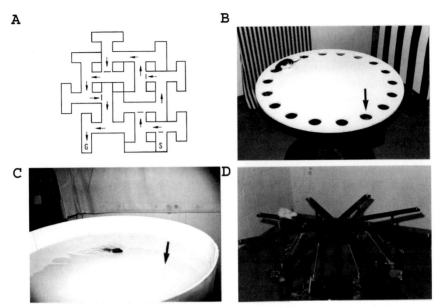

FIGURE 15-1   **(A)** Diagram of the 14-unit T-maze, or Stone maze. The start box is indicated by an *S*, and the arrows show the direction in which the animals must travel to find the goal box (*G*) correctly. Guillotine doors are used to divide the maze into five segments (indicated by lines at choice points) so that the animals are not allowed to go backward through the maze. Motivation for performance can either be food or water reward at the goal or foot shock. (After Ingram, 1988). **(B)** Photograph of the Barnes maze, or circular platform task, illustrating the rat making an "error" by looking into a hole that is not over the dark escape chamber. The arrow points to the correct location of the hole over the goal, which the rat must find on the basis of the features of the environment distal to the platform. **(C)** Photograph of a rat swimming in the spatial version of the Morris water maze. The water is made opaque so that the submerged plat-form is not visible to the animal. The rat must use distal cues in the room to locate the escape platform. The arrow indicates the location of the hidden platform. **(D)** Photograph of a radial eight-arm maze, with a rat prepared for and attached to electrophysiological recording equipment. Food reward is available at the ends of each of the arms in cups, and on any given trial the cup is baited only once (Gallagher and Bostock, 1985).

1979; Barnes et al., 1980; Barnes and McNaughton, 1985; Barnes et al., 1989; Markowska et al., 1989; McLay et al., 1999; Greferath et al., 2000). These findings have been replicated by Bach et al. (1999) in the aged mouse using a modified version of the circular platform task that included a control version for determining how well older animals could see and use distal cues.

Perhaps the most widely used test of spatial learning and memory in rodents is the Morris water maze (Morris, 1981). In its most basic form, a large circular tub of water is filled with opaque water, and it is the rat's job to locate a hidden escape platform submerged just beneath the surface of the water (Fig.

15-1C). On each trial, the rat is released into the water at different predetermined start locations and allowed to swim until either the escape platform is located or a certain amount of time has elapsed. Initially, rats will swim around the pool randomly. But after having found the platform over repeated trials, most young animals learn to swim directly to the escape platform. The two dependent variables that are typically measured are the latency to find the hidden platform and the length of the path taken to reach the platform, although other, more complex analyses can be used, such as Gallagher's "corrected integrated path length" or "learning index" (Gallagher et al., 1993). After a certain number of training trials (typically 20–24), a probe trial can be administered in which the platform is removed from the tank. Those animals that have learned the task (i.e., the location of the platform) will spend a considerable amount of time swimming in the area where the platform was previously located, whereas those animals that have not successfully learned the task will continue to swim randomly around the tank.

Numerous laboratories have demonstrated that aged rats are consistently slower at learning the Morris water maze task than are adult rats and show poorer retention of the hidden platform location from test day to test day than do younger animals (e.g., Gage et al., 1984; Biegon et al., 1986; Pelleymounter et al., 1987; Rapp et al., 1987; Markowska et al., 1989; Frick et al., 1995; Barnes et al., 1997b; Shen et al., 1997; Bizon et al., 2004; Nicholson et al., 2004; Tombaugh et al., 2005). When the performance of individual rats is assessed for each trial, both young and aged rats show bimodal performance on early learning trials (Barnes et al., 1997b). This means that on some trials, both age groups sometimes take a short path to the platform, but on others take a long path. Over the course of training, however, the performance of young rats is increasingly unimodal, since most young rats will learn consistently to take the most direct path to the platform. In contrast, the aged rats continue to display relatively bimodal performance, even on the final days of training. This indicates that aged rats are not completely forgetting or not learning the location of the hidden platform; in fact, aged rats do show moderate improvements in performance over the course of training. Rather, aged rats have more difficulty than young rats in consistently retrieving the correct memory for the platform's location.

The Morris water maze can be modified to become a visual discrimination task instead of a spatial task, with the platform made visible above the waterline. A large beacon can also be placed over the platform, making it easier to find. When this nonspatial version of the task is used, aged rats without visual deficits have little difficulty solving the task (e.g., Rapp et al., 1987; Rosenzweig et al., 1997; Shen et al., 1997). Additional studies have demonstrated that aged rats can perform well on maze tasks when alternative strategies for solving the task can be used. For example, Barnes et al. (1980) investigated the kind of strategies aged and adult rats use to solve a T-maze task, altering the task so that it could be performed using (1) a local-cue strategy in which the rat learns to associate a particular arm of the maze with a rubber mat on the track surface; (2) a response strategy in which the rat learns to turn left or right, depending on the sequence of behaviors already performed on the maze; (3) a place strategy in which the rat learns to enter the arm located in a particular area of the room, using extra-maze cues as a coordinate system to make this decision. This study revealed that aged rats were more likely than adult rats to use a response strategy to solve the task, whereas the adult rats were more likely to use a place strategy. This finding, along with those of other laboratories, suggests that aged rats may not use effective spatial strategies to solve these problems (e.g., Barnes et al., 1980; Rapp et al., 1987; Gallagher and Pelleymounter, 1988).

Finally, one additional task that has been used to test spatial learning and memory function is the eight-arm radial maze (Fig. 15-1D), which consists of a central platform with eight arms containing rewards radiating from it. The rat is required to obtain each of the rewards without reentering an arm where reward has already been obtained. Revisiting arms where reward was already obtained is considered an error of spatial working memory. The task can be made more difficult by imposing a delay between the retrieval of the first four and the last four rewards. Numerous laboratories have demonstrated that aged rats are impaired on this task, with and without the imposed delay (e.g., Barnes et al., 1980; J.E. Wallace et al., 1980; de Toledo-Morrell et al., 1984; Beatty et al., 1985; de Toledo-Morrell and Morrell, 1985; Gallagher et al., 1985; van Gool et al., 1985; Geinisman et al., 1986; Caprioli et al., 1991; Mizumori et al., 1996; McLay et al., 1999; Rossi et al., 2005). When this task is modified so that it can be solved using nonspatial strategies (i.e., a local-cue strategy), aged rats perform as well as young rats (Barnes et al., 1987).

B. Spatial Learning and Memory: Human Studies

Age-related declines in navigational skills are well documented in humans (for review see Rosenzweig and Barnes, 2003; Allen et al., 2004; Driscoll et al., 2005). An example from an early study that examined this issue used a large-scale real-world environmental setting to examine the navigational performance of aged (60–80 years) and middle-aged (26–45 years) residents. The task was to recall as many buildings as possible from the downtown area of Orange, California, and to locate them on a grid. To minimize the types of variables that can negatively impact aged subjects' performance (i.e., speed, anxiety), the task was self-paced, and subjects were told that the purpose of the study was to help urban planners design better cities. Although aged subjects had actually lived in the area longer than the younger subjects, they had less complete knowledge of their home environment when asked to recall environmental features and locate them accurately on a grid (Evans et al., 1984). A more recent study by Uttl and Graf (1993) assessed the spatial memory of several hundred young and old visitors (15–74 years of age) to a museum exhibit. In this study, participants were required to recall three pieces of information: (1) the route they traveled through the exhibit, (2) the landmarks they encountered on the route they traveled, and (3) the temporal order of landmarks they encountered. Aged subjects were able to recognize the landmarks they had encountered, but they had difficulty recalling the route they traveled through the exhibit as well as the temporal order of the landmarks they encountered (Uttl and Graf, 1993).

As noted in the previous section, the Morris water maze is one of the most widely used tests of spatial learning and memory for rodents. A task has been developed for humans that parallels the Morris water maze task used for rats. In the human version, subjects are explicitly instructed to learn the location of a landmark relative to extra-maze cues that are located outside of a large arena. Subjects are then required to place the landmark in the correct location after the environment has been manipulated in some way. As observed in aged rats, aged humans have more difficulty learning the correct location of the landmark than do young subjects (Newman and Kaszniak, 2000). These findings closely parallel those seen with aged rats in the Morris water maze or the circular platform task.

Virtual navigation tasks have become increasingly popular in assessing human spatial cognition. For example, Moffat et al. (2001) had subjects "navigate" a computer-generated maze that consisted of a series of interconnected hallways and alleys, some of which led to target goal positions within the virtual environment while others led to dead ends. Performance of the aged subjects was significantly impaired on this task; following the fifth learning trial, 86% of the young participants located the target goal without error, whereas only 24% of the aged subjects performed at this level. In addition to making significantly more spatial memory errors (i.e., traveled to more dead ends), aged participants traveled a greater distance and took longer to complete the task than did younger subjects.

A virtual analog of the Morris water maze, known as the computer-generated arena (C-G arena), has also been developed (Jacobs et al., 1997, 1998) and has been used to assess age-related changes in human spatial navigation (Thomas et al., 1999). For this task, participants are required to search and locate a target hidden within the virtual environment. Compared to younger adults, older adults had difficulty learning the location of the platform and travel a greater distance through the virtual environment in their search for the platform. In addition, on subsequent probe trials, aged subjects spent less time in the quadrant of the virtual environment that had previously contained the platform than did young subjects. This same pattern of results is demonstrated even when environmental support (i.e, a larger platform, more salient distal cues) and more practice trials are allowed prior to testing (Thomas et al., 1999). As noted earlier for the rodent studies, aged rats have particular difficulty with tests that rely on knowledge of the environment. But if the task can be solved using an alternative strategy, aged rats perform as well as do young rats (e.g., Barnes et al., 1980). Young adults in the Thomas et al. (1999) study reported using a spatial strategy to solve the task, whereas aged adults were less likely to report using a spatial strategy. Thus, these findings, together with those described for rodents, support the idea that aged subjects preferentially employ non-spatial strategies to solve spatial tasks when given an opportunity.

IV. INVOLVEMENT OF THE HIPPOCAMPUS IN SPATIAL LEARNING AND MEMORY

A fundamental goal for research in the area of behavioral neuroscience is to determine which brain structures are necessary for which behaviors. Some of the very earliest attempts to find brain structures that participate in memory relied heavily on lesion methods (for review see Olton, 1991). Defining a role for the hippocampus in memory processes employed a number of lesioning techniques, such as knife cut of the fimbria-fornix, aspiration of the hippo-campus, electrolytic lesions, and chemical lesions. Although many of the tech-niques that have been used to lesion the hippocampus are not selective and may damage closely lying brain structures or fibers-of-passage, these approaches have provided experimental evidence for a critical role of the hippocampus in spatial memory. For example, Morris et al. (1982) removed the entire dorsal and ventral hippocampus and a small area of overlying cortex in rats and then assessed the effects of these lesions on spatial learning and memory using the Morris water maze. Compared to control rats that had comparable cortical lesions, the hippocampal lesioned rats took longer to locate the hidden platform on the spatial version of the task. But when the platform was visible, lesioned rats performed no differently from control rats. Moreover, on the probe trial, when the platform is removed from the tank, control rats showed a preference for the quadrant of the tank where the hidden platform was previously located, whereas the lesioned rats did not show this preference. Animals with lesions of the hippocampus and its closely related structures also show severe deficits on other spatial tasks, such as the 14-unit T-maze (Bresnahan et al., 1988), the Y-maze (Aggleton et al., 1986), the radial arm maze (Olton et al., 1978), and the circular platform task (McNaughton et al., 1989). These deficits mimic those observed in the memory-impaired aged rat on the same tasks; the deficits associated with hippocampal lesions affect performance when the task is spatial in nature but not when the task is nonspatial.

Neuronal recordings from the hippocampus have also provided evidence that the hippocampus is involved in spatial learning and memory. When a rat is allowed to explore an environment, pyramidal and granule "place cells" show patterned neuronal activity that is highly correlated with the rat's position in space (i.e., the place field; O'Keefe and Dostrovsky, 1971). When the activity of many place cells are recorded simultaneously, the position of the rat in a given environment can be accurately reconstructed from the cell firing activity alone (Wilson and McNaughton, 1993). There are several changes that occur to the "hippocampal maps" (i.e., distribution of place cell firing) of aged rats that may contribute to why or how spatial learning and memory deficits occur. For example, in young rats, the maintenance or stability of any particular map can remain intact for long periods of time (i.e., months) such that when a rat is returned to a previously explored environment, the hippocampal map for that particular environment is reliably retrieved (Thompson and Best, 1990). For the aged rat, retrieval of the correct map is somewhat less reliable; upon return to a previously encountered environment, the aged rat is more likely than a young rat to show a different population of activated place cells in a familiar environment (i.e., a different place map is retrieved; Barnes et al., 1997b). This finding correlates well with the bimodal performance of aged rats late in training on the spatial version of the Morris water task described earlier.

Several human studies using functional brain imaging have also supported the idea that the hippocampus is critically involved in spatial memory. For example, Maguire et al. (1998) used positron emission tomography (PET) to scan subjects' brains while they navigated a virtual town. Brain activation was measured in a condition in which the subject navigated directly to a goal, compared to a control condition in which the subject was directed via arrows through the town. The right hippocampus was shown to be more active during navigation than during passive traversal of the town. In addition, the degree of hippocampal activation was positively correlated with more accurate navigation to the goal. A recent study conducted by Moffat et al. (2005) identified differences in brain activation for young and aged subjects during navigation of a virtual environment consisting of several rooms and interconnected hallways. Subjects were instructed to locate and remember the location of six objects within the environment, and performance of this task was compared to a control task in which the subjects followed a path through a similar environment. This study demonstrated that aged and adult subjects had different patterns of brain activation during navigation; in particular, activation of the hippocampus and parahippocampal gyrus was greater in young subjects and was positively correlated with performance on the navigation task (Moffat et al., 2005). Taken together, the results from these studies point to the hippocampus (see Figure 15-2) as a structure that may underlie age-related spatial memory. In fact, the hippocampal formation is among the brain structures that show the earliest signs of age-related changes in rodents, monkeys, and non-demented humans and is an important target structure in the etiology of

Alzheimer's disease (e.g., Hyman et al., 1984; Khachaturian, 1985; for a recent review see Hof and Morrison, 2004). The following discussion provides selected examples of work that attempt to correlate changes in the hippocampus in aging rats and humans with their memory performance. The experiments to be highlighted use neuroanatomical (including imaging studies), electrophysiological, and molecular methods for the correlation of brain function with behavior.

A. Neuroanatomical Findings

Early work on the question of whether aging is accompanied by significant neuron death reported that as much as 25–50% of the neurons in neocortical areas and some hippocampal subfields were lost in humans without Alzheimer's disease as well as in nonhuman primates and rodents (Brody, 1955; Ball, 1977; Brizzee et al., 1980; for review see Coleman and Flood, 1987). These studies, however, were not conducted using unbiased sampling methods, and thus the results were based on neuron density in the structure under investigation instead of on total neuron numbers (for review see Morrison and Hof, 1997). More recent studies that have employed unbiased stereological methods[3] have not detected a significant age-related loss of principle cells in the rat (Rapp and Gallagher, 1996; Rasmussen et al., 1996), mouse (Calhoun et al., 1998), nonhuman primate (M.J. West et al., 1993; Peters et al., 1996), or human (M.J. West, 1993). These studies, therefore, have led to the conclusion that significant neuron loss is not an inevitable consequence of the normal aging process, at least with respect to the hippocampus and much of the neocortex.[4] In addition, although the rate of hippocampal neurogenesis declines with increasing age, new neurons do continue to be generated in the aged rat hippocampus (e.g., Kuhn et al., 1996; Kempermann et al., 2002; Heine et al., 2004). Granule cells are also born in the aged mouse dentate gyrus, and the rate of neurogenesis can be significantly increased by exercise (van Praag et al., 2005; Kronenberg et al., 2005). It remains to be determined, however, if the rate of neurogenesis affects memory function; one study has reported that exercise-induced increases in neurogenesis are associated with better performance on the Morris water maze (van Praag et al., 2005), whereas other studies have found no correlation (Bizon and Gallagher, 2003) or a negative correlation (Bizon et al., 2004) in the aged rodent.

---

[3]Stereological methods are used to analyze biological tissue in 3D. This method uses unbiased sampling and estimation methods to determine parameters of volume, surface area, length, and number in the biological tissue being analyzed.

[4]A recent study by D.E. Smith et al. (2004) has reported a 30% reduction in neuron number of area 8A of the dorsal lateral PFC.

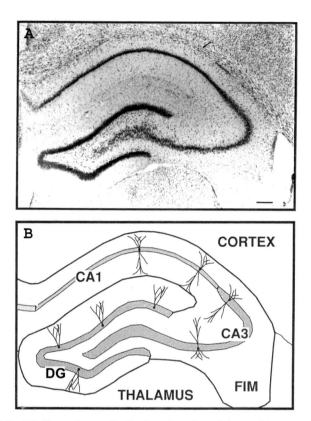

FIGURE 15-2   **(A)** Photomicrograph of a frontal section of the rat hippocampal formation. Calibration bar: 200 µm. **(B)** Outline of the section in part A labeling the subareas of the hippocampal formation. CA1, CA3 represent the hippocampus proper, pyramidal cell fields; DG denotes the dentate gyrus, granule cell fields; FIM, fimbria, a bundle of fibers containing some of the afferents to and the efferent from the hippocampus.

If it is largely true that principle cell numbers are not altered in the aged hippocampus, then what is responsible for the age-related hippocampal-dependent cognitive changes that have been observed? Likely anatomical candidates include the dendrites and axons that connect cells to each other. Early studies that investigated dendritic branching suggested that age-related deterioration occurred in the hippocampus and entorhinal cortex (M.E. Scheibel et al., 1976; A.B. Scheibel, 1979), but these reports were later challenged, primarily on the basis that both demented and nondemented subjects were included in the analysis and nonstereological techniques were used. Subsequent studies demonstrated extensive dendritic branching in CA1, CA3, and the subiculum of both aged and young humans (Flood et al., 1987a; Flood, 1991; Hanks and Flood, 1991), and some studies have also reported increased dendritic branching and length in the DG of aged compared to young adults (Flood et al., 1987b). Increased dendritic branching has also been reported in CA1 pyramidal cells of aged rats (Pyapali and Turner, 1996).

Additional studies have investigated possible changes in cell connectivity by assessing synapse numbers, but these attempts initially produced mixed results (Bondareff and Geinisman, 1976; Geinisman, 1979; Anderson et al., 1983; Scheff et al., 1985), either as an artifact of the method used to count synapses or because the cognitive status of the animals included was not known. More recent studies using unbiased stereological methods and behaviorally character-ized rats have provided a clearer picture. For example, work by Geinisman and his colleagues (1992) demonstrated a decrease in the number of perforated and nonperforated axospinous synapses in the inner and middle molecular layers of the aged dentate gyrus. This finding was specific to aged memory-impaired rats; groups of young rats and aged rats that were not memory impaired were not significantly different from one another. Physiological measures obtained from the dentate gyrus support this finding, showing an age-related decrease in the presynaptic fiber potential elicited by stimulation of the perforant path (e.g., Barnes and McNaughton, 1980a; Foster et al., 1991), indicating the presence of fewer perforant path collateral axons originating from layer II entorhinal cortical cells. In addition, an age-related decrease in the field EPSP has been recorded from dentate granule cells (e.g., Barnes, 1979; Barnes and McNaughton, 1980a; Foster et al., 1991). Both of these findings support the idea that synapse loss is a feature of the aged dentate gyrus.

In area CA1, however, the number of axospinous synapses are not different between aged memory-impaired and unimpaired or young animals (Geinisman et al., 2004). This might seem surprising since the field EPSP in CA1 is reduced in the aged rats (Landfield et al., 1986; Barnes et al., 1992; Deupree et al., 1993). In contrast to the dentate gyrus, the presynaptic fiber potential in CA1 remains unchanged in aged rats (Barnes et al., 1997a; Rosenzweig et al., 1997; Barnes et al., 2000b), suggesting that the decrease in the field EPSP is a result of a loss of functional synaptic contacts rather than a loss of incoming axons. Additional analysis of the synaptic contacts in the CA1 region of aged rats has focused on changes in the size of the postsynaptic density (Nicholson et al., 2004), a region where neurotransmitter receptors, such as those for glutamate, are particularly dense (Peters et al., 1991). The size of the postsynaptic density at axospinous perforated synapses was reduced in memory-impaired aged rats (Nicholson et al., 2004). This finding, taken together with the physiological data, does suggest that a decrease in functional synaptic contacts is a feature associated with aging in this brain region. This raises the possibility that the population of synaptic contacts with smaller postsynaptic densities are the ones that are functionally silent (see Atwood and Wojtowicz, 1999).

Although neuroanatomical measures have been taken from human tissue, there are several methodological issues that impact the results obtained from human tissue samples. For example, the cognitive status of the brain donor may not be known, and issues related to how the brain tissue was handled or preserved prior to analysis is not always consistent. Well-designed structural MRI studies can avoid some of these difficulties and therefore represent a powerful method that can be used to measure potential age-related neuroanatomical changes and to correlate such changes with cognitive function. The relationship between hippocampal size and memory function, however, is not as straightforward as one might first suppose. Studies that have used this technique have reported conflicting results; some studies demonstrate a significant positive correlation between hippocampal volume and memory function, some a significant negative correlation, while others have demonstrated no correlation (for recent reviews see Hedden and Gabrieli, 2005; Van Petten, 2004). There are numerous variables that may contribute to the variability across studies, including the age of the participants in the study, the types and sensitivity of the neuropsychological tests used to assess cognitive function, if or how normalization to head size and/or body size is carried out, and possibly the inclusion of subjects in the early stages of Alzheimer's disease or some other underlying pathology. A recent comprehensive meta-analysis of 33 studies by Van Petten (2004) demonstrated that the relationship between hippocampal volume and episodic memory in healthy aged subjects was weak, due to significant variability across studies. Van Petten did find, however, that the trend was toward volume–memory correlations becoming more positive as the age of the sample under investigation increased (i.e., greater volume is associated with better memory performance). Since the current discrepancy in the literature may be partly due to the inclusion of subjects in the early stages of Alzheimer's disease, assessing hippocampal volume in an organism that does not get Alzheimer's disease is critical. For example, a recent study by Shamy et al. (2006) measured hippocampal volume in behaviorally characterized rhesus macaques. The results of this study indicate that memory-impaired aged monkeys differ by less than 6% from younger monkeys in hippocampal volume. These findings provide additional support for the idea that normal cognitive aging occurs independent of gross structural deterioration in the hippocampus.

B. Electrophysiological Findings

Despite changes at the morphological level, most of the basic electrical properties of the hippocampus remain constant over the life span. Studies investigating resting membrane potential, membrane time constant, threshold to action potential, input resistance, and the amplitude and duration of the Na^+ action potential have not revealed age-associated changes in these properties (for review see Barnes, 1994; Rosenzweig and Barnes, 2003). Some studies, however, have shown an increase in the amplitude of the after-hyperpolarizing potential (AHP) for pyramidal cells of the CA1 subregion (e.g., Landfield and Pitler, 1984; Disterhoft et al., 1996; Thibault and Landfield, 1996), and this change has been correlated with cognitive functioning (Disterhoft et al., 1996; Tombaugh et al., 2005). The larger AHP may be due to changes in Ca^{2+} homeostasis, since the outward K^+ current that underlies the AHP is Ca^{2+} dependent (Thibault and Landfield, 1996). In fact, aged CA1 pyramidal cells have a greater density of L-type Ca^{2+} channels, which may be a major contributing factor to the increased AHP (Thibault and Landfield, 1996). Although the larger AHP predicts slower firing frequencies, no firing rate change has been observed for CA1 pyramidal neurons in vivo. Pyramidal cells of CA3, however, have recently been shown to have elevated firing rates in vivo (I.A. Wilson et al., 2005), suggesting the possibility that the larger AHP in CA1 is an adaptive change that serves to normalize CA1 firing rates (i.e., doesn't allow CA3 to hyperactive CA1).

One particular form of long-term synaptic plasticity, known as *long-term potentiation* (LTP), is thought to be a potential cellular mechanism for long-term memory (Bliss and Gardner-Medwin, 1973; Bliss and Lomo, 1973) and involves the strengthening of connections between neurons. A great deal of work has also been aimed at determining if changes in either the induction or the maintenance of LTP contributes significantly to the memory deficits associated with aging. Although variable results have been reported with respect to LTP induction across age, the differences can be attributed to the specific stimulation parameters used to induce LTP and the region of the hippocampus under investigation. When high-intensity stimulation protocols that are well above threshold for LTP induction are used, there are no age-related deficits in LTP induction at the CA3–CA1 Schaffer collateral synapse (e.g., Landfield and Lynch, 1977; Landfield et al., 1978) or the perforant path–granule cell synapse (e.g., Barnes, 1979). When stimulus parameters are set very close to the LTP induction threshold, however, aged rats consistently show LTP induction deficits at the Schaffer collateral CA1 synapse and the perforant path–granule cell synapse (e.g., Deupree et al., 1993; C.I. Moore et al., 1993; Barnes et al., 2000a).

After LTP has been induced, the durability or maintenance of LTP can also be measured and can be divided into an early phase, lasting one to three hours, and a late phase, lasting for more than 24 hours. New protein synthesis is not required to sustain the early phase of LTP, but new RNA and protein synthesis is required to maintain the late phase. Several studies have demonstrated that in the short term (i.e., over 1 hour), LTP decay rates are not different between aged and young rats (e.g., Landfield and Lynch, 1977; Landfield et al., 1978). Over a longer time course, however, clear age-related maintenance deficits do appear (e.g., Barnes, 1979; Barnes and McNaughton, 1980b; for review see Barnes, 2003). For example, Barnes and McNaughton (1980b) administered LTP-inducing stimuli at 24-hour intervals on 12 consecutive days and then monitored the evoked field response for several weeks. Although there were no age-related differences in the final levels of LTP induction, LTP decayed nearly twice as fast in the aged rats over several weeks.

Several experiments have also demonstrated a significant correlation between LTP and memory deficits in aged animals (Barnes, 1979; Barnes and McNaughton, 1980b; Barnes and McNaughton, 1985; de Toledo-Morrell and Morrell, 1985; Bach et al., 1999). This was first demonstrated by Barnes (1979), who showed that aged rats took longer paths and a greater amount of time to solve the circular platform spatial memory task and also had greater difficulty learning a reversal or change of location for the escape box (Fig. 15-1B). These deficits in spatial memory performance were significantly correlated with the durability of synaptic enhancement in the dentate gyrus both within and between age groups. In subsequent studies, Barnes et al. (1980, 1985) also demonstrated a significant similarity between the rates of acquisition and forgetting of spatial memory and the induction and maintenance of LTP. Aged rats were slower to reach asymptotic performance levels and forgot the problem faster than did young rats.

Similarly, asymptotic LTP levels were reached more slowly and decayed more rapidly in aged rats. In both cases, the rate of decay for the memory of the escape box and LTP was nearly twice as fast for the aged rats. Bach et al. (1999) also found a significant relationship between LTP decay rates in area CA1 in vitro and performance on the Barnes maze in aged but not young mice. These findings are further supported by observations that old animals with spatial memory deficits on the eight-arm radial maze (Fig. 15-1D) also show faster decay rates for hippocampal LTP (de Toledo-Morrell and Morrell, 1985). Taken together, the currently available data indicate that learning and memory deficits observed in aged rats and mice parallel deficits in the induction and maintenance of LTP.

C. Molecular Findings: From Molecules to Memory

The neural signaling events necessary for long-lasting changes to occur in the brain are numerous and will either originate or ultimately converge at the nucleus where information from the entire cell is integrated and transmitted into the transcription of thousands of genes. In turn, this genetic transmission of information can have far-reaching effects on the expression of neuronal proteins, affecting the function not only of the individual neuron but potentially of entire neuronal networks. Long-term memory and LTP are two examples of long-lasting changes in neuronal function that are a result of changes in gene expression. The study of the molecular basis of memory function has resulted in the identification of several molecules now known to be needed for normal cognitive functioning. These molecules include the constitutively expressed regulatory transcription factors, such as CREB, and several downstream immediate-early genes (IEGs), such as c-*fos*, *zif*268, and *Arc*. These molecules are the focus of the following discussion. First, it should be noted that the evaluation of IEG studies is complicated by a number of factors, including: (1) the different methods used to induce their expression (e.g., electrical stimulation that induces seizures or LTP, chemical stimulation with drugs, behavioral stimulation, such as exploration of a novel environment or learning and memory tasks, or no manipulation, so that the constitutive, or resting, activity of these genes can be evaluated); (2) the methods used to evaluate where and to what degree their expression takes place [e.g., *in situ* hybridization or immunohistochemistry to look at the proportions of cells that express a particular mRNA or protein, respectively, reverse transcriptase poly-merase chain reaction (RT-PCR) to evaluate the number of gene copies for a given sample]; (3) the brain regions under investigation (e.g., different regions of the hippocampus, such as the dentate gyrus, cornu ammonis, subcortical or different cortical regions). In spite of the fact that the conclusions drawn are constrained and differ depending on the methods applied, there are several key molecules now known to be involved in memory processes. Moreover, age-associated alterations in the expression of some of these molecules are thought to contribute to the learning and memory deficits observed in the aged organ-ism. For an overview of the cell-signaling cascades that are discussed in the text, see Figure 15-3.

1. cAMP-Response-Element-Binding Protein (CREB)

cAMP-response-element-binding protein (CREB) has been the focus of con-siderable attention in the field of learning and memory, although the exact mechanism through which CREB works is still debated (Mayr and Montminy, 2001; Cha-Molstad et al., 2004). CREB is a common target for many neuronal signaling pathways and can be activated by kinases, such as mitogen–activated kinase (MAPK), signaling pathways involving adenylyl cyclase, cAMP, and also Ca^{2+} (for overview see Nguyen and Woo, 2003; Carlezon et al., 2005). In the nucleus, CREB binds as a dimer to cAMP-response element (CRE) found in the promoter region of numerous genes, and it becomes "active" when it is phosphorylated at serine 133 (p-CREB). CREB may be present in the nucleus in one of two forms, as an activator of gene transcription (CREB-1) or as a repressor of gene transcription (CREB-2). In terms of memory function, overexpression of CREB-1 reduces the number of trials needed to train *Drosphila* on a memory task, whereas overexpression of CREB-2 blocks the formation of long-term memory (Yin et al., 1995). Similarly, results from studies using knockout mice with a targeted disruption of the alpha and delta isoforms of CREB suggest impairments in spatial memory (Bourtchuladze et al., 1994). Blocking CREB expression in rats with antisense oligonucleotides

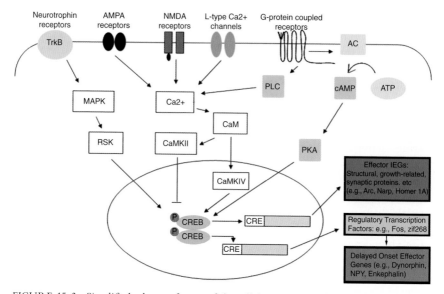

FIGURE 15-3   Simplified scheme of some of the cellular events involved in the regulation of CREB. Neurotransmitters and neurotrophins act at membrane receptors, such as TrkB in the case of neurotropins and AMPA, NMDA, and G-protein-coupled receptors in the case of neurotransmitters such as glutamate and dopamine, to trigger intracellular signaling cascades that culminate in phosphorylation (P) of CREB within the nucleus. Phosphorylation of CREB at serine 133 activates CREB-mediated gene transcription. The cell-signaling events depicted are shown as relatively separate processes, but there is often interaction among the different signaling cascades. The protein products of just a few of the numerous genes targeted by CREB are indicated. *Abbreviations:* AC, adenylyl cyclase; CaM, calmodulin; CaMK, Ca²⁺/calmodulin-dependent kinase II and IV; CRE, cAMP-response element; MAPK, mitogen-activated protein kinase; PDE, phosphodiesterase; PLC, phospholipase C; RSK, MAPK-activated ribosomal S6 kinase; TrkB, neurotrophin tyrosine kinase receptor type 2.

injected directly into the hippocampus also impairs long-term memory for the Morris water maze without impairing initial learning of the task (Guzowski and McGaugh, 1997). The role of CREB has also been investigated in plasticity with regard to its role in LTP. For example, mice with a partial knockout of CREB-1 show normal induction of LTP in area CA1 of the hippocampus; but in these animals, LTP decays to baseline more rapidly than it does for normal control animals (Bourtchuladze et al., 1994).

A number of studies have suggested that there is dysregulation of CREB activity in the aged brain. Monti et al. (2005), reported that basal levels of p-CREB were significantly increased in the hippocampus of aged rats. But when measured 24 hours after a fear-conditioning task, on which aged rats were impaired, the opposite was observed; aged rats had less p-CREB in the hippocampus than did young rats. At least two studies have also investigated potential age-related changes in the number of hippocampal neurons that express CREB. For example, Kudo et al. (2005) showed that a smaller proportion of cells in area CA1 express p-CREB in aged compared to young rats two hours after a contextual fear-conditioning task on which the aged rats were impaired. No age difference was found in the dentate gyrus or area CA3, nor was an age difference observed for levels of unphosphorylated CREB (Kudo et al., 2005). This finding is supported by Ramos et al. (2003), who did not find a significant age effect on the number of CA3 neurons that expressed p-CREB following a spatial working memory task, although these authors did report higher levels of p-CREB in the aged prefrontal cortex. Age-associated changes in CREB-1 but not CREB-2 activity have also been observed. Work by Brightwell et al. (2004) demonstrated a reduction in CREB-1 but not CREB-2 protein in the hippocampus of aged memory-impaired rats compared to nonimpaired aged rats or young rats following transfer training on the Morris water maze. This suggests that the dysregulation of CREB-1 (activator of gene transcription) but not CREB-2 (repressor of gene transcription) may contribute to the spatial memory deficits observed among some aged subjects, most likely as a result of alterations in downstream CREB-dependent gene transcription (e.g., *BDNF* or *Arc*).

2. Immediate-Early Genes

One of the next steps in the molecular events that lead to long-term memory formation involves the transcription of a class of genes known as immediate-early genes (IEGs). As the name suggests, IEGs are among the first group of genes to be expressed following synaptic activity, and they are defined as those genes expressed in the presence of protein synthesis inhibitors (Sheng and Greenberg, 1990). These genes can be broadly classified into (1) genes encoding transcription factors, such as c-*fos* and *zif268* (also known as Krox-24, EGR-1, NGFI-A), that influence cellular activity by regulating the expression of target genes, and (2) genes coding for effector proteins, such as *Arc* (activity-regulated cytoskeletal gene; also known as Arg3.1), *Narp* (neuronal activity-regulated pentraxin), and *BDNF* (brain-derived neurotrophic factor), that directly affect cellular function (Hughes and Dragunow, 1995; Hughes et al., 1999). Correlative studies provided the first evidence that the expression of IEG RNAs and proteins can be increased in hippocampal neurons following training on behavioral tasks such as two-way avoidance (e.g., Nikolaev et al., 1992), brightness discrimination (e.g., Grimm and Tischmeyer, 1997), exposure to a novel environment (Papa et al., 1993; Hess et al, 1995b; C.S. Wallace et al., 1995; Pinaud et al., 2001), odor discrimination tasks (Hess et al., 1995a), and spatial learning tasks (e.g., Guzowski et al., 2000, 2001). In addition, several studies have demonstrated that hippocampal stimulation resulting in LTP also results in the expression of a number of IEGs in the hippocampus (e.g., Cole et al., 1989; Worley et al., 1993; Lanahan et al., 1997; Guzowski et al., 2000; J.L. Lee et al, 2004).

Simply demonstrating that IEGs are induced, however, does not provide evidence that IEGs are necessary or sufficient for maintaining long-lasting changes in neuronal function. In order to demonstrate a more direct role for IEGs in memory function, a number of studies have selectively disrupted the expression of particular IEGs. For example, disrupting the expression of *Arc* and *zif268* using antisense oligonucleotides or knockout technology, respectively, demonstrated that the durability of both long-term spatial memory and LTP are significantly compromised (Guzowski et al., 2000; Jones et al., 2001). These findings in adult animals mimic to some extent those seen in memory-impaired aged animals. Overall, studies like these, which employ strategies to selectively disrupt the expression of specific genes, together with the correlative studies discussed earlier, provide strong evidence that several of the IEGs are required for long-lasting functional changes in the brain.

Studies of IEG expression in the aged brain have focused mainly on potential changes in transcription factors such as *zif268* and c-*fos*. One of the first studies systematically to investigate the transcription factor response of the aged hippocampus used LTP-inducing stimulation to induce gene expression (Worley et al., 1993). The authors report that although aged rats show accelerated rates of decay of the enhanced synaptic response following this treatment, the transcription factor responses do not differ between age groups. A later study that evaluated the expression of a panel of 19 transcription factors and effector IEGs following LTP induction using a reverse Northern strategy, however, found that c-*fos* was elevated in the hippocampus of aged rats compared to young rats (Lanahan et al., 1997). No age-related difference was found for the other 18 genes investigated in this study. The apparent contrast in these results as compared to the findings of Worley et al. (1993) can be explained by the different methods used to measure gene expression.[5] Overall, these findings might be taken to imply that only very subtle differences in gene expression occur during aging or that the LTP induction process itself will not reveal age differences because it results in brain activation that is more synchronous and intense than would occur under normal conditions. If the latter interpretation is correct, then are there different approaches that might be taken to this question? Studies that have used behavioral treatments to induce IEG expression have found age differences for some genes, indicating that this approach may be preferable.

One example of this is provided by Touzani et al. (2003), who investigated behavioral activation of c-*fos* in adult and aged mice following performance

---

[5]The study by Worley et al. (1993) measured the density of the mRNA from autoradiographs following *in situ* hybridization. The study by Lanahan et al. (1997) used a reverse Northern strategy, in which RNA is detected with a hybridization probe and then the amount of RNA present in the sample under investigation is analyzed using gel electrophoresis.

on a radial-arm maze task. As expected, aged mice had difficulty learning the maze task compared to the adult mice, and immunohistochemistry revealed that aged mice had fewer *fos*-positive cells than did adult mice after a six-day retention test in both the CA1 and CA3 regions of the hippocampus but not in dentate granule cells. A second example of age-related changes in behaviorally induced IEG expression comes from the work on *Arc*. As discussed earlier, *Arc* is an effector IEG whose role in long-term plasticity in the brain has been well established (see Steward and Worley, 2002, for review). A study by Small et al. (2004) used a fluorescent *in situ* hybridization protocol with high cellular resolution (known as catFISH; see Guzowski et al., 2005, for a survey of this technique) to map *Arc* expression in the hippocampus of aged and adult rats after they explored a novel environment (Small et al., 2004). This study demonstrated that in aged rats there is a decrease in the proportion of cells expressing *Arc* mRNA in the dentate gyrus but not in other hippocampal subfields. Additional results from this group using real-time RT-PCR to quantify *Arc* mRNA expression in the CA1 region of the hippocampus have revealed that basal levels of *Arc* are lower in aged rats (i.e., in rats without behavioral or any other inducing stimuli); but when *Arc* is induced by exploration of a novel environment, memory-impaired aged rats show a more robust increase in *Arc* expression than do adult rats. Taken together, these results indicate that despite lower basal levels of *Arc* mRNA, the behavioral treatment used to induce *Arc* ultimately brings *Arc* mRNA levels of the aged rats up to levels equivalent to those of adult rats. This finding may indicate a compensatory mechanism in the aged brain, although the avenue through which such a mechanism would work remains undiscovered.

In addition to the studies that have investigated stimulation- or behavior-induced levels of IEGs in the aged brain, there are studies that have sought specifically to characterize basal levels of IEG expression.[6] In the case of *zif*268, Yau et al. (1996) found that its mRNA decreases with age in area CA1 but not in other hippocampal regions, using *in situ* hybridization and densitometry analysis. Using gene microarray technology, Blalock et al. (2003) reported that *zif*268 is one of several IEGs down-regulated in the aged CA1. In addition, Desjardins et al. (1997) observed a decrease in the number of cells that express *zif*268 protein in the aged CA1, using immunohistochemistry and cell count methods. Taken together, the results of these studies demonstrate that basal levels of *zif*268 mRNA and protein are reduced in area CA1 of the aged brain, likely as a result of fewer cells expressing *zif*268 mRNA and protein. In the case of c-*fos*, Desjardins et al. (1997) reported no age difference in the number of cells that express c-*fos* protein. D.R. Smith et al. (2001) found no age-related

---

[6]Although many IEGs are activity dependent and show very low levels of constitutive expression (e.g., *Arc*), some IEGs show relatively high constitutive levels of expression (e.g., *zif*268 and c-*fos*).

change in basal levels of hippocampal c-*fos* mRNA expression using Northern blot analysis. These authors also investigated potential age-associated changes in basal expression of c-*jun* and activator protein-1 (AP-1). AP-1 is a protein complex composed of homodimers of Jun family proteins or heterodimers of Jun and Fos family proteins. If the levels of either c-*fos* or c-*jun* undergo age-associated changes in their expression levels, then the expression and DNA binding activity of AP-1 is likely to also be affected. D.R. Smith et al. (2001) reported that the basal levels of c-*fos* and c-*jun* mRNA were not different between aged and young rats, nor was AP-1 binding activity or composition different between age groups.

Finally, gene microarray studies have been applied to the question of whether there are broad patterns of gene expression disruption during aging (Jiang et al., 2001; C.K. Lee et al., 2000; Blalock et al., 2003; Verbitsky et al., 2004). One of these studies used behaviorally characterized aged and young rats in an attempt to correlate changes in basal levels of gene expression in area CA1 with cognitive decline. The basal expression of many genes were found to be up-regulated, including several associated with CA^{2+} regulation and inflammation, while those found to be down-regulated in aged memory-impaired rats included genes associated with biosynthesis, energy metabolism, and activity-regulated synaptogenesis (Blalock et al., 2003). In addition, the basal levels of two effector IEGS, *Arc* and *Narp*, were down-regulated in area CA1 of aged memory-impaired rats. A recent study has also reported on basal levels of *Arc* protein, demonstrating that *Arc* protein expression is *higher* in the hippocampus of aged compared to adult rats, using a Western blot technique (Monti et al., 2005). Because this is the first study to report on *Arc* protein levels in the aged brain, additional work is needed to determine the functional implications. It is not known if elevated *Arc* expression is a result of a larger proportion of cells expressing *Arc* protein, the same number of cells expressing a larger amount of *Arc* protein, or whether this change is subregion specific.

V. NORMAL BRAIN AGING OUTSIDE THE HIPPOCAMPUS

The hippocampus is not the only brain structure that shows early age-related vulnerability. Of the other brain regions known to show age-related changes, the prefrontal cortex, which is critically involved in working memory and executive functioning (e.g., Divac, 1971; Goldman-Rakic, 1987), is particularly susceptible to advancing age (e.g., Albert, 1997; Bartus et al., 1978; Rapp and Amaral, 1989; Schacter et al., 1996; R.L. West, 1996). Working memory can be readily assessed using a delayed-nonmatching-to-sample (DNMS) task. On one variant of this task, the T-maze alternation task, aged rats show working memory deficits as the delay between trials is increased (e.g., Ando and Ohashi, 1991; Ramos et al., 2003), and aged nonhuman primates also show time-dependent deficits on other variants of the DNMS task (Moss et al., 1988, 1997; Rapp and Amaral, 1989). Working memory function is also affected in the aged human; work by Lyons-Warren et al. (2004) using a computerized spatial delayed-response (SDR) task, demonstrated a significant age effect when delay times were increased.

Executive functioning, which also relies on the PFC, is necessary for processes such as planning, cognitive flexibility, abstract thinking, rule acquisition, and inhibiting inappropriate actions and irrelevant sensory information. In humans, one way to measure executive function is with the Wisconsin card-sorting task. Aged humans have greater difficulty with this task relative to young adults, making more perseverative errors (Rhodes, 2004). Testing executive functioning in animal models is more difficult to accomplish, but analogous executive function tasks have been developed for nonhuman primates. Data from these experiments have confirmed an age-associated deficit, with aged monkeys making more perseverative errors than their younger counterparts (T.L. Moore et al., 2003).

In contrast to the hippocampus, neuron loss may contribute significantly to the cognitive deficits associated with aging and the PFC. Significant neuron loss (~30% reduction) has been found in selected areas of the aged PFC, particularly area 8A of the dorsalateral PFC (D.E. Smith et al., 2004). Longitudinal volumeric analysis of the human PFC has also revealed striking age-related declines in volume in the lateral PFC and orbito-frontal PFC (Raz et al., 2005). Functional imaging studies have also revealed interesting age-associated alterations in the activation of the PFC during memory retrieval tasks. For example, aged adult subjects sometimes show reduced PFC activity as compared to younger adults in areas that show the greatest activation in young subjects. This reduced activation is often correlated with poorer memory performance in the aged subjects. However, in some cases, older people may show equivalent or greater bilateral PFC activation when performing tasks that typically result in unilateral PFC activation for young adults. This pattern of activation is often correlated with better task performance, leading to the hypothesis that it could reflect a compensatory mechanism. A study by Gutchess et al. (2005) suggests that this might be case; functional activation following successful encoding of pictures was lower in the parahippocampus of aged adults as compared to young adults, but activation of the PFC was equivalent for aged and adult subjects. Further, in aged adults, there was a negative correlation between parahippocampal and PFC activation, whereas for young adults the correlation was positive. In a similar vein, Grady et al. (2005) demonstrated a positive correlation between parahippocampal activity and recognition memory performance in young but not aged adults and positive correlations between PFC activation and memory performance in aged but not young adults. Studies like these have important implications for the administration or design of therapeutics aimed at treating age-associated cognitive decline. This was highlighted by Ramos et al. (2003), who demonstrated that treatment strategies aimed at restoring the function of the PFC may be different from those used to restore hippocampal function. Prior to this study, several studies indicated that increasing cAMP/protein kinase A (PKA) activity restored hippocampal function in memory-impaired aged rats and mice (e.g., de Toledo-Morrell et al., 1984; Randt et al., 1982; Barad et al., 1998). Ramos et al. (2003), however, demonstrated that enhancing PKA activity has the opposite effect on the function of the PFC in aged rats and monkeys, leading to poorer working memory performance, whereas PKA inhibition enhances working memory performance. These studies highlight the complex nature of the normal aging process.

VI. CONCLUSIONS

Although most of us will not suffer from a dementing condition as we age, all of us will experience some change in our memory function as we get older. The need for research aimed at identifying those changes in memory that occur during normal nonpathological aging is therefore of great importance. Rather than providing a comprehensive overview of all cognitive changes that occur in aged organisms, age-related changes in spatial memory function were highlighted. The choice of this particular behavior is not meant to imply that other age-associated cognitive changes do not occur or are not equally important, but, rather, changes in this particular cognitive domain are well documented in many species, including humans. This simplifies the task of applying knowledge gained from rodent or other nonhuman studies back to humans. Because the hippocampus is the brain structure that has been most closely associated with spatial cognition, age-associated alterations in the function of this structure were highlighted. It will be important for future work to focus on additional structures (e.g., prefrontal cortex) that also show significant age-dependent changes. Selective region-specific changes occur in the hippocampus, all of which probably contribute to some of the cognitive changes that occur with aging. For example, a change in functional cellular connectivity will alter the efficacy with which neurons can communicate with each other. Similarly, changes at the molecular level may change the ability of neurons to undergo plastic changes that are necessary for normal memory function, such as a change in dendritic spine shape or size or the distribution of glutamate receptors at the postsynaptic membrane. The advantages of gaining a better understanding of the kinds of cognitive changes that occur during normal nonpathological aging, together with an increasingly better understanding of the mechanisms that contribute to these changes, are twofold. First, strategies aimed at preventing or treating age-associated cognitive decline can be developed or improved; second, these investigations will provide information regarding how information processing, and the associated neural mechanisms, function in general.

ACKNOWLEDGMENTS

We would like to thank Michelle Carroll for help with editing and administrative support. Some of the work discussed in this chapter was supported by the McKnight Brain Research Foundation, an Arizona Department of Health Services Grant (AGR2007-37), and a grant from the National Institute on Aging (AG009219 and AG003376).

REFERENCES

Aggleton JP, Hunt PR, Rawlins JN (1986) The effects of hippocampal lesions upon spatial and non-spatial tests of working memory. *Behav Brain Res* 19:133–146.

Albert MS (1997) The aging brain: Normal and abnormal memory. *Phil Trans R Soc Lond B Biol Sci* 352:1703–1709.

Allen GL, Kirasic KC, Rashotte MA, Huan DB (2004) Aging and path integration skill: Kinesthetic and vestibular contributions to wayfinding. *Percept Psychophys* 66:170–179.

Anderson JM, Hubbard BM, Coghill GR, Slidders W (1983) The effect of advanced old age on the neurone content of the cerebral cortex. Observations with an automatic image analyzer point counting method. *J Neurol Sci* 58:235–246.

Ando S, Ohashi Y (1991) Longitudinal study on age-related changes of working and reference memory in the rat. *Neurosci Lett* 128:17–20.

Atwood HL, Wojtowicz JM (1999) Silent synapses in neural plasticity: Current evidence. *Learn Mem* 6:542–571.

Bach ME, Barad M, Son H, Zhuo M, Lu YF, Shih R, Mansuy I, Hawkins RD, Kandel ER (1999) Age-related defects in spatial memory are correlated with defects in the late phase of hippocampal long-term potentiation in vitro and are attenuated by drugs that enhance the cAMP signaling pathway. *Proc Natl Acad Sci USA* 96:5280–5285.

Ball MJ (1977) Neuronal loss, neurofibrillary tangles and granulovacuolar degeneration in the hippocampus with ageing and dementia. A quantitative study. *Acta Neurophathol (Berl)* 37:111–118.

Barad M, Bourtchouladze R, Winder DG, Golan H, Kandel E (1998) Rolipram, a type IV-specific phosphodiesterase inhibitor, facilitates establishment of long-lasting long-term potentiation and memory. *Proceed Natl Acad Sci USA* 95:15020–15025.

Barnes CA (1979) Memory deficits associated with senescence: A neurophysiological and behavioral study in the rat. *J Comp Physiol Psychol* 93:74–104.

Barnes CA (1991) Memory changes with age: Neurobiological correlates. In *Learning and Memory: A Biological View* (Martinez JL and Kesner K, eds.), pp. 259–296. San Diego: Academic Press.

Barnes CA (1994) Normal aging: Regionally specific changes in hippocampal synaptic transmission. *Trends Neurosci* 17:13–18.

Barnes CA (2003) Long-term potentiation and the aging brain. *Philos Trans R Soc Lond B Biol Sci* 358:765–772.

Barnes CA, Eppich C, Rao G (1989) Selective improvement of aged rat short-term spatial memory by 3,4-diaminopyridine. *Neurobiol Aging* 10:337–341.

Barnes CA, Green EJ, Baldwin J, Johnson WE (1987) Behavioral and neurophysiological examples of functional sparing in senescent rat. *Can J Psychol* 41:131–140.

Barnes CA, McNaughton BL (1980a) Physiological compensation for loss of afferent synapses in rat hippocampal granule cells during senescence. *J Physiol (Lond)* 309:473–485.

Barnes CA, McNaughton BL (1980b) Spatial memory and hippocampal synaptic plasticity in senescent and middle-aged rats. In *The Psychobiology of Aging: Problems and Perspectives* (Stein D, ed.), pp. 253–272. New York: Elsevier.

Barnes CA, McNaughton BL (1985) An age comparison of the rates of acquisition and forgetting of spatial information in relation to long-term enhancement of hippocampal synapses. *Behav Neurosci* 99:1040–1048.

Barnes CA, Nadel L, Honig, WK (1980) Spatial memory deficit in senescent rats. *Can J Psychol* 34:29–39.

Barnes CA, Rao G, Foster TC, McNaughton BL (1992) Region-specific age effects on AMPA sensitivity: Electrophysiological evidence for loss of synaptic contacts in hippocampal field CA1. *Hippocampus* 2:457–468.

Barnes CA, Rao G, Houston FP (2000a) LTP induction threshold change in old rats at the per-forant path — granule cell synapse. *Neurobiol Aging* 21:613–620.

Barnes CA, Rao G, Orr G (2000b) Age-related decrease in the Schaffer collateral-evoked EPSP in awake, freely behaving rats. *Neural Plast* 7:167–178.

Barnes CA, Suster MS, Shen J, McNaughton BL (1997b) Multistability of cognitive maps in the hippocampus of old rats. *Nature* 388:272–275.

Barnes CA, Rao G, Shen J (1997a) Age-related decrease in the N-methyl-d-aspartate-R-medi-ated excitatory postsynaptic potential in hippocampal region CA1. Neurobiol Aging 18:445–452.

Bartus RT, Fleming D, Johnson HR (1978) Aging in the rhesus monkey: Debilitating effects on short-term memory. *J Gerontol* 33:858–871.

Beatty WW, Bierley RA, Boyd JG (1985) Preservation of accurate spatial memory in aged rats. *Neurobiol Aging* 6:219–225.

Biegon A, Greenberger V, Segal M (1986) Quantitative histochemistry of brain acetylcholines-terase and learning rate in the aged rat. *Neurobiol Aging* 7:215–217.

Bizon J, Gallagher M (2003) Production of new cells in the rat dentate gyrus over the life span: Relation to cognitive decline. *Eur J Neurosci* 18:215–218

Bizon JL, Lee HJ, Gallagher M (2004) Neurogenesis in a rat model of age-related cognitive decline. *Aging Cell* 3:227–234.

Blalock EM, Chen KC, Sharrow K, Herman JP, Porter NM, Foster TC, Landfield PW (2003) Gene microarrays in hippocampal aging: Statistical profiling identifies novel processes corre-lated with cognitive impairment. *J Neurosci* 23:3807–3819.

Bliss TVP, Gardner-Medwin, AR (1973) Long-lasting potentiation of synaptic transmission in the dentate area of the unanaesthetised rabbit following stimulation of the perforant path. *J Physiol (London)* 232:357–374.

Bliss TVP, Lomo, T (1973) Long-lasting potentiation of synaptic transmission in the dentate area of the anesthetized rabbit following stimulus of perforant path. *J Physiol (London)* 232:331–356.

Bondareff W, Geinisman Y (1976) Loss of synapses in the dentate gyrus of the senescent rat. *Am J Anat* 145:129–136.

Bourlière F (1970) *The Assesment of Biological Age in Man*. Geneva: World Health Organization Pub. No. 37.

Bourtchuladze R, Frenguelli B, Blendy J, Cioffi D, Schutz G, Silva AJ (1994) Deficient long-term memory in mice with a targeted mutation of the cAMP-responsive element-binding protein. *Cell* 79:59–68.

Bresnahan EL, Wiser PR, Ingram DK (1988) Fimbria-fornix lesions in young rats impair acquisi-tion in a 14-unit T-maze similar to prior observed performance deficits in aged rats. *Psycho-biology* 16:243–250.

Brightwell JJ, Gallagher M, Colombo PJ (2004) Hippocampal CREB1 but not CREB2 is decreased in aged rats with spatial memory impairments. *Neurobiol Learn Mem* 81:19–26.

Brizzee KR, Ordy JM, Bartus RT (1980) Localization of cellular changes within multimodal sensory regions in aged monkey brain: Possible implications of age-related cognitive loss. *Neurobiol Aging* 1:45–52.

Brody H (1955) Organization of the cerebral cortex. III. A study of aging in the human cerebral cortex. *J Comp Neurol* 102:511–556.

Calhoun ME, Kurth D, Phinney AL, Long JM, Hengemihle J, Mouton PR, Ingram DK, Jucker M (1998) Hippocampal neuron and synaptophysin-positive bouton number in aging C57BL/6 mice. *Neurobiol Aging* 19:599–606.

Canestrari RE, Jr (1963) Paced and self-paced learning in young and elderly adults. *J Gerontol* 18:165–168.

Caprioli A, Ghirardi O, Giuliani A, Ramacci MT, Angelucci L (1991) Spatial learning and memory in the radial maze: A longitudinal study in rats from 4 to 25 months of age. *Neurobiol Aging* 12:605–607.

Carlezon WA, Jr, Duman RS, Nestler EJ (2005) The many faces of CREB. *Trends Neurosci* 28:436–445.

Cha-Molstad H, Keller DM, Yochum GS, Impey S, Goodman RH (2004) Cell-type-specific binding of the transcription factor CREB to the cAMP-response element. *Proc Natl Acad Sci USA* 101:13572–13577.

Cole AJ, Saffen DW, Baraban JM, Worley PF (1989) Rapid increase of an immediate early gene messenger RNA in hippocampal neurons by synaptic NMDA receptor activation. *Nature* 340:474–476.

Coleman PD, Flood DG (1987) Neuron numbers and dendritic extent in normal aging and Alzheimer's disease. *Neurobiol Aging* 8:521–545.

Costa PT, Jr., McCrae RR (1980) Functional age: A conceptual and empirical critique. In *Epidemiology of Aging* (Haynes SG and Feinleib M, eds.), pp. 23–46. Washington, DC: U.S. Government Printing Office, N.I.H. Pub. No. 80-969.

Craik FIM, McDowd, JM (1987) Age differences in recall and recognition. *J Exp Psychol Learn Mem Cog* 13:474–479.

Crook T, Bartus RT, Ferris SH, Whitehouse P, Cohen GD, Gershon S (1986) Age-associated memory impairment: Proposed diagnostic criteria and measures of clinical change. Report of a National Institute of Mental Health Work Group. *Develop Neuropsychol* 2:261–276.

Crook TH, Ferris SH (1992) Age associated memory impairment. *BMJ* 304:714.

de Toledo-Morrell L, Morrell F (1985) Electrophysiological markers of aging and memory loss in rats. *Ann NY Acad Sci* 444:296–311.

de Toledo-Morrell L, Morrell F, Fleming S, Cohen MM (1984) Pentoxifylline reverses age-related deficits in spatial memory. *Behav Neural Biol* 42:1–8.

Desjardins S, Mayo W, Vallee M, Hancock D, Le Moal M, Simon H, Abrous DN (1997) Effect of aging on the basal expression of c-Fos, c-Jun, and Egr-1 proteins in the hippocampus. *Neurobiol Aging* 18:37–44.

Deupree DL, Bradley J, Turner DA (1993) Age-related alterations in potentiation in the CA1 region in F344 rats. *Neurobiol Aging* 14:249–258.

Disterhoft JF, Thompson LT, Moyer JR, Jr., Mogul DJ (1996) Calcium-dependent afterhyperpolarization and learning in young and aging hippocampus. *Life Sci* 59:413–420.

Divac I (1971) Frontal lobe system and spatial reversal in the rat. *Neuropsychologia* 9:175–183.

Driscoll I, Hamilton DA, Yeo RA, Brooks WM, Sutherland RJ (2005) Virtual navigation in humans: The impact of age, sex, and hormones on place learning. *Horm Behav* 47:326–335.

Evans GW, Brennan PL, Skorpanich MA, Held D (1984) Cognitive mapping and elderly adults: Verbal and location memory for urban landmarks. *J Gerontol* 39:452–457.

Ferris SH, Kluger A (1996) Commentary on age-associated memory impairment, age-related cognitive decline and mild cognitive impairment. *Aging, Neuropsychol Cognition* 3:148–153.

Flood DG (1991) Region-specific stability of dendritic extent in normal aging and regression in Alzheimer's disease. II. Subiculum. *Brain Res* 540:83–95.

Flood DG, Guarnaccia M, Coleman PD (1987a) Dendritic extent in human CA2–3 hippocampal pyramidal neurons in normal aging and senile dementia. *Brain Res* 409:88–96.

Flood DG, Buell SJ, Horwitz GJ, Coleman PD (1987b) Dendritic extent in human dentate gyrus granule cells in normal aging and senile dementia. *Brain Res 402*:205–216.

Foster TC, Barnes CA, Rao G, McNaughton BL (1991) Increase in perforant path quantal size in aged F-344 rats. *Neurobiol Aging 12*:441–448.

Foster TC, Sharrow KM, Masse JR, Norris CM, Kumar A (2001) Calcineurin links Ca^{2+} dysregulation with brain aging. *J Neurosci 21*:4066–4073.

Frick KM, Baxter MG, Markowska AL, Olton DS, Price DL (1995) Age-related spatial reference and working memory deficits assessed in the water maze. *Neurobiol Aging 16*:149–160.

Gage FH, Dunnett SB, Bjorklund A (1984) Spatial learning and motor deficits in aged rats. *Neurobiol Aging 5*:43–48.

Gallagher M, Bostock E, King R (1985) Effects of opiate antagonists on spatial memory in young and aged rats. *Behav Neural Biol 44*:374–385.

Gallagher M, Burwell R, Burchinal M (1993) Severity of spatial learning impairment in aging: Development of a learning index for performance in the Morris water maze. *Behav Neurosci 107*:618–626.

Gallagher M, Pelleymounter MA (1988) Spatial learning deficits in old rats: A model for memory decline in the aged. *Neurobiol Aging 9*:549–556.

Geinisman Y (1979) Loss of axosomatic synapses in the dentate gyrus of aged rats. *Brain Res 168*:485–492.

Geinisman Y, de Toledo-Morrell L, Morrell F (1986) Aged rats need a preserved complement of perforated axospinous synapses per hippocampal neuron to maintain good spatial memory. *Brain Res 398*:266–275.

Geinisman Y, de Toledo-Morrell L, Morrell F, Persina IS, Rossi M (1992) Age-related loss of axospinous synapses formed by two afferent systems in the rat dentate gyrus as revealed by the unbiased stereological dissector technique. *Hippocampus 2*:437–444.

Geinisman Y, Ganeshina O, Yoshida R, Berry RW, Disterhoft JF, Gallagher M (2004) Aging, spatial learning, and total synapse number in the rat CA1 stratum radiatum. *Neurobiol Aging 25*:407–416.

Goldman H, Berman RF, Gershon S, Murphy SL, Altman HJ (1987) Correlation of behavioral and cerebrovascular functions in the aging rat. *Neurobiol Aging 8*:409–416.

Goldman-Rakic PS (1987) Circuitry of the primate prefrontal cortex and the regulation of behavior by representational memory. In *Handbook of Physiology, the Nervous System, Higher Functions of the Brain* (Plum F, ed.), pp. 373–417. Bethesda, MD: American Physiological Society.

Goodrick C (1968) Learning, retention, and extinction of a complex maze habit for mature-young and senescent Wistar albino rats. *J Gerontol 23*:294–304.

Grady CL, McIntosh AR, Craik FI (2005) Task-related activity in prefrontal cortex and its relation to recognition memory performance in young and old adults. *Neuropsychologia 43*:1466–1481.

Greferath U, Bennie A, Kourakis A, Barrett GL (2000) Impaired spatial learning in aged rats is associated with loss of p75-positive neurons in the basal forebrain. *Neuroscience 100*:363–373.

Grimm R, Tischmeyer W (1997) Complex patterns of immediate-early gene induction in rat brain following brightness discrimination training and pseudotraining. *Behav Brain Res 84*:109–166.

Gutchess AH, Welsh RC, Hedden T, Bangert A, Minear M, Liu LL, Park DC (2005) Aging and the neural correlates of successful picture encoding: Frontal activations compensate for decreased medial-temporal activity. *J Cogn Neurosci 17*:84–96.

Guzowski JF, Lyford GL, Stevenson GD, Houston FP, McGaugh JL, Worley PF, Barnes CA (2000) Inhibition of activity-dependent Arc protein expression in the rat hippocampus impairs the maintenance of long-term potentiation and the consolidation of long term memory. *J Neurosci 20*:3993–4001.

Guzowski JF, McGaugh JL (1997) Antisense oligodeoxynucleotide-mediated disruption of hippocampal cAMP response element binding protein levels impairs consolidation of memory for water maze training. Proc Natl Acad Sci USA 94:2693–2698.

Guzowski JF, Setlow B, Wagner EK, McGaugh JL (2001) Experience-dependent gene expression in the rat hippocampus after spatial learning: A comparison of the immediate-early genes Arc, c-fos, and zif268. J Neurosci 21:5089–5098.

Guzowski JF, Timlin JA, Roysam B, McNaughton BL, Worley PF, Barnes CA (2005) Mapping behaviorally relevant neural circuits with immediate-early gene expression. Curr Opin Neurobiol 15:599–606.

Hanks SD, Flood DG (1991) Region-specific stability of dendritic extent in normal human aging and regression in Alzheimer's disease. I. CA1 of hippocampus. Brain Res 540:63–82.

Harwood E, Naylor GFK (1969) Recall and recognition in elderly and young subjects. Australian J Psychol 21:251–257.

Hedden T, Gabrieli JD (2004) Insights into the ageing mind: A view from cognitive neuroscience. Nat Rev Neurosci 5:87–96.

Hedden T, Gabrieli JD (2005) Healthy and pathological processes in adult development: new evidence from neuroimaging of the aging brain. Curr Opin Neurol 18:740–747.

Heine VM, Maslam S, Joels M, Lucassen PJ (2004) Prominent decline of newborn cell proliferation, differentiation, and apoptosis in the aging dentate gyrus, in absence of an age-related hypothalamus-pituitary-adrenal axis activation. Neurobiol Aging 25:361–375.

Hess US, Gary L, Gall CM (1995a) Changes in c-fos mRNA expression in rat brain during odor discrimination learning: Differential involvement of hippocampal subfields CA1 and CA3. J Neurosci 15:4786–4795.

Hess US, Lynch G, Gall CM (1995b) Regional patterns of c-fos mRNA expression in rat hippocampus following exploration of a novel environment versus performance of a well-learned discrimination. J Neurosci 15:7796–7809.

Hof PR, Morrison JH (2004) The aging brain: Morphomolecular senescence of cortical circuits. Trends Neurosci 27:607–613.

Hofer SM, Sliwinski MJ (2001) Understanding aging. An evaluation of research designs for assessing the interdependence of ageing-related changes. Gerontology 47:341–352.

Hughes P, Dragunow M (1995) Induction of immediate-early genes and the control of neurotransmitter-regulated gene expression within the nervous system. Pharmacol Rev 47:133–178.

Hughes PE, Alexi T, Walton M, Williams CE, Dragunow M, Clark RG, Gluckman PD (1999) Activity- and injury-dependent expression of inducible transcription factors, growth factors and apoptosis-related genes within the central nervous systems. Prog Neurobiol 57:421–450.

Hyman BT, Van Hoesen GW, Damasio AR, Barnes CL (1984) Alzheimer's disease: Cell-specific pathology isolates the hippocampal formation. Science 225:1168–1170.

Ingram D (1985) Analysis of age-related impairments in learning and memory in rodent models. Ann NY Acad Sci 444:312–331.

Ingram DK (1983) Toward the behavioral assessment of biological aging in the laboratory mouse: concepts, terminology, and objectives. Exp Aging Res 9:225–238.

Ingram DK (1998) Complex maze learning in rodents as models of age-related memory impairment. Neurobiol Aging 9:475–485.

Ingram DK (2001) Rodent models of age-related memory impairments. In Functional Neurobiology of Aging (Hof PR and Mobbs CV, eds.), pp. 373–386, San Diego: Academic Press.

Jacobs WJ, Laurance HE, Thomas KGF (1997) Place learning in virtual space I: Acquisition, overshadowing, and transfer. Learn Motiv 28:521–541.

Jacobs WJ, Thomas KGF, Laurance HE Nadel L (1998) Place learning in virtual space II: Topographical relations as one dimension of stimulus control. Learn Motiv 29: 288–308.

Jiang CH, Tsien JZ, Schultz PG, Hu Y (2001) The effects of aging on gene expression in the hypothalamus and cortex of mice. *Proc Natl Acad Sci USA* 98:1930–1934.

Jones MW, Errington ML, French PJ, Fine A, Bliss TV, Garel S, Charnay P, Bozon B, Laroche S, Davis S (2001) A requirement for the immediate early gene Zif268 in the expression of late LTP and long-term memories. *Nat Neurosci* 4:289–296.

Kempermann G, Gast D, Gage FH (2002) Neuroplasticity in old age: Sustained fivefold induction of hippocampal neurogenesis by long-term environmental enrichment. *Ann Neurol* 52:135–143.

Khachaturian ZS (1985) Diagnosis of Alzheimer's disease. *Arch Neurol* 42:1097–1105.

Klein QW, Michel ME (1977) A morphometric study of the neocortex of young adult and old maze-differentiated rats. *Mechanisms Aging Devel* 6:441–452.

Kronenberg G, Bick-Sander A, Bunk E, Wolf C, Ehninger D, Kempermann G (2006) Physical exercise prevents age-related decline in precursor cell activity in the mouse dentate gyrus. *Neurobiol Aging* 27:1505–1513.

Kudo K, Wati H, Qiao C, Arita J, Kanba S (2005) Age-related disturbance of memory and CREB phosphorylation in CA1 area of hippocampal of rats. *Brain Res* 1054:30–37.

Kuhn HG, Dickinson-Anson H, Gage FH (1996) Neurogenesis in the dentate gyrus of the adult rat: Age-related decrease of neuronal progenitor proliferation. *J Neurosci* 16:2027–2033.

Lanahan A, Lyford G, Stevenson GS, Worley PF, Barnes CA (1997) Selective alteration of LTP-induced transcriptional response in hippocampus of aged, memory-impaired rats. *J Neurosci* 17:2876–2885.

Landfield PW, Lynch G (1977) Impaired monosynaptic potentiation in in vitro hippocampal slices from aged, memory-deficient rats. *J Gerontol* 32:523–533.

Landfield PW, McGaugh JL, Lynch G (1978) Impaired synaptic potentiation processes in the hippocampus of aged, memory-deficient rats. *Brain Res* 150:85–101.

Landfield PW, Pitler TA (1984) Prolonged Ca^{2+}-dependent after hyperpolarizations in hippocampal neurons of aged rats. *Science* 226:1089–1092.

Landfield PW, Pitler TA, Applegate MD (1986) The effects of high Mg^{2+}-to-Ca^{2+} ratios on frequency potentiation in hippocampal slices of young and aged rats. *J Neurophysiol* 56:797–811.

Lawton AH (1965) The historical developments in the biological aspects of aging and the aged. *Gerontologist* 5:25–32.

Lee CK, Weindruch R, Prolla TA (2000) Gene-expression profile of the ageing brain in mice. *Nat Genet* 25:294–297.

Lee JL, Everitt BJ, Thomas KL (2004) Independent cellular processes for hippocampal memory consolidation and reconsolidation. *Science* 304:839–843.

Lohninger S, Strasser A, Bubna-Littitz H (2001) The effect of l-carnitine on T-maze learning ability in aged rats. *Arch Gerontol Geriatr* 32:245–253.

Lyons-Warren A, Lillie R, Hershey T (2004) Short- and long-term spatial delayed response performance across the life span. *Dev Neuropsychol* 26:661–678

Maguire EA, Burgess N, Donnett JG, Frackowiak RS, Frith CD, O'Keefe J (1998) Knowing where and getting there: A human navigation network. *Science* 280:921–924.

Markowska AL, Stone WS, Ingram DK, Reynolds J, Gold PE, Conti LH, Pontecorvo MJ, Wenk GL, Olton DS (1989) Individual differences in aging: Behavioral and neurobiological correlates. *Neurobiol Aging* 10:31–43.

Mayr B, Montminy M (2001) Transcriptional regulation by the phosphorylation-dependent factor CREB. *Nat Rev Mol Cell Biol* 2:599–609.

McLay RN, Freeman SM, Harlan RE, Kastin AJ, Zadina JE (1999) Tests used to assess the cognitive abilities of aged rats: Their relation to each other and to hippocampal morphology and neurotrophin expression. *Gerontology* 45:143–155.

McNaughton BL, Barnes CA, Meltzer J, Sutherland RJ (1989) Hippocampal granule cells are necessary for normal spatial learning but not for spatially-selective pyramidal cell discharge. *Exp Brain Res* 76:485–496.

Mizumori SJ, Lavoie AM, Kalyani A (1996) Redistribution of spatial representation in the hippocampus of aged rats performing a spatial memory task. *Behav Neurosci* 110:1006–1016.

Moffat SD, Elkins W, Resnick SM (2006) Age differences in the neural systems supporting human allocentric spatial navigation. *Neurobiol Aging* 27:965–972.

Moffat SD, Zonderman AB, Resnick SM (2001) Age differences in spatial memory in a virtual environment navigation task. *Neurobiol Aging* 22:787–796.

Monge R, Hultsch D (1971) Paired-associate learning as a function of adult age and the length of the anticipation and inspection intervals. *J Gerontol* 26:157–162.

Monti B, Berteotti C, Contestabile A (2005) Dysregulation of memory-related proteins in the hippocampus of aged rats and their relation with cognitive impairment. *Hippocampus* 15:1041–1049.

Moore CI, Browning MD, Rose GM (1993) Hippocampal plasticity induced by primed burst, but not long-term potentiation, stimulation is impaired in area CA1 of aged Fischer 344 rats. *Hippocampus* 3:57–66.

Moore TL, Killiany RJ, Herndon JG, Rosene DL, Moss MB (2003) Impairment in abstraction and set shifting in aged rhesus monkeys. *Neurobiol Aging* 24:125–134.

Morris R (1981) Spatial localization does not require the presence of local cues. *Learn Motiv* 12:239–261.

Morris RG, Garrud P, Rawlins JN, O'Keefe J (1982) Place navigation impaired in rats with hippocampal lesions. *Nature* 297:681–683.

Morrison JH, Hof PR (1997) Life and death of neurons in the aging brain. *Science* 278:412–419.

Moss MB, Rosene DL, Peters A (1998) Effects of aging on visual recognition memory in the rhesus monkey. *Neurobiol Aging* 9:495–502.

Moss MB, Killiany RJ, Lai ZC, Rosene DL, Herndon JG (1997) Recognition memory span in rhesus monkeys of advanced age. *Neurobiol Aging* 18:13–19.

Newman M, Kaszniak A (2000) Spatial memory and aging: Performance on a human analog of the Morris water maze. *Aging Neuropsychol Cogn* 7:86–93.

Nguyen PV, Woo NH (2003) Regulation of hippocampal synaptic plasticity by cyclic AMP-dependent protein kinases. *Prog Neurobiol* 71:401–437.

Nicholson DA, Yoshida R, Berry RW, Gallagher M, Geinisman Y (2004) Reduction in size of perforated postsynaptic densities in hippocampal axospinous synapses and age-related spatial learning impairments. *J Neurosci* 24:7648–7653.

Nikolaev E, Kaminska B, Tischmeyer W, Matthies H, Kaczmarek L (1992) Induction of expression of genes encoding transcription factors in the rat brain elicited by behavioral training. *Brain Res Bull* 28:479–484.

O'Keefe J, Dostrovsky J (1971) The hippocampus as a spatial map. Preliminary evidence from unit activity in the freely moving rat. *Brain Res* 34:171–175.

Olton DS (1991) Experimental strategies to identify the neurobiological bases of memory: Lesions. In *Learning and Memory: A Biological View* (Martinez JL and Kesner K, eds.), pp. 441–446. San Diego: Academic Press.

Olton DS, Walker JA, Gage FH (1978) Hippocampal connections and spatial discrimination. *Brain Res* 139:295–308.

Papa M, Pellicano MP, Welzl H, Sadile AG (1993) Distributed changes in c-Fos and c-Jun immunoreactivity in the rat brain associated with arousal and habituation to novelty. *Brain Res Bull* 32:509–515.

Park DC, Lautenschlager G, Hedden T, Davidson NS, Smith AD, Smith PK (2002) Models of visuospatial and verbal memory across the adult life span. *Psychol Aging* 17:299–320.

Pelleymounter MA, Smith MY, Gallagher M (1987) Spatial learning impairments in aged rats trained with a salient configuration of stimuli. *Psychobiology* 15:248–254.

Perls T (2004) Dementia-free centenarians. *Exp Gerontol* 39:1587–1593.

Peters A, Rosene DL, Moss MB, Kemper TL, Abraham CR, Tigges J, Albert MS (1996) Neurobiological bases of age-related cognitive decline in the rhesus monkey. *J Neuropathol Exp Neurol* 55:861–874.

Peters AA, Palay SL, Webster HD (1991) *The Fine Structure of the Nervous System: Neurons and Their Supporting Cell*, 3rd ed. New York: Oxford University Press.

Pinaud R, Penner MR, Robertson HA, Currie RW (2001) Upregulation of the immediate early gene arc in the brains of rats exposed to environmental enrichment: Implications for molecular plasticity. *Brain Res Mol Brain Res* 91:50–56.

Pyapali GK, Turner DA (1996) Increased dendritic extent in hippocampal CA1 neurons from aged F344 rats. *Neurobiol Aging* 17:601–611.

Ramos BP, Birnbaum SG, Lindenmayer I, Newton SS, Duman RS, Arnsten AF (2003) Dysregulation of protein kinase a signaling in the aged prefrontal cortex: New strategy for treating age-related cognitive decline. *Neuron* 40:835–845.

Randt CT, Judge ME, Bonnet KA, Quartmain D (1982) Brain cyclic AMP and memory in mice. *Pharmacol Biochem Behav* 17:677–680.

Rapp PR, Amaral DG (1989) Evidence for task-dependent memory dysfunction in the aged monkey. *J Neurosci* 9:3568–3576.

Rapp PR, Gallagher M (1996) Preserved neuron number in the hippocampus of aged rats with spatial learning deficits. *Proc Natl Acad Sci USA* 93:9926–9930.

Rapp PR, Rosenberg RA, Gallagher M (1987) An evaluation of spatial information processing in aged rats. *Behav Neurosci* 101:3–12.

Rasmussen T, Schliemann T, Sorensen JC, Zimmer J, West MJ (1996) Memory impaired aged rats: No loss of principal hippocampal and subicular neurons. *Neurobiol Aging* 17:1443–1447.

Raz N, Lindenberger U, Rodrigue KM, Kennedy KM, Head D, Williamson A, Dahle C, Gerstorf D, Acker JD (2005) Regional brain changes in aging healthy adults: General trends, individual differences and modifiers. *Cereb Cortex* 15:1676–1689.

Rhodes MG (2004) Age-related differences in performance on the Wisconsin card sorting test: A meta-analytic review. *Psychol Aging* 19:482–494.

Rosenzweig ES, Barnes CA (2003) Impact of aging on hippocampal function: Plasticity, network dynamics, and cognition. *Prog Neurobiol* 69:143–179.

Rosenzweig ES, Rao G, McNaughton BL, Barnes CA (1997) Role of temporal summation in age-related long-term potentiation-induction deficits. *Hippocampus* 7:549–558.

Rossi MA, Mash DC, deToledo-Morrell L (2005) Spatial memory in aged rats is related to PKCgamma-dependent G-protein coupling of the M1 receptor. *Neurobiol Aging* 26:53–68.

Salthouse TA, Nesselroade JR (2002) An examination of the Hofer and Sliwinski evaluation. *Gerontology* 48:18–21.

Schacter DL, Savage CR, Alpert NM, Rauch SL, Albert MS (1996) The role of the hippocampus and frontal cortex in age-related memory changes: A PET study. *Neuroreport* 7:1165–1169.

Schaie KW (1996) *Intellectual Development in Adulthood: The Seattle Longitudinal Study*. Cambridge: Cambridge, UK: University Press.

Scheff SW, Anderson KJ, DeKosky ST (1985) Strain comparison of synaptic density in hippocampal CA1 of aged rats. *Neurobiol Aging* 6:29–34.

Scheibel AB (1979) The hippocampus: Organizational patterns in health and senescence. *Mech Aging Dev* 9:89–102.

Scheibel ME, Lindsay RD, Tomiyasu U, Scheibel AB (1976) Progressive dendritic changes in the aging human limbic system. *Exp Neurol* 53:420–430.

Schonfield D, Robertson BA (1966) Memory storage and aging. *Can J Psychol 20*: 228–236.

Shamy JL, Buonocore MH, Makaron LM, Amaral DG, Barnes CA, Rapp PR (2006) Hippo-campal volume is preserved and fails to predict recognition memory impairment in aged rhesus monkeys (Macaca mulatta). *Neurobiol Aging* (in press).

Shen J, Barnes CA, McNaughton BL, Skaggs WE, Weaver KL (1997) The effect of aging on experience-dependent plasticity of hippocampal place cells. *J Neurosci 17*:6769–6782.

Sheng M, Greenberg ME (1990) The regulation and function of c-Fos and other immediate-early genes in the nervous system. *Neuron 4*:477–485.

Skalicky M, Bubna-Littitz H, Hofecker G (1984) The influence of persistent crowding on the age changes in behavioral parameters and survival characteristics of rats. *Mech Aging Dev 28*:325–336.

Small SA, Chawla MK, Buonocore M, Rapp PR, Barnes CA (2004) Imaging correlates of brain function in monkeys and rats isolates a hippocampal subregion differentially vulnerable to aging. *Proc Natl Acad Sci USA 101*:7181–7186.

Smith DE, Rapp PR, McKay HM, Roberts JA, Tuszynski MH (2004) Memory impairment in aged primates is associated with focal death of cortical neurons and atrophy of subcortical neurons. *J Neurosci 24*:4373–4381.

Smith DR, Hoyt EC, Gallagher M, Schwabe RF, Lund PK (2001) Effect of age and cognitive status on basal level AP-1 activity in rat hippocampus. *Neurobiol Aging 22*:773–786.

Steward O, Worley P (2002) Local synthesis of proteins at synaptic sites on dendrites: Role in synaptic plasticity and memory consolidation? *Neurobiol Learn Mem 78*:508–527.

Stone CP (1929a) The age factor in animal learning: I. Rats in the problem box and the maze. *Gen Psychol Monog 5*:1–130.

Stone CP (1929b) The age factor in animal learning: II. Rats in a multiple light discrimination box and a difficult maze. *Gen Psychol Monog 6*:125–201.

Thibault O, Landfield PW (1996) Increase in single L-type calcium channels in hippocampal neurons during aging. *Science 272*:1017–1020.

Thomas KGF, Laurance HE, Luczak SE, Jacobs WJ (1999) Age-related changes in a human cognitive mapping system: Data from a computer-generated environment. *CyberPsychol Beh 2*: 545–566.

Thompson LT, Best PJ (1990) Long-term stability of place-field activity of single units recorded from the dorsal hippocampus of freely behaving rats. *Brain Res 509*:299–308.

Tombaugh GC, Rowe WB, Rose GM (2005) The slow afterhyperpolarization in hippocampal CA1 neurons covaries with spatial learning ability in aged Fisher 344 rats. *J Neurosci 25*:2609–2616.

Touzani K, Marighetto A, Jaffard R (2003) Fos imaging reveals ageing-related changes in hip-pocampal response to radial maze discrimination testing in mice. *Eur J Neurosci 17*:628–640.

Uttl B, Graf P (1993) Episodic spatial memory in adulthood. *Psychol Aging 8*:257–273.

van Gool WA, Mirmiran M, van Haaren F (1985) Spatial memory and visual evoked potentials in young and old rats after housing in an enriched environment. *Behav Neural Biol 44*:454–469.

Van Petten C (2004) Relationship between hippocampal volume and memory ability in healthy individuals across the lifespan: A review and meta-analysis. *Neuropsychologia 42*:1394–1413.

van Praag H, Shubert T, Zhao C, Gage FH (2005) Exercise enhances learning and hippocampal neurogenesis in aged mice. *J Neurosci 25*:8680–8685.

Verbitsky M, Yonan AL, Malleret G, Kandel ER, Gilliam TC, Pavlidis P (2004) Altered hip-pocampal transcript profile accompanies an age-related spatial memory deficit in mice. *Learn Mem 11*:253–260.

Verzar-McDougall E (1957) Studies in learning and memory in ageing rats. *Gerontologia 35*:355–363.

Wallace CS, Withers GS, Weiler IJ, George JM, Clayton DF, Greenough WT (1995) Correspondence between sites of NGFI-A induction and sites of morphological plasticity following exposure to environmental complexity. *Brain Res Mol Brain Res 32*:211–220.

Wallace JE, Krauter EE, Campbell BA (1980) Animal models of declining memory in the aged: Short-term and spatial memory in the aged rat. *J Gerontol 35*:355–363.

West MJ (1993) Regionally specific loss of neurons in the aging human hippocampus. *Neurobiol Aging 14*:287–293.

West MJ, Amaral DJ, Rapp PR (1993) Preserved hippocampal cell number in aged monkeys with recognition memory deficits. *Soc Neurosci Abst 19*:599.

West RL (1996) An application of prefrontal cortex function theory to cognitive aging. *Psychol Bull 120*:272–292.

Wilson IA, Ikonen S, Gallagher M, Eichenbaum H, Tanila H (2005) Age-associated alterations of hippocampal place cells are subregion specific. *J Neurosci 25*:6877–6886.

Wilson MA, McNaughton BL (1993) Dynamics of the hippocampal ensemble code for space. *Science 261*:1055–1058.

Worley PF, Bhat RV, Baraban JM, Erickson CA, McNaughton BL, Barnes, CA (1993) Thresholds for synpatic activation of transcription factors in hippocampus: Correlation with long-term enhancement. *J Neurosci 13*:4776–4786.

Yau JL, Olsson T, Morris RG, Noble J, Seckl JR (1996) Decreased NGFI-A gene expression in the hippocampus of cognitively impaired aged rats. *Brain Res Mol Brain Res 42*:354–357.

Yin JC, Del Vecchio M, Zhou H, Tully T (1995) CREB as a memory modulator: Induced expression of a CREB2 activator isoform enhances long-term memory in *Drosophila*. *Cell 81*:107–115.

CHAPTER 16

Neurodegenerative Diseases and Memory: A Treatment Approach

Gary L. Wenk

Department of Psychology, Ohio State University, Columbus, OH 43210

I. INTRODUCTION

Cognitive dysfunction associated with neurodegenerative diseases can occur for many reasons that are not directly associated with any underlying disease process, such as associated with expression of dementia, hormone imbalance, dietary deficiency or excess, drug-induced or heavy metal toxicity, head injury, inherited disorders of the CNS, sleep deprivation, and prolonged stress. All of these causes of cognitive decline are the focus of study. However, because most neurodegenerative diseases are associated with aging, much research has focused on the contribution of normal aging and the changes in vulnerability of selected neurotransmitter systems.

Neuronal dysfunction with aging, particularly with regard to specific neurotransmitter abnormalities, contributes substantially to the development of cognitive symptoms associated with Alzheimer's disease (AD). Due to the widespread incidence of AD in the aging populations of the world and to the potential financial and societal impacts of finding an effective therapy, the study of AD has motivated and guided the investigation of numerous other age-associated neurodegenerative disorders. The potential impact of targeting neuronal dysfunction in AD has been advanced by considerable progress in identifying the normal and pathologic mechanisms underlying the degenerative processes that lead to the degeneration of vulnerable neurotransmitter systems within the brain.

Senile plaques and neurofibrillary tangles are two important diagnostic features of AD. Their distribution is not generalized throughout the brain but, instead, demonstrates an intriguing, and thus far unexplained, regional vulnerability. Whether the vulnerability and dysfunction of selected neural systems is a direct result of the degenerative process or simply due to bystander injury is an important question, because the answer can be used to guide the design of an effective therapeutic approach. This chapter reviews the changes that have been observed in the two most vulnerable neural systems associated with AD, the most common and possibly best-studied neurodegenerative disease, and presents the approaches that have been taken to alleviate the symptoms that result from the underlying pathological processes.

II. THE CHOLINERGIC SYSTEM

Acetylcholinergic (cholinergic) neurons in the basal forebrain innervate the cortex and hippocampus and are involved in normal learning and memory and attention (Olton et al., 1988; Wenk, 2003). As a consequence of their vulnerability to aging and the underlying degenerative processes associated with AD, these basal forebrain acetylcholine-releasing neurons may become dysfunctional during the early stages of the disease process (Whitehouse et al., 1981; Davis et al., 1999). The extent to which this neurotransmitter system is impaired may correlate with the severity of selected cognitive symptoms associated with dementia. For example, dysfunction of cholinergic input to the cortex may contribute to a deficit in attentional abilities; alterations in the projection to the central nucleus of the amygdala may underlie emotional changes; and the dysfunction of cholinergic inputs to the hippocampus clearly underlies the presence of amnesia (Olton et al., 1991, 1988; Davis et al., 1999). A deficit in cholinergic biomarkers, including a decline in the level of cholinergic synthetic enzyme, choline acetyltransferase activity, transmitter production and release, and decline in the level of the principal catabolic enzyme, acetylcholinesterase, are commonly reported biochemical changes within the brains of patients with AD (Whitehouse et al., 1981; Davis et al., 1999).

It is important to recognize that the loss of these biomarkers does not herald the death of the cholinergic neuron; an injured neuron will often reduce the production of its luxury systems related to neurotransmitter function in preference to biochemical processes that are essential for recovery. The persistence of these cholinergic neurons offers an opportunity to rescue them from continued degeneration. As such, experimental manipulation of the functional integrity of cholinergic neurons in the basal forebrain of young rats has been used as an animal model for this component of AD pathology (Wenk et al., 1994; Wenk and Willard, 1999). Moreover, drug therapies designed to attenuate memory deficits associated with AD have focused on alleviating these impairments in the cholinergic synaptic function.

III. TREATMENT APPROACH: ACETYLCHOLINESTERASE INHIBITORS

Treatment has focused on reducing the memory impairments by counteracting cholinergic deficits through inhibition of the catabolic enzyme acetylcholinesterase (McGeer et al., 1984; Daiello et al., 2005). The currently available inhibitors of this enzyme, including donepezil, rivastigmine, and galantamine, function by blocking the destruction of released acetylcholine and increasing the life span of the neurotransmitter molecule in the synaptic cleft. This compensates for the reduced production and release of synaptic acetylcholine. However, there is a potential flaw with this therapeutic approach; for example, these enzymatic inhibitors rely on the production and release of acetylcholine from functional cholinergic terminals, and long-term treatment with these acetylcholinesterase inhibitors does not rescue or slow the eventual degeneration of cholinergic cells. Therefore, these drugs rely on the enhancement of synaptic function within a neural system that is actively degenerating, for reasons that are unknown. However, for some AD patients, acetylcholinesterase inhibitors may stabilize cognitive and behavioral function, producing a modest symptomatic benefit that varies between patients, depending on a number of factors, such as age, gender, and the presence or absence of specific genetic mutations (Doody et al., 2003). Approximately 20% of patients given an acetylcholinesterase inhibitor demonstrate a cognitive stabilizing effect that can last up to two years; these findings are consistent with the hypothesis of a potential modifying effect by these drugs on the disease process.

IV. THE GLUTAMATERGIC SYSTEM

Glutamate is the brain's principal excitatory neurotransmitter, with many physiologic and pathological roles that become important at different stages of life. It is responsible for interneuronal communication in local circuits within virtually every region of the brain and spinal cord and, in many instances, also for communication between distant brain regions. Glutamatergic neurotransmission is critical to normal learning and memory, in part because of its documented role in long-term potentiation and other forms of synaptic plasticity. When the activity of glutamate neurons becomes excessive or the stimulation of its primary receptors becomes dysfunctional, there are well-characterized pathological consequences, particularly associated with age-related neurodegenerative diseases. Accumulating evidence suggests that the anomalous glutamatergic activity associated with AD may be due to a postsynaptic receptor defect (Rogawski and Wenk, 2003). This defect may be related to an inappropriately timed or sustained glutamate activation of a specific type of glutamate receptor that responds to the agonist N-nethyl-D-aspartate (NMDA), leading to neuronal injury and death (Francis, 2003). Abnormal NMDA receptor function may underlie the selective vulnerability and loss of specific neural systems, ultimately producing cognitive deficits. NMDA receptors are present at many excitatory synapses. They do not contribute to ongoing synaptic transmission because of the unique property that ambient Mg^{2+} in the extracellular environment blocks them. Mg^{2+} ions enter the channel and are able transiently to occlude cation flux through the channel by binding to a site deep inside the pore of the channel. At the normal resting potential (e.g., -60 mV) the transmembrane electric field (negative on the inside of the cell) favors entry of Mg^{2+} into the pore of the receptor so that the channel is blocked. Under such resting conditions, NMDA receptors do not conduct ions. However, with sufficient postsynaptic depolarization within the region of the channel, the neuronal membrane surrounding the channel sufficiently depolarizes so that Mg^{2+} is no longer strongly attracted into the pore. Under such depolarized conditions, NMDA receptors activated by synaptically released glutamate are able to allow the influx of sodium and calcium ions and contribute to postsynaptic excitation and activation of second messenger systems (Albin and Greenamyre, 1992).

The calcium that enters through the NMDA receptors can act as a messenger for various cellular processes through the activation of calcium-dependent protein kinases. Importantly, the calcium ion entry through NMDA receptors during synaptic excitation is believed to be the critical event that underlies a form of synaptic plasticity called *long-term potentiation* (LTP). Under the pathological conditions associated with AD, the postsynaptic neuronal membrane near the NMDA channels is chronically depolarized, relieving the Mg^{2+} block of NMDA receptors. Under these conditions, subsequent activation of NMDA receptors by ordinary glutamatergic synaptic activity could permit a continuous "leak" of calcium ions into neurons, theoretically overwhelming the endogenous mechanisms that regulate calcium ion homeostasis (Choi et al., 1987; Albin and Greenamyre, 1992). Oxidative stress may also result in impaired energy production, possibly leading to impaired function of the membrane ion pumps required for maintenance of the resting potential. In any of these situations, excessive calcium ion influx through NMDA receptors could mediate glutamatergic excitotoxicity by activating a host of calcium-dependent signaling pathways, ultimately leading to neuronal degeneration. In addition, calcium entry through NMDA receptors stimulates nitric oxide production through closely associated neuronal nitric oxide synthase. Nitric oxide can react with a superoxide anion to form peroxynitrite, which disintegrates into extremely toxic hydroxyl free radicals that can damage cells in a variety of ways. Therefore, neurons that express NMDA receptors would become selectively vulnerable to normal glutamatergic stimulation. Accordingly, there are a number of reports demonstrating that NMDA receptors are depleted in selected regions of the AD brain (Francis, 2003). For example, acetylcholine neurons express NMDA receptors; this could contribute to the vulnerability of these neurons during the early phases of AD.

V. ROLE OF NEUROINFLAMMATION IN NEURODEGENERATION

Chronic neuroinflammation is a key factor underlying neuronal death and the pathophysiological development of AD (Mrak et al., 1995; Akiyama et al., 2000). Regional inflammatory changes are closely related to the cognitive manifestations of AD (Akiyama et al., 2000). In the brains of AD patients, the entorhinal and frontal cortex inflammatory markers, such as activated microglia, demonstrate a higher correlation with synapse loss than does the number of neurofibrillary tangles (DiPatre and Gelman, 1997) or the degree of deposition of Aβ. The brain of an AD patient expresses a significant and well-organized cascade of immunological changes, and these changes occur very early in the progression of the disease and underlie the progression of atrophy and regional degeneration (Cagnin et al., 2001). Memory impairments in the early phases of AD coincide with the development of inflammation within neuronal populations and regions known to be vulnerable in the brains of AD (Davis et al., 1999). The processes underlying the commencement of the inflammation are not completely understood but lead to a cascade of self-propagating cellular events, including blockade of glutamate uptake by glia (Rothwell et al., 1997), increased release of prostaglandins (Katsuura et al., 1989), and enhanced release of glutamate (Hanisch et al., 1997; Emerit et al., 2004). The consequences of the inflammation can relieve the magnesium ion blockade of voltage-gated NMDA channels, increase nitric oxide levels, both leading to calcium ion flux dysregulation, impaired mitochondrial respiration, oxidative stress, a decline in energy production and membrane depolarization (Chao et al., 1995; Emerit et al., 2004), and initiation of the processes out-lined earlier leading to cell death (Albin and Greenamyre, 1992; Chao and Hu, 1994).

The consequences of long-term, low-level brain inflammation may lead to a destabilization of neuronal calcium ion homeostasis and further alter intracellular signal-transduction cascades (Barry et al., 2005). The amplitude of the calcium ion entry through NMDA channels, the kinetics of calcium ion release from intracellular stores, the decay in its free cytoplasmic levels, and the spatiotemporal pattern of activation of NMDA channels distributed around the neural networks within the hippocampus are the principle means by which calcium ion signals are deciphered into a meaningful biological response that can lead to the consolidation of a new memory. NMDA receptor dysregulation is likely to be most evident within brain regions showing the highest degrees of inflammation.

The general neuronal dysfunction that develops as a consequence could impair the mechanisms underlying synaptic plasticity, such as LTP, ultimately leading to memory impairments and neurodegeneration. Therefore, taken together, these findings suggest that neuroinflammation plays an intermediary role in the neurodegenerative processes during the early phases of AD, ultimately contributing to the neuropathology that develops in later stages of the disease (Eikelenboom et al., 1998; Mrak et al., 1995; Wenk et al., 2000b). The mechanisms outlined in this hypothesis predict that an NMDA channel antagonist could enter the pore of the channel and prevent the influx of excessive amounts of calcium ion and attenuate the consequences of the calcium flux dysregulation.

Chronic neuroinflammation may be responsible for the selective vulnerability of neurons in AD. Using an animal model of chronic brain inflammation (Hauss-Wegrzyniak et al., 1998), we have systematically examined, and then selectively inhibited, each step in the cascade shown next that leads to excessive stimulation of NMDA receptors and cell death (Wenk et al., 2000b; Wenk and Willard, 1999; Willard et al., 2000). Many of the components of this cascade are represented by this equation:

$$\text{LPS} \rightarrow \textbf{Cytokines} \rightarrow \textbf{Prostaglandins} \rightarrow \textbf{[Glutamate]}_{ext} \rightarrow$$
$$\textbf{NMDA(R1)} \rightarrow \textbf{Ca}^{++} \textbf{ influx, NOS} \rightarrow \textbf{NO} \rightarrow$$
$$\textbf{Dysfunction or Cell Death}$$

In the current animal model, we used an infusion of lipopolysaccharide (LPS) to generate the inflammation. However, the inflammatory processes may be initiated naturally in the human brain by a diverse array of stimuli that are deposited in response to underlying genetic mutations or inappropriately processed proteins (Akiyama et al., 2000; Francis, 2003). Following the infusion of LPS, activated microglia can indirectly potentiate glutamate-mediated neurotoxicity via the production of prostaglandins, nitric oxide (Morimoto et al., 2002), and cytokines (Bernardino et al., 2005). The inflammatory processes produce a dysregulation in calcium influx via NMDA receptors that could produce multiple unstable conditions, for example, an elevation in intracellular levels of calcium in a larger-than-usual proportion of neurons or a dramatic increase in the number of neurons, with a disruption in neuroplasticity or leading to cell death (Soliven and Albert, 1992). Given the critical role of NMDA receptors in this cascade, it is not surprising that chronic neuroinflammation leads to a significant decline in the number of NMDA(R1) receptors (Rosi et al., 2004). As predicted by this hypothesis, the greatest receptor loss occurred in those regions of the hippocampus that also had the greatest concentration of activated microglia. Therefore, these results are consistent with the hypothesis that a loss of NMDA receptors in hippocampal regions in response to the presence of chronic neuroinflammation may contribute to the cognitive deficits observed in AD during the earliest phases of the disease (Akiyama et al., 2000; Eikelenboom et al., 1998).

VI. EFFECT OF NEUROINFLAMMATION ON CHOLINERGIC FUNCTION

Neuroinflammation within the basal forebrain selectively destroys acetylcholine neurons in a time-, but not dose-, dependent manner (Wenk et al., 2000a, 2000b; Willard et al., 1999, 2000). Medial septal cholinergic neurons that innervate the hippocampus may also be selectively vulnerable to immune-mediated processes (Kalman et al., 1997). In addition, the level of cholinergic biomarkers was significantly reduced within the septal cholinergic neurons of transgenic mice that express elevated levels of the cytokine tumor necrosis factor-alpha (TNFα; Aloe et al., 1999). Therefore, the entire forebrain cholinergic system may be vulnerable to elevated levels of inflammatory proteins, particularly TNFα (Wenk et al., 2000b; Willard et al., 1999, 2000). Stimulation of TNFα receptors may induce cell death by "silencing of survival signals" via the inhibition of insulin-like growth factor-1-mediated signaling within neurons (Venters et al., 2000). TNFα can also inhibit glutamate reuptake into astrocytes and may potentiate glutamate receptor–mediated toxicity (Soliven and Albert, 1992; Chao and Hu, 1994; Chao et al., 1995; Probert et al., 1997) within the basal forebrain, a region that is vulnerable to excess glutamatergic function (Kim and Ko, 1998).

Although the mechanism underlying the degeneration of basal forebrain cholinergic cells is unknown, the specificity of these effects on cholinergic neurons was initially suggested by a study that isolated antibodies from the sera of AD patients that selectively recognized and destroyed basal forebrain cholinergic cells when injected into a rat brain (Foley et al., 1988). In addition, head trauma in humans is a significant risk factor for AD (Rasmusson et al., 1995) and is associated with increased levels of inflammatory proteins (Griffin et al., 1994) and a decline in the number of basal forebrain cholinergic neurons (Murdoch et al., 1998). In vitro studies also indicate that brain inflammation may selectively destroy basal forebrain cholinergic neurons (McMillian et al., 1995).

VII. TREATMENT APPROACH: ANTI-INFLAMMATORY THERAPY

Current treatments for AD as well as many other neurodegenerative diseases typically provide only moderate symptomatic benefits and do not modify the progression of disease. The inflammatory process represents a target for potential disease-modifying drugs that may influence multiple critical steps in the pathogenesis of AD, such as glial activation and production of cytokines and complement proteins (Andersen et al., 1995; Stewart et al., 1997). Results from clinical trials have increased insight into the potential mechanisms of action of multiple nonsteroidal anti-inflammatory drugs (NSAIDs). For example, some NSAIDs may lower the amyloid burden without affecting other important physiological pathways. Epidemiological evidence provides strong support of the chronic use of NSAIDs for reducing the risk of AD (Andersen et al., 1995) and slowing the cognitive decline (Breitner et al., 1994; McGeer et al., 1990; McGeer and McGeer, 1998; Wyss-Coray and Mucke, 2000). Investigators have found a significant association between exposure to NSAIDs for more than two years and AD risk reduction (Breitner et al., 1994). Recent studies suggest that although anti-inflammatory agents do not appear to slow progression of dementia, they may have a preventative influence on the development of AD pathology (Breitner et al., 1994; Wenk et al., 2000b; Marchetti and Abbracchio, 2005). However, multiple clinical trials have elicited mixed (Rogers et al., 1993; In 'T Veld et al., 2001), albeit mostly negative, results (McGeer et al., 1990). Additional work to evaluate their potential protective properties further is warranted.

VIII. TREATMENT APPROACH: GLUTAMATE CHANNEL ANTAGONISM

Although the basis of the vulnerability of cholinergic neurons in AD is not understood, one possibility is that the degeneration of these neurons might be due to inappropriate activation of the NMDA receptors they express (Wenk et al., 1994, 1995, 1997; Muir, 1997). Indeed, infusion of NMDA or glutamate receptor agonists such as quinolinic acid into the rodent basal forebrain region is associated with a loss of cholinergic neurons, as demonstrated by a decrease in the release of acetylcholine in the projection regions and a decline in the activity of choline acetyltransferase. Furthermore, the loss of these neurons, due either to the infusion of NMDA or the presence of chronic neuroinflammation, is associated with impaired spatial memory. Antagonism of the glutamatergic channel function has become a potential target for novel pharmacotherapies. For example, memantine is an uncompetitive, low-affinity NMDA-receptor antagonist. It is believed to have an impact on the abnormal glutamatergic process associated with AD by selectively blocking the excitotoxic effects of atypical glutamate transmission while allowing normal physiological function to occur (Rogawski and Wenk, 2003). Memantine binds deep inside the NMDA receptor channel, not at the glutamate binding site. This characteristic plus its low affinity for the channel enables memantine to discriminately block abnormal glutamate activity. Once glutamate binds to its receptor, memantine falls away. Through this mechanism of action, memantine remotely modulates the receptor, preventing excessive flow but allowing normal function. Theoretically, in the presence of this drug, abnormal glutamate activity that leads to neuronal loss is prevented, but physiological activation that produces learning and memory is preserved. Acetylcholine neurons rescued from excitotoxicity will then be available for further treatment using an acetylcholinesterase inhibitor, as described earlier. Moreover, if memantine can prevent abnormal glutamate neurotransmission, it may provide neuroprotection in both the early stages of many different neurodegenerative diseases when toxicity is generated, e.g., due to the presence of brain inflammation, as well as later in the disease process, when symptoms are more apparent. To the extent that similar mechanisms contribute to cell death in other neurodegenerative diseases, memantine could theoretically slow their progression as well (Möbius, 2003).

Alzheimer's disease is acknowledged to be a disease of multifactorial pathology expressed through a range of cognitive, behavioral, and functional symptomatology. These characteristics, together with its progressive nature, suggest that treatment with a combination of drugs may maximize response to therapy. A reasonable prediction, then, is that memantine and an acetylcholinesterase inhibitor should work together effectively to optimize pharmacotherapy for AD patients (Wenk et al., 2000a; Daiello et al., 2005; Doody et al., 2003). Recent clinical trial results support this prediction.

IX. ROLE OF OXIDATIVE STRESS AND MITOCHONDRIA FAILURE IN NEURODEGENERATION

Deficits in energy metabolism associated with aging play an important role in the vulnerability of neurons and in neurodegenerative diseases, such as AD (Beal, 1995; Blass et al., 2000; Emeritt et al., 2004). A defect in energy production would make neurons that express glutamatergic receptors more vulnerable to elevated or normal levels of endogenous glutamate for the following reasons. Decreased levels of intracellular ATP would lead to a partial, and chronic, membrane depolarization, the relief of the voltage-dependent Mg^{2+} blockade at NMDA receptors, and a persistent increase in the influx of calcium ions into the cells. Ultimately, the accumulation of intracellular calcium ions following the activation of NMDA receptors by glutamate would lead to neuronal death. Oxidative stress or impaired intracellular calcium buffering may also result in impaired energy production, possibly leading to impaired function of the membrane ion pumps required for maintenance of the resting potential. In any of these situations, excessive calcium ion influx through NMDA receptors could activate a host of calcium ion–dependent signaling pathways and stimulate nitric oxide production through closely associated neuronal nitric oxide synthase. Nitric oxide can react with a superoxide anion to form peroxynitrite, which disintegrates into extremely toxic hydroxyl free radicals that can further impair mitochondrial function and energy production. Intracellular calcium may become concentrated within the post-synaptic mitochondria, further contributing to the impaired energy production within the region of the NMDA channels (Peng and Greenamyre, 1998; Duchen, 2000).

Mitochondrial dysfunction coupled with activation of glutamatergic receptors could underlie enhanced cholinergic vulnerability associated with aging and AD. These results suggest that under conditions that lead to a mitochondrial energy deficit, such as that produced by exposure to 3-nitropropionic acid, activation of NMDA receptors can lead to the death of the neuron (Wenk et al., 1996). In addition, mitochondrial dysfunction might have a much greater and earlier impact on the integrity of cholinergic neurons, in part due to their dependence on normal mitochondrial function for the production of acetyl coenzyme A, a precursor to the synthesis of acetylcholine on normal mitochondrial function. These findings are consistent with the hypothesis that mitochondrial failure and neurochemical processes involving NMDA receptor overactivation and oxidative stress play a role in neurodegeneration in vulnerable brain regions (Barnham et al., 2004).

X. NEURODEGENERATIVE DISEASES ASSOCIATED WITH β-AMYLOID

AD is characterized by progressive deterioration of cognition and memory and disturbed emotional reactivity caused by dysfunction and degeneration of neurons in the limbic system and cerebral cortex. Affected brain areas typically contain extracellular neuritic plaques composed of fibrillar Aβ deposits and intracellular neurofibrillary tangles composed of paired helical filaments of hyperphosphorylated tau. The deposition of Aβ is a key element leading to the neuronal loss seen in the AD brain. The gradual accumulation of Aβ in the interstitial fluid of the brain may provide a focus for the subsequent deposition of other proteins and the release of inflammatory proteins by activated glia; these conditions may promote the transformation of diffuse filamentous Aβ deposits into a possibly more neurotoxic form (Schubert et al., 1998). The accumulation of toxic fibrillar Aβ in the surrounding neuropil may be a principle stimulus for the activation of resident microglia to secrete cytokines and reactive oxygen species (Meda et al., 1999). Neuronal homeostasis may become disrupted and eventually induce cell death (Cowburn et al., 1997). Aβ also induces oxidative stress and perturbs neuronal ion homeostasis by promoting membrane lipid peroxidation, which can impair the function of membrane-bound ion, glucose, and amino acid (including glutamate) transport proteins. In addition to producing oxidative stress and affecting Ca^{2+} homeostasis, Aβ may increase the vulnerability of neurons to glutamate, leading to glutamate excitotoxicity and the opportunity for NMDA channel antagonists to reduce this vulnerability. Aβ can chronically depolarize neurons through its action on the metabotropic glutamate receptor, mGluR1 (Blanchard et al., 2002), and partially relieve the voltage-dependent Mg^{2+} blockade of NMDA channels. Under these conditions, subsequent activation of NMDA receptors by ordinary glutamatergic synaptic activity could permit a continuous entry of calcium ion into neurons, overwhelming the endogenous mechanisms that regulate calcium homeostasis. Therefore, due to the continuing presence of Aβ, neurons that express NMDA receptors would become selectively vulnerable to normal glutamatergic stimulation. This is similar to the situation described earlier due to the presence of chronic brain inflammation.

XI. β-AMYLOID: TREATMENT APPROACHES

Aβ can interact with NMDA receptors and enhance NMDA receptor–mediated excitotoxicity. For example, radioligand-binding experiments in rat cortical membranes suggest that Aβ selectively binds to the glutamate and glycine binding sites of the NMDA receptor, and not to non–NMDA glutamate receptor subtypes (Cowburn et al., 1997). Mature cultured murine cortical neurons and fetal human cerebral cortical cell cultures exposed to Aβ were more susceptible to excitotoxic injury by glutamate or NMDA as compared to neurons that were not exposed to Aβ (Kim and Ko, 1998; Mattson et al., 1992). Given the role of the NMDA channel in the vulnerability of neurons, it was not surprising that a chronic infusion of memantine reduced local neuronal cell loss produced by intrahippocampal injection of Aβ (Miguel-Hidalgo et al., 2002). Indeed, a brief exposure of cultured cortical neurons to memantine, which would produce only a transient block of NMDA receptors, inhibited the toxicity of Aβ for up to 48 hours (Tremblay et al., 2000). The relevance of this brief effect of memantine with regard to chronic therapy in AD remains to be investigated.

In transgenic mice that demonstrate an accelerated amyloid deposition in hippocampus and cortex that is associated with dystrophic neurites and reactive astrocytes (Price et al., 1998) and suppressed gene expression underlying consolidation (Dickey et al., 2004), memantine improved performance in both

T-maze and Morris water maze paradigms for spatial working memory and spatial long-term memory, respectively (Tanila et al., 2003). In cultured human neuroblastoma cells, treatment with memantine for 24–48 hours provided evidence that the drug may enhance amyloid degradation (Chen et al., 2002). It remains to be determined whether memantine can produce a similar disease-modifying effect in the AD brain. To the extent that similar mechanisms contribute to cell death in AD and age-related neurodegenerative diseases, selective NMDA channel antagonists could theoretically slow their progression as well.

In spite of the enormous amount of research effort that has focused on AD and other neurodegenerative disorders, the underlying pathophysiological pro-cesses are not understood in sufficient detail to guide the design of effective drug therapies. For example, the relationship between the presence of specific genes associated with AD and the appearance of amyloid deposits has focused attention on the molecular pathways involved in amyloid clearance and strate-gies to prevent amyloid production and aggregation. The current strategy is to try to reduce Aβ production from amyloid precursor protein (APP). APP is a transmembrane protein that is normally metabolized via one of two well-studied pathways associated with the enzymes α-secretase, β-secretase, or γ-secretase. APP processing via the α-secretase pathway avoids the formation of Aβ. APP processing via the sequential activities of the β- and γ-secretases leads directly to the production of the toxic form of Aβ and its subsequent deposi-tion into senile plaques. Therapeutic strategies have been tested using these secretases as targets; for example, drugs have been produced that can lead to the stimulation of α-secretase, inhibition of β-secretase, or inhibition of the γ-secretase (Citron, 2004). Pharmaceutical companies have investigated the effects on γ-secretase inhibitors on the progression of the pathology and dementia associated with AD (Citron, 2004). Although research using these drugs on transgenic animal models of AD has produced promising results, there are significant drawbacks due to lack of substrate specificity. The problem is that the APP processing molecules would be only one of many potential substrates that could be influenced by γ-secretase inhibitors (De Strooper et al., 1999).

XII. NMDA RECEPTOR FUNCTION IN NEURODEGENERATIVE DISEASES ASSOCIATED WITH TAU PROTEINS

The evidence just discussed indicates a clear negative effect of Aβ on the func-tion of NMDA receptors. In addition, NMDA receptor function can alter the expression and functional state of tau. Tau is a microtubule-associated protein that promotes microtubule polymerization and stabilization. Hyperphosphorylated tau accumulates in paired helical neurofilaments to form neurofibrillary tangles in the brains of patients with AD. A potential link between glutamate-induced excitotocity and tau was first demonstrated by studies using cultured rat hippocampal neurons; glutamate-induced neurodegeneration was associated with immunostaining that was specific for the presence of neurofibrillary tangles (Mattson, 1990). Acute or chronic NMDA-induced excitotoxicity in neuronal cultures can also significantly enhance tau production (Pizzi et al., 1995; Sindou et al., 1992) and selectively increases phosphorylated tau (Couratier et al., 1996, 1997). Given the potentially significant role of neuro-fibrillary tangle formation in the clinical progression of AD dementia (Bierer et al., 1995) and that augmented tau phosphorylation can be prevented by an NMDA receptor antagonist (Couratier et al., 1996), it is likely that NMDA receptor–dependent influences on tau phosphorylation could promote the evolution of AD pathology and dementia. The abnormal hyperphosphorylation of tau may be related to the impaired activity of protein phosphatase (PP)-2A; treatment with the NMDA antagonist memantine restored PP-2A function and reduced the accumulation of tau in a rat hippocampal slice preparation (Li et al., 2004). Taken together, these findings suggest that NMDA receptors play a critical role in the progress of neuropathology associated with tauopathies and that drugs similar to memantine might be useful for the treatment.

XIII. TREATING NEURODEGENERATIVE DISEASE SYMPTOMS WITH GINKGO BILOBA

Negative results for therapies targeting those cognitive impairments described earlier are often underrepresented in the scientific literature. This fact underscores the need for caution when assessing the cognitive enhancing or neuro-protective effects of any drug; this is particularly true for herbal medications that target the brain. Today, extracts of the Ginkgo biloba tree are perhaps the herbal treatment most widely used specifically to augment cognitive functions, particularly memory impairments associated with normal aging and neurodegenerative diseases. Ginkgo has been used by humans in different cultures for centuries and, at the very least, is a relatively safe, although likely inert, treatment. The few published studies that show an enhancement of learning and memory in rodents are difficult to evaluate due to insufficient information about experimental procedures. Close examination of the dozens of investigations of the cognitive effects of ginkgo in humans quickly leads one to conclude that the drug produces only mild beneficial effects on various aspects of cognitive functioning (Gold et al., 2003). The majority of studies have involved subjects with significant cognitive impairments, typically a diagnosis of early- to middle-stage Alzheimer's disease, or, more recently, in healthy normal elderly subjects who are typically classified as having mild to moderate cognitive impairment. In general, these studies have found quite limited, but sometimes statistically significant, improvements in performance on various standardized tests of cognitive function after chronic treatment with ginkgo, as compared with placebo. The improvements were usually revealed using tests requiring various cognitive abilities, such as attention, short-term memory, and choice reaction time. In healthy elderly subjects, the available data do not as yet allow any conclusion to be made with certainty. None of the potential effects of ginkgo in humans reported to date are clearly attributable to direct effects of the drug on memory processes. In each case, indirect effects of the drug on memory via effects on other cognitive processes (such as arousal and attention) are probable (Gold et al., 2003).

Ginkgo may enhance the uptake of choline into acetylcholine-releasing neurons (Kristofikova and Klaschka, 1997). Increased availability of choline might be able to enhance the production of acetylcholine within these neurons. Similar therapeutic approaches have led to the use of diets enriched with choline. The production and release of acetylcholine also requires the availability of other precursors, which can be derived from dietary glucose. Consumption of ginkgo extract can increase glucose utilization and acetylcholine production in the frontal and parietal cortex and cerebellum, areas of the brain that are important for processing sensory information and movement and for attentional abilities (Kunkel, 1993). However, the problem is that even though these neurons might synthesize and store more acetylcholine presynaptically, they do not necessarily increase their probability of release of acetylcholine. Thus, the ultimate impact of increased acetylcholine formation on cognitive performance is unclear.

A comparison of the efficacy of ginkgo with acetylcholinesterase inhibitors (Oken et al., 1998) showed that the mean extent of improvement resulting from gingko treatment was approximately 10–20%, a value roughly comparable to the magnitude of improvement often seen in clinical trials with acetylcholinesterase inhibitors. A direct comparison in rats of gingko versus acetylcholinesterase inhibitors showed clearly greater efficacy of the enzyme inhibitors. A recent six-week randomized, double-blind, placebo-controlled, parallel-group trial involving over 200 healthy elderly patients examined the effects of ginkgo using standardized neuropsychological tests of verbal and nonverbal learning and memory, attention and concentration, naming and expressive language abilities (Solomon et al., 2002). There were no significant benefits due to the gingko therapy on any outcome measure. These data suggest that when taken following the manufacturer's instructions, ginkgo provides no measurable benefit in memory or related cognitive function to adults with healthy cognitive function.

Overall, there is simply too little data to base a clear recommendation regarding the benefits of ginkgo extracts, or any other herbal or pharmaceutical agent, on learning and memory or other cognitive functions. Many years of experience with investigations of new drugs have demonstrated that the initial positive results from studies involving a small number of subjects tend to disappear when the drugs are tested on larger numbers of subjects from diverse populations.

XIV. TREATMENT APPROACH OF THE FUTURE: NEUROPROTECTION

Anti-inflammatory drugs, secretase inhibitors, and NMDA antagonists offer the potential of neuroprotection against neurodegeneration involving inflammation, amyloid deposition, and glutamate excitotoxicity. Theoretically, treatments focused on these processes should slow the progression of AD. In AD, glutamatergic dysfunction, the consequences of chronic inflammation, prolonged oxidative stress, Aβ and tau production, and deposition not only contribute to the cell death but also interact with each other, leading to exaggerated pathology through positive feedback mechanisms. However, the principle problem in identifying the neuroprotective actions of these treatment approaches is related to the long treatment durations (1–3 years) that are required to demonstrate true neuroprotection from the neurodegenerative processes that are thought to underlie AD. Drug toxicity and side-effect profiles also become more important in the elderly, increasing further problems with long-term treatment. Clinical trials aimed at showing neuroprotection requires both placebo control groups and a relatively long washout period from the drug to ensure that testing is done when drug is not present in the brain. There are clear ethical concerns about such clinical trial design. That is, is it appropriate to withhold treatment to test a scientific theory? Any drug that exhibits potential neuroprotective properties will require long follow-up periods to allow beneficial effects to be clearly documented. In addition, neuroprotective trials will need to utilize drugs that target different aspects of the known neurodegenerative changes discussed earlier. The best clinical outcome may be achieved by a combination of therapies, particularly if the intent is to demonstrate synergistic effects over time. Combination drug therapies may require that the dose of each drug be reduced in order to limit the drug toxicity; unfortunately, combination therapy will likely add to the overall cost and complexity of trial design. Given our earlier discussions, a combination therapy would consist of agents targeting chronic neuroinflammation, oxidative stress, Aβ, and tau. Thus far, there have been few attempts to show neuroprotective activity in AD. Numerous epidemiological studies strongly indicate that people taking anti-inflammatory drugs have a significantly reduced prevalence of AD or a slower cognitive decline. Although interventional studies have not been successful, anti-inflammatory treatment is likely to be more effective when administered many years prior to the onset of symptomatic AD, a time when brain inflammation is more prominent (McGeer and McGeer, 1998; Mackenzie and Munoz, 1998; Wyss-Coray and Mucke, 2000; Akiyama et al., 2000; Gahtan and Overmier, 1999; Wenk et al., 2000b).

XV. SUMMARY

This chapter has presented evidence that demonstrates that no pharmacological treatment available today has been demonstrated to slow the consequences of human aging on the decline in cognitive function associated with neuro-degenerative disease. Studies of the neuropathological changes associated with specific diseases, primarily AD, have improved our understanding of the neural processes that might be targeted for manipulation; unfortunately, the initial results from clinical trials on humans have not been encouraging. Currently, if one desires to attenuate the consequences of aging and possibly slow the decline in learning and memory abilities, only one intervention, the consumption of a low-calorie, nutritionally balanced diet, can effectively increase longevity and prolong good health (Lane et al., 2002). Therefore, *bon raisonnable appétit!*

REFERENCES

Albin RL and Greenamyre JT (1992) Alternative excitotoxic hypothesis. *Neurology* 42: 733–738.

Akiyama H, Barger S, Barnum S, Bradt B, Bauer J, Cooper NR, Eikelenboom P, Emmerling M, Fiebich B, Finch CE, Frautschy S, Griffin WST, Hampel H, Landreth G, McGeer PL, Mrak R, MacKenzie I, O'Banion K, Pachter J, Pasinetti G, Plata-Salaman C, Rogers J, Rydel R, Shen Y, Streit W, Strohmeyer R, Tooyoma I, Van Muiswinkel FL, Veerhuis R, Walker D, Webster S, Wegrzyniak B, Wenk G, and Wyss-Coray A (2000) Inflammation in Alzheimer's disease. *Neurobiol Aging* 21: 383–421.

Aloe L, Fiore M, Probert L, Turrini P, and Tirassa P (1999) Overexpression of tumor necrosis factor-alpha in the brain of transgenic mice differentially alters nerve growth factor levels and choline acetyltransferase activity. *Cytokine* 11: 45–54.

Andersen K, Launer LJ, Ott A, Hoes AW, Breteler MMB, and Hoffman A (1995) Do nonsteroidal anti-inflammatory drugs decrease the risk of Alzheimer's disease? *Neurology* 45: 1441–1445.

Barnham KJ, Masters CL, and Bush AI (2004) Neurodegenerative diseases and oxidative stress. *Nature Rev Drug Disc* 3: 205–214.

Barry CE, Nolan Y, and Clarke RM (2005) Activation of c-Jun-N-terminal kinase is critical in mediating LPS-induced changes in rat hippocampus. *J Neurochem* 93: 221–231.

Beal MF (1995) Aging, energy, and oxidative stress in neurodegenerative diseases. *Ann Neurol* 38: 357–366.

Bernardino L, Xapelli S, and Silva AP (2005) Modulator effects of interleukin-1b and tumor necrosis factor-a on AMPA-induced excitotoxicity in mouse organotypic hippocampal slice cultures. *J Neurosci* 25: 6734–6744.

Bierer LM, Haroutunian V, and Gabriel S (1995) Neurochemical correlates of dementia severity in Alzheimer's disease: Relative importance of the cholinergic deficits. *J Neurochem 64*: 749–760.

Blanchard BJ, Thomas VL, and Ingram VM (2002) Mechanism of membrane depolarization caused by the Alzheimer Ab1-42 peptide. *Biochem Biophys Res Commun 293*: 1197–1203.

Blass JP, Sheu RK-F, and Gibson GE (2000) Inherent abnormalities in energy metabolism in Alzheimer disease: Interaction with cerebrovascular compromise. *Ann NY Acad Sci 903*(1): 204–221.

Breitner JCS, Gau BA, Welsh KA, Plassman BL, McDonald WM, Helmas MJ, and Anthony JC (1994) Inverse association of anti-inflammatory treatments and Alzheimer's disease. *Neurology 44*: 227–232.

Cagnin A, Brooks DJ, and Kennedy AM (2001) In vivo measurement of activated microglia in dementia. *Lancet 358*: 461–467.

Chao C and Hu S (1994) Tumor necrosis factor alpha potentiates glutamate neurotoxicity in human fetal brain cell cultures. *Dev Neurosci 16*: 172–179.

Chao CC, Hu S, Ehrlich L, and Peterson PK (1995) Interleukin-1 and tumor necrosis factor alpha synergistically mediate neurotoxicity: Involvement of nitric oxide and N-methyl-D-aspartate receptors. *Brain Beh Immunol 9*: 355–365.

Chen D, Alley GM, Ge Y-W, Farlow MR, Banerjee PK, and Lahiri DK (2002) Memantine and the processing of the beta-amyloid precursor protein. Program No. 296.3 (abstract). Washington, DC: Society for Neuroscience.

Choi DW, Maulucci-Gedde M, and Kriegstein AR (1987) Glutamate neurotoxicity in cortical cell culture. *J Neurosci 7*: 357–368.

Citron M (2004) Strategies for disease modification in Alzheimer's disease. *Nat Rev Neurosci 5*: 677–685.

Couratier P, Lesort M, Sindou P, Esclaire F, Yardin C, and Hugon J (1997) Modifications of neuronal phosphorylated tau immunoreactivity induced by NMDA toxicity. *Mol Chem Neuropathol 27*: 259–273.

Couratier P, Lesort M, Terro F, Dussartre C, and Hugon J (1996) NMDA antagonist blockade of AT8 tau immunoreactive changes in neuronal cultures. *Fundam Clin Pharmacol 10*: 344–349.

Cowburn RF, Wiehager B, Trief E, Li-Li M, and Sundstrom E (1997) Effects of beta-amyloid-(25–35) peptides on radioligand binding to excitatory amino acid receptors and voltage-dependent calcium channels: Evidence for a selective affinity for the glutamate and glycine recognition sites of the NMDA receptor. *Neurochem Res 22*: 1437–1442.

Daiello LA, Galvin JE, and Wenk GL (2005) A case-based approach to management of Alzheimer's disease across the disease continuum. *US Pharmacist 30* (suppl 2): 2–5.

Davis KL, Mohs RC, Marin D, Purohit DP, Perl DP, Lantz M, Austin G, and Haroutunian V (1999) Cholinergic markers in elderly patients with early signs of Alzheimer disease. *JAMA 281*: 1401–1406.

De Strooper B, Annaert W, Cupers P, Saftig P, Craessaerts K, Mumm JS, Schroeter EH, Schrijvers V, Wolfe MS, Ray WJ, Goate A, and Kopan R (1999) A presenilin-1-dependent γ-secretase-like protease mediates release of Notch intracellular domain. *Nature 398*: 518–522.

Dickey CA, Gordon MN, and Mason JE (2004) Amyloid suppresses induction of genes critical for memory consolidation in APP+PS1 transgenic mice. *J Neurochem 88*: 434–442.

DiPatre PL and Gelman BB (1997) Microglial cell activation in aging and Alzheimer disease: Partial linkage with neurofibrillary tangle burden in the hippocampus. *J Neuropathol Exp Neurol 56*: 143–149.

Doody RS, Mintzer JE, Sano M, Wenk GL, and Grossberg GT (2003) Alzheimer's disease: Emerging noncholinergic treatments. *Geriatrics* (suppl): 3–11.

Duchen MR (2000) Mitochondria and calcium: From cell signalling to cell death. *J Physiol 529*: 27–68.

Eikelenboom P, Roxemuller JM, and van Muiswinkel FL (1998) Inflammation and Alzheimer's disease: Relationships between pathogenic mechanisms and clinical expression. *Exp Neurol 154*: 89–98.

Emerit J, Edeas M, and Bricaire F (2004) Neurodegenerative diseases and oxidative stress. *Biomed Pharmacother 58*: 39–46.

Foley P, Bradford HF, Dochart M, Fillet H, Luine VN, McEwen B, Buch G, Winblad B, and Hardy J (1988) Evidence for the presence of antibodies to cholinergic neurons in the serum of patients with Alzheimer disease. *J Neurol 235*: 466–471.

Francis PT (2003) Glutamatergic systems in Alzheimer's disease. *Int J Ger Psychiat 18*: S15–S21.

Gahtan E and Overmier JB (1999) Inflammatory pathogenesis in Alzheimer's disease: Biological mechanisms and cognitive sequeli. *Neurosci Biobehav Rev 23*: 615–633.

Gold PE, Cahill L, and Wenk GL (2003) The lowdown on Ginkgo biloba. *Sci Am* (April): 86–91.

Griffin DE, Wesselingh SL, and McArthur JC (1994) Elevated central nervous system prostaglandins in human immunodeficiency virus-associated dementia. *Ann Neurol 35*: 592–597.

Hanisch U-K, Neuhaus J, Rowe W, Van Rossum D, Möller M, Kettenmann H, and Quirion R (1997) Neurotoxic consequences of central long-term administration of interleukin-2 in rats. *Neuroscience 79*: 799–818.

Hauss-Wegrzyniak B, Dobrzanski P, Stoehr JD, and Wenk GL (1998) Chronic neuroinflammation in rats reproduces components of the neurobiology of Alzheimer's disease. *Brain Res 780*: 294–303.

In 'T Veld BA, Ruitenberg A, Hofman A, Launer LJ, van Duijin CM, Stijnen T, Breteler MMB, and Stricker BHC (2001) Nonsteroidal anti-inflammatory drugs and the risk of Alzheimer's disease. *New Engl J Med 346*: 1515–1521.

Kalman J, Engelhardt JI, Le WD, Xie W, Kovacs I, Kasa P, and Appel SH (1997) Experimental immune-mediated damage of septal cholinergic neurons. *J Neuroimmunol 77*: 63–74.

Katsuura G, Gottschall PE, Dahl RR, and Arimura A (1989) Interleukin-1 beta increases prostaglandin E2 in rat astrocyte cultures: Modulatory effect of neuropeptides. *Endocrinology 124*: 3125–3127.

Kim WK and Ko KH (1998) Potentiation of *N*-methyl-D-aspartate-mediated neurotoxicity by immunostimulated murine microglia. *J Neurosci Res 54*: 17–26.

Kristofikova Z and Klaschka J (1997) In vitro effect of Ginkgo biloba extract (EGb 761) on the activity of presynaptic cholinergic nerve terminals in rat hippocampus. *Dement Geriatr Cogn Disord 8*: 43–48.

Kunkel H (1993) EEG profile of three different extractions of Ginkgo biloba. *Neuropsychobiol 27*: 40–45.

Lane MA, Ingram DK, and Roth GS (2002) The serious search for an anti-aging pill. *Scientific American 287*: 19–37.

Li L, Sengupta A, Haque N, Grundke-Iqbal I, and Iqbal K (2004) Memantine inhibits and reverses the Alzheimer-type abnormal hyperphosphorylation of tau and associated neurodegeneration. *FEBS Lett 566*: 261–269.

Mackenzie IR and Munoz DG (1998) Nonsteroidal anti-inflammatory drug use and Alzheimer-type pathology in aging. *Neurology 50*: 986–990.

Marchetti B and Abbracchio MP (2005) To be or not to be (inflamed) — is that the question in anti-inflammatory drug therapy of neurodegenerative diseases? *TIPS 26*: 517–525.

Mattson MP (1990) Antigenic changes similar to those seen in neurofibrillary tangles are elicited by glutamate and Ca^{2+} influx in cultured hippocampal neurons. *Neuron 4*: 105–117.

Mattson MP, Cheng B, Davis D, Bryant K, Lieberburg I, and Rydel RE (1992) b-Amyloid peptides destabilize calcium homeostasis and render human cortical neurons vulnerable to excitotoxicity. *J Neurosci* 12: 376–389.

McGeer EG and McGeer PL (1998) The importance of inflammatory mechanisms in Alzheimer disease. *Exp. Gerontol.* 33: 371–378.

McGeer PL, McGeer EG, Rogers J, and Sibley J (1990) Anti-inflammatory drugs and Alzheimer's disease. *Lancet* 335: 1037.

McGeer PL, McGeer EG, Suzuki J, Dolman CE, and Nagai T (1984) Aging, Alzheimer's disease and the cholinergic system of the basal forebrain. *Neurology* 34: 741–745.

McMillian M, Kong L-Y, Sawin SM, Wilson B, Das K, Hudson P, Hong J-S, and Bing G (1995) Selective killing of cholinergic neurons by microglial activation in basal forebrain mixed neuronal/glial cultures. *Biochem Biophys Res Comm* 215: 572–577.

Meda L, Baron P, Prat E, Scarpini E, Scarlato G, Cassatella MA, and Rossi F (1999) Proinflammatory profile of cytokine production by human monocytes and murine microglia stimulated with beta-amyloid. *J Neuroimmunol* 93: 45–52.

Miguel-Hidalgo JJ, Alvarez XA, Cacabelos R, and Quack G (2002) Neuroprotection by memantine against neurodegeneration induced by b-amyloid(1–40). *Brain Res* 958: 210–221.

Möbius HJ (2003) Memantine: Update on the current evidence. *Int J Geriatr Psychiat* 18: S47–S54.

Morimoto E, Murasugi M, and Oda T (2002) Acute neuroinflammation exacerbates excitotoxicity in rat hippocampus in vivo. *Exp. Neurol.* 177: 95–104.

Mrak RE, Sheng JG, and Griffin WST (1995) Glial cytokines in Alzheimer's disease: Review and pathogenic implications. *Human Pathol* 26: 816–823.

Muir JL (1997) Acetylcholine, aging, and Alzheimer's disease. *Phamacol Biochem Behav* 56: 687–696.

Murdoch I, Perry EK, Court JA, Graham DI, and Dewar D (1998) Cortical cholinergic dysfunction after head injury. *J Neurotrauma* 15: 295–305.

Oken BS, Storzbach DM, and Kaye JA (1998) The efficacy of Gingko biloba on cognitive function in Alzheimer's disease. *Arch Neurol* 55: 1409–1415.

Olton DS, Wenk GL, Church RM, and Meck WH (1988) Attention and the frontal cortical cortex as examined by simultaneous temporal processing. *Neuropsychologia* 26: 307–318.

Olton DS, Wenk GL, and Markowska AL (1991) Basal forebrain, memory, and attention. In R Richardson (ed.), *Activation to Acquisition: Functional Aspects of the Basal Forebrain Cholinergic System*. pp. 247–262, Boston: Birkhauser.

Peng TI and Greenamyre JT (1998) Privileged access to mitochondria of calcium influx through N-methyl-D-aspartate receptors. *Molecul Pharmacol* 53: 974–980.

Pizzi M, Valerio A, and Arrighi V (1995) Inhibition of glutamate-induced neurotoxicity by a tau antisense oligonucleotide in primary culture of rat cerebellar granule cells. *Eur J Neurosci* 7: 1603–1613.

Price DL, Tanzi RE, Borchelt DR, and Sisodia SS (1998) Alzheimer's disease: Genetic studies and transgenic models. *Annu Rev Genet* 32: 461–493.

Probert L, Akassaglou K, Kassiotis G, Pasparakis M, Alexopoulou L, and Kollias G (1997) TNFα transgenic and knockout models of CNS inflammation and degeneration. *J Neuroimmunol* 72: 137–141.

Rasmusson DX, Brandt J, Martin DB, and Folstein MF (1995) Head injury as a risk factor in Alzheimer's disease. *Brain Injury* 9: 213–219.

Rogawski M and Wenk GL (2003) The neuropharmacological basis for memantine in the treatment of Alzheimer's disease. *CNS Drug Rev* 9: 275–308.

Rogers J, Kirby LC, Hempelman SR, Berry DL, McGeer PL, Kaszniak AW, Zalinski J, Cofeild M, Mansukhani L, Wilson P, and Kogan F (1993) Clinical trial of indomethacin in Alzheimer's disease. *Neurology* 43: 1609–1611.

Rosi S, Ramirez-Amaya V, Hauss-Wegrzyniak B, and Wenk GL (2004) Chronic brain inflammation leads to a decline in hippocampal NMDA R1 receptors. *J Neuroinflam 1*: 12–18.

Rothwell NJ, Allan S, and Toulmond S (1997) The role of interleukin-1 in acute neurodegeneration and stroke: Pathophysiological and therapeutic implications. *J Clin Invest 100*: 2648–2652.

Schubert P, Ogata T, Miyazaki H, Marchini C, Ferroni S, and Rudolphi K (1998) Pathological immuno-reactions of glial cells in Alzheimer's disease and possible sites of interference. *J Neural Transmis 54*: 167–174.

Sindou P, Couratier P, Barthe D, and Hugon J (1992) A dose-dependent increase of tau immunostaining is produced by glutamate toxicity in primary neuronal cultures. *Brain Res 572*: 242–246.

Soliven B and Albert J (1992) Tumor necrosis factor modulates Ca^{2+} currents in cultured sympathetic neurons. *J Neurosci 12*: 2665–2671.

Solomon PR, Adams F, Silver A, Zimmer J, and DeVeaux R (2002) Ginkgo for memory enhancement: A randomized controlled trial. *JAMA 288*: 835–840.

Stewart WF, Kawas C, Corrada M, and Metter J (1997) Risk of Alzheimer's disease and duration of NSAID use. *Neurology 48*: 626–632.

Tanila H, Minkeviciene R, and Banerjee PK (2003) Behavioral effects of subchronic memantine treatment in APP/PS1 double mutant mice modeling Alzheimer's disease. *J Neurochem 85*: 48.

Tremblay R, Chakravarthy B, Hewitt K, Tauskela J, Morley P, Atkinson T, and Durkin JP (2000) Transient NMDA receptor inactivation provides long-term protection cultured cortical neurons from a variety of death signals. *J Neurosci 20*: 7183–7192.

Venters HD, Dantzer RR, and Kelley KW (2000) A new concept in neurodegeneration: TNFα is a silencer of survival signals. *TINS 23*: 175–180.

Wenk GL (2003) Neurotransmitters. In L Nadel (ed.), *Encyclopedia of Cognitive Science*, pp. 2414–2421, London: Nature Publishing Group, Macmillan.

Wenk GL, Danysz W, and Mobley SL (1994) Investigations of neurotoxicity and neuroprotection within the nucleus basalis of the rat. *Brain Res 655*: 7–11.

Wenk GL, Danysz W, and Mobley SL (1995) MK-801, memantine and amantadine show neuroprotective activity in the nucleus basalis magnocellularis. *Eur J Pharmacol 293*: 267–270.

Wenk GL, Danysz W, and Roice DD (1996) The effects of mitochondrial failure upon cholinergic toxicity in the nucleus basalis. *Neuroreport 7*: 1453–1456.

Wenk GL, Hauss-Wegrzyniak B, and Willard LB (2000b) Pathological and biochemical studies of chronic neuroinflammation may lead to therapies for Alzheimer's disease. In P Patterson, C Kordon, and Y Christen (eds.), *Research and Perspectives in Neurosciences: Neuro-Immune Neurodegenerative and Psychiatric Disorders and Neural Injury.* pp. 73–77, Heidelberg: Springer-Verlag.

Wenk GL, Quack G, Möbius H-J, and Danysz W (2000a) No interaction of memantine with acetylcholinesterase inhibitors approved for clinical use. *Life Sciences 66*: 1079–1083.

Wenk GL and Willard LB (1999) The neural mechanisms underlying cholinergic cell death within the basal forebrain. *Int J Dev Neurosci 16*: 729–735.

Wenk GL, Zajaczkowski W, and Danysz W (1997) Neuroprotection of acetylcholinergic basal forebrain neurons by memantine and neurokinin B. *Behav Brain Res 83*: 129–133.

Whitehouse PJ, Price DL, Clark AW, Coyle JT, and DeLong MR (1981) Alzheimer disease: Evidence for selective loss of cholinergic neurons in the nucleus basalis. *Ann Neurol 10*: 122–126.

Willard LB, Hauss-Wegrzyniak B, Danysz W, and Wenk GL (2000) The cytotoxicity of chronic neuroinflammation upon basal forebrain cholinergic neurons of rats can be attenuated by glutamatergic antagonism or cyclooxygenage-2 inhibition. *Exp Brain Res 134*: 58–65.

Willard LB, Hauss-Wegrzyniak B, and Wenk GL (1999) The pathological and biochemical consequences of acute and chronic neuroinflammation within the basal forebrain of rats. *Neuroscience 88*: 193–200.

Wyss-Coray T and Mucke L (2000) Ibuprofen, inflammation and Alzheimer disease. *Nature Medicine 6*: 973–974.

CHAPTER 17

Enhancement of Learning and Memory Performance: Modality-Specific Mechanisms of Action

Stephen C. Heinrichs

Department of Psychology, Boston College, Chestnut Hill, MA, 02467

Many of us are anxious to find ways to improve our memories; none of us have to deal with the problem of how to forget. In S's case, however, precisely the reverse was true. The big question for him, and the most troublesome, was how he could learn to forget.

From Luria, 1968, p. 67.

I. INTRODUCTION

The first edition of *Neurobiology of Learning and Memory* described the beneficial impact in several animal species of environmental enrichment and administration of mnemonic drugs on subsequent performance of a variety of learning and memory-intensive tasks. More recently, a great deal of attention has been focused on drug, nutriceutical, and lifestyle interventions capable of enhancing learning and memory capacity (McDaniel et al., 2003; Nyberg et al., 2003; Barch, 2004). At the present time, a concerted effort is under way to commercialize these basic neuroscientific discoveries (Marshal, 2004). One overarching research goal that has been articulated for the future is to couple the explanations of learning and memory processes at the systems and cellular levels of analysis (Weeber and Sweatt, 2002; Morris et al., 2003; Dash et al., 2004). This chapter attempts to rise to this challenge by employing cognitive, neuro-anatomical, physiological, and molecular terminology to characterize several nervous system functions that are thought to be capable of enhancing learning and memory.

The main organizing theme for the chapter is represented by three cognitive function labels vital to any discussion of underlying synaptic plasticity (hence-forth *learning/memory*): (1) the awareness and acquisition of new stimulus/ response associations, episodic events, etc. (*attention/encoding*); (2) the deliberate recording and persistence of learned information (*storage/consolidation*); and (3) the recovery and expression of remembered information (*retrieval/recall*). These three labels are useful for dividing up the temporal stages in the learning/ memory sequence (early, intermediate, and late) as well as for dissociating brain sites and nervous system signaling pathways involved in mobilizing changes necessary for one or more stages of the sequence (Martinez et al., 1991; Baddeley, 1995; Tonegawa et al., 2003). Thus, while the review of relevant literature is limited largely to recent research and theoretical advances appearing since the 1998 publication of the first edition, coverage is compre-hensive, in the sense that the cognitive and neural mechanisms described herein can be thought of as least common denominators through which any valid learning/memory manipulation would likely exert its enhancing effect on performance (see Fig. 17-1).

Memory-fitness strategies to combat learning/memory loss represent a vitally important topic that has been addressed in book-length form by researchers as well as clinicians in the learning/memory field (Einstein and McDaniel, 2004; Small, 2004). However, this chapter presents an account of learning/memory function in normal, unimpaired organisms based on the assumption that the unperturbed nervous system provides the best possible context in which to establish rigorous evidence of cognitive enhancer efficacy. Morris and colleagues argue, for example, that distinct memory processes described in later sections of this chapter cannot be characterized easily by studying organisms with permanent brain damage or irreversible brain lesions (Morris et al., 2003). Even the interpretation of transient memory-impairing treatments such as protein synthesis inhibition is viewed as problematic by some investigators (Silva et al., 1998). Yet the overriding goal, as articulated elegantly by Sweatt, is to "focus on the essential, defining characteristic of the mechanism at the heart of memory" (Sweatt, 2003). Thus, the present focus on learning/memory enhancement in the normal, intact nervous system will be sharpened by not addressing the copious literature related to learning/ memory impairment, amnestic drugs/traumas, or aging-related decline in cognitive capacity (Barch, 2004; Leonard et al., 2004; Lupien et al., 2005).

The credibility of manipulations claimed to enhance learning/memory performance as well as the appropriateness of expending scientific capital and effort in this pursuit are open to question (Gerlai, 2003). It is certainly true that effective clinical treatment of dementia accompanied by memory loss represents a pressing health care goal. However, this aim is completely different in character from the use of cognitive enhancers as lifestyle supplements (Bohn et al., 2003) intended to increase performance in "intact intellects" (individuals with no identifiable pathology) in an ethically and legally dubious manner (Whitehouse et al., 1997; Farah et al., 2004; Mehlman, 2004). This is true in spite of the fact that the neurobiological mechanisms of action would presumably be shared in the therapeutic and lifestyle applications of cognitive enhancers; this circumstance further strengthens the present focus on enhancement mechanisms themselves while leaving the implementation and application of this knowledge to the reader. Moreover, in order to provide examples of nonpharmacological, noninvasive means of achieving cognitive enhancement, three stress-related, physical exercise, and state-dependent mechanisms (a.k.a. mneumonotechniques) reported to benefit performance of learning/memory tasks are also provided in this chapter as a supplement to information on pharmacological enhancement strategies (a.k.a. pro-cognitive drugs).

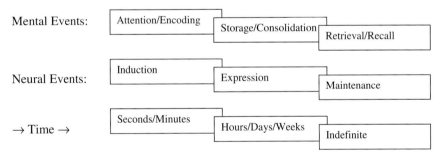

FIGURE 17-1   Schematic diagram relating the time course of mental and neural events underlying various inferred learning/memory processes. The first row labels essential components of the modal model of learning/memory performance. The second row labels presumed states of hippocampal long-term potentiation. The third row reflects evidence in multiple species for the timing of the various components of synaptic plasticity (Clayton, 2000; Dudai, 2004; Horn, 2004). The boxes overlap, and the time labels are coarse in order to reflect the lack of precise transitional boarders between the various entities. Indeed, the contribution of multiple learning/memory processes at the time of performance assessment may contribute to conflicting results in the literature (Barch, 2004). While the horizontal progression from left to right is very well worked out for each of the three rows, the vertical, column-wise linkages are only tentative and await further experimental work.

II. MECHANISMS OF ATTENTION/ ENCODING ENHANCEMENT

While therapeutic enhancement of cognitive function was once a topic for science fiction (Keyes, 1959), the neuroscience and clinical literature of the past few years provides ample documentation of basic brain mechanisms for and potential cognitive enhancement efficacy of pharmacological strategies tested in tasks sensitive to learning and memory performance (Buccafusco and Terry, 2000). The process of attention serves as an initial filter for discriminating novel stimuli during circumstances when organisms are required to shift from one perceptual dimension to another or to perform a reversal-learning sort of task (Dalley et al., 2004). The flexibility and accuracy of identifying target stimuli in a multiple-choice format is one attentional task that is dependent on prefrontal cortex function (Dalley et al., 2004). These processes reflect the earliest stages of mental activity necessary for successful learning/memory. Experimental studies have delineated later-occurring temporal phases of memory and synaptic plasticity including one short-lived form that is established soon after exposure to a novel stimulus and can be established in the absence of new mRNA and protein synthesis (Kelleher et al., 2004). This so-called *working memory* can be operationally defined as that store necessary for performing the current trial of a memory experiment but not for future trials, and like attention is also dependent on a functional prefrontal cortex (Dalley et al., 2004). It is critically important to note that treatments administered before training (i.e., initial learning) are capable of impacting all subsequent learning/memory stages (Martinez, 1986). However, little careful research is available to specify the exact point in time at which a particular cognitive enhancer mechanism is activated by an experimental treatment. Thus, learning/memory stage classifications for particular treatments listed later were assisted wherever possible according to neurobiological correlates that provide a supplemental index of the temporal characteristics of plasticity. Some pharmacological and behavioral strategies for the enhancement of attention to novel stimuli and the encoding of new memories (Fig. 17-2) are listed next.

A. Neuropharmacological Enhancements for Attention/Encoding

1. Dopamine Receptor Modulators

Discussion of pharmacological enhancement of learning/memory capacity ought to begin with a consideration of the utility of dopamine receptor modulators, given the enormous and long-standing influence of psychostimulant drug use to benefit attention and working memory (Goldman-Rakic, 1998; Leonard et al., 2004). Administration in monkeys of low doses of dopamine

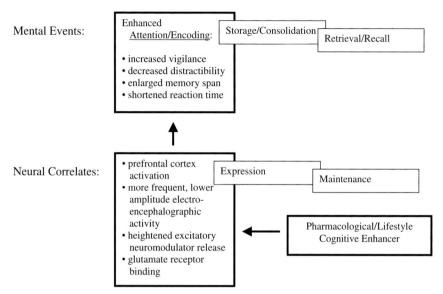

FIGURE 17-2   Schematic diagram relating the mental and neurobiological correlates of attention and encoding enhancement. Note that later-occurring mental and neural event labels are retained since there are likely proactive effects of attention/encoding enhancement on storage/consolidation and retrieval/recall of the same memory.

receptor agonists that result in an optimization of dopamine activation can improve performance of a working-memory task (Williams and Goldman-Rakic, 1995). This finding generalizes to healthy human participants administered nonselective dopamine receptor agonists, such as amphetamine and methylphenidate, which act to increase accuracy or shorten response latency in working-memory tasks (Barch, 2004). One study employed brain imaging to examine physiological correlates of the effects of dextroamphetamine on working-memory performance in healthy controls (Mattay et al., 2000). The catecholamine psychostimulant increased prefrontal cortex activation during a working-memory task performed at mnemonic capacity. However, dextroamphetamine improved performance only in participants who had relatively low working-memory capacity at baseline (Mattay et al., 2000). The degree of task-related enhancement of dorsolateral prefrontal cortex activity in response to amphetamine administration was associated with larger improvement in accuracy. Methylphenidate administration in healthy adults is also reported to enhance performance in a spatial pointing task, although marked individual differences are also reported (Leonard et al., 2004). Rather than direct mediation of the learning/memory trace, the nonspecific mechanism of cognitive enhancement via dopaminergic activation is presumed to be elevation of reward expectancy (Rossetti and Carboni, 2005). Thus, the hypothesis that dopamine-augmenting agents can improve working memory is supported at the present time, and current research is focused on identifying dopamine receptor–selective agonists suitable for human administration (Barch, 2004).

2. Glutamate Receptor Modulators

A well-accepted model for a likely neurobiological mechanism involved in encoding at the earliest stages of learning involves two classes of transmitter receptors colocalized at excitatory, glutamatergic synapses (Horn, 2004). AMPA-type receptors generate depolarizing currents needed to disinhibit voltage-sensitive NMDA-type receptors, which then admit calcium into the dendrites on the postsynaptic side of the synapse (Lynch, 2002). One view is that increases in AMPA receptor number, recycling, or efficiency contribute to enhanced postsynaptic currents that define long-term potentiation, but only for a brief period of time (perhaps 30 minutes or less), following task acquisition/encoding.

So-called *ampakines* are allosteric modulators of AMPA receptors that enhance and prolong synaptic currents generated by release of glutamate from axon terminals (Lynch, 2002). Positive modulation of AMPA receptors can potentially enhance cognition by, first, offsetting losses of glutamatergic synapses; second, by promoting synaptic plasticity; and third, by increasing the production of trophic factors (Lynch, 2004). Ampakines affect only those AMPA receptors activated by endogenously released glutamate and thus only target active circuits. This functional selectivity is bolstered by the absence of ampakine sites of action outside of the central nervous system. The advent of small molecules that selectively enhance AMPA receptors in the brain has made it possible to test these hypotheses. For instance, a small-molecule ampakine capable of facilitating glutamate release in vitro is reported to improve significantly performance of a delayed matching-to-sample task by young adult rhesus monkeys, perhaps by accelerating memory encoding (Buccafusco et al., 2004). Ongoing clinical work (Marshal, 2004) reveals that young adult participants exhibit small-to-moderate improvements in tests of visual association, odor recognition, and visuospatial maze acquisition (Lynch, 2002).

3. Adenosine Receptor Antagonists

Caffeine is the most widely consumed central-nervous-system stimulant (Nehlig et al., 1992). Methylxanthines including caffeine act as adenosine receptor antagonists to activate noradrenaline neurons and alter the local release of dopamine. Behavioral measurements indicate a general improvement in the efficiency of information processing after caffeine consumption, while electro-encephalographic data support the general belief that caffeine acts as a stimulant (Lorist and Tops, 2003). Studies using event-related potentials to measure the timing and amplitude of cortical reactivity to stimulus presentation indicate that caffeine has an effect on attention, which is independent of specific stimulus characteristics. Thus, the effects of caffeine on learning/memory performance are likely related to methylxanthine enhancement of arousal and vigilance.

One study tested this potential link between consumption of caffeinated beverages and arousal (Ryan et al., 2002). Memory performance depends on the time of day, with performance being optimal early in the morning and declining during the late afternoon hours, and it is possible to examine whether this decline is ameliorated by the adenosine receptor antagonist caffeine (Ryan et al., 2002). Adults over the age of 65 who considered themselves "morning types" were tested twice using a list-learning test requiring both free recall and recognition, once in the morning and once in the late afternoon. Participants who ingested decaffeinated coffee showed a significant decline in memory performance from morning to afternoon. In contrast, those who ingested caffeine showed no decline in performance from morning to afternoon. The results suggest that time-of-day effects may be mediated by nonspecific changes in level of arousal (Ryan et al., 2002). This finding is not surprising from an epidemiological point of view, although it may be difficult to disentangle learning/memory modulatory effects of caffeine from self-medicating behaviors in a human population that is increasingly caffeine dependent or at least well experienced with the psychoactive efficacy of caffeine (Nehlig, 1999).

4. Nicotinic Receptor Agonists

The nicotinic acetylcholine receptor is a ligand-gated, Ca^{2+}-permeable channel that facilitates neurotransmitter release at presynaptic sites in the central nervous system (Hejmadi et al., 2003). Of particular interest for the study of synaptic plasticity, Ca^{2+} entry through nicotinic receptors on hippocampal mossy fiber terminals elicits bursts of excitatory postsynaptic currents. Nicotinic receptor activation also promotes a Ca^{2+}-dependent second-messenger cascade (i.e., protein kinase, ERK/MAPK, and CREB) that participates in long-term potentiation. Given the broad-spectrum transmitter and second-messenger modulatory actions of nicotine (Hejmadi et al., 2003), it is not possible to discern if some or all of these events participate in cognitive enhancement. However, nicotine-like ligands exert in several species a wide range of behavioral effects, including improvements in a variety of cognitive tasks, whereas nicotine receptor antagonists impair these functions (Rezvani and Levin, 2001). Nicotinic agonists are reported to be effective in reducing distractibility (i.e. facilitating attention to the task at the earliest possible stage of learning/ memory) in young adult animals (Buccafusco and Terry, 2000). Administration in mice of a β4-nicotinic receptor agonist is reported to be dose-dependently effective in enhancing working memory in a delayed nonmatching-to-place task using an eight-arm radial maze (Bontempi et al., 2003). The weight of evidence suggests that the nicotinic mechanism of action is sufficient to produce enhancement at the earliest processing stages of attention/encoding, although intermediate- and late-stage facilitation of storage/consolidation via facilitation of protein synthesis could also play an important modulatory role as time progresses.

In clinical studies, nicotine is known to increase cortical arousal, as measured electroencephalographically, which is thought to be associated closely with the quality of attentional efficiency. In order to elucidate the neural correlates of cognitive effects of nicotine, one study examined behavioral performance and blood oxygenation-dependent regional brain activity, using functional magnetic resonance imaging, during a working-memory task in healthy nonsmoking males after the administration of nicotine or saline (Kumari et al., 2003). Nicotine, compared to placebo, improved accuracy and shortened response latency under heavy memory load conditions. Nicotine activated the anterior cingulate, superior frontal cortex, and superior parietal cortices and midbrain tectum in all active conditions and the parahippocampal gyrus, cerebellum, and medial occipital lobe during a rest period. These observations point to altered neuronal activity in a distributed neural network associated with attention and arousal systems as a mechanism mediating nicotine enhancement of attention and working memory in humans (Kumari et al., 2003). Moreover, the hemispheric lateralization of activation as a function of nicotine dependence suggests that chronic exposure to nicotine or withdrawal from nicotine affects cognitive strategies used to perform a working-memory task (Ernst et al., 2001).

5. Neurosteroids

Neurosteroids, synthesized in the central and peripheral nervous systems from cholesterol or steroidal precursors (Baulieu et al., 2001), can alter rapidly neuronal excitability by nongenomic modulation of GABA and glutamate neurotransmission (Paul and Purdy, 1992). Two neurosteroids, pregnenolone sulfate (PREGS) and dehydroepiandrosterone (DHEAS), act as antagonists at GABA-A receptors and positively modulate NMDA receptor responses (Bergeron et al., 1996; Maurice et al., 1997). Evidence suggests a role for PREGS and DHEAS in improving performance of hippocampally mediated memory tasks, such as spatial recognition (Pallares et al., 1998; Darnaudery et al., 2000), Y-maze alternation (Mathis et al., 1996; Akwa et al., 2001), visual discrimination go/no-go (Meziane et al., 1996), and motivated lever-press learning (Mathis et al., 1996). PREGS and DHEAS are both effective at increasing learning and retention when administered pre- and posttraining, but not when administered just prior to retention testing in a passive-avoidance paradigm (Reddy and Kulkarni, 1998).

Several mechanisms could contribute to the promnestic actions of PREGS. One possibility is that PREGS enhances central cholinergic function, based on the observation that administration of PREGS in the nucleus basalis magnocellularis, the main source of cortical cholinergic innervation, improves memory performance of young rats (Pallares et al., 1998). Additionally, central administration of PREGS increases extracellular acetylcholine concentrations in the hippocampus (Vallee et al., 1997; Darnaudery et al., 2000). PREGS enhances NMDA-activated currents and inhibits GABA-mediated currents in cultured rat hippocampal neurons (Bowlby, 1993). These in vitro results are consistent with neuronal excitatory and convulsant effects of PREGS in vivo (Majewska et al., 1989). In addition, PREGS could influence NMDA and GABA-A receptor functions by a nonspecific action such as altering membrane fluidity (Nilsson et al., 1998). Pretraining efficacy in learning/memory contexts as well as glutamatergic mediation of neurosteroid actions are thus consistent with the classification of neurosteroids as attention/encoding enhancers.

6. Estrogen

The estrogens are a family of steroid hormones that regulate and sustain female sexual development and reproductive function. Besides affecting the hypothalamus and other brain areas related to reproduction, ovarian steroids have widespread effects throughout the brain, on serotonin pathways, catecholaminergic neurons, the basal forebrain cholinergic system, and the hippocampal formation. Ovarian hormones regulate synapse turnover in the CA1 region of the hippocampus during the four- to five-day estrous cycle of the female rat (Woolley et al., 1990). Formation of new excitatory synapses is induced by estradiol, involves NMDA receptors, and is mediated by acetylcholine (Daniel and Dohanich, 2001). It is also likely that estrogens locally regulate events at the sites of synaptic contact in the excitatory pyramidal neurons where the synapses form (McEwen, 2002). Estrogen interacts with the rat cholinergic system in numerous ways, such as enhancing cortical cholinergic enervation and preserving synaptic density following excitotoxic lesions in the basal forebrain (Horvath et al., 2002). Hippocampal long-term potentiation is facilitated by increased levels of circulating estrogen, as evidenced by the finding that cyclical changes in endogenous estrogen levels can augment long-term potentiation (Cordoba, Montoya, and Carrer, 1997; Good et al., 1999).

In rats, intrahippocampal infusions of estradiol potentiate acetylcholine- and glutamate-mediated memory retention in an avoidance-learning task (Farr et al., 2000). Estrogen administration results in improved performance in avoidance (Singh et al., 1994) and Morris water maze tasks (O'Neal et al., 1996). In intact male and female mice, chronic estrogen treatment improves radial arm maze working-memory performance (Heikkinen et al., 2002).

Additionally, estrogen-mediated improvement in radial maze working-memory is dependent on acetylcholine acting through M2 muscarinic receptors to increase NMDA receptor binding in the hippocampus (Daniel and Dohanich, 2001). These studies support the hypothesis that estrogen is a regulator of learning-related mechanisms. Further, evidence suggests that estrogen increases NMDA receptor activity, and this is likely a further mechanism through which it enhances long-term potentiation (Gureviciene et al., 2003). Broad involvement of estrogen in modulating synaptic plasticity is consistent with classification as an attention/encoding enhancer.

B. Everyday Attention/Encoding Enhancer: Stress–Cognition Axis

An introductory psychology textbook could characterize the adaptive relationship between stressor exposure and learning/memory functions by stating in so many words that "one who learns to run away lives to learn another day!" Specialist researchers also highlight the efficacy of emotionally salient stimuli in evoking long-lasting or particularly intense memories (Cahill, 2003). Moreover, the physiological and psychological consequences of acute and chronic activation of the hypothalamus, amygdala, and pituitary/adrenocortical glands are well-known modulators of learning/memory performance across the life span (Lupien et al., 2005). Noteworthy for the present discussion is the ability of these mechanisms to facilitate learning/memory of particularly salient, affectively significant events (McGaugh, 2003).

1. Hypothalamo-Pituitary-Adrenocortical Peptides and Steroids

Stress and behavioral plasticity are interrelated; levels of alertness correspond to success in performance of a learning task in what can be described as the "stress–cognition" axis (Heinrichs, 1999). Stressors, circulating stress-related hormones, and central nervous system releasing factors that facilitate the pituitary-adrenocortical cascade all modulate learning/memory processes. One study characterized the action of an acute immobilization stressor on hippocampus-dependent learning and synaptic plasticity in the mouse hippocampus (Blank et al., 2003). Acute stress facilitated long-term potentiation of population spikes and enhanced context-dependent fear conditioning. Due to the involvement of hippocampus and induction of long-term potentiation, which are characterstic of early-stage learning/memory processing, stressor exposure can be classified as an attention/encoding enhancer. However, acute stress-induced enhancement of long-delay retrieval performance has also been demonstrated following exposure to a species-typical social stressor (Fig. 17-3).

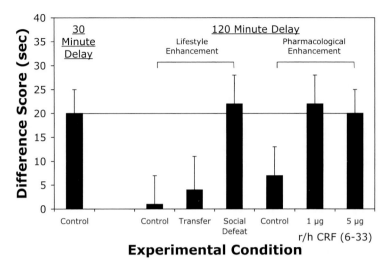

FIGURE 17-3 Social defeat and stress neuropeptide activation enhance social recognition memory. Female rats taken directly from the home cage, transferred to a holding cage, exposed previously to a social defeat stressor, or administered 0-, 1-, or 5-μg doses of the CRF-binding protein ligand inhibitor were allowed to investigate a juvenile conspecific. The duration of adult exploration of the juvenile on first presentation relative to the second presentation of the same juvenile either 30 minutes later (short delay) or 120 minutes later (long delay) was used to compute a difference-score measure (mean ± SEM) of social recognition memory. The attenuation of poor performance in the long-delay condition by prior exposure to the social defeat stressor or two doses of CRF-binding protein ligand inhibitor was used as the index of memory enhancement (Heinrichs, 2003). Cognitive enhancers are often evaluated using long training-to-test intervals (i.e., at the mnemonic limit), as in the present case, so that the degree of baseline performance restoration can be assessed (Buccafusco and Terry, 2000).

The glucocorticoids are a group of adrenocortical steroid hormones whose metabolic effects include stimulation of gluconeogenesis, increased catabolism of proteins, mobilization of free fatty acids, and potent inhibition of the inflammatory response. In addition, the effect on learning capacity of chronic activation of the hypothalamo–pituitary–adrenal axis has been characterized using long-term peripheral administration of glucocorticoids in mice, rats, monkeys, and humans (Lupien et al., 2005). In particular, glucocorticoid administration alters acquisition of a previously unlearned task in a dose-related, inverted-U-shaped fashion (de Kloet et al., 2002). The amygdala, which expresses high levels of adrenal steroid receptors, is a malleable brain structure that is important for certain types of learning and memory (McEwen and Chattarji, 2004). Repeated stress promotes behavioral changes, such as enhancement of fear and aggression, that can be associated with this brain structure. At a cellular level, fear conditioning elicits growth and remodeling of dendrites in the lateral amygdala (McEwen and Chattarji, 2004). Thus, short-term exposure to physiological levels of exogenous glucocorticoids could be expected to enhance performance in a learning and memory context, and this hypothesis is supported by animal and human clinical studies (Buchanan and Lovallo, 2001; McGaugh and Roozendaal, 2002). One prediction from the correlational link between the level of arousal and performance of learned behaviors is that intrinsic overactivation and/or long-term stimulation of neurobiological and endocrine substrates of the stress response would have the effect of producing learning and memory deficits, and this corollary hypothesis is also supported by available data (Lupien et al., 2005).

Corticotropin-releasing factor (CRF) is recognized widely as part of a neuropeptide system whose activation is a necessary component of the biological response to stressor exposure (Heinrichs and De Souza, 2001). Evidence supports a physiological role for CRF systems in information-processing functions of the central nervous system. First, steady-state levels of endogenous CRF family neuropeptide receptor agonists appear sufficient to modulate learning/memory functions, since pharmacological dissociation of CRF and a related neuropeptide, urocortin, from their binding protein enhances performance (Fig. 17-3) in appetitively and aversively motivated memory tasks (Behan et al., 1995a; Heinrichs et al., 1997; Eckart et al., 1999; Zorrilla et al., 2001). In addition, central CRF administration exerts electrophysiological and neurochemical activation of hippocampal circuits relevant for learning/memory processes in several species (Bonaz and Rivest, 1998; Wang et al., 1998; Fuchs et al., 2001). These findings suggest that CRF activation is sufficient to ensure that the early states of learning/memory plasticity are set in motion (Fig. 17-4). This conclusion is supported by results indicating that CRF, adrenocorticotropic hormone, and glucocorticoids continue to be significant modulators of learning and memory processes when either the organism or the treatment itself is rendered incapable of pituitary-adrenocortical activation (Honour and White, 1988). Thus, peptides and steroid hormones of the HPA axis, such as CRF and glucocorticoids, are presumed to be the neurochemical modulators of enhanced long-term memory for stressful or emotionally arousing experiences (Roozendaal, 2002). The pharmacological (Behan et al., 1995b), neurobiological (Radulovic et al., 2000), and clinical (Bernardi et al., 2000) evidence necessary to support this claim convincingly is only now being assembled.

III. MECHANISMS OF STORAGE/ CONSOLIDATION ENHANCEMENT

The process of postacquisition stabilization of long-term memory, labeled *consolidation*, is still germain, in spite of the age-old vintage of this concept (McGaugh, 2000; Dudai, 2004). The term *systems consolidation* can be defined as the process by which memory becomes independent of the hippocampus,

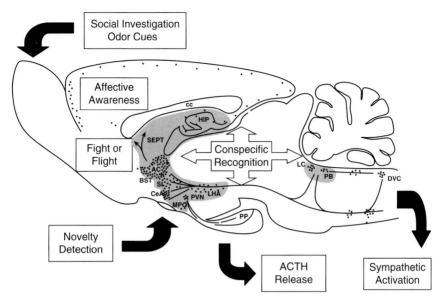

FIGURE 17-4   Schematic diagram of CRF/urocortin brain activity in which the rat detects a conspecific olfactory cue that is recognized as a familiar/unfamiliar juvenile during social investigation while performing the social recognition memory task. CRF/urocortin-mediated efferent responses to threat could include hypophysiotropic release of ACTH, amygdalo-medullary increases in heart rate, septo-hippocampal avoidance learning, and septo-amygdalar anxiogenic-like behavior (Heinrichs and Koob, 2004). AC, anterior commissure; ACTH, adrenocorticotropic hormone; BST, bed nucleus of the stria terminalis; cc, corpus callosum; CeA, central nucleus of amygdala; DVC, dorsal vagal complex; HIP, hippocampus; LC, locus coeruleus; LHA, lateral hypothalamus; MPO, medial preoptic area; PP, posterior pituitary; PVN, paraventricular nucleus; SEPT, septum; SI, substantia innominata.

whereas *cellular consolidation* is defined as the transition of memory traces from protein synthesis and gene expression–dependent states to independence (Dash et al., 2004; Dudai, 2004). The hallmarks of the consolidation process are (1) relocation from short-term hippocampal memory stores to distributed neocortical networks and (2) the gradual, time-dependent process of laying down long-term memories, in which the most recent memories are the most fragile (Sara, 2000). The intermediacy of storage/consolidation can be further delineated by stipulating that this learning/memory stage does not employ any sensorimotor faculties or rely on short-term memory, as does the attention/encoding stage described earlier (Dudai, 2004). The mechanism for storage is hypothesized to be the transit of synthesized proteins via axonal transport to, or induction of gene products in, extra-hippocampal synapse assemblies that are distinguished from all other potential neural assemblies by prior synaptic activity in the tagged locations. Activation of cell-signaling cascades and

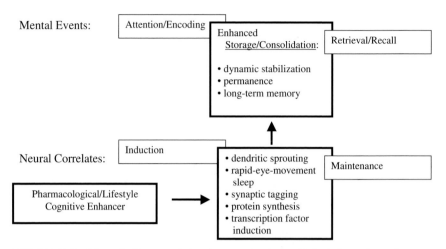

FIGURE 17-5   Schematic diagram relating the mental and neurobiological correlates of storage and consolidation enhancement. Note that earlier- and later-occurring mental and neural event labels are retained, since there are likely proactive, carryover effects of storage/consolidation enhancement on retrieval/recall of the same memory as well as carryover effects of original storage/consolidation on subsequently occurring learning/memory events.

phosphorylated transcription factors, termed the *genomic action potential* (Clayton, 2000), appear to be proximal mediators of cellular consolidation processes that can be identified in relevant hippocampal, limbic, and cortical regions following conditioning in animals. The elegance of this dynamic, distributed information-storage mechanism provides a fitting resolution to Lashley's experimental quandary in which a localized "engram" proved difficult to isolate using progressive cortical ablation methodologies (Lashley, 1958). Some pharmacological and behavioral strategies for enhancing storage and consolidation of memories (Fig. 17-5) are listed next.

A. Neuropharmacological Enhancements for Storage/Consolidation

The use of treatments administered shortly after training to enhance memory provides a highly effective method of influencing memory consolidation without necessarily affecting either attention/encoding or retrieval/recall (McGaugh, 2000). For example, daily pretraining administration in rats of the general stimulant strychnine increases the rate of maze learning in rats (McGaugh and Krivanek, 1970). The possibility that apparent increases in learning/memory were attributable to drug-induced increases in running speed or other performance changes characteristic of acute drug intoxication (including potential enhancement of early-state attention/encoding processes) were ruled out by the finding that posttraining administration of strychnine also facilitated learning performance during the next day's trial, when rats were tested in a drug-free state.

1. Adrenergic Neurotransmission

Epinephrine and drugs that activate adrenergic receptors enhance memory for many kinds of training experiences (McGaugh, 2000). For example, β-adrenergic agonists infused after training either systemically or directly into the basolateral amygdala enhance learning/memory performance. The literature also supports the ability of α_2 noradrenergic receptor agonists to facilitate learning/memory performance. For instance, administration of the α_2 noradrenergic receptor agonist clonidine in healthy adults improves performance in a spatial pointing task in a dose- and practice-dependent manner (Coull et al., 1997). Pharmacological enhancement via noradrenergic receptor agonist treatment is likely mediated by activation of a locus coeruleus to forebrain noradrenergic pathway associated with increased accuracy of response to task-relevant stimuli (Aston-Jones et al., 2000). Levels of noradrenaline in prefrontal cortex increase when rats perform correctly in a delayed variant of a spatial working-memory task (Rossetti and Carboni, 2005). Tellingly, noradrenaline levels were markedly enhanced in animals trained to alternate as compared with rats that acquired the spatial information about the location of food in the maze but were untrained to make a choice to obtain the reward. One study investigated the effect of enhanced noradrenergic activity on memory consolidation in humans (Southwick et al., 2002). Thirty participants viewed a series of slides that depicted an emotionally arousing story. Multiple blood samples were drawn for determining plasma levels of the noradrenaline metabolite MHPG. One week later, participants completed an unannounced memory test for the slides. Linear regression revealed a significant effect of MHPG on memory score for the group as a whole. These findings strengthen support for the hypothesis that enhanced memory for emotionally arousing events in humans depends critically on postlearning adrenergic modulation (Southwick et al., 2002). A more general conclusion is that noradrenergic neurotransmission is required for active maintenance and flexible manipulation of learned information during successful goal-seeking behavior. The classification of adrenergic modulation as a storage/consolidation enhancer is based on nonhippocampal distribution of noradrenergic circuitry in the brain as well as the relatively large literature describing posttraining efficacy of noradrenergic receptor ligands.

2. Trophic Factors

Nerve growth factor (NGF) is a multimeric protein, the beta subunit of which is required for the proper development and maintenance of the sensory neurons of the dorsal root ganglion and of the postganglionic sympathetic neurons. NGF-induced facilitation of learning/memory performance is supported by efficacy of NGF treatment in a variety of animal species and testing contexts. In a classical fear-conditioning task, endogenous NGF is reported to increase one week after training (i.e., during the consolidation phase), while infusion of antisense for the NGF receptor (TrkA) in the hippocampus one week post-training can block this retention (Woolf et al., 2001). In developing CD-1 mice, a single intracerebroventricular administration of NGF at postnatal day 15 resulted in adultlike spatial novelty discrimination in males but not females tested at postnatal day 18, although increased choline acetyltransferase activity was observed in both sexes as a result of NGF treatment (Calamandrei et al., 2002). The link between NGF and cholinergic activity is supported by the observation that the effects of NGF on recent memory in the delayed non-matching-to-position task correlate with changes in the cholinergic system, including increased size of cholinergic neurons and a change in the terminal fields of these same neurons (Gustilo et al., 1999). Taken together, these results suggest that central administration of exogenous NGF can facilitate learning/memory consolidation while remodeling cholinergic brain areas thought to subserve synaptic plasticity.

Brain-derived neurotrophic factor (BDNF), an NGF-related neurotrophin with high affinity for the TrkB receptor, is known to have numerous roles in learning and memory and contributes to the process of hippocampal long-term potentiation (Tyler et al., 2002; Yamada et al., 2002). Both long-term potentiation and spatial learning are associated with increased phosphorylation of TrkB (BDNF receptor) and extracellullar signal-regulated kinase (ERK) in the dentate gyrus following administration of BDNF (Gooney et al., 2002). Although it is still unclear whether BDNF exerts housekeeping functions to maintain neuronal functioning, BDNF appears to play an important role in long-term potentiation induction and modulation (Kovalchuk et al., 2002; Messaoudi et al., 2002). The specific mechanism of BDNF-mediated long-term potentiation, which is induced postsynaptically (Kovalchuk et al., 2002), suggests that BDNF interacts directly with NMDA receptors to increase their activity (Mizuno et al., 2003). Mnemonic effects of BDNF are found in rodents, in primates, where the peptide is up-regulated in the inferior temporal cortex during visual pair-association learning (Tokuyama et al., 2000), and in day-old chicks, where BDNF antisense administration impairs memory consolidation in a one-trial inhibitory-avoidance paradigm (Johnston and Rose, 2001). Lee and colleagues argue for a specific role of BDNF in long-term learning/memory consolidation, based on the ability of BDNF antisense administration into dorsal hippocampus to impair performance in a contextual fear-conditioning task (Lee et al., 2004). In contrast, BDNF knockdown had no effect on the encoding of associative memory because short-term memory was normal three hours after training. Thus, the weight of evidence for NGF and BDNF efficacy when administered posttraining, often with a substantial delay, and a general neural modulatory role independent of the hippocampus allows classification of trophic factors as putative storage/consolidation enhancers.

3. Cholinergic Neurotransmission

For more than 20 years, the ability of drugs, such as physostigmine, that enhance synaptic levels of acetylcholine to facilitate learning/memory recall has been recognized (Deutsch, 1983; Robbins et al., 1997). One report describes a randomized, double-blind, parallel group, placebo-controlled study to test the effects of the acetylcholinesterase inhibitor donepezil on aircraft pilot performance in healthy, middle-aged, licensed pilots (Yesavage et al., 2002). After 30 days of treatment, the donepezil-treated group showed greater ability to retain the capacity to perform a set of complex simulator tasks than the placebo group. Thus, donepezil appears to have beneficial effects on retention of training of complex aviation skills in nondemented, older adults. The hypothesis that enhancement of cholinergic transmission facilitates learning/memory processes (typically referred to as the *cholinergic hypothesis* when referencing therapeutic approaches for dementing disorders) is supported by a broad efficacy of cholinomimetic agents testing in learning/memory contexts (Buccafusco and Terry, 2000). Cholinergic enhancement also facilitates visual attention by increasing activity in extrastriate and prefrontal cortices (Furey et al., 2000; Bentley et al., 2004); see also the earlier section on "Nicotinic Receptor Agonists" as attention/encoding enhancers. High levels of acetylcholine present during active encoding of new information in the hippocampus suggest a role for cholinergic neurotransmission in learning/memory consolidation (Hasselmo, 1999).

B. Everyday Storage/Consolidation Enhancer: Exercise/Activity

Studies of adult animals indicate that metabolic and neurochemical functions improve with aerobic fitness. For example, the effects of physical activity on hippocampal cholinergic function, parietal cortical cholinergic function, and spatial memory have been examined in rats (Fordyce and Farrar, 1991). Three weeks prior to the end of the 14-week chronic treadmill-running protocol, a group of chronic-run rats and their nonrun controls were tested on a stringent version of a place-learning-set task. Chronic-run rats exhibited enhanced performance on the spatial task by significantly reduced second-trial latencies and elevated first- and second-trial proximity ratio scores. Chronic-run rats tested for spatial memory also showed enhanced hippocampal high-affinity choline uptake and muscarinic receptor binding (Fordyce and Farrar, 1991). An additional study from this group supported the hypotheses of enhanced hippo-campal PKC activity in spatial learning and enhancement of spatial learning performance in rodents by physical activity (Fordyce and Wehner, 1993). These data indicate that chronic physical activity improves spatial learning performance, which is correlated with enhancement of hippocampal plasticity.

A review of studies assessing the effects of acute bouts of physical activity on cognitive performance in healthy adults reveals that submaximal aerobic exercise performed for periods up to 60 minutes in duration facilitates specific aspects of information processing (Tomporowski, 2003). One study followed 124 previously sedentary adults over a six-month period of either aerobic exercise (walking) or anaerobic exercise (stretching and toning) (Kramer et al., 1999). Participants who received aerobic training showed substantial improvements in performance on tasks requiring executive control (e.g., task switching and the ability to ignore task-irrelevant stimuli) relative to the anaerobic comparison group. In support of the role of tropic factors in synaptic plasticity, long-lasting expression of BDNF and TrkB in cerebellum, motor cortex, and hippocampus can be induced by exercise and complex acrobatic training (Klintsova et al., 2004). Moreover, voluntary wheel-running exercise can reverse a high-fat-diet-induced decrement in BDNF and its downstream plasticity effectors (Molteni et al., 2004). Thus, physical activity appears to enhance learning/memory performance in a cholinergic and neurotrophic factor–dependent manner, consistent with a storage/consolidation mechanism of action.

IV. MECHANISMS OF RETRIEVAL/ RECALL ENHANCEMENT

A brainwide distributed network orchestrates the recall and retrieval of explicit memory (i.e., memory of facts and events). The network was initially identified in humans and is being investigated systematically in molecular/genetic, single-unit, lesion, and imaging studies in animals (Miyashita, 2004). The unique association between environmental stimuli and context depends on neural acti-vation in the medial temporal lobe (i.e., hippocampus and associated regions), whereas memory traces representing repeated associations reside in domain-specific regions in the temporal cortex. These regions are reactivated during remembering and contribute to the contents of a memory (Fig. 17-6). Note that the functional neuroanatomy of human memory retrieval based on brain-imaging studies is unexpected, based on evidence derived from brain-lesion studies (Fletcher et al., 1997), hence supporting the focus on the healthy, normal nervous system in the present chapter and highlighting the need for convergent neuropharmacological evidence described next.

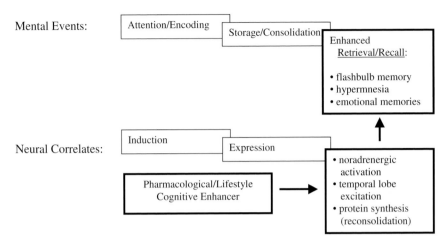

FIGURE 17-6 Schematic diagram relating the mental and neurobiological correlates of retention and recall enhancement. Note that earlier-occurring mental and neural event labels are retained, since there are likely carryover effects of original retrieval/recall enhancement on subsequently occurring learning/memory events.

A. Neuropharmacological Enhancement of Retrieval/Recall

There are relatively few pharmacological studies of direct effects of drugs on memory retrieval (Sara, 2000). Agents reported to exert retrieval enhancement include strychnine, cocaine, nootropics such as piracetam, and nicotine (three of these four agents were discussed in previous sections of this chapter). Efficacy can be selective in the case of amphetamine, which facilitates retrieval of a forgotten maze task, without affecting task performance when administered prior to or immediately following acquisition trials. The hypothesized mechanism of action for psychostimulant drug–induced facilitation of retrieval performance is increased arousal or vigilance (White and Salinas, 1998). Some pharmacological and behavioral strategies for enhancement of retrieval or recall of learned information (Fig. 17-6) are listed next. Note that coming as it does at the end of a long sequence of learning/memory stages, retrieval/recall enhancement is increasingly difficult to dissociate from earlier activated learning/memory mechanisms — hence a relatively diminished number of retrieval/recall enhancers are discussed and in a more guarded fashion at that.

1. Serotonin Neurotransmission

Substantial evidence indicates that serotonin receptors are involved in the regulation of acetylcholine release in brain regions important to mnemonic processes, and they may thus be exploited pharmacologically as targets for memory improvement (Terry et al., 1996). In a series of studies, potent serotonin receptor agonists have been evaluated for potential memory-enhancing effects in macaque monkeys trained to perform a delayed matching-to-sample task (Terry et al., 1998, 2004). For example, serotonin 5HT3 receptor antagonists and 5HT4 receptor agonists appear to be efficacious cognitive enhancers (Terry, 2004). Improvements above baseline are typically observed in medium- and long-delay conditions, suggesting that the mechanism of serotonin receptor stimulation is selective enhancement of retention under conditions of poor performance.

B. Everyday Storage/Consolidation Enhancer: State-Dependent Retrieval

Drug-induced-state–dependent learning as well as similar effects on memory retrieval exercised by physiological states have been known since 1830 (Overton, 1991). The main finding is that memories can be retrieved only when a common, particular emotional or neurohumoral state is established both at the time of acquisition and at the time of expression (Izquierdo, 1989). In so-called *state-dependent* retrieval, awareness of a discriminable physiological state is efficacious in enhancing retention of information learned previously while in the same physiological state. Pharmacological cues such as psychostimulant, sedative, and opiate drugs are quite effective in enhancing retention of an arbitrary learning task if present at the time of retention testing, relative to the poor retention demonstrated during retention testing in the absence of the drug. For example, morphine-induced state-dependent learning has been investigated in animal models. Pretraining administration of morphine dose-dependently decreased the learning of a one-trial passive-avoidance task, whereas pretest administration of morphine-induced state-dependent retrieval of the memory acquired following pretraining administration of morphine (Zarrindast and Rezayof, 2004). Moreover, successful retrieval appears to become increasingly morphine dependent with repetition (Bruins, Slot, and Colpaert, 2003). Endogenous stress-related hormones, such as ACTH, vasopressin, and epinephrine, are effective in creating an easily discriminable internal state that can exert state-dependent retrieval benefits on performance (a.k.a. mood-congruent effect), although higher-order state variables, such as social rank and aerobic exercise, are also reported to be effective retention enhancers (Miles and Hardman, 1998; Barnard and Luo, 2002). Thus, state-dependent learning is a phenomenon in which the retrieval of newly acquired information is possible only if the organism is in the same sensory context and physiological state as during the encoding phase.

A cellular mechanism for the extensive behavioral and pharmacological characterization of the state-dependent-learning phenomenon has recently been proposed. In this model, state-dependent learning is imposed in cortical neurons that exhibit acetylcholine-induced functional plasticity (Shulz et al., 2000). This was demonstrated using neurons of rat somatosensory cortex tuned to the temporal frequency of whisker deflections. Pairing whisker stimulation with acetylcholine applied iontophoretically yielded selective lasting modification of responses, the expression of which depended on the presence of exogenous acetylcholine. Administration of acetylcholine during testing revealed frequency-specific changes in response that were not expressed when tested without acetylcholine or when the muscarinic antagonist atropine was applied concomitantly (Shulz et al., 2000). These results suggest that learning/memory recall can be controlled by the cortical release of acetylcholine.

V. CONCLUSIONS AND FUTURE DIRECTIONS

An increasing number of structurally heterogeneous compounds, which may act via very different neuronal mechanisms, have been proposed to facilitate attention and acquisition, storage, and retrieval of information (Sarter, 1991). The cognitive enhancers just described fall into two general classifications of action: (1) direct mediators, such as glutamatergic agents, which are capable of adjusting the machinery of synaptic plasticity in which the mediator participates directly, and (2) modulators, such as adrenergic agents, which modify diffuse networks of neurons to adjust up or down the induction or expression of synaptic plasticity in an indirect manner without comprising a necessary component of the machinery of that plasticity (Clayton, 2000). It should be noted that a host of other putative cognitive-enhancement mechanisms that were omitted from this chapter likely exert efficacy via common upstream learning/memory modulator systems in a nonspecific fashion. For example, one could rightly suppose that direct neuropharmacological actions of treatments that regulate energy balance, such as neuropeptide Y and insulin, exert their long-term effects on cognitive function by indirect stimulation of glucose availability (Wenk, 1989). It would *not* be appropriate to conclude, however, that learning/memory performance in a reaction-time task was enhanced or impaired due to ergogenic or paralytic effects, respectively, of a treatment on muscle reactivity. This problem of specificity has been called one of the most difficult in learning and memory research (Martinez et al., 1983; Lombardi and Weingartner, 1995; Barco et al., 2003) and can be pursued in the future by defining the number and identity of unique mechanisms in local brain areas for optimizing learning and memory capability separate from other cognitive functions (Lombardi and Weingartner, 1995). Systematic coupling of functional efficacy with known neurobiological mechanisms for cognitive enhancement is one strategy employed in this chapter to work toward this goal (see Table 17-1).

TABLE 17-1   Prominent Anatomical, Physiological, Neurochemical, and Molecular Mediators of Early-, Intermediate-, and Late-Stage Learning/Memory Enhancement

Substrate	Attention/Encoding	Storage/Consolidation	Retrieval/Recall
Anatomical	Prefrontal cortex	Basolateral amygdala	Medial temporal cortex
Physiological	Presynaptic facilitation	Long-term potentiation	Metabolic activation
Neurochemical	Nicotinic receptors	NMDA/AMPA receptors	Noradrenergic receptors
Molecular	Immediate-early genes	CREB	MAPK

Of particular concern to the applied scientist is the process of rational target selection from among the many candidate learning/memory mediators/modulators known presently to be efficacious (Martinez et al., 2004). The isolation of a particular target as an essential link in the synaptic plasticity/gene expression/protein synthesis cascade may not confer on the mechanism utility for safe and selective cognitive enhancement. For example, the transcription factor CREB is argued *not* to be a particularly viable candidate for cognitive enhancement, since this signaling system is essential to many nonlearning/memory-related functions of the nervous system that would be impacted by general administration of a CREB-related drug (Barco et al., 2003). Similarly, it is unlikely that treatments that impact learning/memory performance as modulators of motivation or arousal, for example, could prevail over the inevitable pharmacological-side-effect profile that would accompany exposure to a candidate drug such as amphetamine (White and Salinas, 1998). The most promising cognition-enhancement strategies at the present time are replacement therapies, in which a depleted hormone or neurochemical is restored, estrogen therapy in postmenopausal women or cholinergic enhancement in Alzheimer's disease, for example (Resnick and Maki, 2001; Terry and Buccafusco, 2003). To the consternation of neuroscientists everywhere, the most productive and carefree strategies for selective and physiologically optimal cognitive enhancement in normal-functioning individuals may ultimately turn out to be environmental enrichment, physical exercise, or dietary changes in lifestyle.

One important utility for neurobiological mechanisms of synaptic plasticity is to quantify the magnitude of potential increase in learning/memory capacity that is or can be expected via mechanisms of cognitive enhancement. One pessimistic view is that existing enhancement strategies exert modest effect sizes on the order of 10–20% improvement (Buccafusco and Terry, 2000) and that this is the maximum expected enhancement using a pharmacological approach given inherent nonspecific drug actions (White and Salinas, 1998). On the other hand, Lisman argues for a graded state of synaptic activation that accommodates a range of values from "silent" through "disinhibited" to "potentiated" and describes how temporal parameters within a long-term-potentiation experiment can be adjusted to achieve higher levels of activation (Lisman, 2003). Indeed, at the cellular consolidation level of analysis, synaptic plasticity can be viewed as a mathematical certainty, based on the concentration of interacting synaptic proteins (Clayton, 2000). While synaptic activation can be considered as a necessary prerequisite in order for learning/memory substrates of the nervous system to operate, a further and more rigorous extension of this criterion would require that the degree of activation match the span, persistence, or accuracy of learning/memory performance. In contrast, a mismatch of perpetual or overly robust synaptic plasticity in the absence of incoming afferent information may produce high noise-to-signal ratios, which are associated with deleterious consequences in animal and insect learning/memory models (Barco et al., 2003) or "cognitive chaos" in humans (McGaugh, 2003).

Can reserve learning/memory capacity present in the nervous system and exploitation of this reserve via effective cognitive enhancement be considered adaptive from an evolutionary point of view? From a neuroethological perspective, the ability to conserve information about changing environmental or predatory threats for an extended period of time following a particular incident can be tremendously beneficial for the survival of an organism (Clayton, 2000). On the other hand, enhancements in the ability to recall autobiographical events, sometime referred to as *hypermnesia* (Bluck et al., 1999), can be considered obtrusive or pathogenic when inhibitory mechanisms break down (McNaughton and Wickens, 2003; Osman et al., 2004). The clinical case study of Luria's patient presented at the beginning of this chapter's introduction illustrates this point. One example of a potential pathological consequence of overly efficient recall is the flood of traumatic memories symptomatic of post-traumatic stress disorder (Layton and Krikorian, 2002). Similarly, it may be prudent to remain mindful of the inverted-U-shaped curve that typically governs the relationship between dose of a particular pharmacological cognitive enhancer and performance, lest an overused learning/memory-improvement strategy begins to have adverse consequences (Martinez et al., 1991; Barch, 2004; Lupien et al., 2005).

The search for drugs that enhance learning/memory requires the development and refinement of behavioral tests for animals (Olton and Wenk, 1990; Barnes, 2003; Morris et al., 2003). These tests must be able to identify potentially therapeutic treatments and reject ineffective enhancement strategies. Therefore, a coherent conceptual and experimental framework is needed to organize future research in this area (Brown et al., 2000). Unfortunately, previous preclinical research strategies appear to have focused on the demonstration of enhancing effects in a wide variety of tests of uncertain validity, rather than on determination of the specific psychological and neurobiological processes affected by putative cognition enhancers (McDaniel et al., 2003). For example, some sort of noxious stimulus, such as electric shock delivered unexpectedly to rodent paw pads or escape from a water bath, is typically used to motivate learning in the widely used avoidance-conditioning (Uvnas-Moberg et al., 2000) and Morris-maze (Meijer et al., 2005) contexts, in spite of the fact that shock exposure and forced swimming produce an unconditioned affective state that can confound interpretation of learning performance in the task (Penka et al., 2004). A further disincentive for employing alarming and traumatic stimuli in the conditioning environment is provided by behavioral and cognitive neuroscience studies demonstrating that the affective salience of stimuli can bias encoding and retrieval of learned information in an automatic manner (Ochsner, 2000; Cardinal et al., 2002). Thus, future efforts require explicit identification of the goals of the research, the cognitive process, the neural systems and cellular gene products involved in the phenomenon, the selectivity and sensitivity of tasks that measure the process, and the validity of the behavioral tasks as a model to predict the effects of the cognitive-enhancement strategy in humans.

In conclusion, the present learning/memory classification scheme, based on evidence from cognitive psychology, has proven to be a useful rubric for distinguishing temporally and conceptually a variety of neurobiological events that presumably underlie synaptic plasticity. The linkage is not one to one, however, because events characterized in this chapter as influencing primarily the process of attention (e.g., general arousal) can certainly carry over to impact later consolidation and retrieval stages; an example of this nonspecific involvement at several steps in the process is state-dependent learning, in which, by definition, coordinated activation at two separate learning/memory stages is required. Similarly, changes in gene expression occasioned by some original learning/memory event can certainly impact encoding of any later-occurring event. The complexity of multifactorial enhancement will only increase with the emergence of mnemonic drugs acting at multiple targets in the nervous system (Buccafusco and Youdim, 2004).

VI. SUMMARY

1. Improvement in normal learning/memory performance using a so-called *enhancement neurology* approach is an achievable, if ethically dubious, goal.

2. Memory-fitness strategies to combat learning/memory loss represent a vitally important topic for public health.

3. An encoding/consolidation/retrieval classification scheme is a useful rubric for distinguishing temporally and conceptually a variety of neurobiological events that presumably underlie enhancement of synaptic plasticity.

4. Early-stage attentional and working-memory functions appear to benefit from optimization of dopaminergic, glutamatergic, nicotinic, and adenosinergic neurotransmission.

5. Neuro-, gonadal, and adrenal steroid systems appear to modulate early-stage encoding via indirect actions on other neurochemical and stress neuropeptide learning/memory mediators.

6. Intermediate-stage consolidation and learning/memory storage functions appear to benefit from optimization of adrenergic, trophic factor, and cholinergic neurotransmission.

7. Physical activity appears to enhance learning/memory performance in a cholinergic and neurotrophic factor–dependent manner consistent with a storage/consolidation mechanism of action.

8. Late-state memory recall and retrieval functions appear to benefit from optimization of serotonergic neurotransmission and from reexposure to subjective states present at the time of initial learning.

9. Cognitive enhancers include direct mediators, such as glutamatergic agents, which adjust the machinery of synaptic plasticity in which the mediator participates directly, and modulators, such as adrenergic agents, which modify diffuse networks of neurons to adjust synaptic plasticity in an indirect manner.

10. Given the difficulty in achieving specificity of pharmacological action, the most productive and carefree strategies for selective and physiologically optimal cognitive enhancement in normal-functioning individuals may ultimately turn out to be lifestyle changes.

ACKNOWLEDGMENTS

I thank Melanie Leussis for her assistance in researching and preparing this manuscript.

GLOSSARY

Acetylcholine — A neurotransmitter involved in learning/memory functions.

Ampakine — Drug that modulates glutamatergic neurotransmission in order to facilitate learning/memory.

Cholinesterase inhibitor — Drug that increases synaptic availability of acetylcholine via inhibition of cholinesterase enzymes.

Cognition — Mental function involving memory, language abilities, visual and spatial skills, intelligence, and reasoning.

Dentrites — Short, branching extensions of neurons that receive impulses from other neurons when stimulated by neurotransmitters released from presynaptic terminals.

Hippocampus — A seahorse-shaped temporal lobe structure of the brain that provides a model system for synaptic plasticity.

Long-term memory — Persistent record of experience that has been organized and rehearsed.

Mediation (of learning/memory) — Direct alteration in the specific substrate of learning/memory encoding, consolidation, or retrieval.

Modulation (of learning/memory) — Nonspecific, potentially bidirectional change in performance achieved indirectly by altering the strength or expression of a separate and independent learning/memory mechanism.

Nutraceutical — Natural dietary supplements not regulated by the Food and Drug Administration that are purported to sharpen mental functions and counteract the aging process.

Short-term memory — Transient record of experience lasting only a few seconds/minutes.

U-shaped curve — Hyperbolic line plot of drug dose/stimulus intensity relative to treatment efficacy reveals zeniths at low- and high-dose/intensity extremes and an intermediate nadir to form the shape of the letter U.

REFERENCES

Akwa, Y., N. Ladurelle, D.F. Covey, and E.E. Baulieu (2001). The synthetic enantiomer of pregnenolone sulfate is very active on memory in rats and mice, even more so than its physiological neurosteroid counterpart: Distinct mechanisms? *Proc Nat Acad Sci USA* 98(24): 14033–14037.

Aston-Jones, G., J. Rajkowski, and J. Cohen (2000). Locus coeruleus and regulation of behavioral flexibility and attention. *Prog Brain Res* 126: 165–182.

Baddeley, A.D. (1995). The psychology of memory. *Handbook of Memory Disorders*, A.D. Baddeley, B.A. Wiilson, and F.N. Watts, eds. pp. 3–25, Wiley: New York.

Barch, D.M. (2004). Pharmacological manipulation of human working memory. *Psychopharmacol (Berl)* 174(1): 126–135.

Barco, A., C. Pittenger, and E.R. Kandel (2003). CREB, memory enhancement and the treatment of memory disorders: Promises, pitfalls and prospects. *Expert Opin Ther Targets* 7(1): 101–114.

Barnard, C.J., and N. Luo (2002). Acquisition of dominance status affects maze learning in mice. *Behav Processes* 60(1): 53–59.

Barnes, C.A. (2003). Long-term potentiation and the ageing brain. *Philos Trans R Soc Lond B Biol Sci* 358(1432): 765–772.

Baulieu, E.E., P. Robel, and M. Schumacher (2001). Neurosteroids: Beginning of the story. *Int Rev Neurobiol* 46: 1–32.

Behan, D.P., E.B. De Souza, P.J. Lowry, E. Potter, P. Sawchenko, and W.W. Vale (1995a). Corticotropin releasing factor (CRF) binding protein: A novel regulator of CRF and related peptides. *Frontiers Neuroendocrinol* 16: 362–382.

Behan, D.P., S.C. Heinrichs, J.C. Troncoso, X.J. Liu, C.H. Kawas, N. Ling, and E.B. De Souza (1995b). Displacement of corticotropin releasing factor from its binding protein as a possible treatment for Alzheimer's disease. *Nature* 378(6554): 284–287.

Bentley, P., M. Husain, and R.J. Dolan (2004). Effects of cholinergic enhancement on visual stimulation, spatial attention, and spatial working memory. *Neuron* 41(6): 969–982.

Bergeron, R., C. de Montigny, and G. Debonnel (1996). Potentiation of neuronal NMDA response induced by dehydroepiandrosterone and its suppression by progesterone: Effects mediated via sigma receptors. *J Neurosci* 16(3): 1193–1202.

Bernardi, F., A. Lanzone, R.M. Cento, R.S. Spada, I. Pezzani, A.D. Genazzani, S. Luisi, M. Luisi, F. Petraglia, and A.R. Genazzani (2000). Allopregnanolone and dehydroepiandrosterone response to corticotropin-releasing factor in patients suffering from Alzheimer's disease and vascular dementia. *Eur J Endocrinol* 142(5): 466–471.

Blank, T., I. Nijholt, S. Vollstaedt, and J. Spiess (2003). The corticotropin-releasing factor receptor 1 antagonist CP-154,526 reverses stress-induced learning deficits in mice. *Behav Brain Res* 138(2): 207–213.

Bluck, S., L.J. Levine, and T.M. Laulhere (1999). Autobiographical remembering and hypermnesia: A comparison of older and younger adults. *Psychol Aging* 14(4): 671–682.

Bohn, A.M., M. Khodaee, and T.L. Schwenk (2003). Ephedrine and other stimulants as ergogenic aids. *Curr Sports Med Rep* 2(4): 220–225.

Bonaz, B., and S. Rivest (1998). Effect of a chronic stress on CRF neuronal activity and expression of its type 1 receptor in the rat brain. *Am J Physiol* 275(5 Part 2): R1438–R1449.

Bontempi, B., K.T. Whelan, V.B. Risbrough, G.K. Lloyd, and F. Menzaghi (2003). Cognitive enhancing properties and tolerability of cholinergic agents in mice: A comparative study of nicotine, donepezil, and SIB-1553A, a subtype-selective ligand for nicotinic acetylcholine receptors. *Neuropsychopharmacology* 28(7): 1235–1246.

Bowlby, M.R. (1993). Pregnenolone sulfate potentiation of N-methyl-D-aspartate receptor channels in hippocampal neurons. *Mol Pharmacol* 43: 813–819.

Brown, R.E., L. Stanford, and H.M. Schellinck (2000). Developing standardized behavioral tests for knockout and mutant mice. *ILAR J* 41(3): 163–174.

Bruins Slot, L.A., and F.C. Colpaert (2003). A persistent opioid-addiction state of memory. *Behav Pharmacol* 14(2): 167–171.

Buccafusco, J., and M.B.H. Youdim (2004). Drugs with multiple CNS targets. In *Cognitive Enhancing Drugs*, J. Buccafusco, ed. pp. 179–198, Birkhauser: Basil, Switzerland.

Buccafusco, J.J., and A.V. Terry, Jr. (2000). Multiple central nervous system targets for eliciting beneficial effects on memory and cognition. *J Pharmacol Exp Ther* 295(2): 438–446.

Buccafusco, J.J., T. Weiser, K. Winter, K. Klinder, and A.V. Terry (2004). The effects of IDRA 21, a positive modulator of the AMPA receptor, on delayed matching performance by young and aged rhesus monkeys. *Neuropharmacology* 46(1): 10–22.

Buchanan, T.W., and W.R. Lovallo (2001). Enhanced memory for emotional material following stress-level cortisol treatment in humans. *Psychoneuroendocrinology* 26(3): 307–317.

Cahill, L. (2003). Similar neural mechanisms for emotion-induced memory impairment and enhancement. *Proc Natl Acad Sci USA* 100(23): 13123–13124.

Calamandrei, G., A. Valanzano, and L. Ricceri (2002). NGF induces appearance of adult-like response to spatial novelty in 18-day male mice. *Behav Brain Res* 136(1): 289–298.

Cardinal, R.N., J.A. Parkinson, J. Hall, and B.J. Everitt (2002). Emotion and motivation: The role of the amygdala, ventral striatum, and prefrontal cortex. *Neurosci Biobehav Rev* 26(3): 321–352.

Clayton, D.F. (2000). The genomic action potential. *Neurobiol Learn Mem* 74(3): 185–216.

Cordoba Montoya, D.A., and H.F. Carrer (1997). Estrogen facilitates induction of long-term potentiation in the hippocampus of awake rats. *Brain Res* 778(2): 430–438.

Coull, J.T., C.D. Frith, R.J. Dolan, R.S. Frackowiak, and P.M. Grasby (1997). The neural correlates of the noradrenergic modulation of human attention, arousal and learning. *Eur J Neurosci* 9(3): 589–598.

Dalley, J.W., R.N. Cardinal, and T.W. Robbins (2004). Prefrontal executive and cognitive functions in rodents: Neural and neurochemical substrates. *Neurosci Biobehav Rev* 28(7): 771–784.

Daniel, J.M., and G.P. Dohanich (2001). Acetylcholine mediates the estrogen-induced increase in NMDA receptor binding in CA1 of the hippocampus and the associated improvement in working memory. *J Neurosci* 21(17): 6949–6956.

Darnaudery, M., M. Koehl, P.V. Piazza, M. Le Moal, and W. Mayo (2000). Pregnenolone sulfate increases hippocampal acetylcholine release and spatial recognition. *Brain Res.* 852(1): 173–179.

Dash, P.K., A.E. Hebert, and J.D. Runyan (2004). A unified theory for systems and cellular memory consolidation. *Brain Res Brain Res Rev* 45(1): 30–37.

de Kloet, E.R., J. Grootendorst, A.M. Karssen, and M.S. Oitzl (2002). Gene × environment interaction and cognitive performance: Animal studies on the role of corticosterone. *Neurobiol Learn Mem* 78(3): 570–577.

Deutsch, J.A. (1983). The cholinergic synapse and the site of memory. In *The Physiological Basis of Memory*, J.A. Deutsch, ed. pp. 367–386, New York: Academic Press.

Dudai, Y. (2004). The neurobiology of consolidations, or, how stable is the engram? *Annu Rev Psychol* 55: 51–86.

Eckart, K., J. Radulovic, M. Radulovic, O. Jahn, T. Blank, O. Stiedl, and J. Spiess (1999). Actions of CRF and its analogs. *Curr Med Chem* 6(11): 1035–1053.

Einstein, G., and M.A. McDaniel (2004). *Memory Fitness: A Guide for Successful Aging.* New Haven, CT: Yale University Press.

Ernst, M., S.J. Heishman, L. Spurgeon, and E.D. London (2001). Smoking history and nicotine effects on cognitive performance. *Neuropsychopharmacology* 25(3): 313–319.

Farah, M.J., J. Illes, R. Cook-Deegan, H. Gardner, E. Kandel, P. King, E. Parens, B. Sahakian, and P.R. Wolpe (2004). Neurocognitive enhancement: What can we do and what should we do? *Nat Rev Neurosci* 5(5): 421–425.

Farr, S.A., W.A. Banks, and J.E. Morley (2000). Estradiol potentiates acetylcholine and glutamate-mediated post-trial memory processing in the hippocampus. *Brain Res* 864(2): 263–269.

Fletcher, P.C., C.D. Frith, and M.D. Rugg (1997). The functional neuroanatomy of episodic memory. *Trends Neurosci* 20(5): 213–218.

Fordyce, D.E., and R.P. Farrar (1991). Physical activity effects on hippocampal and parietal cortical cholinergic function and spatial learning in F344 rats. *Behav Brain Res* 43(2): 115–123.

Fordyce, D.E., and J.M. Wehner (1993). Physical activity enhances spatial learning performance with an associated alteration in hippocampal protein kinase C activity in C57BL/6 and DBA/2 mice. *Brain Res* 619(1–2): 111–119.

Fuchs, E., G. Flügge, F. Ohl, P. Lucassen, G.K. Vollmann-Honsdorf, and T.M. Michaelis (2001). Psychosocial stress, glucocorticoids, and the structural alterations in the tree shrew hippocampus. *Physiol. Behav.* 73: 285–291.

Furey, M.L., P. Pietrini, and J.V. Haxby (2000). Cholinergic enhancement and increased selectivity of perceptual processing during working memory. *Science* 290(5500): 2315–2319.

Gerlai, R. (2003). Memory enhancement: The progress and our fears. *Genes Brain Behav* 2(4): 187–188; discussion 189–190.

Goldman-Rakic, P.S. (1998). The cortical dopamine system: Role in memory and cognition. *Adv Pharmacol* 42: 707–711.

Good, M., M. Day, and J.L. Muir (1999). Cyclical changes in endogenous levels of oestrogen modulate the induction of LTD and LTP in the hippocampal CA1 region. *Eur J Neurosci* 11(12): 4476–4480.

Gooney, M., K. Shaw, A. Kelly, S.M. O'Mara, and M.A. Lynch (2002). Long-term potentiation and spatial learning are associated with increased phosphorylation of TrkB and extracellular signal-regulated kinase (ERK) in the dentate gyrus: Evidence for a role for brain-derived neurotrophic factor. *Behav Neurosci* 116(3): 455–463.

Gureviciene, I., J. Puolivali, R. Pussinen, J. Wang, H. Tanila, and A. Ylinen (2003). Estrogen treatment alleviates NMDA-antagonist-induced hippocampal LTP blockade and cognitive deficits in ovariectomized mice. *Neurobiol Learn Mem* 79(1): 72–80.

Gustilo, M.C., A.L. Markowska, S.J. Breckler, C.A. Fleischman, D.L. Price, and V.E. Koliatsos (1999). Evidence that nerve growth factor influences recent memory through structural changes in septohippocampal cholinergic neurons. *J Comp Neurol* 405(4): 491–507.

Hasselmo, M.E. (1999). Neuromodulation: Acetylcholine and memory consolidation. *Trends Cogn Sci* 3(9): 351–359.

Heikkinen, T., J. Puolivali, L. Liu, A. Rissanen, and H. Tanila (2002). Effects of ovariectomy and estrogen treatment on learning and hippocampal neurotransmitters in mice. *Horm Behav* 41(1): 22–32.

Heinrichs, S.C. (1999). Stress-axis, coping and dementia: Gene-manipulation studies. *Trends Pharmacolog Sci* 20: 311–315.

Heinrichs, S.C. (2003). Modulation of social learning in rats by brain corticotropin-releasing factor. *Brain Res.* 994: 107–114.

Heinrichs, S.C., and E.B. De Souza (2001). Corticotropin-releasing factor in brain: Executive gating of neuroendocrine and functional outflow. *Handbook of Physiology*. In B.S. McEwen, ed. pp. 125–137, New York: Oxford University Press. Volume 7: Coping with the environment: *Neural and endocrine mechanisms*.

Heinrichs, S.C., and G.F. Koob (2004). Corticotropin-releasing factor in brain: A role in activation, arousal, and affect regulation. *J Pharmacol Exp Ther* 311(2): 427–440.

Heinrichs, S.C., J. Lapsansky, T.W. Lovenberg, E.B. De Souza, and D.T. Chalmers (1997). Corticotropin-releasing factor CRF1, but not CRF2, receptors mediate anxiogenic-like behavior. *Regulatory Peptides* 71: 15–21.

Hejmadi, M.V., F. Dajas-Bailador, S.M. Barns, B. Jones, and S. Wonnacott (2003). Neuroprotection by nicotine against hypoxia-induced apoptosis in cortical cultures involves activation of multiple nicotinic acetylcholine receptor subtypes. *Mol Cell Neurosci* 24(3): 779–786.

Honour, L.C., and M.H. White (1988). Pre- and postnatally administered ACTH, Organon 2766 and CRF facilitate or inhibit active avoidance task performance in young adult mice. *Peptides* 9(4): 745–750.

Horn, G. (2004). Pathways of the past: The imprint of memory. *Nat Rev Neurosci* 5(2): 108–120.

Horvath, K.M., W. Hartig, R. Van der Veen, J.N. Keijser, J. Mulder, M. Ziegert, E.A. Van der Zee, T. Harkany, and P.G. Luiten (2002). 17beta-estradiol enhances cortical cholinergic innervation and preserves synaptic density following excitotoxic lesions to the rat nucleus basalis magnocellularis. *Neuroscience* 110(3): 489–504.

Izquierdo, I. (1989). Different forms of post-training memory processing. *Behav Neural Biol* 51(2): 171–202.

Johnston, A.N., and S.P. Rose (2001). Memory consolidation in day-old chicks requires BDNF but not NGF or NT-3; an antisense study. *Brain Res Mol Brain Res* 88(1–2): 26–36.

Kelleher 3rd, R.J., A. Govindarajan, and S. Tonegawa (2004). Translational regulatory mechanisms in persistent forms of synaptic plasticity. *Neuron* 44(1): 59–73.

Keyes, D. (1959). *Flowers for Algernon*. New York: Bantam Books.

Klintsova, A.Y., E. Dickson, R. Yoshida, and W.T. Greenough (2004). Altered expression of BDNF and its high-affinity receptor TrkB in response to complex motor learning and moderate exercise. *Brain Res* 1028(1): 92–104.

Kovalchuk, Y., E. Hanse, K.W. Kafitz, and A. Konnerth (2002). Postsynaptic induction of BDNF-mediated long-term potentiation. *Science* 295(5560): 1729–1734.

Kramer, A.F., S. Hahn, N.J. Cohen, M.T. Banich, E. McAuley, C.R. Harrison, J. Chason, E. Vakil, L. Bardell, R.A. Boileau, and A. Colcombe (1999). Aging, fitness and neurocognitive function. *Nature* 400(6743): 418–419.

Kumari, V., J.A. Gray, D.H. ffytche, M.T. Mitterschiffthaler, M. Das, E. Zachariah, G.N. Vythe-lingum, S.C. Williams, A. Simmons, and T. Sharma (2003). Cognitive effects of nicotine in humans: An fMRI study. *Neuroimage* 19(3): 1002–1013.

Lashley, K.S. (1958). Cerebral organization and behavior. *Res Publ Assoc Res Nerv Ment Dis* 36: 1–4; discussion 14–18.

Layton, B., and R. Krikorian (2002). Memory mechanisms in posttraumatic stress disorder. *J Neuropsychiatry Clin Neurosci* 14(3): 254–261.

Lee, J.L., B.J. Everitt, and K.L. Thomas (2004). Independent cellular processes for hippocampal memory consolidation and reconsolidation. *Science* 304(5672): 839–843.

Leonard, B.E., D. McCartan, J. White, and D.J. King (2004). Methylphenidate: A review of its neuropharmacological, neuropsychological and adverse clinical effects. *Hum Psychopharmacol* 19(3): 151–180.

Lisman, J. (2003). Long-term potentiation: Outstanding questions and attempted synthesis. *Philos Trans R Soc Lond B Biol Sci* 358(1432): 829–842.

Lombardi, W.J., and H. Weingartner (1995). Pharmacoloigcal treatment of impaired memory function. *Handbook of Memory Disorders*, A.D. Baddeley, B.A. Wilson, and F.N. Watts, eds. pp. 577–601, New York: Wiley.

Lorist, M.M., and M. Tops (2003). Caffeine, fatigue, and cognition. *Brain Cogn* 53(1): 82–94.

Lupien, S.J., A. Fiocco, N. Wan, F. Maheu, C. Lord, T. Schramek, and M.T. Tu (2005). Stress hormones and human memory function across the lifespan. *Psychoneuroendocrinology* 30(3): 225–242.

Luria, A.R. (1968). *A Little Book About a Vast Memory: The Mind of a Mnemonist.* Cambridge, MA: Harvard University Press.

Lynch, G. (2002). Memory enhancement: The search for mechanism-based drugs. *Nat Neurosci* 5 Suppl: 1035–1038.

Lynch, G. (2004). AMPA receptor modulators as cognitive enhancers. *Curr Opin Pharmacol* 4(1): 4–11.

Majewska, M.D., Bluet, M. Pajot, T.P. Robel, and E.E. Baulieu (1989). Pregnenolone sulfate antagonizes barbituate-induced hypnosis. *Pharmacol Biochem Behav* 33(701): 701–703.

Marshal, E. (2004). A star-studded search for memory-enhancing drugs. *Science.* 304: 36–38.

Martinez, J.L., Jr. (ed.) (1986). Memory: Drugs and hormones. In *Learning and Memory: A Biological View.* San Diego: Academic Press.

Martinez, J.L., Jr., R.A. Jensen, and J.L. McGaugh (1983). Facilitation of memory consolidation. In *The Physiological Basis of Memory*, J.A. Deutsch, ed. pp. 49–70, New York: Academic Press.

Martinez, J.L., G. Schulteis, and S.B. Weinberger (1991). How to increase and decrease the strength of memory traces: The effects of drugs and hormones. In *Learning and Memory: A Biological View*, J.L. Martinez, Jr., and R.P. Kesner, eds. pp. 149–198, New York: Academic Press.

Martinez, J.L. Jr., K.A. Thompson, M. McFadyen-Leussis, and S.C. Heinrichs (2004). Peptide and steroid hormone receptors as drug targets for enhancement of learning and memory performance. In *Cognitive-Enhancing Drugs*, J. Buccafusco, ed. Basel Switzerland: Birkhauser.

Mathis, C., E. Vogel, B. Cagniard, F. Criscuolo, and A. Ungerer (1996). The neurosteroid pregnenolone sulfate blocks deficits induced by a competitive NMDA antagonist in active avoidance and lever-press learning tasks in mice. *Neuropharmacology* 35(8): 1057–1064.

Mattay, V.S., J.H. Callicott, A. Bertolino, I. Heaton, J.A. Frank, R. Coppola, K.F. Berman, T.E. Goldberg, and D.R. Weinberger (2000). Effects of dextroamphetamine on cognitive performance and cortical activation. *Neuroimage* 12(3): 268–275.

Maurice, T., J.L. Junien, and A. Privat (1997). Dehydroepiandrosterone sulfate attenuates dizocilpine-induced learning impairment in mice via sigma 1-receptors. *Behav Brain Res* 83(1–2): 159–164.

McDaniel, M.A., S.F. Maier, and G.O. Einstein (2003). "Brain-specific" nutrients: A memory cure? *Nutrition* 19(11–12): 957–975.

McEwen, B. (2002). Estrogen actions throughout the brain. *Recent Prog Horm Res* 57: 357–384.

McEwen, B.S., and S. Chattarji (2004). Molecular mechanisms of neuroplasticity and pharmacological implications: The example of tianeptine. *Eur Neuropsychopharmacol* 14(Suppl 5): S497–S502.

McGaugh, J.L. (2000). Memory — a century of consolidation. *Science* 287(5451): 248–251.

McGaugh, J.L. (2003). *Memory and Emotion: The Making of Lasting Memories*. New York: Columbia University Press.

McGaugh, J.L., and J.A. Krivanek (1970). Strychnine effects on discrimination learning in mice: Effects of dose and time of administration. *Physiol Behav* 5(12): 1437–1442.

McGaugh, J.L., and B. Roozendaal (2002). Role of adrenal stress hormones in forming lasting memories in the brain. *Curr Opin Neurobiol* 12(2): 205–210.

McNaughton, N., and J. Wickens (2003). Hebb, pandemonium and catastrophic hypermnesia: The hippocampus as a suppressor of inappropriate associations. *Cortex* 39(4–5): 1139–1163.

Mehlman, M.J. (2004). Cognition-enhancing drugs. *Milbank Q* 82(3): 483–506, table of contents.

Meijer, O.C., B. Topic, P.J. Steenbergen, G. Jocham, J.P. Huston, and M.S. Oitzl (2005). Correlations between hypothalamus-pituitary-adrenal axis parameters depend on age and learning capacity. *Endocrinology* 146(3): 1372–1381.

Messaoudi, E., S.W. Ying, T. Kanhema, S.D. Croll, and C.R. Bramham (2002). Brain-derived neurotrophic factor triggers transcription-dependent, late-phase long-term potentiation in vivo. *J Neurosci* 22(17): 7453–7461.

Meziane, H., C. Mathis, S.M. Paul, and A. Ungerer (1996). The neurosteroid pregnenolone sulfate reduces learning deficits induced by scopolamine and has promnestic effects in mice performing an appetitive learning task. *Psychopharmacology* 126(4): 323–330.

Miles, C., and E. Hardman (1998). State-dependent memory produced by aerobic exercise. *Ergonomics* 41(1): 20–28.

Miyashita, Y. (2004). Cognitive memory: Cellular and network machineries and their top-down control. *Science* 306(5695): 435–440.

Mizuno, M., K. Yamada, J. He, A. Nakajima, and T. Nabeshima (2003). Involvement of BDNF receptor TrkB in spatial memory formation. *Learn Mem* 10(2): 108–115.

Molteni, R., A. Wu, S. Vaynman, Z. Ying, R.J. Barnard, and F. Gomez-Pinilla (2004). Exercise reverses the harmful effects of consumption of a high-fat diet on synaptic and behavioral plasticity associated to the action of brain-derived neurotrophic factor. *Neuroscience* 123(2): 429–440.

Morris, R.G., E.I. Moser, G. Riedel, S.J. Martin, J. Sandin, M. Day, and C. O'Carroll (2003). Elements of a neurobiological theory of the hippocampus: The role of activity-dependent synaptic plasticity in memory. *Philos Trans R Soc Lond B Biol Sci* 358(1432): 773–786.

Nehlig, A. (1999). Are we dependent upon coffee and caffeine? A review on human and animal data. *Neurosci Biobehav Rev* 23(4): 563–576.

Nehlig, A., J.L. Daval, and G. Debry (1992). Caffeine and the central nervous system: Mechanisms of action, biochemical, metabolic and psychostimulant effects. *Brain Res Brain Res Rev* 17(2): 139–170.

Nilsson, K.R., C.F. Zorumski, and D.F. Covey (1998). Neurosteroid analogues. 6. The synthesis and GABAA receptor pharmacology of enantiomers of dehydroepiandrosterone sulfate, pregnenolone sulfate, and (3alpha,5beta)-3-hydroxypregnan-20-one sulfate. *J Med Chem* 41(14): 2604–2613.

Nyberg, L., J. Sandblom, S. Jones, A.S. Neely, K.M. Petersson, M. Ingvar, and L. Backman (2003). Neural correlates of training-related memory improvement in adulthood and aging. *Proc Natl Acad Sci USA* 100(23): 13728–13733.

Ochsner, K.N. (2000). Are affective events richly recollected or simply familiar? The experience and process of recognizing feelings past. *J Exp Psychol Gen* 129(2): 242–261.

Olton, D.S., and L. Wenk (1990). The development of behavioral tests to assess the effects of cognitive enhancers. *Pharmacopsychiatry* 23(Suppl 2): 65–69.

O'Neal, M.F., L.W. Means, M.C. Poole, and R.J. Hamm (1996). Estrogen affects performance of ovariectomized rats in a two-choice water-escape working memory task. *Psychoneuroendocrinology* 21(1): 51–65.

Osman, S., M. Cooper, A. Hackmann, and D. Veale (2004). Spontaneously occurring images and early memories in people with body dysmorphic disorder. *Memory* 12(4): 428–436.

Overton, D.A. (1991). Historical context of state dependent learning and discriminative drug effects. *Behav Pharmacol* 2(4 and 5): 253–264.

Pallares, M., M. Darnaudery, J. Day, M. Le Moal and W. Mayo (1998). The neurosteroid pregnenolone sulfate infused into the nucleus basalis increases both acetylcholine release in the frontal cortex or amygdala and spatial memory. *Neuroscience* 87(3): 551–558.

Paul, S.M., and R.H. Purdy (1992). Neuroactive steroids. *FASEB J* 6: 2311.

Penka, L.L., T.L. Bond, and S.C. Heinrichs (2004). Nonspecific effect of fear conditioning and specific effect of social defeat on social recognition memory performance in female rats. *Stress* 7(1): 63–72.

Radulovic, J., A. Fischer, U. Katerkamp, and J. Spiess (2000). Role of regional neurotransmitter receptors in corticotropin-releasing factor (CRF)-mediated modulation of fear conditioning. *Neuropharmacology* 39(4): 707–710.

Reddy, D.S., and S.K. Kulkarni (1998). The effects of neurosteroids on acquisition and retention of a modified passive-avoidance learning task in mice. *Brain Res* 791(1–2): 108–116.

Resnick, S.M., and P.M. Maki (2001). Effects of hormone replacement therapy on cognitive and brain aging. *Ann NY Acad Sci* 949: 203–214.

Rezvani, A.H., and E.D. Levin (2001). Cognitive effects of nicotine. *Biol Psychiatry* 49(3): 258–267.

Robbins, T.W., G. McAlonan, J.L. Muir, and B.J. Everitt (1997). Cognitive enhancers in theory and practice: Studies of the cholinergic hypothesis of cognitive deficits in Alzheimer's disease. *Behav Brain Res* 83(1–2): 15–23.

Roozendaal, B. (2002). Stress and memory: Opposing effects of glucocorticoids on memory consolidation and memory retrieval. *Neurobiol Learn Mem* 78(3): 578–595.

Rossetti, Z.L., and S. Carboni (2005). Noradrenaline and dopamine elevations in the rat prefrontal cortex in spatial working memory. *J Neurosci* 25(9): 2322–2329.

Ryan, L., C. Hatfield, and M. Hofstetter (2002). Caffeine reduces time-of-day effects on memory performance in older adults. *Psychol Sci* 13(1): 68–71.

Sara, S.J. (2000). Retrieval and reconsolidation: Toward a neurobiology of remembering. *Learn Mem* 7(2): 73–84.

Sarter, M. (1991). Taking stock of cognition enhancers. *Trends Pharmacol Sci* 12(12): 456–461.

Shulz, D.E., R. Sosnik, V. Ego, S. Haidarliu, and E. Ahissar (2000). A neuronal analogue of state-dependent learning. *Nature* 403(6769): 549–553.

Silva, A.J., J.H. Kogan, P.W. Frankland, and S. Kida (1998). CREB and memory. *Annu Rev Neurosci* 21: 127–148.

Singh, M., E.M. Meyer, W.J. Millard, and J.W. Simpkins (1994). Ovarian steroid deprivation results in a reversible learning impairment and compromised cholinergic function in female Sprague-Dawley rats. *Brain Res* 644(2): 305–312.

Small, G. (2004). *The Memory Prescription*. New York: Hyperion.

Southwick, S.M., M. Davis, B. Horner, L. Cahill, C.A. Morgan 3rd, P.E. Gold, J.D. Bremner, and D.C. Charney (2002). Relationship of enhanced norepinephrine activity during memory consolidation to enhanced long-term memory in humans. *Am J Psychiatry* 159(8): 1420–1422.

Sweatt, J.D. (2003). *Mechanisms of Memory*. Amsterdam: Elsevier Academic Press.

Terry, A.V. (2004). Drugs that target sertonergic receptors. In *Cognitive-Enhancing Drugs*, J. Buccafusco, ed. pp. 79–88, Birkhauser: Basel, Switzerland.

Terry, A.V., Jr., and J.J. Buccafusco (2003). The cholinergic hypothesis of age and Alzheimer's disease-related cognitive deficits: Recent challenges and their implications for novel drug development. *J Pharmacol Exp Ther* 306(3): 821–827.

Terry, A.V., Jr., J.J. Buccafusco, and G.D. Bartoszyk (2005). Selective serotonin 5-HT(2A) receptor antagonist EMD 281014 improves delayed matching performance in young and aged rhesus monkeys. *Psychopharmacoogy (Berl)* 179(4): 725–732.

Terry, A.V., Jr., J.J. Buccafusco, W.J. Jackson, M.A. Prendergast, D.J. Fontana, E.H. Wong, D.W. Bonhaus, P. Weller, and R.M. Eglen (1998). Enhanced delayed matching performance in younger and older macaques administered the 5-HT4 receptor agonist, RS 17017. *Psychopharmacology (Berl)* 135(4): 407–415.

Terry, A.V., Jr., J.J. Buccafusco, M.A. Prendergast, W.J. Jackson, D.L. Fontana, E.H. Wong, R.L. Whiting, and R.M. Eglen (1996). The 5-HT3 receptor antagonist, RS-56812, enhances delayed matching performance in monkeys. *Neuroreport* 8(1): 49–54.

Tokuyama, W., H. Okuno, T. Hashimoto, Y. Xin Li, and Y. Miyashita (2000). BDNF upregulation during declarative memory formation in monkey inferior temporal cortex. *Nature neuroscience* 3(11): 1134–1142.

Tomporowski, P.D. (2003). Effects of acute bouts of exercise on cognition. *Acta Psychol (Amst)* 112(3): 297–324.

Tonegawa, S., K. Nakazawa, and M.A. Wilson (2003). Genetic neuroscience of mammalian learning and memory. *Philos Trans R Soc Lond B Biol Sci* 358(1432): 787–795.

Tyler, W.J., M. Alonso, C.R. Bramham, and L.D. Pozzo-Miller (2002). From acquisition to consolidation: On the role of brain-derived neurotrophic factor signaling in hippocampal-dependent learning. *Learn Mem* 9(5): 224–237.

Uvnas-Moberg, K., M. Eklund, V. Hillegaart, and S. Ahlenius (2000). Improved conditioned avoidance learning by oxytocin administration in high-emotional male Sprague-Dawley rats. *Regulatory Peptides* 88(1–3): 27–32.

Vallee, M., W. Mayo, M. Darnaudery, C. Corpechot, J. Young, M. Koehl, M. Le Moal, E.E. Baulieu, P. Robel, and H. Simon (1997). Neurosteroids: Deficient cognitive performance in aged rats depends on low pregnenolone sulfate levels in the hippocampus. *Proc Nat Acad Sci USA* 94(26): 14865–14870.

Wang, H.L., M.J. Wayner, C.Y. Chai, and E.H.Y. Lee (1998). Corticotropin-releasing factor produces a long-lasting enhancement of synaptic efficacy in the hippocampus. *Eur J Neurosci* 10: 3428–3437.

Weeber, E.J., and J.D. Sweatt (2002). Molecular neurobiology of human cognition. *Neuron* 33(6): 845–848.

Wenk, G.L. (1989). An hypothesis on the role of glucose in the mechanism of action of cognitive enhancers. *Psychopharmacol (Berl)* 99(4): 431–438.

White, N.M., and J.A. Salinas (1998). Pharmacological approaches to the study of learning and memory. In *Neurobiology of Learning and Memory*, J. Martinez and R. Kesner, eds. pp. 143–176, San Diego: Academic Press.

Whitehouse, P.J., E. Juengst, M. Mehlman, and T.H. Murray (1997). Enhancing cognition in the intellectually intact. *Hastings Cent Rep* 27(3): 14–22.

Williams, G.V., and P.S. Goldman-Rakic (1995). Modulation of memory fields by dopamine D1 receptors in prefrontal cortex. *Nature* 376(6541): 572–575.

Woolf, N.J., A.M. Milov, E.S. Schweitzer, and A. Roghani (2001). Elevation of nerve growth factor and antisense knockdown of TrkA receptor during contextual memory consolidation. *J Neurosci* 21(3): 1047–1055.

Woolley, C.S., E. Gould, M. Frankfurt, and B.S. McEwen (1990). Naturally occurring fluctuation in dendritic spine density on adult hippocampal pyramidal neurons. *J Neurosci* 10(12): 4035–4039.

Yamada, K., M. Mizuno, and T. Nabeshima (2002). Role for brain-derived neurotrophic factor in learning and memory. *Life Sci* 70(7): 735–744.

Yesavage, J.A., M.S. Mumenthaler, J.L. Taylor, L. Friedman, R. O'Hara, J. Sheikh, J. Tinklenberg, and P.J. Whitehouse (2002). Donepezil and flight simulator performance: Effects on retention of complex skills. *Neurology* 59(1): 123–125.

Zarrindast, M.R., and A. Rezayof (2004). Morphine state-dependent learning: Sensitization and interactions with dopamine receptors. *Eur J Pharmacol* 497(2): 197–204.

Zorrilla, E.P., G. Schulteis, N. Ling, G.F. Koob, and E.B. De Souza (2001). Performance-enhancing effects of CRF-BP ligand inhibitors. *Neuroreport* 12(6): 1231–1234.

Index

Page numbers followed by *f* or *t* indicate entries in figures or tables, respectively.

α/β neurons, of mushroom body, 115–116, 115*f*
AC. *See* Adenylate cyclase
Acetylcholine (ACh), 29
  in Alzheimer's disease, 520–521, 525
  in context processing, 164
  estrogen and, 254–255, 549–550
  in fear-potentiated startle, 394–396
  in mushroom body, 116
  in reversal learning, 371–372
  in state-dependent retrieval, 560–561
  in striatum, 357, 358, 364–365, 371–372
Acetylcholine inhibitors, 29
Acetylcholine receptor agonists
  for attention/encoding enhancement, 547–548
  for storage/consolidation enhancement, 557
Acetylcholinesterase (AChE), 14, 16
Acetylcholinesterase inhibitors
  for Alzheimer's disease, 521, 527, 532
  for storage/consolidation enhancement, 557
ACh. *See* Acetylcholine
AChE. *See* Acetylcholinesterase
Acoustic startle reflex, fear-potentiated, 385, 388*f*. *See also* Fear-potentiated startle
ACTH. *See* Adrenocorticotropic hormone
Action potential, genomic, 553–554
Activator protein-1 (AP-1), aging and, 506
Activity, and storage/consolidation enhancement, 557–558
Addiction. *See* Drug abuse/addiction
Adenosine receptor antagonists, for attention/encoding enhancement, 546–547
Adenosine triphosphate (ATP)
  conversion to cAMP, 112
  in myelination, 69
Adenylate cyclase (AC)
  estrogen and, 256
  genetics of pathway, 112–114, 113*f*

Adenylate cyclase inhibitors, 29
Adf1 transcription factor, 115, 118, 119
Adrenal hormones, 243–250, 551–552. *See also specific hormones*
Adrenergic receptor agonists, for storage/consolidation enhancement, 555
Adrenocorticotropic hormone (ACTH), 249, 560
Adult brain
  estrogen effects on, 251
  neurogenesis in, 77–80, 495
  plasticity in, 76–80
Affect attribute, 273–274
After-hyperpolarizing potential (AHP), aging and, 498–599
Age/aging, 41–42, 483–508
  and CREB expression, 500–503
  definitions of, 484–485
  demographic shift in United States, 483–484
  and executive functioning, 486, 506–508
  and hippocampus, 493–506
    electrophysiological findings in, 498–500
    molecular findings in, 500–506
    neuroanatomical findings in, 495–498, 496*f*
  and immediate early genes, 504–506
  and long-term potentiation, 499–500, 505
  methods for studying effects of, 484–487
  and neurodegenerative diseases, 519. *See also* Alzheimer's disease
  and neurogenesis, 77–80
  and plasticity, 17–18
  and prefrontal cortex, 506–508
  and progesterone effects, 259
  and short-term memory, 486, 506–508
  and spatial learning/memory, 487–506
    human models of, 491–493
    rodent models of, 487–491, 489*f*